Deterministic and Stochastic Models of
AIDS Epidemics and HIV Infections
with Intervention

Deterministic and Stochastic Models of AIDS Epidemics and HIV Infections with Intervention

WAI-YUAN TAN

University of Memphis, USA

HULIN WU

University of Rochester, USA

World Scientific

NEW JERSEY • LONDON • SINGAPORE • BEIJING • SHANGHAI • HONG KONG • TAIPEI • CHENNAI

Published by

World Scientific Publishing Co. Pte. Ltd.

5 Toh Tuck Link, Singapore 596224

USA office: 27 Warren Street, Suite 401-402, Hackensack, NJ 07601

UK office: 57 Shelton Street, Covent Garden, London WC2H 9HE

British Library Cataloguing-in-Publication Data
A catalogue record for this book is available from the British Library.

DETERMINISTIC AND STOCHASTIC MODELS OF AIDS EPIDEMICS AND HIV INFECTIONS WITH INTERVENTION

ISBN 981-256-139-0

Typeset by Stallion Press
Email: enquiries@stallionpress.com

Printed in Singapore by World Scientific Printers (S) Pte Ltd

CONTENTS

CHAPTER 1

MATHEMATICAL MODELS FOR HIV TRANSMISSION AMONG INJECTING DRUG USERS

Vincenzo Capasso and Daniela Morale

In this paper, we propose a survey of mathematical models for the spread of HIV/AIDS through shared drug injection equipment in groups of injecting drug users. We extend the expression for the force of infection by including the mechanism of exchange of a drug injecting equipment (DIE) in a friendship group. The classical models based on the law of mass action for direct transmission are obtained as limiting cases. We analyze models for a homogeneous population; for multiple stages of infectivity of HIV-infected individuals, and for stratified populations and multiple stages of infectivity by providing threshold theorems and possible stability results for nontrivial endemic states.

Keywords: HIV/AIDS, transmission models, injecting drug users, multistage-multigroup models.

1. Introduction

A large amount of literature is available about the mathematical modeling of HIV/AIDS epidemics. Most of the proposed models may be seen as extensions of classical epidemic models, by the inclusion of various modifications in order to take into account relevant peculiarities of the mechanism of transmission of HIV. Among those, we may mention the substantial variability of the infection rates for different subpopulations at risk (homosexuals, heterosexuals, intravenous drug users, etc.); the long incubation period before the exhibition of symptoms; the variability of the infection with respect to the evolution of the infection in each individual; etc. As a consequence, the mathematical modeling of the dynamics of HIV/AIDS infections cannot avoid including "structures" in the population. In order to reduce the level of complexity, most of the authors have proposed to start from simplified models in order to obtain some first information about the macroscopic qualitative behavior of the system.

Table 1. Data of the diffusion of HIV epidemic in the world at the end of 2001 (Data source: UNAIDS-OMS).

Region	HIV/AIDS-infected	New HIV infections	Main means of transmission
Subsaharian Africa	28.1 millions	3.4 millions	hetero
North Africa, Middle West	440,000	80,000	hetero — DIE
South and South East Asia	6.1 millions	800,000	hetero — DIE
Eastern Asia and Pacific Islands	1 million	270,000	DIE — hetero — homo
Latin America	1.4 millions	130,000	homo — TD — hetero
Caraibi	420,000	60,000	hetero — homo
Eastern Europe, Central Asia	1 million	250,000	DIE
Western Europe	560,000	30,000	homo — DIE
North America	940,000	45,000	homo — DIE — hetero
Australia, New Zealand	15,000	500	homo

Case notification data from the World Health Organization (WHO) show that in many countries, sharing drug injection equipment (DIE) is one of the main causes of spread of HIV and AIDS (cf. Table 1). For example in Italy, it is shown that even though the number of AIDS/HIV infections due to sexual transmission in the last few years is increasing, drug users still represent the main reservoir (about 60%) of the infection (Notices of the Italian National Institute of Health); (Blythe & Castillo–Chavez, 1989) therefore modeling the transmission in this subpopulation can help to understand the whole phenomena and to suggest some prevention policies for the entire population. Of course, a more realistic model of the spread of HIV and AIDS should also include the interaction between the syringe exchange and the (hetero and homo) sexual transmission, but for the sake of simplicity, hereby we shall ignore this ingredient.

In this paper we present a review of the results obtained by the authors in an attempt to introduce in classical models AIDS as a force of infection that takes into account the specific mechanism of syringe exchange in friendship groups of drug addicts.

In countries (like USA) where there is no free access to injection equipment, sharing occurs mainly through shooting galleries, where DIE's are (sequentially) rented to drug users; a model for this mechanism of transmission was proposed by Kaplan (1989). On the other hand, in many other countries (like Italy) there is free access to DIE's, so that shooting galleries need not exist. In such countries interaction among drug users occurs more likely through syringe sharing in friendship groups; this is the reason why it is necessary to provide an expression for the force of infection specific of this pattern of contacts.

Starting from the papers of Jacquez *et al.* (1988) and Castillo–Chavez *et al.* (1989) based on classical SIR, SEIR, etc. models with vital dynamics, the present authors proposed the derivation of explicit expressions for the "force" of infection acting on any susceptible individual, by means of stochastic models of the kinetics of HIV transmission due to sharing of drug injection equipment within a friendship group of drug addicts (Capasso *et al.*, 1994; 1995; 1998). The kinetics involves both model hypotheses on the behavioral of the addicts, as well as the Poisson distributions of the grouping and social/medical parameters as the level of infectivity or the probability of disinfecting the syringe.

The possibility of explicitly writing the force of infection in terms of the behavioral parameters of individuals leads to a consequent explicit dependence of the the basic reproductive ratio on such parameters. Since mathematical analysis shows that the basic reproductive ratio plays the role of threshold parameter with respect to the extinction of an emerging epidemic, control policies may then be proposed in order to bring the system below threshold.

2. One Population Model

We consider an SIR model to describe the transmission of HIV through drug injection equipment (DIE) in a population of injecting drug users (IDU). This means that we consider a system of three non-linear ordinary differential equations, for the time evolution of the susceptibles (S), the HIV infectives (I) and the recovered individuals (A), i.e. those individuals in the population that exhibit clear symptoms of the AIDS disease. The main feature of the model is that the force of infection is obtained by averaging the probability of exposure of a susceptible individual to an infectious equipment with respect to the group size, and the order of exchange of the DIE. Models of this kind, in which a continuum deterministic model is obtained at the macroscale by averaging at the scale of individuals (microscale), are known as "hybrid models".

The model at the population scale is still given by

$$\begin{aligned} \frac{dS}{dt} &= \Lambda - \mu S - g(S,I)S \\ \frac{dI}{dt} &= g(S,I)S - \mu I - \kappa I \\ \frac{dA}{dt} &= \kappa I - \mu A - \delta A \end{aligned} \tag{1}$$

where Λ is the constant recruitment rate of susceptibles in the population, κ is the transfer rate from I to A, μ is the natural death rate specific of the IDU population, δ is the differential death rate for individuals with AIDS, and $g(S, I)$ denotes the force of infection.

The first important assumption is that only individuals in S and I classes are considered to be "active" and participate in syringe sharing. Thus, the active population size is

$$T = S + I$$

and the fraction of infected individuals (that is the probability that a "sharer" is infectious) is

$$\pi = \frac{I}{T} = \frac{I}{S+I}. \tag{2}$$

From system (1), we derive that the total active population $T(t)$ evolves according to

$$\frac{dT}{dt} = \Lambda - \mu T - \kappa I\,,$$

independent of the force of infection $g(S, I)$; therefore

$$\Lambda - (\mu + \kappa)T \le \frac{dT}{dt} \le \Lambda - \mu T\,.$$

Hence, the following inequalities hold

$$T(t) - \frac{\Lambda}{\mu} \le \left[T(0) - \frac{\Lambda}{\mu}\right] e^{-\mu t}; \quad T(t) - \frac{\Lambda}{\mu + \kappa} \ge \left[T(0) - \frac{\Lambda}{\mu + k}\right] e^{-(\mu+k)t}\,.$$

Proposition 1: (*Capasso*, 1993) *For* $\mu > 0$, *if* $T(0) \le \Lambda/\mu$, *then* $T(t) \le \Lambda/\mu$, *for any* $t \ge 0$. *In particular the set*

$$\Omega := \left\{ (S, I) \in \mathbb{R}^2_+ \left| \frac{\Lambda}{\mu + \kappa} \le S + I \le \frac{\Lambda}{\mu} \right.\right\}$$

is a compact, convex and invariant set for System (1).

2.1. *Force of infection and social rules*

In order to provide an explicit expression for the force of infection $g(S, I)$, we have imported some basic ideas from Kaplan's model (1989) with some modifications that take into account the friendship grouping of drug addicts. We assume the following:

(i) All individuals participate in friendship grouping at random. The group size is a random variable following a Poisson distribution with parameter $\nu > 0$. Each group gets a new (or anyway uninfectious) DIE at the first injection; then the syringe is shared in sequence among all members of the group; after that it is thrown away.

(ii) Grouping occurs according to a Poisson process with parameter $\lambda > 0$.

(iii) A DIE becomes infectious if it is used by an infected addict. When an infectious DIE is used by an uninfected addict, the act of injecting will "rid" the equipment with probability θ. Thus, any susceptible who uses an infectious DIE is considered exposed to HIV, but he may also disinfect the same equipment for the next drug user.

(iv) Given exposure to HIV, an addict becomes infected (and immediately infectious) with probability α. Syringe sharing is the only means by which addicts may become infected.

According to these assumptions, the force of infection is given by

$$g(S, I) = \lambda\alpha\beta(S, I)\,, \tag{3}$$

where $\beta(S, I)$ is the probability of exposure to HIV, i.e. the probability that an addict encounters an infectious DIE. In order to determine it, we proceed as follows. Let E be the event that an addict encounters an infectious equipment, G the random size of the group, and $R(k)$ the order position of the addict in sharing the DIE in his group of size k. So we have

$$\beta(S, I) = P(E) = \sum_{k=2}^{\infty} P(E|G = k)P(G = k) \tag{4}$$

$$= \sum_{k=1}^{\infty} P(G = k) \sum_{j=1}^{k} P(E|G = k, R(k) = j)P(R(k) = j)\,. \tag{5}$$

Only groups with size greater or equal than 2 are relevant for the transmission; due to the hypothesis (i), we have

$$P(G = k) = \frac{\nu^{k-2}e^{-\nu}}{(k-2)!}\,, \quad k = 2, 3, \ldots \tag{6}$$

and the average size of the group is $\bar{\nu} = \nu + 2$. Furthermore, each position within the group is equally probable, so that, if the group size is k,

$$P(R(k) = j) = \frac{1}{k}\,. \tag{7}$$

Thus we only need to calculate the probability $P(E|G = k, R(k) = j)$ of encountering an infectious equipment, being the jth in a group of k

individuals

$$P(E|G = k, R(k) = j) = P\left(\bigcup_{i=1}^{j-1} \{I_i \cap U_i\}\right), \tag{8}$$

where I_i is the event that the ith addict is infective, and

$$U_i = \bigcap_{h=i+1}^{j-1} \{\text{the } h\text{th is uninfective and he has not disinfected the DIE}\}. \tag{9}$$

It is clear the $P(I_i) = I/(I+S) = \pi$; furthermore because of the independence of events in (8) and (9), from assumption (iii) the probability (8) is

$$P(E|G = k, R(k) = j) = \sum_{i=1}^{j-1} \pi \left[(1-\pi)^{j-1-i}(1-\theta)^{j-1-i}\right]. \tag{10}$$

From (5) to (10), one has following expression

$$\beta(S, I) = \frac{I}{\theta S + I}\left\{1 - \frac{(\nu - 1)(S + I) + \left[\left(1 - \nu(1-\theta)\right)S + I\right]e^{-\nu\frac{\theta S + I}{S + I}}}{\nu^2(\theta S + I)}\right\}. \tag{11}$$

Actually, β depends upon S and I only through the fraction of infected individuals π. In fact, $\beta(S, I)$ may be rewritten as

$$\beta(S, I) = \tilde{\beta}(\pi(S, I)), \tag{12}$$

where

$$\tilde{\beta}(\pi) = \frac{\pi}{1 - (1-\theta)(1-\pi)} \times \left\{1 - \frac{\nu - 1 + \left[1 - \nu(1-\theta)(1-\pi)\right]e^{-\nu[1-(1-\theta)(1-\pi)]}}{\nu^2\left[1 - (1-\theta)(1-\pi)\right]}\right\}. \tag{13}$$

The expression (13) and, as a consequence, the force of infection depends only on the fraction of infectives of the active population and on some explicit social parameter. From the modeling point of view, this is a nice result, since one might act on these parameters via a social education program to reduce the force of infection.

2.2. *Qualitative analysis*

Let $\mathcal{R} = \mathbb{R}^2$ be the admissible region for the system (1). The first two equations of the system (1) are independent of the variable of A (recovered individuals), so that we may study the evolution of susceptibles and infectives first

$$\frac{dS}{dt} = F_1(S,I) = \Lambda - \mu S - g(S,I)S \tag{14}$$

$$\frac{dI}{dt} = F_2(S,I) = g(S,I)S - \mu I - \kappa I\,. \tag{15}$$

For $I \neq 0$, define

$$\beta^*(S,I) = \frac{S+I}{I}\beta(S,I)\,,$$

thus

$$F_2(S,I) = IG_2(S,I)\,,$$

where

$$G_2(S,I) = \lambda\alpha\beta^*(S,I)\frac{S}{S+I} - (\mu+\kappa)\,.$$

It is not difficult to show that

$$\frac{\partial}{\partial S}\beta(S,I) \le 0\,, \quad \frac{\partial}{\partial I}\beta(S,I) \ge 0\,, \qquad \forall (S,I) \in \mathcal{R}\,; \tag{16}$$

$$\frac{\partial}{\partial S}\beta^*(S,I) \ge 0\,, \quad \frac{\partial}{\partial I}\beta^*(S,I) \le 0\,, \qquad \forall (S,I) \in \mathcal{R}\,. \tag{17}$$

The behavior (16) is shown in Fig. 1.

As a consequence

$$\frac{\partial}{\partial S}F_1(S,I) \le 0\,, \quad \frac{\partial}{\partial I}F_1(S,I) \ge 0\,, \qquad \forall (S,I) \in \mathcal{R}\,; \tag{18}$$

$$\frac{\partial}{\partial S}G_2(S,I) \ge 0\,, \quad \frac{\partial}{\partial I}G_2(S,I) \le 0\,, \qquad \forall (S,I) \in \mathcal{R}\,. \tag{19}$$

Furthermore, $\partial F_1/\partial I = \partial G_2/\partial I = 0$ if and only if $S = 0$.

The quantity

$$R_0 = \frac{\lambda\alpha}{(\mu+\kappa)\theta}\left\{1 - \frac{\nu - 1 + \left[1 - \nu(1-\theta)\right]e^{-\nu\theta}}{\nu^2\theta}\right\} \tag{20}$$

is known as the basic reproduction number, i.e. the number of additional infectives generated by each infective individual in a large population of susceptibles.

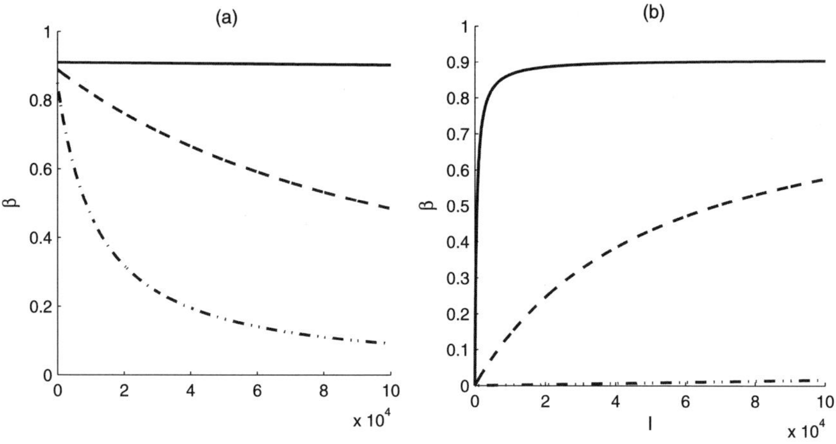

Fig. 1. Behavior of the probability of exposure to HIV, $\beta(S, I)$: (a) dependence on S, for different values of I : 6000 (dot-dash), 60,000 (dash), 6,000,000 (solid); and (b) dependence on I, for different values of S : 10,000 (dash), 100,000 (dot-dash), 1,000,000 (solid).

The following result holds:

Theorem 2: (a) If $R_0 \leq 1$, the disease-free equilibrium $z^* = (\Lambda/\mu, 0)$ is the only equilibrium solution of System (15). Moreover, it is globally asymptotically stable in the admissible region $\mathcal{R}$.

(b) If $R_0 > 1$, the solution z^* is unstable and System (15) admits another equilibrium $z^{**} = (S^{**}, I^{**})$ where $S^{**} \in (\Lambda/(\mu+\kappa), \Lambda/\mu)$ and $I^{**} = (\Lambda - \mu S^{**})/(\mu + \kappa)$. Moreover, z^{**} is globally asymptotically stable (GAS) in $\mathcal{R}$.

Proof: System (15) admits the equilibrium $z^* = (\Lambda/\mu, 0)$. We have two phase-portraits according to the size of R_0. Note that

$$R_0 \leq 1 \Leftrightarrow G_2(S, 0) \leq 0\,. \tag{21}$$

(a) First we show that the only equilibrium is the disease-free state z^*. Indeed, the admissible region $\mathcal{R}$ does not contain the isocline $dI/dt = 0$, since $dI/dt = 0$ if and only if $I = 0$ and $G_2(S, I) = 0$. But G_2 is decreasing in I for (19); as a consequence of (21) we have $G_2(S, I) < 0$ in $\mathcal{R}$. Furthermore, by means of the contracting rectangle technique (Britton, 1986) one can easily prove that z^* is GAS.

(b) In this case, both isoclines are in $\mathcal{R}$. An analysis of the corresponding phase-portrait, along with (18) and (19), allows to determine the existence of an endemic equilibrium $z^{**} = (S^{**}, I^{**})$, where $\Lambda/(\mu+\kappa) < S^{**} < \Lambda/\mu$ and $I^{**} = (\Lambda - \mu S^{**})/(\mu + \kappa)$.

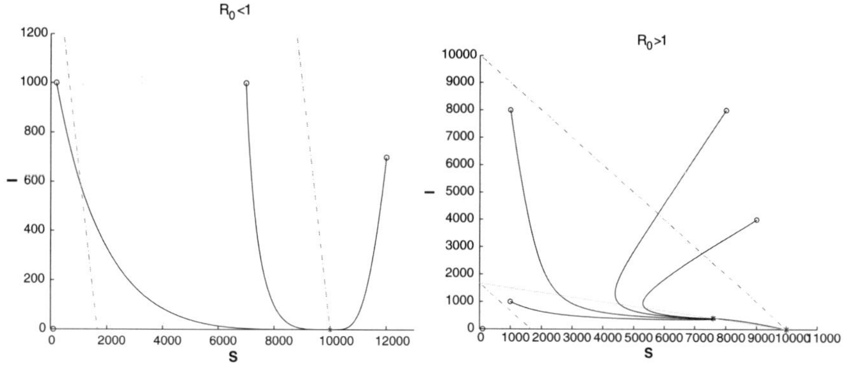

Fig. 2. Phase-portraits of the 1 population model: (S, I) trajectories with the initial conditions (o), and the invariant region (between the two dashed lines); equilibrium points are marked with *. Left: $R_0 < 1$. Right: $R_0 > 1$.

The Jacobi matrix of system (15) at z^{**} has a positive determinant and a negative trace, so that both eigenvalues have negative real parts (Capasso, 1993). Therefore z^* is unstable and z^{**} is locally asymptotically stable. It is possible to rule out the existence of periodic solutions by means of the Bendixon–Dulac criterion (Capasso, 1993; Waltman, 1986), by observing that

$$\frac{\partial}{\partial S}\left(\frac{1}{I}F_1(S,I)\right) + \frac{\partial}{\partial I}\left(\frac{1}{I}F_2(S,I)\right) < 0$$

in the interior of $\mathcal{R}$. Thus, the endemic state z^{**} is GAS in the interior of the admissible region $\mathcal{R}$. □

3. Multipopulation Models

When one considers the spread of an infection, one should take into account not only the different means of infection, but also, not less important, possible "structures" within the active population. For example many of the social parameters we included in the model may depend upon geographical, social, age, level of syringe-sharing and/or other heterogeneities. In particular the probability α of infection might depend upon age, while the Poisson grouping parameter λ might depend upon geographical and/or social factors, as the rate κ of infectives who develop AIDS symptoms may depend on age and/or social factors.

In order to take into account such kind of heterogeneities, the total population is divided into m subpopulations $T_i, i = 1, \ldots, m$. For each of them, S_i denotes the number of susceptibles, I_i the number of infectives,

A_i the number of individuals with AIDS and $T_i = S_i + I_i$ is the number of active individuals.

In such a case, the force of infection acting on each susceptible of the ith group is given by

$$\lambda_i \alpha_i \beta_i(S, I)\,,$$

where $S = (S_1, \ldots, S_m)^T$ and $I = (I_1, \ldots, I_m)^T$.

Hence, model (1) can be extended to the multi-population case as follows

$$\begin{aligned}
\frac{dS_i}{dt} &= \Lambda_i - \mu_i S_i - \lambda_i \alpha_i \beta_i(S, I) S_i \\
\frac{dI_i}{dt} &= \lambda_i \alpha_i \beta_i(S, I) S_i - \mu_i I_i - \kappa_i I_i \quad i = 1, \ldots, m \\
\frac{dA_i}{dt} &= k_i I_i - \mu_i A_i - \delta_i A_i
\end{aligned} \tag{22}$$

In such a model, in order to obtain an explicit expression for $\beta_i(S, I)$, it is necessary to specify the contact rates among individuals of different subpopulations.

We model this interaction among populations via the so called "mixing" matrix $P = (\rho_{ij})_{i,j=1,\ldots,m}$, where ρ_{ij} is the fraction of "partners" (in syringe-sharing) of a member of the ith population, coming from the jth population.

3.1. *The force of infection*

The general case of multiple populations is more difficult to handle unless we assume $\nu_i = \nu$ and $\theta_i = \theta$, independent of i. This could also be reasonable since, for example, the ability of flushing the needle could be more a physiological factor, common to the active population. In such a case, the probability of exposure to HIV is given by

$$\beta_i(S, I) = \tilde{\beta}(\pi_i(S, I))\,,$$

where $\tilde{\beta}(\pi)$ is again given by

$$\begin{aligned}
\tilde{\beta}(\pi) &= \frac{\pi}{1 - (1-\theta)(1-\pi)} \\
&\quad \times \left\{1 - \frac{\nu - 1 + [1 - \nu(1-\theta)(1-\pi)]\, e^{-\nu[1-(1-\theta)(1-\pi)]}}{\nu^2\,[1 - (1-\theta)(1-\pi)]}\right\} \\
&= \pi \sum_{k=2}^{\infty} P_k(\pi) \frac{e^{-\nu} \nu^{2+h}}{\nu(\nu+2)!}\,,
\end{aligned} \tag{23}$$

while the quantity $\pi_i(S, I)$ is the probability that a DIE encounters an infected individual in the group that hosts the susceptible addict of the ith subpopulation and is given by

$$\pi_i(S, I) = \sum_{j=1}^{m} \rho_{ij} \frac{I_j}{T_j}.$$

The expression of P_k in (23) is given by

$$P_k(\pi) = (k-1) + (k-2)(1-\theta)(1-\pi) + (k-3)(1-\theta)^2(1-\pi)^2 + \cdots + \\ + 2(1-\theta)^{k-3}(1-\pi)^{k-3} + (1-\theta)^{k-2}(1-\pi)^{k-2}.$$

Therefore, the force of infection acting on each susceptible in the ith subpopulation is

$$\lambda_i \alpha_i \tilde{\beta}(\pi_i(S, I)).$$

3.2. *Interaction among different populations*

The mixing probabilities ρ_{ij} must satisfy the following properties (Blythe, 1989):

(i) ρ_{ij} are probabilities

$$0 \le \rho_{ij} \le 1, \quad i, j = 1, \ldots, m$$
$$\sum_{j=1}^{m} \rho_{ij} = 1, \qquad i = 1, \ldots, m$$

(ii) the number of sharing encounters between T_i and T_j individuals must equal the number of sharing between T_j and T_i individuals

$$\lambda_i T_i \rho_{ij} = \lambda_j T_j \rho_{ji}, \quad i, j = 1, \ldots, m$$

(iii) there are mixing only among active individuals

$$\lambda_i T_i \lambda_j T_j = 0 \Rightarrow \rho_{ij} = \rho_{ji} = 0. \tag{24}$$

Busenberg and Castillo–Chavez (1989) gave an explicit formula for the construction of arbitrary mixing matrices that represent all forms of mixing as deviations from random mixing. An "affinity" matrix $\Phi = (\Phi_{ij})_{i,j=1,\ldots,n}$ is introduced, where Φ_{ij} measures the preference of i addicts for j addicts. The following theorem holds.

Theorem 3: Representation Theorem For each mixing matrix P at any time, a symmetric affinity matrix Φ can be found such that

$$\rho_{ij} = \bar{p}_j \left[\frac{R_i R_j}{\sum_{k=1}^n \bar{p}_k R_k} + \Phi_{ij} \right],$$

where

$$\bar{p}_j = \frac{\lambda_j T_j}{\sum_{k=1}^n \lambda_k T_k};$$

and

$$R_j = 1 - \sum_{k=1}^n \bar{p}_k \Phi_{jk}.$$

The converse also holds; that is, every symmetric matrix Φ defines a mixing matrix P.

The structure of the matrix $P := (\rho_{ij})_{i,j=1,\ldots,n}$ defines different patterns of contact between groups. Let consider the following two cases:

Case I: Restricted mixing (Jacquez, 1988): all contacts are restricted to within group contacts;

$$\rho_{ij} = 1, \quad i = j = 1, \ldots, n; \qquad P = I_{n \times n}.$$

In this case, we have

$$\beta_i(S, I) = \beta_i(S_i, I_i)$$

and, since each subpopulation is independent of the others, the spread of HIV can be analyzed as in the "one population" model.

Case II: Proportionate (or random) mixing (Capasso, 1993; Capasso *et al.*, 1994; Hethcote *et al.*, 1987; Lin, 1991): the fraction of contacts of group $i = 1, \ldots, n$ with group j is equal to the fraction of contacts due to group j $(i, j = 1, \ldots, n)$, with respect to the the number of contacts due to all groups $\sum_{k=1}^n \lambda_k T_k$, we have in this case

$$\rho_{ij} = \frac{\lambda_j T_j}{\sum_{k=1}^n \lambda_k T_k}, \quad i, j = 1, \ldots, n. \tag{25}$$

This means that the affinity matrix Φ is such that Φ_{ij} is constant for each i and j. In this case $\pi_i(S, I)$ is independent of i and is given by

$$\pi_i(S, I) = \sum_{j=1}^m \frac{\lambda_j T_j}{\sum_{k=1}^n \lambda_k T_k} \frac{I_j}{T_j} = \frac{\sum_{j=1}^m \lambda_j I_j}{\sum_{k=1}^n \lambda_k T_k} := \pi(S, I); \tag{26}$$

in such a case $\beta_i(S,I)$ is independent of i and the following formula holds

$$\beta_i(S,I) = \tilde{\beta}(\pi(S,I))\,. \tag{27}$$

3.3. *Qualitative analysis*

We study system (22) in the special case of random mixing (26) and $\mu_i = \mu, (i = 1,\ldots,m)$; in addition, since the first $2m$ equations in (22) are independent of $A = (A_1,\ldots,A_m)^T$, we can restrict the analysis to these $2m$ equations, without loss of generality. So, we study the system

$$\frac{dS_i}{dt} = \Lambda_i - \mu S_i - \lambda_i\alpha_i\tilde{\beta}(\pi(S,I))S_i \tag{28}$$

$$\frac{dI_i}{dt} = \lambda_i\alpha_i\tilde{\beta}(\pi(S,I))S_i - \mu I_i - \kappa_i I_i \tag{29}$$

for $i = 1,\ldots,m$ where $\pi(S,I)$ is defined in (26).

For this case it is possible to establish a threshold theorem about the existence of a nontrivial endemic state. We shall follow the approach by Castillo–Chavez *et al.* (1989) and Lin (1991; 1993); in accordance with them we use μ as the bifurcation parameter.

System (29) always admits a trivial disease-free equilibrium $(S^*,I^*) = \left(\frac{\Lambda_1}{\mu},\ldots,\frac{\Lambda_m}{\mu},0,\ldots,0\right)$. We want to find now the condition that has to be satisfied for the existence of a non trivial endemic equilibrium.

Consider the following shift of variables, so that the origin becomes the trivial equilibrium:

$$v_i = \frac{\Lambda_i}{\mu} - S_i - I_i\,,$$

so that system (29) assumes the form

$$\begin{aligned} \frac{dv_i}{dt} &= f_{1i} = -\mu v_i + \kappa_i I_i \\ \frac{dI_i}{dt} &= f_{2i} = \tfrac{\lambda_i\alpha_i}{\mu}\tilde{\beta}(\pi(v,I))(\Lambda_i - \mu v_i - \mu I_i) - \mu I_i - \kappa_i I_i\,, \end{aligned} \tag{30}$$

for $i = 1,\ldots,m$, where

$$\pi(v,I) = \mu\frac{\sum_{i=1}^n \lambda_i I_i}{\sum_{h=1}^n \lambda_h(\Lambda_h - \mu v_h)}\,. \tag{31}$$

Proposition 4: *$\bar{\beta} = \tilde{\beta}/\pi$ is a decreasing function of π, while $\tilde{\beta}$ is an increasing function of π.*

Proof: From (23), we have

$$\frac{\partial}{\partial\pi}(\bar\beta) = \sum_k \frac{\partial}{\partial\pi}(P_k)\frac{e^{-\nu}\nu^{2+k}}{\nu(\nu+2)!} - \sum_h \frac{e^{-\nu}\nu^{2+k}}{\nu(\nu+2)!}\left[(k-2)(1-\theta)+2(k-3)(1-\theta)^2(1-\pi)+\cdots + 2(k-3)(1-\theta)^{k-3}(1-\pi)^{k-4}+(k-2)(1-\theta)^{k-2}(1-\pi)^{k-3}\right] < 0\,. \tag{32}$$

Besides

$$\frac{\partial\tilde\beta}{\partial\pi} = \sum_k \left(\pi\frac{\partial P_k}{\partial\pi}+P_k\right)\frac{e^{-\nu}\nu^{2+k}}{\nu(\nu+2)!} > 0\,. \tag{33}$$

Indeed,

$$\pi\frac{\partial P_k}{\partial\pi}+P_k = [(k-1)-(k-2)(1-\theta)]+2(1-\theta)[(k-2)-(k-3) \times(1-\theta)](1-\pi)+\cdots+(k-3)(1-\theta)^{k-3}[2-(1-\theta)] \times(1-\pi)^{k-3}+(k-2)(1-\theta)^{k-2}[2-(1-\theta)](1-\pi)^{k-2} > 0\,.$$

□

As already said, we study (30) instead of (29).

Proposition 5: *The origin* $v^* = (0,0)$ *is an equilibrium of* (30).

In order to study the stability property of v^*, we need to calculate the Jacobian matrix. We need the following relations:

$$\begin{aligned}
b_0 &= \tilde\beta'(0) = \frac{1}{\theta}\left\{1-\frac{\nu-1+[1-\nu(1-\theta)]\,e^{-\nu\theta}}{\nu^2\theta}\right\};\\
b_0^* &= \frac{d}{d\pi}\left(\frac{\tilde\beta(\pi)}{\pi}\right)\Bigg|_{\pi=0};\\
\kappa_0 &= \frac{\sum_{k=1}^m \lambda_k\Lambda_k}{b_0};\\
K &= diag(\kappa_i);\\
L_1 &= (\alpha_i\lambda_i\lambda_j)_{i,j=1,\dots,n};\\
L &= \frac{1}{\kappa_0}diag(\Lambda_i)L_1;\\
A(\mu) &= L-(\mu E+K);
\end{aligned} \tag{34}$$

where E is the $n\times n$ identity matrix.

Proposition 6: *The Jacobi matrix of System* (30) *at the origin is*

$$J_f(v^*) = \begin{pmatrix} -\mu E & K \\ 0 & A(\mu) \end{pmatrix}.$$

Proof: By direct calculations, from (31), we have the following partial derivatives

$$\frac{\partial \pi}{\partial v_i} = -\frac{\mu^2 \lambda_i (\sum_j \lambda_j I_j)}{(\sum_h \lambda_h (\Lambda_h - \mu v_h))^2}, \quad \frac{\partial \pi}{\partial I_i} = \frac{\mu \lambda_i}{\sum_h \lambda_h (\Lambda_h - \mu v_h)},$$

so that

$$\frac{\partial \pi}{\partial v_i}(0) = 0; \quad \frac{\partial \pi}{\partial I_i}(0) = \frac{\mu \lambda_i}{\sum_h \lambda_h \Lambda_h};$$
$$\frac{\partial \beta}{\partial v_i}(0) = 0; \quad \frac{\partial \beta}{\partial I_i}(0) = \frac{\mu \lambda_i}{\sum_h \lambda_h \Lambda_h b_0};$$

$$\frac{\partial f_{1i}}{\partial v_i}(0) = -\mu; \quad \frac{\partial f_{1i}}{\partial I_i}(0) = \kappa_i;$$
$$\frac{\partial f_{1i}}{\partial v_j}(0) = 0; \quad \frac{\partial f_{1i}}{\partial I_j}(0) = 0;$$
$$\frac{\partial f_{2i}}{\partial v_i}(0) = 0; \quad \frac{\partial f_{2i}}{\partial v_j}(0) = 0;$$

$$\begin{aligned} \frac{\partial f_{2i}}{\partial I_i}(0) &= \frac{\alpha_i \lambda_i}{\mu} \frac{\partial \beta}{\partial I_i}(0) \Lambda_i - \lambda_i \beta(0) - (\mu + \kappa_i) \\ &= \frac{\alpha_i \lambda_i^2 \Lambda_i}{\sum_h \lambda_h \Lambda_h} b_0 - (\mu + \kappa_i) \\ \frac{\partial f_{2i}}{\partial I_j}(0) &= \frac{\alpha_i \lambda_i}{\mu} \frac{\partial \beta}{\partial I_j}(0) \Lambda_i = \frac{\alpha_i \lambda_i \lambda_j}{\sum_h \lambda_h \Lambda_h} b_0; \end{aligned}$$

The proof is concluded. □

Note that L, and therefore $L - K$, is an irreducible quasimonotone matrix if we assume L_1 to be so, so that the Perron–Frobenius theory may be applied (Capasso, 1993). From now on, we suppose L to be irreducible.

Let $\mu_0 = \rho(L - K)$ denote the spectral radius of $L - K$. $A(\mu)$ decides the stability of the no-disease equilibrium: as a matter of fact, the disease free equilibrium is locally asymptotically stable if $\mu > \mu_0$ and unstable if $\mu < \mu_0$.

We define the function

$$h(\mu_0) = \sum_{j=1}^{m} x_j I_{0j} I^*_{0j} \,,$$

where I_0 and I_0^* are the eigenvectors of $H-K$ and $(H-K)^T$ corresponding to the eigenvalue μ_0 and

$$x_i = \frac{\mu_0 + k_i}{\Lambda_i} \left[\frac{\kappa_0(\mu_0+\kappa_i)I_{0i}}{\Lambda_i} - \frac{\sum_{j=1}^{m} \lambda_j I_{0j}(b_0^*\mu_0 + b_0\kappa_j)}{b_0^2} \right] .$$

The following theorem holds (Huang, 1988):

Theorem 7: Let system (30) be such that the matrix (35) is irreducible. Let $\mu_0 = \rho(L-K)$ denote the spectral radius of $L-K$. Let $h(\mu_0)$ defined by (3.3).

If $h(\mu_0) \neq 0$, then μ_0 is a bifurcation point.

If $h(\mu_0) > 0$ $(h(\mu_0) < 0)$, there exist $\epsilon > 0$ and continuously differentiable functions S and I mapping $(\mu_0 - \epsilon, \mu_0]$ $([\mu_0, \mu_0 + \epsilon)) \to R^m$ such that $(S(\mu_0), I(\mu_0)) = \left(\frac{\Lambda}{\mu_0}, 0\right)$ and $(S(\mu), I(\mu))$ is a positive endemic equilibrium of System (30).

Moreover, $(S(\mu), I(\mu))$ is locally asymptotically stable for $\mu \in (\mu_0 - \epsilon, \mu_0)$ it is unstable if $\mu \in (\mu_0, \mu_0 + \epsilon)$.

Hence we have shown the in the case only one trivial equilibrium exists, it is locally asymptotically stable; otherwise this equilibrium is unstable and another equilibrium (endemic) exists. In the next session, we study the uniqueness and the stability of the positive equilibrium.

3.4. *About the uniqueness of the positive endemic equilibrium*

Since $A(\mu)$ is a quasimonotone matrix, from Perron–Frobenius's theory it follows that if $s(A(\mu)) < 0$, then the origin is LAS. Our main goal is to show that if $A(\mu))$ is non negative, then only one positive equilibrium exists (Lin, 1991). We want to underline that from the characterization of a matrix with all eingenvalues with negative real part follows that if $A(\mu)$ is non negative, then $s(A(\mu)) > 0$, i.e. the origin is unstable (Capasso, 1993).

Since our system is not quasimonotone, we introduce a quasimonotone auxiliary system, which has the same number of positive equilibria of (30).

Suppose $\frac{dv_i}{dt} = 0$. It follows that $v_i = \frac{\kappa_i}{\mu} I_i$. Substituting in the second equation of (30), we obtain the following auxiliary system

$$\begin{aligned}\frac{dI_i}{dt} &= \lambda_i \alpha_i \frac{\sum_{i=1}^{n} \lambda_i I_i}{\sum_{h=1}^{n} \lambda_h (\Lambda_h - \mu v_h)} \bar{\beta}(\pi(S, I))(\Lambda_i - (\kappa_i + \mu) I_i) - (\mu + \kappa_i) I_i \\ &= f_i \,. \end{aligned} \tag{35}$$

Proposition 8: *The number of positive endemic equilibria of* (30) *is equal to the number of equilibria of* (35) *in the region*

$$G = \left\{ 0 < I_i < \frac{\Lambda_i}{\kappa_i + \mu}; i = 1, \dots, n \right\} .$$

Proposition 9: *System* (35) *is quasimonotone.*

Proof: From Proposition 4 and the proof of Proposition 6,

$$\frac{\partial f_i}{\partial I_J} = \frac{\lambda_i \alpha_i}{\mu} (\Lambda_i - (\kappa_i + \mu) I_i) \frac{\partial \tilde{\beta}}{\partial \pi} \frac{\partial \pi}{\partial I_J} > 0 \,.$$ □

There is a correspondence between the Jacobian of (30) and the Jacobian of (35), i.e.

Proposition 10: *If* $\begin{pmatrix} -\mu E & K \\ B & A(\mu) \end{pmatrix}$ *is the Jacobian of* (30) *at an equilibrium* $(v^*, I^*)^T$, *then* $-\mu^{-1} BK + A$ *is the Jacobian of* (35) *at the corresponding equilibrium* I^*. *Therefore* $A(\mu)$ *is the Jacobian of* (35) *at the origin.*

For a proof we refer to Lin (1991).

Proposition 11: *Let* I^* *be a positive equilibrium of* (35) *in* G. *If* $A(\mu) I^* > 0$, *then* I^* *is a sink.*

Proof: Our main task is to show that $Df(I^*)I^* < 0$, so that, by applying the theory of quasimonotone matrices, $s(Df(I^*) < 0$, i.e. I^* is a sink (Hirsch, 1982; 1985; Smith, 1988; 1995).

The Jacobian of (35) at I^* is

$$
\begin{aligned}
&Df(I^*)\\
&= \bar{\beta}(I^*)\left[\frac{\sum_{i=1}^n \lambda_i I_i^*}{\sum_h \lambda_h(\Lambda_h - \kappa_h I_h^*)^2}\right.\\
&\quad\times\begin{pmatrix} \lambda_1^2\alpha_1\kappa_1(\Lambda_1-(\mu+\kappa_1)I_1^*) & \cdots & \lambda_1\alpha_1\lambda_n\kappa_n(\Lambda_1-(\mu+\kappa_1)I_1^*)\\ & \vdots & \\ \lambda_n\alpha_n\lambda_1\kappa_1(\Lambda_n-(\mu+\kappa_n)I_n^*) & \cdots & \lambda_n^2\alpha_n\kappa_n(\Lambda_n-(\mu+\kappa_n)I_n^*)\end{pmatrix}\\
&\quad+\frac{1}{\sum_h \lambda_h(\Lambda_h-\kappa_h I_h^*)}\\
&\quad\times\begin{pmatrix} \lambda_1^2\alpha_1(\Lambda_1-(\mu+\kappa_1)I_1^*) & \cdots & \lambda_1\alpha_1\lambda_n(\Lambda_1-(\mu+\kappa_1)I_1^*)\\ & \vdots & \\ \lambda_n\alpha_n\lambda_1(\Lambda_n-(\mu+\kappa_n)I_n^*) & \cdots & \lambda_n^2\alpha_n(\Lambda_n-(\mu+\kappa_n)I_n^*)\end{pmatrix}\\
&\quad\left.-\frac{\sum_i \lambda_i I_i^*}{\sum_h \lambda_h(\Lambda_h-k_h I_h^*)}\begin{pmatrix}\lambda_1\alpha_1(\mu+\kappa_1) & \cdots & 0\\ & \vdots & \\ 0 & \cdots & \lambda_n\alpha_n(\mu+\kappa_n)\end{pmatrix}+\right]\\
&\quad+\pi(I^*)\frac{\partial\bar{\beta}}{\partial\pi}(I^*)\cdot C\,,
\end{aligned}
$$

where C is the sum of the first two matrices in square brackets. It is clear from Proposition (4) that

$$A = \pi(I^*)\frac{\partial\bar{\beta}}{\partial\pi}(I^*)\cdot C < 0\,.$$

Denote $\bar{\beta}(I^*)$ with b_{I^*}. Then the jth component of $Df(I^*)I^*$ is

$$
\begin{aligned}
&A_j - \frac{b_{I^*}\sum_h \lambda_h I_h^*}{\sum_h \lambda_h(\Lambda_h-\kappa_h I_h^*)}\left[\lambda_j\alpha_j\Lambda_j\frac{\sum_h\lambda_h I_h^*}{\sum_h\lambda_h\Lambda_h} - \frac{(\mu+\kappa_j)}{b_{I^*}}I_j^*\right]\\
&\quad\le A_j - \frac{b_{I^*}\sum_h \lambda_h I_h^*}{\sum_h \lambda_h(\Lambda_h-\kappa_h I_h^*)}\left[\lambda_j\alpha_j\Lambda_j\frac{\sum_h\lambda_h I_h^*}{\sum_h\lambda_h\Lambda_h} - \frac{(\mu+\kappa_j)}{b_0}I_j^*\right]\\
&\quad= A_j - \frac{b_{I^*}b_0\sum_h \lambda_h I_h^*}{\sum_h \lambda_h(\Lambda_h-\kappa_h I_h^*)}\left[A(\mu)I^*\right]_j < 0.
\end{aligned}
$$

The last inequality is due to the hypotheses. □

Remember that we have supposed L to be irreducible. Now we are able to provide the main result of the paper.

Theorem 12: If $A(\mu)$ is nonnegative, then system (30) has a unique positive equilibrium.

Proof: Since $A(\mu)$ is non negative, then there is almost one eigenvalue with positive real part. Hence $s(A(\mu)) > 0$, i.e. the trivial steady state is unstable. The quasimonotonicity of System (35) implies the existence of a positive endemic equilibrium (Smith, 1988). We need to show uniqueness. For each positive equilibrium I^* in G, the hypothesis $A(\mu)$ non negative implies $A(\mu)I^* > 0$. Then, for Proposition (11), I^* is a sink. Suppose that there are two positive equilibria I^* and I^{**} ($I^* < I^{**}$), then, since they are sinks, there is an unstable equilibrium $\bar{I}$ such that $I^* < \bar{I} < I^{**}$ (Smith, 1988). That is a contradiction since $\bar{I}$ is a sink, being positive. So System (35) has only one positive equilibrium. Because of Proposition (8), the proof is concluded. □

4. The Multistage Model

We wish to point out that in the model introduced above, it has been assumed not only homogeneous mixing, but also that an infectious individual has always the same infectiousness, independently of the time at which he has got the infection.

Actually this assumption is not realistic. In fact several studies show that for any individual the infectiousness of HIV infected individuals is not constant during the incubation period: it reaches its maximum in the period of primary infection, then there is a long period of low infectivity in the asymptomatic phase and finally it grows again when an infected individual is in the stage leading to AIDS.

In order to simulate a gamma-distributed incubation period, a mathematical model has been developed with m stages of infection, corresponding to the stages related to CD4 lymphocyte counts, used by Longini

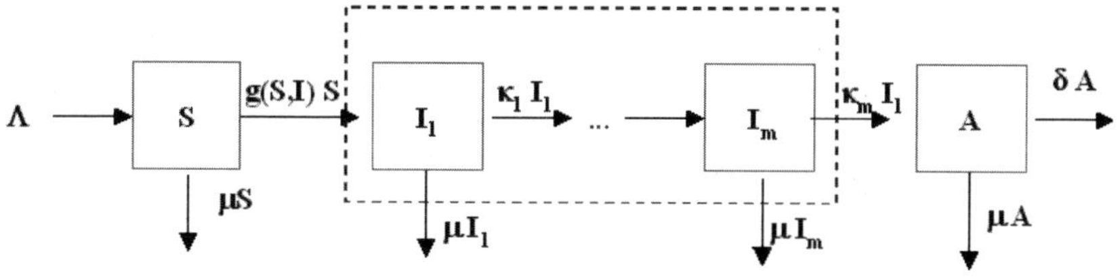

Fig. 3. Scheme for the multistage model.

et al. (1989). The different stages of infectivity are modeled as a cascade of a finite number of stages connected by a linear decay; the decay rates correspond to the parameters of exponentially distributed permanence times in the different stages.

Analogous assumptions were made by Jacquez *et al.* (1988) and Lewis and Grrenhalgh (2001), where anyway the typical force of infection, based on the law of mass action, was used. Here we consider only one population.

Starting from model (1), the I class is splitted into $m \geq 1$ subclasses $I_j, j = 1, \ldots, m$. The model is given by the following system of differential equations:

$$\begin{cases} \dfrac{dS}{dt} = \Lambda - \mu S - g(S, I)S \\ \dfrac{dI_1}{dt} = g(S, I)S - (\kappa_1 + \mu)I_1 \\ \dfrac{dI_j}{dt} = \kappa_{j-1}I_{j-1} - (\kappa_j + \mu)I_j \quad j = 2, \ldots, m \\ \dfrac{dA}{dt} = \kappa_m I_m - (\mu + \delta)A \end{cases} \tag{36}$$

where, again, Λ is the constant recruitment rate of susceptibles in the population, μ is the natural mortality rate specific of the IDU population, δ is the differential mortality rate due to AIDS, κ_j is the transfer rate from the jth to the $(j + 1)$th stage and $g(S, I)$ denotes the force of infection acting on each susceptible drug user.

To obtain an explicit expression for $g(S, I)$ regarding the syringe sharing in a friendship group we make the same assumptions (i), (ii) and (iv) as in the single population model. In order to take into account the variable infectiousness with respect to the different stages, assumption (iii) is modified as follows:

(iii) If a DIE is used by an I_j addict, it becomes infectious with probability α_j. If an infectious DIE is used by an I_j (S) addict, the act of injecting will "rid" the equipment with probability θ_j (θ_0). In other words, when a DIE is used by an infective in the jth stage, it becomes infectious with probability α_j, it becomes uninfectious with probability θ_j and keeps its HIV status with probability $(1 - \alpha_j)(1 - \theta_j)$.

As in the single population model, the force of infection is given by

$$g(S, I_1, \ldots, I_m) = \lambda\alpha\beta(S, I_1, \ldots, I_m)$$

where, again, $\beta(S, I)$ denotes the probability that a drug user encounters an infectious syringe (in this case, $I = (I_1, \dots, I_m)^T$). In order to determine β we can proceed as follows. If the addict is the first user in the group, the syringe is certainly uninfectious; thus

$$P(E|G = k, R = 1) = 0$$

(E denotes the event that the drug user injects with an infectious equipment, G the group size and R the order position in the friendship group).

The second user encounters an infectious DIE only if the first user was infected and has left the syringe infectious; this happens with probability

$$P(E|G = k, R = 2) = \sum_{j=1}^{m} \alpha_j \frac{I_j}{T} =: \pi(S, I) =: \pi$$

$\big(T$ denotes the total "active" population $T = S + \sum I_j\big)$.

The probability that the first couple of users leaves the syringe infectious is

$$P(E|G = k, R = 3) = \sum_{i=0}^{m} \sum_{j=0}^{m} \left[\alpha_j + (1 - \alpha_j)\alpha_i(1 - \theta_j)\right] \frac{I_i}{T} \frac{I_j}{T} = \pi(1 + \tau)$$

where

$$\tau := \tau(S, I) := (1 - \theta_0)\frac{S}{T} + \sum_{j=1}^{m} (1 - \alpha_j)(1 - \theta_j)\frac{I_j}{T}$$

with $I_0 = S$ and $\alpha_0 = 0$.

The general case of the jth user is a straightforward consequence:

$$P(E|G = k, R = j) = \pi(1 + \tau + \dots + \tau^{j-2}) \quad j \geq 2\,.$$

Now, since

$$\begin{aligned}\beta(S, I) = P(E) &= \sum_{k=2}^{\infty} P(E|G = k) P(G = k) \\ &= \sum_{k=2}^{\infty} \frac{\nu^{k-2} e^{-\nu}}{(k - 2)!} \frac{1}{k} \sum_{j=1}^{k} P(E|G = k, R = j)\,,\end{aligned}$$

we are able to provide an explicit expression for the force of infection:

$$g(S, I) = \frac{\lambda\alpha\pi}{1 - \tau}\left\{1 - \frac{\nu - 1 + (1 - \nu\tau)e^{-\nu(1-\tau)}}{\nu^2(1 - \tau)}\right\}.$$

(since λ and α appear always together in the formulae, from now on we shall denote the product $\lambda\alpha$ simply by λ). Again one may show that the force of infection is a decreasing function of S and an increasing function of each I_j, $(j = 1, \dots, m)$.

System (36) always admits a disease-free equilibrium $\left(\frac{\Lambda}{\mu}, 0, \dots, 0\right)$; we make a shift of coordinates $S = \frac{\Lambda}{\mu} - X$ to move such an equilibrium into the origin. With respect to these coordinates, the model is represented by the system

$$
\begin{aligned}
\frac{dX}{dt} &= -\mu X - g(X, I)\left(\frac{\Lambda}{\mu} - X\right) \\
\frac{dI_1}{dt} &= g(X, I)\left(\frac{\Lambda}{\mu} - X\right) - (\kappa_1) + \mu)I_1 \\
\frac{dI_j}{dt} &= \kappa_{j-1} I_{j-1} - (\kappa_j + \mu) I_j \quad j = 2, \dots, m\,.
\end{aligned}
\tag{37}
$$

If we denote by x the vector $(X, I_1, \dots, I_m)^T$, System (37) can be rewritten as

$$
\frac{dx}{dt} = Ax + Q(x)
$$

where

$$
A := \left(\begin{array}{c|c} -\mu & * \\ \hline 0 & A_1 \end{array}\right);
$$

$$
A_1 := \begin{pmatrix}
\lambda b_0 \alpha_1 - (\kappa_1 + \mu) & \lambda b_0 \alpha_2 & \dots & \dots & \lambda b_0 \alpha_m \\
\kappa_1 & -(\kappa_2 + \mu) & 0 & \dots & 0 \\
0 & \kappa_2 & -(\kappa_3 + \mu) & 0 & 0 \\
\dots & \dots & \dots & \dots & \dots \\
\dots & \dots & \dots & \dots & \dots \\
0 & \dots & 0 & \kappa_{m-1} & -(\kappa_m + \mu)
\end{pmatrix};
$$

and

$$
b_0 = \left(\frac{\beta(\pi(x), \tau(x))}{\pi(x)}\right)\bigg|_{x=0};
$$

$$
Q(x) := \begin{pmatrix}
-g(X, I)X - \frac{\lambda b_0 \mu}{\Lambda} \sum \alpha_j I_j + o(x) \\
-g(X, I)X - \frac{\lambda b_0 \mu}{\Lambda} \sum \alpha_j I_j + o(x) \\
0 \\
\vdots \\
0
\end{pmatrix}.
$$

Due to the specific structure of the matrix A, the following theorem holds:

Theorem 13: Let $r := \rho(A_1)$ be the spectral radius of A_1.

(a) If $r < 0$ then $x^* = 0$ is GAS in $\mathbb{R}^n_+$;
(b) If $r > 0$ then x^* is unstable and there exists a unique nontrivial equilibrium.

Proof: (a) A_1 is an irreducible quasi-monotone matrix. Since $Q(x)$ is $o(x)$, x^* is LAS for Perron–Frobenius Theorem. As in Jacquez *et al.* (1988), to show that x^* is GAS, we can use the Lyapunov function

$$V(X, I) := (0, v)^T(X, I) = v^T I$$

where $v \in R^m$ is the eigenvector of A_1 corresponding to r; since $r < 0$, $I \geq 0$, $Q(x) \leq 0$ and $v > 0$, then

$$\dot{V}(X, I) = v^T A_1 I + (0, v)Q(X, I) = rv^T I + (0, v)Q(X, I) \leq 0$$

and the global asymptotical stability follows by Lyapunov theorems.

(b) Since $r > 0$ is a real eigenvalue of A_1, the equilibrium is unstable. For the research of nontrivial equilibria, we use the equivalent system

$$\begin{cases} B(I_1) = 0 \\ S = \dfrac{\Lambda - (\kappa_1 + \mu)I_1}{\mu} \\ I_j = \dfrac{\kappa_1\kappa_2 \cdots \kappa_{j-1}}{(\kappa_2 + \mu) \cdots (\kappa_j + \mu)} I_1 := \bar{\kappa}_j I_1 \quad j = 2, \ldots, m \end{cases}$$

where

$$B(I_1) : = \lambda \sum_{j=1}^{m} \alpha_j \bar{\kappa}_j \frac{\Lambda - (\kappa_1 + \mu)I_1}{\Lambda - (\kappa_1 + \mu)I_1 + \mu\tilde{\kappa}I_1} \times \frac{1}{1 - \tau}\left\{1 - \frac{\nu - 1 + (1 - \nu\tau)e^{-\nu(1-\tau)}}{\nu^2(1 - \tau)}\right\} - (\kappa_1 + \mu)$$

and

$$\tilde{\kappa} := \sum_{j=1}^{m} \bar{\kappa}_j .$$

The problem now is to find the possible solutions of the equation $B(I_1) = 0$ in the set $\{I_1 \in \mathbb{R} | \ 0 < I_1 < \Lambda/(\kappa_1 + \mu)\}$. By direct calculations, it is possible to prove that B is a decreasing function of I_1; in fact

$$\lim_{I_1 \to 0} B(I_1) = \lambda b_0 \sum_j \alpha_j \bar{\kappa}_j - (\kappa_1 + \mu) = r > 0$$

$$\lim_{I_1 \to \frac{\Lambda}{\kappa_1 + \mu}} B(I_1) = -(\kappa_1 + \mu) < 0\,.$$

So, there exists a unique I_1, included in the interval $(0, \Lambda/(\kappa_1 + \mu))$, at which the function B vanishes. □

The proof of the Theorem 13 allows us to determine the expression of the threshold parameter

$$R_0 := \lambda b_0 \bar{\alpha}$$

where

$$\bar{\alpha} = \frac{\sum_{i=1}^{m} \prod_{j<i} \kappa_j \prod_{j>i} (\kappa_j + \mu)}{\prod_{j=1}^{m} (\kappa_j + \mu)} = \sum_{i=1}^{m} \alpha_i \frac{\kappa_1 \cdots \kappa_{i-1}}{(\kappa_1 + \mu) \cdots (\kappa_i + \mu)}\,.$$

R_0 is the basic reproduction number. From the proof it appears that $R_0 > 1$ if and only if $r > 0$ and $R_0 < 1$ if and only if $r < 0$.

In Fig. 4, numerical simulations of the trends of susceptibles and infectives in the case $R_0 > 1$ are shown. It is clear from Fig. 4(b) that the system tends to an endemic equilibrium. The higher curve represent the

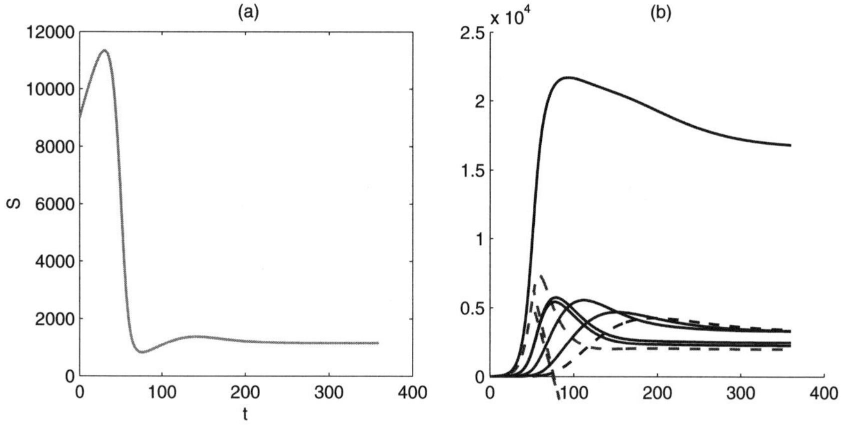

Fig. 4. Susceptibles (a) and infectives (b) trends in the case of multiple stages of infectivity.

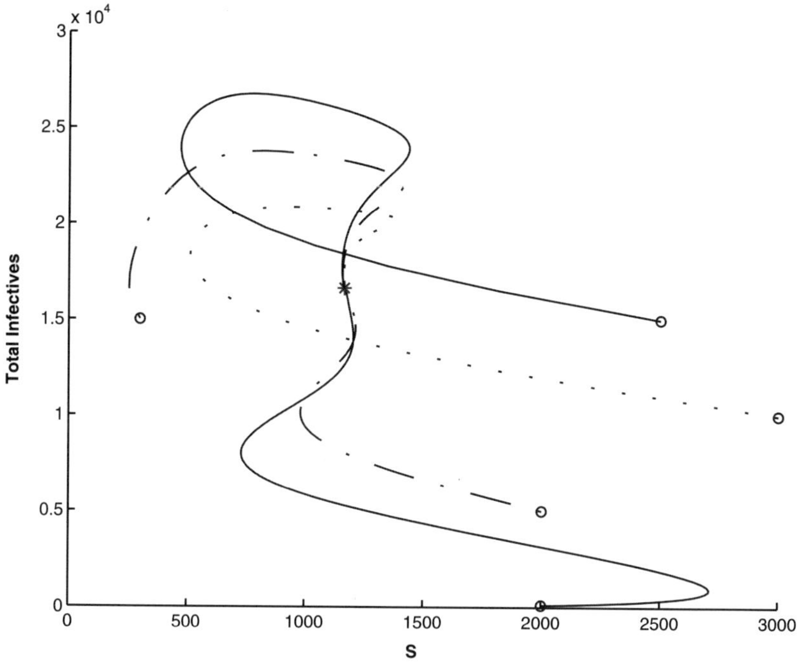

Fig. 5. Phase portrait for the susceptibles and total infectives in the case of $R_0 > 1$.

total number of infectives, the dashed one the infectives at the first stages. Furthermore we may suppose that the equilibrium is asymptotically stable, as highlighted by Fig. 5.

5. Multipopulation models with multiple stages of infection

A further extension of our model takes into account both heterogeneities and the multistage structure of the infectivity of HIV infected individuals a multipopulation-multiple stage model. For the ith subpopulation, S_i denotes the number of susceptibles, I_{ij} the number of infectives in the jth stage, I_i the total number of infectives, A_i the number of individuals with full blown AIDS and $T_i = S_i + I_i$ is the number of active individuals.

In such a case, the force of infection acting on each susceptible of the ith group will be of the form $\lambda_i \alpha_i \beta_i(S, I)$ where $S = (S_1, \ldots, S_n)^T$ and $I = (I_1, \ldots, I_n)^T$.

So, model (36) can be extended to the multipopulation case as follows

$$\begin{aligned}
\frac{dS_i}{dt} &= \Lambda_i - \mu_i S_i - \lambda_i \gamma_i \beta_i(S,I) S_i \\
\frac{dI_{i1}}{dt} &= \lambda_i \gamma_i \beta_i(S,I) S_i - (\kappa_{i1} + \mu_i) I_{i1} \quad i = 1,\dots,n \\
\frac{dI_{ij}}{dt} &= \kappa_{i(j-1)} I_{i(j-1)} - (\kappa_{ij} + \mu_i) I_{ij} \quad j = 2,\dots,m \\
\frac{dA_i}{dt} &= \kappa_{im} I_{im} - \mu_i A_i - \delta_i A_i.
\end{aligned} \tag{38}$$

By the same assumptions made in Sections 3 and 4, we obtain the following probability of exposure to HIV:

$$\beta_i(S,I) = \tilde{\beta}(\pi_i(S,I), \tau_i(S,I))$$

where

$$\begin{aligned}
\tilde{\beta}(\pi,\tau) &= \frac{\pi}{1-\tau} \sum_{k=2}^{\infty} \left[1 - \frac{1-\tau^k}{k(1-\tau)}\right] \frac{e^{-\nu}\nu^{k-2}}{(k-2)!} \\
&= \frac{\pi}{1-\tau} \left\{1 - \frac{\nu - 1 + (1-\nu\tau)e^{-\nu(1-\tau)}}{\nu^2(1-\tau)}\right\};
\end{aligned}$$

and

$$\pi_i = \sum_{j=1}^{n} \rho_{ij} \hat{\pi}_j \qquad \tau_i = \sum_{j=1}^{n} \rho_{ij} \hat{\tau}_j$$

with

$$\hat{\pi}_j = \sum_{r=0}^{m} \alpha_r \frac{I_{jr}}{T_j} \qquad \hat{\tau}_j = \sum_{r=0}^{m} (1-\alpha_r)\,(1-\theta_r) \frac{I_{jr}}{T_j}$$

($I_{j0} = S$).

Therefore, the force of infection acting on each susceptible in the ith subpopulation is

$$g_i(S,I) = \lambda_i \alpha_i \tilde{\beta}(\pi_i(S,I), \tau_i(S,I)).$$

For the expression of the mixing matrix we use the explicit formula given by Busenberg and Castillo–Chavez in their "Representation Theorem", as in Section 3.

If we sum up the first $m+1$ equations in System (38) for any $i = 1,\dots,n$, we get the evolution equation for the total active population in the ith group

$$\frac{d}{dt} T_i = -\mu_i T_i + \Lambda_i - \kappa_{im} I_{im} \quad i = 1,\dots,n.$$

System (38) satisfies the following properties:

(a) For $\mu_i > 0$ if $T_i(0) \leq \frac{\Lambda_i}{\mu_i}$, then $T_i(t) \leq \frac{\Lambda_i}{\mu_i}$, for all $t \geq 0$. Furthermore, if $\kappa_i = \max_j \kappa_{ij}$, the set

$$\Omega := \Bigg\{ (S_1, \ldots, S_n, I_{11}, \ldots, I_{nm}) \in \mathbb{R}_+^{n+nm} \Bigg| \\ \frac{\Lambda_i}{\mu_i + \kappa_i} \leq T_i \leq \frac{\Lambda_i}{\mu_i} \quad i = 1, \ldots, n \Bigg\}$$

is a compact, convex and invariant set for System (38).

(b) For $\mu_i > 0$, System (38) admits the disease-free equilibrium

$$S_i^* = \frac{\Lambda_i}{\mu_i} \quad I_{i1}^* = \cdots = I_{im}^* = A_i^* = 0 \qquad i = 1, \ldots, n\,.$$

In conclusion we may underly how the knowledge of the force of infection directly related to some individual social parameter it is crucial for the diffusion of HIV epidemic among drug addict. Indeed, in Fig. 6 the trend of the HIV incidence rate (solid line) in Italy between 1984 and 2001: the typical behavior of the spread of an epidemic disease is shown; from a comparison with simulation results (cf. Fig. 7) we see how a qualitative similar trend is obtained by suitable reductions in the behavioral parameters (rate of syringe sharing, size of friendship groups and rate of recruitment of new drug users). Indeed, the solid line show the trajectory of the force of

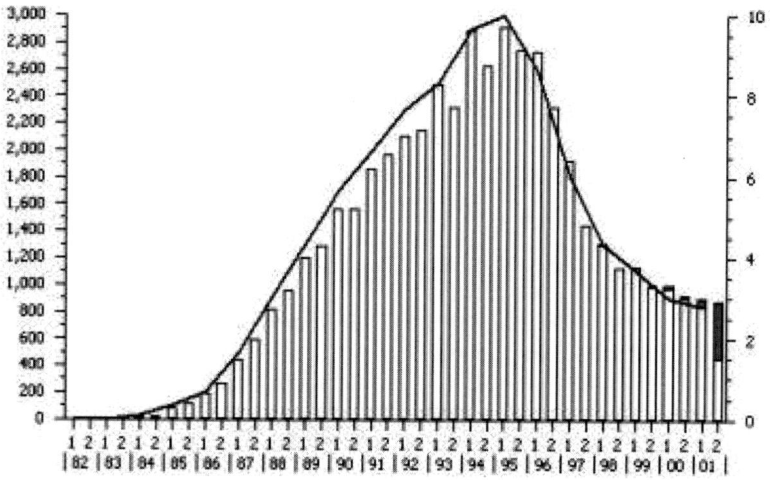

Fig. 6. Incidence in Italy till 2001 (solid line).

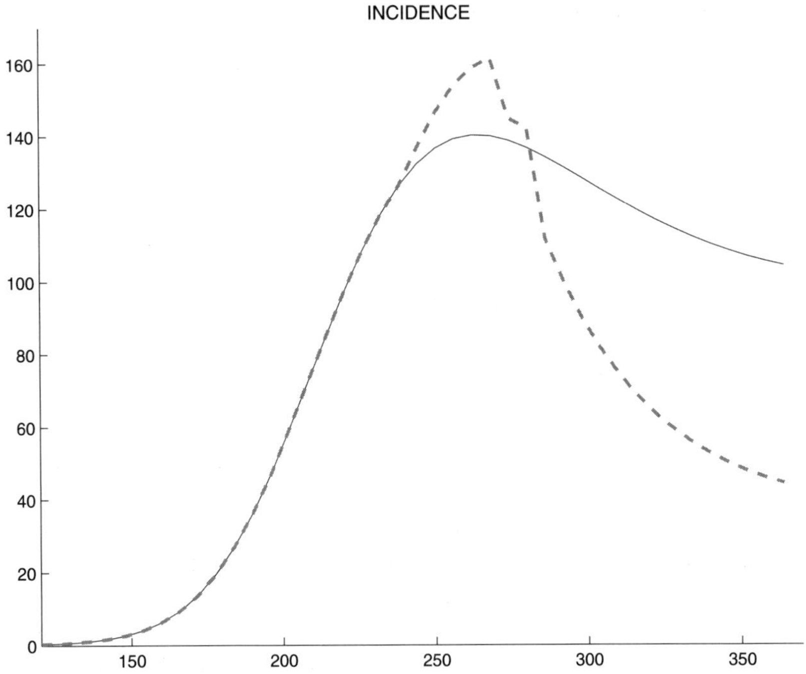

Fig. 7. Simulated incidence with (dashed line) and without (solid line) behavioral changes.

infection without any change in the behavioral parameters, while the dash line show the same trajectory after behavioral changes. Actually, in Italy, we have seen a minor attitude of sharing syringes among drug addicts in the last decade, due above all to the smaller size of the friendship groups.

References

Blythe SP, Castillo–Chavez C (1989) Like-with-like preference and sexual mixing models, *Math Biosci* **96**: 221–238.

Britton NF (1986) *Reaction-Diffusion Equations and their Applications to Biology*, Academic Press, London.

Busenberg S, Castillo–Chavez C (1989) Interaction, pair formation and force of infection terms in sexually transmitted diseases, in Castillo–Chavez C (ed.) *Mathematical and Statistical Approaches to AIDS Epidemiology*, Lecture Notes in Biomathematics, Vol. 83, Springer–Verlag, Heidelberg.

Capasso V (1993) *Mathematical Structures of Epidemic Systems*, Lecture Notes in Biomathematics, Vol. 97, Springer–Verlag, Heidelberg.

Capasso V, Villa M (1994) A multigroup model for HIV transmission among injecting drug users, in Martelli M *et al.* (eds.) *Differential Equations and Applications to Biology and Industry*, pp. 49–56, World Scientific, Singapore.

Capasso V, Sicurello F, Villa M (1995) Mathematical models for HIV transmission in groups of injecting drug users via shared drug equipment, *J Biol Sys* **3**: 747–758.

Capasso V, Villa M, Morale D, Di Somma M, Sicurello F, Nicolosi A (1998) Multistage models of HIV trasmission among injecting drug users via shared drug injection equipment, in Arino O, Axelrod A and Kimmel M (eds.) *Advances in Mathematical Population Dynamics: Molecules, Cells and Man*, pp. 511–528, World Scientific Publishing Company.

Castillo–Chavez C, Cooke KL, Huang W, Levin SA (1989) On the role of long incubation periods in dynamics of acquired immunodeficiency syndrome (AIDS). Part 1: Single population models, *J Math Biol*, **27**: 373–398.

Castillo–Chavez C, Cooke KL, Huang W, Levin SA (1989) On the role of long incubation periods in dynamics of acquired immunodeficiency syndrome (AIDS). Part 2: Multiple group models, in Castillo–Chavez C (ed.) *Mathematical and Statistical Approaches to AIDS Epidemiology*, Lecture Notes in Biomathematics, Vol. 83, Springer–Verlag, Heidelberg.

Hethcote HW, van Ark P (1987) Epidemiological models for hetrogeneous populations: proportionate mixing, parameter estimation and immunization programs, *Math Biosci* **84**: 85–118.

Hirsch MW (1982) Systems of differential equations which are competitive or cooperative, I: Limit sets, *SIAM J Math Anal* **13**: 167–179.

Hirsch MW (1985) Systems of differential equations which are competitive or cooperative, II: Convergence almost everywhere, *SIAM J Math Anal* **16**: 423–439.

Huang W (1990) *Studies in Differential Equations and Applications*, Ph.D. Dissertation, The Claremont Graduate School, Claremont (CA).

Jacquez JA, Simon CP, Koopman JS, Sattenspiel L, Parry T (1988) Modeling and analyzing HIV transmission: the effect of contact patterns, *Math Biosci* **92**: 119–199.

Kaplan EH (1989) Needles that kill: modeling Human Immunodeficiency Virus transmission via shared drug injection equipment in shooting galleries. *Rev Inf Diseases* **11**: 289–298.

Lewis F, Greenhalgh D (2001) Three stages AIDS incubation period: a worst case scenario using addict needle interaction assumptiond, *Math Biosci* **169**: 53–87.

Lin X (1991) On the uniqueness of endemic equilibria of an HIV/AIDS transmission model for a heterogeneous population, *J Math Biol* **29**: 779–790.

Lin X, Hethcote HW, Van den Driessche P (1993) An epidemiological model for HIV/AIDS with proportional recruitment, *Math Biosci* **118**: 181–195.

Longini IM, Jr., Clark WS, Haber M, Horsburgh CR, Jr. (1989) The stages of HIV infection: waiting times and infectious contact patterns, in Castillo–Chavez C (ed.) *Mathematical and Statistical Approaches to AIDS Epidemiology*, Lecture Notes in Biomathematics, Vol. 83, Springer–Verlag, Heidelberg.

Notices of the Italian National Institute of Health, http://www.ministerosalute.it/aids/statistiche/sezStatistiche.jsp.

Smith HL (1988) Systems of ordinary differential equations which generate an order preserving flow. A survey of results, *SIAM Rev* **30**: 87–113.

Smith HL (1995) Monotone Dynamical Systems. An introduction to the theory of competitive and cooperative systems, in *Mathematical Surveys and Monographs*, Vol. 41, Amer. Math. Soc. Providence, Rhode Island.

Waltman PA (1986) *A Second Course in Elementary Differential Equations*, Academic Press, Orlando.

CHAPTER 2

ESTIMATION OF HIV INFECTION AND SEROCONVERSION PROBABILITIES IN IDU AND NON-IDU POPULATIONS BY STATE SPACE MODELS

Wai-Yuan Tan, Li-Jun Zhang and Lih-Yuan Deng

To estimate the probabilities of HIV infection and HIV seroconversion and to compare HIV seroconversions from different populations, in this paper we have developed some statistical models and state space models for HIV infection and seroconversion. By combining these models with the multi-level Gibbs sampling procedures, in this paper we have developed some efficient methods to estimate simultaneously these probabilities and the state variables as well as other unknown parameters. By using the complete likelihood function, we have also developed a generalized likelihood ratio test for comparing several HIV seroconversion distributions. As an illustration, we have applied the models and the methods to some data generated by the cooperative study on HIV under IDU and cocaine crack users by the National Institute of Drug Abuse/NIH. Our results show that there are significant differences in HIV seroconversion and HIV infection between populations of IDU, homosexuals and individuals with both IDU and homosexual behavior. For homosexuals, IDU and homosexuals with IDU, the probability density functions of times to HIV infection and HIV seroconversion are bi-model curves with two peaks and with heavy weights on the right. The average window period is about 2.75 months. Also, there are significant differences between the death and retirement rates of S (susceptible) people and I (infected but not seroconverted) people in all populations.

Keywords: HIV seroconversion, IDU, homosexuals, likelihood ratio test, probability of HIV infection, statistical model, stochastic difference equation, stochastic system model.

1. Introduction

In the US, the major avenues for spreading HIV are IV drug use and homosexual and bisexual behavior. For developing strategies to control the spread of HIV under these conditions, it is imperative to estimate the probabilities of HIV infection and to assess effects of risk variables which affect HIV infection under these conditions. Based on HIV seroconversion data, in this paper we proceed to develop efficient statistical procedures to estimate these probabilities. To achieve these objectives, we will first develop a state space model for HIV seroconversion with the observation model being based on information from observed seroconversion status. By combining this state space model with the multi-level Gibbs sampling method, we will then develop efficient statistical procedures to estimate the probability of HIV infection as well as the state variables and HIV infection prevalence under various conditions including IV drug use.

In Section 2, we illustrate how to develop a stochastic model for the spread of HIV infection and the development of HIV antibodies. In Section 3, we will describe the available data on seroconversion and illustrate how to develop a statistical model for HIV seroconversion; we will derive the complete likelihood function. In Section 4, by combining the stochastic model in Section 2 with the statistical model in Section 3, we will illustrate how to develop a state space model for HIV seroconversion. In this section we will also illustrate how to develop efficient statistical procedures to estimate the probabilities of HIV infection, the unknown parameters, the state variables and the prevalence of HIV infection and HIV seroconversion. As an application of the model and the method developed in previous sections, in Section 5 we will apply the model and the method to some data sets on seroconversion under IV drug use and homosexuals generated by the NIDA Cooperative Agreement Study. We will compare the HIV epidemic patterns between IV drug use and homosexuals. Finally in Section 6, we will discuss the model and the methods and draw some conclusions.

2. A Stochastic Model for HIV Infection and HIV Seroconversion

To develop a stochastic model for the spread of HIV and for HIV seroconversion, consider a large closed population at risk for HIV infection. For developing a stochastic model for the development of HIV antibodies, we will consider three types of people in the population: Susceptible (S), infected but not seroconverted (I), and seroconverted (C). An S person is a person who is susceptible to HIV infection but has not yet contracted HIV virus; an I person is a person who has already contracted the HIV virus but has not yet developed HIV antibodies in the blood; and a C person is a person who has contracted the HIV virus and has developed HIV antibodies in the blood (this is called HIV seroconversion). Thus barring false negative, C people will always be tested positive by ELISA test; similarly, barring false positive I people are always tested negative by ELISA. The transitions are from S to I and from I to C (that is, $S \rightarrow I \rightarrow C$).

Let $\underset{\sim}{X}(t) = \{S(t), I(u,t), u = 0, 1, \ldots, t, C(t)\}$ denote the numbers of S people, I(u) people and new seroconverters at the tth month respectively. Then we are entertaining a high dimensional stochastic process. This is basically a Markov chain with discrete time $\{t = 0, 1, \ldots, \infty\}$. Because the dimension of states increases with time t, the traditional approach is very complicated to be of much use. In this paper we will thus propose an alternative equivalent approach through stochastic difference equations.

2.1. *Stochastic equations for the state variables*

To derive stochastic equations for the state variables, consider the transition during the jth month. Then, $\underset{\sim}{X}(t+1)$ derives from $\underset{\sim}{X}(t)$ through HIV infection of S people and the transition from I people to C people during the tth month stochastically. These stochastic transitions are characterized by the following stochastic variables:

$D_S(t)$ = Number of deaths of S people during the tth month;
$D_I(u,t)$ = Number of deaths of $I(u)$ people during the tth month, $u = 0, \ldots, t$;
$F_S(t)$ = Number of $S \rightarrow I$ transition (i.e. HIV infection of S people) during tth month;
$F_I(u,t)$ = Number of transition $I(u) \rightarrow C$ (i.e. seroconversion of I(u) people) during the tth month, $u = 0, \ldots, t$.

Let $\mu_S(t)$ and $\mu_I(u,t)$ denote the probability of death of S people and $I(u)$ people at the tth month respectively. Let $p_S(t)$ denote the probability that S people contract HIV viruses at the tth month and $\gamma(u,t) = \gamma(u)$ the probability that $I(u)$ people develop HIV antibodies at the tth month, where an $I(u)$ person is an I person who has contracted HIV viruses for a period of u months. As shown in Tan, Lee and Tang (1995), one may practically assume these probabilities as non-random deterministic functions. Thus, as illustrated in Tan (2000, Chapters 3, 6), the conditional probability distribution of $\{F_S(t), D_S(t)\}$ given $S(t)$ is multinomial with parameters $\{S(t); p_S(t), \mu_S(t)\}$, i.e.

$$\{F_S(t), D_S(t)\}|S(t) \sim ML\{S(t); p_S(t), \mu_S(t)\}\,.$$

Similarly, $\{F_I(u,t), D_I(u,t)\}|I(u,t) \sim ML\{I(u,t); \gamma(u,t), \mu_I(u,t)\}$, independently of $\{F_S(t), D_S(t)\}$. Hence, the conditional expected numbers of $\{F_S(t), D_S(t)\}$ are $E[F_S(t)|S(t)] = S(t)p_S(t)$ and $E[D_S(t)|S(t)] = S(t)\mu_S(t)$. Similarly, $E[F_I(u,t)|I(u,t)] = I(u,t)\gamma(u,t)$ and $E[D_I(u,t)|I(u,t)] = I(u,t)\mu_I(u,t)$. Define the random noises $\underset{\sim}{\epsilon}(t) = \{\epsilon_S(t), \epsilon_I(u,t), u = 0, 1, \ldots, t, \epsilon_C(t)\}$ by:

$$\begin{aligned}
\epsilon_S(t+1) &= -[F_S(t) - S(t)p_S(t)] - [D_S(t) - S(t)\mu_S(t)]\,,\\
\epsilon_I(0,t+1) &= [F_S(t) - S(t)p_S(t)]\,,\\
\epsilon_I(u+1,t+1) &= -[F_I(u,t) - I(u,t)\gamma(u,t)] - [D_I(u,t) - I(u,t)\mu_I(u,t)]\,,\\
\epsilon_C(t+1) &= \sum_{u=0}^{t}[F_I(u,t) - I(u,t)\gamma(u,t)]\,.
\end{aligned}$$

When there are no immigration and recruitment, by the conservation law we have the following stochastic equations for the state variables $\{S(t), I(u,t), C(t)\}$:

$$\begin{aligned}
S(t+1) &= S(t) - F_S(t) - D_S(t)\\
&= S(t)\{1 - p_S(t) - \mu_S(t)\} + \epsilon_S(t+1)\,, && (1)\\
I(0,t+1) &= F_S(t) = S(t)p_S(t) + \epsilon_I(0,t+1)\,, && (2)\\
I(u+1,t+1) &= I(u,t) - F_I(u,t) - D_I(u,t)\\
&= I(u,t)\{1 - \gamma(u,t) - \mu_I(u,t)\} + \epsilon_I(u+1,t+1)\,, && (3)\\
C(t+1) &= \sum_{u=0}^{t} F_I(u,t) = \sum_{u=0}^{t} I(u,t)\gamma(u,t) + \epsilon_C(t+1)\,. && (4)
\end{aligned}$$

Let $F(t+1,t)$ be a $(t+4) \times (t+3)$ matrix defined by:

$$F(t+1,t) = \begin{bmatrix}
\phi_S(t) & 0 & 0 & \cdots & 0 & 0\\
p_S(i,t) & 0 & 0 & \cdots & 0 & 0\\
0 & \phi_I(0,0) & 0 & \cdots & 0 & 0\\
0 & 0 & \phi_I(1,t) & \cdots & 0 & 0\\
\vdots & \vdots & \vdots & \ddots & \vdots & \vdots\\
0 & 0 & 0 & \cdots & \phi_I(t,t) & 0\\
0 & \gamma(0,t) & \gamma(1,t) & \cdots & \gamma(k,t) & 0
\end{bmatrix},$$

where $\phi_S(t) = 1 - p_S(t) - \mu_S(t)$ and $\phi_I(i,t) = 1 - \gamma(i,t) - \mu_I(i,t)$

Then, in matrix notation, equations (1)–(4) can be expressed as

$$\underset{\sim}{X}(t+1) = F(t+1,t)\underset{\sim}{X}(t) + \underset{\sim}{\epsilon}(t+1)\,. \tag{5}$$

It can easily be shown that the random noises $\underset{\sim}{\epsilon}(t)$ in equations (1)–(5) have expectation zero. The variances and covariances of these random noises given $\underset{\sim}{X}(t)$ are easily obtained as

$$\begin{aligned}
&Cov\{\epsilon_S(s+1), \epsilon_S(t+1)|S(t), I(u,t)\}\\
&\quad = \delta_{ts} ES(t)\{p_S(t)[1-p_S(t)] + \mu_S(t)[1-\mu_S(t)] - 2p_S(t)\mu_S(t)\}\,,\\
&Cov\{\epsilon_S(s+1), \epsilon_I(u+1,t+1)|S(t), I(u,t)\}\\
&\quad = -\delta_{0u}\delta_{ts} E\{S(t)p_S(t)[1-p_S(t)]\}\,,\\
&Cov\{\epsilon_S(s+1), \epsilon_C(t+1)|S(t), I(u,t)\} = 0\,;\\
&Cov\{\epsilon_I(u+1,s+1), \epsilon_I(v+1,t+1)|S(t), I(u,t)\}\\
&\quad = \delta_{uv}\delta_{ts} EI(u,t)\{\gamma(u,t)[1-\gamma(u,t)] + \mu_I(u,t)[1-\mu_I(u,t)]\\
&\qquad - 2\gamma(u,t)\mu_I(u,t)\}\,,\\
&Cov\{\epsilon_I(u+1,s+1), \epsilon_C(t+1)|S(t), I(u,t)\}\\
&\quad = -\delta_{ts} E\{I(u,t)\gamma(u,t)[1-\gamma(u,t)]\}\,,\\
&Cov\{\epsilon_C(s+1), \epsilon_C(t+1)|S(t), I(u,t)\}\\
&\quad = \delta_{ts} \sum_{u=0}^{t} E\{I(u,t)\gamma(u,t)[1-\gamma(u,t)]\}\,.
\end{aligned}$$

By using the formula $Cov(X,Y) = E\{Cov[(X,Y)|Z]\} + Cov[E(X|Z), E(Y|Z)]$, it can be shown easily that the random noises $\varepsilon_j(t)$ are uncorrelated with the state variables $\{S(t), I(u,t), u = 0,1,\ldots,t\}$. Since the random noises $\varepsilon_j(t)$ are random noises associated with the random transitions during the interval $[t, t+1)$, one may also assume that the random noises $\varepsilon_j(t)$ are uncorrelated with the random noises $\varepsilon_l(\tau)$ for all j and l if $t \neq \tau$.

2.2. *The expected numbers of the state variables*

By using the stochastic equations given by equations (1)–(4), one may readily generate Monte Carlo data to study the probability behavior of the process. Also, by using these stochastic equations, one may derive equations for the mean numbers, the variances and covariances between the state variables as well as other higher cumulants. In particular, the equations for the mean numbers $\{u_S(t) = E[S(t)], u_I(u,t) = E[I(u,t)], u = 0, 1, \ldots, t, u_C(t) = E[C(t)]\}$ are given by:

$$u_S(t+1) = u_S(t)[1 - p_S(t) - \mu_S(t)]\,, \tag{6}$$

$$u_I(0, t+1) = u_S(t)p_S(t)\,, \tag{7}$$

$$u_I(u+1, t+1) = u_I(u,t)[1 - \gamma(u,t) - \mu_I(u,t)]\,, \quad u = 0, 1, \ldots, t\,, \tag{8}$$

$$u_C(t+1) = \sum_{u=0}^{t} u_I(u,t)\gamma(u,t)\,. \tag{9}$$

Denote by $\underset{\sim}{u}(t) = \{u_S(t), u_I(u,t), u = 0, 1, \ldots, t, u_C(t)\}'$. Then the solution of equations (6)–(9) is

$$\underset{\sim}{u}(t+1) = \prod_{j=0}^{t} F(t+1-j, t-j)\underset{\sim}{u}(0), \quad \text{where } \underset{\sim}{u}(0) = \{S(0), I(0,0)\}'.$$

Or, equivalently,

$$u_S(t+1) = u_S(0)\prod_{j=0}^{t}[1 - p_S(j) - \mu_S(j)]\,, \tag{10}$$

$$u_I(0, t+1) = u_S(0)\left\{\prod_{j=0}^{t-1}[1 - p_S(j) - \mu_S(j)]\right\}p_S(t)\,, \tag{11}$$

$$\begin{aligned} u_I(u+1, t+1) &= u_I(0, t-u)\prod_{j=0}^{u}[1 - \gamma(j, t-u+j) - \mu_I(j, t-u+j)] \\ &= u_S(0)\left\{\prod_{j=0}^{t-u-1}[1 - p_S(j) - \mu_S(j)]\right\}p_S(t-u-1) \\ &\quad \times \prod_{j=0}^{u}[1 - \gamma(j, t-u+j) - \mu_I(j, t-u+j)]\,, \\ u &= 0, 1, \ldots, t\,, \end{aligned} \tag{12}$$

$$u_C(t+1) = \sum_{u=0}^{t} u_I(u,t)\gamma(u,t)\,. \tag{13}$$

In the above equations, notice that the equations are the same equations for defining the deterministic model which assume the $\{S(t), I(u,t), j = 0, 1, \ldots, t, C(t)\}$ as deterministic functions of time, ignoring completely the randomness of these variables. By using these equations, one may derive the least square estimators of the parameters; see Section 4. Some examples of applications are given in Tan and Wu (1998).

2.3. *The probability distribution of the state variables*

Denote by $\boldsymbol{X} = \{\underset{\sim}{X}(1), \ldots, \underset{\sim}{X}(t_M)\}$, where t_M is the last time point (Dec. 1998 in the data). Using the probability distribution given in Section 2.1 and the stochastic equations (1)–(4), the conditional probability density function $P\{\underset{\sim}{X}(t+1)|\underset{\sim}{X}(t)\}$ of $\underset{\sim}{X}(t+1)$ given $\underset{\sim}{X}(t)$ can readily be derived as:

$$
\begin{aligned}
&P\{\underset{\sim}{X}(t+1)|\underset{\sim}{X}(t)\} \\
&\quad = \binom{S(t)}{I(0,t+1)}[p_S(t)]^{I(0,t+1)}[1-p_S(t)]^{S(t)-I(0,t+1)} g_S(t) \\
&\qquad \times \sum_{\{\sum_{r=0}^{t} j_r = C(t+1)\}} \prod_{r=0}^{t} \left\{ \binom{I(r,t)}{j_r} [\gamma(r,t)]^{j_r}[1-\gamma(r,t)]^{I(r,t)-j_r} g_I(r,t) \right\}
\end{aligned}
$$

where $\sum_{\{\sum_{r=0}^{t} j_r = C(t+1)\}}$ denotes summation over all nonnegative integers $(j_r, r = 0, \ldots, t)$ satisfying the conditions $\{\sum_{r=0}^{t} j_r = C(t+1), 0 \le j_r \le I(r,t)\}$;

$$
g_S(t) = \binom{S(t)-I(0,t+1)}{a_1(t)} \left(\frac{\mu_S(t)}{1-p(S,t)}\right)^{a_1(t)} \left(1 - \frac{\mu_S(t)}{1-p(S,t)}\right)^{a_2(t)}
$$

with $a_1(t) = Max(0, S(t)-S(t+1)-I(0,t+1))$ and $a_2(t) = Min(S(t+1), S(t)-I(0,t+1))$ and where

$$
g_I(r,t) = \binom{I(r,t)-j_r}{b_1(r,t)} \left(\frac{\mu(r,t)}{1-\gamma(r,t)}\right)^{b_1(r,t)} \left(1 - \frac{\mu(r,t)}{1-\gamma(r,t)}\right)^{b_2(r,t)}
$$

with $b_1(r,t) = Max(0, I(r,t) - I(r+1,t+1) - j_r)$ and $b_2(r,t) = Min(I(r+1,t+1), I(r,t)-j_r)$.

From this, one may readily derive the conditional probability density of $\boldsymbol{X}$ given $\underset{\sim}{X}(0)$ as

$$
P\{\boldsymbol{X}|\underset{\sim}{X}(0)\} = \prod_{t=0}^{t_M-1} P\{\underset{\sim}{X}(t+1)|\underset{\sim}{X}(t)\}. \tag{14}
$$

It is reasonable to assume $\mu_S(t) = \mu_S, \mu_I(u,t) = \mu_I(u) = \mu_I$. Also, because the time between HIV infection and HIV seroconversion is very short, one may assume $\gamma(u,t) = \gamma$. Thus, the parameters in the above distribution are $\Theta = \{\Theta_1 = [p_S(t), t = 1, \ldots, t_M, \gamma], \Theta_2 = [\mu_S(t) = \mu_S, \mu_I(u,t) = \mu_I]\}$. Notice that some of the parameters are time-dependent and are numerous; but these problems can be handled by penalized likelihood or smoothing and additional information from the stochastic system model and prior information, see Tan (2000, Chapter 4).

3. Statistical Models and Data for HIV Seroconversion

Statistical models for HIV seroconversion are based on times to HIV seroconversion for S people in the population. This has been referred to as time to event models by Tan (2000, Chapter 4). The data for these models are usually seroconversion status of individuals in some random samples from the population. In this section, we will describe some statistical models and some data to make inferences on HIV seroconversion.

3.1. *The time to event models for HIV seroconversion*

To describe the statistical model, let $t_0 = 0$ be the time the HIV epidemic starts. That is, at time 0 all people in the population are S people who do not carry the HIV virus; but to start the HIV epidemic, some HIV virus have been introduced into the population. Let T_C denote the time from the S state to the onset of seroconversion, T_I the time from the S state to the onset of the I state and T_W the time from the initiation of the I state to the onset of seroconversion. Then $T_C = T_I + T_W$. If T_I and T_W are independently distributed of each other, then T_C is the convolution of T_I and T_W.

To characterize the probability distributions for these random times, let $h(j)$ $(j \geq 0)$ and $f_I(i)$ $(i \geq 0)$ denote the probability density functions (pdf) of T_C and T_I respectively and $g(i,j) = g(j-i)$ the pdf of T_W given the initiation of the I stage at time i. Denoted by $\lambda(j)$, the probability of seroconversion at the jth month and let $S_c(j)$ denote the cumulative distribution function (cdf) of seroconversion. Then

$$f_I(i) = p_S(i) \prod_{r=0}^{i-1} [1 - p_S(r)], \quad i = 0, \ldots, \infty\,;$$

$$g(j) = \gamma(j) \prod_{r=1}^{j-1} [1 - \gamma(r)] = (1-\gamma)^{j-1}\gamma, \quad j = 0, \ldots, \infty\,;$$

$$h(j) = S_C(j-1)\lambda(j), \quad j = 1, \ldots, \infty\,,$$

where $S_C(j) = \prod_{r=1}^{j}[1 - \lambda(r)]$ is the survival function of HIV sero-conversion.

Under the assumption that T_I and T_W are independently distributed of each other,

$$h(t) = \sum_{j=0}^{t} f_I(j)g(j,t)\,. \tag{15}$$

Equation (15) is the basic formulation to link HIV infection and transition from the I state to the C state with HIV seroconversion. Obviously, the parameters $\lambda(j)$ are uniquely determined by the parameters $p_S(i)$ and $\gamma(r) = \gamma$. Notice that with $\gamma(r) = \gamma$, the probability density $g(i,j) = g(j-i)$ of the window period is $g(j) = (1-\gamma)^{j-1}\gamma, j = 1, \ldots, \infty$. That is, $g(j)$ is the pdf of a geometric distribution with parameter γ.

3.2. *The data and a statistic model for seroconversion*

For studying HIV infection under IV drug use and/or homosexual behavior, the data available are usually data on seroconversion status of individuals in the respective population. In this paper we will use some data generated by the Cooperative Agreement Study during the period 1992–1998 by the National Institute on Drug Abuse of NIH (NIDA/NIH) (Study No. 3023). The objective of this study was to monitor risk factors, risk behaviors, and rate of HIV sero-prevalence and sero-incidence among out-of-treatment, multi-ethnic/racial injection drug users and crack cocaine users. This study has generated a huge data set on IDU and HIV infection under IDU, involving 31,088 individuals in 23 US cities, Puerto Rico and Brazil, and more than 80 risk variables. For HIV infection and seroconversion for each participant, the study has generated information on seroconversion for each individual.

Data collection in the NIDA Cooperative Agreement began in 1992 and continued through the first half of 1998 over a period of six years. During these six years, people entered the study randomly and were tested for HIV status by ELISA and Southern Blot. Thus for each individual, the data provide the date of testing and testing result. To illustrate the type of data available on HIV seroconversion, consider four individuals who entered the study and were tested at times $s_j, j = 1, 2, 3, 4$ respectively. Suppose that the first individual was tested positive at s_1, then the HIV seroconversion time for this person is $\leq s_1$ (this is called left truncation). Suppose that the second person was tested negative at s_2 and was not followed; then the HIV seroconversion time for this person is $> s_2$ (this is called right censoring at s_2). Suppose that the third person was tested negative at s_3 and was followed up at t_3 and tested positive at t_3; then the HIV seroconversion time for this person is between s_3 and t_3 (this is called interval censoring). Suppose that the fourth person was tested negative at s_4 and was followed up at t_4 and tested negative at t_4; then the HIV seroconversion time for this person is $> t_4$ (this is right censoring at t_4). From this it followed that the seroconversion times of the 31,088 individuals fall into one of the three groups: Left truncation (C_1), right censoring (C_2) and interval censoring (C_3). Let n_i denote the number of individuals in C_1 with seroconversion time $\leq i$ $(i > 0)$, m_j the number of individuals in C_2 with seroconversion time $> j$ $(j > 0)$ and $k(s,t)$ the number of individuals in C_3 with seroconversion time in $(s,t]$ $(t > s > 0)$. Then, the data is given by $\boldsymbol{Y} = \{n_i, m_i, i = 1, \ldots, t_C, k(s,t), 0 < s < t \leq t_C\}$, where t_C is the maximum seroconversion time. Let $F_C(j) = \sum_{i=1}^{j} h(i)$ be the cumulative distribution function (cdf) of HIV seroconversion time so that $S_C(j) = 1 - F_C(j)$ is the survival function of HIV seroconversion time. Denote by $\Lambda = \{\lambda_j, j = 1, \ldots, t_C\}$. The likelihood function $(L(\Lambda|\boldsymbol{Y}) = P\{\boldsymbol{Y}|\Lambda\})$ of Λ given $\boldsymbol{Y}$ is:

$$L(\Lambda|\boldsymbol{Y}) \propto \prod_{i=1}^{t_C} [F_C(i)]^{n_i} [S_C(i)]^{m_i} \prod_{s=1}^{t_C-1} \prod_{t=s+1}^{t_C} [F_C(t) - F_C(s)]^{k(s,t)} . \quad (16)$$

3.3. *Statistical inferences on HIV seroconversion*

To make inference on HIV seroconversion incidence and hence the seroconversion distribution by using the likelihood function in (16), there are two major difficulties: (a) The likelihood function in (16) is very complicated. (b) The number of parameters is very large. As illustrated in detail in Tan *et al.* (1998) and in Tan (Chapter 4), the first difficulty can readily be resolved by introducing the EM algorithm whereas the second difficulty can be resolved by introducing a smoothing step or a penalized likelihood method. From the traditional sampling theory approach, one of the efficient method arrived by these procedures for estimating HIV seroconversion is the EMS algorithm (i.e. Expectation, Maximization and Smoothing algorithm); see Tan (2000, Chapter 4), and Tan *et al.* (1998). These algorithms have been discussed in detail in Tan *et al.* and in Tan (2000, Chapter 4). As an alternative method, in the next section we will use the Bayesian approach to derive Bayesian estimates of the unknown parameters and the state variables on HIV seroconversion.

3.4. *The Bayesian approach for estimating seroconversion*

The Bayesian approach for making inference about Λ is based on the posterior density $P\{\Lambda|\boldsymbol{Y}\}$ of Λ given data $\boldsymbol{Y}$. Because the posterior density $P\{\Lambda|\boldsymbol{Y}\}$ is very complicated, we propose an indirect approach to generate samples from $P\{\Lambda|\boldsymbol{Y}\}$ by combining a complete likelihood with the prior and by coupling with the multi-level Gibbs sampling procedure. The complete likelihood $L_C\{\Lambda|\boldsymbol{Y},\boldsymbol{V}\}$ is the joint density $P\{\boldsymbol{Y},\boldsymbol{V}|\Lambda\}$ of $\boldsymbol{Y}$ and $\boldsymbol{V}$ given Λ, where $\boldsymbol{V}$ are some intermediate un-observable variables. For HIV seroconversion, these intermediate random variables are: (a) $n_i(j) =$ the number of individuals with seroconversion time during $(j-1,j]$ among the n_i individuals with seroconversion time in $(0,i]$ in C_1, (b) $m_i(j) =$ the number of individuals with seroconversion time during $(j-1,j]$ among the m_i individuals with seroconversion time in $(i,t_C]$ in C_2 and (c) $k_j(s,t) =$ the number of individuals with seroconversion time during $(j-1,j]$ among the $k(s,t)$ individuals with seroconversion time in $(s,t]$ in C_3. These intermediate random variables $\{n_i(j), m_i(j), k_j(s,t)\}$ are not observed but

$$\sum_{j=1}^{i} n_i(j) = n_i\,, \quad \sum_{j=i+1}^{t_C} m_i(j) = m_i\,, \quad \sum_{j=s+1}^{t} k_j(s,t) = k(s,t)\,.$$

Notice that with $\boldsymbol{N} = \{\underset{\sim}{n}_i, i=1,\ldots,t_C\}, \boldsymbol{M} = \{\underset{\sim}{m}_i, i=1,\ldots,t_C\}$ and $\boldsymbol{K} = \{\underset{\sim}{k}(s,t), s=1,\ldots,t_C-1, t=s+1,\ldots,t_C\}$, $\boldsymbol{V} = \{\boldsymbol{N},\boldsymbol{M},\boldsymbol{K}\}$.

To derive the Bayesian estimates of Λ, notice that given Λ, the conditional probability densities of $\underset{\sim}{n}_i = \{n_i(j), j=1,\ldots,i\}'$ given n_i, of $\underset{\sim}{m}_i = \{m_i(j), j=i+1,\ldots,t_C\}'$ given m_i and of $\underset{\sim}{k}(s,t) = \{k_j(s,t), j = s+1,\ldots,t\}'$ given $k(s,t)$ are given respectively by:

$$\underset{\sim}{n}_i|n_i \sim Multinomial\left\{n_i; \frac{h(j)}{F_C(i)}, j=1,\ldots,i\right\} \tag{17}$$

$$\underset{\sim}{m}_i|m_i \sim Multinomial\left\{m_i; \frac{h(j)}{S_C(i)}, j=i+1,\ldots,t_C\right\} \tag{18}$$

$$\underset{\sim}{k}(s,t)|k(s,t) \sim Multinomial\left\{k(s,t); \frac{h(j)}{S_C(s)-S_C(t)}, j=s+1,\ldots,t\right\}. \tag{19}$$

Define $\delta_j(s,t)$ by:

$$\begin{aligned}\delta_j(s,t) &= 1 \quad \text{if } s<j\le t, s=1,\ldots,t_C-1, t=s+1,\ldots,t_C\\ &= 0 \quad \text{otherwise.}\end{aligned}$$

Put $r(j) = \sum_{i=j}^{t_C} n_i(j) + \sum_{i=1}^{j-1} m_i(j) + \sum_{s=1}^{t_C-1} \sum_{t=s+1}^{t_C} k_j(s,t)\delta_j(s,t)$ and $R(j) = \sum_{i=j+1}^{t_C} r(i)$. Using the distribution results in (17)–(19), the complete likelihood function $L_C(\Lambda|\boldsymbol{V},\boldsymbol{Y}) = P\{\boldsymbol{V},\boldsymbol{Y}|\Lambda\}$ of Λ given $\{\boldsymbol{V},\boldsymbol{Y}\}$ is:

$$L_C(\Lambda|\boldsymbol{V},\boldsymbol{Y}) = \prod_{j=1}^{t_C} [h(j)]^{r(j)} = \prod_{j=1}^{t_C} \{\lambda(j) S_C(j-1)\}^{r(j)}$$
$$= \prod_{j=1}^{t_C} [\lambda(j)]^{r(j)} [1-\lambda(j)]^{R(j)} . \tag{20}$$

Let $P(\Lambda)$ be the prior distribution of HIV seroconversion incidence given by:

$$P(\Lambda) \propto \prod_{j=1}^{t_C} [\lambda(j)]^{a_j-1} [1-\lambda(j)]^{b_j-1} , \tag{21}$$

where $\{a_j > 0, b_j > 0\}$ are the hyper-parameters which can be estimated by previous studies; see Tan (2000, Chapter 6; 2002, Chapter 9).

The above prior is the conjugate prior as defined in Raiffer and Schlafer (1961). If $a_j = b_j = 1$, the above prior is the non-informative uniform prior representing that the prior information is vague and imprecise. Notice that if one adopts the non-informative uniform prior, then numerically the Bayesian estimates are approximately the same as the classical sampling theory estimates although the probability concepts of the two approaches are very different.

Using the prior in (21), the posterior density of Λ given $\{\boldsymbol{Y},\boldsymbol{V}\}$ is

$$P\{\Lambda|\boldsymbol{Y},\boldsymbol{V}\} = \prod_{j=1}^{t_C} [\lambda(j)]^{\hat{r}(j)+a_j-1} [1-\lambda(j)]^{\hat{R}(j)+b_j-1} . \tag{22}$$

Using the above results, the multi-level Gibbs sampling procedure for estimating Λ and $\boldsymbol{V}$ given $\boldsymbol{Y}$ as described in Tan (2000, Chapter 6; 2002, Chapter 9) is given by:

(1) Given Λ and $\boldsymbol{Y}$, generate $\boldsymbol{V} = \{\boldsymbol{N},\boldsymbol{M},\boldsymbol{K}\}$ from the conditional densities given by (17)–(19). Denote the generated sample by $\boldsymbol{V}^{(*)}$. Then $\boldsymbol{V}^{(*)}$ is a random sample of size one from $P\{\boldsymbol{V}|\boldsymbol{Y},\Lambda\}$.
(2) On substituting $\boldsymbol{V}^{(*)}$ for $\boldsymbol{V}$, generate Λ from the conditional posterior density $P\{\Lambda|\boldsymbol{V}^{(*)},\boldsymbol{Y}\}$ given by equation (22).
(3) With Λ being generated from Step 2 above, go back to Step 1 and repeat the above (1)–(2) loop until convergence.

At convergence, one then generates a random sample of $\boldsymbol{V}$ from the conditional distribution $P\{\boldsymbol{V}|\boldsymbol{Y}\}$ of $\boldsymbol{V} = \{\boldsymbol{N},\boldsymbol{M},\boldsymbol{K}\}$ given $\boldsymbol{Y}$ independent of Λ and a random sample of Λ from the posterior density $P\{\Lambda|\boldsymbol{Y}\}$ of Λ given $\boldsymbol{Y}$, independent of $\boldsymbol{V}$. The convergence of these procedures are proved by using the basic theory of homogeneous Markov chains as given in Chapter 3 of Tan (2002).

Repeat the above procedures we generate a random sample of size n of $\boldsymbol{V}$ from $P\{\boldsymbol{V}|\boldsymbol{Y}\}$, independent of Λ and a random sample of size m of Λ from $P\{\Lambda|\boldsymbol{Y}\}$, independent of $\boldsymbol{V}$. If m is very large, the sample means and the sample variances and covariances of the samples of Λ provide a close approximation of the posterior means and posterior variances and covariances of Λ, independently of $\boldsymbol{V}$. Similarly, if n is vary large, the sample means and sample variances and covariances of $\boldsymbol{V}$ provide a close approximation of the conditional means and conditional variances and covariances of $\boldsymbol{V}$ from $P\{\boldsymbol{V}|\boldsymbol{Y}\}$, independently of Λ. Thus one may use these sample means as the estimates of $\boldsymbol{V}$ and Λ respectively and use the sample variances and covariances as the estimates of the variances and covariances of these estimates respectively.

The above procedures for deriving the Bayesian estimates of the parameters are analogous and comparable to the EM algorithm in the sampling theory approach if one uses a non-informative uniform prior. Because the number of parameters is very large, as in the sampling theory approach, one may also introduce the smooth step to smooth the above Bayesian estimates to derive the final estimates. Under the non-informative uniform prior, numerically this will yield approximately identical estimates as the EMS method in the sampling theory approach; but it is to be noted that the probability concept between the Bayesian approach and the sampling theory approach are very different.

The Bayesian estimates of the parameters Λ using the above procedures are derived by combining information from two sources: (i) The data via the likelihood. (ii) The prior information of the parameters. It has ignored information from the stochastic system model. One may think of the smoothing step as a step to recover the latter information; in this sense the estimates by the EMS method or the Bayesian method with smoothing may be analogous to the estimates by using state space models; see Section 4.

3.5. *A likelihood ratio test for comparing several HIV seroconversion distributions*

To develop optimal intervention procedures, often one would need to compare several HIV seroconversion distributions. For example, a practical question is: Is the HIV seroconversion under some interventions different from those without interventions? Specifically, suppose that there are k populations at risk for HIV infection. Let $\{\lambda_i(j), j = 1, \ldots, t_C\}$ denote the HIV seroconversion incidence from the ith ($i = 1, \ldots, k$) population. The problem is then to test the statistical hypothesis $H_0 : \lambda_i(j) = \lambda(j), j = 1, \ldots, t_C, i = 1, \ldots, k$ versus H_1 : Negation. In this section we will proceed to develop a likelihood ratio test for testing this hypothesis.

To derive procedures to test H_0, let $\{\bar{n}_i(j) = n_i \frac{h(j)}{F_C(i)}, \bar{m}_i(j) = m_i \frac{h(j)}{S_C(i)}, \bar{k}_j(s,t) = k(s,t) \frac{h(j)}{S_C(s) - S_C(t)}\}$. Let $\{\hat{n}_i(j), \hat{m}_i(j), \hat{k}_j(s,t)\}$ be derived from $\{\bar{n}_i(j), \bar{m}_i(j), \bar{k}_j(s,t)\}$ respectively by replacing the $\{h(j), F_C(j), S_C(s,t)\}$ by its MLE respectively. Put $\bar{\boldsymbol{V}} = \{\bar{\boldsymbol{N}}, \bar{\boldsymbol{M}}, \bar{\boldsymbol{K}}\}$, where $\bar{\boldsymbol{N}} = \{\bar{n}_i(j), j = 0, 1, \ldots, i, i = 1, \ldots, t_C\}$, $\bar{\boldsymbol{M}} = \{\bar{m}_i(j), j = 0, 1, \ldots, i, i = 1, \ldots, t_C\}, \bar{\boldsymbol{K}} = \{\bar{k}(s,t), s = 0, 1, \ldots, t_C - 1, t = s+1, \ldots, t_C\}$. Similarly, define $\hat{\boldsymbol{V}} = \{\hat{\boldsymbol{N}}, \hat{\boldsymbol{M}}, \hat{\boldsymbol{K}}\}$, where $\hat{\boldsymbol{V}}$ is the $\{\hat{n}_i(j), \hat{m}_i(j), \hat{k}_j(s,t)\}$. Then we have an approximation $L_A(\Lambda|\hat{\boldsymbol{V}}, \boldsymbol{Y})$ to the likelihood of Λ given $\boldsymbol{Y}$:

$$\begin{aligned} L_A(\Lambda|\hat{\boldsymbol{V}}, \boldsymbol{Y}) &\cong \prod_{j=1}^{t_C} [h(j)]^{\hat{r}(j)} = \prod_{j=1}^{t_C} \left\{ \lambda(j) \prod_{i=1}^{j-1} [1 - \lambda(i)] \right\}^{\hat{r}(j)} \\ &= \prod_{j=1}^{t_C} [\lambda(j)]^{\hat{r}(j)} [1 - \lambda(j)]^{\hat{R}(j)} \end{aligned} \tag{23}$$

where $\hat{r}(j) = \sum_{i=j}^{t_C} \hat{n}_i(j) + \sum_{i=1}^{j-1} \hat{m}_i(j) + \sum_{s=1}^{t_C - 1} \sum_{t=s+1}^{t_C} \hat{k}_j(s,t) \delta_j(s,t)$ and $\hat{R}(j) = \sum_{i=j+1}^{t_C} \hat{r}(i)$.

We will use the above approximation as the likelihood function for developing procedures to test H_0. This is motivated by the following observations:

(1) By using the distribution results in (17)–(19), the conditional summation of $L_C\{\Lambda|\boldsymbol{V}, \boldsymbol{Y}\}$ over non-negative integers $\{n_i(j), m_i(j), k_j(s,t)\}$ under the constraints $\sum_{j=1}^{i} n_i(j) = n_i$, $\sum_{j=i+1}^{t_C} m_i(j) = m_i$, $\sum_{j=s+1}^{t} k_j(s,t) = k(s,t)$ is the likelihood function $L\{\Lambda|\boldsymbol{Y}\}$ in equation (20). (2) The partial derivatives of the log of $L(\Lambda|\boldsymbol{Y})$ equal to those of the log of $L(\Lambda|\bar{\boldsymbol{V}}, \boldsymbol{Y})$ respectively, That is,

$$\frac{\partial}{\partial \lambda} \log L(\Lambda|\boldsymbol{Y}) = \frac{\partial}{\partial \lambda} \log L(\Lambda|\bar{\boldsymbol{V}}, \boldsymbol{Y}) \,.$$

Let $L_i(\Lambda_i|\boldsymbol{Y}_i)$ denote the approximate likelihood function from the ith population, where $\boldsymbol{Y}_i$ denotes the HIV seroconversion data from the ith population and $\Lambda_i = \{\lambda_i(j), j = 1, \ldots, t_C\}$. Then,

$$L_i(\Lambda_i|\boldsymbol{Y}_i) = \prod_{j=1}^{t_C} [\lambda_i(j)]^{\hat{r}_i(j)}[1 - \lambda(j)]^{\hat{R}_i(j)} \tag{24}$$

where the $\hat{r}_i(j)$ are the $\hat{r}(j)$'s from the ith population and $\hat{R}_i(j) = \sum_{i=j+1}^{t_C} \hat{r}_i(j)$.

The joint likelihood function of $\{\Lambda_i, i = 1, \ldots, k\}$ given $\{\boldsymbol{Y}_i, i = 1, \ldots, k\}$ is $L = \prod_{i=1}^{k} L_i(\Lambda_i|\boldsymbol{Y}_i)$. Under H_0, L becomes

$$L = \prod_{j=1}^{t_C} [\lambda(j)]^{r_S(j)}[1 - \lambda(j)]^{R_S(j)} \tag{25}$$

where $r_S(j) = \sum_{i=1}^{k} \hat{r}_i(j)$ and $R_S(j) = \sum_{i=1}^{k} \hat{R}_i(j)$.

Obviously, $\hat{L}_i = \underset{\Lambda_i}{Max}\ L_i(\Lambda_i|\boldsymbol{Y}_i)$ and $\hat{L} = \underset{\Lambda_i}{Max}\ L$ are given respectively by:

$$\hat{L}_i = \prod_{j=1}^{t_C} \left[\frac{\hat{r}_i(j)}{\hat{r}_i(j) + \hat{R}_i(j)}\right]^{\hat{r}_i(j)} \left[\frac{\hat{R}_i(j)}{\hat{r}_i(j) + \hat{R}_i(j)}\right]^{\hat{R}_i(j)},$$

$$\hat{L} = \prod_{j=1}^{t_C} \left[\frac{r_S(j)}{r_S(j) + R_S(j)}\right]^{r_S(j)} \left[\frac{R_S(j)}{r_S(j) + R_S(j)}\right]^{R_S(j)}.$$

Let $\hat{\omega} = \hat{L}/\{\prod_{i=1}^{k} \hat{L}_i\}$. Then the generalized likelihood ratio test procedure is to reject H_0 if $0 < \hat{\omega} < K_0$ for some constant $K_0 \leq 1$. When the sample size in each population is very large, approximately $-2\log\hat{\omega} \sim \chi^2_{T_C(k-1)}$ under H_0, where $\chi^2_{T_C(k-1)}$ denotes a central chi-square distribution with degrees of freedom $t_C(k-1)$. Thus, the p-value is $p = P\{\chi^2_{T_C(k-1)} > -2\log\hat{\omega}\}$, where $\hat{\omega}$ is the observed likelihood ratio. One rejects H_0 at level α if $p < \alpha$.

3.6. *Estimation of HIV infection*

As shown in Section 3.1, the HIV seroconversion time is a convolution of HIV infection and the window period. This is analogous to the AIDS onset time which is a convolution of HIV infection time and the HIV incubation period (from HIV infection to the AIDS onset). Hence, as shown in Tan (2000, Chapter 5), the above statistical model and methods are not identifiable in the sense that one can not estimate the HIV infection and the distribution of the window period simultaneously unless additional information can be derived from sources other than data. Hence, in the event that there are no other information than the data and the likelihood, one needs to assume the probability distribution of window period as known if one wishes to estimate the HIV infection probability; similarly, in the event that there are no other information than the data and the likelihood, one needs to assume the HIV infection probabilities as known if one wishes to estimate the probability distribution of the window period.

For making statistical inferences, notice that besides information from the data via the likelihood, there are two additional sources of information: (i) The information from the stochastic system and the dynamic AIDS epidemiology and (ii) possible prior information from previous experiences. Since the state space modeling approach coupled with the multi-level Gibbs sampling method is a natural avenue to combine information from these two sources with those from the data and the likelihood, in this paper, we will resolve these problems by introducing state space models, the multi-level Gibbs sampling method and the generalized Bayesian procedure. This is discussed in detail in Section 4 below.

4. A State Space Model for HIV Seroconversion

Based on observed information on HIV seroconversion over time, in this paper we develop a state space model for the HIV seroconversion. For this state space model, the state variables are $\boldsymbol{X} = \{\underset{\sim}{X}(t), t = 1, \ldots, t_C\}$ where $\underset{\sim}{X}(t) = \{S(t), I(u,t), u = 0, 1, \ldots, t, C(t)\}$ and t_C is the maximum of HIV seroconversion. Thus the stochastic system model is represented by the stochastic equations (1)–(6) given above for these variables and the probability distributions of the variables in Section 2. Based on seroconversion data, the observation model is given by the statistical model based on the complete and the approximate likelihood functions given by equations (20) and (23) respectively.

4.1. *The stochastic system model and the probability distribution of state variables*

The probability distribution of the state variables is given in Section 2. For implementing the multi-level Gibbs sampling procedure to estimate the unknown parameters and the state variables, we define the un-observed state variables $\underset{\sim}{U}(t) = \{D_S(t), D_I(u,t)\}$ and put $\boldsymbol{U} = \{\underset{\sim}{U}(t), t = 0, 1, \ldots, t_C - 1\}$. Using the distribution result in Section 3.2, it is easily shown that the conditional density of $\underset{\sim}{U}(t)$ given $\underset{\sim}{X}(t)$ is

$$P\{\underset{\sim}{U}(t)|\underset{\sim}{X}(t)\} = C_1(t){\mu_S}^{D_S(t)}(1-\mu_S)^{S(t)-D_S(t)}{\mu_I}^{\sum_{u=0}^{t} D_I(u,t)} \times (1-\mu_I)^{\sum_{u=0}^{t}[I(u,t)-D_I(u,t)]}, \quad (26)$$

where $C_1(t) = \binom{S(t)}{D_S(t)} \prod_{u=0}^{t} \binom{I(u,t)}{D_I(u,t)}$.

The conditional density of $\underset{\sim}{X}(t+1)$ given $\{\underset{\sim}{U}(t), \underset{\sim}{X}(t)\}$ is

$$\begin{aligned} P\{\underset{\sim}{X}(t+1)|\underset{\sim}{U}(t), \underset{\sim}{X}(t)\} &= C_X(t)\left[\frac{p_S(t)}{1-\mu_S}\right]^{A(t)}\left[1-\frac{p_S(t)}{1-\mu_S}\right]^{S(t+1)} \\ &\times \left[\frac{\gamma}{1-\mu_I}\right]^{\sum_{u=0}^{t} B(u,t)}\left[1-\frac{\gamma}{1-\mu_I}\right]^{\sum_{u=0}^{t} I(u+1,t+1)}, \end{aligned} \quad (27)$$

where $C_X(t) = \binom{S(t)-D_S(t)}{A(t)}\{\prod_{u=0}^{t}\binom{I(u,t)-D_I(u,t)}{B(u,t)}\}$, $\{A(t) = S(t) - S(t+1) - D_S(t), B(u,t) = I(u,t) - I(u+1,t+1) - D_I(u,t)$.

The joint density of $\{\boldsymbol{X}, \boldsymbol{U}\}$ given $\Theta = \{p_S(i), i = 1, \ldots, t_C, \mu_S, \mu_I, \gamma\}$ is

$$\begin{aligned} P\{\boldsymbol{X}, \boldsymbol{U}|\Theta\} = P\{\underset{\sim}{X}(0)|\Theta\} \prod_{t=1}^{t_M} P\{\underset{\sim}{X}(t)|\underset{\sim}{U}(t-1), \underset{\sim}{X}(t-1)\} \\ \times P\{\underset{\sim}{U}(t-1)|\underset{\sim}{X}(t-1)\}, \end{aligned} \quad (28)$$

4.2. *The observation model and the probability distribution of the number of the observed seroconvertors*

For HIV seroconversion, the observation model is a statistic model described in Section 2. Thus, the observation model is represented by the complete likelihood and the approximate likelihood given in (20) and (23) respectively. To extract information on $\{p_S(t), \gamma\}$ from the data, we let $u_i(j)$ denote the number of individuals with HIV infection during $(i-1, i]$ among the $r(j)$ individuals with seroconversion in $(j-1, j]$. Then $\sum_{i=1}^{j} u_i(j) = r(j)$ and the probability distribution of $\underset{\sim}{u}_j = \{u_i(j), i = 1, \ldots, j\}$ given $r(j)$ is

$$\{u_i(j), i = 1, \ldots, j\}|r(j) \sim ML\{r(j), f_I(i)g(j-i)/h(j), i = 1, \ldots, j\}.$$

Put $P\{\underset{\sim}{u}_j|\boldsymbol{U}, \boldsymbol{Y}, \Theta\} = \{r(j)!/\prod_{i=1}^{j} u_i(j)!\} \prod_{i=1}^{j}\{f_I(i)g(j-i)/h(j)\}^{u_i(j)}$ and $\boldsymbol{R} = \{\underset{\sim}{u}_j, j = 1, \ldots, t_C\}$. Then the density of $\boldsymbol{R}$ given $\{\boldsymbol{U}, \boldsymbol{Y}, \Theta\}$ is $P\{\boldsymbol{R}|\boldsymbol{U}, \boldsymbol{Y}, \Theta\} = \prod_{j=1}^{t_C} P\{\underset{\sim}{u}_j|\boldsymbol{U}, \boldsymbol{Y}, \Theta\}$.

Denote $\{u_i = \sum_{j=i} u_i(j), W_i = \sum_{j=i+1}^{t_C} u_j\}$ $\{v_k = \sum_{j=k+1}^{t_C} u_{j-k}(j), V_R = \sum_{k=1}^{t_C} v_k, V_g = \sum_{k=1}^{t_C}\sum_{j=k+1}^{t_C} v_j\}$, then the joint density $P\{\boldsymbol{R}, \boldsymbol{U}, \boldsymbol{Y}|\Theta\}$ of $\{\boldsymbol{R}, \boldsymbol{U}, \boldsymbol{Y}\}$ given Θ is

$$\begin{aligned} P\{\boldsymbol{R}, \boldsymbol{U}, \boldsymbol{Y}|\Theta\} &= P\{\boldsymbol{R}|\boldsymbol{V}, \boldsymbol{Y}, \Theta\}P\{\boldsymbol{U}, \boldsymbol{Y}|\Theta\} \\ &= C_R \prod_{j=1}^{t_C}\prod_{i=1}^{j}\{f_I(i)g(j-i)\}^{u_i(j)} \\ &= C_R \prod_{i=1}^{t_C}[f_I(i)]^{u_i} \prod_{k=1}^{t_C} g(k)^{v_k} \\ &= C_R\gamma^{V_R}(1-\gamma)^{V_g}\prod_{i=1}^{t_C}\{[p_S(i)]^{u_i}[1-p_S(i)]^{W_i}\}, \end{aligned} \tag{29}$$

where $C_K = \frac{M!}{\prod_{j=1}^{t_C}\prod_{i=1}^{j}[u_i(j)]!}$ with $M = \sum_{j=1}^{t_C}\sum_{i=1}^{j} u_i(j)$.

4.3. *The contribution to the observed number of seroconverters by the data*

To extract the observed number of individuals with seroconversion in $(j-1, j]$ from the data, consider a individual with seroconversion in $(a_i, b_i]$. Then this individual (say the ith individual) is left truncated at b_i if $a_i = 0$ and $b_i < t_C$, right censored at a_i if $a_i > 0$ and $b_i = t_C$ and interval censored in $(a_i, b_i]$ if $a_i > 0$ and $b_i < t_C$. Define $y_j(i) = 0$ if $r_{1i} = Max\{j-1, a_i\} \geq r_{2i} = Min\{j, b_i\}$ and define $y_j(i)$ by the formula below if $r_{1i} < r_{2i}$:

$$y_j(i) = \frac{S_C(r_{1i}) - S_C(r_{2i})}{S_C(a_i) - S_C(b_i)}.$$

Then, as shown in Tan *et al.* (1998) and Tan (2000, Chapter 4), $y_j(i)$ is the contribution to the number of individuals with seroconversion in $(j-1, j]$ from this individual so that $y_j = \sum_i y_j(i)$ is contribution to the total number of individuals with seroconversion in $(j-1, j]$ from the data.

Let c_j be the number of seroconvertors generated by the model in Section 2 starting with n individuals, where n is the sample size in the data. Then, one may assume that the $e_j = \frac{y_j - c_j}{\sqrt{c_j}}$ are independently distributed as normal with means 0 and variance σ^2. That is, the density of y_j is

$$f_Y\{y_j|c_j\} = \{2\pi\, c_j\sigma^2\}^{-1/2} exp\left\{-\frac{1}{2c_j\sigma^2}(y_j - c_j)^2\right\}. \tag{30}$$

In Section 5, these distribution results will be used to estimate the unknown parameters via multi-level Gibbs sampling procedures.

4.4. The conditional posterior distribution

Put $\Omega = \{\Theta, \sigma^2\}$. Following Box and Tiao [4], we assume $P\{\Omega\} = P\{\Theta\}P(\sigma^2)$ with $P(\sigma^2) \propto \sigma^{-2}$. Further, following Raiffa and Schlaiffer (1961), we will assume a conjugate prior for Θ as

$$\begin{aligned} P\{\Theta\} \propto \{&{\mu_S}^{a_S-1}(1-\mu_S)^{b_S-1}{\mu_I}^{a_I-1}(1-\mu_I)^{b_I-1} \\ &\times\gamma^{a_G-1}(1-\gamma)^{b_G-1} \\ &\times\prod_{i=1}^{t_C}[p_S(i)]^{c_1(i)-1}[1-p_S(i)]^{c_2(i)-1}\,. \end{aligned} \quad (31)$$

where the hyper-parameters $\{a_S, b_S, a_I.b_I, a_G, b_G, c_j(i), i = 1, \ldots, t_C, j = 1, 2\}$ are positive real numbers.

The hyper-parameters in the above prior can be determined by results from previous studies if such studies have been conducted. In the event that the prior information is vague and imprecise, one may assume a non-informative uniform prior by letting all the hyper-parameters equal to one.

Let $D_S = \sum_{t=1}^{t_C} D_S(t), S_T = \sum_{t=0}^{t_C-1} S(t+1), I_T = \sum_{t=0}^{t_C-1}\sum_{u=0}^{t} I(u+1, t+1), B = \sum_{t=1}^{t_C}\sum_{u=0}^{t} B(u,t), D_I = \sum_{t=1}^{t_C}\sum_{u=0}^{t} D_I(u,t)$.

Because the $r(j)$ and $R(j)$ are uniquely determined by $\boldsymbol{Y}$, the conditional posterior density of Θ given $\{\boldsymbol{X}, \boldsymbol{U}, \boldsymbol{R}, \boldsymbol{Y}\}$ is

$$\begin{aligned} &P\{\Theta|\boldsymbol{X}, \boldsymbol{U}, \boldsymbol{R}, \boldsymbol{Y}\} \\ &\quad\propto \{{\mu_S}^{D_S+a_S-1}(1-\mu_S)^{S_T+b_S-1}{\mu_I}^{D_I+a_I-1} \\ &\qquad\times(1-\mu_I)^{I_T+b_I-1}\gamma^{B+V_R+a_G-1}(1-\gamma)^{V_g+b_G-1}\left(1-\frac{\gamma}{1-\mu_I}\right)^{I_T} \\ &\qquad\times(\sigma^2)^{-\frac{t_C}{2}-1}exp\left\{-\frac{t_C\hat{\sigma}^2}{2\sigma^2}\right\}\prod_{i=1}^{t_C}[p_S(i)]^{A(i)+u_i+c_1(i)-1} \\ &\qquad\times[1-p_S(i)]^{W_i+c_2(i)-1}\left[1-\frac{p_S(i)}{1-\mu_S}\right]^{S(i+1)}, \end{aligned} \quad (32)$$

where $\hat{\sigma}^2 = \frac{1}{t_C}\sum_{j=1}^{t_C}\frac{1}{c_j}(y_j - c_j)^2$.

From the distribution in (32), one can easily derive the probability distribution of each of the parameters $\{\mu_S, \mu_I, \gamma, \sigma^2, p_S(i), i = 1, \ldots, t_C\}$ given $\{\boldsymbol{X}, \boldsymbol{U}, \boldsymbol{R}, \boldsymbol{Y}\}$. This shows that all parameters are identifiable if Θ is estimable; see Remark 1.

Remark 1: In the state space model, $\boldsymbol{X}$ and $\boldsymbol{U}$ are the state variables which are time dependent. The classical state space modeling approach (i.e. the Kalman filter) is to estimate and predict the state variables assuming the parameters as known. This is achieved through the stochastic system model which provide information about the system via the mechanism of the system and the observation model which is a statistical model based on data. The number of data set is very limited, far less than the number of state variables to be estimated; but due to additional information from the system besides information from data, this is made possible. Notice that the observation model is basically a statistical model which is the model used by statisticians to estimate parameters. (The information provided by the observation model is very limited and is not adequate even to estimate all parameters, much less state variables.) Now given the values of state variables (i.e. $\boldsymbol{X}, \boldsymbol{U}$), one can readily estimate the unknown parameters by using the stochastic system model and the observation model. Thus, in state space models, one can readily bring these two aspects together by Gibbs sampling method to estimate simultaneously the state variables and the unknown parameters. This is the approach used by Tan and Ye (2000a,b) to estimate the state variables (i.e. the numbers of S people, HIV infected people and AIDS cases) as well as the unknown parameters (the HIV infection incidence, the HIV incubation incidence, the death rates and the immigration rates) in the Swiss homosexual and IV drug population and in the San Francisco homosexual population. A formal proof of the convergence of this approach is given in Tan, Zhang and Xiong (2004).

5. Simultaneous Estimation of Unknown Parameters and State Variables

For the above model, the important state variables are the numbers of I people and C people and prevalence of HIV infection and seroconversion. The unknown parameters are $\Omega = \{\Theta, \sigma^2\} = \{p_S(t), \gamma(u,t) = \gamma, \mu_S(t) = \mu_S, \mu_I(t) = \mu_I, \sigma^2\}$.

The multi-level Gibbs sampling procedure is a Gibbs sampling method applied to multi-variate data as proposed by Sheppard (1994) and Chen and Liu (1998). The details of this procedure as applied to problems of AIDS and cancers have been given in Tan (2000, Chapter 6; 2002, Chapter 9).

Using the above results, one can readily estimate simultaneously the state variables and the unknown parameters Ω by using the multi-level Gibbs sampling procedure; see Remark 1. This procedure for estimating Ω

and the state variables given $\boldsymbol{Y}$ is described in Tan (2000, Chapter 6; 2002, Chapter 9) and is given by the following loops:

(1) Combining a large sample from $P\{\boldsymbol{U}, \boldsymbol{X}|\Theta\}$ with the densities of y_j in (30) through the weighted Bootstrap method due to Smith and Gelfand (1992), we generate $\{\boldsymbol{U}, \boldsymbol{X}\}$ (denote the generated sample $\{\boldsymbol{U}^{(*)}, \boldsymbol{X}^{(*)}\}$) from $P\{\boldsymbol{U}, \boldsymbol{X}|\Theta, \boldsymbol{Y}\}$ although the latter density is unknown. Because given $\boldsymbol{Y}$, $\{\boldsymbol{X}, \boldsymbol{U}\}$ are independent of $\boldsymbol{R}$, $\{\boldsymbol{U}^{(*)}, \boldsymbol{X}^{(*)}\}$) is also a random sample from the conditional density of $\{\boldsymbol{U}, \boldsymbol{X}\}$ given $\{\Theta, \boldsymbol{Y}, \boldsymbol{R}\}$. An illustration of the weighted bootstrap method in the present situation is given in Appendix A in Tan, Zhang and Xiong (2004).
(2) Using the distributions in (17)–(19), generate $\boldsymbol{R}$ given $\{\boldsymbol{Y}, \Theta\}$ and denote it by $\boldsymbol{R}^{(*)}$. Because given $\boldsymbol{Y}$, $\{\boldsymbol{X}, \boldsymbol{U}\}$ are independent of $\boldsymbol{R}$, $\boldsymbol{R}^{(*)}$ is also a random sample from the conditional density of $\boldsymbol{R}$ given $\{\Theta, \boldsymbol{Y}, \boldsymbol{X}, \boldsymbol{U}\}$.
(3) On substituting $\{\boldsymbol{U}^{(*)}, \boldsymbol{X}^{(*)}, \boldsymbol{R}^{(*)}\}$, which are generated by the above steps, generate Θ from the conditional density $P\{\Theta|\boldsymbol{X}^{(*)}, \boldsymbol{U}^{(*)}, \boldsymbol{R}^{(*)}, \boldsymbol{Y}\}$ given by equation (32) above.
(4) With Θ generated from Step 3 above and with $\boldsymbol{R}^{(*)}$ generated from Step 2, go back to Step 1 and repeat the above (1)–(2) loop until convergence.

At convergence, one generates a random sample of $\boldsymbol{X}, \boldsymbol{U}$ from the conditional distribution $P\{\boldsymbol{X}, \boldsymbol{U}|\boldsymbol{Y}\}$ of $\{\boldsymbol{X}, \boldsymbol{U}\}$ given $\{\boldsymbol{Y}\}$, independent of $\{\boldsymbol{R}, \Theta\}$ and a random sample of Θ from the distribution $P\{\Theta|\boldsymbol{Y}\}$ of Θ given $\boldsymbol{Y}$, independent of $\{\boldsymbol{X}, \boldsymbol{U}, \boldsymbol{R}\}$. Repeat these procedures we then generate a random sample of size n of $(\boldsymbol{X}, \boldsymbol{U})$ and a random sample of size m of Θ. One may then use the sample means to derive the estimates of $(\boldsymbol{X}, \boldsymbol{U})$ and Θ and use the sample variances as the variances of these estimates. The convergence of these procedures are proved by using the basic theory of homogeneous Markov chains as given in Chapter 3 of Tan (2002) and is given in Appendix B of Tan, Zhang and Xiong (2004).

6. An Illustrative Example

In this section we will apply the above models and methods to the NIDA/NIH data to estimate the probabilities of time to HIV infection and time to HIV seroconversion in populations of IDU drug users and homosexuals. This data set has been described in some detail in Section 2. For this data set, the sample size for the populations of homosexuals, IDU and homosexuals with IDU are given by 12,069, 5177, and 2404 respectively. Among the homosexuals, 195 have age $<$ 20 years old at the testing time and 11,874 have age $\geq$ 20 years old at the testing time.

To apply the models and methods, we let $t = 0$ be the time point to start the HIV epidemic. That is, at $t = 0$ every one in the population are HIV non-carriers; but at $t = 0$ there are some HIV viruses in the populations to infect S people. Thus, because HIV infection in IV drug users is caused by IV drug use, we let the first time to use IV drug of an IV drug user be the starting time point of that IV drug user. Because the average age that people start sexual behavior is between 13 and 18 years old, for those homosexuals with age less than 20 years old at the testing time, we let the time of 15 years old as the starting time point; for those homosexuals with age greater than 20 years old at the testing time and because on the average it takes about 5 years for homosexual S people to get HIV infected (Tan 2000, Chapter 6), we chose January 1985 as the starting time point to start the HIV epidemic. We note in passing, however, that as shown in

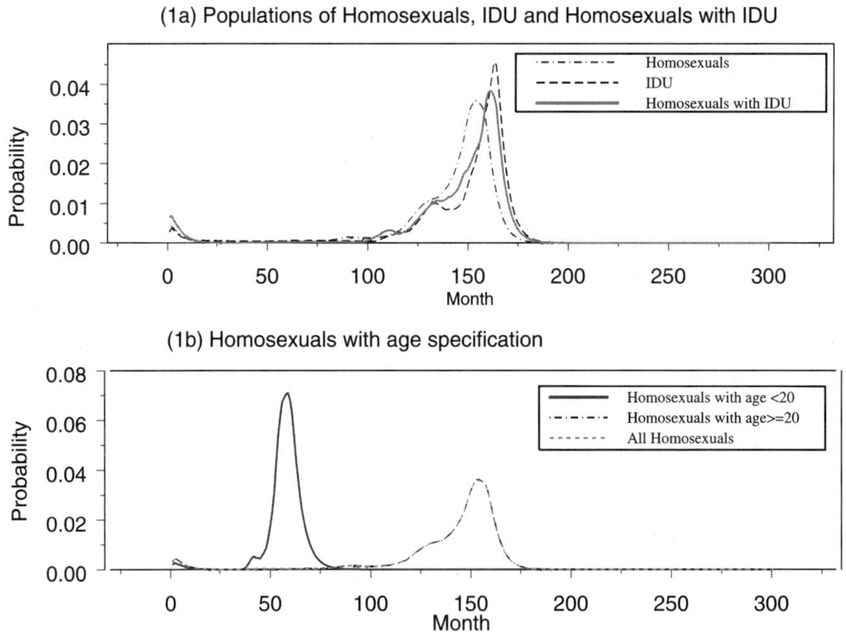

Fig. 1. Plots showing the probability density function of time to seroconversion.

Tan (2000, Chapter 6), the choice of starting time point for HIV infection has little impacts on the pattern of the HIV infection curve.

Under the above specification we have applied the methods in Sections 3 and 4 to the NIDA/NIH data to estimate the unknown parameters and the numbers of HIV infected people and HIV seroconverters monthly. The estimate of σ^2 is 0.688 ± 0.264. The probability densities of times to HIV seroconversion in populations of homosexuals, IDU and homosexuals with IDU are plotted in Fig. 1. Plotted in Fig. 2 are the probability densities of times to HIV infection and times to seroconversion in various populations respectively. Given in Table 1 are the means and variances of the HIV infection and seroconversion densities in various populations. Given in Table 2 are the estimates of the death and retirement probabilities per month as well as the monthly transition probabilities from $I \to C$ in various populations. From these we have obtained the following results:

(1) The likelihood ratio tests show that there are significant differences between seroconversion incidence of homosexuals, IDU and homosexuals with IDU. The observed $-2\log\hat{\omega}$ statistics for IDU versus homosexuals, IDU versus homosexuals with age $\geq$ 20 years at testing time, IDU versus

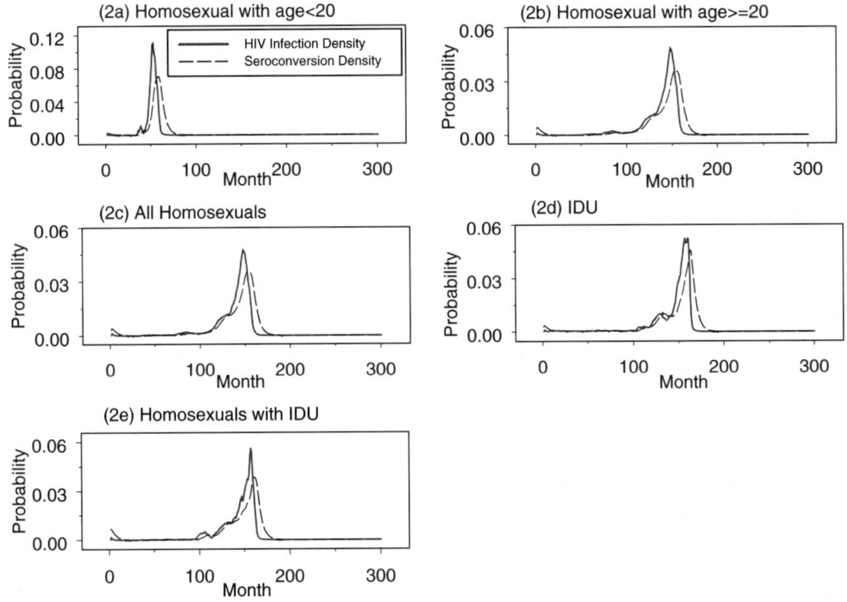

Fig. 2. Plots showing the probability densities of time to HIV infection and HIV seroconversion in populations of homosexuals, IDU and homosexuals with IDU.

Table 1. The means and variances (in month) of time to HIV infection and seroconversion of homosexuals, IDU and homosexuals with IDU.

Populations	HIV Infection		HIV Seroconversion	
	Mean	Variance	Mean	Variance
IDU	142.430	698.630	145.279	1132.378
Homosexuals with Age < 20	51.335	25.770	54.016	92.850
Homosexuals with Age ≥ 20	137.253	393.309	139.899	908.169
All Homosexuals	136.830	429.025	139.407	892.304
Homosexuals with IDU	142.185	423.012	145.100	1248.493

Table 2. The estimated probability of death and retirement (per month) of homosexuals, IDU and homosexuals with IDU.

Populations	Normal people	Infected people	γ Value
IDU	6.935E-06 $\pm$ 7.960E-06	4.589E-03 $\pm$ 2.319E-03	0.351 $\pm$ 0.028
Homosexuals with Age < 20	1.00E-07 $\pm$ 1.458E-10	5.337E-03 $\pm$ 1.243E-03	0.373 $\pm$ 0.070
Homosexuals with Age ≥ 20	5.496E-06 $\pm$ 5.712E-06	4.689E-03 $\pm$ 1.552E-03	0.378 $\pm$ 0.023
All Homosexuals	4.758E-06 $\pm$ 4.049E-06	4.663E-03 $\pm$ 4.355E-04	0.388 $\pm$ 0.019
Homosexuals with IDU	1.772E-05 $\pm$ 2.453E-05	4.977E-03 $\pm$ 2.7780E-03	0.343 $\pm$ 0.033

homosexuals with age $<$ 20 years old at testing time and IDU versus individuals of homosexuals with IDU are given by (4359.500, 4610.312, 9111.172, 1270.719) respectively. (Notice that under H_0, for large sample size the statistic $-2\log\hat{\omega}$ is distributed as a central chi-square variate with degrees of freedom $t_C = 300$ so that all p-values are less than 0.0001.) Also, there are significant differences in HIV seroconversion between those homosexuals with age $\geq$ 20 and those with age $<$ 20 with $-2\log\hat{\lambda} = 10492.000$ and with p-value less than 0.0001. As discussed in Section 7, the latter results might be an artifact and consequences of the HIV epidemic and the sampling scheme for collecting data.

(2) From Figs. 1 and 2, we observe that the probability densities of times to HIV infection and seroconversion for homosexuals, IDU and homosexuals with IDU appear to be mixtures and with heavy weight on the right tail. For those homosexuals with age $\geq$ 20 years old at the testing time (to be referred to as older homosexuals) and homosexuals with

age < 20 years old at the testing time (to be referred to as younger homosexuals), the densities appear roughly to be curves with a single mode. Also, the curve of the older homosexuals is almost the same as that of all homosexuals and is shifted considerably to the left than the younger homosexuals. The mean times to HIV infection and HIV seroconversion for younger homosexuals and older homosexuals are given by $(51.335, 54.016)$ and $(137.253, 139.899)$ months respectively. This indicates that on the average it takes a much longer time for older homosexuals than younger homosexuals to be infected by HIV. As discussed in Section 7, this might be an artifact and the consequences of the sampling scheme for collecting data.

(3) From Figs. 1 and 2, we observe that the probability densities of times to HIV infection for homosexuals, IDU and for homosexuals with IDU appear to be vague mixtures with two peaks and with heavy weight on the right tail. Further, the curves of the IDU shifted more to the left than those of homosexuals.

(4) From results in Table 1, it appears that the mean time to HIV infection in IDU are greater than the corresponding times of homosexuals and homosexuals with IDU. It follows that on average, it takes much longer for the IDU to contract HIV than the homosexuals and homosexuals with IDU. In all cases, the variances of times to HIV infection and seroconversion in IDU are much greater than those from homosexuals.

(5) From Table 2, it is observed that the death and retirement probabilities from HIV infected individuals are significantly greater than those from susceptible people in all populations respectively. These results are consistent with the findings by Tan and Ye (2000b) in the Swiss populations of IDU and homosexuals. Comparing these rates from different populations in Table 2, however, there do not seem to have any significant differences between different populations.

(6) From Table 2, it is observed that there are no significant differences in the window period incidence (i.e. γ) between different populations. This indicates that the window periods from various populations are independently and identically distributed as a geometric distribution with parameter γ. The estimated mean values (i.e. $1/\gamma$) for IDU, younger homosexuals, older homosexuals, all homosexuals and homosexuals with IDU are given by $\{2.849, 2.681, 2.646, 2.577, 2.915\}$ respectively. Thus, after HIV infection, it takes about 2.75 months to develop HIV antibodies for HIV-infected individuals.

7. Conclusions and Discussion

To assess impact of IV drug use and homosexual behavior, one would need to estimate the HIV seroconversion incidence and to compare HIV seroconversions from several different populations such as IDU and homosexuals. The data available are usually HIV seroconversion status of individuals in the population. These data are subject to left truncation, right censoring and interval censoring. Based on these data sets, in this paper we have developed Bayesian estimates of the seroconversion incidence and hence the probability density of time to HIV seroconversion by using the multi-level Gibbs sampling method. Also, we have developed a likelihood ratio testing procedure to compare HIV seroconversion incidence from several different populations. We have applied these models and methods to some HIV seroconversion data under IDU and under homosexuals generated by a cooperative agreement study supported by NIDA/NIH. The results indicate that there are significant differences in times to HIV seroconversion and hence HIV infection between populations of homosexuals, IDU and homosexuals with IDU. The probability densities of times to HIV seroconversion are basically mixtures with two peaks; further, the density curve of IDU shifts more to the left of those from homosexuals. The mean time to HIV seroconversion in month of IDU are greater than those from homosexuals and homosexuals with IDU. Thus it takes a longer time for IDU to become seroconverted than homosexuals. This can help explain why the AIDS incidence under IV drug use in US is only about 50% of that of homosexuals in US; see CDC surveillance report (1997). Contradicting the findings by Tan *et al.* (1998) in the San Francisco homosexual population, the results from the NIDA/NIH data suggested that there were significant differences in HIV seroconversion between homosexuals with age < 20 years old at the testing time and homosexuals with age ≥ 20 years old at the testing time. From the following two observations, it appears that this difference might not be real but is an artifact of the sampling scheme generating the NIDA/NIH data: (i) The sample of the NIDA/NIH data had excluded AIDS cases and were taken over the period between 1992 and 1998 which has passed the primary infection period 1980–1990; furthermore the majority of individuals (over 98%) in the sample have age ≥ 20 old years at the testing time. (There are only 195 individuals among the 12,096 homosexuals who have age < 20 years old at the testing time.) Further, the number of individuals who tested positive at the testing time is 970 among the 11,874 homosexuals who have age ≥ 20 at the testing time. Because it takes about

10 years on average to develop AIDS after HIV infection and AIDS cases are excluded from the data, most individuals (over 90%) in homosexuals with age $\geq$ 20 years at the testing time have low sexual activity level. On the other hand, homosexuals with age $<$ 20 years old at the testing time are expected to be homosexuals with median or high sexual activity levels. (ii) As shown in Tan *et al.* (1998), there are significant differences in HIV seroconversion between individuals with high or median sexual activity level and individuals with low sexual activity level. Comparing the seroconversion density of homosexuals in Figs. 1(b) and 2 with that from the San Francisco homosexual population, clearly the density of homosexuals with age $\geq$ 20 at testing time as given in Fig. 2 matches well with the seroconversion density of homosexuals around the second peak of the curve from the San Francisco population; similarly, the density of homosexuals with age $<$ 20 at the testing time as given in Fig. 1(b) matches well with the seroconversion density of homosexuals around the first peak of the curve from San Francisco population. The apparent discrepancy between the vague mixture of the over-all seroconversion density of homosexuals and that of the San Francisco homosexual population is that more than 90% of homosexuals in the sample have age $\geq$ 20 years old at the testing time who had tested negative and hence are individuals with low sexual activity level.

To estimate the probability of HIV infection of susceptible people in IDU and/or homosexuals, we notice that based on data on HIV seroconversion status only, the problem is not identifiable in the sense that it is not possible to estimate these probabilities and the distribution of window period simultaneously. To resolve this difficulty, in this paper we have developed a state space model for HIV infection and HIV seroconversion. By coupling this state space model with the multi-level Gibbs sampling method, we have developed a generalized Bayesian procedure to estimate simultaneously the unknown parameters under IV drug use and homosexuals. These unknown parameters include the probabilities of HIV infection of S people in populations of IDU and homosexuals, the incidence of the window period, the death and retirement rates of S people and HIV-infected people. We have applied these models and methods to some HIV seroconversion data under IDU and under homosexual behavior generated by a cooperative agreement study supported by NIDA/NIH. We have obtained the following results:

(1) For the probability distribution of the window period, it appears that there are no significant differences between homosexuals, IDU and homosexuals with IDU and are independent of the age of individuals. It takes on the average about 2.75 months for an I person to develop HIV

antibodies. This is close to the estimate of 3 months for the mean of window period by Horsburgh *et al.* (1989).

(2) It appears that there are significant differences in the probabilities of death and retirement between S people and I people in all populations; however, there are no significant differences in these probabilities between IDU and homosexuals and are time independent. This is consistent with the findings by Tan and Ye (2000b) in the Swiss populations of IDU and homosexuals.

(3) The probability densities of times to HIV infection are quite similar to the probability densities of times to HIV seroconversions respectively except that the densities of times to HIV infection have shifted to the right by about 2.8 months. Thus, the probability densities of times to HIV infection are mixtures for homosexuals, IDU and homosexuals with IDU. For homosexuals with age $\geq$ 20 years old at the testing time, the density is a curve with a single peak which matches that of homosexuals with low sexual activity levels whereas that for homosexuals with age $<$ 20 years old at the testing time, the density is a curve with a single peak which matches that of homosexuals with high or median sexual activity levels. These results are similar to the findings in the San Francisco homosexual population by Tan and Ye (2000a) and in the Swiss populations of IDU and homosexuals by Tan and Ye (2000b); but there are significant differences between results in Tan and Ye (2000a, b) and those given in Fig. 2. As discussed above, these differences might be an artifact and consequences of the sampling scheme for generating data in the NIDA/NIH study.

To develop a state space model for HIV serocnversion based on HIV seroconversion status, we have used the data through the likelihood in (16) and the estimated number y_j of seroconverters in the jth month. We have assumed the $e_j = (y_j - c_j)/\sqrt{c_j}$'s as independent normal variate. To check this assumption, we have carried out a Q-Q plot of estimates of e_j as given in Fig. 3. From Fig. 3, it is clear that except at a few extreme points, the estimated residuals appear to fit well with the normal curve. This indicates that one can in fact assume the e_j as normal variables with means 0 and variance σ^2.

From the above demonstration, it appears that we have proposed a useful procedure via the application of the state space models to estimate simultaneously the unknown parameters and the state variables in HIV epidemic. To make this approach applicable to a wide range of problems, however, many problems need to be examined: First, in many practical situations, one would need to extend the model to deal with the problem of

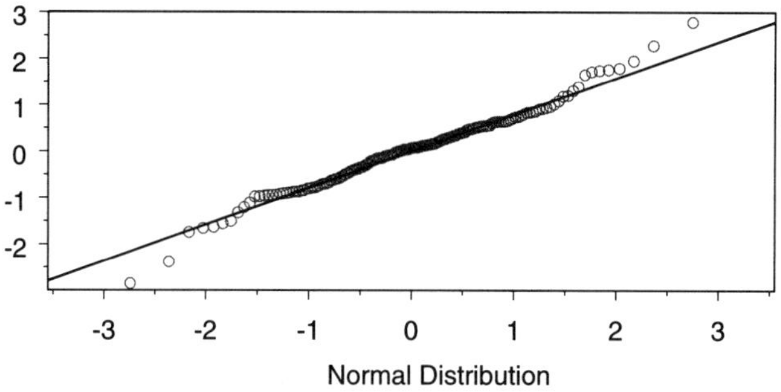

Fig. 3. Q-Q Plot for residuals (Kolmogorov–Smirnov Test Statistic is 0.0533 and P-Value = 0.1500).

immigration and recruitment. Second, in many cases, AIDS incidence data and other data may also be available; it would be necessary to incorporate these additional data sets in the observation model to estimate the HIV incubation and prevalence of AIDS. Third, in some cases, other risk variables than age may also be available. In these cases, it may be necessary to study effects of risk variables on HIV infection and HIV seroconversion in the HIV seroconversion data. We will address these problems in our future research; we will not go any further here.

Acknowledgements

The authors wish to acknowledge the help of Dr. S. Strauss from National Development and Research Institute, Inc for sending us the data.

References

Box GEP, Tiao GC (1973) *Bayesian Inference in Statistical Analysis.* Addison–Wesley, Reading, Mass.

CDC (1997) *Surveillance Report of HIV/AIDS*, June, Atlanta, GA.

Horsburgh CR Jr, Qu CY, Jason IM *et al.* (1989) Duration of human immuodeficiency virus infection before detection of antibody. *Lancet* **2**: 637–640.

Liu JS, Chen R (1998) Sequential Monte Carlo method for dynamic systems. *J Am Stat Assoc* **93**: 1032–1044.

Raiffa H, Schlaiffer R (1961) *Applied Statistical Decision Theory*, The M.I.T. Press, Cambridge.

Shephard N (1994) Partial non-Gaussian state space. *Biometrika* **81**: 115–131.

Smith AFM, Gelfand AE (1992) Bayesian statistics without tears: A sampling-resampling perspective. *Am Stat* **46**: 84–88.

Tan WY (2000) *Stochastic Modeling of AIDS Epidemiology and HIV Pathogenesis*, World Scientific, Singapore.

Tan WY (2002) *Stochastic Models With Applications to Genetics, Cancers, AIDS and Other Biomedical Systems*, World Scientific, Singapore.

Tan WY, Lee SR, Tang SC (1996) Characterization of the HIV incubation distributions and some comparative studies. *Stat in Med* **15**: 197–220.

Tan WY, Tang SC, Lee SR (1995) Effects of randomness of risk factors on the HIV epidemic in homo-sexual populations. *SIAM J Appl Math* **55**: 1697–1723.

Tan WY, Tang SC, Lee SR (1998) Estimation of HIV seroconversion and effects of age in San Francisco homosexual populations. *J Appl Stat* **25**: 85–102.

Tan WY, Wu H (1998) Stochastic modeling of the dynamic of $CD4^+$ T cell infection by HIV and some monte carlo studies. *Math Bioscience* **147**: 173–205.

Tan WY, Ye ZZ (2000) Estimation of HIV infection and HIV incubation via state space models. *Math Biosci* **167**: 31–50.

Tan WY, Ye ZZ (2000) Simultaneous estimation of HIV infection, HIV incubation, immigration rates and death rates as well as the numbers of susceptible people, infected people and AIDS cases. *Comm Stat (Theory & Methods)* **29**: 1059–1088.

Tan WY, Zhang P, Xiong X (2004) A state space model for HIV pathogenesis under anti-viral drugs and applications, in *Deterministic and Stochastic Models of AIDS and HIV with Intervention*, World Scientific, Singapore.

CHAPTER 3

A BAYESIAN MONTE CARLO INTEGRATION STRATEGY FOR CONNECTING STOCHASTIC MODELS OF HIV/AIDS WITH DATA

Charles J. Mode

Department of Mathematics
Drexel University
Philadelphia, PA 19104

Whenever a Bayesian procedure that involves the numerical evaluation of multi-dimensional integrals is contemplated, a decision as to whether proceed with a computer implementation of the procedure will often depend on the ease with which the software may be written. The ease with which the software may be written will, in turn, depend not only programming language chosen but also a programmer's knowledge of the language. Because the author has had rather extensive experience in using the programming language, APL, which, among other things, is known for ease with which succinct code may be written to process complex arrays, it was known at the outset that the software to program Bayesian structure outlined in this paper could written and validated with relative ease. Furthermore, it seems likely that any other programming language with array processing capabilities could also be used to write code the implement the ideas presented in this paper. Consequently, a decision was made to organize the ideas in a precise mathematical form as a first step to writing computer code in any programming language chosen by an investigator. One of the outstanding problems in using the Bayesian paradigm to estimate unknown parameters and make statistical inferences, using stochastic models of HIV/AIDS epidemics, was that of developing a methodology for drawing sample from the joint conditional posterior distribution of the parameters and the stochastic process, given the data. Chapter 6 in the recent book by (Tan, 2000) may be consulted for technical details. Among other things, this paper contains a novel procedure, depending on the ease with which large arrays may be processed in a computer, for drawing random samples from the posterior of the parameters and the process, given a sample of data.

Keywords: Stochastic models of HIV/AIDS; Bayesian methods; Monte Carlo integration; weighted bootstrap; selecting maximal values of the likelihood function.

1. Introduction

Stochastic models of epidemics often have complicated likelihood functions that are very difficult to compute given a set of data. Consequently, there is a need to develop strategies for coping with difficult computational problems that arise in connection with interfacing stochastic models of epidemics with data. In this note, attention will be focused on a Monte Carlo integration strategy that yields an approach for computing samples from the joint conditional distribution of the parameters and the process, given the data, within a Bayesian paradigm. In Bayesian terminology, this conditional distribution is usually referred to as the posterior distribution.

A reading of Chapter 6 in the recent book and interesting book by (Tan, 2000) motivated the thought processes underlying the development of these notes. In this chapter, Tan presented a methodology for estimating parameters in a model of HIV/AIDS epidemic in a population of male homosexuals, using incidence data from San Francisco and Switzerland. The stochastic structure used by this author was a chain binomial model and was linear in that probabilities of infection did not depend on the state of the population but they were allowed to change with time. The method used to estimate the unknown parameters fell within the Bayesian paradigm and employed such methods as Gibbs sampling and a weighted boot strap, when the posterior distribution of the process, given the data, was considered.

As this chapter was studied, it also became apparent that the proposed Bayesian methods were not adequate for estimating the many unknown parameters of the non-linear models of HIV/AIDS epidemics considered in Chapters 10, 11 and 12 of (Mode and Sleeman, 2000), which do depend on the state of the population at any time. Consequently, a decision was made to undertake an in depth development of the methods proposed by Tan in which more attention would be paid to the mathematical fundamentals underlying the suggested computer intensive methods.

In this regard, the paper by (Smith and Gelfand, 1992) provided inspiration, even though at the time of publication, as the title suggests, it was considered primarily as a didactic exercise. But with availability of fast personal computers with increasing memory capacity, it seems reasonable that the Monte Carlo integration strategies suggested by these authors for coping with difficult integration problems be brought into the research arena on stochastic models of infectious diseases. An interesting book on Bayesian methodology that focuses attention on tractable likelihood functions is that of (Tanner, 1996). A more recent book, containing an extensive account of Monte Carlo simulation methods in statistics, is that of (Roberts and Casella, 2000).

2. Basic Bayesian Concepts

To set the stage for the development of Monte Carlo integration strategy, it will be helpful to cast the problem at a more abstract level than that considered in Chapter 6 of (Tan, 2000). Let $g(\theta)$ be the prior *p.d.f.* of the parameters, where θ is a vector taking values in some parameter space Θ_{d_1} of dimension $d_1 \geq 1$, and let $g(x \mid \theta)$ be the conditional *p.d.f.* of the process, given a value $\theta \in \Theta_{d_1}$, where x is a vector taking values in the set $\mathfrak{X}_{d_2}$ of all possible realizations of the process of dimension $d_2 \geq 1$. Next, let $g(y \mid \theta, x)$ be the conditional density of the data, given vectors $\theta \in \Theta_{d_1}$ and $x \in \mathfrak{X}_{d_2}$, where the data vector $y \in \mathfrak{D}_{d_3}$, the set of all possible data points of some dimension $d_3 \geq 1$. Then, the joint density function of the parameters, a sample of realizations of the process and the data is, by definition,

$$g(\theta, x, y) = g(\theta)g(x \mid \theta)g(y \mid \theta, x). \tag{2.1}$$

The marginal *p.d.f.* of the data is, by definition,

$$g_3(y) = \int_{\Theta_{d_1}} \int_{\mathfrak{X}_{d_2}} g(\theta, x, y)\, \mu_1(d\theta)\, \mu_2(dx), \tag{2.2}$$

where $\mu_1(d\theta)$ and $\mu_2(dx)$ are measures chosen in a way that is suitable for the model under consideration. For example, $\mu_1(d\theta)$ will usually be a product Lebesque measure, because the set Θ_{d_1} is frequently chosen as a set of d_1-dimensional vectors of real numbers; whereas, because realizations of the process usually are vectors of non-negative integers, $\mu_2(dx)$ will be a counting measure.

Let A and B be subsets of Θ_{d_1} and $\mathfrak{X}_{d_2}$, respectively. Technically, these sets will be measurable, but the details will be omitted. When working within the Bayesian paradigm, a capability for computing conditional probabilities of the form

$$P[\theta \in A,\ x \in B \mid y] \tag{2.3}$$

is required. Formally, this conditional probability has the formula

$$P[\theta \in A, x \in B \mid y] = \frac{\int_A \int_B g(\theta, x, y)\, \mu_1(d\theta)\, \mu_2(dx)}{\int_{\Theta_{d_1}} \int_{\mathfrak{X}_{d_2}} g(\theta, x, y)\, \mu_1(d\theta)\, \mu_2(dx)}. \tag{2.4}$$

The conditional *p.d.f.* of θ and x, given the data y, is

$$f(\theta, x \mid y) = \frac{g(\theta, x, y)}{\int_{\Theta_{d_1}} \int_{\mathfrak{X}_{d_2}} g(\theta, x, y)\, \mu_1(d\theta)\mu_2(dx)}. \tag{2.5}$$

In general, multidimensional integrals that appear in the numerator and denominator of (2.4) are very difficult to compute in practical situations, which often mitigates against working within the Bayesian paradigm. However, as will be demonstrated, it is possible to design Monte Carlo integration strategies such that these multidimensional integrals can be estimated.

3. A Monte Carlo Integration Strategy

Let

$$\{(\theta_i, x_i) \mid i = 1, 2, \ldots, N\} \tag{3.1}$$

denote a sample of independent realizations of the parameter vector θ and the process x. That is, for each $i = 1, 2, \ldots, N$, θ_i is a sample from $g(\theta)$ and, given θ_i, x_i is a realization of the process sampled from $g(x \mid \theta_i)$. For every $i = 1, 2, \ldots, N$, let

$$w_i(\theta_i, x_i) = g(y \mid \theta_i, x_i). \tag{3.2}$$

Then, given the data vector y,

$$\{w_i(\theta_i, x_i) \mid i = 1, 2, \ldots\} \tag{3.3}$$

is, by assumption, a sequence of conditionally independent random variables.

Let $I_{A\times B}(\theta, x)$ be the indicator function of the product set $A \times B$. That is $I_{A\times B}(\theta, x) = 1$ if $(\theta, x) \in A \times B$ and $I_{A\times B}(\theta, x) = 0$ if $(\theta, x) \notin A \times B$. For every $i = 1, 2, \ldots, N$ the conditional expectation of $w_i(\theta_i, x_i)I_{A\times B}(\theta, x)$, given the data y, is

$$E[w_i(\theta_i, x_i) I_{A\times B}(\theta_i, x_i) \mid y] = \int_A \int_B g(\theta) g(x \mid \theta) g(y \mid \theta, x)\, \mu_1(d\theta)\, \mu_2(dx). \tag{3.4}$$

Similarly, for every $i = 1, 2, \ldots$, the conditional expectation of $w_i(\theta_i, x_i)$, given the data y, over the entire product space $\Theta_{d_1} \times \mathfrak{X}_{d_3}$ is

$$E[w_i(\theta_i, x_i) \mid y] = \int_{\Theta_{d_1}} \int_{\mathfrak{X}_{d_2}} g(\theta) g(x \mid \theta) g(y \mid \theta, x)\, \mu_1(d\theta)\, \mu_2(dx). \tag{3.5}$$

Observe that (3.4) and (3.5) are, respectively, the numerator and the denominator of (2.4).

Next let

$$R_{A\times B}(N) = \frac{1}{N} \sum_{i=1}^{N} w_i(\theta_i, x_i) I_{A\times B}(\theta_i, x_i) \tag{3.6}$$

and let

$$R_{\Theta_{d_1}\times \mathfrak{X}_{d_2}}(N) = \frac{1}{N} \sum_{i=1}^{N} w_i(\theta_i, x_i). \tag{3.7}$$

Then, from the ratio

$$\frac{R_{A\times B}(N)}{R_{\Theta_{d_1}\times \mathfrak{X}_{d_2}}(N)}. \tag{3.8}$$

By the strong law of large numbers, it follows that the limit

$$\lim_{N\uparrow\infty} \frac{R_{A\times B}(N)}{R_{\Theta_{d_1}\times \mathfrak{X}_{d_2}}(N)} = P[\theta \in A, x \in B \mid y] = \frac{\displaystyle\int_A \int_B g(\theta, x, y)\, \mu_1(d\theta)\, \mu_2(dx)}{\displaystyle\int_{\Theta_{d_1}} \int_{\mathfrak{X}_{d_2}} g(\theta, x, y)\, \mu_1(d\theta)\, \mu_2(dx)} \tag{3.9}$$

holds with probability one for any measurable product set $A \times B$.

It important to observe that in applications of the ratio in (3.9) to approximate the posterior conditional probability, $P[\theta \in A,\ x \in B \mid y]$, the numerical evaluation of the densities $g(\theta)$ and $g(x \mid \theta)$ would not be necessary in a computational procedure. All this is needed are procedures for simulating realizations of random samples from these distributions. This observation is particularly important for the case of the conditional density of the process, $g(x \mid \theta)$, given θ, because, as can be seen from consulting Chapters 10, 11 and 12 in Mode and Sleeman (2000), specific forms of this function would be very difficult to derive and compute.

Equation (3.9) has two easy corollaries. For example, suppose one is interested in the marginal conditional probability $P[\theta \in A \mid y]$. That is, given the data, one wishes to draw inferences about the conditional distribution of the parameters ignoring the process. Then,

$$P[\theta \in A \mid y] = \frac{\int_A \int_{\mathfrak{X}_{d_2}} g(\theta, x, y)\, \mu_1(d\theta)\, \mu_2(dx)}{\int_{\Theta_{d_1}} \int_{\mathfrak{X}_{d_2}} g(\theta, x, y)\, \mu_1(d\theta)\, \mu_2(dx)}, \tag{3.10}$$

where A is any measurable subset of Θ_{d_1}. Similarly, the conditional distribution of the process, given the data y, is

$$P[x \in B \mid y] = \frac{\int_{\Theta_{d_1}} \int_B g(\theta, x, y)\, \mu_1(d\theta)\, \mu_2(dx)}{\int_{\Theta_{d_1}} \int_{\mathfrak{X}_{d_2}} g(\theta, x, y)\, \mu_1(d\theta)\, \mu_2(dx)}, \tag{3.11}$$

where B is any measurable subset of $\mathfrak{X}_{d_2}$. In a subsequent section, applications of these formulas for the conditional marginal distributions will be discussed.

4. On the Conditional Likelihood Function of the Data Given a Point in the Parameter Space and a Realization of the Process

Available data on epidemics of infectious diseases usually consists of a time series of relatively short duration consisting of a few days, weeks, months or years. Suppose the vector y of observations has the form

$$y = (y_1, y_2, \ldots, y_{t_1}), \tag{4.1}$$

where t_1 is the last time interval for which data are available. In many data sets, these numbers would be the incidence data, i.e. the number of new cases of an infectious disease reported at the end of a set of t_1 time intervals. Given a realizations of the parameter vector θ_i in the ith sample, let x_i be a realization of the process. And, let

$$(x_{i1}, x_{i2}, \ldots, x_{it_1}) \tag{4.2}$$

denote a sub-vector of the realization x_i corresponding the data vector in (4.1). That is, the vector in (4.2) is the simulated data, given the parameter point θ_i.

Then, a useful choice for the conditional likelihood function of the data, given the ith simulated point (θ_i, x_i), is

$$w_i(\theta_i, x_i) = g(y \mid \theta_i, x_i) = \frac{1}{(2\pi)^{\frac{t_1}{2}}} \exp\left[-\frac{1}{2}\sum_{j=1}^{t_1}(y_j - x_{ij})^2 \right]. \tag{4.3}$$

Observe that the sum in the exponent on the right is the square of the Euclidean distance of the simulated data from the observed data. Moreover, the likelihood function is a decreasing function of this distance and is maximized when the distance is zero. From the probabilistic point of view, the conditional likelihood function is that of t_1 independent normal random variables, whose conditional mean vector is that in (4.2) and with common variance 1. One could, of course, introduce a variance σ_i^2 to be estimated by the method of maximum likelihood for each simulated pair (θ_i, x_i), but this seems unnecessary, because, as will be demonstrated, when the whole set of N is simulated points is considered, choices will be made based on the largest values of the weights in (4.3).

A possible difficulty that may arise in applying (4.3) is that the sum of squares in the exponent could be so large as to cause a computer over flow problem, resulting in a termination of a computer run and a incomplete experiment. One way around this potential difficulty is to introduce a variance σ_i^2, compute its maximum likelihood estimate and evaluate the likelihood function in (4.3) at this estimate. This procedure would result in the weight

$$w_i(\theta_i, x_i) = \frac{t_1^{\frac{t_1}{2}}}{\left(2\pi \sum_{j=1}^{t_1}(y_j - x_{ij})^2\right)^{\frac{t_1}{2}}} \exp\left[-\frac{t_1}{2}\right]. \tag{4.4}$$

Observe that this weight is also a decreasing function of the square of Euclidean distance of the simulated data from the observed data and could be used in the procedures discussed in subsequent sections.

5. A Weighted Boot Strap Method for Resampling the Posterior Distribution

From (3.9) it follows that for any measurable product set $A \times B$

$$\frac{R_{A\times B}(N)}{R_{\Theta_{d_1}\times \mathfrak{X}_{d_2}}(N)} = \frac{\sum_{i=1}^{N} w_i(\theta_i, x_i) I_{A\times B}(\theta_i, x_i)}{\sum_{i=1}^{N} w_i(\theta_i, x_i)} \approx P[\theta \in A, x \in B \mid y] \tag{5.1}$$

for N sufficiently large. The issue of deciding when N is sufficiently large, so that the approximation of the right is valid, will be addressed subsequently. A weighted boot strap method for sampling the set of simulated data points consists of sampling from the multinomial distribution, whose probabilities are given by

$$p_i(\theta_i, x_i) = \frac{w_i(\theta_i, x_i)}{\sum_{i=1}^{N} w_i(\theta_i, x_i)}, \tag{5.2}$$

where $i = 1, 2, \ldots, N$.

As a first step in describing this resampling procedure, let $\mathfrak{A}$ denote a $N \times (d_1 + d_2)$ matrix such that the ith row in this matrix is the simulated data point (θ_i, x_i) for $i = 1, 2, \ldots, N$. The next step in the conception of this resampling procedure is to draw a sample of size 1 from the multinomial distribution in (5.2). The computer output from this simulated draw will be a multinomial indicator function

$$\Phi = (\varphi_i \mid i = 1, 2, \ldots, N) \tag{5.3}$$

where for every i, $\varphi_i = 0$ or $\varphi_i = 1$ and

$$\sum_{i=1}^{N} \varphi_i = 1. \tag{5.4}$$

Observe that all the $\varphi's$ are zero except one, which indicates which one among the N disjoint events occurred. For example, suppose it is the νth. Then, $\varphi_\nu = 1$ and $\varphi_i = 0$ for all $i \neq \nu$.

The next device to introduce into the resampling procedure is that of programming the computer to select the appropriate index, given that the νth event occurred. To this end, it is useful to let

$$\mathfrak{I} = (i \mid 1, 2, \ldots, N) \tag{5.5}$$

denote a vector of indices, corresponding the rows of the matrix $\mathfrak{A}$. If the νth event occurred, then the inner product of these vectors is

$$(\Phi, \mathfrak{I}) = \sum_{i=1}^{N} i\varphi_i = \nu. \tag{5.6}$$

Hence, the first simulated data point selected in the resampling procedure would be

$$\left(\theta_\nu^{(1)}, x_\nu^{(1)}\right) \leftarrow \mathfrak{A}[\nu;], \tag{5.7}$$

where the symbol on the right indicates that the νth row of the matrix $\mathfrak{A}$ was selected.

This resampling procedure may be continued step-wise to produce a sub-sample of simulated data

$$\left\{\left(\theta_{\nu_j}^{(j)}, x_{\nu_j}^{(j)}\right) \mid j = 1, 2, \ldots, n\right\}, \tag{5.8}$$

where $n < N$. Observe that the set of indices

$$\{\nu_j \mid j = 1, 2, \ldots, n\} \tag{5.9}$$

may not be distinct elements of the index vector $\mathfrak{I} = (i \mid i = 1, 2, \ldots, N)$, indicating that some rows of the matrix $\mathfrak{A}$ may have been selected more than once. There is a ready explanation for this possibility, because the resampling procedure would tend to select events that were among those with the larger probabilities in (5.2). These larger probabilities would also reflect better fits of the simulated data to the actual data, see (4.3) and (4.4).

According to (3.10), if N is sufficiently large, then the set of parameter vectors

$$\left\{\left(\theta_{\nu_j}^{(j)}\right) \mid j = 1, 2, \ldots, n\right\} \tag{5.10}$$

is a random sample from the conditional distribution of the parameters, given a data vector y. The next step in the analysis of the set of parameter vectors in (5.10) would be that computing statistical summaries of each element in the sample of parameter vectors of dimension d_1. Among the summary statistics for each element of a parameter vector that could be informative regarding the variation in the parameter values would be the min, max, and median as well as the mean and standard deviation. If the dimension of the parameter space is $d_1 = 2$, then a two-dimensional scatter plot of the points in the data set in (5.10) could also be informative. It may also be of interest to compute a covariance and correlation matrix from the sample of parameter points, when $d_1 \geq 2$.

Similarly, according to (3.11), the set of realizations of the process

$$\left\{x_{\nu_j}^{(j)} \mid j = 1, 2, \ldots, n\right\} \tag{5.11}$$

is a random sample from the conditional distribution of the process, given a data vector y. Each vector in this sample is a realization of the process as a function of time. An informative way of statistically summarizing this sample of realizations is to compute the order statistics at each time point. Then, given an array of order statistics, it would be possible to compute estimated quantile trajectories of the process. Among the quantile trajectories that may be informative would be the min, max and median.

If desired, the mean and standard deviation trajectories could also be computed. After the desired trajectories of the process have been computed, a simultaneous plot of the estimated quantile trajectories would provide an informative summarization of the probable behavior of the process, given a data vector y.

To gain some insights into how the predictions of the process may change, given the data, it may also be informative to compute the estimated quantiles of the entire set

$$\{x_i \mid i = 1, 2, \ldots, N\} \tag{5.12}$$

of realizations of the process. A plot of these quantile trajections could then be informative when compared with plots of the quantile trajectories computed from the sample in (5.11).

The procedure outlined above for drawing n samples from the multinomial distribution in (5.2) one at a time can easily be implemented in APL, using the each operator and nested arrays. However, because the algorithm used in drawing a sample of size 1 from the a multinomial distribution entails a recursive procedure, N loops will be required for each sample of size 1. When N is large, $n \times N$ loops would be required to complete a run, using the procedure described above. This could lead to long waiting times for the computer to complete a resampling experiment. Consequently, it is of interest to consider another approach.

The recursive algorithm just mentioned may also be used to compute a sample of size n for the distribution in (5.2), entailing only N loops. Let the vector

$$(z_i \mid i = 1, 2, \ldots, N) \tag{5.13}$$

denote a sample from the multinomial distribution in (5.2) resulting from this method of sampling. Then,

$$\sum_{i=1}^{N} z_i = n. \tag{5.14}$$

In this approach, the next step would be to examine every element of the vector of non-negative integers in (5.14) as follows.

If $z_1 = 0$, then go to z_2. If $z_1 \geq 1$, then create an array of z_1 repeats of the index 1. For example, if $z_1 = 2$, this array would be 1 1. Then, go to z_2 and repeat the procedure just described for z_1. By continuing recursively in this way, an array of the form

$$\{R_i \mid i = 1, 2, \ldots, N\} \tag{5.15}$$

could be created such that if $z_i = 0$, then the sub-array R_i is empty, but if $z_i \geq 1$, then the sub-array R_i would contain z_i repeats of the index i. For any non-empty array R_i, the ith row of the matrix $\mathfrak{A}$ would be sampled $z_i \geq 1$ times. This procedure could also be easily implemented in APL or, indeed, in any programming language of an investigator's choice. Although it is known that applying the algorithm for sampling from a multinomial distribution one sample at a time is very efficient, the procedure just described may be more computationally efficient.

Another advantage of programming in APL is that many nice properties are built into the language, which often simplifies the writing of code. For example, in APL the syntax for creating the arrays in (5.15) is the simple expression $R_i \leftarrow z_i \rho i$. If $z_i = 0$, then the array R_i is empty, but if $z_i \geq 1$, then R_i is an array with the index i repeated z_i times. Thus, in writing code to generate the array in (5.15), it would not be necessary to test whether $z_i = 0$ or $z_1 \geq 1$ for every i, if the each operator were used. Briefly, the each operator is a short hand notation for instructing the computer to perform some operation in each component of a nested array.

6. Resampling the Posterior Distribution Based on the Largest Probabilities

Because the multinomial probabilities in (5.2) are random variables, it makes sense to compute their order statistics and draw a sample from the posterior distribution based on the largest probabilities in this distribution. In a sense, this selection of a sample would be based on a goodness-of-fit criterion, because the larger probabilities would correspond to the better the fits of the simulated data to the observed sample points.

Let

$$\{p_{r_i}(\theta_{r_i}, x_{r_i}) \mid i = 1, 2, \ldots, N\} \tag{6.1}$$

be the order statistics of the probabilities in (5.2). By definition, the number r_1 is a member of the index set $\mathfrak{I} = \{i \mid i = 1, 2, \ldots, N\}$ such that $p_{r_1}(\theta_{r_1}, x_{r_1})$ is the smallest probability in (5.2) and so on for the other ranks in (6.1). Observe that the set of ranks

$$\{r_i \mid i = 1, 2, \ldots, N\} \tag{6.2}$$

is a permutation of the index set $\mathfrak{I}$.

Given this set of ranks, a number of interesting ways of selecting a sub-sample from the posterior distribution becomes apparent. For example, the rank r_N corresponds to the largest probability in (5.2). Consequently, the row

$$(\theta_{r_N}, x_{r_N}) \leftarrow \mathfrak{A}[r_N;] \tag{6.3}$$

of the matrix $\mathfrak{A}$ of simulated data would provide the "best" fit to the observed data. However, in the face of uncertainty and variation, it would seem wise not to base statistical inferences on only a point estimate.

Therefore, one may wish to consider a sub-sample based on a chosen subset of the largest probabilities to get some idea of the variability among the parameters and realizations of the process among those simulated sample points that gave "good" fits to the data. For some $n < N$, let

$$\{r_i \mid i = n, n+1, \ldots, N\} \tag{6.4}$$

be the set of ranks corresponding to the larger probabilities. Then, by selecting the rows of the matrix $\mathfrak{A}$ corresponding to the set of ranks in (6.4) would result in a sub-sample of simulated points that would be the "better" fits to the observed data. The number n could be chosen such that, for example, the sample of the largest probabilities would represent some percentage of the entire sample of size N. This percentage could be chosen as 25 or any other percentage deemed appropriate. Given this sub-sample of simulated data points, the procedures used to obtain informative statistical summaries could be those suggested in the previous section. From the computational point of view, this procedure for selecting a sub-sample from the posterior would be more efficient computationally, because fewer loops in a program would be required to do the calculations.

7. A Criterion for Selecting Sample Size

By the law of large numbers, the sequence of means

$$\bar{W}_n = \frac{1}{n}\sum_{i=1}^{n} w_i(\theta_i, x_i), \quad n = 1, 2, \ldots \tag{7.1}$$

will converge with probability one to a constant as $n \uparrow \infty$, see (3.6) and (3.9). Therefore, an approach to choosing the simulated sample size N is to continue sampling until some number N is reached such that $\bar{W}_n$ is constant within some preselected number of decimal points for all n in the neighborhood of the integer N. It seems plausible that constancy to within one to three decimals would suffice for many practical situations.

As a further test as to whether convergence is actually occurring, one may also wish to investigate the convergence of the sequence of means

$$\bar{W}_n(A \times B) = \frac{1}{n}\sum_{i=1}^{n} w_i(\theta_i, x_i) I_{A\times B}(\theta_i, x_i), \quad n = 1, 2, \ldots \tag{7.2}$$

for a few selected product sets $A \times B$. Because, the parameter space of a model Θ_{d_1} is usually much simpler than space of the process $\mathfrak{X}_{d_2}$, in testing for convergence in (7.2) it would be expedient to choose a product set of the form $A \times \mathfrak{X}_{d_2}$, where A is a simple sub-set of Θ_{d_1}. For example, if $d_1 = 1$, then the set A could be chosen as $A = \{\theta_1 \mid \theta_1 \leq \theta_0\}$ for some $\theta_0 \in \Theta_1$. Similar remarks for parameter spaces of dimension $d_1 > 1$ could also be made. It should also be mentioned, that when the programming language APL is used, the implementation of the ideas outlined in this section would be quite straight forward.

It is interesting to note that in a chapter their book, devoted to issues of convergence when using either the Metropolis–Hastings or Gibbs sampling algorithms (Roberts and Casella, 2000), recommend using a sequence of means similar to those in (7.1) to check convergence. The rationale underlying this recommendation is that in both algorithms the sampling procedure determines a transition function of a Markov chain with a stationary distribution that is the joint conditional distribution of the parameters, given the data. By applying the well-known ergodic theorem for Markov chains with a stationary transition function, it follows that sequence of means will converge with probability one to an expectation determined by the stationary distribution.

8. Strategies for Confronting Issues of Computer Performance

From the foregoing discussion, it is clear that the computer implementation of the ideas just outlined could result in long computer run times. At the same time, however, the computer implementation of these ideas seems plausible for many models of interest. For example, even in models with parameter spaces of high dimension, attention could be focused on those parameters such that slight changes in their values have profound effects on the behavior of the process. For the case of models of infectious diseases, probabilities that a susceptible person becomes infected per contact with an infectious individual are examples of such parameters. In such cases, given a set of background parameters such that the process is less sensitive to changes in their values, attention could be focused on probabilities of infection per contact, which could lead to parameter spaces of low dimension, say $d_1 = 1, 2, 3$. Indeed it is recommended that preliminary sensitivity analyses be run to determine what parameters have the greatest effects on the evolution of the process.

Another factor that contributes to the plausibility of implementing the ideas described above is that, for the case of an infectious disease, time series of incidence data are relatively short so that realizations of the process would not have to be computed for long periods of time. This factor could be an important one in reaching a decision whether a computer implementation of the ideas discussed above could be used to analyze a set of incidence data.

As shown in Mode and Sleeman (2000) and elsewhere, see particularly Chapters 10, 11, and 12, deterministic models may be embedded in a stochastic process in a systematic way. Furthermore, the computation of trajectories of deterministic model requires relatively little computer time. For example, the computing of a trajectory of an embedded deterministic model, which usually entails a recursive evaluation of a set of non-linear difference equations, can be accomplished in seconds or at most minutes. Whereas, the computation of a sample of realizations of the process may take a half hour or up to few or many hours, depending on the fineness of the time scale and the length of the projection period. Therefore, the efficiency with which computer experiments with the embedded deterministic model can be conducted suggests that a prudent course to follow would be that conducting exploratory experiments of the parameter space with the deteministic model before preceding with confirmatory experiments with the process by computing a sufficiently large sample of its realizations.

To this end, let

$$\{(\theta_i, \widehat{x}_i) \mid i = 1, 2, \ldots, N\} \tag{8.1}$$

denote a sample resulting from the computations of a set of N trajectories of the embedded deterministic model. That is, for each $i = 1, 2, \ldots, N$, θ_i is a sample from $g(\theta)$ and, given θ_i, $\hat{x}_i$ is a computed trajectory of the embedded deterministic model. Given the sample in (8.1), its analysis could proceed by the methods outlined in the foregoing sections. Observe, that with respect to the parameter space Θ_{d_1}, a realization of a trajectory $\hat{x}$ of the embedded deterministic model is a random element.

When working with the embedded deterministic model, at least two cases arise. The first case is that in which a model is essentially linear so that the embedded deterministic model is actually the mean function of the process. In such cases, it would make sense to fit the mean to the data, because if one wishes to predict the sample functions of the process, the best predictor is the mean function in the sense of least squares. On the other hand, if a model has non-linearities resulting from its construction, then the embedded deterministic model is not the mean function of the process. For further information on models of this types, see Mode and Sleeman (2000), and Chapters 10–12. For this class of models, the trajectories computed from the embedded deterministic model may not be good measures of central tendency for the sample functions of the process. In such cases, it will be necessary to run confirmatory experiment with the process to assess the validity of estimates of parameters based on using the embedded deteministic model in exploratory experiments.

9. On Choosing Prior Distributions of the Parameters

When purely analytical methods are emphasized, most Bayesian analyses entail the use of conjugate prior distributions for the parameters. Briefly, given a statistical model in the exponential class of distributions, a class of models that constitute the vast majority of models used in the statistical analysis of data, conjugate prior distributions for the parameters are often chosen to expedite the integrations needed to carry out a Bayesian analysis. Indeed, in some cases, the required form of the posterior distribution of the data can be expressed in a simple closed form. However, even when parametrized form of conjugate priors are considered, parameter values are often chosen such that a prior distribution reduces to a uniform distribution on some interval. A good example of this situation is when a Beta prior is chosen for a Bernoulli or binomial distribution.

An advantage of the computer intensive methodology outlined in this note is that there is no need to choose particular forms of prior distributions for the parameters. In fact, in preliminary experiments with the embedded deterministic model may suggest those regions of a parameter space that are most sensitive and seem to lead to the most realistic scenarios for an epidemic to develop. Given this preliminary information, uniform priors may be placed of those regions of the parameter space that seem most relevant. To simply matters, for the case of multidimensional parameter space, one could assume independent uniform priors of some intervals of most interest. After a sample from the posterior distribution has been computed, it may be checked statistically for correlations among the parameters.

It is also relevant to observe that this approach to selecting prior distributions is simple and, in some cases, may eliminate the need for developing sampling algorithms that are necessary when using the Metropolis–Hastings or a Gibbs sampler algorithm. In the case of the Metropolis–Hastings algorithm, this procedure does not seem applicable, because of computational difficulties that arise when attempting to compute the likelihood function of the process. Moreover, for similar reasons, it would be very difficult to derive formulas for the conditional distributions making up a Gibbs sampler algorithm. Hence, neither of these widely used algorithms seem applicable to the class of models under consideration.

References

1. Mode CJ, Sleeman CK (2000) *Stochastic Processes in Epidemiology, HIV/AIDS, Other Infectious Diseases, and Computers*, World Scientific.
2. Robert CP, Casella G (2000) *Monte Carlo Statistical Methods*, Springer, New York, Berlin, Heidelberg.
3. Smith AFM, Gelfand AE (1991) Bayesian statistics without tears: a sampling-resampling perspective, *Amer Statist* **46**: 84–88.
4. Tan WY (2000) *Stochastic Modeling of AIDS Epidemiology and HIV Pathogenesis*, World Scientific, Singapore.
5. Tanner MA (1996) *Tools for Statistical Inference — Methods for the Exploration of Posterior Distribution and Likelihood Functions*, Springer-Verlag, New York, Berlin, Heidelberg.

CHAPTER 4

A CLASS OF METHODS FOR HIV CONTACT TRACING IN CUBA: IMPLICATIONS FOR INTERVENTION AND TREATMENT

Ying-Hen Hsieh*, Hector de Arazoza[a], Rachid Lounes[b], and Jose Joanes[c]

A class of four linear and nonlinear differential equations models is given to describe the detection of HIV-positive individuals in Cuba through random screening and contact tracing. The basic reproduction number is obtained for each of the four models. Cuban HIV data from 1986 to 2002 are used to fit the models for the purpose of comparison. We also use the models to gauge the difference in detection time through random screening and contact tracing. Remarks on the implications for intervention measures and treatment of people living with HIV in Cuba are also given.

Keywords: HIV, AIDS, Cuba, contact tracing, mathematical model, basic reproduction number, intervention measures, HAART.

1. Introduction

The first Acquired Immunodeficiency Syndrome (AIDS) case was diagnosed in Cuba in April of 1986. This signaled the starting point of the Human Immunodeficiency virus (HIV/AIDS) epidemic in the country, although some HIV-positive persons had been detected at the end of 1985. At the early stages the Cuban Government had initiated several preventive measures to try to contain the possible outbreak of the epidemic (Granich *et al.*, 1995; Pèrez Avila *et al.*, 1996; Swanson *et al.*, 1995). Among these measures

*Department of Applied Mathematics, Nacional Chung Hsing University, Taichung, Taiwan 402
[a]University of Havana, Dept. Ecuaciones Diferenciales, Facultad Matemática y Computación, San Lázaro y L, C. de La Habana, Cuba. arazoza@matcom.uh.cu
[b]Université René Descartes, Laboratoire de Statistique Médicale, équipe MAP5 VMR CNRS 8145, 45 rue des Saints-Pères 75270 Paris cedex 06, France. lounes@math-info.univ-paris5.fr
[c]Dept. of Epidemiology, Ministry of Public Health, Havana, Cuba

was a total ban on the import of blood, and blood byproducts. Once the first cases were confirmed, a program based on the experience with other sexually transmitted diseases was started. In 1983, the Ministry of Health in Cuba set up a national committee on AIDS and had screened eight million people by June 1990, with all provinces in Cuba started building AIDS sanatoriums. By 1993, day-care hospitals started treating patients replacing the sanatorium system (Koike, 2002). The AIDS program also had, among other measures, the tracing of sexual contacts of known HIV-positive (HIV+) persons, to prevent the spreading HIV. When a person is detected as living with HIV, an epidemiological interview is carried out by the Epidemiology Department of his municipality or by his family doctor as part of the Partner Notification Program. After this interview, the Epidemiology Department tries to locate the sexual partners of the person through the network of the Health System. The person living with HIV usually does not participate in this process, though they normally help in notifying their present partners. Trying to locate the sexual partners is a very complex job and one that, in some cases, takes a lot of time. This task is one of high level of priority for the Health System, and it is something that is under constant supervision to try to determine how effective it is in the prevention of the spread of HIV. All data used in this work are from the time period of 1986–2002.

The number of AIDS cases in Cuba at the end of 2002 is 2090 with 448 females and 1642 males. Of the males, 83.5% are homo-bisexuals (we consider the group of homo-bisexuals to be formed by homosexuals and bisexuals). There have been 1057 deaths due to AIDS. Through the Health System HIV/AIDS program a total of 4517 HIV-positive individuals have been found, including 942 females and 3575 males. Of the HIV-positive males, 85.7% are homo-bisexuals. As of December of 2003, number of people in Cuba receiving HAART treatment is 1287 which constitutes all those in need according to CDC AIDS definition case of patients with cd4 of less than 200 cels/mm^3. However, in the near future, it is planned that those with cd4 around 350 or less (or high viral load) will also be treated. In fact, currently there are some cases (around 5%) with cd4 over 200 or only having a high viral load, are already being treated but the authors have not received the exact numbers from the sources (Ministry of Public Health, 2003).

From the table, we can see the epidemic is very low-prevalent. Indeed, with a population of around 11 million, Cuba has a cumulative incidence rate for AIDS of 190 per million (11.2 per million per year). One of the characteristics of the Cuban Program for the HIV/AIDS epidemic is that there is an active search of HIV-positive persons through the sexual

Table 1. New HIV+, AIDS cases and deaths due to AIDS by year in Cuba, 1986–2002.

Year	HIV+	AIDS	Death due to AIDS
1986	99	5	2
1987	75	11	4
1988	93	14	6
1989	121	13	5
1990	140	28	23
1991	183	37	17
1992	175	71	33
1993	102	82	59
1994	122	102	62
1995	124	116	82
1996	234	99	93
1997	363	129	99
1998	362	150	99
1999	493	176	123
2000	545	258	143
2001	642	392	117
2002	644	407	90
Total	4517	2090	1057

contacts of known HIV-infected persons. As a result, 29.1% of all HIV-positive persons detected have been found through contact tracing. The rest of the infected persons are found through a "blind" screening, a search of HIV-positive individuals by serotesting of blood donors, pregnant women, persons with other sexually transmitted diseases, etc. at clinics. Non-parametric estimation of the mean time for the health authority to find a sexual partner notified by a detected HIV-positive person through the Contact Tracing Program has been found to be 54.3 months, with a standard deviation of 0.631 (Fig. 1).

Contact tracing has been used as a method to control endemic contagious diseases (Hethcote *et al.*, 1982; 1984). While there is still a debate about contact tracing for the HIV infection (April *et al.*, 1995; Rutherford *et al.*, 1988) the resurgence of infectious tuberculosis and outbreaks of drug-resistant tuberculosis secondary to HIV-induced immunodepression is forcing many public health departments to reexamine this policy (Altman, 1997; CDC-MMWR, 1991). A model of the HIV epidemic allowing for contact tracing would help evaluate the effect of this method of control on the size of the HIV epidemic, and give some idea as to the effectiveness of the Health System in finding them.

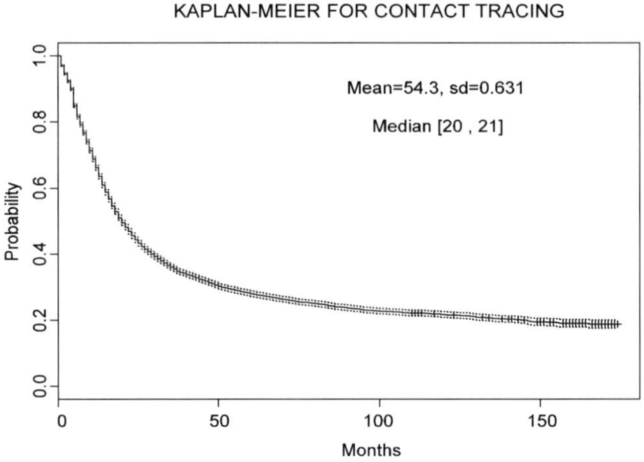

Fig. 1. Estimated time for contact tracing using declared sexual partners from 1986–2001 using Kaplan–Meier method.

Our objective is to model the contact tracing aspect of the HIV detection system, to try to obtain some information that could be useful to the Health System in Cuba in evaluating the way the program is working, and to ascertain its usefulness in terms of intervention and treatment of HIV. Other models have been used to study the effect of contact tracing with this objective in mind (Lounes *et al.*, 1999; Arazoza *et al.*, 2000). However, these were essentially linear models. We will now introduce non-linearity to model contact tracing. We will also discuss the implications of our results for the purpose of intervention and treatment of HIV/AIDS in Cuba.

2. The Models

As we noted, the Cuban Program to control the HIV/AIDS epidemic is based on the active search of persons infected with HIV, long before they show any signs of AIDS. Our objective is not to model how new infections by HIV are generated, but how the HIV infected persons are detected. We will consider the following variables:

(i) $X(t)$ is the number of HIV-infected persons that do not know they are infected at time t;
(ii) $Y(t)$ is the number of HIV-infected persons that know they are infected at time t; and
(iii) $Z(t)$ is the number of persons with AIDS at time t.

As the detection system has several search methods, we will separate the individuals in $Y(t)$ into two classes:

- $Y_1(t)$ is the number of HIV-infected persons that know they are infected at time t and were detected in a random type search; and
- $Y_2(t)$ the number of HIV-infected persons that know they are infected at time t and were detected through contact tracing.

Evidently, $Y(t) = Y_1(t) + Y_2(t)$ for all t. With the following constant coefficients:

(i) N — sexually active population.
(ii) α — the rate of recruitment of new HIV-infected persons infected by X.
(iii) α' — the rate of recruitment of new HIV-infected persons infected by Y.
(iv) k_1 — the rate at which the unknown HIV-infected persons are detected by the system, independently of other HIV-positive persons (through "random" screening).
(v) k_2 — the rate at which unknown HIV-infected persons are detected by the system through contact tracing.
(vi) β — the rate at which the undetected HIV-positive persons develop AIDS, reciprocal of the mean incubation.
(vii) β' — the rate at which the detected HIV-positive persons develop AIDS, the reciprocal of the mean time it takes to go from Y to Z.
(viii) μ — the mortality rate of the sexually active population.
(ix) μ' — the mortality rate of the population with AIDS.

The model dynamics is described by the following system:

$$\begin{aligned}
\frac{dX}{dt} &= \alpha NX + \alpha' NY - (k_1 + \mu + \beta)X - f(k_2, X, Y)\,, \qquad (1)\\
\frac{dY_1}{dt} &= k_1 X - (\mu + \beta')Y_1\,,\\
\frac{dY_2}{dt} &= f(k_2, X, Y) - (\mu + \beta')Y_2\,,\\
\frac{dZ}{dt} &= \beta X + \beta' Y - \mu' Z\,.
\end{aligned}$$

We consider the system only in the region $\boldsymbol{D} = \{X \geq 0, Y \geq 0, Z \geq 0\}$. It is clear that $\boldsymbol{D}$ is positively invariant under the flow induced by (1). First, we make following three remarks regarding the system in Eq. (1):

(i) In (1), there are two ways individuals can move from the unknown HIV-infected class (X) to the known HIV-infected class (Y). One way is through the term $f(k_2, X, Y)$. This is the term we utilize to model contact tracing. In other words, the individual is found through his contacts with persons that are known to live with HIV. The other way they can be detected is through the screening term $k_1 X$ which models all the other "random" ways of searching for HIV-positives. (For other HIV models using constant and nonlinear screening terms, see Hsieh (1991), Velasco–Hernandez *et al.* (1994), de Arazoza *et al.* (2003).) It is important to note that $1/k_1$ can be viewed as the mean time from infection to detection for the persons found through means other than contact tracing.

(ii) The term $f(k_2, X, Y)$ models contact tracing, the way it is given in the model indicates that the process is one that goes on for a long time, because it involves all the individuals in the class Y and this is confirmed numerically by the result that the mean time to find a contact is 54.3 months (Fig. 1). If we consider the numerical result that the mean time from detection to AIDS is 86.8 months (Fig. 2) we can see that, an average, infected contacts of an HIV-positive person are found for more than 32 months before that person who is living with HIV develops AIDS. To consider more than one class of known HIV-positive persons (before onset of AIDS) in the model, i.e. one class where contact are found and another where contacts are no longer found, would complicate the model unnecessarily for it is not clear that this complication would give more information on the dynamics of the epidemic. Of course variations are high; some persons have very few contacts and are easy to locate, while others may have a large number of contacts, of which some may be impossible to locate. Furthermore, some persons have a lot of "casual" contacts and they do not remember enough information on these contacts to make it possible to find them. Others may have fewer contacts and possess a better knowledge of the contact's full name and/or addresses that make it more likely for the Health System to find them. Some contacts would refuse being tested

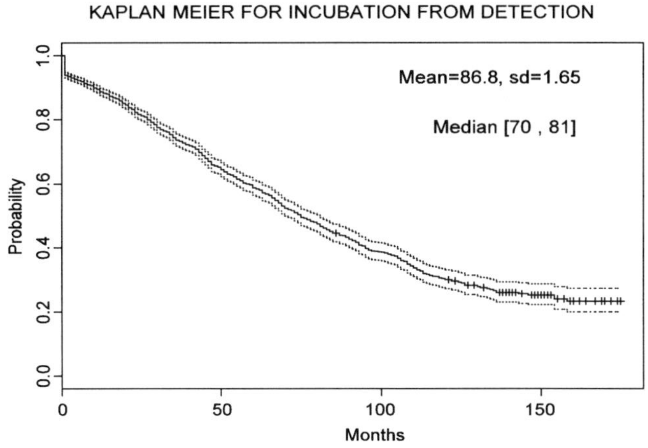

Fig. 2. Estimated time from detection to AIDS-defined illness (ADI) for 4517 HIV/AIDS patients in Cuba from 1986–2002 using Kaplan–Meier method.

for HIV, even if they are found. In general, of the more than 15,000 contacts, 80% have been found and tested. We will attempt, as a first approximation, to find the value of k_2 and to ascertain the general effect of this contact tracing on the time it takes for a person living with HIV to be detected.

(iii) We assume that the known HIV infected persons are infectious, but at a much lower rate than those that do not know they are infected. In this sense α' will be taken as a fraction of α.

(iv) The passage to AIDS is modeled in a linear way. This could be modeled in a more general way, but for the Cuban case the best fit to an incubation curve is still an exponential. This can be seen in Fig. 3, which gives us the cumulative hazard function for the time to AIDS which is a straight line and this corresponds to an exponential model.

Several possibilities arise for the contact tracing term $f(k_2, X, Y)$. In this work, we will consider the following four models:

(i) $k_2 X$
(ii) $k_2 Y$
(iii) $k_2 XY$
(iv) $k_2 \dfrac{XY}{X + Y}$

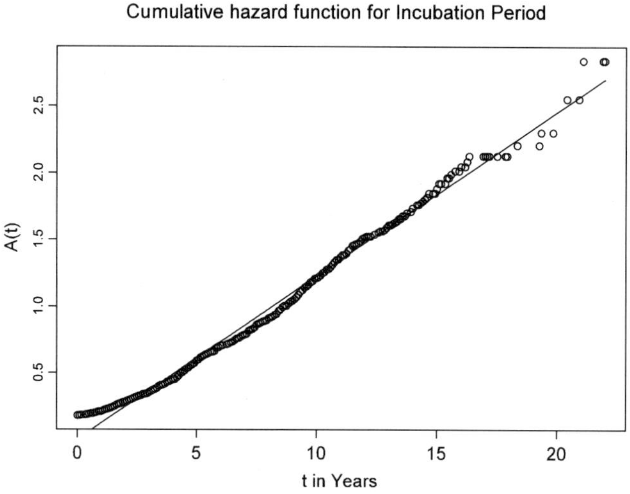

Fig. 3. Cumulative hazard function for time to AIDS in Cuba, 1986–2002.

The first two models are linear models, while the last two are non-linear. In order to make full comparison, we will give some analytical results for each of the models and compute the basic reproduction number for each one. Furthermore, we will fit the models to the Cuban contact tracing data to see which model explains the data most satisfactorily.

2.1. The k_2X model

In this case the system is:

$$\begin{aligned} \frac{dX}{dt} &= (\lambda - k - \mu - \beta)X + \lambda' Y\,, \\ \frac{dY}{dt} &= kX - (\mu + \beta')Y\,, \\ \frac{dZ}{dt} &= \beta X + \beta' Y - \mu' Z\,, \end{aligned} \tag{2}$$

where $k = k_1 + k_2$.

It is a linear model and the basic reproduction number is:

$$R_0 = \frac{\lambda}{k + \mu + \beta} + \frac{\lambda'}{\mu + \beta'} \frac{k}{k + \mu + \beta}\,.$$

If $R_0 > 1$, then all trajectories go to infinity, otherwise the Disease-free Equilibrium (DFE) at (0,0) is the only equilibrium and is globally asymptotically stable.

2.2. The k_2Y model

For this case the system is:

$$\begin{aligned}\frac{dX}{dt} &= \lambda X + \lambda' Y - (k_1 + \mu + \beta)X - k_2 Y\,,\\ \frac{dY}{dt} &= k_1 X + k_2 Y - (\mu + \beta')Y\,,\\ \frac{dZ}{dt} &= \beta X + \beta' Y - \mu' Z\,.\end{aligned} \tag{3}$$

It is also a linear model and the basic reproduction number is:

$$R_0 = \frac{\lambda}{k_1 + \mu + \beta} + \frac{k_1 \lambda'}{(k_1 + \mu + \beta)(\mu + \beta')} + \frac{k_2}{k_2 + \mu + \beta}\frac{\mu + \beta - \lambda}{\mu + \beta'}\,.$$

Again, if $R_0 > 1$, then all trajectories go to infinity, otherwise (0, 0) is unique and globally asymptotically stable.

2.3. The k_2XY model

This model is similar to one studied by de Arazoza and Lounes (2002), but it only considers the variable Y. The system is

$$\begin{aligned}\frac{dX}{dt} &= \lambda X + \lambda' Y - (k_1 + \mu + \beta)X - k_2 XY\,,\\ \frac{dY}{dt} &= k_1 X + k_2 XY - (\mu + \beta')Y\,,\\ \frac{dZ}{dt} &= \beta X + \beta' Y - \mu' Z\,.\end{aligned} \tag{4}$$

In the region $\boldsymbol{D} = \{X \geq 0,\ Y \geq 0,\ Z \geq 0\}$, the system has two equilibria, one is the DFE at $P_0 = (0,\ 0,\ 0)$, and the other is a unique endemic equilibrium $P^* = (X^*, Y^*, Z^*)$ at

$$X^* = \frac{\sigma\gamma + \lambda' k_1}{k_2(\sigma + k_1)}\,, \quad Y^* = \frac{\sigma\gamma + \lambda k_1}{k_2(\gamma - \lambda')}\,, \quad Z^* = \frac{\beta X^* + \beta' Y^*}{\mu'}\,, \tag{5}$$

with $\sigma = \lambda - k_1 - \beta - \mu$, $\gamma = \beta' + \mu$.

The basic reproduction number for the system is

$$R_0 = \frac{\lambda}{k_1 + \mu + \beta} + \frac{\lambda'}{\mu + \beta'}\frac{k_1}{k_1 + \mu + \beta}\,.$$

If $R_0 < 1$, and the endemic equilibrium P^* is feasible (i.e., $\gamma > \lambda'$), then the DFE P_0 is stable and its basin of attraction consists of a triangle formed by the axes and the line of slope

$$\frac{\sigma + k_1}{\sigma(\lambda') - \gamma}\left\{\frac{\lambda' k_1}{\gamma - \lambda'} + \lambda_1\right\}$$

that passes through P^*, where λ_1 is the negative eigenvalue of the Jacobian matrix at P^*.

If $R_0 > 1$, then P_0 is unstable and P^* is globally asymptotically stable in the region $\boldsymbol{D}$.

If $R_0 = 1$, P^* and P_0 coincide, 0 is a simple eigenvalue for the Jacobian at P_0 and the other eigenvalue is $\sigma - \gamma$ which is negative. Therefore P_0 is globally asymptotically stable in the region $\boldsymbol{D}$.

2.4. *The $k_2XY/(X+Y)$ model*

The system considered here is:

$$\begin{aligned}
\frac{dX}{dt} &= \lambda X + \lambda' Y - (k_1 + \mu + \beta)X - k_2\frac{XY}{X+Y}\,,\\
\frac{dY}{dt} &= k_1 X + k_2\frac{XY}{X+Y} - (\mu + \beta')Y\,,\\
\frac{dZ}{dt} &= \beta X + \beta' Y - \mu' Z\,.
\end{aligned} \tag{6}$$

The basic reproduction number of the system is

$$\begin{aligned}
R_0 = &\frac{\lambda}{(\mu+\beta) + k_1 + \sigma} + \frac{k_1}{(\mu+\beta) + k_1 + \sigma}\frac{\lambda}{(\mu+\beta')}\\
&- \sigma\frac{(\lambda - (\mu+\beta) + k_1)}{(\mu+\beta')((\mu+\beta) + k_1 + \sigma)}\,.
\end{aligned}$$

As before, we will consider the system formed by the first two equations in Eq. (6). Let $x = \frac{X}{X+Y}, y = \frac{Y}{X+Y}$ be the respective proportions of unknown and known HIV-positives in the HIV-positive population. Since $x + y = 1$, we have

$$\begin{aligned}
x' &= \lambda x + \lambda' y - ((\mu+\beta) + k_1)x - k_2 xy - x[\lambda x + \lambda' y - (\mu+\beta)x\\
&\quad -(\mu+\beta')y]\\
&= [\lambda x + \lambda' y - (\mu+\beta)x](1-x) - k_1 x - k_2 xy + (\mu+\beta')xy\\
&= (\lambda' - \lambda + k_2 + \beta - \beta')x^2 + (\lambda - 2\lambda' - k_1 - k_2 + \beta' - \beta)x + \lambda'\,,\\
y' &= k_1 x + k_2 xy - (\mu+\beta)y - y[\lambda x + \lambda' y - (\mu+\beta)x - (\mu+\beta')y]\\
&= k_1 x + k_2 xy - (\mu+\beta')y(1-y) - y[\lambda x - \lambda' y - (\mu+\beta)x]\,.
\end{aligned}$$

Here $x' = 0 \Rightarrow -[\lambda x + \lambda'(1-x) - (\mu+\beta)x](1-x) + k_1 x + k_2 x(1-x) - (\mu+\beta)x(1-x) = 0$. Equivalently $(\lambda - \gamma\lambda - k_2)x^2 - (\lambda - 2\gamma\lambda - k_1 - k_2)x - \gamma\lambda = 0$ has unique positive solution x^* between 0 and 1. Also let y^* be the corresponding solution for $y' = 0$. x^* is globally asymptotically stable on the one-dimensional line between 0 and 1, which means that the straight line $X(t)/Y(t) = x^*/y^*$ on the positive XY-quadrant is an asymptotic line for all trajectories, either going to (0, 0) or infinity.

3. Fitting the Models to Cuban Data

We fit the models to the data for the known HIV-positives and AIDS cases in Cuba. We will use the following values for the parameters which were estimated from the HIV data in Cuba:

$X(0) \in [200, 230]$, the number of unknown HIV-positives in 1986 estimated from the number of HIV-positives who were detected after 1986 but were found to be already infected in 1986,

$Y(0) = 94$, number of HIV positives who were known to be alive at the end of 1986,

$Z(0) = 3$, number of AIDS cases who were alive at the end of 1986,

$\mu = 0.0053$, yearly mortality rate for the HIV-positive cases for 1991–2000, (S.D. $= 0.0030$), computed from the number of death for HIV infected persons not related to AIDS,

$\mu' \in [0.66, 0.85]$, obtained from the 95% confidence interval for the median of the survival time to AIDS (1987–2000),

$\lambda = \alpha N = 0.5744$, the infection rate of the undetected HIV-positives obtained from the value of the parameter λ in the model developed in (Lounes *et al.*, 1999), (S.D. $= 0.0096$),

$\beta = 0.1135$, from the incubation period $(1/\beta)$ estimated from 1218 persons whose probable date of infection has been determined during the observation period 1987–2000, (S.D. $= 0.0031$),

$\beta' = 0.1350$, from the mean time from detection to AIDS $(1/\beta')$ (1987–2000), (S.D. $= 0.0026$),

We take λ', the infection rate of the known HIV-positives, to be a fraction of λ; $\lambda' = r\lambda$ and consider $r \in (0, 0.1)$.

We fit the models to the data to obtain values for k_1 and k_2 by minimizing a relative error function. As traditional optimization methods failed to work properly we used a genetic algorithm approach and a random search for local minima (Arazoza *et al.*, 2000). To compute standard errors for the parameters, 300 fitting runs were made using different values for the known parameters taken randomly from their confidence interval.

The following table gives the mean values found for k_1 and k_2.

4. Discussion and Concluding Remarks

The question now is which model is the best one in the sense of best fit for the data. In other words, which of the four models offers the best model for contact tracing? By comparing the mean errors of the four models in Table 1, the order of suitability of the four models in the sense of the smallest mean errors is:

(i) $k_2 X$
(ii) $k_2 \dfrac{XY}{X+Y}$
(iii) $k_2 Y$
(iv) $k_2 XY$

However, we note that the contact tracing term $k_2 \frac{XY}{X+Y}$ in Model 4 can be approximated by Model 1 (i.e. $k_2 X$) when $Y >> X$ and by Model 2 (i.e. $k_2 Y$) when $X >> Y$. In other words, Model 4 approximates whichever that is the smaller of the two linear models whenever one contact tracing term is much larger than the other. Moreover, $k_2 \frac{XY}{X+Y}$ is smaller than the corresponding contact tracing terms in either Model $k_2 X$ or Model $k_2 Y$. Hence Model 4 $\left(k_2 \frac{XY}{X+Y}\right)$, as a model for the contact tracing, offers a conservative compromise between the two extremes of contact tracing in the linear models and should be the best from the theoretical point of view. Estimates of the unknown HIV-positive population in Cuba (de Arazoza *et al.*, 2003; Hsieh *et al.*, 2002; 2001), though not a negligible number, have shown that, in recent years, approximately two thirds of the HIV-positive

Table 2. Parameters.

Model	Mean k_1	sd k_1	Mean k_2	sd k_2	Mean error	sd Error
$k_2 X$	0.2423	0.0229	0.1232	0.0110	16.6823	1.2322
$k_2 Y$	0.2786	0.0280	0.0984	0.0036	18.5417	1.0827
$k_2 \dfrac{XY}{X+Y}$	0.2547	0.0242	0.2457	0.0133	17.1199	1.1790
$k_2 XY$	0.3031	0.0254	0.00024	0.000038	20.3743	0.9922

persons in Cuba have been detected. Hence realistically Model 4 is probably more appropriate than either Models 1 or 2. The simple "mass action" contact tracing term in Model 3 (k_2XY) gives the "worst" fit and should be discarded.

We also made an ANOVA test on the error for the 4 models, yielding the following result (see Table 3).

That is, the errors for the 4 groups of models are different and testing one against the other gives the order that we have given in Table 2.

To further utilize our result, we compute the mean detection time for random screening to detect an HIV-positive, $(1-p)/k_1$, for Models 1, 2 and 4, and the mean detection time for contact tracing to detect one HIV-positive person, p/k_2, for Model 1 only. Here $p = 0.291$ is the proportion of known HIV-positive persons detected through contact tracing. The results are given in Table 4 below.

The result indicates that, using Model 1, the contact tracing program shortens the time of detection for an HIV-positive person by 6.8 months. The other models cannot be used to draw any similar conclusions.

Finally, we consider the implications of contact tracing for the intervention measures and treatment of HIV in Cuba. The intervention program in Cuba has gone through several phases, the most recent being the establishment of anonymous testing sites for HIV. Currently there are three anonymous testing sites functioning in Havana (which reports more than 50% of HIV-positive people nationwide) and there are immediate plans to

Table 3. ANOVA analysis for fitting errors for the 4 models.

SOURCE	**SS**	**df**	**MS**	**F**	**Prob > F**
COLUMNS	2493	3	831.255	656.45	0
ERROR	1514.49	1196	1.266	—	—
TOTAL	4008.25	1199	—	—	—

Table 4. Mean detection time by random screening and contact tracing, when applicable.

Model	Mean detection time by random screening (A)	Mean detection time by contact tracing (B)	Difference (A − B)
1. k_2X	35.1 months	28.3 months	6.8 months
2. $k_2\dfrac{XY}{X+Y}$	33.4 months	NA	NA
3. k_2Y	30.1 months	NA	NA

have at least one in each of the Central and Eastern regions in addition to the ones in Havana. The long-term plan is to have at least one site in every province. These sites also have pre- and post-counseling services to report contacts and reduce future contacts. Around 80% of people testing positive in anonymous sites eventually decide to adhere to confidential HIV reporting system (normal system). Typically once a person knows his or her HIV sero-positive status, change in sexual behavior occurs even without having a good educational background. Only a minority of people keep on having risky behaviors after knowing his/her positive serological status. Hence early diagnosis through random screening, contact tracing, and, more recently, anonymous testing has been instrumental in keeping the HIV prevalence in Cuba at a low level (Joanes Fiol, 2003).

The detection of HIV-positive persons subsequently made their treatment possible. The first therapeutic methods employed in Cuba, appearing first in 1986, consisted of the use of domestically produced immune-modulators. Treatments making use of transfer factor and the recombinant interferon alpha were conducted with satisfactory results. In 1987, AZT therapy was introduced into Cuba's healthcare system. Donations made by individuals and international organizations made it possible in 1996 to offer triple AIDS therapy to 100 Cubans afflicted with the disease. Though more substantial donations were made in the years to follow, due to the recent increase in the HIV-infected population size (Hsieh *et al.*, 2001) the need of the populations were, of yet, not entirely met (Lantero Abreu, 2003).

In 1997, when the domestic production of these pharmaceuticals entered its research phase, the Cuban government paid the international prices to acquire all of the medication needed to offer triple therapy (HAART) to mothers and children who were HIV positive. From 2001 onward, a wider variety of domestically manufactured anti-retroviral agents became increasingly available in Cuba, resulting in a 100% level of coverage by the end of 2002. With the advances in therapeutic treatment, the early detection and diagnosis of HIV-positive persons through contact tracing, as evident from our modeling, has taken an increasingly important role in improving the quality of life for those living with HIV/AIDS.

Acknowledgments

The authors would like to thank Drs. Jose Joanes Fiol and Maria Isela Lantero Abreu of National Aids Programme, Ministry of Public Health of Cuba, for providing the up-to-date information on the intervention and treatment of HIV in Cuba. YHH was supported by National Science Council (Taiwan) under grant NSC92-2115-M-005-002. H. de Arazoza received support from Equipe Probabilités, Statistique et Modélisation Laboratoire de Mathématique Université Paris-Sud, France and from Université René Descartes, Laboratoire de Statistique Médicale, équipe MAP5 FRE CNRS 2428. Part of the work was carried while H. de Arazoza was a guest researcher at the Dpt. Tecnología Electrónica, E.T.S. Ingeniería Telecomunicación, Universidad de Málaga, Spain, and while H. de Arazoza visited National Chung Hsing University under a grant from NSC (Taiwan).

References

Altman L (1997) Sex, Privacy and Tracking the HIV Infection, *New York Times*, November 4th.

April K, Thévoz F (1995) Le Contrôle de l'entourage ("Contact Tracing") a été négligé dans le cas des infections par le VIH. *RevueMédicale de la Suisse Romande*, **115**: 337–340.

Arazoza H, Lounes R, Hoang T, Interian Y (2000) Modeling HIV Epidemic under Contact Tracing — The Cuban Case, *J Theor Med* **2**: 267–274.

Center for Disease Control and Prevention (1991) Transmission of multidrug resistant tuberculosis from an HIV positive client in a residential substance-abuse treatment facility — Michigan, MMWR, **40**(8): 129.

de Arazoza H, Lounes R (2002) A non linear model for a sexually transmitted disease with contact tracing, *Math Med Biol* **19**(3).

de Arazoza H, Lounes R, Pérez J, Hoang T (2003) What percentage of the Cuban HIV-AIDS Epidemic is known?, *Revista del Instituto de Medicina Tropical de Cuba*, **55**(1).

Granich R, Jacobs B, Mermin J, Pont A (1995) Cuba's national AIDS program — The first decade, *West J Med* **163**: 139–144.

Hethcote HW, Yorke JA (1984) *Gonorrhea Transmission Dynamics and Control*, in Lecture Notes in Biomathematics, Vol. 56, Springer Verlag.

Hethcote HW, Yorke JA, Nold A (1982) Gonorrhea modeling: Comparison of control methods, *Math Biosci* **58**: 93–109.

Hsieh YH (1991) Modeling the effect of screening in HIV transmission dynamics, in Busenberg S and Martelli M (eds.) *Proceedings of International Conference on Differential Equations and its Applications to Mathematical Biology, Claremont, 1990* pp. 99–120 Lecture Notes in Biomathematics, Vol. 92, Berlin Heidelberg New York, Springer.

Hsieh YH, Arazoza H, Lee SM, Chen CWS (2002) Estimating the number of HIV-infected Cubans by sexual contact, *Int J of Epidemiology* **31**: 679–683.

Hsieh YH, Lee SM, Chen CWS, Arazoza H (2001) On the recent sharp increase of HIV infections in Cuba, *AIDS* **15**(3): 426–428.

Joanes Fiol J (2003) National Aids Programme, Ministry of Public Health, Cuba (personal communication).

Koike S (2002) Assessment of the Cuban approach to AIDS and HIV. *Nippon Koshu Eisei Zasshi* **49**(12): 1268–1277 (in Japanese).

Lantero Abreu MI, Head of National Aids Programme, Ministry of Public Health, Cuba (personal communication).

Lounes R, de Arazoza H (1999) A two-type model for the Cuban national programme on HIV/AIDS *IMA. J Math Appl Med Biol* **16**: 143–154.

Ministry of Public Health (2003) HIV/AIDS/STD Weekly report, National Department of Epidemiology, Ministry of Public Health, Cuba, December 6th.

Pèrez Avila J, Peña Torres R, Joanes Fiol J, Lantero Abreu M, Arazoza Rodriguez H (1996) HIV control in Cuba, *Biomed Pharmacother* **50**: 216–219.

Rutherford G, Woo J (1988) Contact tracing and the control of human immunodeficiency virus, *J Amer Med Assoc* **259**: 3609–3610.

Swanson JM, Gill AE, Wald K, Swanson KA (1995) Comprehensive care and the sanatorium: Cuba's response to HIV/AIDS. *JANAC* **6**(1): 33–41.

Velasco-Hernandez JX, Hsieh YH (1994) Modeling the effect of treatment and behavioral change in HIV transmission dynamics, *J Math Biol* **32**: 233–249.

CHAPTER 5

SIMULTANEOUS INFERENCES OF HIV VACCINE EFFECTS ON VIRAL LOAD, CD4 CELL COUNTS, AND ANTIRETROVIRAL THERAPY INITIATION IN PHASE 3 TRIALS

Peter B. Gilbert[1] and Yanqing Sun[2]

[1]*Statistical Center for HIV/AIDS Research and Prevention*
Fred Hutchinson Cancer
Research Center, Seattle, Washington 98109, U.S.A.
e-mail: pgilbert@scharp.org

[2]*Department of Mathematics and Statistics*
University of North Carolina at Charlotte
9201 University City Boulevard, Charlotte
North Carolina 28223, U.S.A.
e-mail: yasun@uncc.edu

There is an urgent need for a preventive HIV vaccine. Several candidate vaccines currently under development are designed to suppress HIV viral load and slow HIV progression post acquisition of HIV. In randomized, blinded efficacy trials of such a vaccine, concerns about the unreliability of viral load and CD4 cell counts as surrogate endpoints for AIDS-defining events and HIV transmission to others, and the use of antiretroviral therapies (ARTs) by infected trial participants, complicate the choice of post infection endpoints and analytic approaches for judging the vaccine's effectiveness. This chapter reviews some endpoint and analytic issues involved in assessing post infection vaccine effects, and evaluates in simulations (based on VaxGen's Phase 3 trial in Thailand) the use of time-to-event composite endpoints defined as the first event of ART initiation or biomarker failure (either high viral load or low CD4 count). Methods are presented for hypothesis testing and for constructing confidence bands about efficacy parameters corresponding to the composite endpoints; the tests and confidence bands are simultaneous over multiple clinically relevant biomarker failure thresholds. The simulation experiments support that simultaneous inferences of the viral load/ART and CD4/ART composite endpoints is a useful component of an analysis plan to evaluate post-infection vaccine effects.

Keywords: Gaussian multipliers; HIV vaccine trial; simultaneous hypothesis testing; Kaplan–Meier estimator; simultaneous confidence intervals.

1. Introduction

A preventive HIV vaccine (administered to HIV uninfected persons) would help curb the HIV epidemic if it had one or more of the following effects: (1) to lower susceptibility of vaccinated individuals (vaccinees) to acquire HIV infection; (2) to slow HIV disease progression in vaccinees who become infected; and (3) to decrease secondary transmission of HIV from vaccinees who become infected to others. Two Phase III trials have been conducted between 1998 and 2003 to evaluate efficacy of an HIV vaccine candidate, the first in North America/Netherlands and the second in Bangkok, and the primary objective of each trial was to assess vaccine effect (1) (Harro *et al.*, 2004; rgp120 HIV Vaccine Study Group, 2005). The trials were randomized, doubled-blinded, and placebo-controlled, and (1) was assessed by comparing the incidence rate of HIV infection between the vaccine and placebo study arms over a 36 month follow-up period. Participants who became HIV infected during the trials were followed for an additional 24 months on a Month 0, 0.5, 1, 2, 4, 8, 12, 16, 20, 24 post infection diagnosis schedule, and at each of these visits data were collected on plasma HIV RNA level (viral load), CD4 cell count, antiretroviral therapy (ART) initiation, and HIV-related illnesses. In each trial, effect (2) was assessed by comparing the incidence of HIV-related clinical outcomes between the vaccine and placebo arms, both in the entire randomized cohort and in the subgroup of HIV infected subjects. Effect (3) was not assessed in the North America/Netherlands (domestic) trial because sexual partners of infected trials participants were not enrolled, whereas in the Thai trial, by comparing the rate of HIV transmission to sexual partners between infected vaccine and infected placebo recipients.

Both efficacy trials were powered to address the primary study question (1). Based on 368 endpoints (HIV infections), the domestic trial had greater than 90% power to detect vaccine efficacy $> 30\%$ if the true vaccine efficacy (defined as the % reduction in the hazard rate of infection for vaccine versus placebo) exceeded 60%, and based on 211 endpoints, the Thai trial had greater than 90% power to detect vaccine efficacy $> 30\%$ if the true vaccine efficacy exceeded 70%. In contrast, both trials had low power for identifying a vaccine effect to slow the time until an AIDS-defining or related illness, because there were only 77 events in the domestic trial and 70 events in the Thai trial. A follow-up period longer than 2 years post infection diagnosis would be needed to assess (2) with high power. In addition, in the Thai trial there was low power to assess (3), because only a small number of sexual partners of infected trial participants were enrolled (Dale Hu, personal communication).

Although (2) and (3) were not directly assessable with high power within the two efficacy trials, they are critically important objectives because most effective licensed vaccines work through post-infection effects (Fine, 1993; Murphy and Chanock, 1996; Clemens, Naficy and Rao, 1997; Halloran, Struchiner and Longini, 1997; Clements–Mann, 1998), mathematical models show that post infection vaccine effects can have a great impact on the epidemic through indirect effects including herd immunity (Anderson and May, 1991; Quinn *et al.*, 2000; Gray *et al.*, 2001), and because there are a whole series of HIV vaccine candidates under development that are designed to control viral load post infection through induction of cytotoxic T cell and T helper responses (e.g. Shen and Siliciano, 2000; Nabel, 2001; Graham, 2002; Shiver *et al.*, 2002; HVTN, 2004; IAVI, 2004). Multiple such "T cell" vaccines have shown promise in simian-HIV (SHIV) macaque challenge studies, with a demonstrated ability to suppress SHIV viral load post infection by a factor of 2 to 3 $\log_{10}$. One HIV vaccine efficacy trial is ongoing in Thailand, of a prime-boost regimen (prime: recombinant canarypox T cell vaccine; boost: VaxGen's recombinant envelope protein subunit vaccine), and several newer T cell candidate vaccines are expected to enter efficacy trials by 2007 (HVTN, 2004; IAVI, 2004). This recent direction of HIV vaccine research underscores the importance of assessing HIV vaccine effects (2) and (3).

To license a vaccine that protects primarily through post HIV acquisition mechanisms, it will likely ultimately be necessary to directly demonstrate vaccine effects (2) and (3), through long term follow-up of infected trial participants for AIDS-defining illnesses and other clinical events, Phase 4 epidemiologic surveillance studies (cf. Orenstein, Bernier and Hinman, 1988), augmented partners efficacy trials (Longini, Susmita and Halloran, 1996; Halloran, Struchiner and Longini, 1997), and/or cluster randomized trials (Donner, 1981; Halloran, Struchiner and Longini, 1997; Hayes, 1998). However, given the urgent need to introduce even a modestly protective HIV vaccine to public health practice, it may be unacceptable to wait for definitive evidence of clinical vaccine effects (2) and (3) before licensing a vaccine. Instead, a vaccine showing strong effects to control viral load, maintain CD4 cell counts, and delay ART initiation over a 2–4 year period post infection diagnosis may warrant provisional licensure (or accelerated approval).

The two leading biomarker endpoints on which to base provisional licensure decisions are viral load and CD4 cell counts, as both markers have been shown to be prognostic for progression to AIDS and death (e.g. Mellors *et al.*, 1997; HIV Surrogate Marker Collaborative Group, 2000), and viral load has also been shown to correlate with infectiousness among heterosexual couples (Quinn *et al.*, 2000; Gray *et al.*, 2001). Furthermore, these biomarker endpoints have been used as a central basis for licensing antiretroviral therapies for HIV-infected persons (cf. DHHS Guidelines, 2002). Within an HIV vaccine efficacy trial, a pattern of lower viral loads and higher CD4 counts in vaccine relative to placebo recipients may suggest effects (2) and (3). These biomarkers must be used carefully in vaccine trials, because there are numerous examples in medical research in which treatment effects on surrogate endpoints did not translate into treatment effects on clinical endpoints (e.g. Fleming, 1992; Fleming and DeMets, 1996).

Some efficacy trial participants who become HIV-infected will receive the ethically mandated antiretroviral therapy (ART) (UNAIDS, 2001; DHHS Guidelines, 2002), which will suppress viral load and maintain CD4 cell counts in most recipients. The potent impact of ART on the biomarkers may overwhelm and mask any possible effect of vaccine on these markers. To address this problem, all biomarker measurements made after the initiation of ART can be excluded from the analysis, and inferences made on pre-ART values. This restriction raises the thorny methodologic problem that the censoring of biomarker trajectories by ART initiation is dependent on the trajectories themselves. In fact, all current and planned efficacy trials use standardized ART initiation guidelines in which infected participants are provided therapy when their CD4 counts drop below a certain value, implying a form of structural dependent censoring.

Attempts to correct for the dependent censoring by ART initiation can be made with linear mixed effects (lme) models (cf. Verbeke and Molenberghs, 2000) or with estimating equations methods. With an lme model, if the probability of ART initiation over time depends solely on covariates included in the model and on observed pre-ART responses, then unbiased inferences can be made on pre-ART responses. With estimating equations, inverse probability of censoring weighted (IPCW) equations can potentially be used to make unbiased inferences (cf. Rotnitzky and Robins, 1997). However, if standardized ART guidelines are used and are perfectly adhered to, the censoring (ART initiation) can be predicted perfectly by covariates, in which case the IPCW estimating equations contain infinite weights and the approach breaks down. Given the strongly dependent censoring and the large extent of censoring (44% and 34% of infected participants started ART within 24 months in VaxGen's domestic and Thai trials, respectively), it is difficult to have assurance about the validity of results obtained by methods that attempt to correct for the dependent censoring.

Because of the aforementioned concerns, Gilbert *et al.* (2003) proposed the use of two relatively simple time-to-event composite endpoints for basing provisional licensure decisions that obviate the dependent censoring problem. Specifically, the first endpoint is the event of viral failure (viral load exceeds a pre-specified failure threshold x_{vl} copies/mL) or ART initiation, whichever occurs first, and the second endpoint is the event of CD4 failure (CD4 count drops below a pre-specified failure threshold x_{cd4} copies/mL) or ART initiation, whichever happens first. The composite endpoints measure the magnitude of vaccine effect through the choice of failure threshold x_{vl} or x_{cd4}, and measure the durability of vaccine effect by registering endpoints during a sufficiently long period after HIV detection. These composite endpoints can be validly analyzed using standard survival analysis techniques such as Kaplan–Meier estimators. Furthermore, it is scientifically justifiable to include ART initiation into the endpoint definition, because ART initiation exposes a patient to drug toxicities and risk for drug resistance and loss of future therapy options (Hirsch *et al.*, 2000; DHHS Guidelines, 2002), and because under certain ART initiation guidelines, the event of ART initiation is coincident or highly correlated with biomarker failure. For example, if ART is provided when CD4 $<$ 200 copies/mL, then the composite endpoint defined by CD4 $<$ 200 or starting ART has straightforward interpretation as CD4 failure with no interference by ART.

In the first HIV vaccine efficacy trial, the main assessment of post infection vaccine effects was based on the composite viral load/ART initiation endpoint with endpoints captured out to $\tau = 12$ months. Because the failure threshold level x_{vl} below which the vaccine would be able to suppress viral load was unknown, the analysis plan specified studying the composite endpoint for four thresholds: $x_{vl} = 1500$, 10,000, 20,000, 55,000. Given x_{vl}, the vaccine efficacy to prevent the composite endpoint by time τ post-infection diagnosis, $VE(\tau, x_{vl})$, was defined as one minus the ratio (vaccine/placebo) of probabilities of the composite endpoint occurring by

τ months post-infection detection. $VE(\tau, x_{vl})$ is interpreted as the percent reduction by vaccine in the rate of the composite endpoint by τ months. The 1500 copies/mL failure threshold was selected based on the heterosexual HIV-discordant partners study in Uganda in which 0 of 51 HIV + partners with viral load < 1500 copies/mL transmitted HIV to their partners (Quinn *et al.*, 2000). The other three thresholds were selected based on an incidence cohort of men who have sex with men in the Multicenter AIDS Cohort Study (MACS), for whom the 3-year risk of progression to AIDS was 64.4% for those with viral load $> 55,000$ copies/mL, 36.4% for those with viral load between 20,000 and 55,000 copies/mL, and 4.1% for those with viral load $< 20,000$ copies/mL.

Motivated by the dataset and analysis plan of VaxGen's first trial, Gilbert and Sun (2005) developed a procedure for constructing simultaneous confidence bands for $VE(\tau, x_{vl})$ over the four thresholds x_{vl}, as well as for x_{vl} spanning continuously over the range 1500 to 55,000 copies/mL, and they performed a simulation study to assess the properties of the procedure. A limitation of the simulation study was that estimation of $VE(\cdot)$ was only studied over $\tau = 12$ months because the trial's analysis plan specified this time-span. However, vaccine suppression of viremia out to one year may be too short in duration to warrant initiation of provisional licensure proceedings, because 12 months post-infection is quite distant from AIDS, and the early efficacy may wane due to the emergence of HIV vaccine resistance mutations (e.g. Lukashov *et al.*, 2002). SHIV challenge studies in Rhesus macaques underscore this possibility: in multiple experiments initial profound suppression of viremia by a T cell vaccine was completely lost 6–24 months post SHIV acquisition, and HIV CTL epitope escape mutations were implicated in the failures (Barouch *et al.*, 2002; 2003). Because of the concern for escape, it may be important to assess the composite endpoint over at least 2 to 4 years post-infection detection to support provisional licensure, and in this chapter results of new simulations over this time-range are presented. In addition, Gilbert and Sun (2004) did not study the CD4/ART initiation composite endpoint, because this endpoint was not specified in the trial's analysis plan. However, international ART initiation guidelines (World Health Organization, 2003) currently in use in several countries including Botswana, South Africa, Thailand, and Uganda are based on CD4 thresholds but do not use viral load data. In addition, some cohort studies suggest that CD4 counts decline more rapidly in some international settings compared to domestic settings (Morgan *et al.*, 2002), which implies that ongoing and planned trials may have relatively high power to assess CD4 failure-based endpoints. Furthermore, MACS data suggest that CD4 counts in the range 200–350 are more prognostic for AIDS than high viral load values (Hughes *et al.*, 1998; Lyles *et al.*, 2000; HIV Surrogate Marker Collaborative Group, 2000). These considerations motivate use of the CD4/ART composite endpoint in some international trials, and motivate a simulation study to examine its use.

The remainder of this chapter is organized as follows. The procedure developed by Gilbert and Sun (2004) for constructing simultaneous confidence bands for $VE(\tau, x_{vl})$ is reviewed in Section 2, and is adapted to apply for simultaneous inferences on $VE(\tau, x_{cd4})$. Corresponding hypothesis testing methods and graphical diagnostic techniques are also developed. The procedures are studied in efficacy trial simulations over 48 months of follow-up post-infection diagnosis in Section 3. Section 4 illustrates application of the methods to a simulated vaccine trial dataset based on VaxGen's trial in Thailand. The chapter concludes with discussion on limitations of the composite endpoints and on other endpoints that might be considered for measuring post-infection vaccine effects.

2. Simultaneous Inferences for $VE(\tau, X_{vl})$ and $VE(\tau, X_{cd4})$

Work related to the problem addressed here includes methodology for constructing confidence bands for a functional of two survival curves (Parzen, Wei and Ying, 1997) or of two cause-specific cumulative incidence functions (McKeague, Gilbert and Kanki, 2001). These procedures approximated the distribution of interest using the Gaussian multipliers technique introduced by Lin *et al.* (1993); we also apply this technique.

2.1. *Notation and vaccine efficacy parameters of interest*

Suppose n trial participants become HIV-infected, with $n_1(n_2)$ in the vaccine (placebo) arm, and $n_k/n \to \rho_k \in, (0,1)$. For τ, a fixed time point post-infection detection, and x_{vl}, a fixed viral failure threshold, define the viral load/ART efficacy parameters as

$$VE^{vl}(\tau, x_{vl}) = 1 - F_1^{vl}(\tau, x_{vl})/F_2^{vl}(\tau, x_{vl}),$$

where $F_1^{vl}(\tau, x_{vl})$ $(F_2^{vl}(\tau, x_{vl}))$ is the probability of viral load exceeding x_{vl} copies/mL or ART initiation occurring by τ months post-infection diagnosis for an infected vaccine (placebo) recipient.

The CD4/ART composite efficacy parameter, $VE^{cd4}(\tau, x_{cd4})$, is defined in the same way except that the distributions are for CD4 count $< x_{cd4}$ or

ART initiation by τ. For the n_1 infected vaccinees (group $k = 1$) and n_2 infected placebo recipients (group $k = 2$), let $T_{k1}, \ldots, T_{kn_k}$ be the possibly right-censored times between infection diagnosis and ART initiation, $Y_{k1}^{vl}(t), \ldots, Y_{kn_k}^{vl}(t)$ be the viral loads at time t, and $Y_{k1}^{cd4}(t), \ldots, Y_{kn_k}^{cd4}(t)$ be the CD4 counts at time t. Within each group the triples $\{Y_{ki}^{vl}(t), Y_{ki}^{cd4}(t), T_{ki}\}$, $i = 1, \ldots, n_k$, are independent, identically distributed (iid), and the two samples are independent. Further assume for each $k = 1, 2$ that $F_k^{vl}(t, x_{vl})$ is continuous on $[0, \tau] \times [x_L^{vl}, x_U^{vl}]$ and $F_k^{cd4}(t, x_{cd4})$ is continuous on $[0, \tau] \times [x_L^{cd4}, x_U^{cd4}]$, where $F_k^{vl}(\tau, x_L^{vl}) < 1$, $F_2(\tau, x_U^{vl}) > 0$ and $F_k^{cd4}(\tau, x_U^{cd4}) < 1$, $F_2(\tau, x_L^{cd4}) > 0$. We summarize how to construct simultaneous confidence bands for $VE^{vl}(\tau, x_{vl})$ with x_{vl} varying over the prespecified range $x_{vl} \in, [x_L^{vl}, x_U^{vl}]$, and for $VE^{cd4}(\tau, x_{cd4})$ with x_{c4} varying over $[x_L^{cd4}, x_U^{cd4}]$.

For subject i in group k, the time to viral load/ART composite endpoint can be written as

$$\tilde{T}_{ki}^{vl}(x_{vl}) = \min\left\{\inf\left\{t\colon \sup_{0 \le s \le t} Y_{ki}^{vl}(s) \ge x_{vl}\right\}, T_{ki}\right\}.$$

The time to CD4/ART composite endpoint is

$$\tilde{T}_{ki}^{cd4}(x_{cd4}) = \min\left\{\inf\left\{t\colon \sup_{0 \le s \le t} Y_{ki}^{cd4}(s) \le x_{cd4}\right\}, T_{ki}\right\}.$$

For fixed threshold x_{vl} and x_{cd4},

$$F_k^{vl}(\tau, x_{vl}) = P\left\{\tilde{T}_{ki}^{vl}(x_{vl}) < \tau\right\} = 1 - P\left\{\sup_{0 \le s \le \tau} Y_{ki}^{vl}(s) < x_{vl}, T_{ki} \ge \tau\right\}$$

and

$$\begin{aligned} F_k^{cd4}(\tau, x_{cd4}) &= P\left\{\tilde{T}_{ki}^{cd4}(x_{cd4}) < \tau\right\} \\ &= 1 - P\left\{\sup_{0 \le s \le \tau} Y_{ki}^{cd4}(s) > x_{cd4}, T_{ki} \ge \tau\right\}. \end{aligned}$$

2.2. *Point and simultaneous confidence interval estimates for vaccine efficacy*

Let C_{ki} be the censoring (drop-out or study completion) time for subject i in group k. Let $\tilde{X}_{ki}^{vl}(x_{vl}) = \min\{\tilde{T}_{ki}^{vl}(x_{vl}), C_{ki}\}$, and censoring indicator $\delta_{ki}^{vl}(x_{vl}) = I(\tilde{T}_{ki}^{vl}(x_{vl}) \le C_{ki})$. Let $\tilde{X}_{ki}^{cd4}(x_{cd4}) = \min\{\tilde{T}_{ki}^{cd4}(x_{cd4}), C_{ki}\}$, and censoring indicator $\delta_{ki}^{cd4}(x_{cd4}) = I(\tilde{T}_{ki}^{cd4}(x_{cd4}) \le C_{ki})$. We assume independent censoring within each group: $\tilde{T}_{ki}^{vl}(x_{vl})$ and $\tilde{T}_{ki}^{cd4}(x_{cd4})$ independent of C_{ki}. The constructions of the confidence intervals for $VE^{vl}(\tau, x_{vl})$ and $VE^{cd4}(\tau, x_{cd4})$ based on possible censored composite endpoints follow the same procedure. We shall drop the indices that distinguish the functions for the two types of composite endpoints in the following derivations. The possible censored composite endpoints at a failure threshold x are given as $\{\tilde{X}_{ki}(x), \delta_{ki}(x), i = 1, \ldots, n_k, k = 1, 2\}$ and $F_k(\tau, x) = P\{\tilde{T}_{ki}(x) < \tau\}$ is the cumulative distribution at τ of a composite endpoint with the failure threshold x.

Let $\hat{F}_k(\tau, x) = 1 - \hat{S}_k(\tau, x)$, where $\hat{S}_k(\tau, x)$ is the Kaplan–Meier estimator of $S_k(\tau, x) = 1 - F_k(\tau, x)$ based on possible censored composite endpoints. Then $VE(\tau; x)$ is estimated by $\widehat{VE}(\tau; x) = 1 - \hat{F}_1(\tau; x)/\hat{F}_2(\tau; x)$. Also let $\hat{\Lambda}_k(\tau, x)$ be the Nelson-Aalen estimator for the cumulative hazard function $\Lambda_k(\tau, x) = -\log S_k(\tau, x)$. Set $N_{ki}(t, x) = I(\tilde{X}_{ki}(x) \leq t, \delta_{ki}(x) = 1)$, $R_{ki}(t, x) = I(\tilde{X}_{ki}(x) \geq t)$, $R_k(t, x) = \sum_{i=1}^{n_k} R_{ki}(t, x)$, $\hat{\Lambda}_k(t, x) = \sum_{i=1}^{n_k} \int_0^t \frac{dN_{ki}(s,x)}{R_k(s,x)}$, $M_{ki}(t, x) = N_{ki}(t, x) - \int_0^t R_{ki}(s, x) d\Lambda_k(s, x)$, and $r_k(t, x) = P\{\tilde{X}_{ki}(x) \geq t\}$. Then, the Martingale representation

$$\sqrt{n_k}(\hat{F}_k(\tau, x) - F_k(\tau, x)) = S_k(\tau, x) \int_0^{\tau} \frac{\sqrt{n_k} \sum_{i=1}^{n_k} dM_{ki}(s, x)}{r_k(s, x)} + o_p(1) \tag{2.1}$$

holds uniformly for $x \in [x_L, x_U]$, and converges in distribution to a mean-zero Gaussian variable with variance $\sigma_k^2(\tau, x) = S_k^2(\tau, x) \int_0^{\tau} d\Lambda_k(s, x)/r_k(s, x)$ that is estimated consistently by $\hat{\sigma}_k^2(\tau, x) = n_k \hat{S}_k^2(\tau, x) \int_0^{\tau} d\hat{\Lambda}_k(s, x)/R_k(s, x)$ (see Gilbert and Sun, 2005). Define the process

$$U(\tau, x) = \sqrt{n} \left(\frac{\hat{F}_1(\tau, x)}{\hat{F}_2(\tau, x)} - \frac{F_1(\tau, x)}{F_2(\tau, x)} \right). \tag{2.2}$$

Then it follows that

$$\begin{aligned} U(\tau, x) = \sqrt{n} \bigg(& \frac{1}{F_2(\tau, x)} (\hat{F}_1(\tau, x) - F_1(\tau, x)) \\ & - \frac{F_1(\tau, x)}{(F_2(\tau, x))^2} (\hat{F}_2(\tau, x) - F_2(\tau, x)) \bigg) + o_p(1) \end{aligned} \tag{2.3}$$

holds uniformly in $x \in [x_L, x_U]$.

By the central limit theorem, for fixed x, $U(\tau, x)$ converges in distribution to a mean-zero Gaussian variable with variance

$$\sigma^2(\tau, x) = \rho_1^{-1} (F_2(\tau, x))^{-2} \sigma_1^2(\tau, x) + \rho_2^{-1} (F_1(\tau, x))^2 (F_2((\tau, x))^{-4} \sigma_2^2(\tau, x),$$

which is estimated by $\hat{\sigma}^2(\tau, x)$ obtained by replacing ρ_k with n_k/n, $F_k(\tau, x)$ with $\hat{F}_k(\tau, x)$, and $\sigma^2(\tau, x)$ with $\hat{\sigma}^2(\tau, x)$. Let $VE(\tau, x) = 1 - \hat{F}_1(\tau, x)/\hat{F}_2(\tau, x)$. Large sample $100(1-\alpha)\%$ pointwise confidence bands for $VE(\tau, x)$ at x are given by

$$\widehat{VE}(\tau, x) \pm n^{-1/2} z_{\alpha/2} \hat{\sigma}(\tau, x) , \tag{2.4}$$

where $z_{\alpha/2}$ is the upper $\alpha/2$ quantile of a standard normal distribution.

From (2.1) and (2.3), the process $U(\tau, x)$ can be written as

$$\begin{aligned} U(\tau, x) = (n/n_1)^{1/2} (F_2(\tau, x))^{-1} S_1(\tau, x) \int_0^\tau \frac{n_1^{-1/2} \sum_{i=1}^{n_1} dM_{1i}(s, x)}{r_1(s, x)} \\ - (n/n_2)^{1/2} \frac{F_1(\tau, x)}{(F_2(\tau, x))^2} S_2(\tau, x) \\ \times \int_o^\tau \frac{n_2^{-1/2} \sum_{i=1}^{n_2} dM_{2i}(s, x)}{r_2(s, x)} + o_p(1) . \end{aligned} \tag{2.5}$$

Let Z_{1i}, Z_{2j}, $i = 1, \ldots, n_1$, $j = 1, \ldots, n_2$, be iid $N(0, 1)$ variables, and put

$$\begin{aligned} U^*(\tau, x) = (n/n_1)^{1/2} (\hat{F}_2(\tau, x))^{-1} \hat{S}_1(\tau, x) \int_0^\tau \frac{\sqrt{n_1} \sum_{i=1}^{n_1} Z_{1i} d\hat{M}_{1i}(s, x)}{R_1(s, x)} \\ - (n/n_2)^{1/2} \frac{\hat{F}_1(\tau, x)}{(\hat{F}_2(\tau, x))^2} \hat{S}_2(\tau, x) \int_0^\tau \frac{\sqrt{n_2} \sum_{i=1}^{n_2} Z_{2i} d\hat{M}_{2i}(s, x)}{R_2(s, x)} , \end{aligned} \tag{2.6}$$

where $\hat{M}_{ki}(t, x) = N_{ki}(t, x) - \int_0^\tau R_{ki}(s, x) d\hat{\Lambda}_k(s, x)$. The process $U(\tau, x)$ converges weakly to a mean zero Gaussian process for $x \in [x_L, x_U]$ and conditional on the observed data, $U^*(\tau, x)$ converges weakly to the same Gaussian process as $U(\tau, x)$.

Consider a subset $R_0 \subset [x_L, x_U]$. Let $c_{\alpha/2}$ be the asymptotic $1-\alpha$ quantile of $\sup_{x \in R_0} |U(\tau, x)/\hat{\sigma}(\tau, x)|$, and let $U_b^*(\tau, x)$, $b = 1, \ldots, B$, be B independent copies of $U^*(\tau, x)$, realized by conditioning on the observed data and generating iid $N(0, 1)$ variates $\{Z_{1i}, Z_{2j}, i = 1, \ldots, n_1, j = 1, \ldots, n_2\}$. The quantile $c_{\alpha/2}$ is estimated consistently by the $1 - \alpha$ quantile of $\{\sup_{x \in R_0} |U_b^*(\tau, x)/\hat{\sigma}(\tau, x)|, b = 1, \ldots, B\}$. Then, as proved in the Appendix of Gilbert and Sun (2005), asymptotic $100(1-\alpha)\%$ symmetric simultaneous confidence bands for $VE(\tau, x)$ over $x \in R_0 \subset [x_L, x_U]$ are given by

$$\widehat{VE}(\tau, x) \pm n^{-1/2} c_{\alpha/2} \hat{\sigma}(\tau, x) . \tag{2.7}$$

2.3. *Hypothesis testing for vaccine efficacy*

Let VE_0 be a specified level of vaccine efficacy. We consider testing the following 2-sided and 1-sided hypotheses on vaccine efficacy, respectively, at a given time $\tau = 24$ or 48:

$$\begin{aligned} H_0^2 &: VE(\tau, x) = 0 \ \forall x \in R_0 \text{ versus} \\ H_a^2 &: VE(\tau, x) \neq 0 \text{ for some } x \in R_0; \text{ and} \end{aligned} \tag{2.8}$$

$$\begin{aligned} H_0^1 &: VE(\tau, x) \leq VE_0 \ \forall x \in R_0 \text{ versus} \\ H_a^1 &: VE(\tau, x) > VE_0 \text{ for some } x \in R_0 \end{aligned} \tag{2.9}$$

where R_0 is a set of virologic failure thresholds or CD4 failure thresholds.

The statistic for testing H_0^2 versus H_a^2 is given by

$$S = \sup_{x \in R_0} \left| \sqrt{n} VE(\tau, x)/\hat{\sigma}(\tau, x) \right| .$$

The null hypothesis H_0^2 is rejected when $S > s_\alpha$, where $s_\alpha = z_{\alpha/2}$ if R_0 contains a single threshold value and s_α is estimated by the $1 - \alpha$ quantile of $\{S_b^* = \sup_{x \in R_0} |U_b^*(\tau, x)/\hat{\sigma}(\tau, x)|, b = 1, \ldots, B\}$ if R_0 contains more than one threshold value. A p-value of the two-sided test can be calculated as the percentage of the S_b^*'s that exceed S.

The statistic for evaluating H_0^1 versus H_a^1 is given by

$$T = \sup_{x \in R_0} \sqrt{n}(\widehat{VE}(\tau, x) - VE_0)/\hat{\sigma}(\tau, x) .$$

The null hypothesis H_0^1 is rejected when $T > c_\alpha$, where $c_\alpha = z_\alpha$ if R_0 contains a single threshold value and c_α is estimated by the $1 - \alpha$ quantile of $\{\sup_{x \in R_0}(-U_b^*(\tau, x)/\hat{\sigma}(\tau, x)|, b = 1, \ldots, B\}$ if R_0 contains more than one threshold value.

A graphical method for checking H_0^1 versus H_a^1 comes along with the proposed formal numerical test procedure by plotting the observed test process $W(\tau, x) = \sqrt{n}(\widehat{VE}(\tau, x) - VE_0)/\hat{\sigma}(\tau, x)$, $x \in R_0$, along with a number of trajectories, say 20, simulated from $W^*(\tau, x) = (-U_b^*(\tau, x)/\hat{\sigma}(\tau, x)$, $x \in R_0$. If the path of $W(\tau, x)$ falls above the trajectories from $W^*(\tau, x)$, it is an indication of statistical evidence to support H_a^1, and lack of evidence otherwise. A similar diagnostic procedure is obtained for H_0^2 versus H_a^2 by setting $VE_0 = 0$. In this case, if the path of $W(\tau, x)$ falls above or below the trajectories from $W^*(\tau, x)$, it is an indication of statistical evidence to support H_a^2. Some simulated examples of such graphical diagnoses are given in the next section.

3. Simulations of a Phase 3 HIV Vaccine Trial

We simulate viral load, CD4 count, and ART initiation data in a similar manner as Gilbert and Sun (2005), with the following main differences: (i) Data are simulated over 4 years post-infection diagnosis instead of 1 year; (ii) the probability of ART initiation over time depends only on CD4 counts but not on both CD4 counts and viral load; (iii) lower CD4 counts are generated to better represent some international trial populations; (iv) an inverse correlation among intra-subject viral loads and CD4 counts is induced; and (v) larger vaccine effects (1 or 1.2 $\log_{10}$ mean lower viral loads in vaccine than placebo arm, compared to 0.33 or 0.5 $\log_{10}$ mean shifts in the earlier work). Change (i) was made to study longer range vaccine effects on the composite endpoints, and change (ii) was made to reflect the current reality that international ART guidelines provide ART when CD4 < 200 or $350\,\text{cells/mm}^3$ but usually do not use viral load information; Gilbert and Sun (2005) designed the simulations to reflect the U.S. recommendations to start ART either when viral load $> 55{,}000\,\text{copies/mL}$ or CD4 $< 350\,\text{copies/mL}$ (DHHS Guidelines, 2002). Thus, due to changes (ii) and (iii) the simulations presented here are directly relevant for ongoing and planned international efficacy trials. Change (iv) was made to reflect the reality of an inverse correlation of viral load and CD4 counts (in Gilbert and Sun (2005) this issue was less important because the CD4/ART composite endpoint was not studied).

Regarding change (v), in Gilbert and Sun (2005) the simulations were used to study size and power of the test corresponding to the confidence bands for detecting $VE^{vl}(\tau; x_{vl}) \neq 0$ for at least one x_{vl}, whereas in the current simulations, we study the test for detecting $VE^{vl}(\tau; x_{vl}) > 0.2$ (or 0.1) for at least one x_{vl}. Testing a 0.2 (or 0.1) null instead of a 0.0 null is important because a vaccine showing only modest effects on viral load would arguably be insufficiently promising to warrant provisional licensure, due to uncertainties in whether the surrogate effects translate into clinical effects, the limited data on durability of vaccine efficacy, and possible biases in the result stemming from post-randomization selection bias. A 0.2 efficacy cutoff is selected based on mathematical models (Anderson and Garnett, 1996) and on numerical experiments suggesting that use of a vaccine with $VE^{vl}(\cdot)$ greater than 0.2 over at least 2 years is likely to provide appreciable beneficial effects on progression and secondary transmission (Gilbert *et al.*, 2003).

For the CD4/ART efficacy parameter, we use the less stringent $VE^{cd4}(t, x_{cd4}) = 0$ null hypothesis. This decision is based on the fact that

CD4 $< 200\,\text{cells/mm}^3$ is an AIDS-defining event, so that the CD4/ART efficacy parameter is more directly linked to clinical efficacy than the viral load/ART efficacy parameter.

3.1. *Simulation model set-up*

The basis of the simulation model is the data from VaxGen's efficacy trial in Thailand, with an addendum of 24 months of follow-up beyond the 24 months of follow-up used in the real trial:

(i) $n_1 = 106$ infected vaccine recipients and $n_2 = 105$ infected placebo recipients.
(ii) Infected subjects are followed for 48 months after HIV diagnosis.
(iii) 30% random dropout before the composite endpoint by 48 months for each group.
(iv) Viral loads and CD4 counts are measured from samples drawn at 15 times on or near scheduled visit times t_j, $1 \leq j \leq 15$ at months 0.5, 1, 2, 4, 8, 12, 16, 20, 24, 28, 32, 36, 40, 44, 48 post-infection diagnosis. For subject i in group k, the jth visit time t_{kij} is generated from $N(t_j, \sigma_j^2)$, where $\sigma_1 = 0.05$, $\sigma_2 = 0.06$, $\sigma_3 = 0.10$ and $\sigma_j = 0.12$ for $j = 4, \ldots, 15$.
(v) For subject i in the placebo group, the viral loads ($\log_{10}$ transformed) are generated from a linear mixed effects (lme) model,

$$Y_{2i}^{vl}(t) = \beta_0^{vl} + \beta_1^{vl}t + \beta_2^{vl}t^2 + r_{0i}^{vl} + r_{1i}^{vl}t + r_{2i}^{vl}t^2 + \varepsilon_i^{vl}(t)\,, \tag{3.1}$$

where $\left(\beta_0^{vl}, \beta_1^{vl}, \beta_2^{vl}\right)^T = (4.624433, -0.049451, 0.001169)^T$ are fixed effects and $\left(r_{0i}^{vl}, r_{1i}^{vl}, r_{2i}^{vl}\right)^T$ are random effects with a multivariate normal distribution with mean zero and $Std\left(r_{0i}^{vl}\right) = 0.655587$, $Std\left(r_{1i}^{vl}\right) = 0.089251$, $Std\left(r_{2i}^{vl}\right) = 0.002135$, $Corr\left(r_{0i}^{vl}, r_{1i}^{vl}\right) = -0.425$, $Corr\left(r_{0i}^{vl}, r_{2i}^{vl}\right) = -0.400$ and $Corr\left(r_{1i}^{vl}, r_{2i}^{vl}\right) = -0.980$. The iid measurement errors $\varepsilon_i^{vl}(t_{kij})$ have mean zero and standard deviation 0.3834.
(vi) For subject i in the placebo group, the CD4 cell counts are generated from a lme model,

$$\begin{aligned} Y_{2i}^{cd4}(t) = \beta_0^{cd4} + \beta_1^{cd4}t + \beta_2^{cd4}t^2 - 33\left\{Y_{2i}^{vl}(t) - \left[\beta_0^{vl} + \beta_1^{vl}t + \beta_2^{vl}t^2\right]\right\} \\ + r_{0i}^{cd4} + r_{1i}^{cd4}t + r_{2i}^{cd4}t^2 + \varepsilon_i^{cd4}(t)\,, \end{aligned} \tag{3.2}$$

where $\beta_0^{cd4} = 513.4167$, $\beta_1^{cd4} = -8.7049$, $\beta_2^{cd4} = 0.1291$ and $(r_{0i}^{cd4}, r_{1i}^{cd4}, r_{2i}^{cd4})^T$ has a multi-normal distribution with mean zero and $Std\left(r_{0i}^{cd4}\right) =$

203.983, $Std\left(r_{1i}^{cd4}\right) = 13.065$, $Std\left(r_{2i}^{cd4}\right) = 0.380$, $Corr\left(r_{0i}^{cd4}, r_{1i}^{cd4}\right) = -0.534$, $Corr\left(r_{0i}^{cd4}, r_{2i}^{cd4}\right) = 0.495$ and $Corr\left(r_{1i}^{cd4}, r_{2i}^{cd4}\right) = -0.951$. The iid measurement errors $\varepsilon_i^{cd4}(t_{kij})$ have mean zero and standard deviation 96.499. The term with -33 in front induces the inverse correlation between viral loads and CD4 counts; for example if subject i has viral load 1 $\log_{10}$ higher than the mean value at time t, then his or her CD4 cell count is lowered by 33 cells.

For the vaccine group, the viral loads and CD4 cell counts are simulated in three ways:

(i) **null model** (denoted by NULL) where the viral load processes follow (3.1) and the CD4 processes follow (3.2).
(ii) **constant mean shift model** (denoted by CONS) with mean viral load shift $s_{vl} = 1.0$ or 1.2 ($\log_{10}$) at all 15 time points, lower in vaccine than placebo. When there is a $s_{vl}(\log_{10})$ mean viral load shift, the mean CD4 count shifts at Months 0.5, 1, 2, 4, 8, 12, 16, 20, 24, 32, 36, 40, 44, and 48 are given by $s_{vl} \times (0, 6, 12, 25, 50, 75, 100, 125, 150, 175, 200, 225, 250, 275, 300)$, higher in vaccine than placebo. We denote CONS(1) and CONS(2) for the cases with $s_{vl} = 1.0$ and $s_{vl} = 1.2$, respectively.
(iii) **non-constant mean shift model** (denoted by NCONS) with mean viral load shift s_{vl} lower at Months 0.5, 1, 2, 4, 8, and 12, mean $0.5s_{vl}$ lower at Months 16, 20, 24, 32, and 0 lower at Months 36, 40, 44, and 48. The corresponding mean CD4 count shifts are given by $s_{vl}\times(0, 6, 12, 25, 50, 75, 100)$ at Months 0.5, 1, 2, 4, 8, 12, and $100s_{vl}$ at later time points, higher in vaccine than placebo. We denote NCONS(1) and NCONS(2) for the cases with $s_{vl} = 1.0$ and $s_{vl} = 1.2$, respectively.

In the NCONS scenario the vaccine suppresses viral load for a year, but then vaccine resistance gradually emerges, and suppression is completely lost by 3 years.

Once the biomarker processes for an individual are generated, the time to ART initiation is simulated assuming that the trial has a policy to provide ART when CD4 counts decline $\leq 200\,\text{cells/mm}^3$ or clinical symptoms present; this policy is used in the ongoing Phase 3 trial in Thailand. We assume that no participants receive ART prior to either of these conditions; this was the case for the completed VaxGen trial in Thailand. We also assume that not all subjects reaching an ART criterion start ART (e.g. due to participant refusal, problems with drug distribution, etc.). Accordingly, at each visit time we set the probability of starting ART within the next month (for visits at Months 0.5, 1, 2) and within the next two months (for visits at Months 4, 8, 12, 16, 20, 24, 28, 32, 36, 40, 44) as a function of the current CD4 count. In particular, the probabilities of starting ART at a visit during the subsequent 1 or 2 month interval are set to be:

CD4 Count	Probability of ART initiation
CD4 $\leq$ 200	0.7
200 $<$ CD4 $\leq$ 350	0.05
CD4 $>$ 350	0.02

The probabilities 0.05 and 0.02 in the last two rows are greater than 0 to represent low rates of clinical symptoms at these CD4 levels.

3.2. *Empirical sizes and powers*

We conduct a simulation study to test for vaccine efficacy with the hypotheses (2.8) and (2.9) for both the viral load/ART and CD4/ART composite endpoints.

For the viral load/ART composite endpoint, we test (2.9) with $VE_0 = 0.2$ at $\tau = 24$ and $VE_0 = 0.1$ at $\tau = 48$ months. We consider the hypothesis testing problems over the following sets of values for x_{vl}: $R_1^{vl} = [10{,}000, 100{,}000]$; $R_2^{vl} = \{10{,}000, 20{,}000, 55{,}000, 100{,}000\}$; $R_3^{vl} = \{10{,}000\}$; $R_4^{vl} = \{20{,}000\}$; $R_5^{vl} = \{55{,}000\}$; $R_6^{vl} = \{100{,}000\}$.

For the CD4/ART composite endpoint, we test (2.9) with $VE_0 = 0.0$ at $\tau = 24$ and 48 months. We consider the hypothesis testing problems over the following sets of values for x_{cd4}: $R_1^{cd4} = [200, 500]$; $R_2^{cd4} = \{200, 350, 500\}$; $R_3^{cd4} = \{200\}$; $R_4^{cd4} = \{350\}$; $R_5^{cd4} = \{500\}$.

Empirical sizes and powers are estimated using 1000 generated datasets and $B = 1000$. The nominal level is set to be $\alpha = 0.05$.

Table 1 shows empirical sizes and powers for testing vaccine efficacy with the viral load/ART composite endpoint at month 24 and 48 for the one-sided hypothesis (2.9). The empirical sizes of the one-sided test are all zero under the model NULL for which the true vaccine efficacy across all thresholds and times τ is zero. The test has strong power for vaccine efficacy greater than 0.2 at 24 months under model CONS(1) and CONS(2) except for the threshold set $R_3^{vl} = \{10{,}000\}$ under CONS(1). The test also shows reasonable power under NCONS(1) and NCONS(2) except for the threshold set R_3^{vl}. The power is greater under NCONS(2) than under

Table 1. Empirical sizes and powers × 100 for the one-sided test of H_0^1 with the viral load/ART composite endpoint.

Month	Model	R_1^{vl}	R_2^{vl}	R_3^{vl}	R_4^{vl}	R_5^{vl}	R_6^{vl}
		H_0:	$VE \leq 0.2$	vs.	H_a:	$VE > 0.2$	
24	NULL	0	0	0	0	0	0
	CONS(1)	92.4	96.5	40.0	90.8	97.8	95.8
	CONS(2)	99.2	99.4	92.1	99.6	99.7	98.8
	NCONS(1)	59.8	73.2	5.7	48.2	81.9	77.3
	NCONS(2)	84.8	91.3	39.8	85.0	94.2	90.4
		H_0:	$VE \leq 0.1$	vs.	H_a:	$VE > 0.1$	
48	NULL	0	0	0	0	0	0
	CONS(1)	77.3	89.4	7.8	57.0	91.9	93.6
	CONS(2)	93.3	98.0	52.0	89.7	98.3	98.7
	NCONS(1)	0.2	2.2	0	0	1.6	4.8
	NCONS(2)	0.9	3.2	0	0	3.4	6.8

NCONS(1). The reason for lack of power with R_3^{vl} is that most subjects in both of the vaccine and placebo groups fail virologically at the viral failure threshold value of 10,000. This implies that the vaccine is not powerful enough to suppress viral load below 10,000 copies/mL under CONS(1). The vaccine efficacy diminishes by 48 months. The test clearly has good power for vaccine efficacy greater than 0.1 at 48 months under CONS(1) and CONS(2). The simulation results indicate vaccine efficacy less than 0.1 for the NCONS(1) and NCONS(2) models at 48 months. Although the true vaccine efficacy under these models is difficult to compute and is unknown, the fact that the data were simulated with identical viral loads and CD4 counts in the vaccine and placebo groups after 36 months suggests that the methods correctly ascertain low vaccine efficacy at 48 months.

Table 2 lists empirical sizes and powers of the two-sided test of (2.8) with the viral load/ART composite endpoint. The empirical sizes under NULL provide an assessment of the coverage probabilities of the pointwise and simultaneous confidence bands of vaccine efficacy. The coverage probabilities are close to the nominal level 95% for R_2^{vl}, R_4^{vl}, R_5^{vl} and R_6^{vl} and greater than 98% for $R_1^{vl} = [10{,}000, 100{,}000]$ and $R_3^{vl} = \{10{,}000\}$. As we have just mentioned, most subjects in both the vaccine and placebo groups fail virologically at a viral failure threshold value close to 10,000. The sizes of the risk sets $R_1(s, x)$ and $R_2(s, x)$ in (2.6) are rather small, resulting in larger variation and hence greater coverage probabilities.

Tables 3 and 4 show empirical sizes and powers for testing vaccine efficacy with the CD4/ART composite endpoint at months 24 and 48 for

Table 2. Empirical sizes and powers × 100 for the two-sided test of H_0^2 with the viral load/ART composite endpoint.

Month	Model	R_1^{vl}	R_2^{vl}	R_3^{vl}	R_4^{vl}	R_5^{vl}	R_6^{vl}
		H_0:	$VE = 0$	vs.	H_a:	$VE \neq 0$	
24	NULL	1.6	4.9	1.4	3.8	4.6	4.2
	CONS(1)	100	100	100	100	100	99.9
	CONS(2)	100	100	100	100	100	100
	NCONS(1)	100	100	99.8	100	99.9	99.3
	NCONS(2)	100	100	100	100	100	99.7
		H_0:	$VE = 0$	vs.	H_a:	$VE \neq 0$	
48	NULL	1.4	5.3	0.1	0.2	4.2	4.7
	CONS(1)	99.5	99.8	93.1	99.0	99.5	99.4
	CONS(2)	100	100	99.2	99.9	100	100
	NCONS(1)	27.0	42.1	0.05	25.6	39.7	38.6
	NCONS(2)	35.6	52.1	10.1	34.5	48.9	45.4

Table 3. Empirical sizes and powers × 100 for the one-sided test of H_0^1 with the CD4/ART composite endpoint.

Month	Model	R_1^{cd4}	R_2^{cd4}	R_3^{cd4}	R_4^{cd4}	R_5^{cd4}
		H_0:	$VE \leq 0$	vs.	H_a:	$VE > 0$
24	NULL	3.7	6.9	7.4	6.0	5.7
	CONS(1)	54.6	66.6	53.4	63.3	50.1
	CONS(2)	68.2	79.2	61.3	74.4	65.9
	NCONS(1)	47.3	59.9	47.9	52.5	44.9
	NCONS(2)	59.0	69.1	53.6	66.3	53.7
		H_0:	$VE \leq 0$	vs.	H_a:	$VE > 0$
48	NULL	2.8	7.3	7.2	6.1	4.5
	CONS(1)	78.0	86.0	80.8	82.1	62.7
	CONS(2)	88.9	94.8	89.9	91.8	77.7
	NCONS(1)	37.0	51.9	45.4	44.6	36.1
	NCONS(2)	47.3	61.7	52.8	55.8	41.7

(2.9) and (2.8), respectively. Table 3 lists empirical sizes and powers of the one-sided test with $VE_0 = 0.0$. The empirical sizes are close to the nominal level 0.05. The test has reasonable power under the alternative models CONS(1) and CONS(2). The power is greater at 48 months than at 24 months under CONS(1) and CONS(2), undoubtedly due to the accrual of additional events. The power decreases for NCONS(1) and NCONS(2) with smaller power at 48 months than at 24 months. As presented in Table 4, for the two-sided tests the empirical sizes under the NULL model are close to the nominal level 0.05. Similar patterns on the power are observed in Table 4 as in Table 3.

Table 4. Empirical sizes and powers × 100 for the two-sided test of H_0^2 with the CD4/ART composite endpoint.

Month	Model	R_1^{cd4}	R_2^{cd4}	R_3^{cd4}	R_4^{cd4}	R_5^{cd4}
		H_0:	$VE = 0$	vs.	H_a:	$VE \neq 0$
24	NULL	2.8	5.1	4.5	3.9	4.7
	CONS(1)	47.1	57.1	43.5	52.3	39.3
	CONS(2)	62.5	71.0	51.0	64.8	54.0
	NCONS(1)	40.4	50.7	37.9	42.6	32.5
	NCONS(2)	51.4	60.4	43.5	53.5	43.3
		H_0:	$VE = 0$	vs.	H_a:	$VE \neq 0$
48	NULL	2.0	5.9	5.0	4.9	3.6
	CONS(1)	71.4	80.8	73.4	72.9	50.2
	CONS(2)	85.5	91.7	85.8	86.6	67.2
	NCONS(1)	30.4	42.0	33.6	33.7	24.1
	NCONS(2)	42.2	51.8	41.6	44.1	30.3

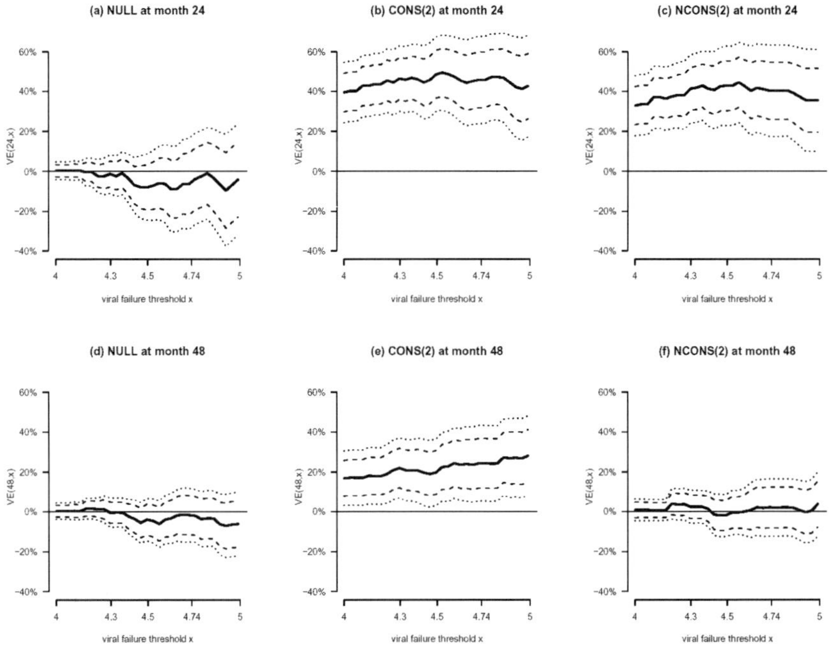

Fig. 1. (a)–(c) show point estimates (solid lines) and 95% pointwise (dashed lines) and simultaneous (dotted lines) confidence intervals for $VE^{vl}(24, x_{vl})$ with x_{vl} ranging between 10,000 and 100,000 copies/mL, for a single vaccine trial dataset simulated under the (a) NULL, (b) CONS(2), and (c) NCONS(2) models, respectively. (d)–(f) show the corresponding analyses of $VE^{vl}(48, x_{cd4})$.

4. Example

A goal of this chapter is to evaluate use of the composite endpoint efficacy parameters when 4 years of follow-up post-infection diagnosis are available. However, no real data are yet available for this assessment. Accordingly, using the simulation model of Section 3 that is based on the real efficacy trial in Thailand, we generate three datasets to illustrate application of the methods, one dataset each under the NULL, CONS(2), and NCONS(2) models.

Figure 1 shows simultaneous confidence bands for vaccine efficacy $VE^{vl}(\tau, x_{vl})$ at months $\tau = 24$ and 48 based on the viral load/ART composite endpoint for the NULL, CONS(2) and NCONS(2) datasets, whereas Figure 2 shows the corresponding plots for $VE^{cd4}(\tau, x_{cd4})$ based on the CD4/ART composite endpoint. For the NULL dataset, all of the estimated efficacy parameters are near zero at all failure thresholds, with confidence

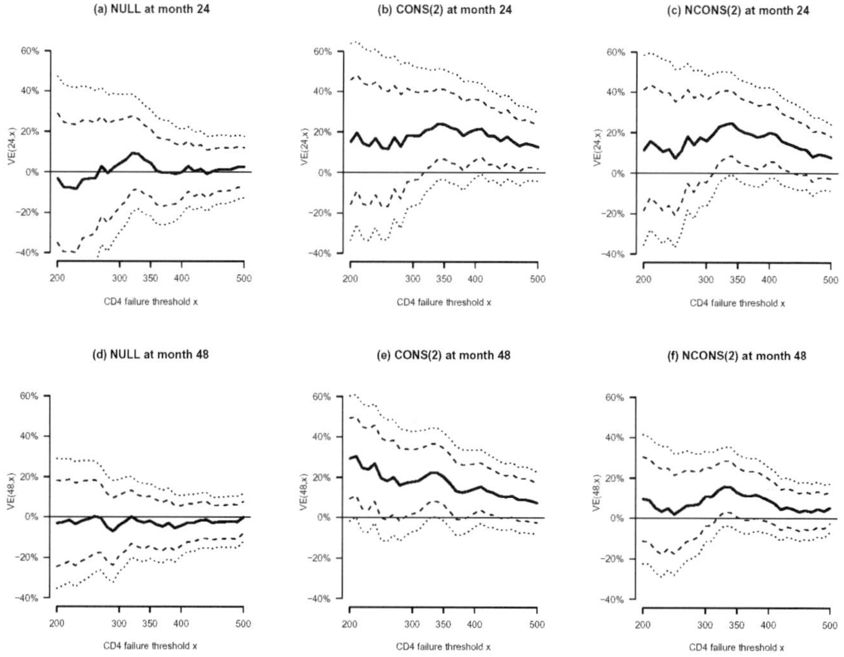

Fig. 2. (a)–(c) show point estimates (solid lines) and 95% pointwise (dashed lines) and simultaneous (dotted lines) confidence intervals for $VE^{cd4}(24, x_{cd4})$ with x_{cd4} ranging between 200 and 500 cells/mm^3, for the vaccine trial datasets also used for Fig. 1 that were simulated under the (a) NULL, (b) CONS(2), and (c) NCONS(2) models, respectively. (d)–(f) show the corresponding analyses of $VE^{cd4}(48, x_{cd4})$.

bands quite narrow and always including 0; therefore the methods correctly identify the vaccine as non-effective. For the viral load/ART endpoint, the confidence bands widen as the failure threshold increases, while for the CD4/ART endpoint, the confidence bands narrow as the failure threshold increases. This result occurs because power is greater for a greater number of events. Precision is greatest for the lowest viral threshold 10,000 copies/mL and for the highest CD4 threshold 500 cells/mm^3.

For the CONS(2) dataset, for which the vaccine has a persistent 1.2log10 mean-drop in viral load, the estimate of $VE^{vl}(24, x_{vl})$ is about 0.40 at all thresholds x_{vl}, with approximately constant-width confidence bands 0.25 – 0.55, and the estimate of $VE^{cd4}(24, x_{cd4})$ is about 0.20 with wider confidence bands -0.20 – 0.45 at $x_{cd4} = 200$ to 0.03 – 0.25 at $x_{cd4} = 500$. Thus, at 24 months the vaccine effect on viral load is greater than the effect on CD4 cell count. At 48 months the vaccine effect on both the viral load/ART

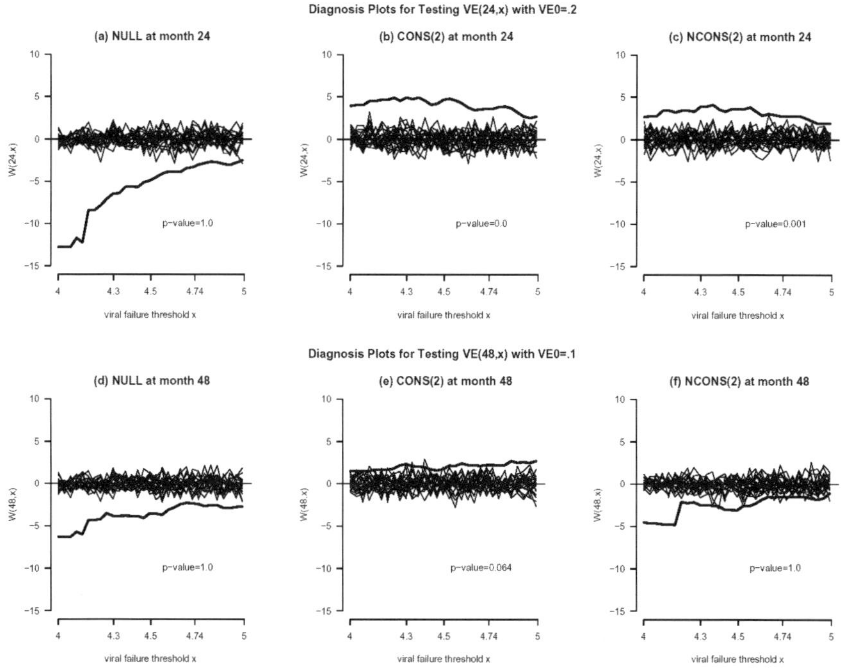

Fig. 3. (a)–(c) show the test process $W(24, x_{vl})$ (thick lines) and 20 trajectories from $W^*(24, x_{vl})$ (thin lines) for the vaccine trial datasets simulated under the (a) NULL, (b) CONS(2), and (c) NCONS(2) models, respectively. (d)–(f) show the corresponding test processes $W(48, x_{vl})$ and $W^*(48, x_{vl})$.

and CD4/ART efficacy parameters is diminished, likely due to the additional ART initiation between 24 and 48 months that occurs in both groups.

For the NCONS(2) dataset, the methods do very well to identify that the vaccine is efficacious to suppress viral load for up to 24 months, but the efficacy is completely lost by 48 months (Fig. 1). The analysis of the CD4/ART parameter does not clearly identify this effect (Fig. 2), suggesting that under the simulation model we used the methods are more sensitive for the viral load-based efficacy parameter. This result can be explained by the fact that an intervention can impact viral load immediately after infection, whereas longer term follow-up is needed to observe a substantial effect to sustain CD4 cell counts, because CD4 counts decline gradually after HIV infection.

The test processes $W(\tau, x)$ for testing vaccine efficacy greater than 0.2 at 24 months and greater than 0.1 at 48 months with the viral load/ART composite endpoint are plotted in Fig. 3, along with 20 trajectories from

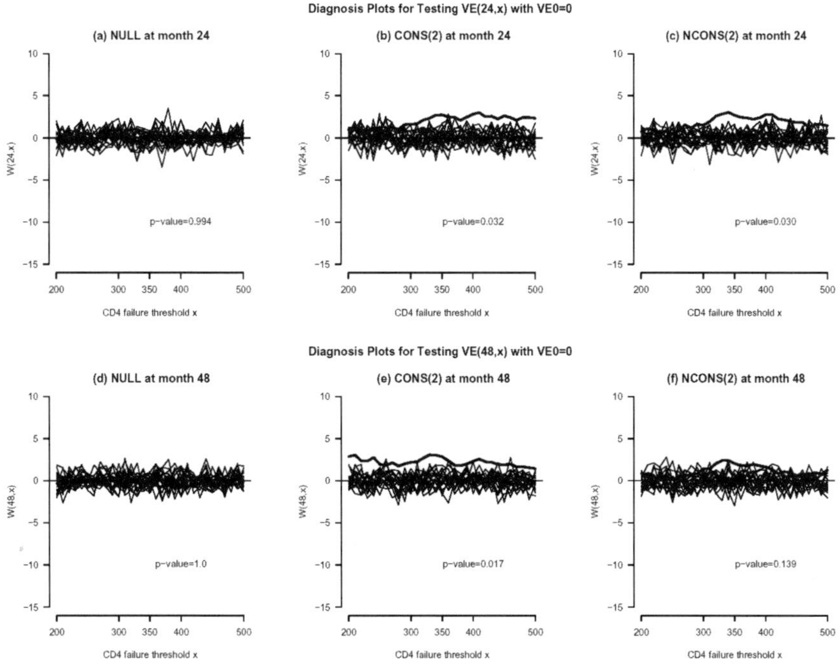

Fig. 4. (a)–(c) show the test process $W(24, x_{cd4})$ (thick lines) and 20 trajectories from $W^*(24, x_{cd4})$ (thin lines) for the vaccine trial datasets simulated under the (a) NULL, (b) CONS(2), and (c) NCONS(2) models, respectively. (d)–(f) show the corresponding test processes $W(48, x_{cd4})$ and $W^*(48, x_{cd4})$.

$W^*(\tau, x)$. Estimated one-sided p-values based on $B = 1000$ are also shown on the plots. Figures 3(b), (c) and (e) indicate statistical evidence against the corresponding null hypotheses while Figs. 3(a), (d) and (f) do not provide such evidence. The diagnostic plots for testing vaccine efficacy greater than zero for the CD4/ART composite endpoint are given in Fig. 4. Other than Figs. 4(a), (d), and (f), the plots provide some statistical evidence against the null hypothesis.

5. Discussion

In this chapter, methods for simultaneous inferences on HIV vaccine efficacy parameters based on viral failure or ART initiation, and on CD4 failure or ART initiation, have been developed. Simulation experiments demonstrated that the methods perform well for datasets expected from forthcoming efficacy trials, with reliably accurate coverage probabilities. The methods have been used in assessments of post-infection vaccine effects in the world's first two HIV vaccine efficacy trials, and are anticipated to be applied to the ongoing efficacy trial in Thailand and to planned efficacy trials. The approach to assessing vaccine effects on HIV progression based on the composite endpoints has a number of issues and limitations, some of which we now discuss.

First, can the composite efficacy parameters be assessed with sufficient power and precision? We found that with at least a 1 $\log_{10}$ drop in viral load by vaccine, there is high power to study the viral load/ART initiation endpoint by 24 or 48 months, and the simultaneous confidence bands about the efficacy parameters are quite narrow (Fig. 1). The method can correctly distinguish between a vaccine with durable efficacy out to 48 months versus a vaccine with transient efficacy that wanes due to emergence of HIV vaccine resistance (likely due to development of CTL epitope or T helper epitope escape mutations; Barouch *et al.*, 2002, 2003; Lukashov *et al.*, 2002). Power and precision for the composite CD4/ART endpoint was considerably less than for the viral load/ART endpoint (confidence bands about twice as wide), due to the lower number of CD4 failure events and to the longer time-frame required for vaccine effects on CD4 cells to become apparent (Fig. 2). The result of relatively low precision for the CD4/ART parameter may also be sensitive to the simulation model we used, which specified large variability of the CD4 cell counts (standard deviation of random intercept approximately 200 cells/mm^3, based on VaxGen's efficacy trial in Thailand). For forthcoming efficacy trials the relative precision for assessment of the viral- and CD4-based efficacy parameters is uncertain, and will depend on the population under study.

Second, are there other surrogate efficacy parameters that would provide superior assessments of vaccine efficacy on HIV progression? The time-to-event composite parameters only capture failure status, and efficacy parameters based on the quantitative level of viral load or CD4 count could be assessed with greater power. In addition to the level of viral load or CD4 count, the rate of HIV clearance during primary infection (estimated with a viral dynamics model) is a candidate surrogate endpoint, since rapid clearance of viremia has been shown to correlate with progression to AIDS (Blattner *et al.*, 2004). A central motivation of the composite parameters is to measure durability of vaccine effects in a way that accommodates ART initiation in a simple way that permits valid inference by standard time-to-event methods; in contrast, analysis of continuous parameters requires careful handling of dependent censoring by ART initiation. As mentioned in the Introduction and as others have discussed (cf. Gilbert, Bosch and Hudgens, 2003; Hudgens, Gilbert and Self, 2004), ART initiation could be accommodated by rank-based analysis (i.e. assign ART initiation as the worst rank), or pre-ART biomarker values could be assessed with adjustment for dependent censoring by ART using likelihood-based linear mixed effects models or inverse probability of censoring weighted estimating equations methods (e.g. Rotnitzky and Robins, 1997). Further research is needed into the relative utility of time-to-event post-infection composite endpoints versus quantitative biomarker-based endpoints.

Third, the analysis of post-infection vaccine effects that condition on infection are susceptible to post-randomization selection bias (Halloran and Struchiner, 1995; Hudgens, Hoering and Self, 2003; Gilbert, Bosch and Hudgens, 2003). Because of this issue, we believe that post-infection vaccine effects should generally be studied for the entire randomized cohort as well as for the infected subcohort. For continuous endpoints, "severity of infection" or "burden of illness" scores or ranks could be assessed in all randomized vaccinees versus all randomized placebo recipients as proposed by Chang, Guess and Heyse (1994). For the composite endpoints focused on here, Gilbert and Sun (2005) argued that the analyses should be done both for all randomized subjects and for the infected subcohort. An advantage of the composite endpoints is that they can be assessed straightforwardly in the entire study cohort based on time from randomization, and the corresponding efficacy parameters have clear interpretations as vaccine reductions in the time to clinically significant HIV infection.

Fourth, the composite endpoint parameters are surrogate endpoints, and their accuracy for capturing clinical endpoints is unknown and difficult to validate. This uncertainty argues the importance of assessing the composite endpoints for as long as possible after infection diagnosis, because more durable surrogate effects are more likely to translate into clinical effects. The uncertainty also underscores that vaccine effects on clinical endpoints must be assessed directly, and that more research is needed to understand the relationship between vaccine effects on the surrogate and clinical endpoints. It appears that meta-analysis will be required to validate surrogate endpoints (cf. Daniels and Hughes, 1997), and more research is needed to understand how many vaccine trials of what sizes are sufficient for this purpose. Clinical endpoints of interest include AIDS-defining illnesses and other clinical events, for example defined based on the World Health Organization's staging system for classifying HIV progression. Morgan *et al.* (2002) collected data on rates of WHO stage 2, 3, 4 clinical events among HIV-infected persons in Uganda, which are useful for power calculations on the ability to detect vaccine effects on clinical endpoints. More data on rates of various clinical endpoints are needed in other geographic regions (such as Southern Africa) where HIV vaccine efficacy trials are being planned.

Finally, emergence of vaccine resistance post acquisition of HIV is a major concern, and new methods are needed for detecting vaccine resistance based on analysis of the genetic sequences that infect trial participants and on immune responses induced by infected vaccinees. The event of vaccine resistance could be defined as loss of virologic suppression, but it is unclear how to best implicate particular mutations in the HIV genome as possibly causative of resistance. Barouch *et al.* (2002) provided a fairly definitive analysis for monkey challenge studies and methodologic techniques are needed for human trials.

Acknowledgments

This research was supported by NIH grant 1RO1 AI054165-01 (Gilbert) and NSF grant DMS-0304922 (Sun).

References

Albert JM, Ioannidis JPA, Reichelderfer P *et al.* (1998) Statistical issues for HIV surrogate endpoints: point/counterpoint, *Stat Med* **17**: 2435–2462.

Anderson RM, Garnett GP (1996) Low-efficacy HIV vaccines: potential for community-based intervention programs, *Lancet* **348**: 1010–1013.

Anderson RM, May RM (1991) *Infectious Diseases of Humans: Dynamics and Control.* Oxford: Oxford University Press.

Barouch DH, Kunstman J, Kuroda MJ *et al.* (2002) Eventual AIDS vaccine failure in a rhesus monkey by viral escape from cytotoxic T lymphocytes, *Nature* **415**: 335–339.

Barouch DH, Kunstman J, Glowczwskie J *et al.* (2003) Viral escape from dominant simian immunodeficiency virus epitope-specific cytotoxic T lymphocytes in DNA-vaccinated rhesus monkeys, *J Virol* **77**: 7367–7375.

Bilias Y, Gu M, Ying Z (1997) Towards a general asymptotic theory for Cox model with staggered entry, *Ann Stat* **25**: 662–682.

Blattner WA, Oursler KA, Cleghorn F, Charurat M, Sill A, Bartholomew C, Jack N, O'Brien T, Edwards J, Tomaras G, Weinhold K, Greenberg M (2004) Rapid clearance of virus after acute HIV-1 infection: correlates of risk of AIDS, *J Infec Dis* **189**: 1793–1801.

Chang MN, Guess HA, Heyse JF (1994) Reduction in burden of illness: a new efficacy measure for prevention trials, *Stat Med* **13**: 1807–1814.

Clemens JD, Naficy A, Rao MR (1997) Long-term evaluation of vaccine protection: methodological issues for phase 3 trials and phase 4 studies. in: Levine MM, Woodrow GC, Kaper JB, Cobon GS (eds.) *New Generation Vaccines* pp 47–67, New York: Marcel Dekker, Inc.

Clements-Mann ML (1998) Lessons for AIDS vaccine development from non-AIDS vaccines, *AIDS Res Hum Retrov* **14**(Supp.3): S197–S203.

Daniels MJ, Hughes MD (1997) Meta-analysis for the evaluation of potential surrogate markers, *Stat Med* **16**: 1965–1982.

DHHS Guidelines (2002) 2002 Panel on clinical practices for treatment of HIV infection. Department of Health and Human Services. Guidelines for the use of antiretroviral agents in HIV-infected adults and adolescents, Feb 4, 2002, http://www.aidsinfo.nih.gov/guidelines/

Fleming TR (1992) Evaluating therapeutic interventions (with discussion and rejoinder), *Stat Sci* **7**: 428–456.

Fleming TR, DeMets DL (1996) Surrogate endpoints in clinical trials: are we being misled? *Ann Intern Med* **125**: 605–613.

Gilbert P, De Gruttola V, Hudgens M, Self S, Hammer S, Corey L (2003) What constitutes efficacy for an HIV vaccine that ameliorates viremia: issues involving surrogate endpoints in Phase III trials, *J Infect Dis* **188**: 179–183.

Gilbert P, Sun Y (2005) Failure time analysis of HIV vaccine effects on viral load and antiretroviral therapy initiation. *Biostatistics*, in press.

Gray RH, Wawer MJ, Brookmeyer R *et al.* (2001) Probability of HIV-1 transmission per coital act in monogamous, heterosexual, HIV-1 discordant couples in Rakai, Uganda, *Lancet* **357**: 1149–1153.

Halloran ME, Struchiner CJ, Longini IM (1997) Study designs for evaluating different efficacy and effectiveness aspects of vaccines, *Amer J Epidem* **146**: 789–803.

Harro CD, Judson FN, Gorse GJ, Mayer KH, Kostman JR, Brown SJ, Koblin B, Marmor M, Bartholow BN (2004) Popovic V for the VAX004 study group recruitment and baseline epidemiologic profile of participants in the first phase 3 HIV vaccine efficacy trials. *J Acq Immunodef Syndr*, in press.

Hayes R (1998) Design of human immunodeficiency virus intervention trials in developing countries, *J Roy Stat Soc Series A* **161**(2): 251–263.

Hirsch MS, Brun-Vezinet F, D'Aquila RT *et al.* (2000) Antiretroviral drug resistance testing in adult HIV-1 infection: recommendations of an International AIDS society-USA Panel, *J Amer Med Assoc* **283**: 2417–2426.

Hughes MD, Daniels MJ, Fischl MA, Kim S, Schooley RT (1998) CD4 cell count as a surrogate endpoint in HIV clinical trials: a meta-analysis of studies of the AIDS Clinical Trials Group, *AIDS* **12**: 1823–1832.

HIV Surrogate Marker Collaborative Group (2000) Human immunodeficiency virus type 1 RNA level and CD4 count as prognostic markers and surrogate endpoints: a meta-analysis. *AIDS Res Hum Retrov* **16**: 1123–1133.

Hudgens MG, Hoering A, Self SG (2003) On the analysis of viral load endpoints in HIV vaccine trials, *Stat Med* **22**: 2281–2298.

Hudgens M, Gilbert P, Self S (2004) Statistical issues for endpoints in vaccine trials, *Stat Meth Med Res*, in press.

HVTN (HIV Vaccine Trials Network) (2004) The pipeline project, http://www.hvtn.org/

IAVI (International AIDS Vaccine Initiative) (2004) State of current AIDS vaccine research, http://www.iavi.org/

Lin DY, Wei LJ, Ying Z (1993) Checking the Cox model with cumulative sums of martingale-based residuals, *Biometrika* **80**: 557–72.

Longini IM, Susmita D, Halloran ME (1996) Measuring vaccine efficacy for both susceptibility to infection and reduction in infectiousness for prophylactic HIV-1 vaccines, *J Acq Imm Defic Syndr Hum Retrovir* **13**: 440–447.

Lukashov VV, Goudsmit J, Paxton WA (2002) The genetic diversity of HIV-1 and its implications for vaccine development, in: Wong-Staal F, Gallo RC (eds.) *AIDS Vaccine Research*, pp 93–120, New York: Marcel Dekker.

Lyles RH, Munoz A, Yamashita TE *et al.* (2000) Natural history of human immunodeficiency virus type 1 viremia after seroconversion and proximal to AIDS in a large cohort of homosexual men. Multicenter AIDS cohort study, *J Infec Dis* **181**: 872–880.

McKeague IW, Gilbert PB, Kanki PJ (2001) Omnibus tests for comparison of competing risks with adjustment for covariate effects, *Biometrics* **57**: 818–828.

Mellors JW, Munoz A, Giorgi JV (1997) Plasma viral load and CD4+ lymphocytes as prognostic markers of HIV-1 infection, *Ann Intern Med* **126**: 946–954.

Morgan D, Mahe C, Mayanja B, Whitworth JAG (2002) Progression to symptomatic disease in people infected with HIV-1 in rural Uganda: prospective cohort study, *Brit Med J* **324**: 193–197.

Murphy BR, Chanock RM (1996) Immunization against virus disease, in: Fields BN, Knipe DM, Howley PM, Chanock RM, Melnick JL, Monath TP, Roizman B, Straus SE (eds.) *Fields Virology* pp 467–497, Philadelphia: Lippincott–Raven.

Nabel GJ (2001) Challenges and opportunities for development of an AIDS vaccine, *Nature* **410**: 1002–1007.

Orenstein WA, Bernier RH, Hinman AR (1998) Assessing vaccine efficacy in the field, *Epidemiol Rev* **10**: 212–241.

Parzen MI, Wei LJ, Ying Z (1999) Simultaneous confidence intervals for the difference of two survival functions, *Scand J Stat* **24**: 309–314.

Quinn TC, Wawer MJ, Sewankambo N *et al.* (2000) Viral load and heterosexual transmission of human immunodeficiency virus type 1, *New Engl J Med* **342**: 921–929.

rgp120 HIV Vaccine Study Group (2004) Placebo-controlled trial of a recombinant glycoprotein 120 vaccine to prevent HIV infection, *J Infect Dis*, in press.

Rotnitzky A, Robins J (1997) Analysis of semi-parametric regression models with non-ignorable nonresponse, *Biometrics* **16**: 81–102.

Shen X, Siliciano RF (2000) AIDS: preventing AIDS but not HIV-1 infection with a DNA vaccine, *Science* **290**: 463–465.

Shiver JW, Fu T-M, Chen L *et al.* (2002) Replication-incompetent adenoviral vaccine vector elicits effective anti-immunodeficiency virus immunity, *Nature* **415**: 331–335.

UNAIDS (2001) Joint United Nations Programme on HIV/AIDS, 2001. Symposium: ethical considerations in HIV preventive vaccine research: examining the 18 UNAIDS guidance points, *Devel World Bioeth* **1**: 121–134.

Verbeke G, Molenberghs G (2000) *Linear Mixed Models for Longitudinal Data.* New York: Springer.

World Health Organization (2003) Scaling up antiretroviral therapy in resource-limited settings: Treatment guidelines for a public health approach, http://www.who.int3by5publicationsguidelinesenarv_guidelines.pdf.

CHAPTER 6

A REVIEW OF MATHEMATICAL MODELS FOR HIV/AIDS VACCINATION

Shu-Fang Hsu Schmitz

This chapter first summarizes results of recent vaccine trials. Then we present structures and results of selected mathematical models for prophylactic (non-live-attenuated and live-attenuated) and therapeutic HIV/AIDS vaccines. Finally, important modeling considerations and discussions are provided. The purpose of this chapter is not to emphasize the usefulness of mathematical modeling, but to provide modelers a good collection of various types of deterministic models and discussions concerning differences among the models.

Keywords: HIV, AIDS, vaccine, mathematical model.

1. Introduction

1.1. *Current situation of HIV/AIDS epidemic*

Around the world more than 3 million people died of HIV/AIDS and 5 million acquired HIV in 2003, and 40 million are currently living with HIV/AIDS (UNAIDS 2003). This disease has become a big social and medical burden in many countries. Various treatments have been developed with good success in slowing down disease progression. However, the high cost makes treatments unaffordable for most patients in developing countries. Moreover, long-term therapy is associated with several harmful side-effects and drug-resistance. Finally, no cure can be expected with such treatments.

As AIDS is an uncurable disease, prevention is the best approach to protecting populations at risk from infection. Vaccination has been a successful, cost-effective, preventive intervention to control or eradicate several infectious diseases, e.g. smallpox, poliomyelitis, measles, mumps, influenza, hepatitis B, tuberculosis, etc. (Plotkin and Mortimer, 1994; Blower *et al.*, 2003). Therefore, it is a natural hope to develop effective vaccines against HIV/AIDS.

1.2. *HIV/AIDS vaccines*

Depending on development strategies, vaccines can be broadly classified into two groups. One group consists of live-attenuated vaccines (e.g. smallpox, poliomyelitis, measles, mumps) and the other group is developed by other strategies. Live-attenuated vaccines have the potential advantages of high protective efficacy, low cost and simple immunization schedules, but have the risk of being infected by the attenuated strains and possibly resulting in serious disease or death (Blower *et al.*, 2001). Due to safety concerns of live-attenuated vaccines, most candidate HIV/AIDS vaccines are in the group of non-live-attenuated vaccines.

The ultimate goal of HIV/AIDS vaccines is to prevent infection, i.e. prophylactic (preventive) vaccines for healthy individuals. To supplement or replace current therapies, some vaccines are being developed for treatment purpose, i.e. therapeutic vaccines to prevent or postpone progression to AIDS, with the potential to reduce emergence of drug-resistant HIV strains and to reduce side effect on patients.

1.3. *The purpose and structure of this chapter*

Clinical vaccine trials reveal the potential effects of candidate vaccines at the individual level. Before a vaccine can be applied to the general population, not only limited in clinical trial participants, its effects at the population level should be carefully evaluated. This is especially important when the vaccines are imperfect. Currently there is no vaccine ready for general use. Hard data of vaccine effects at the population level is thus not available. Under this situation, mathematical modeling can be a very useful tool for decision making on public health strategies. Various mathematical models with different considerations have been proposed in the literature. Blower and colleagues have provided good reviews of various models (McLean and Blower, 1995; Blower *et al.*, 2003). The purpose of this chapter is not to emphasize the usefulness of mathematical modeling to the medical researchers. Instead, it is intended as a summarizing reference for modelers, with focus on the collection of various types of deterministic models and on the recognition of differences among them.

The subsequent sections are organized as follows: Section 2 summarizes results of recent vaccine trials, so we know what we can expect in the near future; Sections 3 and 4 present model structure and results for prophylactic non-live-attenuated vaccines and live-attenuated vaccines, respectively; Section 5 is devoted to models of therapeutic vaccines; Section 6 describes models that investigate combined interventions including vaccination;

finally important modeling considerations and discussions are provided in Section 7.

2. Results of Recent HIV Vaccine Trials

2.1. *Prophylactic vaccines*

The development of efficacious prophylactic HIV/AIDS vaccines progresses very slowly due to lack of immune correlates, unidentified immunogens that induce broad and long-lasting immunity, genetic diversity of the virus, and limitations of animal models (Nabel, 2001). The aim of candidate vaccines was to induce neutralizing antibodies in the first wave, to stimulate CD8+ T-cell response in the second wave, and to achieve both responses in the current wave (Esparza and Osmanov, 2003). It is likely that a vaccine to prevent HIV/AIDS will eventually be developed (Stratov *et al.*, 2004). Several candidate HIV/AIDS vaccines have been investigated in different stages of clinical trials since 1987. Only very recently results of large-scale efficacy trials become available and summarized below.

A phase III efficacy trial comparing a gp120 (non-live-attenuated) vaccine with placebo in gay men was conducted in the United States, Puerto Rico, Canada and the Netherlands. The data show no difference in HIV infection rates between vaccine and placebo recipients. However, in a subgroup analysis, the vaccine protected 67% of the black, Asian, and mixed-race participants (Cohen, 2003a).

Another phase III trial of a similar vaccine was conducted in injecting drug users in Thailand. The preliminary results also show no difference in HIV infection rates between vaccine and placebo recipients. However, this vaccine will be used as a booster shot to another vaccine in an ongoing large-scale trial in Thailand (Cohen, 2003b).

2.2. *Therapeutic vaccines*

Therapeutic vaccines seem to have more success. An oral immunogenic preparation, V-1 Immunitor, was developed in Thailand. It could boost CD4 and CD8 lymphocyte counts, decrease viral load, and reverse AIDS associated wasting in HIV-infected patients (Jirathitikal and Bourinbaiar, 2002). In the subsequent trial, V-1 Immunitor was administered to terminally ill, end-stage AIDS patients and resulted in prolonged survival and return to normal life (Metadilogkul *et al.*, 2002). The benefits of V-1 Immunitor were also demonstrated in a retrospective study with 650 HIV positive patients (Jirathitikal *et al.*, 2004).

In addition, the prophylactic potential of V-1 Immunitor was investigated in a phase II, placebo-controlled trial conducted in Thailand. Uninfected individuals receiving V-1 Immunitor had significantly higher CD4 and CD8 cell counts than those receiving placebo. This finding suggests that CD4 and CD8 counts may become an easily measurable immune correlate of the efficacy of HIV/AIDS vaccines and V-1 Immunitor has the potential as a prophylactic vaccine (Jirathitikal *et al.*, 2003).

3. Models for Prophylactic, Non-Live-Attenuated Vaccines

Due to difficulties in vaccine development as mentioned in the Introduction, the candidate prophylactic vaccines are unlikely to be perfect and might fail in different ways. Mathematical models help to evaluate the potential of imperfect candidate vaccines under different situations.

3.1. *Simple one-sex models*

McLean and Blower (1993) consider three facets of vaccine failure: "take" (ε), the fraction that the vaccine has an effect on the vaccinated individuals; "degree" (ψ), reduction fraction in probability of infection upon exposure among vaccinated individuals who take the vaccine; and "duration" ($1/\omega$) of vaccine protection. They evaluate the vaccine impact on the epidemic in a homosexual population, with individuals classified as susceptible, vaccinated, infectious, or AIDS patients (assumed not sexually active). Vaccination is applied only to a fraction (p) of new (susceptible) recruits and only individuals who "take" the vaccine are included in vaccinated class. No revaccination is considered for those whose vaccine protection has waned. No differences in infectiousness and progression rate are assumed between infectious individuals coming from the susceptible class and those from the vaccinated class. Random mixing is assumed for sexually active individuals across different classes, i.e. disease status and vaccination status do not influence pair formation. Individuals leave the population at a natural (not due to AIDS death) removal rate (μ).

The severity of the epidemic before introduction of any intervention can be summarized by the **basic** reproductive number (R_0), defined as the number of new HIV infection cases resulting from a single case in an otherwise disease-free population (McDonald, 1952; May and Anderson, 1987). If $R_0 < 1$, then the epidemic will die out; but if $R_0 > 1$, then the disease spread will continue. One outcome measure for vaccine effects is the **effective** reproductive number (R), similarly defined as the number of new

HIV infection cases resulting from a single case in an otherwise disease-free population in which a program of mass vaccination is in place. For the model described above, the relationship between R and R_0 is given by

$$R = R_0(1 - p\,\phi),$$

where ϕ is the summary measure of vaccine imperfections, called vaccine impact, and defined as:

$$\phi = \varepsilon\,\psi\,\mu/(\mu + \omega).$$

The critical vaccination coverage (p_c) required to eradicate the disease for a given R_0 and a given ϕ can be calculated by letting $R = 1$, more specifically,

$$p_c = (1/\phi)(1 - 1/R_0).$$

If $p_c > 1$, then it is not possible to eradicate the disease by such vaccines.

Similarly, the critical vaccination impact (ϕ_c) required to eradicate the disease for a given R_0 and a given p can be calculated by letting $R = 1$, more specifically,

$$\phi_c = (1/p)(1 - 1/R_0).$$

Blower and McLean (1994) applied the model above to the homosexual male population in San Francisco. The R_0 is estimated to be between 2 and 5, i.e. a severe epidemic. The results show that vaccination alone is unlikely to eradicate the disease in this population. They also investigate the effects of change in risky behavior induced by the vaccination campaign. The relative level of risky behavior is represented by the product of the average transmission efficiency of HIV per sexual partnership (β, reflecting the degree of condom use) and the average number of receptive anal sex partners experienced per unit time (c). If the vaccination campaign induces increase in risky behavior in all sexually active classes, then the severity of the epidemic could even become worse.

Following McLean and Blower (1993), Anderson and Garnett (1996) considered two additional parameters to reflect the potential therapeutic effect of the vaccine: ratio q of the infectiousness and ratio r of the incubation period, in infected vaccinated individuals relative to that of unvaccinated individuals. The critical vaccination coverage is now given by

$$p_c = (1 - 1/R_0)/\{(\phi/\psi)[1 - (1 - \psi)r\,q]\}.$$

They also conclude that it is difficult to eliminate the infection unless the vaccine is moderately to highly effective in preventing infection and the vaccine protection duration is long.

3.2. *A one-sex model with preferred mixing*

Anderson *et al.* (1995) investigate the potential impact of vaccines with low to moderate efficacy in two models. The first model is a one-sex model with sexually active individuals classified by sexual activity (rate of sexual partner change) and disease/vaccination status (susceptible, infected, or vaccinated). Vaccination is applied to susceptible individuals and new recruits at different rates, with the possibility of revaccination. The mixing pattern between individuals of different sexual activity classes follows a preferred mixing matrix, which can range from random mixing (no specific preference for partner's sexual activity class) to complete assortative mixing (individuals choose partners only of the same sexual activity class) (Garnett and Anderson 1993a; 1993b). It is assumed that disease status has no influence on mixing matrix. The outcome measure is the possibility of blocking HIV transmission. Using parameters reflecting high transmission areas, the model results show that it will be difficult to block transmission for any mixing pattern.

3.3. *A two-sex age-structured model*

The second model of Anderson *et al.* (1995) is a two-sex model including sexually active and inactive individuals. The basic setting is similar to that in the first model, but with additional grouping according to sex and age. In addition to heterosexual transmission, vertical transmission is also included (details in Anderson *et al.*, 1991 and Garnett and Anderson, 1994). Since the simple model above already indicates that blocking transmission will be difficult, different outcome measures are considered here, i.e. HIV seroprevalence and number of AIDS cases prevented 20 years after the introduction of immunization. Using parameters reflecting a HIV epidemic in an urban center in sub-Saharan Africa, the model results show that vaccines with efficacy of 50% or less and protection duration of 5 years can prevent many cases of HIV infection and associated mortality if the vaccination coverage can be maintained to a sufficiently high level, which might be costly. Hence, proceeding with phase III efficacy trials of such vaccines may be appropriate in high HIV transmission regions.

3.4. *One-sex models with genetic heterogeneity*

Several mutant genes in HIV co-receptors are associated with reduction in susceptibility to HIV-1 or/and rate of progression to AIDS (Samson *et al.*, 1996; Dean *et al.*, 1996; Smith *et al.*, 1997; Quillent *et al.*, 1998; Winkler *et al.*, 1998). Some of these genes have non-ignorable allele frequencies in general populations. Their effects on the HIV/AIDS dynamics and the related interventions may be significant. Taking the genetic heterogeneity into account, Hsu Schmitz (2000) investigates effects of treatment or/and vaccination on HIV transmission in a homosexual population. Uninfected individuals are classified by degree of natural resistance to HIV-1 infection due to mutant genes (no resistance, partial resistance, or complete resistance) and vaccination status (protected or not) into 5 groups. Those with complete natural resistance will not be vaccinated and never be infected. New recruits enter the 3 susceptible groups at certain proportions depending on the frequency of the mutant alleles. Vaccination is applied only to a proportion of new (susceptible) recruits, i.e. no revaccination, and only individuals who "take" the vaccine are included in the vaccinated groups. The vaccine does not provide therapeutic effects in subsequent infection. Infected individuals are classified by progression rate (rapid, normal, or slow progressor) and treatment status (treated or not) into 6 groups. Newly infected individuals enter different infected groups at certain proportions depending on the genotype. Those with the favorable mutant genes have a higher chance to become slow progressors than those without. Treatment (chemotherapy or immunotherapy) is administered only to a proportion of newly infected individuals and helps reduce transmission rate and progression rate by certain factors. Partnerships are formed by random mixing. The outcome measures are related to the basic reproductive numbers calculated under different situations, together with the critical vaccination proportion and critical treatment proportion for disease eradication.

The model is applied to the homosexual population in San Francisco with focus on the CCR-5 $\Delta 32$ mutation. Vaccination alone is helpful in slowing down the disease spread, but only highly efficacious and long-lasting vaccines have the potential to eradicate the epidemic.

3.5. *A complex one-sex model for cost-effectiveness*

Focusing on cost-effectiveness instead of possibility of disease eradication, Edwards *et al.* (1998) analyzed the total health benefits and costs of a wide range of possible vaccination programs in a population of homosexual men in San Francisco. Individuals are classified by disease stage, screening status and vaccination status into 8 groups: uninfected, vaccinated uninfected, unidentified infected asymptomatic, vaccinated unidentified infected asymptomatic, identified infected asymptomatic, vaccinated identified infected asymptomatic, infected symptomatic, and AIDS. New recruits enter the 4 unvaccinated, non-AIDS groups at constant rates. The preventive vaccine is applied to uninfected individuals and unidentified infected asymptomatic individuals. When the vaccine protection wanes, individuals can be revaccinated. The therapeutic vaccine is administered to identified infected asymptomatic individuals and may prolong the duration of the asymptomatic period. Partnerships are formed by random mixing. Factors considered in the cost-effectiveness analysis include type of vaccines (preventive or therapeutic), vaccine characteristics (efficacy, protection duration and cost), infectivity (i.e. infectiousness) induced by therapeutic vaccines, epidemic stage (early stage with rapid growing or late stage with slow growing) and risky behavior change (condom usage). Two outcome measures of vaccine programs are evaluated: 1) total discounted economic (direct and indirect) costs of the vaccination program; and 2) total discounted quality-adjusted life-years (QALYs), with a year of life with HIV infection being considered less desirable than that without and a year of life with asymptomatic infection being more desirable than that with symptomatic infection. The differences in the two outcome measures between a population with and a population without a vaccination program are calculated.

The results of the analysis for preventive vaccines suggest that the outcome depends on risky behavior induced by vaccination and epidemic stage. If condom usage is decreased by 25%, then the vaccination program would be cost-effective only when the preventive vaccine has a relatively high efficacy and a relatively long protection duration. If condom usage is increased by 25%, then any vaccination program would be cost-effective. In the early epidemic, the vaccines are more cost-effective than in the late epidemic.

3.6. *A complex one-sex model for intravenous drug users*

For intravenous drug users (IDUs), the model considerations are quite different. Bogard and Kuntz (2002) present a model to investigate the impact of a partially effective vaccine on a population of IDUs in Bangkok. Individuals are classified by vaccination status (unvaccinated, vaccinated and not susceptible, or vaccinated and susceptible), level of injection-related risky behavior (low or high), and disease status (uninfected, unidentified asymptomatic infected, identified asymptomatic infected, symptomatic infected, or AIDS) into 14 groups. The injection risky behavior is characterized by contact rate (number of "partners", i.e. different persons with whom needles are shared, each year), total number of injections per year, and the probability that a needle was shared (i.e. used by someone else immediately before injection). The infectivity of an infected needle is described by per contact infectivity, which is defined as the probability of viral transmission per infected needle contact. The probability of viral transmission per infected partner is then a function of per contact infectivity, the probability that a needle was shared, contact rate, and the number of injections per year. Contacts are formed by "restricted" mixing, i.e. individuals share needles only with other individuals in the same risk group. Revaccination is allowed. The vaccines are assumed not to have therapeutic effects in vaccinated infected individuals.

The outcome measure here is HIV prevalence after 40 years. The model results are highly sensitive to changes in risky behavior among susceptible individuals, the probability that a needle was shared, the number of injections per year, and vaccine protection duration. Even though the vaccines are not perfect, they provide benefit in most situations. Unfavorable risky behavior change decreases the benefit of vaccination, but has a limited effect on the outcome when vaccine efficacy is high, e.g. 75%.

4. Models for Prophylactic, Live-Attenuated Vaccines

Live-attenuated HIV vaccines (LAHVs) contain a specific HIV strain with reduced activity that is different from the wild-type strains. Individuals vaccinated with a LAHV have the risk of developing AIDS caused by the attenuated strain. To predict the tradeoff between efficacy and safety of LAHVs, Blower *et al.* (2001) present a one-sex model with individuals classified into 5 groups: susceptible, infected with the wild-type strain, infected with the vaccine strain (through vaccination or transmission), infected first with the vaccine strain and then also the wild-type strain, and AIDS patients (not sexually active). The considered LAHV is assumed to have two beneficial effects: 1) a degree effect, i.e. it induces some degree of protection against infection with wild-type strain; and 2) a therapeutic effect, i.e. it reduces the progression rate in dually infected individuals. The vaccine is applied only to a proportion of new (initially uninfected) recruits. The dually infected individuals can only transmit the wild-type strain. Random mixing is assumed for sexually active individuals across different classes, i.e. disease and vaccination status does not influence pair formation.

Using the model described above together with demographic parameters for Zimbabwe and Thailand, simulations are performed for 1000 sets of vaccine parameters. The vaccine *efficacy* is evaluated by annual AIDS death rate due to wild-type strain and by the possibility of eradicating the wild-type strain. The vaccine *safety* is evaluated by annual AIDS death rate due to vaccine strain and specifically quantified here as the percentage of vaccinated individuals who progress to AIDS in 25 years as a result of a LAHV. The higher value the vaccine safety parameter is, the higher the risk of progression, i.e. the less safe the vaccine. Considering both efficacy and safety, the net effect of the vaccination campaigns at the population level is specifically assessed by total AIDS death rate (AIDS deaths due to either strain).

For both Zimbabwe and Thailand, the wild-type HIV would be eradicated by mass vaccination campaigns after several decades and the prevalence of the vaccine strain would reach an endemic steady state. Hence, AIDS deaths due to wild-type would decrease with time and AIDS deaths due to vaccine strain would increase with time. Fortunately the vaccine strain could be eradicated by stopping vaccination because the associated basic reproductive number is less than unity. For Zimbabwe where the transmission rate and current annual AIDS death rate (1987 per 100,000 individuals) are high, the net effect of vaccination is always very beneficial, with total annual AIDS death rate decreasing substantially over time for all 1000 simulations. In contrast, for Thailand where the transmission rate and the current annual AIDS death rate (159 per 100,000 individuals) are not too high, the net effect is often detrimental, with total AIDS death rate increasing over time in the majority of the simulations, especially when the vaccine safety is not good enough. The same vaccine can be beneficial in Zimbabwe, but detrimental in Thailand. This suggests that the transmission rate determines the tradeoff between a LAHV's efficacy and safety characteristics. A threshold transmission rate (TTR) can be identified. If the transmission rate is greater than the TTR, then the efficacy effect of a LAHV outweighs the safety effect. On the other hand, if the transmission rate is smaller than the TTR, then the safety effect outweighs the efficacy effect.

5. Models for Therapeutic Vaccines

Treatments (including therapeutic vaccines) are likely to slow down the progression rate, hence increase the incubation period (from infection to AIDS symptoms). This is certainly a benefit for patients. However, the effects of treatment on infectiousness are not clear. Due to prolonged incubation period, there will be more infected individuals in the population; hence the chance to pair with an infected individual is higher. Whether treatments give also benefits at the population level is not obvious. Mathematical models help to project the possible outcome in the population.

5.1. *A simple one-sex model*

Anderson, Gupta and May (1991) present a simple model to investigate the potential of chemotherapy or immunotherapy (i.e. therapeutic vaccines) to control of HIV epidemic in a homosexual population. Individuals are classified into 4 groups: unvaccinated susceptible, vaccinated susceptible, untreated infected, treated (with chemotherapy and/or immunotherapy) infected. The two infected groups include symptomatic AIDS patients. The immunization is applied to a proportion of new (susceptible) recruits and at a constant per capita rate to susceptible individuals. The vaccine does not provide protection against HIV infection, but with life-long therapeutic effects. Chemotherapy is administered at a given per capita rate to infected individuals. The chemotherapy and immunotherapy help to slow down the progression rate, hence prolong the incubation period. Random mixing is assumed across different classes, i.e. disease/treatment status does not influence pair formation.

Three outcome measures for the impact of treatment on the population are examined: 1) ratio of population size at equilibrium in the presence of treatment (N^*) over the size in the absence of treatment (N_A), with the ratio > 1 indicating that the treatment is beneficial for the population, while ratio < 1 indicating that it is detrimental; 2) net death rate due to AIDS; and 3) ratio of infected population size in the presence of treatment over the size in the absence of treatment. If the treatment does not decrease infectiousness, then it could be detrimental for the population when the severity of the initial epidemic is not too high and incubation period is increased by a few years. The detrimental effects on the community may be negated if the treated infected individuals reduce their sexual activity to certain extend. When the severity of the initial epidemic is very high, treatments will be always beneficial. These results draw attention to two important aspects: 1) treatment should be linked with counseling in order to avoid increase in risky behaviors; and 2) clinical trials should assess change in infectiousness induced by the treatment.

5.2. *One-sex models with genetic heterogeneity*

The model by Hsu Schmitz (2000) described above for prophylactic vaccines also investigates effects of treatments. Although not specifically for immunotherapy, the results for the general sense of treatments may still provide useful hints. For the homosexual population in San Francisco, it is unlikely to eradicate the disease by treatment alone. The treatment effect on the epidemic at the population level is sensitive to the reduction factor for transmission rate induced by the treatment. Individuals with partial resistance are more likely to be slow progressors once infected. The critical treatment proportion is very sensitive to the transmission rate of slow progressors. This reflects the indirect genetic influence on the epidemic and disease control.

5.3. *A complex one-sex model for cost-effectiveness*

The model by Edwards *et al.* (1998) described above for prophylactic vaccines is also applied to therapeutic vaccines. The results of the analysis for therapeutic vaccines in a population of homosexual men in San Francisco suggest that the outcome depends largely on changes in infectivity and risky behavior induced by vaccination. If condom use is decreased by 25% (increase in risky behavior), then the vaccination program would be cost effective only when the therapeutic vaccine can sufficiently reduce infectivity and prolong incubation time. If condom use is increased by 25%, then any vaccination program would be cost effective. Therapeutic vaccines that reduce infectivity by 25% or less are more efficient in the late than in the early epidemic. Therapeutic vaccines that reduce infectivity by 75% or more and prolong 2 years of life are efficient in both epidemic stages, but can save more money in the early epidemic.

Based on the same model, Owens *et al.* (1998) raise another point for consideration. Preventive vaccines reduce HIV infection incidence and prevalence. In contract, therapeutic vaccines that do not affect infectivity may increase slightly the incidence and prevalence; however, they can still provide substantial benefit in term of net gain in QALY.

6. Models with Combined Interventions

Results from several models mentioned above demonstrate that it is unlikely to eliminate the disease by a single intervention. Adding vaccination and new interventions to existing disease control programs might be a promising approach.

6.1. *Additivity*

Anderson and Garnett (1996) introduce a measure of additivity for combinations of different interventions. This measure (A) is calculated as the reduction in prevalence of HIV when two interventions are combined, divided by the sum of the reduction in prevalence when the two interventions are used separately. $A = 1$ reflects simple additivity, $A > 1$ synergism (i.e. more than additivity), and $A < 1$ antagonism (i.e. less than additivity). In their examples, combining vaccination with other interventions gives at least additive effect and hence seems to be cost effective.

6.2. *One-sex models with genetic heterogeneity*

The model by Hsu Schmitz (2000) described above for prophylactic vaccines also consider the combination with treatments. For the homosexual population in San Francisco, adding vaccination to an existing treatment program makes the elimination of transmission more feasible.

Following the same approach, Del Valle *et al.* (2004) present a model taking additional education intervention into account. Individuals of the 3 unvaccinated susceptible groups are educated at certain rate. The infection rates of the 3 resulting educated groups are reduced by certain factor. The education effect is assumed permanent, therefore there is no need to re-educate susceptible individuals. For the homosexual population in San Francisco, education appears to be an effective control measure even with only a small education rate. Combining education programs with treatment and vaccination makes it more feasible to eradicate the epidemic.

7. Discussions

Important characteristics of selected models are listed in Table 1 and commented below.

- AIDS patients are included in some models and excluded in some others. If interest is focused on disease transmission by sexual act, then AIDS patients can be ignored when they are assumed sexually inactive. If AIDS deaths and costs related to AIDS patients are of interest, then these patients should be included in the model. Note that in Model 9, AIDS patients are included in the infected group instead of a separate group.
- Different outcome measures have been proposed. The choice depends on vaccine type and feasibility. For non-live-attenuated vaccines, the main concern is efficacy; while for live-attenuated vaccines, additional safety measures should be considered. Some outcome measures might be feasible in some countries, but not in other countries. It is possible that conclusions based on different outcome measures differ from each other for the same vaccination programs. Although it is a favorable approach to consider simultaneously various outcome measures, sometimes it might be difficult to reach an overall recommendation.

Table 1. Overview of characteristics of selected models.

Model nr.	Ref. nr.	Incl. AIDS	Outcome measures	Mixing pattern	Behavior change
1	25	Yes	R_0 and p_c	Random	All sexually active groups
2	4	No	Disease eradication	Preferred	—
3	4	Yes	HIV prevalence, number of AIDS cases prevented in 20 years	Preferred	—
4	18	No	R_0	Random	—
5	11	No	R_0	Random	Due to education in unvaccinated susceptible groups
6	7	Yes	HIV prevalence after 40 years	Restricted	A proportion change between high- and low-risk groups
7	5	Yes	Efficacy: Annual AIDS death rate and eradication of the wild-type strain Safety: Annual AIDS death rate due to vaccine	Random	—
8a	12	Yes	Cost & QALY	Random	—
8b	12	Yes	Cost & QALY	Random	—
9	2	Yes	Ratio of equilibrium population sizes, net AIDS death rate, ratio of infected population sizes	Random	Treated infected

References

1 — Anderson *et al.* (1991)
2 — Anderson *et al.* (1995)
3 — Blower *et al.* (2001)
4 — Bogard and Kuntz (2002)
5 — Del Valle *et al.* (2004)
6 — Edwards *et al.* (1998)
7 — Hsu Schmitz (2000)
8 — McLean and Blower (1993).

- Several models use the basic (or effective) reproductive number R_0 as or to derive outcome measures. Kribs–Zaleta and Velasco–Hernández (2000) raise caution to the possibility of backward bifurcation, with which one may fail to control the disease even the $R_0 < 1$.
- Mixing pattern can also play an important role in the epidemic and decision making for vaccination programs. Random mixing is a good approximation if the status of the stratifying factors is not known to the potential partners.

Table 1. Overview of characteristics of selected models (cont.).

Model nr.	Vaccine type	Vaccination target	Distinguish vaccinated	Revaccination	Therap. effects
1	Prophyl.	New recruits	Only "take"	No	No
2	Prophyl.	New recruits & susceptibles	Only "take"	Yes	No
3	Prophyl.	New recruits & susceptibles	Only "take"	Yes	No
4	Prophyl.	New recruits	Only "take"	No	No
5	Prophyl.	New recruits	Only "take"	No	No
6	Prophyl.	Susceptibles	"Take" & "no take"	No	No
7	Prophyl., LAHV	New recruits	Only "take"	No	Yes
8a	Prophyl.	Uninfected & unidentified	Only "take"	Yes	No
8b	Therap.	Identified & symptomatic	Only "take"	No	Yes
9	Therap.	New recruits & susceptibles	Only "take"	No	Yes

- Vaccination-induced behavior changes are considered in various subject groups and reflected by different parameters. Both favorable and unfavorable changes are possible.
- Prophylactic vaccines are targeted to new recruits and/or susceptible individuals. In Model 8a is also applied to unidentified asymptomatic infected individuals, which is very likely in practice if the screening program is too expensive or not effective to identify asymptomatic-infected individuals.
- Therapeutic vaccines are naturally targeted to infected individuals. In Model 9, it is given to susceptible individuals.
- Usually only individuals who are vaccinated and "take" the vaccine are included in the "vaccinated" group. In Model 6 vaccinated individuals are stratified as "take" or "no take". The "no take" group is distinct from

the unvaccinated susceptible group. The purpose of such distinction is to allow different behavior. If vaccinated individuals cannot be informed about their "take" status soon after vaccination, those "no take" individuals may change their risky behavior after vaccination even though they are not protected by the vaccine. Such vaccination-induced changes are not expected for other groups in this model, but they are in Model 1.

- Revaccination is not considered for therapeutic vaccine models and for some prophylactic vaccine models. The necessity of revaccination will depend on the protection duration of the prophylactic vaccines considered.
- Therapeutic effects for prophylactic vaccines are only considered in Model 7 with live-attenuated vaccines. In fact, vaccines with both the prophylactic and therapeutic effects are quite possible, e.g. V-1 Immunitor (Jirathitikal *et al.*, 2003).

Not presented in Table 1 is whether the model is one-sex or two-sex. Among the listed models, only Model 3 uses a two-sex model. For applications to homosexual populations or IDUs, it is the natural approach, but for some cases one-sex model, instead of two-sex model, is used for its simplicity and tractability.

Due to space limitations, some other models are not included in this chapter; for instance, stochastic models (Gray *et al.*, 2003) and models involving HIV subtypes and cross-immunity (Porco and Blower, 1998). We still hope that this chapter has provided a useful overview of important aspects for mathematical modeling of HIV/AIDS vaccination.

References

Anderson RM and Garnett GP (1996) Low-efficacy HIV vaccines: potential for community-based intervention programmes, *Lancet* **348**: 1010–1013.

Anderson RM, Gupta S and May RM (1991) Potential of community-wide chemotherapy or immunotherapy to control the spread of HIV-1, *Nature Lond.* **350**: 356–359.

Anderson RM, May RM, Boily M-G, Garnett GP and Rowley JT (1991) The spread of HIV-1 in Africa: sexual contact patterns and the predicted demographic impact of AIDS, *Nature Lond.* **352**: 581–589.

Anderson RM, Swinton J and Garnett GP (1995) Potential impact of low efficacy HIV-1 vaccines in populations with high rates of infection, *Proc. R. Soc. London Ser. B* **261**: 147–151.

Blower SM, Koelle K, Kirschner DE and Mills J (2001) Live attenuated HIV vaccines: predicting the tradeoff between efficacy and safety, *Proc. Natl. Acad. Sci. USA* **98**: 3618–3623.

Blower SM, Schwartz EJ and Mills J (2003) Forecasting the future of HIV epidemics: the impact of antiretroviral therapies and imperfect vaccines, *AIDS Rev.* **5**: 113–125.

Bogard E and Kuntz KM (2002) The impact of a partially effective HIV vaccine on a population of intravenous drug users in Bangkok, Thailand: a dynamic model, *J. Acquir. Immune Defic. Syndr.* **29**: 132–141.

Cohen J (2003a) AIDS vaccine trial produces disappointment and confusion, *Science* **299**: 1290–1291.

Cohen J (2003b) AIDS vaccine still alive as booster after second failure in Thailand, *Science* **302**: 1309–1310.

Dean M, Carrington M, Winkler C, *et al.* (1996) Genetic restriction of HIV-1 infection and progression to AIDS by a deletion allele of the CKR5 structural gene, *Science* **273**: 1856–1862.

Del Valle S, Evangelista AM, Velasco MC, Kribs–Zaleta CM and Hsu Schmitz S-F (2004) Effects of education, vaccination and treatment on HIV transmission in homosexuals with genetic heterogeneity, *Math. Biosci.* **187**: 111–133.

Edwards DM, Shachter RD and Owens DK (1998) A dynamic HIV-transmission model for evaluating the costs and benefits of vaccine programs, *Interfaces* **28**: 144–166.

Esparza J and Osmanov S (2003) HIV vaccines: a global perspective, *Curr. Mol. Med.* **3**: 183–193.

Garnett GP and Anderson RM (1993a) Factors controlling the spread of HIV in heterosexual communities in developing countries: patterns of mixing between different age and sexually active classes, *Phil. Trans. R. Soc. Lond. B* **342**: 137–159.

Garnett GP and Anderson RM (1993b) Contact tracing and the estimation of sexual mixing patterns: the epidemiology of gonococcal infection, *Sex. Transm. Dis.* **20**: 82–91.

Garnett GP and Anderson RM (1994) Balancing sexual partnerships in an age and activity stratified model of HIV transmission in heterosexual populations, *IMA J. Math. App. Math. Biol.* **11**: 161–192.

Gray RH, Li X, Wawer MJ, Gange SJ, Serwadda D, Sewankambo NK, Moore R, Wabwire–Mangen F, Lutalo T, Quinn TC (2003) Rakai Project Group. Stochastic simulation of the impact of antiretroviral therapy and HIV vaccines on HIV transmission; Rakai, Uganda, *AIDS* **17**: 1941–1951.

Hsu Schmitz S-F (2000) Effects of treatment or/and vaccination on HIV transmission in homosexuals with genetic heterogeneity, *Math. Biosci.* **167**: 1–18.

Jirathitikal V and Bourinbaiar AS (2002) Safety and efficacy of an oral HIV vaccine (V-1 Immunitor) in AIDS patients at various stages of the disease, *HIV Clin. Trials* **3**: 21–26.

Jirathitikal V, Metadilogkul O and Bourinbaiar AS (2004) Increased body weight and improved quality of life in AIDS patients following V-1 Immunitor administration, *Eur. J. Clin. Nutrition* **58**: 110–115.

Jirathitikal V, Sooksathan P, Metadilogkul O and Bourinbaiar AS (2003) V-1 Immunitor: oral therapeutic AIDS vaccine with prophylactic potential, *Vaccine* **21**: 624–628.

Kribs–Zaleta CM and Velasco–Hernández JX (2000) A simple vaccination model with multiple endemic states, *Math. Biosci.* **164**: 183–201.

May RM and Anderson RM (1987) Transmission dynamics of HIV infection, *Nature Lond.* **326**: 137–142.

McDonald G (1952) The analysis of equilibrium in malaria, *Trop. Dis. Bull.* **49**: 813–829.

McLean AR and Blower SM (1993) Imperfect vaccines and herd immunity to HIV, *Proc. R. Soc. London Ser. B* **253**: 9–13.

McLean AR and Blower SM (1995) Modelling HIV vaccination, *Trends Microbiol.* **3**: 458–463.

Metadilogkul O, Jirathitikal V and Bourinbaiar AS (2002) Survival of end-stage patients receiving V-1 Immunitor, *HIV Clin. Trials* **3**: 258–259.

Nabel GJ (2001) Challenges and opportunities for development of an AIDS vaccine, *Nature* **410**: 1002–1007.

Owens DK, Edwards DM and Shachter RD (1998) Population effects of preventive and therapeutic HIV vaccines in early- and late-stage epidemics, *AIDS* **12**: 1057–1066.

Plotkin S and Mortimer E (eds), *Vaccine*, 2nd edition. WB Saunders Co 1994.

Porco TC and Blower SM (1998) Designing HIV vaccination policies: subtypes and cross-immunity, *Interfaces* **3**: 167–190.

Quillent C, Oberlin E, Braun J, *et al.* (1998) HIV-1-resistance phenotype conferred by combination of two separate inherited mutations of CCR5 gene, *Lancet* **351**: 14–18.

Samson M, Libert F, Doranz BJ, *et al.* (1996) Resistance to HIV-1 infection in caucasian individuals bearing mutant alleles of the CCR-5 chemokine receptor gene, *Nature* **382**: 722–725.

Smith MW, Dean M, Carrington M, *et al.* (1997) Contrasting genetic influence of CCR2 and CCR5 variants on HIV-1 infection and disease progression, *Science* **277**: 959–965.

Stratov I, DeRose R, Purcell DF and Kent SJ (2004) Vaccines and vaccine strategies against HIV, *Curr. Drug Targets* **5**: 71–88.

UNAIDS. *AIDS Epidemic Update 2003.*

Winkler C, Modi W, Smith MW, *et al.* (1998) Genetic restriction of AIDS pathogenesis by an SDF-1 chemokine gene variant, *Science* **279**: 389–393.

CHAPTER 7

EFFECTS OF AIDS VACCINE ON SUB-POPULATIONS OF CD4$^{(+)}$ T CELLS, CD8$^{(+)}$ T CELLS AND B CELLS UNDER HIV INFECTION

Wai-Yuan Tan, Ping Zhang and Xiaoping Xiong

In this chapter, we have developed stochastic models for subsets of CD4$^{(+)}$ T cells, CD8$^{(+)}$ T cells and B cells under various conditions, including AIDS vaccination and HIV infection. We have used these models to generate some Monte Carlo data to assess both the prophylactic effects and the therapeutic effects of AIDS vaccines. Our Monte Carlo results indicated that in all cases, the immune system under AIDS vaccination will reach a steady state condition in at most a few months. At this steady state condition, the total number of HIV was kept below 100/mL whereas the total number of CD4 T cells is 670/mL in all cases. These results are consistent and comparable with the findings of clinical trials of AIDS vaccines on rhesus monkeys by Amara *et al.* (2001) and Barouch *et al.* (2000). These results seemed to indicate that the AIDS vaccines were very effective as both prophylactic and therapeutic agents if they can stimulate cytotoxic responses and humoral responses although it cannot completely prevent HIV infection. The Monte Carlo results also implied that the prophylactic effects of AIDS vaccines derive mainly from the cytotoxic immune responses through CTL cells (the primed CD8$^{(+)}$ T cells) although the humoral responses through HIV antibodies might also be important albeit with a lesser degree; on the other hand, as therapeutic agents, the humoral vaccines appeared to be more important than the cytotoxic vaccines although the difference may not be very big.

Keywords: Cytotoxic response; humoral response; Monte Carlo studies; prophylactic effects; stochastic models; subsets of CD4 T cells; CD8 T cells and B cells; therapeutic effects.

1. Introduction

When an individual is infected by an outside invader (bacteria, virus or HIV), the immune system of the body mounts a vigorous defense against the infection. In this immune response, the central players are the CD4$^{(+)}$ T cells (to be referred to as T_4 cells), the CD8$^{(+)}$ T cells (to be referred to as T_8 cells) and the B cells. These cells are constantly produced by the precursor stem cells in the bone marrow and thymus and moved to other parts of the body through the blood stream. Under infection by outside invaders, the T_4 cells are activated to proliferate; the activated T_4 cells secret Il-2 to activate T_8 cells and B cells. The activated T_8 cells yield CD8 primed T cells which will eradicate the infected cells (the cytotoxic response) whereas the activated B cells yield plasma B cells to produce neutralizing anti-bodies to neutralize the invaders (the humoral response).

To reveal basic mechanisms of HIV pathogenesis and to assess effects and impacts of intervention such as treatment and AIDS vaccines in HIV-infected individuals, it is necessary and important to develop biologically based stochastic models of HIV pathogenesis involving these cells. In this paper we will proceed to develop biologically based stochastic models for HIV pathogenesis involving subsets of T_4 cells, T_8 cells and B cells under various conditions including AIDS vaccination and HIV infection.

Because it is crucial to be familiar with the basic biology of immune response to develop these models, in Section 2, we will present some basic biological aspects of the immune response. Based on these basic biological mechanisms, in Section 3, we will develop stochastic models for these subsets in both unvaccinated and vaccinated normal individuals. In Section 4, we will extend the results of Section 3 to cases under HIV infection in both unvaccinated and vaccinated individuals. To assess impacts of AIDS vaccines on HIV pathogenesis, in Section 5, we will generate and analyze some computer Monte Carlo samples by using the above stochastic models. Finally in Section 6, we will provide some conclusions and discuss the basic implications and possible applications of the models.

2. Some Biological Background

To develop stochastic models of HIV pathogenesis involving subsets of T_4 cells, T_8 cells and B cells, it is necessary to be familiar with some basic biological mechanism of these cells. In this section we thus give a brief introduction of the following relevant biological mechanisms; for more detail, we refer the readers to the books (DeVita *et al.*, 1997; Fauci and Pantaleo, 1997; Jilek *et al.*, 1992; Levy, 1998; Paul, 1998; Tan, 2000) and the papers (Bajaria *et al.*, 2002; 2004; Bajaria and Kirschner, 2004; Benoist and Mathis, 1999; Brememmann, 1995; Coffin, 1995; Cohen *et al.*, 1998; DeFranco, 1999; Essunger and Perelson, 1994; Fauci, 1996; Hraba and

Dolezal, 1995; Murry *et al.*, 1998; Nowak and McMichael, 1995; Pantaleo *et al.*, 1993; Perelson and Nelson, 1999; Perelson *et al.*, 1993; Phillip, 1996; Seder and Mosmann, 1999; Stilinakis *et al.*, 1997; Tan and Wu, 1998; Tan and Ye, 2000; Weiss, 1999).

i) Virgin T_4 cells (denoted by T_v cells), virgin T_8 cells (denoted by P_v cells) and virgin B cells (denoted by B_v cells) are produced by precursor stem cells in the bone marrow and matured in the thymus. Through the blood stream these matured cells move to other tissues. Since the number of precursor stem cells are very large whereas the probability for each stem cell to produce the T_v cells, P_v cells and B_v cells are very small during small time intervals, one may assume Poisson processes to generate these cells with rates $s_i(t)$ at time t(subscript $i = v$ for T_v cells, $i = p$ for P_v cells and $i = b$ for B_v cells); see Tan and Wu (1998) and Tan and Ye (2000). Further, free HIV can also infect the precursor stem cells; thus one may take the $s_i(t)$ as decreasing functions of the numbers of free HIV. One such function used by Perelson *et al.* (1993) and Kirschner and Perelson (1995) is $s_i(t) = s_i/[\alpha_i + V(t)], i = v, p, b$; where s_i is the rate in uninfected individuals, α_i the saturation constant and $V(t)$ the total number of HIV per unit micro-liter (ml) of the body.

ii) The T_v cells can be activated by pathogens, or AIDS vaccine or HIV infection to become activated CD4$^{(+)}$ T_4 cells (denote by T_a cells) by acquiring interleukin I. The proliferation rate of T_a cells may further be enhanced by acquiring interleukin II receptor to become effector T_4 cells (denoted by T_e cells). The T_a cells and T_e cells may lose its proliferation capability to become memory T_4 cells (denoted by T_m cells) whereas the T_m cells may be reactivated to become T_a cells by pathogens or AIDS vaccine. The T_v and T_m cells are resting T_4 cells and it takes months for these cells to recycle; on the other hand, as shown by Benoist and Mathis (1999) the T_a and T_e cells are activated cells with T_a cells recycling in days and the T_e cells recycling in hours. Murry *et al.* (1998) noted that when the T_v and T_m cells are activated, it would generate 216 T_a cells whereas Essunger and Perelson (1994) assumed that when the T_a cells divide, each will immediately give 2 T_m cells. Since it takes about 7–8 cell divisions for a single cell to generate 216 cells, we postulate that after T_v and T_m cells being activated and becoming T_a cells, then the T_a either divide for increasing T_a cells, or lose its proliferation becoming T_m cells. Similarly, because the turnover rate of T_e cells is very high (in hours), we assume that the T_e cells would divide with higher rate to accumulate T_e cells but many of these cells would shed its interleukins to become T_m cells. A schematic presentation of the dynamic network of CD4 subsets in normal individuals is given in Fig. 1.

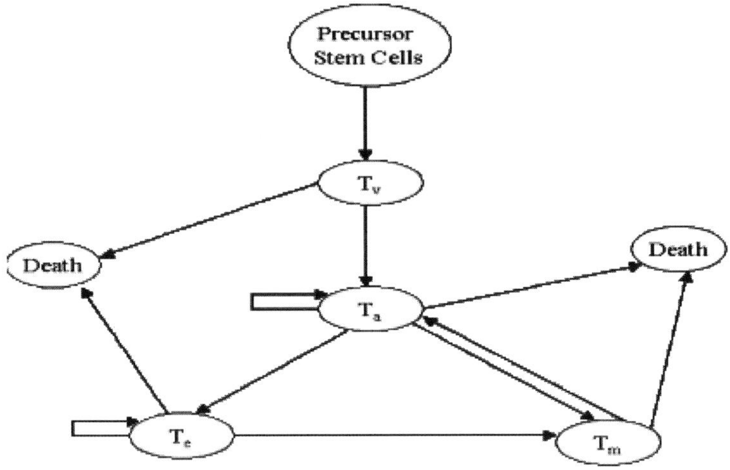

Fig. 1. Plot showing the subset of T_4 cells and the dynamic network.

iii) An unique feature of HIV pathogenesis is that the HIV infect mainly $CD4^{(+)}$ T cells, although it can also infect many other cells such as macrophage albeit at a lesser level. Upon HIV infection, as shown by Bajaria and Kirschner (2004) the dendritic cells present the HIV proteins on its surface and form a complex with T_v cells to activate these cells. The HIV can infect not only all types of T_4 cells but also attack the precursor stem cells. The HIV may also infect T_8 cells and B cells but the level of infection for these cells is very low to give any appreciative direct effects on HIV replication (Bajaria *et al.*, 2000; 2004; Bajaria and Kirschner, 2004).

When the T_v cells and the T_m cells are infected by HIV, they become abortively HIV-infected T_4 cells to be referred to as $T_v^{(*)}$ cells and $T_m^{(*)}$ cells respectively (Bajaria *et al.*, 2002; 2004; Remark 1). In these cells the RNA HIV molecules remain in the cytoplasm and have not yet been integrated and converted to provirus; hence these cells will not generate new HIV unless they are activated to became productively HIV-infected CD4 T cells. These cells are unstable so that within a few weeks most of these cells would either return to uninfected CD4 T cells by shedding the RNA HIV molecules, or be eliminated by apoptosis through a homing process (Bajaria *et al.*, 2002; 2004); with a very small probability a small proportion may also be activated to become productively HIV-infected CD4 T cells. These cells normally do not express HIV antigens on its membrane so that they would not be recognized by primed CD8 cells and /or natural killer cells. Hence the death rates of these cells are much smaller than those of productively HIV-infected CD4 T cells although they are greater than those of un-infected T_4 cells due to apoptosis of these cells (Bajaria *et al.*, 2002; 2004). Notice that when these cells are activated to become productively HIV-infected T_4 cells, some may shed its HIV to become T_a cells.

iv) When the T_a and T_e cells are infected by HIV, they become productively HIV-infected T_4 cells to be referred to as $T_a^{(*)}$ cells and $T_e^{(*)}$ cells respectively. When these cells lose their proliferation capabilities by shedding the interleukins, they become memory productively HIV-infected T_4 cells ($T_m^{(**)}$ cells), which have been referred to as latently HIV-infected CD4 T cells by Bajaria *et al.* (2004); see Remark 1. In these cells, the HIV have been integrated into the genome of the cells to become proviruses. Thus, when these cells are activated and dividing, the proviruses would generate HIV RNA and HIV proteins to form new HIV. These newly generated HIV bud out of the cells to become new free HIV which can infect new T_4 cells; when this happens, normally the cells would die. (We postulate that the $T_m^{(**)}$ cells would generate free HIV only after they are activated to become $T_a^{(*)}$ cells.) Because the $T_a^{(*)}$ cells and $T_e^{(*)}$ cells present the HIV proteins on the surface of its membrane, these cells would normally be recognized and then killed by primed T_8 cells (P_c cells) and/or natural killing cells; however, this latter type of death do not generate infectious free HIV in the blood. Because of this, we postulate two types of deaths for productively HIV-infected CD4 T cells; one type of death would generate free HIV to infect new T_4 cells, and the other does not generate HIV and occurs through elimination by primed CD8 T cells or natural killer cells.

v) The activated T_4 cells (T_a cells and T_e cells) secret cytokines to further activate P_v cells to become activated T_8 cells (denoted by P_a cells) and to activate B_v cells to become activated B cells (denoted by B_a cells). Hence, either acting directly or through activation of T_4 cells, the AIDS vaccine or HIV infection can enhance activation of P_v cells and B_v cells to produce P_a cells and B_a cells respectively.

vi) The P_v cells are produced by precursor stem cells in the bone marrow and thymus with rate $s_p(t)$ (about 2/3 of $s_v(t)$; see Hraba and Dolezal, 1995) but can die with rates $\nu_{P_v}(t)$ and activated with rate $k_{va}(t)$ by vaccine and/or T_a cells to become P_a cells. The P_a cells have a certain probability ($p_C(t)$) to transform with rate $\omega_m(t)$ into P_m cells or to transform with rate $\omega_p(t)$ into P_p cells; P_p cells are the CD8 T cells that kill HIV-infected cells and have been referred to as cytotoxic T lymphocytes (CTL) cells. The P_m cells may be reactivated with rate $r_{ma}(t)$ to become P_a cells. The

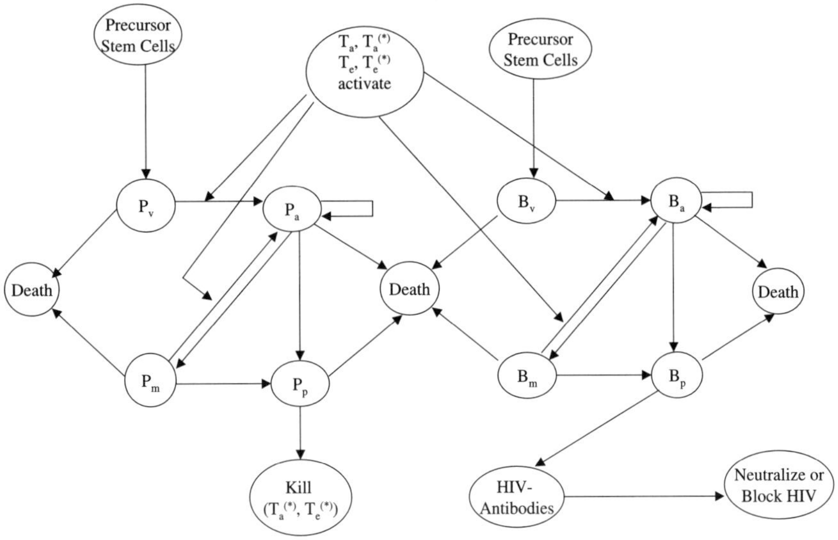

Fig. 2. Plot showing the subsets of CD8$^{(+)}$ T cells and B cells and the dynamic network.

P_a cells may lose its proliferation to either become P_m cells or differentiate to become P_p cells. The P_m cells are resting cells but can be reactivated to become P_a cells; on the other hand, the P_p cells are end cells which function as a killer cells to kill productively HIV-infected T_4 cells (i.e. the $T_a^{(*)}$ cells and $T_e^{(*)}$ cells). This is the cellular cytotoxic response of CD8$^{(+)}$ T_8 cells. A schematic presentation of this dynamic network is given in Fig. 2.

vii) As illustrated in Prikrylova, Jilek and Waniewski (1992), the B_v cells are produced by precursor stem cells in the bone marrow and thymus with rate $s_b(t)$; these cells can die with rate $\nu_{B_v}(t)$; or be activated with rate $q_{va}(t)$ by vaccine and/or T_a cells to become B_a cells. B_a cells have a certain probability ($p_B(t)$) to stop proliferation either to transform with rate $\gamma_m(t)$ into B_m cells or to differentiate with rate $\gamma_p(t)$ to become B_p cells which produce antibodies. The B_m cells may be reactivated with rate $q_{ma}(t)$ to become B_a cells or be differentiated with rate $q_{mp}(t)$ to become antibody-producing plasma B cells. (This is the humoral response.) A schematic presentation of this dynamic network is given in Fig. 2. From the above biological mechanism, the T_a cells can be generated from the $T_v, T_m, T_v^{(*)}$ and $T_m^{(*)}$ cells and by cell division; these cells are decreased by death, by infection by HIV, by converting to T_e cells and T_m cells. The $T_a^{(*)}$ cells are increased through HIV infection of T_a cells, through activation of $T_v^{(*)}, T_m^{(*)}$ cells and $T_m^{(**)}$ cells; they are decreased by death and cell division

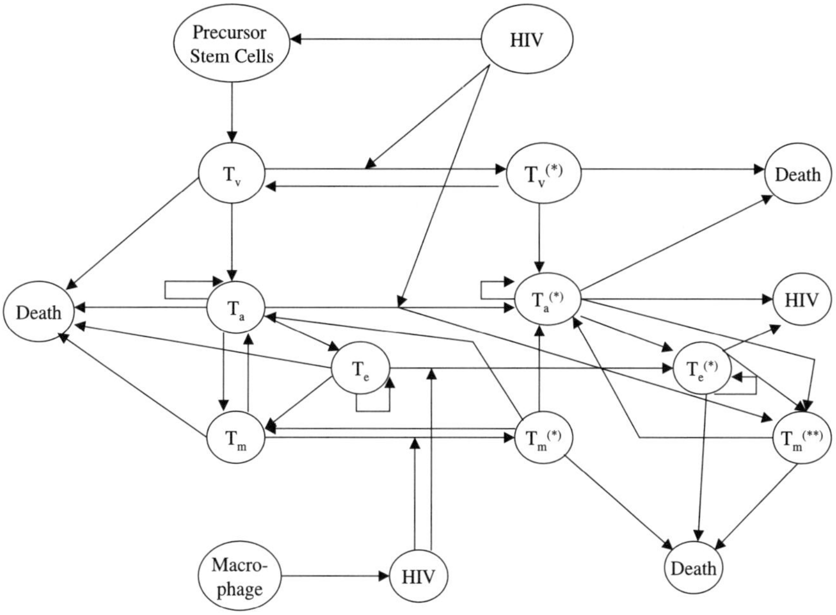

Fig. 3. Plot showing the subset of HIV-infected T4 cells and the dynamic network.

(see Remark 2), by converting to $T_m^{(**)}$ memory cells and $T_e^{(*)}$ cells. The T_v cells are generated by precursor stem cells or converted from $T_v^{(*)}$ cells by shedding of HIV; T_v cells are decreased by death, by activation to T_a cells and by infection by HIV to become $T_v^{(*)}$ cells. The network of these processes under HIV infection are described in Fig. 3.

Remark 1: As described in O'Hara (2004), whereas in most of the so-called latently HIV-infected CD4 T cells the HIV molecules have not yet been integrated into the genome; in a few of these cells, however, the HIV do have been integrated into the cell genome to become pro-viruses. The former cells are equivalent to the abortive cells defined by Bajaria *et al.* (2002) and are the $\{T_v^{(*)}, T_m^{(*)}\}$ cells in this paper; on the other hand, the latter cells are the latently HIV-infected cells in Bajaria *et al.*'s definition and are the $T_m^{(**)}$ cells in this paper.

Remark 2: When a productively HIV-infected cell is activated and then divide, the proviruses in the cell produce virus RNA and virus proteins and virus enzymes; by the action of protease these virus RNA and proteins and enzymes are coupled together to form new infectious HIV viruses which bud out of the cells to become free HIV. When this happens, the cell would normally die. Thus, in HIV pathogenesis, cell division of productively HIV-infected CD4 T cells leads to death. The productively HIV-infected CD4 T cells can also be killed by primed CD8 T cells or natural killer cells; but this type of death will not generate free HIV.

3. Stochastic Models of Immune Response Under Vaccination

AIDS vaccines can be considered as antigens with no infectivity (Baltimore and Heilman, 1998) and hence no harm. Hence, AIDS vaccines can promote activation of T_v and T_m cells into T_a cells which secrete cytokines to activate CD8 T cells and B T cells. The activated CD8 T cells will generate CD8 primed T cells, which would kill infected CD4 T cells (cytotoxic effects), whereas the activated B cells generate plasma B cells, which produce antibodies to neutralize HIV (humoral effects). For uninfected normal individuals, AIDS vaccines can then protect or reduce the chance of infection by HIV, to be referred to as the prophylactic effect. On the other hand, for HIV-infected individuals, AIDS vaccines can enhance the immune response against HIV, referred to as the therapeutic effect of AIDS vaccines. To assess these effects of AIDS vaccines, in this section, we will develop stochastic models involving subsets of CD4 T cells, B cells and CD8 T cells under vaccination for normal uninfected individuals. In Section 4, we will extend these models to subsets of CD4 T cells for HIV-infected individuals.

3.1. *Stochastic models of subset of CD4 T cells in uninfected individuals*

Let $\underset{\sim}{X}_1(t) = \{T_v(t), T_a(t), T_e(t), T_m(t)\}'$ denote the numbers of $\{T_v, T_a, T_e, T_m\}'$ cells at time t respectively. Then $\underset{\sim}{X}_1(t)$ is a Markov chain with continuous time $t > 0$. Thus, $\underset{\sim}{X}_1(t+\Delta t)$ derives from $\underset{\sim}{X}_1(t)$ through stochastic transitions independent of past history. From the basic biological mechanisms in Section 2 and the mechanistic-dynamic network as described in Fig. 1, these stochastic transitions are characterized by the following random variables:

- $S_v(t)$ = Number of T_v cells generated by the precursor stem cells in the bone marrow and thymus with rate $s_v(t)$ during $[t, t+\Delta t)$;
- $\{F_{va}(t), F_{am}(t), F_{ae}(t), F_{ma}(t), F_{em}(t)\}$ = Numbers of $\{T_v \to T_a, T_a \to T_m, T_a \to T_e, T_m \to T_a, T_e \to T_m\}$ with rates $\{k_{va}(t), k_{am}(t), k_{ae}(t), k_{ma}(t), k_{em}(t)\}$ during $[t, t+\Delta t)$ respectively;

- $D_i(t)$ = Number of death of T_i $(i = v, a, e, m)$ cells with rates $\{\mu_i(t), i = v, a, e, m\}$ during $[t, t + \Delta t)$ respectively;
- $B_{Ti}(t)$ $(i = a, e)$ = Number of Cell Division of T_i $(i = a, e)$ cells with rates $\{b_i(t), i = a, e\}$ during $[t, t + \Delta t)$ respectively.

To the order of $o(\Delta t)$, $S_v(t)$ is a Poisson random variable with mean $s_v(t)\Delta t$ and the conditional probability distributions of these random variables given the state variables are basically multinomial distributions. That is,

$$\begin{aligned}
S_v(t) &\sim \text{Poisson}\{s_v(t)\Delta t\};\\
\{F_{va}(t), D_v(t)\}|T_v(t) &\sim \text{Multinomial}\{T_v(t); k_{va}(t)\Delta t,\\
&\quad \mu_v(t)\Delta t\};\\
\{F_{am}(t), F_{ae}(t), B_{Ta}(t), D_a(t)\}|T_a(t) &\sim \text{Multinomial}\{T_a(t); k_{am}(t)\Delta t,\\
&\quad k_{ae}(t)\Delta t, b_a(t)\Delta t, \mu_a(t)\Delta t\};\\
\{F_{em}(t), B_{Te}(t), D_e(t)\}|T_e(t) &\sim \text{Multinomial}\{T_e(t); k_{em}(t)\Delta t,\\
&\quad b_e(t)\Delta t, \mu_e(t)\Delta t\};\\
\{F_{ma}(t), D_m(t)\}|T_m(t) &\sim \text{Multinomial}\{T_m(t); k_{ma}(t)\Delta t,\\
&\quad \mu_m(t)\Delta t\}.
\end{aligned}$$

By the conservation law and by subtracting from the random variables its conditional mean values respectively, we obtain the following stochastic equations for the state variables:

$$\begin{aligned}
dT_v(t) &= T_v(t + \triangle t) - T_v(t) = S_v(t) - F_{va}(t) - D_v(t)\\
&= \{s_v(t) - T_v(t)[k_{va}(t) + \mu_v(t)]\}\triangle t + e_v(t)\triangle t, \qquad (1)\\
dT_a(t) &= T_a(t + \triangle t) - T_a(t)\\
&= F_{va}(t) + F_{ma}(t) + B_{Ta}(t) - F_{am}(t) - F_{ae}(t) - D_a(t)\\
&= \{T_v(t)k_{va}(t) + T_m(t)k_{ma}(t) + T_a(t)b_a(t) - T_a(t)[\mu_a(t)\\
&\quad + k_{am}(t) + k_{ae}(t)]\}\triangle t + e_a(t)\triangle t, \qquad (2)\\
dT_e(t) &= T_e(t + \triangle t) - T_e(t) = F_{ae}(t) + B_{Te}(t) - F_{em}(t) - D_e(t)\\
&= \{[T_a(t)k_{ae}(t) + T_e(t)b_e(t)] - T_e(t)[\mu_e(t)\\
&\quad + k_{em}(t)]\}\triangle t + e_e(t)\triangle t, \qquad (3)\\
dT_m(t) &= T_m(t + \triangle t) - T_m(t) = F_{am}(t) + F_{em}(t) - F_{ma}(t)\\
&\quad - D_m(t) = \{[T_a(t)k_{am}(t) + T_e(t)k_{em}(t)]\\
&\quad - T_m(t)[\mu_m(t) + k_{ma}(t)]\}\triangle t + e_m(t)\triangle t. \qquad (4)
\end{aligned}$$

Table 1. Parameter values for CD4 T cells in uninfected individuals.

Variables	Parameters	Parameter values
$s_v(t) = s_v$	Rate of generation of T cells	8.076/day mm^3
T_{max}	Maximum attainable T cells	1500/mm^3
$b_a(t) = b_a$	Rate of proilferation of activated T cells	$0.3 * \left(1 - \frac{TOT(t)}{T_{max}}\right)$
$b_e(t) = b_e$	Rate of proilferation of effector T cells of effector T cells	$3. * \left(1 - \frac{TOT(t)}{T_{max}}\right)$
$\mu_v(t) = \mu_v$	Death rate of virgin T cells	0.025/day
$\mu_a(t) = \mu_a$	Death rate of activated T cells	0.85/day
$\mu_e(t) = \mu_e$	Death rate of effector T cells	0.95/day
$\mu_m(t) = \mu_m$	Death rate of memory T cells	0.03/day
$k_{am}(t) = k_{am}$	Rate of $T_a \rightarrow T_m$	0.026/day
$k_{ae}(t) = k_{ae}$	Rate of $T_a \rightarrow T_e$	2.5/day
$k_{em}(t) = k_{em}$	Rate of $T_e \rightarrow T_m$	0.5/day
Without vaccine:		
$k_{va}(t) = k_{va}$	Rate of $T_v \rightarrow T_a$	0.0025/day
$k_{ma}(t) = k_{ma}$	Rate of $T_m \rightarrow T_a$	0.0065/day
Under vaccine:		
$k_{va}(t) = k_{va}$	Rate of $T_v \rightarrow T_a$	0.015/day
$k_{ma}(t) = k_{ma}$	Rate of $T_m \rightarrow T_a$	0.95/day

For assessing effects of vaccine, one may assume that $k_{va}(t) = \kappa_0 + \kappa_{va} Vac(t)/(\theta_V + Vac(t))$, and $k_{ma}(t) = \kappa_0 + \kappa_{ma} Vac(t)/(\theta_V + Vac(t)$ with $Vac(t)$ being the dose of vaccine at time t and with θ_V being the saturation constant. As in Perelson *et al.* (1993), one may also assume logistic growth for T_a and T_e cells. That is, for $(j = a, e)$, we assume $b_j(t) = q_j(t)\{1 - T_{tot}(t)/T_{max}\}$ with $T_{tot}(t) = T_v(t) + T_a(t) + T_m(t) + T_e(t)$ and T_{max} the maximum population size of all T cells. The random noises in the above equations are derived by subtracting from the random variables their conditional expected values respectively. These random noises have expected values zero and are un-correlated with the state variables $\{T_v(t), T_a(t), T_e(t), T_m(t)\}$. It can also be shown that the random noises from different time intervals are independent of one another. For generating Monte Carlo data we give some parameter values in Table 1 for the relevant parameters.

3.2. *Stochastic models of subset of CD8 T cells*

Let $\underset{\sim}{Z}(t) = \{P_v(t), P_a(t), P_p(t), P_m(t)\}$ denote the numbers of P_v cells, P_a cells, P_p cells and P_m cells at time t, respectively. Then, as $\underset{\sim}{X}_1(t)$, $\underset{\sim}{Z}(t)$ is also a 4-dimensional Markov chain with continuous time $t > 0$ so that

$\underset{\sim}{Z}(t+\Delta t)$ derives from $\underset{\sim}{Z}(t)$ by stochastic transitions independent of past history. These stochastic transitions are specified schematically by the dynamic network in Fig. 2. To model these stochastic transitions, let $S_p(t)$ denote the number of P_v cells generated by the precursor stem cells in the bone marrow and thymus during $[t, t+\Delta t)$; $\left\{F_{va}^{(p)}(t), F_{am}^{(p)}(t), F_{ap}^{(p)}(t), F_{ma}^{(p)}(t)\right\}$ the numbers of transitions of $\{P_v \to P_a, P_a \to P_m, P_a \to P_p, P_m \to P_a, P_m \to P_p\}$ during $[t, t+\Delta t)$ respectively. Denote by $\{D_{P_i}(t), i = v, a, p, m\}$ the numbers of death of P_i $(i = v, a, p, m)$ cells during $[t, t+\Delta t)$ respectively and let $B_{Pa}^{(p)}$ be the number of cell division of P_a cells during $[t, t+\Delta t)$. Then, as in the case of CD4 T cells, $S_p(t)$ is a Poisson random variable with mean $s_p(t)\Delta t$ and the conditional probability distributions of other random variables given the state variables are basically multinomial distributions. This is summarized in Table 2. Using these probability distributions, $\underset{\sim}{Z}(t)$ is modeled by the following stochastic differential equations:

$$\begin{aligned} dP_v(t) &= P_v(t+\triangle t) - P_v(t) = S_p(t) - F_{va}^{(p)}(t) - D_{P_v}(t) \\ &= \{s_p(t) - P_v(t)[\alpha_{va}(t) + \mu_{P_v}(t)]\}\triangle t + e_{P_v}(t)\triangle t, \end{aligned} \tag{5}$$

$$\begin{aligned} dP_a(t) &= P_a(t+\triangle t) - P_a(t) = F_{va}^{(p)}(t) + F_{ma}^{(p)}(t) + B_{Pa}^{(p)}(t) \\ &\quad - F_{am}^{(p)}(t) - F_{ap}^{(p)}(t) - D_{P_a}(t) = \{P_v(t)\alpha_{va}(t) \\ &\quad + P_a(t)(1 - p_C(t))q_a(t) + P_m(t)\alpha_{ma}(t) - P_a(t)\mu_{P_a}(t) \\ &\quad - P_a(t)p_C(t)[\omega_m(t) + \omega_p(t)]\}\triangle t + e_{P_a}(t)\triangle t, \end{aligned} \tag{6}$$

$$\begin{aligned} dP_m(t) &= P_m(t+\triangle t) - P_m(t) = F_{am}^{(p)}(t) - F_{ma}^{(p)}(t) - F_{mp}^{(p)}(t) \\ &\quad - D_{P_m}(t) = \{P_a(t)p_C(t)\omega_m(t) - P_m(t)[\alpha_{ma}(t) \\ &\quad + \alpha_{mp}(t) + \mu_{P_m}(t)]\}\triangle t + e_{P_m}(t)\triangle t, \end{aligned} \tag{7}$$

$$\begin{aligned} dP_p(t) &= P_p(t+\triangle t) - P_p(t) = F_{ap}^{(p)}(t) + F_{mp}^{(p)}(t) - D_{P_p}(t) \\ &= \{P_a(t)p_C(t)\omega_p(t) + P_m(t)\alpha_{mp}(t) \\ &\quad - P_p(t)\mu_{P_p}(t)\}\triangle t + e_{P_p}(t)\triangle t. \end{aligned} \tag{8}$$

Note that in general the activation rates $\{\alpha_{va}(t), \alpha_{ma}(t)\}$ are functions of $\{T_a(t), Vac(t)\}$. Also, the random noises in the above equations have expected values zero and are un-correlated with the state variables $\{P_v(t), P_a(t), P_p(t), P_m(t)\}$. For generating Monte Carlo studies, some parameter values are given in Table 2.

3.3. *Stochastic models of subset of B cells*

As illustrated in Prikrylova, Jilek and Waniewski (1992) and in the dynamic network in Fig. 2, virgin B cells (B_v cells; also referred to as precursor

Table 2. Parameter values for CD8 T cells in uninfected individuals.

(a) Variables

Variables	Parameters	Parameter values
$s_p(t) = s_b$	Rate of generation of T_8 cells from precursor stem cells	8.076/day mm^3
T_{max}	Maximum attainable T_8 cells	1000/mm^3
$b_p(t) = b_p$	Rate of proliferation of activated T_8 cells	$0.26 * \left(1 - \frac{TOT(t)}{T_{max}}\right)$
$\mu_{P_v}(t) = \mu_{P_v}$	Death rate of virgin T_8 cells	0.02/day
$\mu_{P_a} = \mu_{P_a}$	Death rate of activated T_8 cells	0.085/day
$\mu_{P_m} = \mu_{P_m}$	Death rate of memory T_8 cells	0.015/day
$\mu_{P_p}(t) = \mu_{P_p}$	Death rate of primed T_8 cells	0.013/day
$\alpha_{am}(t) = \alpha_{am}$	Rate of $P_a \to P_m$	0.015/day
$\alpha_{ap}(t) = \alpha_{ap}$	Rate of $P_a \to P_p$	0.042/day
$\alpha_{mp}(t) = \alpha_{mp}$	Rate of $P_m \to P_p$	0.0015/day
Without vaccine		
$\alpha_{va}(t) = \alpha_{va}$	Rate of $P_v \to P_a$	0.0045/day
$\alpha_{ma}(t) = \alpha_{ma}$	Rate of $P_m \to P_a$	0.0065/day
Under vaccine		
$\alpha_{va}(t) = \alpha_{va}$	Rate of $P_v \to P_a$	0.055/day
$\alpha_{ma}(t) = \alpha_{ma}$	Rate of $P_m \to P_a$	0.068/day

(b) Distribution results

$$S_p(t) \sim \text{Poisson}\{s_p(t)\Delta t\};$$
$$\left\{F_{va}^{(p)}(t), D_{P_v}(t)\right\}|P_v(t) \sim \text{Multinomial}\{P_v(t); \alpha_{va}(t)\Delta t, \mu_{P_v}(t)\Delta t\};$$
$$\left\{F_{am}^{(p)}(t), F_{ap}^{(p)}(t), D_{P_a}(t)\right\}|P_a(t) \sim \text{Multinomial}\{P_a(t); p_C(t)\omega_m(t)\Delta t = \alpha_{am}(t)\Delta t, p_C(t)\omega_p(t)\Delta t = \alpha_{ap}(t)\Delta t, \mu_{P_a}(t)\Delta t\};$$
$$\left\{F_{ma}^{(p)}(t), F_{mp}^{(p)}(t), D_{P_m}(t)\right\}|P_m(t) \sim \text{Multinomial}\{P_m(t); \alpha_{ma}(t)\Delta t, \alpha_{mp}(t)\Delta t, \mu_{P_m}(t)\Delta t\};$$
$$\{D_{P_p}(t)\}|P_p(t) \sim \text{Binomial}\{P_p(t); \mu_{P_p}(t)\Delta t\};$$

B cells) are produced by precursor stem cells in the bone marrow and thymus with rate $s(t)$; these cells can die with rate $\mu_{B_v}(t)$; or be activated with rate $\beta_{va}(t)$ by vaccine and/or T_a cells to become activated B cells (B_a cells; also referred to as proliferating B cells). B_a cells have certain probability ($p_B(t)$) to stop proliferation either to transform with rate $\gamma_m(t)$ into memory B cells (B_m cells) or to differentiate with rate $\gamma_p(t)$ to become plasma B cells (B_p cells) which produce antibodies. The B_m cells may be reactivated with rate $\beta_{ma}(t)$ to become B_a cells, or be differentiated with rate $\beta_{mp}(t)$ to become antibody-producing plasma B cells.

Let $\underset{\sim}{W}(t) = \{B_v(t), B_a(t), B_m(t), B_p(t)\}'$ be the numbers of B_v cells, B_a cells, B_m cells and B_p cells at time t, respectively. Then $\underset{\sim}{W}(t)$ is a Markov chain with continuous time $t > 0$. Hence $\underset{\sim}{W}(t + \Delta t)$ derives from $\underset{\sim}{W}(t)$ through stochastic transitions and is independent of past history. To model these stochastic transitions, let $S_b(t)$ denote the number of B_v cells generated by the precursor stem cells in the bone marrow and thymus during $[t, t+\Delta t)$; $\left\{F_{va}^{(b)}(t), F_{am}^{(b)}(t), F_{ap}^{(b)}(t), F_{ma}^{(b)}(t), F_{mp}(t)\right\}$ the numbers of transitions of $\{B_v \to B_a, B_a \to B_m, B_a \to B_p, B_m \to B_a, B_m \to B_p\}$ during $[t, t + \Delta t)$ respectively.

Denoted by $D_{B_i}(t)$ the number of death of B_i $(i = v, a, p, m)$ cells during $[t, t+\Delta t)$ respectively and let $B_{Ba}(t)$ be the number of cell division of B_a cells during $[t, t + \Delta t)$. Then, as in the case of CD4 T cells, $S_b(t)$ is a Poisson random variable with mean $s_b(t)\Delta t$ and the conditional probability distributions of other random variables given the state variables are basically multinomial distributions. This is summarized in Table 3. Using these probability distributions, $\underset{\sim}{W}(t)$ is modeled by the following stochastic differential equations:

$$\begin{aligned} dB_v(t) &= B_v(t + \triangle t) - B_v(t) = S_b(t) - F_{va}^{(b)}(t) - D_{B_v}(t) \\ &= \{s_b(t) - B_v(t)[\beta_{va}(t) + \mu_{B_v}(t)]\}\triangle t + e_{B_v}(t)\triangle t, \quad (9) \\ dB_a(t) &= B_a(t + \triangle t) - B_a(t) = F_{va}^{(b)}(t) + F_{ma}^{(b)}(t) + B_{Ba}(t) \\ &\quad - F_{am}^{(b)}(t) - F_{ap}^{(b)}(t) - D_{B_a}(t) = \{B_v(t)\beta_{va}(t) \\ &\quad + B_m(t)\beta_{ma}(t) + B_a(t)[1 - p_B(t)]b_B(t) - B_a(t)\mu_{B_a}(t) \\ &\quad - B_a(t)p_B(t)[\gamma_m(t) + \gamma_p(t)]\}\triangle t + e_{B_a}(t)\triangle t, \quad (10) \\ dB_m(t) &= B_m(t + \triangle t) - B_m(t) = F_{am}^{(b)}(t) - F_{ma}^{(b)}(t) - F_{mp}^{(b)}(t) \\ &\quad - D_{B_m}(t) = \{B_a(t)p_B(t)\gamma_m(t) - B_m(t)[\beta_{ma}(t) \\ &\quad + \mu_{B_m}(t) + \beta_{mp}(t)]\}\triangle t + e_{B_m}(t)\triangle t, \quad (11) \\ dB_p(t) &= B_p(t + \triangle t) - B_p(t) = F_{ap}^{(b)}(t) + F_{mp}^{(b)}(t) - D_{B_p}(t) \\ &= \{B_a(t)p_B(t)\gamma_p(t) + B_m(t)\beta_{mp}(t) \\ &\quad - B_p(t)\mu_{B_p}(t)\}\triangle t + e_{B_p}(t)\triangle t, \quad (12) \end{aligned}$$

Note that in general, the activation rates $\{\beta_{va}(t), \beta_{ma}(t)\}$ are functions of $\{T_a(t), Vac(t)\}$. Also, the random noises in the above equations have expected values zero and are un-correlated with the state variables $\{B_v(t), B_a(t), B_p(t), B_m(t)\}$. For generating Monte Carlo studies, some parameter values are given in Table 3.

Table 3. Parameter values for B cells in uninfected individuals

(a) Variables

Variables	Parameters	Parameter values
$s_b(t) = s_b$	Rate of generation of B cells from precursor stem cells	8.076/day mm^3
T_{max}	Maximum attainable B cells	1000/mm^3
$b_b(t) = b_b$	Rate of proliferation of activated B cells	$0.15 * \left(1 - \frac{TOT(t)}{T_{max}}\right)$
$\mu_{B_v}(t) = \mu_{B_v}$	Death rate of virgin B cells	0.025/day
$\mu_{B_a} = \mu_{B_a}$	Death rate of activated B cells	0.15/day
$\mu_{B_m} = \mu_{P_m}$	Death rate of memory B cells	0.063/day
$\mu_{B_p}(t) = \mu_{B_p}$	Death rate of plasma B cells	0.035/day
$\beta_{am}(t) = \beta_{am}$	Rate of $B_a \to B_m$	0.065/day
$\beta_{ap}(t) = \beta_{ap}$	Rate of $B_a \to B_p$	0.055/day
$\beta_{mp}(t) = \beta_{mp}$	Rate of $B_m \to B_p$	0.053/day
Without vaccine:		
$\beta_{va}(t) = \beta_{va}$	Rate of $B_v \to B_a$	0.008/day
$\beta_{ma}(t) = \beta_{ma}$	Rate of $B_m \to B_a$	0.0095/day
Under vaccine:		
$\beta_{va}(t) = \beta_{va}$	Rate of $B_v \to B_a$	0.055/day
$\beta_{ma}(t) = \beta_{ma}$	Rate of $B_m \to B_a$	0.068/day

(b) Distribution results

$$S_b(t) \sim \text{Poisson } \{s_b(t)\Delta t\};$$
$$\left\{F_{va}^{(b)}(t), D_{B_v}(t)\right\}|B_v(t) \sim \text{Multinomial } \{B_v(t); \beta_{va}(t)\Delta t, \mu_{B_v}(t)\Delta t\};$$
$$\left\{F_{am}^{(b)}(t), F_{ap}^{(b)}(t), D_{B_a}(t)\right\}|B_a(t) \sim \text{Multinomial } \{B_a(t); p_B(t)\gamma_m(t)\Delta t = \beta_{am}(t)\Delta t, p_B(t)\gamma_p(t)\Delta t = \beta_{ap}(t)\Delta t, \mu_{B_a}(t)\Delta t\};$$
$$\left\{F_{ma}^{(b)}(t), F_{mp}^{(b)}(t), D_{B_m}(t)\right\}|B_m(t) \sim \text{Multinomial } \{B_m(t); \beta_{ma}(t)\Delta t, \beta_{mp}(t)\Delta t, \mu_{B_m}(t)\Delta t\};$$
$$\{D_{B_p}(t)\}|B_p(t) \sim \text{Binomial } \{B_p(t); \mu_{B_p}(t)\Delta t\};$$

4. A Stochastic Model of Immune Response under Vaccination for HIV-Infected Individuals

Because HIV can infect all types of CD4 T cells, under HIV infection we would encounter 5 additional variables to account for HIV-infected CD4 T cells, $\{T_v^{(*)}, T_m^{(*)}, T_a^{(*)}, T_e^{(*)}, T_m^{(**)}\}$. In these cells, $\{T_a^{(*)}, T_e^{(*)}\}$ are productively HIV-infected CD4 T cells, $\{T_v^{(*)}, T_m^{(*)}\}$ the abortively HIV-infected CD4 T cells (Bajaria *et al.*, 2002; 2004) and $T_m^{(**)}$ the latently

HIV-infected CD4 T cells (Bajaria *et al.*, 2002; 2004). (In the literature, the abortively HIV-infected CD4 T cells have also been referred to as latently HIV-infected CD4 T cells, see O'Hara, 2004.) Let $\underset{\sim}{X}_2(t) = \{T_v^{(*)}(t), T_m^{(*)}(t), T_a^{(*)}(t), T_e^{(*)}(t), T_m^{(**)}(t)\}'$ denote the numbers of these cells at time t respectively. With $\underset{\sim}{X}_1(t) = \{T_v(t), T_a(t), T_e(t), T_m(t)\}'$, we are then entertaining a 9-dimensional Markov process $\underset{\sim}{X}(t) = \{\underset{\sim}{X}_1(t)', \underset{\sim}{X}_2(t)'\}'$ with continuous time $t > 0$ for the subsets of CD4 T cells under HIV infection.

To model this process, besides the transition random variables given in Section 3, we define some additional random variables to account for HIV infection of CD4 T cells and for the transition of $\{T_i^{(*)} \to (T_i, T_a, T_a^{(*)}), i = v, m, T_a^{(*)} \to (T_m^{(**)}, T_e^{(*)}), T_e^{(*)} \to T_m^{(**)}, T_m^{(*)} \to T_m^{(**)}, T_m^{(**)} \to T_a^{(*)}\}$. Thus, we let $G_i(t)$ $(i = v, a, m, e)$ be the number of HIV infection of T_i $(i = v, a, m, e)$ by HIV during $[t, t + \Delta t)$ respectively; $\{R_{vv}(t), R_{va}(t)\}$ the number of $\{T_v^{(*)} \to T_v, T_v^{(*)} \to (T_a, T_a^{(*)})\}$ during $[t, t + \Delta t)$ respectively and $\{R_{mm}(t), R_{mm}^{(*)}(t), R_{ma}(t)\}$ the number of $\{T_m^{(*)} \to T_m, T_m^{(*)} \to T_m^{(**)}, T_m^{(*)} \to (T_a, T_a^{(*)})\}$ during $[t, t + \Delta t)$ respectively and $\{R_{ae}(t), R_{am}(t), R_{em}(t)\}$ the numbers of $\{T_a^{(*)} \to T_e^{(*)}, T_a^{(*)} \to T_m^{(**)}, T_e^{(*)} \to T_m^{(**)}\}$ during $[t, t + \Delta t)$ respectively. Among the $\{R_{ia}(t), i = v, m\}$ transitions, let $\{P_{ia}(t), i = v, m\}$ be the number of T_a^* cells respectively. Let $\{D_i^{(*)}(t), i = v, m, D_m^{(**)}(t)\}$ be the number of deaths of $\{T_i^{(*)}, i = v, m, T_m^{(**)}\}$ cells during $[t, t + \Delta t)$. For the $\{T_i^{(*)}, i = a, e\}$ cells, as described in Section 2, we define two types of deaths and let $\{D_{ia}^{(*)}(t), D_{ic}^{(*)}(t), i = a, e\}$ denote the number of deaths through activation and through elimination by primed CD8 T cells during $[t, t + \Delta t)$ respectively. Notice that only $\left(D_{ia}^{(*)}(t), i = a, e\right)$ generate free HIV.

To model these random variables, let $\{k_{ii}^{(*)}(t), k_{ia}^{(*)}(t), i = v, m\}$ be the rates of $\{T_i^{(*)} \to T_i, T_i^{(*)} \to (T_a, T_a^{(*)}), i = v, m\}$ respectively; $\{k_{mm}^{(*)}(t), k_{am}^{(*)}(t), k_{ae}^{(*)}(t), k_{em}^{(*)}(t), k_{ma}^{(**)}(t)\}$ the rates of $\{T_m^{(*)} \to T_m^{(**)}, T_a^{(*)} \to T_m^{(**)}, T_a^{(*)} \to T_e^{(*)}, T_e^{(*)} \to T_m^{(**)}, T_m^{(**)} \to T_a^{(*)}\}$ respectively. Among the $\{R_{ia}(t), i = v, m\}$ transitions, let $\{\delta_{ia}, i = v, m\}$ be the proportion of T_a cells respectively. Let $\{\lambda_i(t), i = v, m, a, e\}$ be the infection rates of $\{T_i, i = v, m, a, e\}$ cells by HIV respectively and $\{\mu_i^{(*)}(t), i = v, m, \mu_m^{(**)}(t), \mu_V(t)\}$ the death rates of $\{T_i^{(*)}, i = v, m, T_m^{(**)}\}$ cells and HIV virus at time t, respectively. For the $\{T_i^{(*)}, i = a, e\}$ cells, let $\{\mu_{ia}(t), \mu_{ic}(t), i = a, e\}$ denote the death rates for $\{D_{ia}^{(*)}(t),$

$D_{ic}^{(*)}(t), i = a, e\}$ respectively. For the generation of free HIV, let $g(t)$ be the number of viruses from macrophage and other long-lived cells, $N(t)$ the number of free HIV released by the activation and death of $T_a^{(*)}$ or $T_e^{(*)}$ cells; $\{\xi_1, \xi_2\}$ the proportions of non-infectious HIV among the free HIV generated by the $T_a^{(*)}$ and $T_e^{(*)}$ cells through activation and division and hence death respectively. Then as given in Table 4, the conditional distributions of these variables given the state variables are basically binomial and multinomial distributions. Using these distribution results and the

Table 4. Parameter values for HIV-infected CD4 T cells.

(a) Variables

Variables	Parameters	Parameter values
$s_v(t) = s_v$	Rate of generation of T cells from precursor stem cells	$\frac{8.076*14}{1+V(t)^{0.35}}$
T_{max}	Maximum attainable T cells	$1500/\text{mm}^3$
$b_a(t) = b_a$	Rate of proliferation of activated T cells	$0.3 * \left(1 - \frac{TOT(t)}{T_{max}}\right)$
$b_e(t) = b_e$	Rate of proliferation of effector T cells	$3 * \left(1 - \frac{TOT(t)}{T_{max}}\right)$
$\mu_v(t) = \mu_v$	Death rate of virgin T cells	0.025/day
$\mu_a(t) = \mu_a$	Death rate of activated T cells	0.85/day
$\mu_e(t) = \mu_e$	Death rate of effector T cells	0.95/day
$\mu_m(t) = \mu_m$	Death rate of memory T cells	0.03/day
$\mu_v^{(*)}(t) = \mu_v^{(*)}$	Death rate of $T_v^{(*)}$ cells	0.33/day
$\mu_m^{(*)}(t) = \mu_m^{(*)}$	Death rate of $T_m^{(*)}$ cells	0.95/day
$\mu_m^{(**)}(t) = \mu_m^{(**)}$	Death rate of $T_m^{(**)}$ cells	1.3/day
$k_{am}(t) = k_{am}$	Rate of $T_a \to T_m$	0.026/day
$k_{ae}(t) = k_{ae}$	Rate of $T_a \to T_e$	0.15/day
$k_{ma}(t) = k_{ma}$	Rate of $T_m \to T_a$	0.85/day
$k_{em}(t) = k_{em}$	Rate of $T_e \to T_m$	0.5/day
$k_{vv}^{(*)}(t) = k_{vv}^{(*)}$	Rate of $T_v^{(*)} \to T_v$	1.5/day
$k_{va}^{(*)}(t) = k_{va}^{(*)}$	Rate of $T_v^{(*)} \to T_a^{(*)}$	0.0105/day
$k_{am}^{(*)}(t) = k_{am}^{(*)}$	Rate of $T_a^{(*)} \to T_m^{(*)}$	0.0835/day
$k_{ae}^{(*)}(t) = k_{ae}^{(*)}$	Rate of $T_a^{(*)} \to T_e^{(*)}$	0.385/day
$k_{em}^{(*)}(t) = k_{em}^{(*)}$	Rate of $T_e^{(*)} \to T_m^{(**)}$	0.65/day
$k_{mm}^{(*)}(t) = k_{mm}^{(*)}$	Rate of $T_m^{(*)} \to T_m$	0.85/day
$k_{mm}^{(**)}(t) = k_{mm}^{(**)}$	Rate of $T_m^{(*)} \to T_m^{(**)}$	0.00058/day
$k_{ma}^{(**)}(t) = k_{ma}^{(**)}$	Rate of $T_m^{(**)} \to T_a^{(*)}$	0.65/day
$\delta_{va}(t) = \delta_{va}$	proportion of $T_v^{(*)} \to T_a$	2.5/day
$\delta_{ma}(t) = \delta_{ma}$	proportion of $T_m^{(*)} \to T_a$	1.5/day
$\lambda_v(t) = \lambda_v$	Rate of $T_v \to T_v^{(*)}$	0.00525/day
$\lambda_a(t) = \lambda_a$	Rate of $T_a \to T_a^{(*)}$	0.00856/day

Table 4. (*Continued*)

(a) Variables		
$\lambda_e(t) = \lambda_e$	Rate of $T_e \to T_e^{(*)}$	0.0093/day
$\lambda_m(t) = \lambda_m$	Rate of $T_m \to T_m^{(*)}$	0.00595/day
$\xi_1(t) = \xi_1$	Proportion of non-infectious HIV among the free HIV generated by the $T_a^{(*)}$ cells through activation and death	0.015/day
$\xi_2(t) = \xi_2$	Proportion of non-infectious HIV among the free HIV generated by the $T_e^{(*)}$ cells through activation and death	0.085/day
$g(t) = g$	Viruses from macrophage	$40/\text{mL}^3/\text{day}$
Without vaccine:		
$k_{va}(t) = k_{va}$	Rate of $T_v \to T_a$	0.02/day
$\mu_{ac}^{(*)}(t) = \mu_{ac}^{(*)}$	Death rate of $T_a^{(*)}$ eliminated by primed T_8	1.5/day
$\mu_{ec}^{(*)}(t) = \mu_{ec}^{(*)}$	Death rate of $T_e^{(*)}$ eliminated by primed T_8	3.5/day
$\mu_{aa}^{(*)}(t) = \mu_{aa}^{(*)}$	Death rate of $T_a^{(*)}$ by activation	$0.25 * \left(1 - \frac{TOT(t)}{T_{max}}\right)$
$\mu_{ea}^{(*)}t) = \mu_{ea}^{(*)}$	Death rate of $T_e^{(*)}$ by activation	$2.2 * \left(1 - \frac{TOT(t)}{T_{max}}\right)$
$N(t)$	Number of virus produced per $T_a^{(*)}$ and $T_e^{(*)}$ $N(t) = 1000 * \left(1 + \frac{8.85t^2}{t^2+17000^2}\right)$	
$\mu_V(t) = \mu_V$	Death rate of HIV	3.5/day
Under vaccine:		
$k_{va}(t) = k_{va}$	Rate of $T_v \to T_a$	0.5/day
$\mu_{aa}^{(*)}(t) = \mu_{aa}^{(*)}$	Death rate of $T_a^{(*)}$ through activation	$0.2 * \left(1 - \frac{TOT(t)}{T_{max}}\right)$
$\mu_{ea}^{(*)}(t) = \mu_{ea}^{(*)}$	Death rate of $T_e^{(*)}$ through activation	$2. * \left(1 - \frac{TOT(t)}{T_{max}}\right)$
$\mu_{ac}^{(*)}(t) = \mu_{ac}^{(*)}$	elimination rate of $T_a^{(*)}$ by primed T_8	$1.5 * (1 + exp(-1.5$ $*log_{10}(PP(t) + 1)))$
$\mu_{ec}^{(*)}(t) = \mu_{ec}^{(*)}$	elimination rate of T_e^* by primed T_8	$5. * (1 + 0.75 * exp(-5.$ $*log_{10}(PP(t) + 1)))$
$\mu_V(t) = \mu_V$	Death rate of virus	$0.4 * (1 + 0.5$ $*log_{10}(BP(t) + 1))$
$N(t)$	Number of virus produced by per $T_a^{(*)}$ and $T_e^{(*)}$	$530 * \left(1 + \frac{1.035*t^2}{t^2+15000^2}\right)$

(b) Distribution results

$S_v(t) \sim$ Poisson $\{s_v(t)\Delta t\}$;

$\{F_{va}(t), G_v(t), D_v(t)\}|[T_v(t), V(t)]$

$\sim$ Multinomial $\left\{T_v(t); k_{va}(t)\Delta t, \lambda_v(t)\frac{V(t)}{1+V(t)^{0.5}}\Delta t, \mu_v(t)\Delta t\right\}$;

Table 4. (*Continued*)

(b) Distribution results
$\left\{R_{vv}(t), R_{va}(t), D_v^{(*)}(t)\right\}\Big\vert T_v^{(*)}(t)$ $\sim$ Multinomial $\left\{T_v^{(*)}(t); k_{vv}^{(*)}(t)\Delta t, k_{va}^{(*)}(t)\Delta t, \mu_v^{(*)}(t)\Delta t\right\}$;
$\{F_{am}(t), F_{ae}(t), G_a(t), D_a(t)\}\vert[T_a(t), V(t)]$ $\sim$ Multinomial $\left\{T_a(t); k_{am}(t)\Delta t, k_{ae}(t)\Delta t, \lambda_a(t)\frac{V(t)}{1+V(t)^{0.75}}\Delta t, \mu_a(t)\Delta t\right\}$;
$\left\{R_{am}(t), R_{ae}(t), D_{aa}^{(*)}(t), D_{ac}^{(*)}(t)\right\}\Big\vert T_a^{(*)}(t)$ $\sim$ Multinomial $\left\{T_a^{(*)}(t); k_{am}^{(*)}(t)\Delta t, k_{ae}^{(*)}(t)\Delta t, \mu_{aa}^{(*)}(t)\Delta t, \mu_{ac}^{(*)}(t)\Delta t\right\}$;
$\left\{F_{em}(t), G_e(t), D_e(t)\right\}\vert[T_e(t), V(t)]$ $\sim$ Multinomial $\left\{T_e(t); k_{em}(t)\Delta t, \lambda_e(t)\frac{V(t)}{1+V(t)^{0.4}}\Delta t, \mu_e(t)\Delta t\right\}$;
$\left\{R_{em}(t), D_{ea}^{(*)}(t), D_{ec}^{(*)}(t)\right\}\Big\vert T_e^{(*)}(t)$ $\sim$ Multinomial $\left\{T_e^{(*)}(t); k_{em}^{(*)}(t)\Delta t, \mu_{ea}^{(*)}(t)\Delta t, \mu_{ec}^{(*)}(t)\Delta t\right\}$;
$\left\{F_{ma}(t), G_m(t), D_m(t)\right\}\vert[T_m(t), V(t)]$ $\sim$ Multinomial $\left\{T_m(t); k_{ma}(t)\Delta t, \lambda_m(t)\frac{V(t)}{1+V(t)^{0.95}}\Delta t, \mu_m(t)\Delta t\right\}$;
$\left\{R_{mm}(t), R_{mm}^{(*)}(t), R_{ma}(t), D_m^{(*)}(t)\right\}\Big\vert T_m^{(*)}(t)$ $\sim$ Multinomial $\left\{T_m^{(*)}(t); k_{mm}^{(*)}(t)\Delta t, k_{mm}^{(**)}(t)\Delta t, k_{ma}^{(*)}(t)\Delta t, \mu_m^{(*)}(t)\Delta t\right\}$;
$\left\{R_{ma}^{(*)}(t), D_m^{(**)}(t)\right\}\Big\vert T_m^{(**)}(t)$ $\sim$ Multinomial $\left\{T_m^{(**)}(t); k_{ma}^{(**)}(t)\Delta t, \mu_m^{(**)}(t)\Delta t\right\}$;
$G(t) = V(t) - G_v(t) - G_a(t) - G_m(t) - G_e(t)$; $\{D_V(t)\}\vert V(t) - G(t) \sim$ Binomial $\{V(t) - G(t); \mu_V(t)\Delta t\}$;

biological mechanisms as given in Section 2 above, we have the following system of stochastic differential equations for the state variables:

$$\begin{aligned} dT_v(t) &= T_v(t+\Delta t) - T_v(t) = S_v(t) + R_{vv}(t) - F_{va}(t) - G_v(t) - D_v(t) \\ &= \left\{s_v(t) + T_v^{(*)}(t)k_{vv}^{(*)}(t) - T_v(t)\left[\frac{V(t)}{1+V(t)^{0.5}}\lambda_v(t)\right.\right. \\ &\quad \left.\left. + k_{va}(t) + \mu_v(t)\right]\right\}\Delta t + \varepsilon_v(t)\Delta t, \end{aligned} \tag{13}$$

$$\begin{aligned} dT_v^{(*)}(t) &= T_v^{(*)}(t+\Delta t) - T_v^{(*)}(t) = G_v(t) - R_{vv}(t) - R_{va}(t) - D_v^{(*)}(t) \\ &= \left\{\frac{V(t)}{1+V(t)^{0.5}}T_v(t)\lambda_v(t) - T_v^{(*)}(t)\left[k_{vv}^{(*)}(t)\right.\right. \\ &\quad \left.\left. + k_{va}^{(*)}(t) + \mu_v^{(*)}(t)\right]\right\}\Delta t + \varepsilon_v^{(*)}(t)\Delta t, \end{aligned} \tag{14}$$

$$
\begin{aligned}
dT_a(t) &= T_a(t+\triangle t) - T_a(t) = F_{va}(t) + F_{ma}(t) + B_{Ta}(t) \\
&\quad + [R_{ma}(t) - P_{ma}(t)] + [R_{va}(t) - P_{va}(t)] - [G_a(t) \\
&\quad + F_{am}(t) + F_{ae}(t) + D_a(t)] = \Big\{ T_v(t)k_{va}(t) \\
&\quad + T_m(t)k_{ma}(t) + T_a(t)b_a(t) + (1-\delta_{ma})T_m^{(*)}(t)k_{ma}^{(*)}(t) \\
&\quad + (1-\delta_{va})T_v^{(*)}(t)k_{va}^{(*)}(t) - T_a(t)\bigg[\frac{V(t)}{1+V(t)^{0.75}}\lambda_a(t) \\
&\quad + k_{am}(t) + k_{ae}(t) + \mu_a(t)\bigg]\Big\}\triangle t + \varepsilon_a(t)\triangle t, \qquad (15)
\end{aligned}
$$

$$
\begin{aligned}
dT_a^{(*)}(t) &= T_a^{(*)}(t+\triangle t) - T_a^{(*)}(t) = G_a(t) + P_{va}(t) + R_{ma}^{(*)}(t) \\
&\quad + P_{ma}(t) - \Big[F_{am}(t) + F_{ae}(t) + D_{aa}^{(*)}(t) + D_{ac}^{(*)}(t)\Big] \\
&= \bigg\{\frac{V(t)}{1+V(t)^{0.75}}T_a(t)\lambda_a(t) + \delta_{va}T_v^{(*)}(t)k_{va}^{(*)}(t) \\
&\quad + T_m^{(**)}(t)k_{ma}^{(**)}(t) + \delta_{ma}T_m^{(*)}(t)k_{ma}^{(*)}(t) - T_a^{(*)}(t) \\
&\quad \times \Big[k_{ae}^{(*)}(t) + k_{am}^{(*)}(t) + \mu_{aa}^{(*)}(t) + \mu_{ac}^{(*)}(t)\Big]\bigg\}\triangle t + \varepsilon_a^{(*)}(t)\triangle t, \qquad (16)
\end{aligned}
$$

$$
\begin{aligned}
dT_e(t) &= T_e(t+\triangle t) - T_e(t) = F_{ae}(t) + B_{Te}(t) - [G_e(t) + F_{em}(t) \\
&\quad + D_e(t)] = \Big\{[T_a(t)k_{ae}(t) + T_e(t)b_e(t)] - T_e(t) \\
&\quad \times \bigg[\frac{V(t)}{1+V(t)^{0.4}}\lambda_e(t) + k_{em}(t) + \mu_e(t)\bigg]\Big\}\triangle t + \varepsilon_e(t)\triangle t, \qquad (17)
\end{aligned}
$$

$$
\begin{aligned}
dT_e^{(*)}(t) &= T_e^{(*)}(t+\triangle t) - T_e^{(*)}(t) = G_e(t) + R_{ae}(t) - \Big[R_{em}(t) + D_{ea}^{(*)}(t) \\
&\quad + D_{ec}^{(*)}(t)\Big] = \bigg\{\frac{V(t)}{1+V(t)^{0.4}}T_e(t)\lambda_e(t) + T_a^{(*)}(t)k_{ae}^{(*)}(t) \\
&\quad - T_e^{(*)}(t)\Big[k_{em}^{(*)}(t) + \mu_{ea}^{(*)}(t) + \mu_{ec}^{(*)}(t)\Big]\bigg\}\triangle t + \varepsilon_e^{(*)}(t)\triangle t, \qquad (18)
\end{aligned}
$$

$$
\begin{aligned}
dT_m(t) &= T_m(t+\triangle t) - T_m(t) = F_{am}(t) + R_{mm}(t) + F_{em}(t) - [G_m(t) \\
&\quad + F_{ma}(t) + D_m(t)] = \Big\{\Big[T_a(t)k_{am}(t) + T_m^{(*)}(t)k_{mm}^{(*)}(t) \\
&\quad + T_e(t)k_{em}(t)\Big] - T_m(t)\bigg[\frac{V(t)}{1+V(t)^{0.95}}\lambda_m(t) + k_{ma}(t) \\
&\quad + \mu_m(t)\bigg]\Big\}\triangle t + \varepsilon_m(t)\triangle t, \qquad (19)
\end{aligned}
$$

$$\begin{aligned} dT_m^{(*)}(t) &= T_m^{(*)}(t+\Delta t) - T_m^{(*)}(t) = G_m(t) - R_{mm}(t) - R_{mm}^{(*)}(t) \\ &\quad - R_{ma}(t) - D_m^{(*)}(t) = \left\{ \frac{V(t)}{1+V(t)^{0.95}} T_m(t)\lambda_m(t) \right. \\ &\quad \left. - T_m^{(*)}(t)\left[k_{mm}^{(*)}(t) + k_{mm}^{(**)}(t) + k_{ma}^{(*)}(t) + \mu_m^{(*)}(t)\right] \right\} \Delta t \\ &\quad + \varepsilon_m^{(*)}(t)\Delta t, \end{aligned} \tag{20}$$

$$\begin{aligned} dT_m^{(**)}(t) &= T_m^{(**)}(t+\Delta t) - T_m^{(**)}(t) = R_{am}(t) + R_{em}(t) - R_{ma}^{(*)}(t) \\ &\quad - D_m^{(**)}(t) = \left\{ T_a^{(*)}(t)k_{am}^{(*)}(t) + T_e^{(*)}(t)k_{em}^{(*)}(t) - T_m^{(**)}(t)k_{ma}^{(**)}(t) \right. \\ &\quad \left. - T_m^{(**)}(t)\mu_m^{(**)}(t) \right\} \Delta t + \varepsilon_m^{(**)}(t)\Delta t, \end{aligned} \tag{21}$$

$$\begin{aligned} dV(t) &= V(t+\Delta t) - V(t) = \left\{ g(t) + N(t)\left[(1-\xi_1)T_a^{(*)}(t)\mu_{aa}^{(*)}(t) \right. \right. \\ &\quad \left. + (1-\xi_2)T_e^{(*)}(t)\mu_{ea}^{(*)}(t)\right] - \left[\frac{V(t)}{1+V(t)^{0.5}} T_v(t)\lambda_v(t) \right. \\ &\quad + \frac{V(t)}{1+V(t)^{0.75}} T_a(t)\lambda_a(t) + \frac{V(t)}{1+V(t)^{0.4}} T_e(t)\lambda_e(t) \\ &\quad \left. \left. + \frac{V(t)}{1+V(t)^{0.95}} T_m(t)\lambda_m(t) + V(t)\mu_V(t)\right] \right\} \Delta t + \varepsilon_V(t)\Delta t. \end{aligned} \tag{22}$$

The random noises in the above equations have expected values zero and are uncorrelated with the state variables. For generating some Monte Carlo studies, some parameter values are given in Table 4.

5. Some Monte Carlo Studies

As shown in Sections 3–4 and in Tables 1–4, the probability distributions of variables for stochastic transitions in the dynamic system of the immune response in human beings are basically binomial and multinomial distributions. By using these probability distributions, in this section, we will use the stochastic models in Sections 3–4 to generate some Monte Carlo studies to assess the prophylactic and therapeutic roles of the AIDS vaccines against HIV infection. For these Monte Carlo studies, the parameters are selected from published papers and are given in Tables 1–4. In this paper we will consider three types of AIDS vaccines: the cytotoxic AIDS vaccines which exert its effects mainly through cytotoxic CD8 T cells (CTL cells), the humoral AIDS vaccines, which exert its effects mainly through plasma B cells and HIV antibodies and the complete AIDS vaccines, which exert its effects through both cytotoxic CD8 T cells and plasma B cells and antibodies. We model the cytotoxic effects of AIDS vaccine by promoting the transitions from $P_v \to P_a$ and from $P_m \to P_a$; similarly we model the humoral effects of AIDS vaccine by promoting the transitions from $B_v \to B_a$ and from $B_m \to B_a$. Thus, if there are no cytotoxic effects, we assume the transition rates for $P_v \to P_a$ and for $P_m \to P_a$ as 0.0045/day and 0.0065/day respectively; if there are cytotoxic effects, then we assume the transition rates for $P_v \to P_a$ and for $P_m \to P_a$ as 0.0455/day and 0.0565/day respectively. Similarly, if there are no humoral effects, we assume the transition rates for $B_v \to B_a$ and for $B_m \to B_a$ as 0.008/day and 0.0095/day respectively; if there are humoral effects, then we assume the transition rates for $B_v \to B_a$ and for $B_m \to B_a$ as 0.015 and 0.018 respectively.

5.1. *The dynamics of immune system in normal individuals*

In unvaccinated normal individuals, the immune system is in the steady state condition. This is illustrated in Fig. 4. Under this condition, most

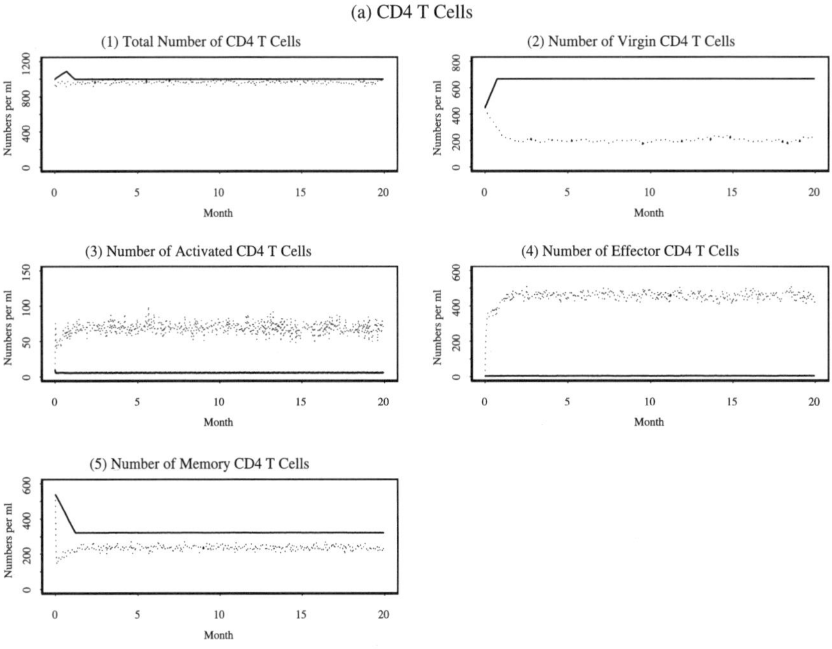

Fig. 4. Plots showing the dynamic of the immune system in vaccinated and unvaccinated normal individuals: solid line, unvaccinated; dotted line, vaccinated.

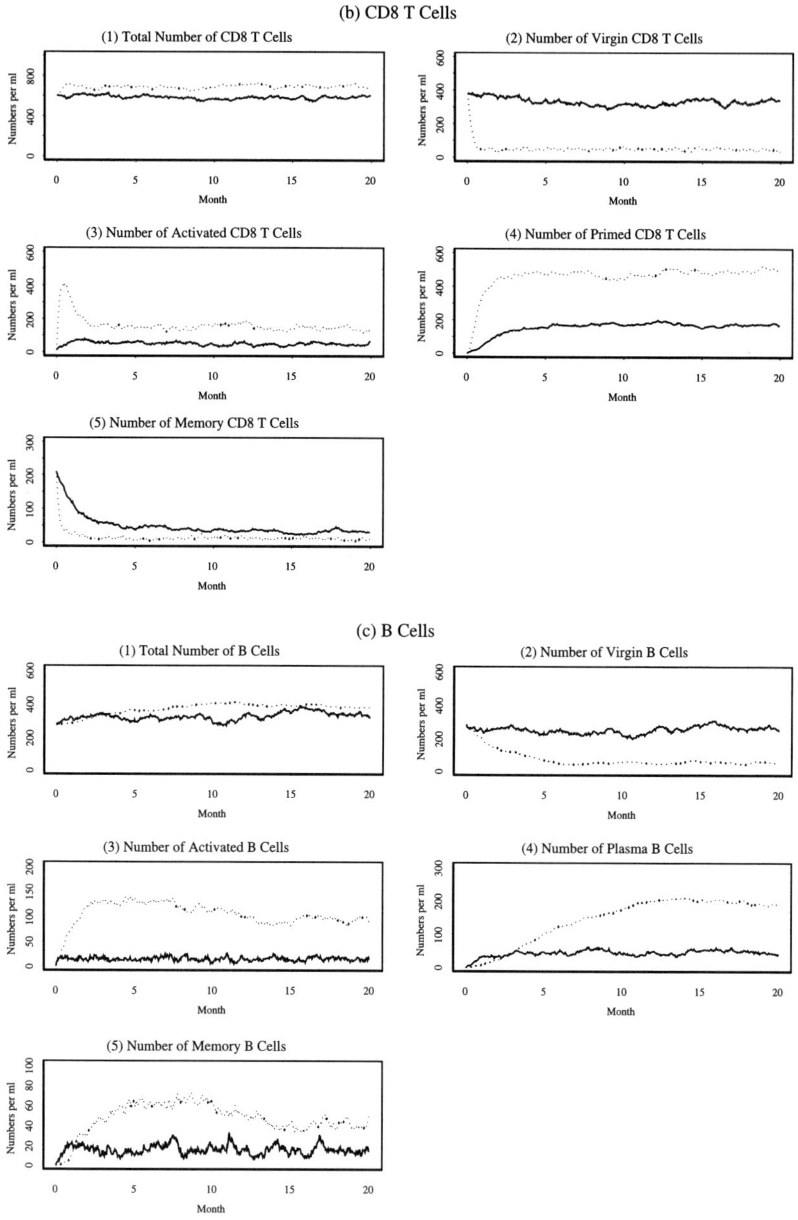

Fig. 4. *(Continued)*

of the CD4 T cells are T_v (660/mL) and T_m (320/mL) cells, whereas the T_a and T_e cells are only about 1% of all CD4 T cells. (The numbers of T_a and T_e cells are 6/mL and 4/mL respectively.) The total number is 1000/mL for a normal person. In the steady state condition, the numbers of $\{P_v, P_m, P_a, P_p\}$ CD8 T cells are about $\{314/\text{mL}, 58/\text{mL}, 42/\text{mL}, 184/\text{mL}\}$ respectively with a total number of CD8 T cells being 598/mL which is about 60% of CD4 T cells; under the steady state condition the numbers of $\{B_v, B_a, B_p, B_m\}$ B cells are $\{203/\text{mL}, 7/\text{mL}, 53/\text{mL}, 24/\text{mL}\}$ respectively with a total number of B cells being 287/mL which is about 28% of CD4 T cells.

When the individuals are vaccinated by an AIDS vaccine and maintained at the level by additional shoots, then the dynamic of the immune system change dramatically leading to a new steady state condition in about 2 months. At this new steady state condition, most of the CD4 T cells are T_e cells. Comparing with unvaccinated individuals, the number of T_a and T_e cells have increased from less than 1% to over 10% and 50% whereas the number of T_v and T_m cells have reduced from 60% to less than 30%. For the CD8 T cells and B cells, at the new steady-state condition, most cells are P_p cells and B_p cells respectively. Comparing with unvaccinated individuals, the number of P_a and P_p cells have increased from less than 20% to over 40% and 80% whereas the number of B_a and B_p cells have increased from approximately 20% to more than 70%; also, the total number of CD8 T cells and B cells have increased about 10% and 30%, from 598/mL to 667/mL for CD8 T cells and from 287/mL to 390/mL for B cells. Since AIDS vaccines can be considered as antigens without infection and harm, these results follow basically from the role that the AIDS vaccines increase significantly the transition rates of stochastic transitions $\{T_v \to T_a, T_m \to T_a, P_v \to P_a, P_m \to P_a, B_v \to B_a, B_m \to B_a\}$. For complete vaccines these results are plotted and illustrated in Fig. 4. For saving printing pages, the plots for cytotoxic vaccines and humoral vaccines are given in Fig. 5.5 in Zhang (2004).

From Fig. 4 and plots in Zhang (2004), it is observed that under the new steady-state condition, the proportions of $\{T_v, T_a, T_e, T_m\}$ CD4 T cells are fluctuating around $\{225/\text{mL}, 70/\text{mL}, 440/\text{mL}, 230/\text{mL}\}$ for complete vaccines; the corresponding proportions for cytotoxic and humoral vaccines are $\{245/\text{mL}, 55/\text{mL}, 370/\text{mL}, 300/\text{mL}\}$ respectively. These results indicate that the AIDS vaccines decrease the numbers of T_v and T_m cells (especially the T_v cells) but increase the numbers of T_a and T_e cells (especially the T_e cells). For the CD8 T cells, at the new steady-state condition under vaccines, the proportions of $\{P_v, P_m, P_a, P_p\}$ CD8 T cells are fluctuating around $\{49/\text{mL}, 146/\text{mL}, 8/\text{mL}, 464/\text{mL}\}$ for complete vaccines, fluctuating around $\{39/\text{mL}, 164/\text{mL}, 17/\text{mL}, 457/\text{mL}\}$ for cytotoxic vaccines and fluctuating around $\{141/\text{mL}, 115/\text{mL}, 26/\text{mL}, 388/\text{mL}\}$ for humoral vaccines. For the B cells, at the new steady-state condition under the vaccines, the proportions of $\{B_v, B_m, B_a, B_p\}$ B cells are fluctuating around $\{50/\text{mL}, 45/\text{mL}, 85/\text{mL}, 220/\text{mL}\}$ for complete vaccines, fluctuating around $\{55/\text{mL}, 35/\text{mL}, 90/\text{mL}, 180/\text{mL}\}$ for cytotoxic vaccines and fluctuating around $\{115/\text{mL}, 35/\text{mL}, 35/\text{mL}, 110/\text{mL}\}$ for humoral vaccines. Interestingly, we observed that the numbers under cytotoxic vaccines are very close to those under complete vaccines respectively, but those under humoral vaccines differ significantly from those under other vaccines. This indicates that the prophylactic effects of AIDS vaccines come mainly from the CTL cells rather than from HIV antibodies.

5.2. *The dynamic of immune system under HIV infection in unvaccinated individuals*

When an individual is infected by HIV, because the HIV infect mainly CD4 T cells, one expects the CD4 T cells to decrease below 200/mL in 10 years. (The individual is classified as an AIDS patient by CDC if the total number of CD4 T cells is below 200/mL in blood; see CDC, 1992.) To illustrate the basic dynamic of HIV, we let the individual be infected by 100/mL HIV and use the stochastic models in Section 4 to generate some Monte Carlo samples for unvaccinated individuals. We assume that initially, the proportion of $\{T_v, T_m, T_a, T_e\}$ are $\{450/\text{mL}, 540/\text{mL}, 10/\text{mL}, 0/\text{mL}\}$ respectively and that there are no HIV-infected CD4 T cells initially. The results are presented schematically in Fig. 5.

As illustrated in Fig. 5(b), the total number of CD4 T cells first decreased to around 600/mL in a week, then decrease slowly but steadily, reaching 245/mL in 5 years and reaching 178/mL in 10 years. On the other hand, after an initial drop, the HIV load increases steadily and reaches a very high level of 25,745/mL in 5 years and reach 39,939/mL in 10 years. These are consistent with many observed results as described in Tan (2000, Chapters 7–8).

From Fig. 5, we observed that the number of $\{T_v, T_m\}$ cells have reduced significantly whereas the $\{T_a, T_e\}$ cells increased significantly. In 5 years, the proportion of $\{T_v, T_a, T_e, T_m, T_v^{(*)}, T_a^{(*)}, T_e^{(*)}, T_m^{(*)}, T_m^{(**)}\}$ cells are $\{97/\text{mL}, 47/\text{mL}, 6/\text{mL}, 45/\text{mL}, 17/\text{mL}, 10/\text{mL}, 7/\text{mL}, 1/\text{mL}, 15/\text{mL}\}$

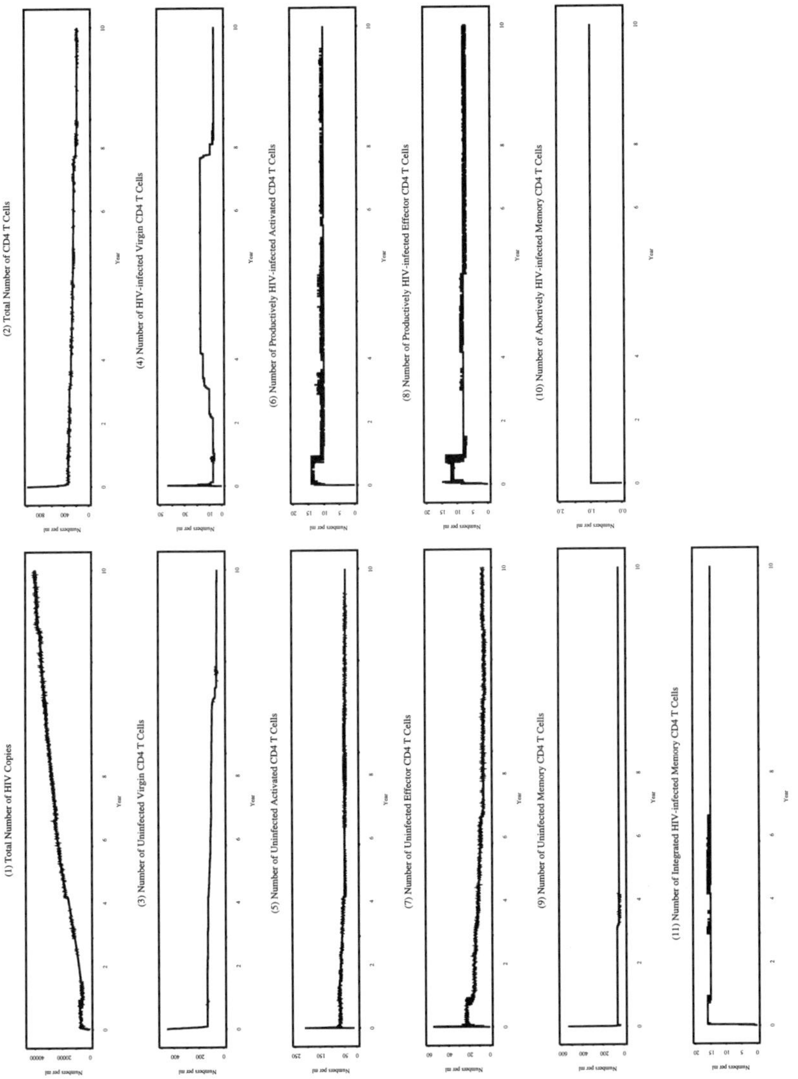

Fig. 5. Plots showing the dynamic in HIV-infected unvaccinated individuals.

respectively. In 10 years, the proportion of $\{T_v, T_a, T_e, T_m, T_v^{(*)}, T_a^{(*)}, T_e^{(*)}, T_m^{(*)}, T_m^{(**)}\}$ cells are {46/mL, 42/mL, 6/mL, 45/mL, 6/mL, 10/mL, 7/mL, 1/mL, 15/mL} respectively. Notice that, in 10 years, the proportion of productively HIV-infected T_4 cells (e.g. $T_a^{(*)}, T_e^{(*)}$ cells), abortively HIV-infected T_4 cells (e.g. $T_v^{(*)}, T_m^{(*)}$ cells) and latently HIV-infected T_4 cells (e.g. $T_m^{(**)}$ cells) are 17/178 (about 0.095), 7/178 (about 0.0412) and 15/178 (about 0.0882) respectively. These results are in consistent with those reported by O'Hara (2004).

5.3. *The prophylactic effects of AIDS vaccines*

One major function of AIDS vaccine is to protect normal individuals from HIV infection through the cytotoxic mechanism and/or through the humoral mechanism of AIDS vaccines. To assess the prophylactic effects of AIDS vaccines, we consider the situation in which an individual is vaccinated by an AIDS vaccine and is maintained at a fixed level by boosting injections. One month after vaccination, at which time the immune system has reached the new steady-state condition, we let the individual be infected by 100/mL, or 1000/mL, or 10,000/mL HIV respectively. In each case, we measure the time period when the HIV loads reduce to the new steady-state level and the HIV concentration at that level. For the case infecting with 10,000/mL HIV, the results are presented schematically in Fig. 6. For saving pages, results of other vaccines are given in Fig. 5.7 in Zhang (2004). Notice that although the number of HIV is undetectable under vaccination, the vaccines cannot prevent HIV infection. These results are consistent and comparable to results from clinical trials of AIDS vaccines on rhesus monkeys by Amara *et al.* (2001) and Barouch *et al.* (2000).

From Fig. 6 and results in Zhang (2004), it is observed that when the individual is infected by 100/mL HIV, it takes about 2 weeks to reach the steady-state level of around 60/mL copies of HIV for complete and cytotoxic vaccines, and takes about 3 to 4 weeks to reach the steady-state level of around 85/mL copies of HIV for humoral vaccines; on the other hand, the total number of CD4 T cells decrease from 1000/mL to about 380/mL in a few days and then increase to reach a steady-state number 670/mL in about 22 days. When the individual is infected by 1000/mL HIV, it takes about 26 days to reach the steady-state level of around 60/mL HIV copies for both complete and cytotoxic vaccines, and takes about 33 days to reach the steady-state level of around 85/mL HIV copies for humoral vaccines; on the other hand, the total number of CD4 T cells decrease from 1000/mL to

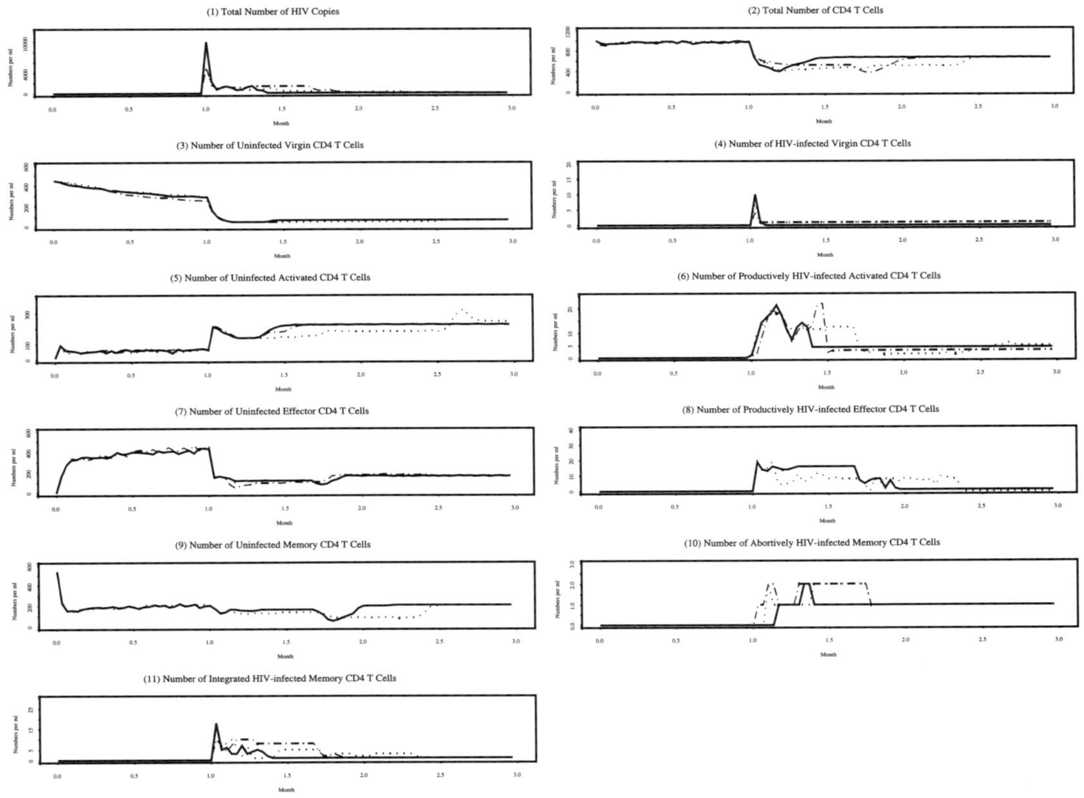

Fig. 6. Plots showing the prophylactic effects of AIDS vaccines in the Monte Carlo study in which the vaccinator is infected by 10,000/mL HIV one month after vaccination. The solid line is for complete vaccine, dashed line for cytotoxic vaccine, and dots for humoral vaccine.

about 370/mL in a few days and then increase to reach a steady-state number 670/mL in about 2 weeks. When the individual is infected by 10,000/mL HIV, it takes about 30 to 35 days to reach the steady-state level of around 90/mL HIV copies for complete and cytotoxic vaccines, and takes about 55 days to reach the steady-state level of around 90/mL HIV copies for humoral vaccines; interestingly, almost exactly as before, the total number of CD4 T cells decrease from 1000/mL to about 400/mL in a few days and then increase to reach a steady-state number of 670/mL cells in about 2 weeks. Notice that at steady-state, the total number of CD4 T cells is the same (670/mL) regardless whether the HIV copies infecting the CD4 T cells is 100/mL, or 1000/mL or 10,000/mL. On the other hand, the HIV loads at the steady-state condition increase slightly by increasing the intensities of HIV at the time of infection although the numbers are less than 100/mL HIV. For the subsets of CD4 T cells, we observed from Fig. 6 and Fig. 5.7 in Zhang (2004) that by the action of AIDS vaccines, the numbers of $\{T_v, T_v^{(*)}, T_m, T_m^{(*)}, T_m^{(**)}, T_a^{(*)}, T_e, T_e^{(*)}\}$ cells were reduced significantly but the number of T_a cells increase significantly to the level of 200/mL cells at the steady-state condition. In all cases, it takes about 2 weeks for all types of cells to reach the steady-state condition. At the steady-state condition, the number of $\{T_v,\ T_m,\ T_a,\ T_e,\ T_a^{(*)},\ T_e^{(*)},\ T_m^{(*)}\}$ is around {70/mL, 200/mL, 160/mL, 200/mL, 5/mL, 1/mL, 1/mL}.

The results in Fig. 6 and in Fig. 5.6 in Zhang (2004) indicated that as prophylactic agents against HIV infection, the cytotoxic vaccines are close to complete vaccines and are more effective than humoral vaccines. The Monte Carlo studies seemed to show that the vaccines should be very effective as a prophylactic agents if it increases CTL (CD8 primed cells) responses and /or humoral responses.

5.4. *The therapeutic effects of AIDS vaccines*

Because the AIDS vaccine increases significantly the transition rates of stochastic transitions $\{T_v \to T_a, T_m \to T_a, P_v \to P_a, P_m \to P_a, B_v \to B_a, B_m \to B_a\}$, one expects that another major function of AIDS vaccines is to boost the immune defense against HIV infection in HIV-infected individuals. To assess this effect of AIDS vaccines, we consider the situation in which an individual is first infected by 100/mL HIV; then after some time period since infection (one month, or one year or 5 years), the individual is vaccinated by an AIDS vaccine and is maintained at a fixed level by boosting injections. Under this experiment, we will use the stochastic models in Sections 3 and 4 to generate Monte Carlo studies. We will check if the vaccines can control the HIV replication and how. For the case of vaccination one year after infection, the results are plotted in Fig. 7. For saving pages, other results are plotted in Fig. 5.8 in Zhang (2004).

From Fig. 7 and Fig. 5.8 in Zhang (2004), we observed that when the AIDS vaccines are applied to HIV-infected individuals, the dynamic of immune system change dramatically and reach eventually a new steady-state condition at which time the HIV loads are $\leq$ 100/mL. The time period to reach this steady-state condition depends on the number of HIV at the time when the AIDS vaccines are used. When the vaccine is used at one month after HIV infection when the HIV has reached 7794/mL, it takes about 10 days to reach the steady-state condition with 60/mL HIV for complete vaccines, and take 30 days for cytotoxic vaccines and 15 days for humoral vaccines reaching the steady-state with 85/mL HIV; interestingly, the total number of CD4 T cells increase from 329/mL at the time of vaccination to the steady-state level 670/mL in about 10 days for all vaccines. Notice that the steady-state number 670/mL CD4 T cells is the same number in Section 5.3. When the vaccine is used one year after HIV infection when the HIV load has reached 8400/mL, the HIV load will reach the steady-state with 85/mL HIV in about 10 days for complete vaccines, about 15 days for humoral vaccines and about a month for cytotoxic vaccines; interestingly, the total number of CD4 T cells increased steadily from below 400/mL at the time of vaccination to reach 670/mL at the steady-state condition for all vaccines in about 2 weeks. This is similar to the results when starting vaccination one month after HIV infection. Similar results were observed when vaccine is used 5 years after HIV infection when the HIV has reached 25,600/mL; in this latter case, the HIV load at the steady-state condition is around 100/mL for all vaccines but the total number of CD4 T cells is still 670/mL at the steady-state condition, and it takes about 2 weeks since vaccination to reach this steady-state condition.

On the subsets of CD4 T cells, the results are similar to the cases in Section 5.3. That is, at the steady-state condition, in almost all cases most cells are $\{T_a$ (200/mL), T_e (200/mL), T_m (200/mL)$\}$ cells; for the HIV-infected CD4 T cells, there are 5 $T_a^{(*)}$ cells, 1 $T_e^{(*)}$ cell and 1 $T_m^{(*)}$ cell.

From results in Figs. 7 and 5.8 in Zhang (2004), we observed that as therapeutic agents, the humoral vaccines are closer to complete vaccines and are slightly better than those of cytotoxic vaccines. This is in contract with the prophylactic effect. Results from the Monte Carlo studies seemed to show that the AIDS vaccines should be very effective as therapeutic

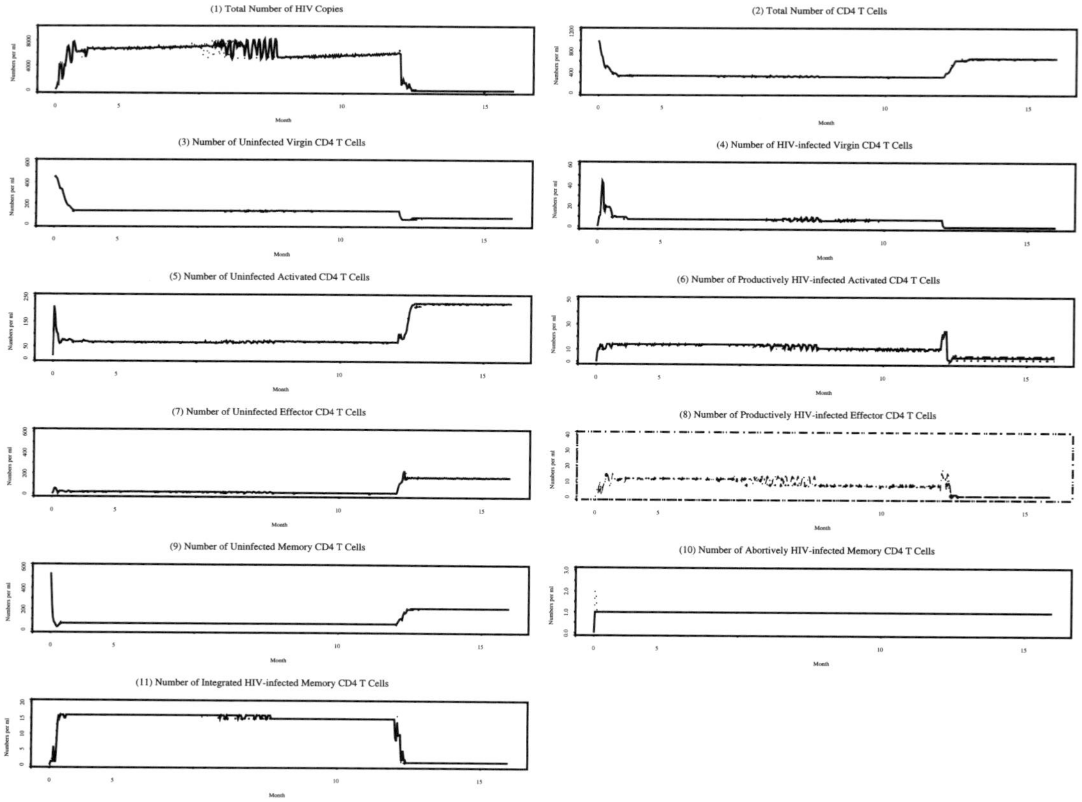

Fig. 7. Plots showing the therapeutic effects of AIDS vaccine in the Monte Carlo study where the individual is infected by HIV one year before vaccination. The solid line is for complete vaccine, dashed line for cytotoxic vaccine, and dots for humoral vaccine.

agents if they can generate effective humoral responses and/or cytotoxic responses.

6. Conclusions and Discussion

To assess effects of AIDS vaccines, in this paper we have developed stochastic models for the subsets of CD4 T cells, CD8 T cells and B cells under various conditions including AIDS vaccination and HIV infection. These are biologically supported models because they are developed by taking into account the basic biological mechanisms as described in Section 2 and in Figs. 1–3. By using these stochastic models, we have generated some Monte Carlo samples to assess effects of AIDS vaccines. Because AIDS vaccines can be considered as antigens without harm and with no infectivity, they function to protect individuals against HIV infection through two basic avenues: the cytotoxic avenue through the primed CD8 T cells (CTL cells) and the humoral avenue through the plasma B cells and HIV antibodies. Through Monte Carlo studies, we have compared the three types of AIDS vaccines. Our results showed that the prophylactic effects of AIDS vaccines derive mainly from cytotoxic avenue although the same effect is achieved by the humoral avenue through a little longer time period; on the other hand, for the therapeutic effects of AIDS vaccines, the humoral avenue appears to be a little better than the cytotoxic avenue.

To assess the prophylactic and therapeutic effects of AIDS vaccines, we considered two experiments in this paper. In one experiment, we first vaccinated the individual and maintained the vaccine level by booster injections; then we let the individual be infected by HIV with different HIV concentrations. This is the design for studying the prophylactic effects of AIDS vaccines. In the second experiment, we let the individual be infected by HIV first; then after some time period (one month, one year, or 5 years) the HIV-infected individual is injected by an AIDS vaccine. This is the experiment for studying the therapeutic effects of AIDS vaccines. In the first experiment, our Monte Carlo studies showed that even when the vaccinated individual is infected by 10,000/mL HIV, in less than 2 months after infection, the HIV loads were reduced to the undetectable level (≤ 100/mL) and maintained at this level afterwards. In the second experiment, our Monte Carlo samples showed that when HIV-infected individuals were vaccinated by an AIDS vaccine, even after 5 years when the number of HIV copies were very high, in about than 2 to 3 weeks the vaccine can bring down the HIV to the undetectable level (≤ 100/mL) and maintained this level afterwards. Interestingly, in all experiments, the total number of CD4 T cells reached 670/mL for all vaccines; furthermore, most of the CD4 T cells are $\{T_a, T_e, T_m\}$ cells and the HIV-infected cells are ≤ 10/mL. The above results are consistent and comparable with results of clinical trials of AIDS vaccines on rhesus monkeys by Amara *et al.* (2001) and Barouch *et al.* (2000).

Vaccines can be broadly classified as live-attenuated vaccines and non-live-attenuated vaccines. The live-attenuated vaccines exert their effects through both cytotoxic and/or humoral responses and hence are in general more effective, but have the potential to infect the vaccinated individuals. On the other hand, the non-live-attenuated vaccines are less effective but safe. Because most candidates of AIDS vaccines such as the gp120 AIDS vaccine are not live-attenuated, the results of clinical trials of these vaccines so far are disappointed and are failures (Cohen, 2003a; 2003b). Our simulation studies seemed to indicate that if any vaccine can induce significant immune response through cytotoxic and/or humoral defense avenues, then the vaccine should be very effective in both prophylactic and therapeutic functions although it cannot completely prevent HIV infection. Thus, one might conclude that if the vaccines induce significant responses through cytotoxic and/or humoral avenues, then the vaccines can control and suppress the HIV replication to undetectable level. In practice, however, the HIV is constantly mutating to new forms to escape the actions of vaccines, thus rending it ineffective. Thus, as expected, if the vaccines can protect against HIV infection over a wide range of HIV types, then AIDS vaccines would be the best avenue to control the HIV epidemic.

In this paper, we are mainly concerned with mathematical models for the immune system under AIDS vaccines and HIV infection. Because AIDS vaccine data are hard to get and AIDS vaccines are not yet available for public use, we have not yet applied the model to actual data sets. When AIDS vaccine data become available in the future, we will apply the models to some real data to estimate the efficacy and usefulness of AIDS vaccines.

Also, for the applications of these models, more researches are definitely needed. Specifically, we need to consider the following topics: (i) we need to apply the above models to assess effects of different AIDS vaccines and to validate the models; (ii) we need to assess how different risk factors affecting effects of AIDS vaccines, (iii) for given AIDS vaccines, we need to compare the vaccines with respect to the robustness against a wide variety of HIV species; and (iv) for given AIDS vaccines, we need to determine the time period for boosting injections to maintain high level of vaccines in the body. We will address these questions in our future research through either Monte Carlo studies and/or real data.

References

Amara RR, Villinger F, Altman JD *et al.* (2001) Control of a mucosal challenge and prevention of AIDS in rhesus macaques by a multiprotein DNA/MVA vaccine, *Science* **292**: 69–74.

Bajaria SH, Webb G, Cloyd M, Kirschner DE (2002) Dynamics of naive and memory CD4+ T lymphocytes in HIV-1 disease progression, *JAIDS* **30**: 41–58.

Bajaria SH, Webb G, Kirschner DE (2004) Predicting differential responses to structured treatment interruptions during HAART. *Bull Math Biol* in press.

Bajaria SH, Kirschner DE (2004) CTL action is determined via interactions with several cell types, in Tan WY (ed.) *Deterministic and Stochastic Models of AIDS and HIV With Intervention*, World Scientific, Singapore.

Baltimore D, Heilman C (1998) HIV vaccine: Prospects and challenges, *Sci Am* **279**: 98–103.

Barouch DH *et al.* (2000) Control of viremia and prevention of clinical AIDS in rhesus monkeys by cytokine-augmented DNA vaccination, *Science* **290**: 486–492.

Benoist C, Mathis D (1999) T-lymphocyte differentiation and biology, in Paul WF (ed.) *Fundamental Immunology*, 4th ed., Chapter 11, pp. 367–410, Lippincott-Raven, Philadelphia, PA.

Bremermann HJ (1995) Mechanism of HIV persistence: implication for vaccines and therapy, *J Acq Immun Def Synd* **9**: 459–483.

CDC (1993) Revised classification system for HIV infection and expanded surveillance case definition for AIDS among adolescents and adults, *MMWR, 1992* **41**: No. RR17.

Coffin JM (1995) HIV population dynamics *in vivo*: implications for genetic variation, pathogenesis and therapy, *Science* **267**: 483–489.

Cohen O, Weissman D, Fauci AS (1998) The immunopathogenesis of HIV infection, in Paul WE (ed.) *Fundamental Immunology*, 4th ed., Chapter 44, pp. 1511–1534, Lippincott–Raven Publishers, Philadelphia.

Cohen J (2003) AIDS vaccine trial produces disappointment and confusion, *Science* **299**: 1290–1291.

Cohen J (2003) AIDS vaccine still alive as booster after second failure in Thailand, *Science* **302**: 1309–1310.

DeFranco AL (1999) B-lymphocyte activation, in Paul WF (ed.) *Fundamental Immunology*, 4th ed., Chapter 7, pp. 225–261, Lippincott–Raven, Philadelphia, PA.

DeVita VT Jr, Hellman S, Rosenberg SA (eds.) (1997) *AIDS*, Fourth edition, Lippincott–Raven, Philadelphia.

Essunger P, Perelson AS (1994) Modeling HIV infection of CD4+ T-cell subpopulations, *J Theor Biol* **170**: 367–391.

Fauci AS (1996) Immunopathogenic mechanisms of HIV infection, *Ann Int Med* **124**: 654–663.

Fauci AS, Pantaleo G (eds.) (1997) *Immunopathogenesis of HIV Infection*, Springer–Verlag, Berlin.

Hraba T, Dolezal J (1995) A mathematical model of $CD8^+$ lymphocyte dynamics in HIV infection, *Folia Biologica (Praba)* **41**: 304–318.

Levy JA (1998) *HIV and Pathogenesis of AIDS* Second Edition ASM Press, Washington, D.C.

Murry JM, Kaufmann G, Kelleher AD, Cooper DA (1998) A model of primary HIV-1 infection, *Math Biosci* **154**: 57–85.

Nossal GJV (1999) Vaccines, in Paul WF (ed.) *Fundamental Immunology*, 4th ed. Chapter 42, pp. 1387–1425, Lippincott–Raven, Philadelphia, PA.

O'Hara K (2004) A model of HIV-1 treatment: the latently infected CD4+ T cells becomes undetectable. This volume, Chapter 13, World Scientific, Singapore.

Pantaleo G, Graziosi C, Fauci AS (1993) The immunopathogenesis of human immunodeficiency virus infection, *New Eng J Med* **328**: 327–335.

Paul WE (ed.) (1998) *Fundamental Immunology* Lippincott–Raven Publishers, New York.

Perelson AS, Kirschner DE, Boer RD (1993) Dynamics of HIV infection of $CD4^{(+)}$ T cells, *Math Biosci* **114**: 81–125.

Perelson AS, Nelson PW (1999) Mathematical analysis of HIV-1 dynamics *in vivo*, *SIAM Rev* **41**: 3–44.

Perelson AS, Neumann AU, Markowitz M, Leonard JM, Ho DD (1996) HIV-1 dynamics *in vivo*: Virion clearance rate, infected cell life-span, and viral generation time, *Science* **271**: 1582–1586.

Phillips AN (1996) Reduction of HIV concentration during acute infection: independence from a specific immune response, *Science* **271**: 497–499.

Prikrylova D, Jilek M, Waniewski J (1992) *Mathematical Models of the Immune Response*, CRC Press, Boca Raton, Florida.

Seder RA, Mosmann TM (1999) Differentiation of effector phenotypes of $CD4^+$ and $CD8^+$ T cells, in Paul WF (ed.) *Fundamental Immunology*, 4th ed. Chapter 26, pp. 879–908, Lippincott–Raven, Philadelphia, PA.

Stilianakis NI, Dietz K, Schenzle D (1997) Analysis of a model for the pathogenesis of AIDS, *Math Biosci* **145**: 27–46.

Tan WY, Wu H (1998) Stochastic modeling of the dynamic of $CD4^+$ T cell infection by HIV and some Monte Carlo studies, *Math Biosci* **147**: 173–205.

Tan WY, Ye ZZ (2000) Assessing effects of different types of HIV and macrophage on the HIV pathogenesis by stochastic models of HIV pathogenesis in HIV-infected individuals, *J Theor Med* **2**: 245-265.

Weiss A (1999) T-lymphocyte activation, in Paul WF (ed.) *Fundamental Immunology*, 4th ed. Chapter 12, pp. 412–447, Lippincott–Raven, Philadelphia, PA.

Zhang P (2004) Stochastic modelling of HIV pathogenesis under therapy and AIDS vaccination, Ph.D. Thesis, Department of Mathematical Sciences, The University of Memphis, Memphis, TN 38152.

CHAPTER 8

DYNAMICAL MODELS FOR THE COURSE OF AN HIV INFECTION

Christel Kamp

Faced with an ongoing HIV virus pandemic, a deeper understanding of the interactions between this virus and an adaptive immune system is as essential to develop efficient counteractive measures. This review introduces models that aim to understand the course of disease as an emergent from the complex network of interactions of virus particles and immune cells at the microscopic scale, though not focusing on the details of the individual interactions. This allows the derivation of constraints on the global dynamics within cellular automata models and an approach based on stochastic processes. Special attention is devoted to a better understanding of the origin of the incubation period distribution. Moreover, vaccination strategies and recent developments in drug design based on fusion inhibition are discussed in the light of the presented theoretical models.

Keywords: HIV; AIDS; incubation period distribution; cellular automata; stochastic process; shape space; percolation.

1. Introduction

Most viral infections lead to disease after a rather short and well-defined incubation period obeying approximately a lognormal distribution which is known as "Sartwell's model" (Sartwell, 1950; 1966). Remarkably different from that, after an acute phase following an HIV infection, patients typically experience a period of latency for several years before reaching the terminal stage of the disease with the onset of AIDS. A lot of work is concerned with making predictions on survivability by fitting empirical data to statistical distributions (Dangerfield and Roberts, 1999). More recent studies go further investigating the origins of the unusual incubation period distribution from an HIV infection to the onset of AIDS (Philippe, 1994; 2000). A broad variety of models describing the interactions between adaptive immune systems and viruses, specifically HIV, has been developed (Perelson and Weisbuch, 1997; Nowak and May, 2000; Kamp, 2003). Nowak *et al.* investigated the time evolution of the number of T-helper cells or $CD4^+$ cells as well as of the viral load during the course of disease after an infection with HIV. Using a differential equation approach, they associate the onset of AIDS with the excession of a viral diversity threshold (Nowak *et al.*, 1991; Nowak and May, 2000). Many attempts have also been made to model host-virus-interactions by means of microscopic computer simulations or cellular automata models which have the advantage to be able to account for stochastic fluctuations in the system. While some approaches focus on virus dynamics in physical space (Zorzenon dos Santos and Coutinho, 2001), most describe viral dynamics in an abstract shape space parameterizing the viral epitopes and complementary immune receptor specificities (Perelson and Weisbuch, 1997).

These approaches likewise try to understand the course of a disease as a feature emergent from the complex network of interactions between immune cells and antigenic intruders (Valera and Coutinho, 1991; Perelson and Weisbuch, 1997; Zorzenon dos Santos and Bernardes, 1998; Brede and Behn, 2001). The adaptive immune response is mainly determined by a diverse set of B- and T-lymphocytes whose mutual interactions are mediated among others by cytokines (Roitt, 1994; Janeway *et al.*, 2001). In response to an antigenic challenge, B- and T-cells get activated if their receptors are complementary to the detected epitopes and a highly efficient immune response ensues that is controlled by a complex regulatory network. Although antigens, and specifically viruses, develop mutants to escape an immune response, they are generally defeated very efficiently by the immune system. In the case of an HIV infection, the above set of interactions is complemented by a characteristic negative feedback from the virus to the immune cells. HIV preferably replicates in T-helper cells (or $CD4^+$-cells), which leads to a continuous depletion of these essential actors within an immune response (Wei *et al.*, 1995; Vergis and Mellors, 2000; Douek *et al.*, 2002). The corresponding degradation of the immune system that leads eventually to its breakdown and the onset of AIDS seems to be an essential feature of HIV dynamics and consequently is considered in most theoretical models.

While a general review on models for HIV dynamics is beyond the scope of this article approaches will be outlined that reproduce the course of disease as an emergent feature of the underlying interactions between immune cells and HIV. Special focus will be devoted to describe the unusual distribution of incubation periods from the HIV infection to the onset of AIDS, i.e. in explanations for the separation of time scales of primary infection and of the period of latency.

2. From Differential Equations to Microscopic Simulations

Hershberg *et al.* (2001) introduce a model for the course of an HIV infection that is implemented as a hybrid model combining differential equation approaches with microscopic simulations to consider stochastic fluctuations when necessary. Their approach extends the perspective from the process of an infection with a single or a few virus strains to the overall set of interactions between multiple virus strains and the whole repertoire of immune cells.

They define a shape space to represent viral strains and their complementary immune receptors by introducing a random lattice with a fixed number of neighbors that are chosen randomly, which is sketched in Fig. 1. Each site i of this lattice is assigned occupation numbers N_{V_i} and N_{C_i} corresponding to the number of virus particles of the shape represented by the lattice site and the number of immune cells with a complementary immune receptor (though not distinguishing between different kinds of lymphocytes). Neighboring sites in the lattice are assumed to correspond to virus strains only differing in a single base mutation. The virus and immune

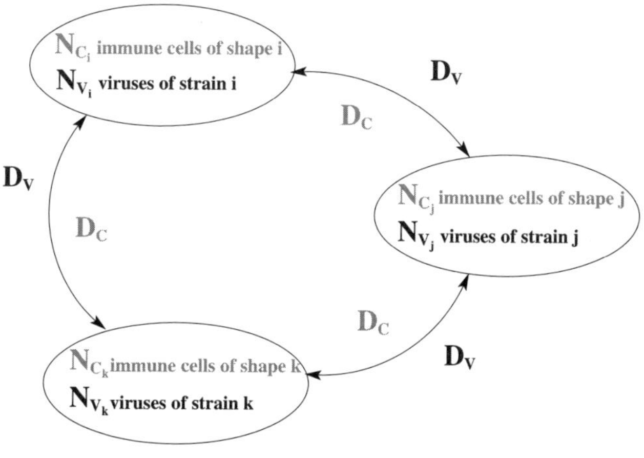

Fig. 1. Each site in the random lattice is characterized by an occupation number of immune cells N_{C_i} and virus particles N_{V_i} as well as their specific growth rates (τ_C, $\tau_V C_{tot}$) and decay rates ($d_C V_{tot}$, d_V). Variability of species is modeled by diffusion at rates D_V and D_C.

cell populations on this lattice are allowed to evolve in time, obeying growth and decay rates dependent on current prevalences of their complementary cells. Immune cells replicate at a rate τ_C if the system also contains their complementary virus epitope whereas each immune cell (irrespective of its receptor shape) decays at a rate of $d_C V_{tot}$, where $V_{tot} = \sum_i N_{V_i}$ is the size of the total viral population. The decay of immune cells reflects the depletion of $CD4^+$ cells by HIV, which affects the functionality of the immune system. In addition, new immune cells with randomly chosen receptor shapes are continuously introduced at a constant rate. Each virus strain in the system grows at a rate $\tau_V C_{tot}$, where $C_{tot} = \sum_i N_{C_i}$ is the total number of immune cells. This is to capture the fact that HIV replicates mainly in immune cells, and more specifically $CD4^+$ cells. A virus strain decays at a rate d_V if it faces a complementary immune receptor. Variation in immune cells and virus strains is modeled by allowing them to diffuse in shape space. The course of an HIV infection is modeled by introducing a restricted number of viral strains at a high concentration to the system and let the system evolve according to the above rules. Considering the HIV and immune cell levels, the system follows the three stages that are commonly observed in an HIV infection, which is shown in Fig. 2. After a short phase of acute infection with a high viral load, for a long time, the system stays in a state of latency before the onset of AIDS with high viral load and strong depletion of immune cells. While the acute phase is observed in all viral infections, the latent phase is characterized within the model by ongoing minor infections by mutant virus particles escaping an immune response and by receptor-unspecific but successive depletion of immune cells as the result of the permanent virus prevalence. The immune system eventually breaks down followed by a vast proliferation of virus particles at the onset of AIDS. Similar to Nowak *et al.* (1991), the authors emphasize the aggregate effect of the successive accumulation of viral strains weakening the immune system until it breaks down. Changing the efficiency of the immune system in these simulations gives rise to the limiting cases of long term latency for a strong immune response and almost immediate onset of AIDS for a weak immune response.

Considering the period of latency as a period of ongoing minor infections allows for an explanation of the hierarchy of the different dynamic time scales observed in the system, i.e. the time of virus replication measured in hours and the whole course of disease typically taking several years. While virus replication takes place on a microscopic time scale of hours, new strains are established by this process within weeks. The overall course

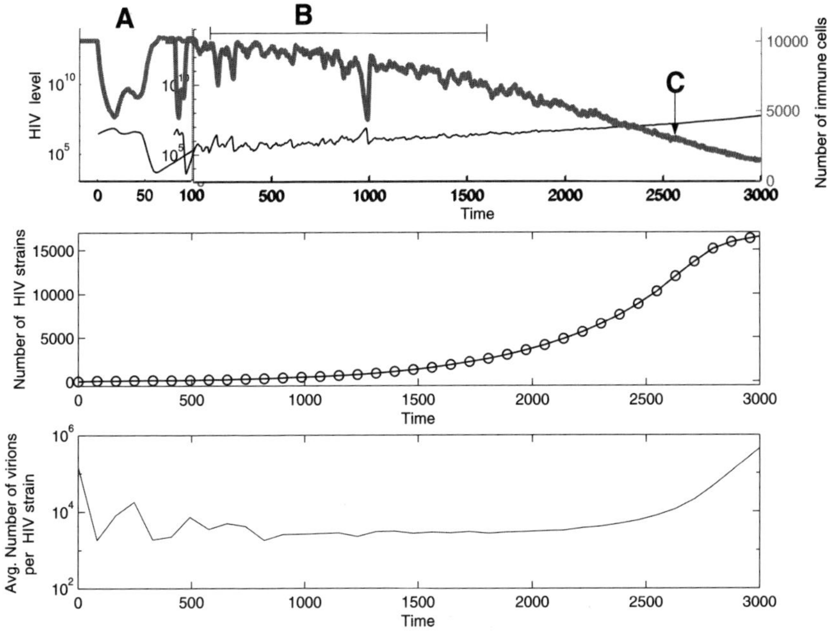

Fig. 2. Typical HIV evolution from the simulation model of Hershberg *et al.*: The top panel shows the viral load (thin line) and the number of immune cells (bold line) during the acute infection (A), the period of latency (B), and at the onset of AIDS (C). The middle and bottom panel show the corresponding number of different HIV strains that are present in the system and the average number of virus particles per strain. This figure has been reproduced from *Physica A*, **298**, 178–190 (2001).

of disease is not determined by the microscopic time scale of hours but by the rate at which new strains are established. Accordingly it takes a few hundred weeks — i.e. several years.

3. Dynamical Percolation and HIV Dynamics

Following the above ideas, an attempt to study a minimal model to describe the course of disease from HIV infection to the onset of AIDS has been made in (Kamp and Bornholdt, 2002). The shape space chosen there is not a random lattice but a sequence space of strings of length n, built up from an alphabet of length λ, modeling the gene sequences that code for viral epitopes and immune receptors. Each string represents a viral strain and its complementary immune receptor. Different from the approach in Section 2, not the exact number of immune cells with given receptor specificity and of virus particles per strain are monitored. Instead, two immune states are introduced corresponding to the presence or absence of a specific immune receptor (1/0). The status of a viral sequence or strain is parametrized by three states using a notion borrowed from so called SIR-models (Hethcote, 2000): A sequence may represent a virus strain that is present in the system ("infected site", I), a viral strain that in principle can be realized within the system but is currently not present ("susceptible site", S), or a site that is inaccessible for a virus because it represents a meaningless genome or is immunized by the presence of an immune sequence ("recovered site" or "removed site", R). The virgin system is assigned a finite fraction ρ_0 of sites that harbor an immune receptor and a finite fraction D_0 of sites that are in principle inaccessible for a virus which leads $R(0) = D_0 + \rho_0 - D_0\rho_0$ for the initial fraction of virally inaccessible sites. Then the system is locally infected by introducing a few viral strains. The predator-prey dynamics between these virus strains and the immune cells is represented by the following dynamic rules that are iterated within computer simulations:

i) Choose a random site.
ii) If this site represents an active immune receptor:
    a) Mutate any bit with probability $1 - q_{is}$.
    b) If mutation has occurred and a new immune strain has been generated and if it matches a site with a virus present, it is assumed to be amplified. The site becomes inaccessible for virus strains (recovered) and is occupied by an immune receptor.
iii) If this site represents a viral strain:
    a) Mutate any bit with probability $1 - q_v$.
    b) If mutation has occurred and a new virus strain has been generated, and if it corresponds to a susceptible site, a new virus strain is established there.

Figure 3 shows an example of some sequences with their immune and viral status as well as the probabilities for mutation connecting them. The dynamics of this system is characterized by the emergence of two regimes with qualitatively different behavior which can be understood against the background of dynamical percolation (Grassberger, 1983; Stauffer and Aharony, 1992) or models of epidemic spreading (Hethcote, 2000). Below a critical point, an infection does not lead to a vast spread of virus strains. The number of strains remains negligible in comparison to the system size.

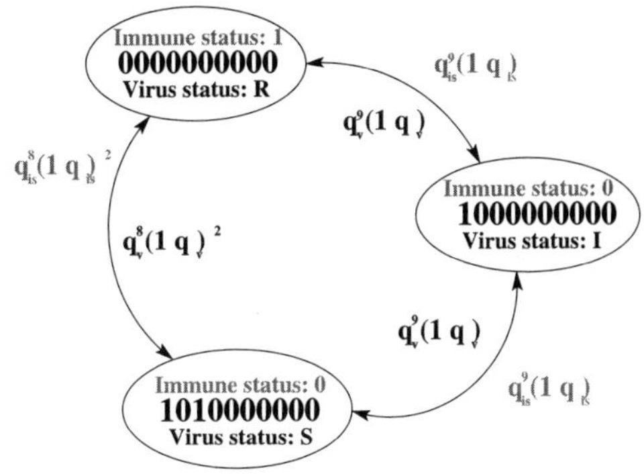

Fig. 3. Each (binary) string ($\lambda = 2$) is assigned an immune status (0/1) and a viral status ($S/I/R$), spread takes place with probability $(1-q_j)^d q_j^{n-d}/(\lambda-1)^d$, where n is the sequence length, d the number of differing positions of source and target sequence, and q_j ($j \in \{v, is\}$) the probability for a correct copy per position for the viral and immune sequence, respectively.

Above a critical point or percolation threshold, which is determined by[1]

$$\frac{1-q_v^n}{1-q_{is}^n} > \frac{\rho_0}{1-R(0)} = \frac{\text{initial immune receptor density}}{\text{initial fraction of virally accessible sites}}, \qquad (1)$$

viral strains seize a finite fraction of sites in sequence space in the case of infinite system size (Kamp and Bornholdt, 2002). If the immune receptor density is too small (i.e. the immune repertoire is not sufficiently diverse) in relation to the virally accessible sites (i.e. the potential number of viral strains) and to the involved mutation rates, a virus can start to spread mutant strains. As one does not generally observe vast spreading of viral genomes during the course of an infection it is reasonable to assume the immune system to operate below the percolation threshold. Nonetheless, it will probably operate near the percolation threshold as an unnecessarily high immune receptor density ρ_0 involves competitive disadvantages.

As a next step, this model is extended to include specific features of HIV dynamics, i.e. the destruction of the immune system by the virus via depletion of $CD4^+$ cells. Accordingly, the basic algorithm is supplemented by an additional rule: Each viral strain meets a random immune clone with a probability that is given by the current density of immunologically covered sites in the system. On encounter, the immune strain is destroyed with probability p and the site becomes virally susceptible if it is in principle

[1]This condition results from a mean field approximation which is a good approximation to the simulation results, cf. Kamp and Bornholdt (2002); Moreno *et al.* (2002).

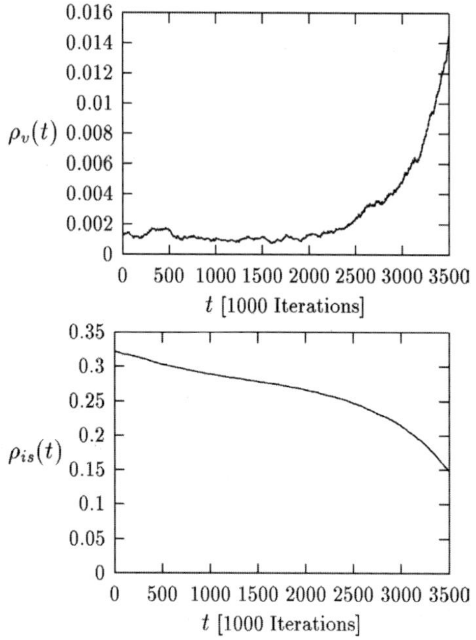

Fig. 4. Density of viral strains $\rho_v(t)$ and immunologically active sites $\rho_{is}(t)$ in sequence space ($D_0 = 0.5$, $\rho_0 = 0.325$, $q_v = q_{is} = 0.95$, $n = 15$, $\lambda = 2$, $p = 0.0001$, the parameter values correspond to subcritical but near critical system as determined by computer simulations ($\rho_0^c = 0.32$)). Note the analogy to the decline in $CD4^+$ cells under HIV infection. This figure has been reproduced from *Proc R Soc Lond B* **269**: 2035–2040 (2002).

accessible for virus strains. For intermediate values of p, a system near but below the percolation threshold shows a fluctuating number of viral strains that drift in sequence space whereas the number of immune strains slowly decreases, which is shown in Fig. 4.

This drift of viral epitopes due to immune pressure can be found in HIV-infected individuals (Wei *et al.*, 1995; Barouch *et al.*, 2002) accompanied with a successive decrease in $CD4^+$ cell counts (Wei *et al.*, 1995; Vergis and Mellors, 2000; Douek *et al.*, 2002), which can be assumed to go along with a loss in diversity of the immune repertoire. If the virus has not become extinct by chance, eventually the number of viral strains sharply increases while the immune system breaks down. These are features that are qualitatively observed in the course of an HIV infection, leading to a breakdown of the immune system and a large diversification of viral strains at the onset of AIDS (Nowak *et al.*, 1991). Within the computer simulation model the

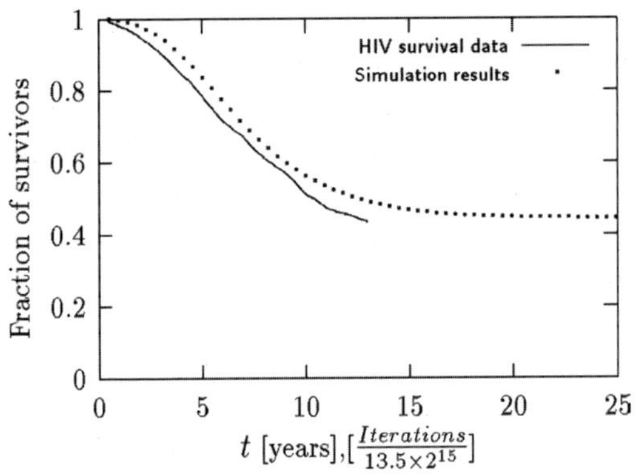

Fig. 5. Comparison of the probability for HIV positives of not yet having developed AIDS with a survival distribution generated by the computer simulations (after adequate renormalization of the time axis, $D_0 = 0.5$, $\rho_0 = 0.325$, $q_v = q_{is} = 0.95$, $n = 15$, $\lambda = 2$, $\rho_v(0) = 0.0012$, $\rho_v^{up} = 0.002$, the parameter values correspond to subcritical but near critical system as determined by computer simulations ($\rho_0^c = 0.32$)). This figure has been reproduced from *Proc R Soc Lond B* **269**: 2035–2040 (2002).

immune system is successively weakened while fighting the ongoing viral attacks and eventually breaks down when the virus begins to percolate sequence space after exceeding a critical viral strain density ρ_v^{up}.

Inspired by this analogy, the survival distribution generated by the computer simulations have been compared with empirical data on the distribution of incubation periods from HIV infection to the onset of AIDS (Robert Koch Institut, 1998) for patients that have not been treated with highly active antiretroviral therapy (Collaborative Group on AIDS Incubation and HIV Survival, 2000; The CASCADE Collaboration, 2000). Note, that the qualitative agreement between the data as shown in Fig. 5 does not depend on the specific set of parameters chosen in this simulation but on the fact the initial system is located in a sub- but near critical regime which is supposed to be the natural state of a healthy immune system within the model.

While the simulation results can already reproduce the empirical data, a deeper understanding can be gained by complementing them by an analytical approach based on stochastic processes.

4. A Stochastic Processes Approach

The computer simulation models introduced in Sections 2 and 3 as well as earlier differential equation models (Nowak *et al.*, 1991) paint a picture in which new strains are continuously generated leading a typical, though fluctuating, number of strains that are dominant at a time. This scenario can lead to two outcomes: typically the number of strains proliferates as soon as it passes a critical barrier; only sometimes the strains die out, which may be due to a strong immune response or due to statistical fluctuations. This fluctuating number of strains during the latent phase of infection is mapped to generalized geometric Brownian motion between two absorbing boundaries in (Kamp and Bornholdt, 2002). In this approach the fraction of realised strains in shape space ρ_v (which is proportional to the number of strains, cf. Fig. 4) is assumed to obey a growth process with a time dependent growth rate $r(t)$ that is superposed by Brownian noise $B_t(0, \sigma^2)$ with mean 0 and variance $\sigma^2 t$ (Øksendal, 1998):

$$\rho_v(t) = \rho_v(0) e^{R(t)+B_t(0,\sigma^2)}$$
$$R(t) = \int_0^t r(t')dt'.$$

In this framework, the onset of AIDS corresponds to the case in which the stochastic process ρ_v reaches the upper boundary (the critical fraction of viral strains in shape space) without having touched the lower (extinction) barrier before. The virus is defeated if absorption takes place first at the lower boundary which corresponds to a fraction of viral strains in shape space that does not allow for the existence of a single strain any more. The incubation period distribution from HIV infection to the onset of AIDS is mapped to the first passage time distribution with respect to the upper boundary $J(b,t)$ (Gardiner, 1983; van Kampen, 1997). This distribution is derived in Kamp and Bornholdt (2002), given a lower absorbing boundary at ρ_v^{low}, an upper absorbing boundary (critical fraction of viral strains) at ρ_v^{up} and an initial fraction of viral strains $\rho_v(0)$:

$$J(b,t) = \frac{F(a,b,\sigma^2 t)}{\sqrt{2\pi\sigma^2 t^3}} e^{-\frac{(b-R(t))^2}{2\sigma^2 t}} \tag{2}$$
$$R(t) = \int_0^t r(t')dt'$$
$$-a = \ln\left(\frac{\rho_v^{low}}{\rho_v(0)}\right) < 0$$
$$b = \ln\left(\frac{\rho_v^{up}}{\rho_v(0)}\right) > 0$$

$$F(a,b,\sigma^2 t) = \frac{e^{\frac{2b(a+b)}{\sigma^2 t}}\left(-a\left(1-e^{-\frac{2b(a+b)}{\sigma^2 t}}\right)+b\left(e^{\frac{2a(a+b)}{\sigma^2 t}}-1\right)\right)}{e^{\frac{2(a+b)^2}{\sigma^2 t}}-1}$$

$$\xrightarrow{a\to\infty} b.$$

In the case of only one absorbing boundary and a constant growth rate $r(t) = \mu$ (i.e. $R(t) = \mu t$), one gets the inverse Gaussian distribution as a well-known solution for this special problem (Feller, 1968; Redner, 2001; Aalen and Gjessing, 2001). The waiting time distribution (2) is fitted to the data of the incubation period distribution generated by the computer simulation model of Section 3 (cf. Fig. 5).[2] It shows that a constant growth rate $r(t) = \mu$ for the fraction of viral strains in shape space $\rho_v(t)$ only corresponds to very aggressive HIV strains which directly lead to a proliferation of strains corresponding to the onset of AIDS. In consequence the simplest time dependent function $r(t) = \mu + \gamma t$ for the growth rate is chosen for better fits of equation (2) to the computer simulation as well as empirical results (Robert Koch Institut, 1998) leading to $\gamma < 0 < \mu$ as shown in Fig. 6. The spread of HIV strains is initially determined by a positive growth rate which is continually diminished due to the influence of the immune system. Nonetheless, in many cases, HIV is able to percolate sequence space if its suppression happens too slowly. This occurs in a

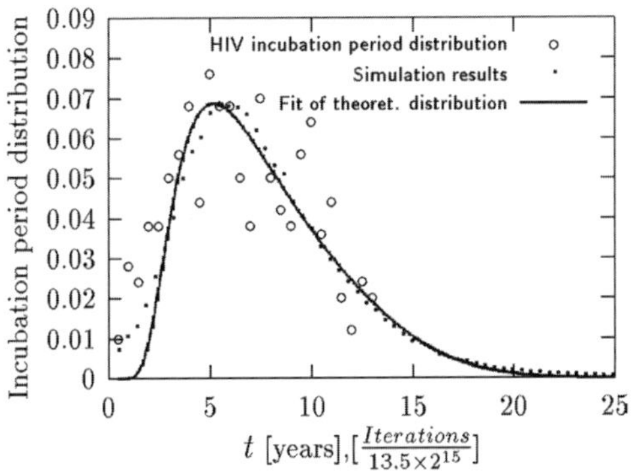

Fig. 6. Comparison of the incubation period distributions corresponding to Fig. 5 with the first passage time distribution with $r(t) = 0.064–0.0092t$, $\sigma^2 = 0.0091$. This figure has been reproduced from *Proc R Soc Lond B* **269**: 2035–2040 (2002).

[2]Parameters have been chosen in accordance with those chosen in the computer simulations: $\rho_v^{low} = 2^{-15}$ $(n = 15)$, $\rho_v(0) = 0.0012$, $\rho_v^{up} = 0.002$, cf. Kamp and Bornholdt (2002).

non-deterministic manner due to stochastic fluctuations leading to the observed incubation period distribution.

5. Conclusions

The above models attempt to describe HIV dynamics as emergent from the complex microdynamics of the virions and immune cells. However, they are different in their focus and techniques and accordingly yield insights into the involved processes from different perspectives. The microscopic simulations of Section 2 draw a detailed picture of the interactions between HIV and the immune system and reproduce very well the qualitative behavior of the course of the disease. In particular, the approach provides an explanation for the separation of time scales of primary infection and the incubation period. The ideas behind this approach are implemented into a minimal model in Section 3. While this approach allows on the one hand for a less detailed description, it on the other hand allows to understand the global behavior of the system in terms of a dynamical percolation transition. It is also capable of qualitatively reproducing the distribution of incubation periods from HIV infection to the onset of AIDS. The computer simulation models are eventually complemented by a stochastic processes model in Section 4. Fitting the dynamically motivated incubation period distribution function to the empirical data results in further insight into the underlying dynamics. The model suggests that the growth rate of the number of viral strains is successively diminished by the immune system — nonetheless, proliferation of viral strains eventually takes place in most cases because the immune system does not take effect sufficiently fast to clear the viral prevalence before it breaks down and the number of viral strains eventually passes a critical barrier. This is in line with findings from (Regoes *et al.*, 2002) showing in a SIV model that it is the continuous presence of the virus rather than temporal peaks in viral load that leads to disease.

However, there is further correspondence between the outlined models and real HIV statistics. Decreasing the aggressiveness of viral strains against the immune system (i.e. decreasing p in Section 3) leads to an increase of waiting times before percolation can be observed as well as an enlarged fraction of cases where viral strains go extinct. This corresponds to a setting in which HIV is obstructed to replicate in $CD4^+$ cells protecting them against depletion. Most HIV strains do not only need the $CD4$ receptor for membrane fusion but also an additional co-receptor named $CCR5$. Individuals homozygous for a deletion allele leading to a non-expression of the $CCR5$ receptor have been shown to be resistant against HIV infection (Lu *et al.*, 1997). In case of a heterozygous individuals and reduced expression of $CCR5$ a slower progression of the disease can be observed together with a higher probability of non-progression (Marmor *et al.*, 2001). The more resistant the $CD4^+$ cells, the longer is the patients' life expectancy and the larger is the probability to survive. In the case of total resistance, patients do not develop AIDS. This behavior is represented within the model of Section 3 by variation of the parameter p, which represents the aggressiveness of the virus. A decrease in p also leads to a stabilization of the system for vanishing p resulting in the dynamics of a "common" infection without a breakdown of the immune system as observed at the onset of AIDS. The concept of fusion inhibition by receptor blocking is also gaining attention in clinical drug trails (Pilcher, 2002; Meanwell and Kadow, 2003) and has recently been approved for the treatment of advanced HIV-1 infection (U.S. Food and Drug Administration, 2003).

Moreover, within the theoretical model of Section 3, vaccination procedures can be interpreted by a local increase in the immune receptor density ρ_0 leading to a stabilization of the system below the percolation threshold. Accordingly it is more difficult for HIV to reach the percolation threshold and to proliferate strains corresponding to an increased probability of prolonged or total survival. Although, the search for an HIV vaccine has been afflicted with as many drawbacks as breakthroughs (Amara *et al.*, 2001; Barouch *et al.*, 2002; Lifson and Martin, 2002; Shiver *et al.*, 2002; Cohen, 2003) the above models nonetheless support this approach and a better understanding how to deal with escape mutants (Nowak and McLean, 1991; Walker and Goulder, 2000; Allen *et al.*, 2000; Goulder *et al.*, 2001; Rambaut *et al.*, 2004; Grenfell *et al.*, 2004) might bring some progress in this area.

More generally, the above models try to derive global dynamics from the local interactions between viruses and immune cells at different levels of abstraction. They are accordingly not capable of explaining HIV dynamics in all detail. But they can expose trends and constraints on the dynamics for further investigation.

Acknowledgements

I would like to thank A. Bunten and M.P.H. Stumpf for a thorough review and many helpful comments to improve this chapter. I also acknowledge U. Hershberg, Y. Louzoun, H. Atlan and S. Solomon for providing Fig. 2 from their publication. This work was supported by a fellowship within the Postdoc-Programme of the German Academic Exchange Service (DAAD).

References

Aalen OO and Gjessing HK (2001) Understanding the shape of the hazard rate: a process point of view, *Stat Sci* **16**: 1–22.

Allen TM *et al.* (2000) Tat-specific cytotoxic T lymphocytes select for SIV escape variants during resolution of primary viraemia, *Nature* **407**: 386–390.

Amara RR *et al.* (2001) Control of mucosal challenge and prevention of AIDS by a multiprotein DNA/MVA vaccine, *Science* **292**: 69–74.

Barouch DH *et al.* (2002) Eventual AIDS vaccine failure in a rhesus monkey by viral escape from cytotoxic T lymphocytes, *Nature* **415**: 335–339.

Brede M and Behn U (2001) Architecture of idiotypic networks: percolation and scaling behavior, *Phys Rev E* **64**: 011908–1–011908–11.

Cohen J (2003) AIDS Vaccine still alive as booster after second failure in Thailand, *Science* **302**: 1309–1310.

Collaborative Group on AIDS Incubation and HIV Survival including the CASCADE EU Concerted Action, (2000) Time from HIV-1 seroconversion to AIDS and death before widespread use of highly-active antiretroviral therapy: a collaborative re-analysis, *Lancet* **355**: 1131–1137.

Dangerfield B and Roberts C (1999) Optimisation as a statistical estimation tool: An example in estimating the AIDS treatment-free incubation period distribution, *Sys Dyn Rev* **15**: 273–291.

Douek DC *et al.* (2002) HIV preferentially infects HIV-specific $CD4^+$ T cells, *Nature* **417**: 95–98.

Feller W (1968) *An Introduction to Probability Theory and Its Applications*, Vol. I, John Wiley and Sons, Inc.

Gardiner CW (1983) *Handbook of Stochastic Methods.* Springer–Verlag.

Goulder PJR *et al.* (2001) Evolution and transmission of stable CTL escape mutations in HIV infection, *Nature* **412**: 334–338.

Grassberger P (1983) On the critical behaviour of the general epidemic process and dynamical percolation, *Math Biosci* **63**: 157–172.

Grenfell BT, Pybus OG, Gog JR, Wood JLN, Daly JM, Mumford JA and Holmes EC (2004) Unifying the epidemiological and evolutionary dynamics of pathogens, *Science* **303**: 327–332.

Hershberg U, Louzoun Y, Atlan H and Solomon S (2001) HIV time hierarchy: winning the war while loosing all the battles, *Physica A* **289**: 178–190.

Hethcote HW (2000) The mathematics of infectious diseases, *SIAM Rev* **42**: 599–653.

Janeway CA, Travers P, Walport M and Shlomchik M (2001) *Immunobiology*, Garland Science.

Kamp C (2003) A quasispecies approach to viral evolution in the context of an adaptive immune system, *Microbes Infect* **5**: 1397–1405.

Kamp C and Bornholdt S (2002) From HIV infection to AIDS: a dynamically induced percolation transiton? *Proc R Soc Lond B* **269**: 2035–2040.

van Kampen NG (1997) *Stochastic Processes in Physics and Chemistry*, Elsevier Science B.V.

Lifson JD and Martin MA (2002) One step forward, one step back, *Nature* **415**: 272–273.

Lu Z-H *et al.* (1997) Evolution of HIV-1 coreceptor usage through interactions with distinct CCR5 and CXCR4 domains, *Proc Natl Acad Sci USA* **94**: 6426–6431.

Marmor M, Sheppard HW, Donnell D, Bozeman S, Celum C, Buchbinder S, Koblin B and Seage GR (2001) Homozygous and heterozygous CCR5-Delta32 genotypes are associated with resistance to HIV infection, *J Acquir Immune Defic Syndr* **27**: 472–481.

Meanwell NA and Kadow JF (2003) Inhibitors of the entry of HIV into the host cells. *Curr Opin Drug Discov Devel* **6**: 451–461.

Moreno Y, Pastor–Satorras R and Vespignani A (2002) Epidemic outbreaks in complex heterogeneous networks, *Eur Phys J B* **26**: 521–529.

Nowak MA and May RM (2000) *Virus Dynamics, Mathematical Principles of Immunology and Virology*, Oxford University Press.

Nowak MA and McLean AR (1991) A mathematical model of vaccination against HIV to prevent the development of AIDS, *Proc R Soc Lond B* **246**: 141–146.

Nowak MA, Anderson RM, McLean AR, Wolfs TFW, Goudsmit J and May R (1991) Antigenic diversity thresholds and the development of AIDS, *Science* **254**: 963–969.

Øksendal B (1998) *Stochastic Differential Equations*, Springer–Verlag.

Perelson AS and Weisbuch G (1997) Immunology for physicists, *Rev Mod Phys* **69**: 1219–1267.

Philippe P (1994) Sartwell's incubation period model revisited in the light of dynamic modeling, *J Clin Epidemiol* **47**: 419–433.

Philippe P (2000) Epidemiology and self-organized critical systems: an analysis in waiting times and disease heterogeneity, *Nonlinear Dyn Psy Life Sci* **4**: 275–295.

Pilcher CD (2002) *T-20 and Beyond: Inhibition of HIV Attachment and Fusion at the 9th CROI*, http://www.natap.org/2002/9retro/day28.htm.

Rambaut A, Posada D, Crandall KA and Holmes EC (2004) The causes and consequences of HIV evolution, *Nature Genet* **5**: 52–61.

Redner S (2001) *A Guide to First-passage Processes*, Cambridge University Press.

Regoes RR, Staprans SI, Feinberg MB and Bonhoeffer S (2002) Contribution of peaks of virus load to Simian Immunodeficiency Virus proliferation, *J Virol* **76**: 2573–2578.

Robert Koch Institut (1998) http://www.rki.de/INFEKT/AIDS_STD/SERO/KONVERT.HTM. Langzeitbeobachtung von Probanden mit bekanntem Zeitpunkt der HIV-Serokonversion, *Epidemiol Bull* **32**: 228–230.

Roitt I (1994) *Ess Immunol*, Blackwell Scientific Publications.

Sartwell PE (1995) The distribution of incubation periods of infectious diseases, *Am J Hyg* 1950; **51**: 310–318, *Am J Epidemiol* **141**: 386–394.

Sartwell PE (1966) The incubation period and the dynamics of infectious disease, *Am J Epidemiol* **83**: 204–216.

Shiver JW *et al.* (2002) Replication-incompetent adenoviral vaccine vector elicits effective anti-immunodeficiency-virus immunity, *Nature* **415**: 331–335.

Stauffer D and Aharony A (1992) *Introduction to Percolation Theory*, Taylor and Francis.

The CASCADE Collaboration (2000) Survival after introduction of HAART in people with known duration of HIV-1 infection, *The Lancet* **355**: 1158–1159.

U.S. Food and Drug Administration (2003) FDA approves first drug in new class of HIV treatments for HIV infected adults and children with advanced disease, http://www.fda.gov/bbs/topics/NEWS/2003/NEW00879.html.

Valera FJ and Coutinho A (1991) Second generation immune networks, *Immunol Today* **12**: 159–166.

Vergis EN and Mellors JW (2000) Natural history of HIV-1 infection, *Infect Dis Clin North Am* **14**: 809–825.

Walker BD and Goulder PJR (2000) Escape from the immune system, *Nature* **407**: 313–314.

Wei X *et al.* (1995) Viral dynamics in human immunodeficiency virus type 1 infection, *Nature* **373**: 117–122.

Zorzenon dos Santos RM and Bernardes AT (1998) Immunization and aging: a learning process in the immune network, *Phys Rev Lett* **81**: 3034–3037.

Zorzenon dos Santos RM and Coutinho S (2001) Dynamics of HIV infection: a cellular automata approach, *Phys Rev Lett* **87**: 168102–1–168102–4.

CHAPTER 9

HOW FAST CAN HIV ESCAPE FROM IMMUNE CONTROL?

W. David Wick and Steven G. Self

HIV can mutate and thereby escape immune recognition — an event that may precipitate AIDS. In a theoretical discussion of escape, it is essential to treat the infected-cell kinetics, and not just the mutations, as a random process. Deterministic modeling of infection dynamics exaggerates the rate-of-escape and obscures scenarios that decrease it. Using stochastic models, we show how extra-Poisson variability in virus production, possibly related to the viral "effective population size" *in vivo*, can lower the escape rate by orders-of-magnitude. If such variation is not operating, our simulations indicate that escape is very fast even if a CTL vaccine lowers viral load two logs. We combine loss-of-fitness and killing-rate data into a formula quantifying the influence of a broad response. Finally, we describe experiments that could reveal mechanisms that delay escape and discuss the relevance of our analysis to CTL vaccine design.

Keywords: HIV evolution *in vivo*; cytotoxic T lymphocytes; escape by mutation; stochastic modeling; effective population number; CTL vaccines.

1. Introduction

Control by the immune-system, whether by antibodies or by cytotoxic T-lymphocytes (CTLs), always suggests the possibility of escape — especially in the case of retroviruses. The reason is the huge virion-production during chronic infection and the characteristically-high mutation rate. For HIV, Wei *et al.* (1995) and Ho *et al.* (1995), investigators in the 1995 trials of triple-combination drug therapy, estimated daily production at tens of billions of virions, produced by 10^7 to 10^8 productively-infected target cells (PITs; mostly CD4+ T cells). Mutations occur at the rate of one nucleotide in 100,000 per replication cycle (Mansky and Temin, 1995) — five orders-of-magnitude higher than for mammalian genomes. The cause is sloppy transcription of viral RNA into DNA by the reverse-transcriptase enzyme encoded by all retroviruses. From these figures, every single-site mutation is likely made each day in infected patients (Coffin, 1995).

Mutant HIV strains that have escaped recognition by particular clones of CTLs have been observed as early after infection as 156 days (Borrow *et al.*, 1997), and as late as ten years (Goulder *et al.*, 1997). In 2000–2001, Amara *et al.* (2001) and Barouch *et al.* (2000) carried out successful trials of vaccines aimed at raising CTLs against SHIV infection in macaques. ("SHIVs" are highly pathogenic chimeras stitched together from SIV and HIV genes. The fast time-course in monkeys — from infection to clinical disease in a year — makes it a practical vaccine test-bed.) The vaccines did not protect the macaques from infection, but the viral count in the vaccinated animals fell to the almost-undetectable level we associate with successful drug treatment. Then, after about a year, one of the vaccinated monkeys developed AIDS and died (Barouch *et al.*, 2002). The monkey had initially mounted a dominant response to a structural protein called GAG that the investigators hoped was conserved; but the virus was able to mutate one nucleotide in the GAG gene and escape CTL recognition.

These observations appear to raise a paradox. If every possible mutation arises every day, and viable escape mutants exist, why should escape from immune recognition take more than one month — the time necessary for a new variant to grow out, assuming growth kinetics similar to the wild strain? If AIDS is the endpoint of an in-host evolutionary process (a by no means universally accepted theory), how can patients remain AIDS-free on average for 10 years?

The key to resolving these paradoxes lies in stochastic modeling. Many authors have modeled viral evolution *in vivo*, separately or together with CTL kinetics, by deterministic rate equations (ordinary differential equations, ODEs) aside perhaps from occurrence of mutations; e.g., Nowak *et al.* (1990), Nowak (1992), Nowak *et al.* (1995), Wodarz and Nowak (2000), Monteiro, Goncalves and Piqueira (2000), Antia *et al.* (2003). Our point is that at least the infection process must be allowed a stochastic evolution. Otherwise, we risk overlooking mechanisms that can greatly extend the time to escape and possibly misunderstanding the cause of events. For example, Nowak *et al.* (1995) and Wodarz and Nowak (2000) constructed high-dimensional ODE models that generate complicated "antigenic oscillations"; but we will argue that the data presented can be understood more simply by means of basic reproductive numbers in a stochastic model (Section 4) without invoking chaotic dynamics. Earlier, Ribeiro and

Bonhoeffer (1999) and we (2000) treated primary HIV infection as a stochastic process, but without detailed immune-system modeling.

In the next sections, we explain why stochastic modeling is essential to understanding escape. Several concepts from classical population biology will prove essential, although we rely for the most part on branching-process arguments and detailed modeling. In subsequent sections we use our models and simulations to predict, under various scenarios, the time to escape from CTL recognition. Yang *et al.* (2003), Jamieson *et al.* (2003), Altman and Feinberg (2004), Leslie *et al.* (2004), and Friedrich *et al.* (2004) have observed that escape is driven by a complicated mix of factors including loss of fitness and epitope recognition. We provide some quantitative insights into this complexity in Section 6. In the last section we discuss experiments relevant to the mechanisms delaying escape and the implications for vaccine design.

2. The Necessity of a Stochastic Infection Model, Part I

We begin with some simple observations about branching-processes that indicate why deterministic modeling would greatly over-estimate escape rates. The wild-type viral strain has a basic reproductive number, R_0(w.t.), and a mutant strain, R_0(mut.); they can be defined in terms of the exponent in a growth-rate formula or, in the traditional language of epidemics, as the average number of infected cells in a second generation given exactly one in the first and no "intervention" — in this case, by the immune system. (The formulas differ by a natural logarithm; see appendix A for details.) These numbers also quantify viral reproductive fitness *in vivo*.

With either definition, a strain can grow out if and only if $R_0 > 1$. In the simplest stochastic (branching-process) model of the PIT population, R_0 takes the form (see appendix A):

$$R_0(\text{w.t.}) = \frac{A}{\delta_{\text{PIT}}}, \quad A = \text{constant}, \tag{1}$$

where A combines various infectivity parameters and δ_{PIT} is natural PIT death rate (inverse lifetime, neglecting immune killing). In the presence of an immune response, we can add a term: $\kappa C, C$ = CTL density and κ a constant, to δ_{PIT} to represent active killing by CTLs. The resulting formula can be rewritten

$$R_e(\text{w.t.}) = \frac{R_0(\text{w.t.})}{1 + aC}, \quad a = \text{constant}, \tag{2}$$

where R_e is now an effective reproductive number. Steady-state conditions before any mutants appear implies

$$R_e(\text{w.t}) = 1\,. \tag{3}$$

Now let the immune system recognize E viral epitopes and consider a mutant strain that has nullified recognition of one; call it a 1-mutant. This means E distinct CTL populations kill PITs that express wild-type epitopes; i.e., $C = C_1 + \cdots + C_E$; but the mutant is only killed by, e.g. $C_2 + \cdots + C_E$. Suppose for simplicity that all CTLs have identical kinetics and killing potential. (We relax this assumption in a later section.) Then the effective reproductive number for the mutant becomes

$$R_e(\text{mut.}) = \frac{R_0(\text{w.t.})}{1 + a(1 - 1/E)C}\,. \tag{4}$$

Using Eqs. (2), (3), and (4), we obtain the following expression for fitness when escaping from one epitope when E are recognized:

$$R_e(\text{mut.}) = \frac{R_0(\text{mut.})}{(1 - 1/E)R_0(\text{w.t.}) + 1/E}\,. \tag{5}$$

This basic, parameter-independent formula is enlightening. For example, suppose that $R_0(\text{mut.}) \approx R_0(\text{w.t.}) = 3.0$ (reasonable for HIV on the basis of its growth rate *in vivo*), and $E = 10$. Then

$$R_e(\text{mut.}) = \frac{1}{1 - \frac{2}{3E}} \approx 1 + \frac{1}{15} \approx 1.067\,. \tag{6}$$

Despite an R_0 identical to the wild-type's, at birth the mutant's effective reproductive number is much closer to one. It lacks fitness because it is still recognized by 90% of the CTLs, which are moreover being continuously stimulated by the wild-type strain.

The cellular immune system can react only after viral densities have changed by many logs; in the mean time, each strain can be modeled separately as a branching process. But branching processes that can grow (called supercritical in the stochastic-process literature) can also go extinct. The formula for the extinction probability in the simplest case is (see Appendix A):

$$P[\text{extinct}] = \frac{1}{R_e}\,. \tag{7}$$

For the survival probability of a mutant strain in our example, we find $P[\text{survive}] = 1 - P[\text{extinct}] \approx 2/3E \approx .07$, i.e. 93% of new mutant strains die out. In an ODE model, provided zero-density is unstable, the mutant strain grows exponentially as soon as it forms. In this example, a deterministic simulation, even implementing the mutations *via* a random number generator, would make the escape rate appear ten times higher than in a stochastic simulation.

Considering the production and mutation figures mentioned earlier, a factor of ten does not resolve the basic timing conundrum. While more epitopes may be recognized — hundreds have been listed in data bases, Korber *et al.* (2001) — considerations from population biology suggest another resolution.

3. The Necessity of a Stochastic Infection Model, Part II

We evaluated the extinction probability in the last section in the simplest branching-process model. One assumption underlying the computation was that virion-production follows a Poisson law. However, biologists have reasons to believe that retroviral production in T cells will exhibit extra-Poisson variation. In test-tube experiments with HIV growing in T-cell lines, 99.9% of the virions produced are non-infectious (Layne *et al.* 1991; Emerman, personal communication). PIT production ranges from 50 to more than 1000 per day, Dimitrov *et al.* (1993), Levy *et al.* (1996). Reproduction in T-cells requires proper integration of the viral genome into the host's, activation of host genes, and other intracellular conditions; thus production may be essentially bimodal, with some PITs making thousands of infectious virions and others virtually none.

Adding variation in a branching process typically increases the chance of extinction. We posited multiple kinds of ITs (infected target cells in the eclipse phase, see Appendix A) and PITs. Let the kth IT-type be formed with probability p_k and the kth PIT-type produce v_k virions per unit time. We adopted values for p_k and v_k keeping R_0 and mean production fixed while yielding a variety of variance-to-mean ratios (VMRs). A Poisson law has a VMR of one. Table 1 lists some survival probabilities for various cases:

Table 1. Survival probabilities.

VMR	P[survive]
32	0.0082
296	0.0029
2690	0.0013
4021	0.0008

Evidently, the effect of incorporating the last-listed survival probability in simulations would be dramatic.

A second line of evidence comes from phylogenetic analysis of the viral quasispecies. In the 1920s, Wright and Fisher introduced the concept of "effective population size" to distinguish different populations by their degree of genetic variation. Recall that the number of PITs in untreated patients during the chronic phase is around 10^8. Call this the "demographic population size", N_d. The effective population size is denoted N_e. An apparent paucity of neutral drift in certain envelope sequences lead Leigh–Brown (1997) to propose that HIV *in vivo* has an N_e as low as 1000 or even 100. Others dispute this finding (Rouzine and Coffin, 1999). Many scenarios can yield a low effective population size, including evolutionary bottlenecks, population fluctuations, and selective sweeps; but non-Poisson production remains a candidate.

In the 1930s, Crow noted that population structure had a large affect on extinction rates and introduced an "extinction population size", which in his examples came out the same as Wright's definition. Geneticists have several definitions of N_e; we adopt here the "inbreeding population size", the reciprocal of the probability that two randomly-selected PITs arose from the same ancestor. Let the probability of a PIT generating n new PITs in its lifetime be p_n, and define the offspring mean, variance, and VMR as usual. In Appendix B we derive the following formula for effective population size, valid for large, fixed N_d (i.e. assuming a steady-state):

$$N_e = \frac{N_d}{\text{VMR}} \,. \tag{8}$$

With the population sizes mentioned before, and attributing population structure entirely to the extra-Poisson hypothesis, we find the extraordinary result: $VMR \approx 10^5$ for HIV *in vivo*. Of course, extra-Poisson variation likely contributes only part of the explanation for a low N_e, if real.

In Appendix C, we derive a formula for the survival probability for the simplest branching process (i.e. ignoring the eclipse phase) with general production law:

$$P[\text{survive}] \approx \frac{2(1 - 1/M)}{\text{VMR}} \,, \tag{9}$$

valid when VMR is large, M (mean offspring number) is near one and a certain condition on a third moment. Adopting the above value for VMR and $M = 1.05$ (recalling the scenario in Section 2), we conclude that deterministic modeling could overestimate the escape rate by a million-fold (but we do not seriously propose that it does).

4. Temporal Considerations

How quickly, and by what mechanism, does the escape mutant replace the existing strain? From one new mutant virus, yielding one new PIT, at least 10^7 must be formed to grow past the wild-type. Using the branching-process definition for R_e (note: all formulas for R agree near one), we put

$$(R_e)^{t/g} = 10^7\,, \tag{10}$$

where g is the generation time (about two days for HIV). In our example from Section 2 with $R_e = 1 + 1/15$,

$$t \approx 15 \times 14 \times \log(10) \approx 483 \text{ days}\,. \tag{11}$$

Here we gain significantly on the time-to-escape problem.

The computation assumes that the mutant's fitness remains low. While the wild-type persists, this is true, but the new strain destabilizes the old one. Initially, $R_e(\text{w.t.}) = 1$, because the wild-type's growth is cancelled by the E sets of CTLs killing it. Once a successful 1-mutant begins to grow out, it adds antigen recognized by $E - 1$ of the CTL families. Proliferation increases in these CTLs and their populations grow. Since the $E - 1$ families still recognize the wild-type, its R_e now drops below one. The wild-type disappears, at some difficult-to-estimate rate. Afterwards, the mutant will be fitter than before and its steady-state level will drift higher than the wild-type's. In the SIV vaccine experiments, after escape the viral load increased a log (Amara *et al.*, 2001; Barouch *et al.*, 2000). Note that neither interspecies competition nor "CTL exhaustion" is required; the replacement is purely a consequence of immune-control. In our view, this mechanism provides a parsimonious explanation for the data presented in Nowak *et al.* (1995).

The process continues until an antigenically-innocent strain appears (which cannot destabilize its predecessors by this mechanism), at which point immune control is lost and (perhaps) AIDS results. Unfortunately, for higher mutants we cannot simply iterate the calculation. For an initially-pure strain will soon evolve into a complex quasispecies: a mixture of 1-mutants, 2-mutants, and so forth, all adding antigen and altering each-other's fitness. Only simulations from detailed models can answer the question posed in the title.

5. Simulations

We used the cellular-immune-system plus viral-infection kinetic model we introduced in an earlier paper (Wick and Self, 2002b), amplified to include mutations. Briefly (for more details see that citation and Appendix D), the immune-half of the model has 18 compartments for each epitope recognized: one for antigen-naïve, resting cells, and eight for naïve, activated cells; with analogous compartments for memory (antigen-experienced) cells. Resting cells become activated with rate proportional to the number of PITs expressing that epitope. We did not model free virions because their lifetime in the body may be as short as 30 minutes (Perelson *et al.*, 1996; Igarashi *et al.*, 1999; Zhang *et al.*, 1999). Activated cells undergo eight divisions (an assumption called "programmed-proliferation", abbreviated PP) before being deleted by apoptosis (programmed cell death) or converted/reverted to resting status. Kaech and Ahmed (2001) proved that the number of divisions is the minimum by adoptive-transfer of virus-specific CTLs into antigen-free, transgenic mice. Antia *et al.* (2003) previously modeled programmed-proliferation of T-cells. Memory cells promote to CTLs four times faster than naïve cells.

Many biologists hold that HIV-specific CTLs are defective in some way (see references in our 2002 paper). To incorporate such a "CTL defect" we assumed that "resting" memory cells did not truly rest (i.e. persist for years without re-exposure to antigen), but rather die in three days if not re-activated. This scenario is supported by clinical observations such as the disappearance of HIV-specific CTLs from patients treated with drugs (Dalod *et al.*, 1998; Gray *et al.*, 1999; Kalams *et al.*, 1999; Ogg *et al.*, 1999) and adoptive-transfer experiments with engineered CTLs (Brodie *et al.*, 1999). In our earlier work (Wick and Self, 2002), we labeled it "defective memory" and contrasted it with other proposals, e.g. defective activation or defective killing. Subsequently, several groups published data which fall on the defective-memory side (Wherry *et al.*, 2003; Addo *et al.*, 2003). Lack of CD4 help for CD8 maturation is a plausible explanation, since HIV-specific CD4s are probably early viral casualties (Altes, Wodarz and Janssen, 2002; Wordarz, 2001; Sun and Bevan, 2003; Bourgeois, Rocha and Tanchot, 2003; Shedlock and Shen, 2003; Day and Walker, 2003; Janssen *et al.*, 2003; Kaech and Ahmed, 2003). Vaccination was assumed to cure the defect.

For infected-cell dynamics we used a simple two-compartment model for each genome instead of the three-compartment model of our (2002). The first compartment represents the "eclipse phase" before virions appear and the second the productive phase. Mutations were introduced at the eclipse phase, at rate μx the number of genomes with one more normal epitope. We did not explicitly model the uninfected target-cell population because we support immune-control over "target-cell scarcity" as the correct theory

Table 2. Parameters in the CTL/Infection model.

Parameter	Meaning	Value (for rates, per day)
α	activation	10^{-10}
β	NR "immigration"	see text
δ_{NR}	NR death rate	0.00017
δ_{MR}	MR death rate	0.00017
δ_{CTL}	CTL death rate	0.333
Rev	reversion/conversion fraction	0.05
n_d	no. of doublings	8
γ_N	naïve cell-cycles/day	2.0
γ_M	memory cell-cycles/day	4.0
η	IT progression/death	1.0
ι	infection	see text
κ	killing by CTLs	10^{-10}
R_0	basic reproductive no.	3.0
φ	memory speed-up factor	7.0
δ_{IT}	IT death rate	0.333
μ	mutation rate per site	10^{-5}

of events leading to the viral set-point (Wick, Self and Corey, 2002). In addition, the virus's primary targets are activated CD4+ T cells; due to immune hyperactivation, these may not fluctuate much even as CD4 counts drop (Wick, 1999a). Table 2 records our parameter choices. We started the process with 10^5 naïve resting cells (see Blattman *et al.*, 2001), no memory cells for unvaccinated subjects and 10^8 resting memory cells for vaccinated.

Because the immune compartments start with many cells (and from the desire to keep simulation run-times reasonable), we used ODEs for that half of the kinetics. For the infected half we used our "switch-over" technique (Wick and Self 2004a; 2004b). Briefly, the algorithm begins by simulating a jump process; when all compartments of a connected component (here, a single genome) grow large enough, it switches to a continuous-state process. The latter, derived from a technique for simulating stochastic differential equations, requires only drawing some Gaussian random variables and performing some linear algebra at each time-step. After switching, the computational labor no longer scales with population size. The technique assures third-order error control at each time-step and performs well even on extinction times.

Figure 1, similar to one in our previous paper (Wick and Self, 2002), shows the immune- and infected-cell dynamics for unvaccinated and vaccinated, neglecting mutation. Note the two-log drop in viral set-point, in agreement with the SIV-vaccine trials.

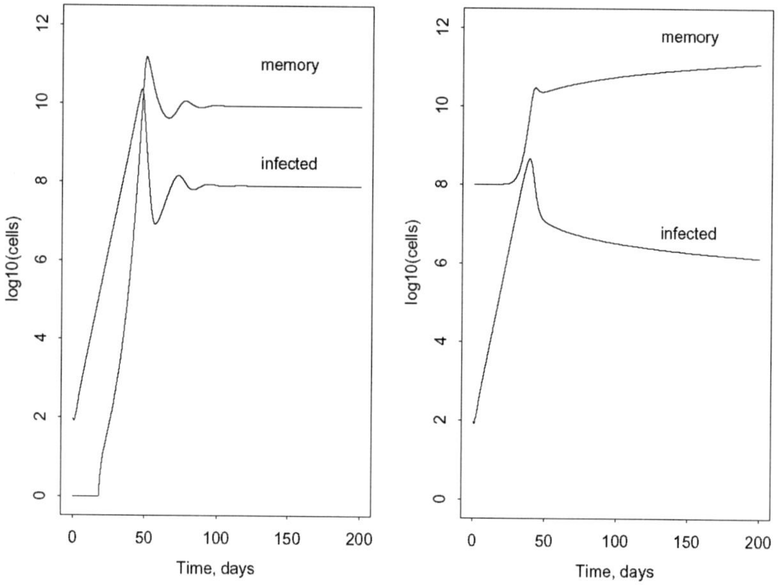

Fig. 1. Steady-states in the programmed-proliferation CTL model plus two-compartment infection model. Left: unvaccinated patient; Right: vaccinated.

Suppose we define "time of escape" to be when the escape mutant (with all epitope-recognition deleted) passes the wild-type.

Figure 2 shows escape from one epitope in 60 days in natural infection. After escape, the mutant grows exponentially (we did not incorporate target-cell limitation in the model).

Figure 3 shows the two-epitope case; the virus escapes in 80 days. Note that, after a 1-mutant grows out, the wild-type is destabilized, as theory predicts.

Figure 4 shows the vaccinated subject, assuming "defective-memory" is fixed; surprisingly, the two-log drop in viral load makes little difference in the timing of escape.

6. Other Biological Scenarios that Could Delay Escape

If the simulations of the previous section appear unrealistic, it is because the virus escapes too fast. For example, the SHIV in the vaccine experiment required 140 days, although SIV escape has been observed in macaques in 4 weeks (O'Conner *et al.*, 2001). Besides scenarios connected to population structure, others exist that could slow the process down.

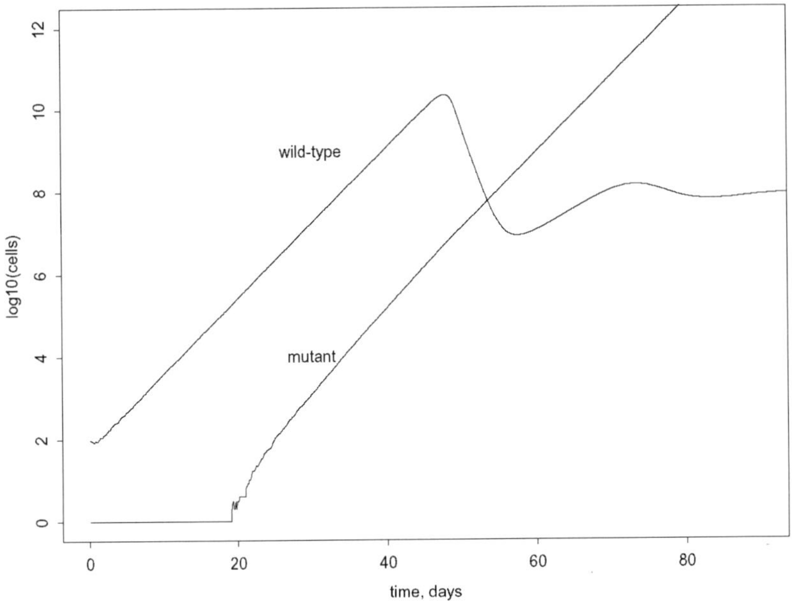

Fig. 2. Escape from recognition of one epitope.

One scenario comes about if the mutant is less fit than the wild-type. It need not be much. From Eq. (5) of Section 2 and $R_0(\text{w.t.}) = 3$,

$$R_e(\text{mut.}) \approx \frac{R_0(\text{mut.})}{R_0(\text{w.t.})} + \frac{2}{3E}, \tag{12}$$

which could be close to or even less than one. Indeed, with $E = 10$ the mutant need only be 8% less fit than the wild-type for this mechanism to prevent it from growing.

More generally, there will be heterogeneity in one-mutant fitnesses and CTL avidities (Yang, Sarkis, Trocha *et al.*, 2003). Equation (5) becomes

$$\begin{aligned} R_e(\text{mut.;j}) &= \frac{R_0(\text{mut.;j})}{(1-\theta_j)R_0(\text{w.t.}) + \theta_j} \\ \theta_j &= \frac{\kappa_j C_j}{\sum_{i=1}^{E} \kappa_i C_i}. \end{aligned} \tag{13}$$

Hence, if

$$R_0(\text{mut.;j}) < (1-\theta_j)R_0(\text{w.t.}) + \theta_j \tag{14}$$

holds for all 1-mutants, none could grow out barring other changes.

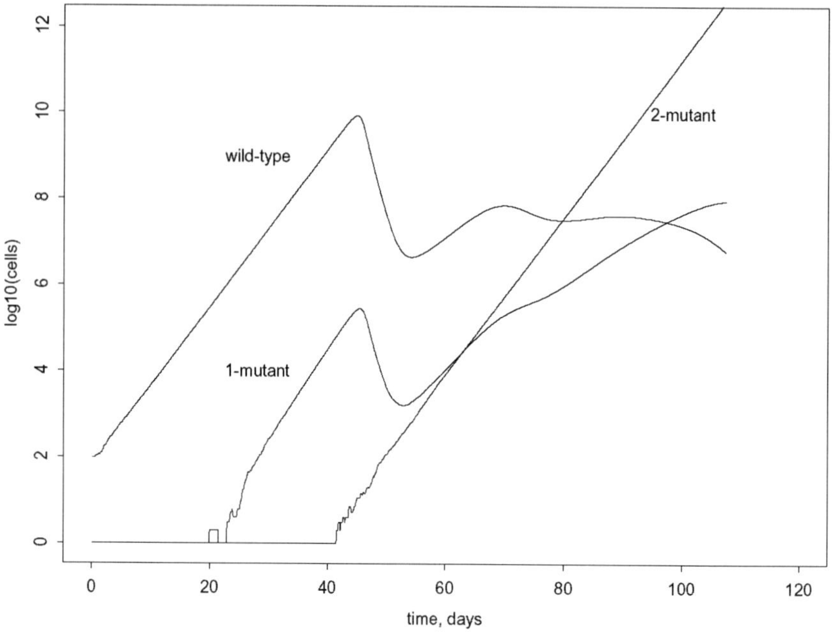

Fig. 3. Escape from two epitopes (unvaccinated patient).

Here is a mechanism that could force the virus to make two or more mutations virtually simultaneously. Surprisingly, simulations revealed that escape may still occur. With three epitopes, $R_0(\text{w.t.}) = 3$, and equal killing rates, the right righthand side of (14) yields a critical value of 2.33. Yet with $R_0(\text{mut.; j}) = 2.29$, additive loss of fitness, and the "vaccinated" conditions, the escape mutant grows out in less than a year (see Fig. 5).

Despite every 1-mutant lineage going extinct, the mutant's density (about 2.5 logs below and tracking the wild-type's) apparently sufficed for escape. At $R_0(\text{mut.; j}) = 2.0$, however, escape was prevented for two years (Fig. 6).

Other interesting conclusions follow from Eq. (13). First, a broad response can be better than a narrow one, even if the CTLs in the narrow response have greater avidity or killing potential: $R_e(\text{mut.; j}) \approx R_0(\text{mut.; j})$ if $\kappa_j C_j \gg \kappa_k C_k$ for all $k \neq j$, while $R_e(\text{mut.; j})$ approaches $R_0(\text{mut.; j})/R_0(\text{w.t.})$ for θ_j small. Second, the virus will likely escape recognition of high-avidity epitopes before low-avidity ones, as O'Conner *et al.* (2002) recently observed for SIV in macaques. With $E = 2$, equal

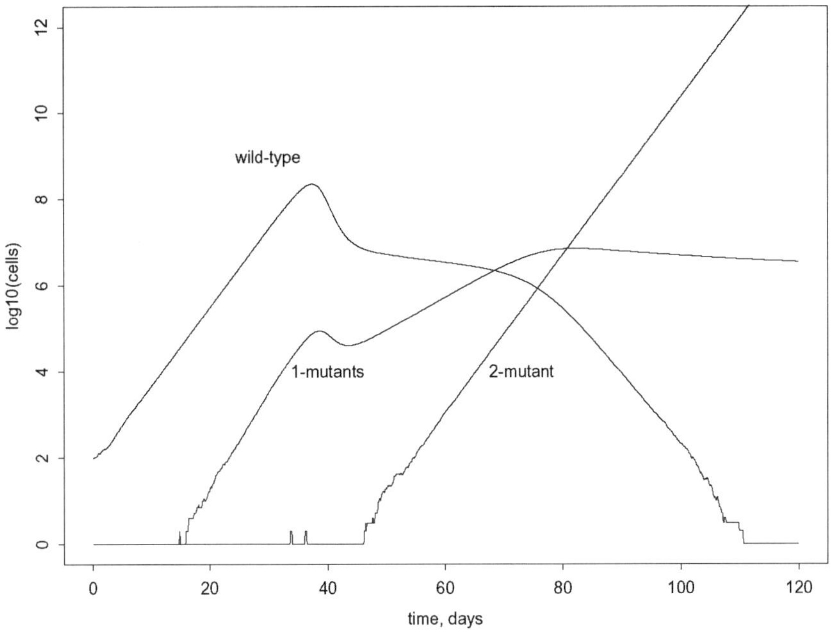

Fig. 4. Escape from two epitopes (vaccinated patient).

killing rates, and $C_1 \gg C_2, \theta_2$ will be small and $R_e(\text{mut.}; 2) \approx R_0(\text{mut.}; 2)/R_0(\text{w.t.}) < 1$, while $\theta_1 \approx 1$ and $R_e(\text{mut.}; 1) \approx R_0(\text{mut.}; 1)$. That is, most attempts at escaping from the second epitope will fizzle out, but not from the first. Note that this explanation does not require that CTLs would have cleared the virus, absent escape (O'Conner, Allen and Watkins, 2001).

A rival to the dynamical explanations considered here invokes the "fitness landscape": the functional dependence of $R_0(\text{mut.})$ on sequence variation. The influence of multiple mutations on fitness could be additive, multiplicative or take some more complicated form. A sequence of neutral, or even slightly deleterious, mutations might be required before a final, adaptive one confers escape. For HIV, Kelleher *et al.* (2001) found that several, possibly compensatory, mutations were required in the p24 GAG protein for escape from CTL recognition. Since neutral mutations do not change fitness, each strain is destined to go extinct; nevertheless the neutral mutants can persist at low density. (In classical theory their frequency is related to the product $N_e\mu$, so the low-N_e question appears again.) The time-to-escape might increase by a huge factor. Figure 7 shows the situation

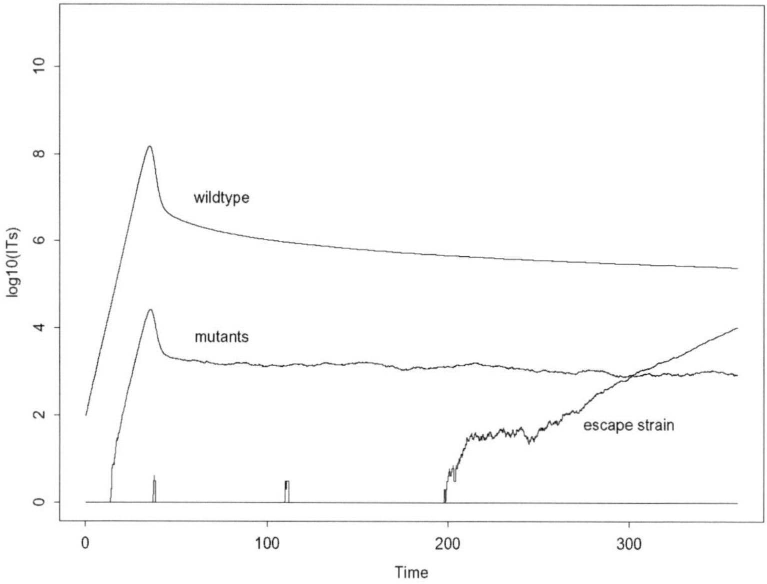

Fig. 5. Escape from three epitopes with LOF just above the critical value (vaccinated patient).

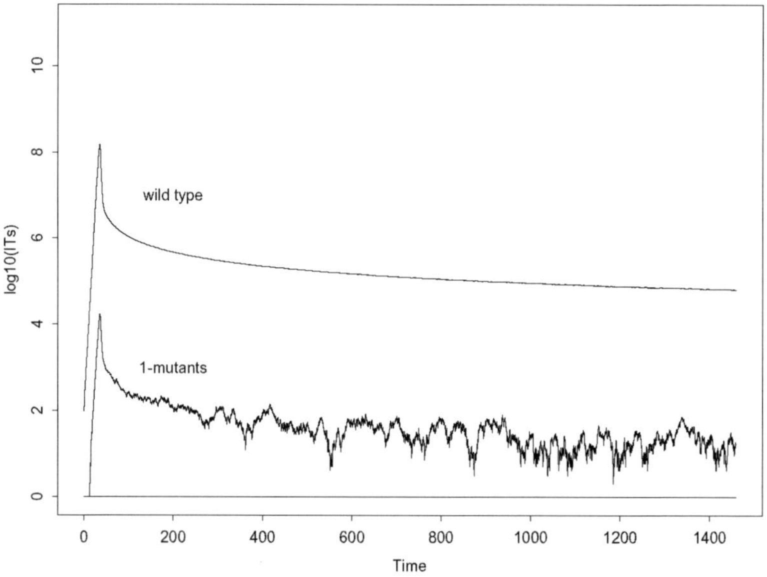

Fig. 6. Lack of escape from three epitopes with LOF further above the critical value.

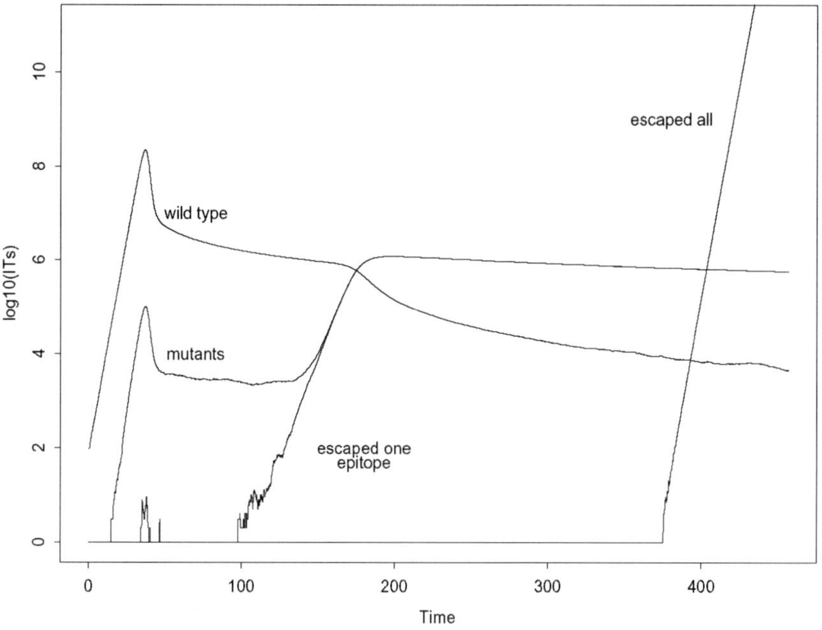

Fig. 7. Escape from two epitopes with two neutral mutations (vaccinated patient).

for two neutral mutations per epitope (both required for escape), $E = 2$, and "vaccinated" conditions.

Escape is delayed relative to Fig. 3 by many months but still occurred in a year. But with the same scenario and $E = 3$, the final escape mutant did not appear in two years (Fig. 8).

7. Can Vaccines Delay Escape?

Besides preventing infection, other conceivable goals of a vaccine include lowing post-infection viral load and preventing or delaying progression to AIDS or death through the possible mechanisms of increasing the number of epitopes recognized and curing a "CTL-defect" (if one exists). Figure 9 shows the time for the virus to delete one epitope when 12 are recognized, in a vaccinated patient, and simulating only the wild-type and single mutants. (Initial conditions were: 12 sets of 10^8 "genuine" resting memory CTLs, each recognizing one epitope, and 100 wild-type ITs.) The 1-mutant quasispecies replaced the wild-type in about 200 days, not a great improvement over Fig. 3.

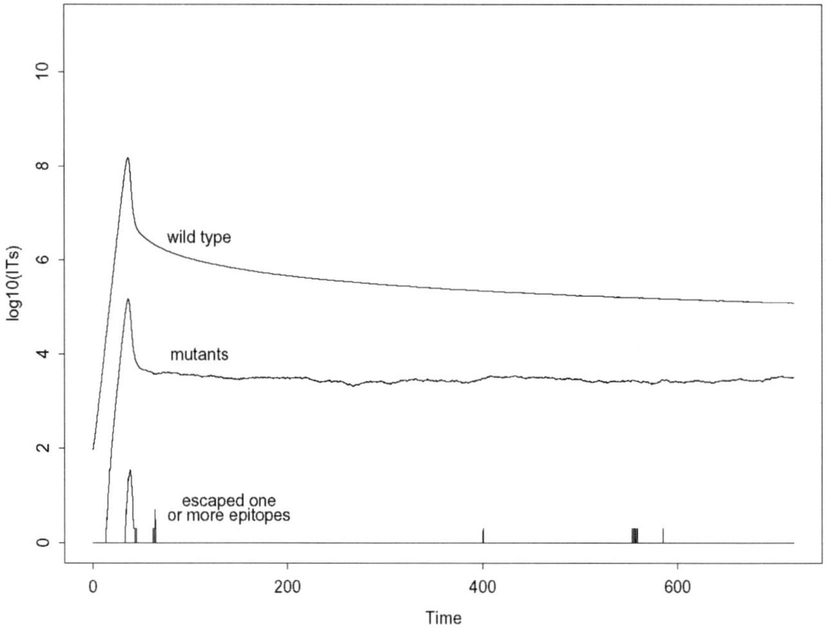

Fig. 8. Lack of escape from three epitopes recognized, with two neutral mutations (vaccinated patient).

How long would the virus take to mutate all epitopes and escape? At present, the numerical explosion in genomes (1024 variants for $E = 10$) deters us from addressing this question *via* stochastic modeling. Therefore we resorted to the following quasi-deterministic calculation. In Appendix A, we derive a formula for the extinction rate of form $P[\text{extinct}] = F(\delta_{PIT})$. We set, for each genome g, $\delta_{PIT}(g) = \delta_{IT} + \kappa C(g)$, where $C(g)$ is the number of CTLs recognizing an epitope expressed by g. Given genome g, let $\rho(g) = (1 - P[\text{extinct}](g))\mu\iota P(g)$, where $P(g)$ denotes the number of PITs of genomes for which g is a 1-mutant and ι is an infection parameter determined by R_0. This factor describes the effective birth-rate of new mutant lines while taking into account the chance of extinction. At each time-step of length h, we sweep through genomes lacking any representatives; if g is one, we draw a uniform random variable u; compute $\rho(g)$; if $u < \exp(-\rho(g)h)$, we add one to the corresponding IT-compartment. Finally, we update all compartments deterministically.

The method faithfully reproduced Fig. 3 (figure not shown). Figure 10 shows the 1-, 2-, and escape mutants for $N = 10$ and an unvaccinated

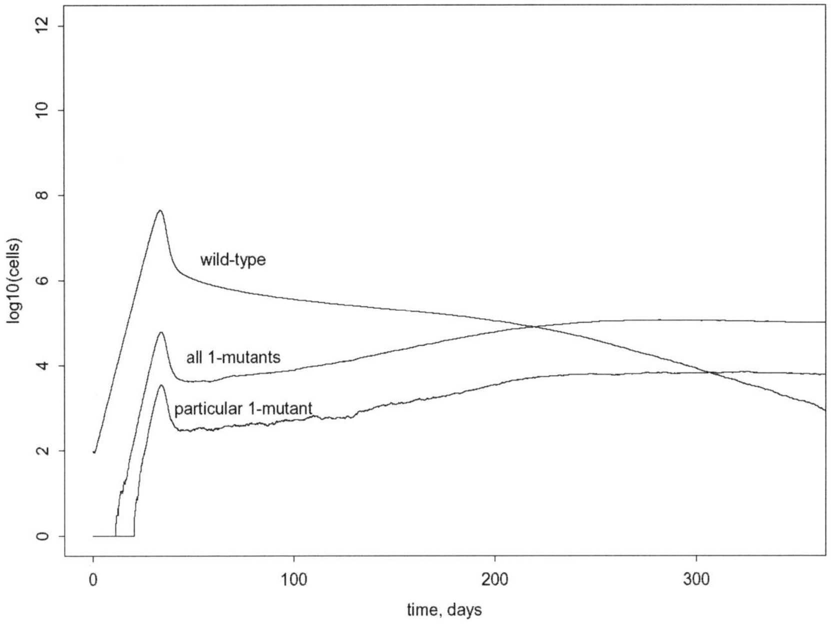

Fig. 9. Generation of one-mutants when 12 are recognized (stochastic simulation).

patient; the virus escapes in six months. For the vaccinated patient the virus escaped only a few weeks later, despite the two-log-suppressed viral load (figure not shown).

8. Discussion

In 2002, an NIH-sponsored HIV vaccine was judged insufficient in generating a CTL response to initiate efficacy evaluation. A phase III clinical trial of a similar product is proceeding and other candidate CTL vaccines are under development. The weighty question arises: what surrogate markers should be accepted as proof of efficacy for a vaccine that does not prevent infection?

Conceivably, the number of CTL epitopes recognized could be as important a marker as post-infection viral load. The question is whether increasing the number of epitopes recognized by five- or ten-fold would slow down escape marginally or dramatically. If CTLs have roughly equivalent kinetics, no extra (non-Poisson) heterogeneity exists, and escape mutants suffer no major loss of viability, our modeling efforts to date support the former

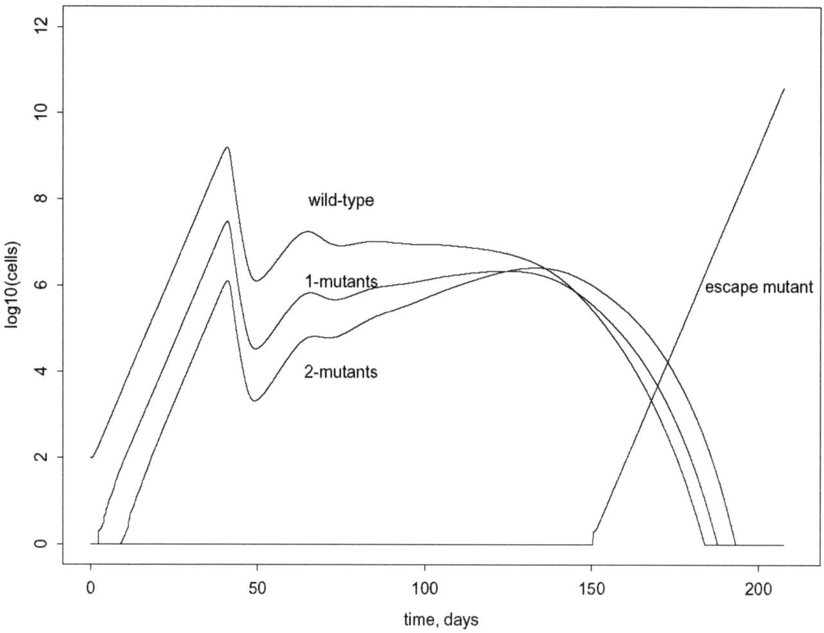

Fig. 10. Escape from 10 epitopes recognized (quasi-deterministic simulation).

conclusion. But, as we remarked in Section 6, even a small loss of fitness in mutants and a broad CTL response could effectively prevent escape.

We have argued that extra-Poisson variation in virion-production could make a large impact on the time to escape from immune control. But other mechanisms can lower N_e besides dynamical heterogeneity. The situation calls for new experiments, which might be carried out *in vitro* or *in vivo*. It would be useful to introduce an artificial "pseudogene" into the HIV genome: a sequence, as long as possible, that is free to accumulate neutral mutations. The construct is inoculated into multiple wells containing T-cell cultures and the pseudogene sequenced at later timepoints. If the infection grows, the effect will appear in the variance of the accumulated mutants, rather than in the mean; for adequate statistical power many cultures would have to be run in parallel. Better would be to maintain cultures in approximate steady-state conditions (by repeated removal of infected cells and infusion of uninfected), since then the difference will appear in the mean. (The extra-Poisson variation will yield a low N_e, resulting in decreased genetic variation.) An experiment in an animal model would of course be preferable, but it may be much more difficult to perform.

To what extent do our immune-response hypotheses affect our time-to-escape results? Any realistic model has to reproduce the observed steady-state CTL cell densities, time-constants, and killing rates, and granted these the impact on infection is fairly stable. The pivotal hypotheses are more likely to be on the infection-side, as we suggested in Section 6: epitope antigenicity, fitness conservation, and virion-production.

On this side there is also a critical parameter: μ, the mutation rate per nucleotide. Yokoyama and Gojobori (1987) first derived a value for μ from cell-free studies of the fidelity of reverse-transcription. Then Mansky and Temin (1995) found it to be 20 times lower *in vivo*. Their figure, 3.4×10^{-5} per b.p. per cycle, is the one usually cited. We re-ran Fig. 3 with $\mu = 10^{-6}$; the escape time increased about 40 days. Another mechanism that can speed up escape is recombination in cells infected by multiple genomes; this important topic should also be addressed in simulation studies.

CTL clones sometimes disappear; what part of this phenomena is due to destabilization by mutants (as in Fig. 3), and what part to "CTL exhaustion"? The latter refers to any special mechanism that deletes or renders anergic CTL clones at high antigen-exposure (Zajac *et al.*, 1998; Wodarz *et al.*, 1998). Pantaleo *et al.* (1997) reported clones disappearing without finding evidence of change in the genetic make-up of the virus. Perhaps the most-avid CTL clones are "exhausted" early and play no further role in immunity. Altfeld *et al.* (2001) noted a broadening of CTL response in untreated patients after primary disease; is this due to expansion of already-existing, lower-avidity clones, appearance of new epitopes, or continued thymogenesis of CTL-precursors?

Although we have made use of N_e and other traditional ideas, we record our qualms here about treating escape within the confines of standard population genetics. See Rouzine *et al.* (2001) for a review of classical theory as it applies in virology. First, classical theory assumes that the demographic population number is fixed (although generalizations to exponentially-growing populations exist). Each new generation derives from randomly sampling N_d offspring of the previous generation. In particular, fixation of a new mutation does not change N_d (or the growth rate if the population is expanding). Second, classical theory assumes neutral or weakly-selective conditions. Third, lacking a detailed population-dynamics model, classical theory cannot evaluate the selection coefficient.

In contrast, we think the essence of immune control of parasites is its dynamic, reactive nature. A selective advantage confers precisely the ability to grow to a higher density, limited only by the immune response. In classical theory, one way to assure a constant N_d is to assume all niches are filled at all times. This assumption might be warranted for Darwin's finches, but in chronic infection HIV does not appear capable of productively-infecting more than a small fraction (typically 0.1 percent) of its target cells. The rate of escape from CTLs shows they exert a strong selection pressure on the virus. Finally, we adopt a detailed model (as in Section 5) except when making general remarks.

Concerning simulation technique, we hope the reader noted the "blips" near the horizontal axis in some of our figures. These were caused by mutant viral lines appearing briefly but going extinct. If we had exclusively used deterministic modeling, each strain would have grown out, greatly exaggerating the escape rate. Moreover, we could not have discussed how extra variability can amplify extinction, as in Section 6. Despite the large numbers involved in immune-system and infection dynamics, sometimes a stochastic treatment is essential.

Acknowledgments

The authors were supported in this research by grants from the NIH and NCI.

References

Addo MM *et al.* (2003) Comprehensive epitope analysis and human immunodeficiency virus-1 (HIV-1)-specific T-cell responses directed against the entire HIV-1 genome demonstrate broadly directed responses but no correlation to viral load, *J Virol* **77**: 2081–2092.

Altman JD and Feinberg MB (2004) HIV escape: there and back again, *Nat Med* **10**.

Amara RR, Villinger F, Altman JD *et al.* (2001) Control of a mucosal challenge and prevention of AIDS in rhesus macaques by a multiprotein DNA/MVA vaccine, *Science* **292**: 69–74.

Antia R, Bergstrom CT, Pilyugin SS, Kaech SM and Ahmed R (2003) Models of CD8+ responses I: what is the antigen-independent proliferation program? *J Theo Biol* **221**: 585–598.

Altes HK, Wodarz D and Janssen VA (2002) The dual role of CD4 T helper cells in the infection dynamics of HIV and their importance for vaccination, *J Theor Biol* **214**: 633–646.

Altfeld M *et al.* (2001) Cellular immune response and viral diversity in individuals treated during acute and early HIV-1 infection, *J Exp Med* **193**(2): 169–180.

Barouch DH *et al.* (2000) Control of viremia and prevention of clinical AIDS in rhesus monkeys by cytokine-augmented DNA vaccination, *Science* **290**: 486–492.

Barouch DH *et al.* (2002) Eventual AIDS vaccine failure in a rhesus monkey by viral escape from cytotoxic T-lymphocytes, *Nature* **415**: 335–339.

Blattman JN, Antia R, Sourdive DJD *et al.* (2002) Estimating the precursor frequency of naïve antigen-specific CD8+ T cells, *J Exp Med* **195**: 657–664.

Borrow P *et al.* (1997) Antiviral pressure exerted by HIV-1-specific cytotoxic T-lymphocytes during primary infection demonstrated by rapid selection of CTL escape virus, *Nat Med* **3**: 205–211.

Brodie SJ, Lewinsohn DA, Patterson BK *et al.* (1999) *Nat Med* **5**: 34–41.

Bourgeois C, Rocha B and Tanchot C (2002) A role for CD40 expression on CD8+ T-cells in the generation of CD8+ T-cell memory, *Science* **297**: 2060–2063.

Coffin JM (1995) HIV population dynamics *in vivo*: implications for genetic variation, pathogenesis, and therapy, *Science* **267**: 483.

Dalod M, Harzic M, Pellegrin I *et al.* (1998) Evolution of cytotoxic T-lymphocyte responses to human immunodeficiency virus type-1 in patients with symptomatic primary infection receiving antiretroviral triple therapy, *J Infect Diseases* **178**: 61–69.

Day CL and Walker BD (2003) Progress in defining CD4 helper cell responses in chronic viral infections, *J Exp Med* **198**: 1773–1777.

Dimitrov DS *et al.* (1993) Quantitation of human immunodeficiency virus type 1 infection kinetics, *J Virol* **67**: 2182–2190.

Friedrich *et al.* (2004) Reversion of CTL escape-variant immunodeficiency viruses *in vivo*, *Nat Med* **10**:

Gray CM, Lawrence J, Schapiro JM *et al.* (1999) Frequency of class I HLA-restricted anti-HIV CD8+ T cells in individuals receiving highly-active antiretroviral therapy (HAART), *J Immunol* **162**: 1780–1788.

Goulder PJR *et al.* (1997) Late escape from an immunodominant cytotoxic T-lymphocyte response associated with progression to AIDS, *Nat Med* **3**: 212–217.

Ho D *et al.* (1995) Rapid turnover of plasma virions and CD4 lymphocytes in HIV-1 infection, *Nature* **373**: 123–126.

Igarashi T, Brown C, Azadegan A, Haigwood N, Dimitrov D, Martin MA and Shibata R (1999) Human immunodeficiency virus type 1 neutralizing antibodies accelerate clearance of cell-free virions from blood plasma, *Nat Med* **5**: 211–216.

Jamieson BD, Yang OO and Hultin L (2003) Epitope escape mutation and decay of human immunodeficiency virus type-1 specific CTL responses, *J Immun* **171**: 5372–5379.

Janssen EM *et al.* (2003) CD4+ cells are required for secondary expansion of memory in CD8+ T lymphocytes, *Nature* **421**: 852–856.

Kaech SM and Ahmed R (2001) Memory CD8+ T cell differentiation: initial antigen encounter triggers a developmental program in naïve cells, *Nat Immunol* **2**: 415–422.

Kaech SM and Ahmed R (2003) CD8 T cells remember with a little help, *Science* **300**: 263–265.

Kalams SA, Goulder PJ, Shea AK *et al.* (1999) Levels of human immunodeficiency virus type 1-specific cytotoxic T-lymphocyte effector and memory responses decline after suppression of viremia with highly-active antiretroviral therapy, *J Virol* **73**: 6721–6728.

Kelleher AD *et al.* (2001) Clustered mutations in HIV-1 gag are consistently required for escape from HLA-B27-restricted cytotoxic T-lymphocyte responses. *J Exp Med* **193**: 375–385.

Korber B, Brander C, Haynes BF, Koup R, Kuiken C, Moore JP, Walker BD and Watkins DI (eds.) (2001) *HIV Molecular Immunology, 2000.* Theoretical Biology and Biophysics Group, Los Alamos National Escape from Immune Control 17 Laboratory, Los Alamos, NM.

Layne SP *et al.* (1992) Factors underlying spontaneous inactivation and susceptibility to neutralization of human immunodeficiency virus, *Virology* **189**: 695–714.

Leigh-Brown AJ (1997) Analysis of HIV-1 *env* gene sequences reveals evidence for a low effective number in the viral population, *Proc Nat Acad Sci USA* **94**: 1862–1865.

Leslie *et al.* (2004) HIV evolution: CTL escape mutation and reversion after transmission, *Nat Med* **10**: (published online Feb. 2004).

Levy JA, Ramachandran B and Barker E (1996) *Science* **271**: 670.

Mansky LM and Temin HM (1995) Lower *in vivo* mutation rate of human immunodeficiency virus type 1 than that predicted by the fidelity of purified reverse transcriptase. *J Virol* **69**: 5087–5094.

Monteiro LHA, Goncalves CHO and Piqueira JRC (2002) A condition for successful escape of a mutant after primary HIV infection, *J Theor Biol* **203**: 399–406.

Myers G, in Myers G, Korber B, Berzosfsky JA, Smith R and Pavlakis GW (eds.) (1992) *Human Retroviruses and AIDS*, Part III, Los Alamos National Laboratory.

Nowak MA *et al.* (1990) The evolutionary dynamics of HIV-1 quasispecies and the development of immunodeficiency disease, *AIDS* **4**: 1095–1103.

Nowak MA (1992) Variability of HIV infections, *J Theor Biol* **155**: 1–20.

Nowak MA *et al.* (1995) Immune responses against multiple epitopes, *J Theor Biol* **175**: 325–353.

Nowak MA, May RM, Phillips RE *et al.* (1995) Antigenic oscillations and shifting immunodominance in HIV-1 infections, *Nature* **375**: 606–611.

O'Conner DH *et al.* (2001) Understanding cytotoxic T-lymphocyte escape during simian immunodeficiency virus infection, *Immunol Rev* **183**: 115–126.

O'Conner DH *et al.* (2002) Acute phase cytotoxic T-lymphocyte escape is a hallmark of simian immunodeficiency virus infection, *Nat Med* **8**: 493–499.

O'Conner D, Allen T, Watkins DI (2001) Vaccination with CTL epitopes that escape: an alternative approach to HIV development? *Immunol Lett* **79**: 77–84.

Ogg GS, Jin X, Bonhoeffer S *et al.* (1999) Decay kinetics of human immunodeficiency virus-specific effector cytotoxic T-lymphocytes after combination antiretroviral therapy, *J Virol* **73**: 797–800.

Pantaleo G *et al.* (1997) Evidence for rapid disappearance of initially expanded CD8+ T cell clones during primary HIV infection, *Proc Nat Acad Sci USA* **94**(18): 9848–9853.

Perelson AS *et al.* (1996) HIV-1 dynamics *in vivo*: virion clearance rate, infected cell lifespan, and viral generation time. *Science* **271**: 1582–1586.

Ribeiro RM and Bonhoeffer S (1999) A stochastic model for primary HIV infection: optimal timing for therapy. *AIDS* **13**: 351–357.

Rouzine IM and Coffin JM (1999) Linkage disequilibrium test implies a large effective population number for HIV *in vivo, Proc Nat Acad USA* **96**: 10758–10763.

Rouzine IM, Rodrigo A and Coffin JM (2001) Transition between stochastic evolution and deterministic evolution in the presence of selection: general theory and application to virology, *Microbiology and Molecular Biology Reviews*, 151–185.

Shedlock DJ and Shen H (2003) Requirement for CD4 help in generating functional CD8 T-cell memory. *Science* **300**: 337–339.

Sun JC and Bevan MJ (2003) Defective CD8 T-cell memory following acute infection without CD4 help. *Science* **300**: 339–342.

Wherry EJ *et al.* (2003) Memory defect in high viral load HIV patients: lineage relationships and protective immunity of memory CD8 T-cell subsets, *Nat Immunol* **4**: 225–234.

Wei X *et al.* (1995) Viral dynamics in human human immunodeficiency virus type 1 infection, *Nature* **373**: 117.

Wick D (1999a) The disappearing CD4+ T cells in HIV infection: a case of overstimulation? *J Theor Biol* **197**: 507–516.

Wick D (1999b) On T-cell dynamics and the hyperactivation theory of AIDS pathogenesis, *Math Biosc* **158**: 127–144.

Wick D and Self SG (2000) Early HIV Infection *in vivo*: branching-process model for studying timing of immune responses and drug therapy, *Math Biosciences* **165**: 115–134.

Wick D, Self SG and Corey L (2002) Do scarce targets or T-killers control primary HIV infection? *J Theor Biol* **214**: 209–214.

Wick D and Self SG (2002b) What's the matter with HIV-directed T-killer cells? *J Theor Biol* **219**: 19–31.

Wick D and Self SG (2004a) On simulating strongly-interacting population stochastic models. *Math Biosci* **187**: 1–20.

Wick D and Self SG (2004b) On simulating strongly-interacting population stochastic models. II. Many-compartment models, *Math Biosci* in press.

Wordarz D (2001) Helper-dependent vs. helper-independent CTL responses in HIV infection: implications for drug therapy and resistance. *J Theor Biol* **213**: 447–459.

Wodarz D, Klenerman P and Nowak MA (1998) Dynamics of cytotoxic T-lymphocyte exhaustion, *Proc Royal Soc London B* **265**: 191–203.
Wodarz D and Nowak MA (2000) Correlates of cytotoxic T-lymphocyte-mediated virus control: implications for immuno-suppressive infections and their treatment. *Philos Trans Royal Soc London B* **355**: 1059–1070.
Yang OO, Sarkis PTN and Ali A (2003) Determinants of HIV-1 mutational escape from cytotoxic T-lymphocytes. *J Exp Med* **197**: 1365–1375.
Yang OO, Sarkis PTN and Trocha A *et al.* (2003) Impact of avidity and specificity on the antiviral efficiency of HIV-1 specific CTL. *J Immunol* **171**: 3718–3724.
Yokoyama S and Gojobori T (1987) Molecular evolution and phylogeny of the human AIDS viruses LAV, HTLV-III and ARV. *J Mol Evol* **24**: 330–336.
Zhang L, Dailey PJ, He T, Gettie A, Bonhoeffer S, Perelson AS and Ho DD (1999) Rapid clearance of simian immunodeficiency virus particles from plasma of rhesus macaques. *J Virol* **73**(1): 855–860.
Zajac AJ, Blattman JN, Murali-Krishna K *et al.* (1998) Viral immune evasion due to persistence of activated T cells without effector function. *J Exp Med* **188**(12): 2205–2213.

Appendices

A. *Infection models*

The simplest infection model with an eclipse phase, the population of which we denote X, and a PIT-phase, which we denote by Y, is, in jump-process language:

$$\begin{array}{lcr} \textit{Type} & \textit{Jump} & \textit{Rate} \\ \text{Births:} & X \to X+1 & \iota Y\,; \\ \text{Deaths:} & X \to X-1 & \delta_{IT} X\,; \\ & Y \to Y-1 & \delta_{PIT} Y\,; \\ \text{Progression:} & X \to X-1 \text{ and } Y \to Y+1 & \eta X\,. \end{array} \tag{15}$$

Here $\delta_{IT}, \delta_{PIT}, \eta$ and ι are positive parameters. See Table 2 for our values for δ_{IT} and η; with an immune response we put

$$\delta_{PIT} = \delta_{IT} + \kappa \sum_{i=1}^{E} 1[\text{epitope } i \text{ is expressed}] C_i\,, \tag{16}$$

where κ, the killing rate, is listed in Table 2, the C_i are populations of CTLs recognizing epitope i, and $1[\,]$ is the indicator function.

The infection rate ι we derived from the basic reproductive number. A growth-condition is most easily derived from the corresponding deterministic model (which describes the same process for large populations):

$$\frac{d}{dt}\begin{pmatrix} X \\ Y \end{pmatrix} = \begin{pmatrix} -(\delta_{\mathrm{IT}} + \eta), & \iota \\ \eta, & -\delta_{PIT} \end{pmatrix} \begin{pmatrix} X \\ Y \end{pmatrix}. \tag{17}$$

For growth, the matrix of the system must have a positive eigenvalue, which yields

$$R_0(\mathrm{ODE}) = \frac{\iota\eta}{\delta_{PIT}(\delta_{IT} + \eta)} > 1\,. \tag{18}$$

The reason for the "ODE" is that (18) is not quite the same as the branching-process definition. Growth in the ODE comes from a factor $\exp(\lambda_+ t)$; for $R_0(\mathrm{ODE})$ near one and $\eta \gg \delta, \lambda_+ \approx (R_0(\mathrm{ODE}) - 1)\delta$. Note that δ is the inverse IT-lifetime; hence if $R_0(\mathrm{br.})$ is the usual branching-process definition (secondary infections caused by one primary case in its lifetime), the relation is: $R_0(\mathrm{ODE}) \approx 1 + \ln R_0(\mathrm{br.})$. Thus some authors assume HIV's basic reproductive number *in vivo* is 6–9, and others 2–3. All definitions agree near unity. We derive ι from (18), with parameters from Table 2 and $\delta_{PIT} = \delta_{IT}$.

Alternatively, R_0 can be derived in the course of computing the extinction probability, as in Wick and Self (2000). We indicate how in connection with the generalization in Section 6 to multiple ITs and PITs. Consider the branching process for the kth cell type:

Type	*Jump*	*Rate*
Births:	$X_k \to X_k + 1$	$\iota p_k \left(\sum_{n=1}^{K} v_n Y_n\right)$;
Deaths:	$X_k \to X_k - 1$	$\delta_{IT} X_k$;
	$Y_k \to Y_k - 1$	$\delta_{PIT} Y_k$;
Progression:	$X_k \to X_k - 1$ and $Y_k \to Y_k + 1$	ηX_k .

(19)

Let r_k, respectively s_k, equal the extinction probability starting with one IT, respectively one PIT, of type k. Either directly or by deriving a PDE for a moment-generating function, one can derive the equations

$$\begin{aligned} 0 &= \delta_{IT}(1 - r_k) + \eta(s_k - r_k)\,; \\ 0 &= \delta_{PIT}(1 - s_k) + \iota v_k s_k \sum_{n=1}^{K} p_n(r_n - 1)\,. \end{aligned} \tag{20}$$

For the one-epitope case, solving (20) yields

$$r = \left(\frac{\iota\delta_{IT} + \delta_{PIT}(\delta_{IT} + \eta)}{\iota(\delta_{IT} + \eta)} \right) ;$$

$$s = \left(\frac{\delta_{PIT}}{\iota} \right) \left(\frac{\delta_{IT} + \eta}{\eta} \right) . \tag{21}$$

The conditions: $r, s < 1$, is easily seen to be equivalent to $R_0 > 1$ with R_0 given by (18), as general branching-theory (for the "super-critical case") predicts.

In the general case, we let the starting distribution be an IT of type k with probability p_k. Rather than extract roots we solved (20) using the iteration:

$$\theta = \sum_{k=1}^{K} p_k(1 - r_k) ;$$

$$s_k = \frac{1}{1 + zv_k\theta} ;$$

$$r_k = \frac{1 + ws_k}{1 + w} , \tag{22}$$

where $z = \iota/\delta_{PIT}$ and $w = \eta/\delta_{IT}$. System (22) converged from $p_k, s_k \equiv 0.9$ in 100 iterations, even with $K = 1000$. To generate different VMR's while keeping R_0 and mean production fixed we adopted:

$$p_k = \frac{\lambda}{k^q} ;$$

$$\lambda = \left(\sum_{k=1}^{K} \frac{1}{k^q} \right)^{-1} ;$$

$$v_k = V k^q . \tag{23}$$

Scale factor V fixes mean production but falls out when calculating the extinction probability. We fixed z using the formula:

$$R_0 = \left(\frac{zw}{1 + w} \right) \left(\sum p_k v_k \right) , \tag{24}$$

which can be derived from the matrix of the ODE system as before, but is more easily obtained by requiring that the iteration have a non-trivial fixed point. Collapsing (22) to $\theta = f(\theta)$, the criterion for an attracting fixed-point θ^* in a one-dimensional map is: $f'(\theta) < 1$. A necessary and sufficient condition for the latter is $1 < f'(0)$, which yields (24).

We used $\eta = 1.0$, $\delta_{IT} = \delta_{PIT} = 0.33$, $q = 2$, $K = 10$–100, and $R_0 = 1 + 1/15$, as for escaping when 10 epitopes are recognized. The figures in Table 1 are rounded to two decimals.

To incorporate multiple genomes, we need only replace index k in (19) by g, with g ranging over the set G of possible genomes of cardinality 2^E, assuming one mutable site per epitope. We set the birth rate into the genome-gIT compartment by:

$$\text{Births:} \quad \iota\mu \sum_{g' \in G} 1[\#(g') = \#(g) + 1] Y_{g'} \,, \tag{25}$$

where $\#(g)$ denotes the number of wild-type epitopes in g. G can be represented as the set $\{0, 1, \ldots, 2^E - 1\}$ with $\#(g)$ the number of 1's in g's binary expansion.

B. *Effective population size*

This section is pure population biology with discrete generations (no ITs and PITs). Let $P_N[\,]$ denote probability in a parental population of size N. Let n_k be the number of offspring of the k-th parent; these are assumed independent with distribution $p_n, n = 0.1, \ldots$. Assuming fixed demographic number, the next generation is formed by choosing N offspring at random from those produced. We wish to compute the probability that two randomly selected offspring have the same parent:

$$\begin{aligned} P_N[\text{same parent}] &= N P_N[\text{given parent}] \\ &= N \sum_{u=0}^{\infty} \sum_{m=0}^{\infty} P_N\left[\text{g.p.} | n_1 = u; \sum_{k=2}^{N} n_k = m\right] \\ &\quad \times P_N[n_1 = u] P_N\left[\sum_{k=2}^{N} n_k = m\right] \\ &= N \sum_{u=0}^{\infty} \sum_{m=0}^{\infty} \left(\frac{u}{u+m}\right)\left(\frac{u-1}{u-1+m}\right) p_m^{*(N-1)} p_u \,. \end{aligned} \tag{26}$$

Here $p^{*(N-1)}$ denotes the $(N-1)$ fold convolution of $p.$ with itself.

Introduce the moment-generating function of p_n:

$$\phi(x) = \sum_{n=0}^{\infty} e^{-nx} p_n \,, \tag{27}$$

into (26) by writing the fractions as exponential integrals; this gives:

$$P_N[\text{same parent}] = N \int_0^{\infty} (e^x - 1) \left\{ \sum_{u=0}^{\infty} p_u u(u-1) e^{-ux} \right\} [\phi(x)]^{N-1} dx \,. \tag{28}$$

Formula (8) follows from (28) by an asymptotic analysis as $N \to \infty$. We display one step:

$$P_N[\text{same parent}] = N[V + M^2 - M] \int_0^{x(N)} dx\, x\, e^{\phi'(0)(N-1)x} dx + \text{remainder}\,, \tag{29}$$

where V = variance, M = mean, and $x(N) \to 0$ will be chosen later. Integrating by parts, the expression gives

$$N[V + M^2 - M] \left\{ \frac{1 - e^{-M(N-1)x(N)}}{M^2(N-1)^2} - \frac{x(N)e^{-M(N-1)x(N)}}{M(N-1)} \right\}. \tag{30}$$

If we choose $x(N) = N^{-0.7}$, the only non-vanishing term in curly brackets is the one without the exponential. Setting $M = 1$ yields Eq. (8) of Section 3; the remainder can be shown to be vanish as well (details on demand to the first author).

C. *Extinction probability with general offspring distribution*

This section refers to the standard continuous-time branching-process (again, without ITs and PITs). The usual technique to compute the extinction probability is to introduce the moment-generating function and compute its fixed point in the unit interval, as explained in Appendix A or any book on branching-processes. The upshot is that $q^* = P[\text{ext.}]$ is the root of

$$\psi(q) = \sum_{n=0}^{\infty} p_n q^{n+1} - 2q + 1 = 0\,. \tag{31}$$

To approximate q^*, use the Taylor's expansion of around $q = 1$ to second order; the root of the resulting quadratic gives

$$\begin{aligned} P[\text{ext.}] &\approx 1 - \frac{2(M-1)}{\sum p_n n^2 + M} \\ &\approx 1 - \frac{2(1 - 1/M)}{\text{VMR}}\,, \end{aligned} \tag{32}$$

for VMR $\gg 1$ and $M \approx 1$. This formula only yields an upper bound, but is accurate provided a certain condition, involving the third moment, holds.

D. *The CTL model*

Our kinetic model for one CTL clone recognizing one epitope is taken from Wick and Self (2002). In jump terminology (Z_{index}: a compartment; *NR*: naïve resting; *MR*: memory resting; n_d: number of permitted doublings):

Type	*Jump*	*Range*	*Rate*
Births:	$Z_i \to Z_i + 1$	MR	β;
Mitosis:	$Z_i \to Z_i - 1$ and	$NR, \ldots, NR + n_d - 1$ and	γ_N;
	$Z_{i+1} \to Z_{i+1} + 2$	$MR, \ldots, MR + n_d - 1$	γ_M;
Deaths:	$Z_i \to Z_i - 1$	$NR, MR,$	δ_{NR}, δ_{MR};
		$NR + n_d, MR + n_d$	δ_{CTL};
Progression:	$Z_i \to Z_i - 1$ and		
	$Z_{MR} \to Z_{MR} + 1$	$MR + n_d$	ρ.

Activation into the cell-cycle of resting, naïve-or-memory CD8s occurs at rate $\upsilon_{NR} = \alpha PITs$ or $\upsilon_{MR} = \alpha\ \varphi\ PITs$, where α is an activation rate constant, φ is a memory speed-up factor, and "PITs" means the sum of all those expressing the particular epitope. We set β ("immigration" of resting, naïve CD8s) $= Z_{NR}(0)\ \delta_{NR}$ so that resting-cells are at steady-state without activation. All E CTL-types have identical kinetics other than activation. Other rate constants are to be found in Table 2.

We simulated the immune half of the process *via* the ODEs defined by

$$dZ_i/dt = \sum_{\text{jumps}} (\text{total rate})\Delta_{\text{jump}} Z_i\,, \tag{33}$$

where "total rate" means Z_i x the per-cell rate above, except for "births" of resting, NR-cells which have rate β; Δ stands for the change made by the jump. We usually used Runge–Kutta, second-order (third-order per timestep) for this half of the process.

CHAPTER 10

CTL ACTION DURING HIV-1 IS DETERMINED VIA INTERACTIONS WITH MULTIPLE CELL TYPES

Seema H. Bajaria and Denise E. Kirschner

During HIV-1 infection, interactions between immune cells and virus yield three distinct stages of infection: high viral levels in acute infection, immune control in the chronic stage, and AIDS, when $CD4^+$ T cells fall to extremely low levels. The immune system consists of many players that have key roles during infection. In particular, $CD8^+$ T cells are important for killing of virally infected cells as well as inhibition of cellular infection and viral production. Activated $CD8^+$ T cells, or cytotoxic T cells (CTLs) have unique functions during HIV-1, most of which are thought to be compromised during HIV-1 infection progression. Controversy exists regarding priming of CTLs, and our work attempts to address the dynamics occurring during HIV-1 infection. To explore the influence of $CD8^+$ T cells as determinants in disease progression and issues relating to their priming and activation, we develop a two-compartment ordinary differential equation model describing cellular interactions that occur during HIV-1 infection. We track $CD4^+$ T cells, $CD8^+$ T cells, dendritic cells, infected cells, and virus, each circulating between blood and lymphatic tissues. Using parameter estimates from literature, we simulate commonly observed infection patterns. Our results indicate that $CD4^+$ T cells as well as dendritic cells likely play a significant role in successful activation of $CD8^+$ T cells into CTLs. Model simulations correlate with clinical data confirming a quantitative relationship between $CD4^+$ T cells and $CD8^+$ T-cell effectiveness.

Keywords: HIV-1; $CD8^+$ T cells; CTLs; T-cell help; dendritic cells; immune activation.

1. Introduction

HIV-1 infection leads to the gradual depletion of $CD4^+$ T cells over a typical adult course of infection in the majority of patients. The mechanisms for this loss remains unclear, and can be attributed to several host and viral factors. Among them is a reduction in thymic production of uninfected cells or infection within the thymus (Douek, 1998; Ye *et al.*, 2002), altered migration of circulation patterns (otherwise known as enhanced homing) (Bajaria *et al.*, 2002), a decrease in T cell life span (enhanced apoptosis) (Hellerstein *et al.*, 1999; McCune *et al.*, 2000), and failure of CD4$^+$ T-cell help to CD8$^+$ T cells, that normally assist them in performing their full cytolytic function (Shedlock and Shen, 2003; Janssen *et al.*, 2003). We have studied altered homing and enhanced apoptosis in previous models (Bajaria *et al.*, 2002; Ye *et al.*, 2002). We now focus on mechanisms related directly to the function of CD8$^+$ T cells.

1.1. *Clinical progression of HIV-1*

During HIV-1 infection, immune cells and virus interact resulting in three distinct stages of infection (Pantaleo *et al.*, 1993). The acute stage is characterized by high levels of virus, low CD4$^+$ T-cell counts, and a rapid increase in CD8$^+$ T-cell counts (Quinn, 1997). The asymptomatic or chronic stage is a function of the host immune system controlling viral growth and spread for an extended period of time, often resulting in undetectable levels of virus, even without the influence of drug therapy. The onset of the AIDS stage occurs when CD4$^+$ T-cell counts fall to extremely low levels, precluding the onset of a variety of opportunistic infections, and eventually death. While this pattern is observed in typical disease progressors, long-term non-progressors (LTNP) are a small percentage (5–10%) of the infected population who, even without the help of antiretroviral therapy, do not progress to AIDS for over 15 years of infection (Zhang *et al.*, 1997).

CD8$^+$ T cells have been shown to be influential in all three infection stages. CD8$^+$ T-cell numbers correlate with both viral load (Koup *et al.*, 1994; Borrow *et al.*, 1994; Ogg *et al.*, 1998) and have a defined correlation with CD4$^+$ T-cell counts (Caruso *et al.*, 1998). Due to the many known activities of CD8$^+$ T cells in antiviral immunity, we explore their specific roles during HIV-1 infection.

1.2. *CD8$^+$ T-cell mechanisms during HIV-1 infection*

CD8$^+$ T cells are important in the control of intracellular pathogens, including viruses such as cytomegalovirus (CMV) and influenza (McMichael *et al.*, 1983; Webby *et al.*, 2003; Ellefsen *et al.*, 2002). These classically termed "killer" cells have various direct and indirect cytotoxic functions during infection. Upon successful differentiation to cytotoxic T cells (CTLs),

$CD8^+$ T cells can perform cytolytic functions such as direct killing of infected cells expressing foreign proteins and MHC I (McMichael and Rowland–Jones, 2001; Yang *et al.*, 1996; Berke, 1995), and noncytolytic functions such as inhibition of virus infection and production (Walker *et al.*, 1986; Levy *et al.*, 1996). The cytolytic pathway begins when CTLs secrete perforin which forms pores in the target cell membrane. Granzymes are then released from CTLs and travel through these pores into target cells, resulting in cell death (McMichael and Rowland–Jones, 2001). A less common mechanism of CTL killing is via Fas ligand on CTLs binding to the Fas receptor on target cells and initiating apoptosis (Katsikis *et al.*, 1995).

In addition to the above mechanisms, CTLs prevent infection and virus production through indirect means during HIV-1 infection. CTLs inhibit viral replication through production of cytokines, such as INF-γ (Meylan *et al.*, 1993; Emilie *et al.*, 1992). CTLs also produce chemokines that compete with HIV-1 for binding with host cell co-receptors (Barker *et al.*, 1998; Appay *et al.*, 2000). HIV-1 entry into host cells requires binding of gp120 on the HIV-1 envelope to the CD4 receptor and a co-receptor on the surface of the host cell. By producing chemokines that are natural ligands for HIV-1 co-receptors, CTLs can outcompete HIV-1 by successfully preventing binding and subsequent entry of HIV-1 into host cells. Two of the most common co-receptors used by HIV-1 *in vivo* are CCR5 and CXCR4. Ligands for CCR5 are the CC-chemokines MIP-1α, MIP-1β, and RANTES (Cocchi *et al.*, 1995; Wagner *et al.*, 1998). High levels of these CC-chemokines are secreted from $CD8^+$ T cells, and have been shown to be inversely correlated with HIV-1 viral load (Ferbas *et al.*, 2000). $CD8^+$ T cells also produce a soluble $CD8^+$ T-cell antiviral factor (CAF) distinct from the CC-chemokines (Barker *et al.*, 1998; Moriuchi *et al.*, 1996; Cocchi *et al.*, 1995; Levy *et al.*, 1996; McMichael and Rowland–Jones, 2001), which may suppress viral transcription through inhibiting the HIV promoter (Chang *et al.*, 2003). There is evidence of co-localization of granzymes and CC-chemokines inside CTLs, implying that the direct killing and inhibitory pathways are linked such that the same CTL may exert both mechanisms during cytotoxic activities (Wagner *et al.*, 1998).

1.3. *Development of successful CTLs*

Differentiation of $CD8^+$ T cells into activated, HIV-1-specific CTLs usually results from the encounter of $CD8^+$ T cells with antigen, commonly found on antigen presenting cells (APCs). However, not all $CD8^+$ T cells that are activated by APCs will become fully differentiated CTLs capable of cytotoxic function (Auphan–Anezin *et al.*, 2003). Differentiation of $CD8^+$ T cells into CTLs that are equipped to perform the functions described above is a result of many interactions involving multiple cell types. There is controversy in the literature as to the importance of $CD4^+$ T-cell help in the priming of the CTL response. Some have found $CD4^+$ T-cell help to be crucial for both priming and maintenance of the CTL response (Wang and Livingstone, 2003), while some have found $CD4^+$ T-cell help to be important only in the memory response (Shedlock and Shen, 2003; Janssen *et al.*, 2003). Furthermore, if $CD4^+$ T-cell help is necessary, there is speculation about what specific interactions take place between $CD4^+$ and $CD8^+$ T cells or whether intermediates are required. We describe below the known functions of dendritic cells and $CD4^+$ T cells in the successful development of CTLs.

1.3.1. *Activation by dendritic cells*

During viral infection, antigen presentation occurs primarily via dendritic cells and macrophages transporting antigen to T-cell rich areas of lymph tissues (LT) (Buseyne *et al.*, 2001). In recent years, dendritic cells (DCs) have been identified as the most efficient APCs during the course of HIV-1 disease progression (Bottomly, 1999). Thus, we focus here on the influence of dendritic cells in initiating the immune response and HIV-1-specific CTL activity.

Prior to antigen stimulation, DCs maintain an immature phenotype in the periphery and filter the environment for foreign antigen, but are inefficient at presenting antigen to T cells (Mellman and Steinman, 2001). Immature DCs internalize antigen upon initial encounter, and migrate to the LT undergoing maturation. Maturation is characterized by a decrease in antigen processing capability and an upregulation of costimulatory molecules CD80 and CD86 as well as increased expression of MHC Class II and B7.2 molecules (Sharpe and Freeman, 2002), all of which allows for proper activation of $CD4^+$ and $CD8^+$ T cells via DCs (Mellman and Steinman, 2001; Bousso and Robey, 2003; Banchereau and Steinman, 1998; Cella *et al.*, 1997). Thus we track two classes of dendritic cells: the immature DCs (IDCs) that capture antigen in the peripheral blood, and mature DCs (MDCs) that present antigen to T cells in the LT.

HIV-1 has the capability to exploit this process of antigen presentation to its advantage (Teleshova *et al.*, 2003). During infection, IDCs bind to gp120 on HIV-1 in the periphery via dendritic cell-specific, intracellular adhesion molecule-grabbing nonintegrin (DC-SIGN) (Curtis *et al.*, 1992; Geijtenbeek *et al.*, 2000). IDCs then internalize the virus through macropinocytosis (Sallusto *et al.*, 1995; Baribaud *et al.*, 2001), which extends the short half-life of the virus to several days and protects its infectious capability (Geijtenbeek *et al.*, 2000; Kwon *et al.*, 2002). IDCs then migrate into LT to present antigen to $CD4^+$ and $CD8^+$ T cells (Geijtenbeek *et al.*, 2000). Among the most significant events that follow is DC activation of resting $CD4^+$ T cells into HIV-1-specific T helper cells. This occurs via interaction of CD40-CD40L or secretion of cytokines (Ridge *et al.*, 1998).

Cell-associated virus carried on dendritic cells can infect both resting and activated $CD4^+$ T cells even more efficiently than free virus (McMichael and Rowland–Jones, 2001; Engering *et al.*, 2002; Albert *et al.*, 1998). Upon contact with a dendritic cell, HIV-1 receptors CD4, CCR5, and CXCR4 on T cells co-localize on the cell surface toward the junction of the dendritic cells with T cells, enhancing the opportunity for successful viral entry (McDonald *et al.*, 2003). There is some evidence that DCs can also become infected, due to their surface expression of both CD4 molecules and the co-receptors CCR5 and CXCR4 (Fong *et al.*, 2002; Patterson *et al.*, 2001). However, this is not thought of as significant in disease progression because the total number of DCs in blood is much less than that of other cell types (Liu, 2001; Grabbe *et al.*, 2000). Additionally, studies have found HIV-1-infected DCs *in vivo* or *in vitro* are significantly less prevalent than infected $CD4^+$ T cells (McIlroy *et al.*, 1995; Haase, 1986). Therefore, infection of DCs is not considered in this study.

Here, we identify the infection route by which virus is carried from the periphery into LT via dendritic cells. We acknowledge the possibility that there are other mechanisms of virus entry into LT but focus on this as the major pathway.

1.3.2. *$CD4^+$ T-cell help to $CD8^+$ T cells*

$CD4^+$ T cells that become activated are termed helper T cells in that they play a significant role in aiding CTLs to perform their effector function. $CD4^+$ T cells have been implicated in both the initial priming (Wang and Livingstone, 2003) as well as long-term memory $CD8^+$ T-cell response (Janssen *et al.*, 2003; Shedlock and Shen, 2003). Thus, since $CD4^+$ T cells decline significantly early in the infection process, the $CD8^+$ T-cell response is adversely affected during both short- and long-term immunity.

HIV-1-specific helper T cells correlate strongly with HIV-1-specific CTLs that have antiviral activity (Kalams and Walker, 1998; Kalams *et al.*, 1999).

The helper function of $CD4^+$ T cells can be attributed to both the release of cytokines as well as direct cell-cell interactions. Activated $CD4^+$ T cells release IL-2, a cytokine that induces T-cell proliferation, thus expanding the populations of both $CD4^+$ and $CD8^+$ T cells. However, $CD4^+$ T cells that have encountered HIV-1 on dendritic cells express various activation markers that can further interact with either dendritic cells or $CD8^+$ T cells. This initiates the stages of $CD8^+$ T-cell activation that lead to fully differentiated CTLs. CD40L, expressed on the surface of activated $CD4^+$ T cells, is important for effector function in immunity (van Essen *et al.*, 1995), particularly during HIV-1 infection (Bennett *et al.*, 1998; Ridge *et al.*, 1998; Schoenberger *et al.*, 1998). This has been shown in experiments in which CD40L inhibition results in a population of unprimed $CD8^+$ T cells (Schoenberger *et al.*, 1998). The interaction of CD40L on activated T cells with CD40 on DCs further stimulates virus-carrying DCs to activate $CD8^+$ T cells (Cella *et al.*, 1996). Once activated, $CD4^+$ T cells can also have direct interaction with $CD8^+$ T cells (Bourgeois *et al.*, 2002). Thus, $CD4^+$ T-cell help to CTLs is ultimately mediated by the interplay between $CD4^+$ T cells and DCs (Caux *et al.*, 1994). The HIV-1 protein Nef can also interfere with the antigen presentation pathway by intersecting CD40-CD40L activation (Pope, 2003; Andrieu *et al.*, 2001).

HIV-1-specific CTL killing and inhibition is crucial even as early as the acute stage of infection (Zajac *et al.*, 1998). Throughout chronic infection, $CD4^+$ T cells lose their ability to dictate functionality of $CD8^+$ T cells towards eliminating the virus, or ensuring specificity of adaptive immunity through a steady decline in $CD4^+$ T-cell numbers (McMichael and Rowland–Jones, 2001; Appay *et al.*, 2000; Lieberman *et al.*, 2001). This effect would be even more pronounced however, in the absence of CTLs specific for HIV-1 activated during acute infection. Even though there may be a significant proportion of $CD8^+$ T cells that are activated, virus can still persist without specific T-cell immunity due to loss of instructional $CD4^+$ T helper cells (Goulder *et al.*, 2000). A higher frequency of HIV-1-specific $CD4^+$ T-cell responses is characteristic of LTNPs as compared to typical progressors (Pitcher *et al.*, 1999). There exists both a quantitative and qualitative correlation between the specificity of the $CD4^+$ T-cell response during HIV-1 infection and the functionality of the $CD8^+$ T-cell response (McMichael and Rowland–Jones, 2001).

1.4. *Failure of CTLs in controlling infection*

There are a number of possibilities for the observed decline in CTL immunity in end-stage infection. In addition to the loss in the number of CTLs, CTL dysfunction could be due to the presence of a high number of activated cells without functionality (Kostense *et al.*, 2002; Trimble and Lieberman, 1998), alterations in cell distributions at infected sites (Pantaleo *et al.*, 1997b), or virus mutation to evade CTLs (Phillips *et al.*, 1991).

CTLs may not necessarily traffic to sites of virus replication (Pantaleo *et al.*, 1997b; Wherry *et al.*, 2003). CCR7 receptors are generally found on molecules that travel to lymphoid tissue (Forster *et al.*, 1999). HIV-1-specific CTLs are characterized as CCR7-, thus lacking the capability for travel to LT, the site of most viral infection and production (Chen *et al.*, 2001). This finding is supplemented by evidence that perforin is not found in the lymph nodes of infected patients in both acute and chronic disease and perforin and granzymes are not co-localized in CTLs found in the LT (Andersson *et al.*, 1999; 2002), both suggesting that CTLs are not in the anatomic location where they are most needed.

HIV-1-specific CTLs are lower in number than CTLs detected during other viral infections such as Epstein–Barr virus or CMV (Gillespie *et al.*, 2000; Hislop *et al.*, 2001). This may be a key reason for better immune control observed in other infections. Additionally, there are functional differences between HIV-1-specific CTLs and for example, CMV-specific CTLs. HIV-1-specific $CD8^+$ T cells express lower levels of perforin compared to CMV-specific cells (Papagno *et al.*, 2002). HIV-1-specific CTLs are also in a less mature state as compared to other CTLs, and their killing and inhibitory capacities are reduced (van Baarle *et al.*, 2002). In this work, we assume activated $CD4^+$ T cells are HIV-1-specific and correlate our simulation results with data on HIV-1-specific $CD4^+$ and $CD8^+$ T cells.

2. Model of HIV-1 Infection

We develop a mathematical model to elucidate the role of $CD8^+$ T cells in various stages of HIV-1 infection as well as to explore priming and differentiation of HIV-1-specific CTLs. The most common clinical markers for progression of HIV-1 infection progression are $CD4^+$ T-cell and viral load measurements from the blood. However, most lymphocytes (98%) reside in lymph tissues and circulate constantly between LT and blood (Haase, 1999). Therefore, we build a two-compartment model of HIV-1 infection in blood and lymph tissues (Fig. 1), similar to our previous work (Bajaria

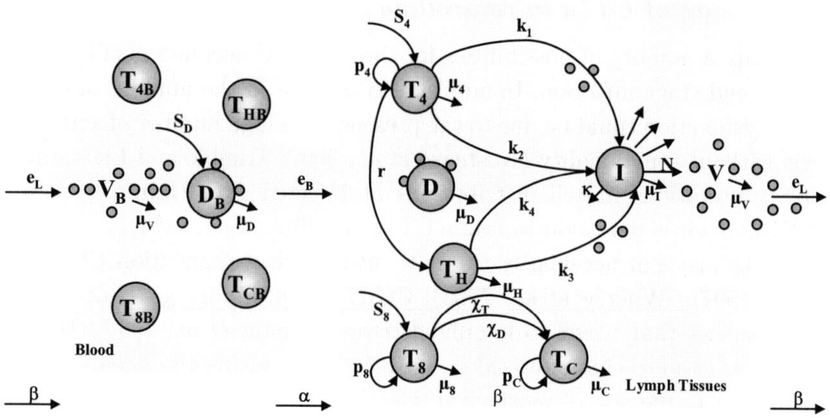

Fig. 1. **Two-compartment HIV-1 infection model.** The model developed here explores mechanisms of HIV-1 infection progression based on the interactions of a single viral strain ($V(t)$) of HIV-1 with resting ($T_4(t)$) and activated ($T_H(t)$) subclasses of CD4$^+$ T cells at rates k_1 and k_3, respectively. This leads to the development of infected cells ($I(t)$), which are also created by infection via virus ($V_B(t)$) on IDCs from the blood ($D_B(t)$) (at rates k_2 and k_4). Upon migration into the LT, IDCs mature into MDCs ($D(t)$) and activate T_4 into T_H at rate r. Infection directly produces actively infected ($I(t)$) cells that are assumed to be present predominantly in the LT compartments, owing to the assumption that infected cells in the blood (containing only 2% of all T cells) constitute a relatively small contribution to overall infection, and that most HIV-1 replication occurs in the LT (Richman, 2000). CD8$^+$ T cells ($T_8(t)$) are activated into CTLs ($T_C(t)$) in the LT due to interaction with either $T_H(t)$ (at rate χ_T) or MDCs ($D(t)$) (at rate χ_D). CTLs can kill infected cells (at rate κ) and inhibit infection of T_4 cells (saturating at c_1 and c_2) and T_H cells (saturating at c_3 and c_4), and inhibit virus production from I (saturating at c_5). CTLs can proliferate in response to IL-2 secreted from T_H cells (at rate p_C). Resting and activated CD4$^+$ and CD8$^+$ T cells proliferate at rates p_4 and p_8 and circulate between blood ($\mathcal{B}$) and LT ($\mathcal{L}$) at rates e_B and e_L scaled by α and β. All initial conditions and parameters are described in Tables 1–3.

et al., 2002). We account for the following populations in both the peripheral blood and lymph tissues: resting and activated CD4$^+$ T cells, resting and activated CD8$^+$ T cells, infected CD4$^+$ T cells, virus, and dendritic cells.

2.1. *Model equations*

Using thirteen nonlinear ordinary differential equations, we tracked dynamics of CD4$^+$ (T_4) and CD8$^+$ (T_8) T cells circulating between blood ($\mathcal{B}$) and lymph tissue ($\mathcal{L}$) compartments. During infection, virus levels are monitored in both the blood ($V_{\mathcal{B}}$) as well as lymph tissues (V), together with infected cells (I) and activated CD4$^+$ (T_H) and CD8$^+$ (T_C) T cells in the lymph tissues, MDCs in the lymph tissues (D) and IDCs in the blood (D_B). Equations for the model system are as follows:

$$\frac{dT_4}{dt} = S_4 - \mu_4 T_4 - \frac{rDT_4}{D+f_1}\left(\frac{f_5}{V+f_5}\right) - k_1\left(\frac{c_1}{T_C+c_1}\right)VT_4 - k_2\left(\frac{c_2}{T_C+c_2}\right)DT_4 + \frac{p_4 T_H T_4}{T_H+f_2} + \alpha e_B T_{4B} - e_L T_4 \tag{1}$$

$$\frac{dT_{4B}}{dt} = \beta e_L T_4 - e_B T_{4B} \tag{2}$$

$$\frac{dT_H}{dt} = \frac{rDT_4}{D+f_1}\left(\frac{f_5}{V+f_5}\right) - \mu_H T_H - k_3\left(\frac{c_3}{T_C+c_3}\right)VT_H - k_4\left(\frac{c_4}{T_C+c_4}\right)DT_H + \alpha e_B T_{HB} - e_L T_H \tag{3}$$

$$\frac{dT_{HB}}{dt} = \beta e_L T_H - e_B T_{HB} \tag{4}$$

$$\frac{dI}{dt} = k_1\left(\frac{c_1}{T_C+c_1}\right)VT_4 + k_2\left(\frac{c_2}{T_C+c_2}\right)DT_4 + k_3\left(\frac{c_3}{T_C+c_3}\right)VT_H + k_4\left(\frac{c_4}{T_C+c_4}\right)DT_H - \mu_I I - \kappa T_C I \tag{5}$$

$$\frac{dV}{dt} = N\left(\frac{c_5}{T_C+c_5}\right)\mu_I I - \mu_V V \tag{6}$$

$$\frac{dV_B}{dt} = \beta e_V V - \mu_V V_B \tag{7}$$

$$\frac{dT_8}{dt} = S_8 - \mu_8 T_8 - \chi_D \frac{T_H}{T_H+f_3} DT_8 - \chi_T T_H T_8 + \frac{p_8 T_H T_8}{T_H+f_4} + \alpha e_B T_{8B} - e_L T_8 \tag{8}$$

$$\frac{dT_{8B}}{dt} = \beta e_L T_8 - e_B T_{8B} \tag{9}$$

$$\frac{dT_C}{dt} = \chi_D \frac{T_H}{T_H+f_3} DT_8 + \chi_T T_H T_8 - \mu_C T_C + p_C \frac{T_H}{T_H+f_6} T_C + \omega\alpha e_B T_{CB} - e_L T_C \tag{10}$$

$$\frac{dT_{CB}}{dt} = \beta e_L T_C - \omega e_B T_{CB} \tag{11}$$

$$\frac{dD}{dt} = \alpha\delta_B \frac{V_B}{V_B+\psi} D_B + \lambda_L \frac{V}{V+\phi_L} D - \mu_D D \tag{12}$$

$$\frac{dD_B}{dt} = S_D - \mu_D D_B + \lambda_B \frac{V_B}{V_B + \phi_B} D_B - \delta_B \frac{V_B}{V_B + \psi} D_B \tag{13}$$

Equations (1) and (8) describe resting CD4$^+$ and CD8$^+$ T-cell populations in the lymph tissues. Their populations depend on influx from the blood at rate e_B (scaled by α for compartmental exchange) and loss due to emigration at rate e_L. Equations (2) and (9) describe resting CD4$^+$ and CD8$^+$ T-cell populations in the blood and have similar circulation terms (scaled by β for entry into the blood). The source of CD4$^+$ (S_4) and CD8$^+$ (S_8) T cells represents new constant input from the thymus while their half-lives are representated by the decay rates μ_4 and μ_8, respectively. Contact between a dendritic cell and a resting CD4$^+$ T-cell can result in either activation into CD4$^+$ T helper cells (T_H) at rate r (saturating at f_1) or infection at rate k_2. Infection can also result directly by meeting virus at rate k_1. Resting cells can also proliferate at rate p_4, dependent on the amount of IL-2 in the environment secreted from T helper cells (T_H), which we assume is a proportion relative to T_H cells. Activation can be hindered by Nef interference with the CD40-CD40L pathway of DC activation (saturating at f_5) (Pope, 2003). Activated CD4$^+$ T cells (T_H) have a death rate (μ_H), and become infected by directly meeting virus (k_3) or through presentation of virus by a dendritic cell (at rate k_4) (Douek *et al.*, 2002). Similar circulation terms for T_H cells describe emigration from and to their counterparts in the blood (T_{HB} in Eq. 4). Infected cells result from any of the pathways described above and are subject to a death rate μ_I. When dendritic cells come into contact with resting CD8$^+$ T cells at rate χ_D, new CTLs are created, an effect that is enhanced in the presence of T_H cells, saturating at rate f_3. CD8$^+$ T cells can also become activated through direct contact with a T_H at rate χ_T because CD8$^+$ T cells transiently express CD40 after activation (Bourgeois *et al.*, 2002). The levels of resting CD8$^+$ T cells are supplemented by proliferation (at rate p_8), also in response to IL-2 secreted from T_H cells. There is evidence that pre-terminally differerentiated CTLs can proliferate (Champagne *et al.*, 2001) in response to IL-2 (Jin *et al.*, 1998) and the majority of HIV-1-specific CD8$^+$ T cells are at an intermediate stage of differentiation (Appay *et al.*, 2002; Champagne *et al.*, 2001), thus we include proliferation of CTLs occurring at rate p_C. This effect saturates at rate f_6, since IL-2 expandable CTL responses are reduced in end-stage HIV-1 infection (Jin *et al.*, 1998). Because there is evidence that HIV-1-specific CTLs lack the capacity to home to the LT during infection (Chen *et al.*, 2001), we include a factor ω as the percentage of CTLs that circulate back into the LT from the blood. IDCs are initially present in homeostasis in blood with source S_D and death rate μ_D. They are recruited into LTs at rate λ_B in response to virus in the blood (V_B), a process which saturates at ϕ_B. Blood IDCs take up virus and migrate to lymph tissues at rate δ_B, which also saturates at ψ. There is also a recruitment rate (λ_L) of MDCs into the lymph tissues in response to viral stimulus or inflammation, an activity which saturates at ϕ_L. Virus (V) is produced in the LT from infected cells (at rate N) and has a short half-life (μ_V). Virus in the blood flows in from the LT (e_V), and is lost to natural death (μ_V).

Direct cell-to-cell contact is necessary for CTLs to eliminate infected cells (Folkvord *et al.*, 2003). We incorporate clearance of infected cells due to CTLs (at rate κ) acting through the perforin/granzyme pathway and/or Fas pathway. As of yet there is little evidence as to what proportion of cells are eliminated by Fas-mediated apoptosis and which by perforin and granzyme killing; only that both result in target cell death. We include these cytolytic activities occurring in the lymph tissues only, concurrent with previous data that most viral replication and production processes occur in the LT (Haase, 1999). Free virus is not a common target for clearance by CTLs and thus the virus equation only includes a loss term due to half-life of the virus. CTLs inhibit both viral infection and virus production. Low CTL counts allow full infectivity and virus production, whereas increased CTL levels impair these infection processes. Thus, the inhibitory effect of CC-chemokines is captured by CTLs decreasing infection rates of CD4$^+$ T cells (k_1 through k_4) and the rate of virus production from infected cells (N), an effect that saturates at a maximum number of CTLs (c_1 through c_5).

2.2. *Parameter estimation*

Table 1 presents initial conditions for the thirteen variables in our model system. In the absence of infection, CD4$^+$ T cells, CD8$^+$ T cells, and dendritic cells exhibit homeostasis. Initial values for infected cell populations and virus are all zero to ensure that initially there is no infection before virus is introduced into the system. We account only for recently emigrated virus-carrying IDCs from the blood, without considering any background level of MDCs in the lymph tissues. Thus the initial condition for MDCs in the LT is zero.

Tables 2 and 3 present parameter values used in our simulations. Infection rates (k_1 through k_4) are estimated using LHS analysis (discussed below) as no data exist on these processes. We estimate, however, that infection of CD4$^+$ T helper cells (at rates k_3 and k_4) highly exceeds that of resting cells (at rates k_1 and k_2) due to high susceptibility of activated

Table 1. Initial conditions for HIV-1 model.

Variable	Definition	Value	Units	Reference
Populations in LT				
$T_4(0)$	Resting CD4$^+$ T cells	2×10^{11}	cells	1
$T_H(0)$	Activated CD4$^+$ T cells	0	cells	
$I(0)$	Infected CD4$^+$ T cells	0	cells	
$V(0)$	Virus	0	viral RNA	
$T_8(0)$	Resting CD8$^+$ T cells	1×10^{11}	cells	1
$T_C(0)$	Activated CD8$^+$ T cells	0	cells	
$D(0)$	Dendritic cells	0	cells	
Populations in blood				
$T_{4B}(0)$	Resting CD4$^+$ T cells	1000	cells/mm^3	1
$T_{8B}(0)$	Activated CD4$^+$ T cells	0	cells/mm^3	
$V_B(0)$	Virus concentration	10	viral RNA/mL	
$T_8(0)$	Resting CD8$^+$ T cells	500	cells/mm^3	1
$T_C(0)$	Activated CD8$^+$ T cells	0	cells/mm^3	
$D_B(0)$	Dendritic cells	20	cells/mm^3	2

[1]Haase, 1999.
[2]Barron *et al.*, 2003.

Table 2. Parameter values for homeostasis.

Parameter	Value	Units	Reference
Scaling between compartments			
Blood to LT (α)	5×10^6	mm^3 or μL	
LT to blood (β)	2×10^{-7}	mm^3 or μL	
Circulation of T cells			
LT to blood (e_L)	0.01	/day	1
Blood to LT (e_B)	0.4	/day	1
Source terms			
CD4$^+$ T cells into LT (S_4)	4×10^8	cells/day	2
CD8$^+$ T cells into LT (S_8)	2×10^8	cells/day	2
Dendritic cells into blood (S_D)	0.1	cells/day	estimated
Death rates			
Resting CD4$^+$ T cells (μ_4)	0.002	/day	3
Resting CD8$^+$ T cells (μ_8)	0.002	/day	3
Dendritic cells (μ_D)	0.005	/day	estimated

[1]Sprent, 1973; Sprent and Basten, 1973.
[2]Haase, 1999.
[3]Richman, 2000.

Table 3. Parameter values for infection.

Parameter	Value	Units
Death rates		
Activated $CD4^+$ T cells (LT) (μ_H)	0.5	/day
Activated $CD8^+$ T cells (LT) (μ_C)	1.5	/day
Infected $CD4^+$ T cells (μ_I)	0.1	/day
Free virus (μ_V)	3.0	/day
Infection rates		
Resting $CD4^+$ T cells by virus (k_1)	2.0×10^{-14}	/day-virion
Resting $CD4^+$ T cells by dendritic cells (k_2)	4.0×10^{-13}	/day-cell
Activated $CD4^+$ T cells by virus (k_3)	4.0×10^{-11}	/day-virion
Activated $CD4^+$ T cells by dendritic cells (k_4)	2.0×10^{-7}	/day-cell
Number of virions produced from infected cells (N)	800	virions/cell
Activation/Differentiation/Proliferation rates		
Activation of $CD4^+$ T cells by dendritic cells (r)	0.01	/day
Differentiation of $CD8^+$ T cells by dendritic cells (χ_D)	1.0×10^{-10}	/day-cell
Differentiation of $CD8^+$ T cells by $CD4^+$ T helper cells (χ_T)	8.0×10^{-11}	/day-cell
Proliferation of $CD4^+$ T cells (p_4)	0.02	/day
Proliferation of $CD8^+$ T cells (p_8)	0.02	/day
Proliferation of CTLs (p_C)	0.5	/day
Half-saturation constants		
$CD4^+$ T cells on activation (f_1)	10^{11}	cells
$CD4^+$ T cells on proliferation (f_2)	10^7	cells
$CD4^+$ T-cell help to $CD8^+$ differentiation (f_3)	10^4	cells
$CD8^+$ T cells on proliferation (f_4)	8.0×10^6	cells
Virus on interference with activation (f_5)	8.0×10^6	cells
Migration of dendritic cells to LT (ψ)	1.0	virus
Recruitment of dendritic cells to LT (ϕ_L)	100.0	virus
Recruitment of dendritic cells to blood (ϕ_B)	10.0	virus
Inhibition constants		
$CD4^+$ T-cell infection by virus (c_1)	10^8	cells
$CD4^+$ T-cell infection by dendritic cells (c_2)	10^6	cells
$CD4^+$ T helper cell infection by virus (c_3)	10^8	cells
$CD4^+$ T helper cell infection by dendritic cells (c_4)	10^7	cells
Virus production from infected cells (c_5)	10^8	cells
Clearance rate of infected cells (κ)	9.0×10^{-10}	/day-cell
Circulation/Migration/Recruitment		
Proportion of CTLs that travel to LT from blood (ω)	0.002	scalar
Circulation of virus (LT to blood) (e_V)	0.8	/day
Migration of blood DC to LT (δ_B)	0.0448	/day
Recruitment of dendritic cells to LT (λ_L)	0.007	/day
Recruitment of dendritic cells to blood (λ_B)	0.04	/day

All parameters were estimated except μ_I (Cavert *et al.*, 1997), μ_V (Perelson *et al.*, 1996; Stafford *et al.*, 2000), and N (Haase, 1999).

cells to infection (Gougeon and Montagnier, 1993). During contact with MDCs, HIV-1 receptors co-localize on the surface of T cells (Douek *et al.*, 2002; McDonald *et al.*, 2003). Additionally, Tat, a gene expressed during HIV-1 infection, exploits immature dendritic cell activities. Tat mediates chemokine upregulation in immature dendritic cells, causing enhanced recruitment of activated T cells (Izmailova *et al.*, 2003). Thirdly, virus that has been taken up by IDCs has a longer half-life than free virus. Lastly, the kinetics of viral transfer from dendritic cells to T cells is much greater than that of free virus to T cells (Gummuluru *et al.*, 2003). This last feature is crucial to infection due to the short half-life of activated cells (Gougeon and Montagnier, 1993). These processes make trans-infection of $CD4^+$ T cells by DCs very efficient, and the infection rate of cell-associated virus is estimated to be much higher than that by free virus (i.e. $k_4 > k_3 > k_2 > k_1$). Similarly, death rates of activated cells are estimated as greater than that of resting cells, with free virus having the shortest half-life (i.e. $\mu_V > \mu_H$ and $\mu_C > \mu_4$ and μ_8). Infected cells are assumed to be productively infected and able to actively produce virus over their lifespan (Haase, 1999). Productively infected cells produce anywhere from 100–1000 virions per cell (Haase, 1999), and we use 800 virions/cell as the baseline value in our model simulations. This parameter has been shown to have a significant impact on the dynamics of infection (Bajaria *et al.*, 2002). Half-saturation constants for the effect of T-cell help to $CD8^+$ T cells (f_3), the strength of CTL inhibition (c_1 through c_5), the impact of dendritic cells on $CD4^+$ T-cell activation (f_1), inhibition of $CD4^+$ T-cell activation by Nef-induced interference (f_5) and the influence of IL-2 on T-cell proliferation (f_2, f_4, f_6) have also been estimated. α and β are included as compartmental parameters scaling exchange between LT and blood compartments, since blood cells are measured per mm^3 of blood and the LT in total cells.

2.2.1. *Uncertainty and sensitivity analysis*

There are several parameters in our model for which no *in vivo* nor human data exists. By comparison with non-human data, *in vitro* experiments, and studies of other pathogens, we can estimate a wide range of possible values for each unknown parameter. Using Latin hypercube sampling (LHS), we find values for those parameters for which reported estimates do not exist (Blower and Dowlatabadi, 1994). This process generates a hypercube through a random combination of parameter values from each estimated range; simulations are then performed for each sample. Output is compared with clinical data on T cells and viral load, and we ultimately choose those parameters for which model simulations exhibit the best fit with known output variables. We further extend this LHS method to test system output sensitivity to various key parameters. By examining the effect each key parameter has on the outcome variable (in this case viral load) using a partial rank correlation (PRC), we can assess the relative influence of each parameter on system dynamics (e.g. Blower and Dowlatabadi, 1994). In fact, those parameters that have significant PRC values have been shown to be bifurcation parameters. Using the LHS/PRC method to study sensitivity and identify bifurcations is a tool to study differences in disease outcomes.

3. Results

Using the equations and parameters described above, we simulate the three distinct stages of typical progression of infection. We then use this positive control to vary parameters related to CD8$^+$ T-cell dynamics and examine effects to the typical HIV-1 course of infection.

3.1. *Healthy control*

We simulate total CD4$^+$ and CD8$^+$ T cells in healthy, uninfected individuals as a negative control. In healthy individuals, total CD4$^+$ T-cell counts in the LT are about 2×10^{11} cells and 1000 cells/μL in the blood (Haase, 1999). The ratio of CD4$^+$ T cells to CD8$^+$ T cells is approximately 2:1 in healthy individuals in both blood and LT (Haase, 1999; Rosenberg *et al.*, 1998). Our model simulations correlate with estimates of 10^{11} CD8$^+$ T-cells in the LT and 500 cells/μL of blood from (Haase, 1999) (data not shown). There is a median 14–20 dendritic cells per milliliter of blood in healthy individuals (Barron *et al.*, 2003). All infected cell populations and dendritic cells in the lymph tissues are set to zero prior to infection.

3.2. *Typical progression of HIV-1*

3.2.1. *CD8$^+$ T cells in acute infection*

The acute stage of HIV-1 infection is characterized by extremely low CD4$^+$ T-cell counts and high levels of virus, often higher than 10^6 copies/mL of blood. Consequently, activation and differentiation of large numbers of CD8$^+$ T cells into CTLs results in a significant number that provide killing and inhibitory activities towards viral infection and production. A strong

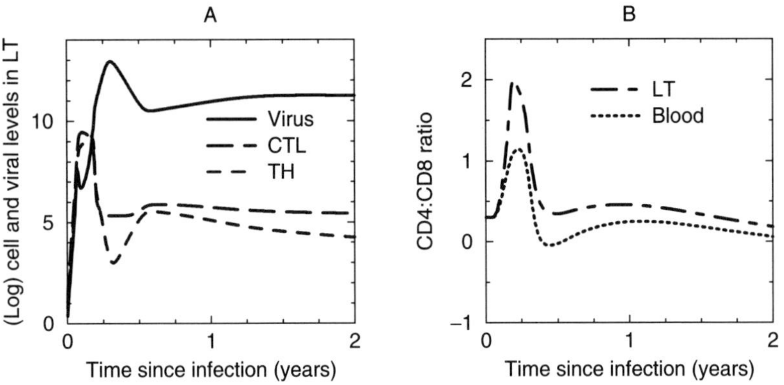

Fig. 2. **Acute stage of HIV-1 infection.** Panel A: Virus (solid line), CTL (long dashed), and TH (short dashed) dynamics over 2 years post-infection in the LT. Panel B: CD4/CD8 ratio in the blood (dotted) and LT (dot-dashed) over 2 years post-infection.

CTL response coincides with resolution of high viremia in the acute stage whereby virus levels decrease approximately 100-fold (McMichael and Rowland–Jones, 2001; Koup *et al.*, 1994; Safrit and Koup, 1995). Additionally, the CD4:CD8 ratio is inverted from 2:1 to 1:2 after seroconversion (Rosenberg *et al.*, 1998; Schacker *et al.*, 1996).

Our model simulations reflect a rise in CTL number concurrent with a rapid increase in viremia during acute infection (Fig. 2, Panel A). $CD4^+$ T-cell help and CTL levels initially follow this increase in virus, while for the remainder of infection progression they correlate inversely with viral load (Ogg *et al.*, 1998). We simulate the ratio inversion of CD4 to CD8 from 2 to 0.5 during acute infection; however our results show that there is a slightly lower CD4:CD8 ratio in the blood with a higher ratio in the LT (Fig. 2, Panel B). This indicates either a higher number of $CD8^+$ T cells or lower numbers of $CD4^+$ T cells in the blood as compared to LT, reflecting the skewed distribution of lymphocytes during massive viral infection.

3.2.2. *$CD8^+$ T cells in chronic and end-stage infection*

After the acute stage of infection, which occurs over the first few months of infection, resting $CD4^+$ T-cell, activated $CD4^+$ T helper cell, and CTL levels increase and virus decreases to a quasi-steady-state level. This resolution of virus to a quasi-steady-state level, referred to as the viral setpoint, occurs for several years and is characteristic of the chronic stage of disease (Ho *et al.*, 1995). Our simulation shows a quasi-steady-state level of virus in the blood and LT following the acute stage of infection that lasts for several years (Fig. 3, Panel A). We also show the typical rebound of resting $CD4^+$ and $CD8^+$ T cells to high levels followed by the gradual $CD4^+$ T-cells decline throughout a typical ten-year course of infection (Fig. 3, Panel B).

Resting $CD8^+$ T cells decline initially due to massive expansion of the CTL population, but recover to normal levels rapidly and only decrease slightly throughout infection progression (Fig. 3, Panel B and McCune *et al.*, 2000; Hellerstein *et al.*, 1999; Kovacs *et al.*, 2001). Because there is little to no helper activity in the blood after the acute stage of infection in data (Betts *et al.*, 2001), we are able to examine dynamics of resting and activated $CD4^+$ and $CD8^+$ T cells only in LT (Fig. 3, Panels C and D).

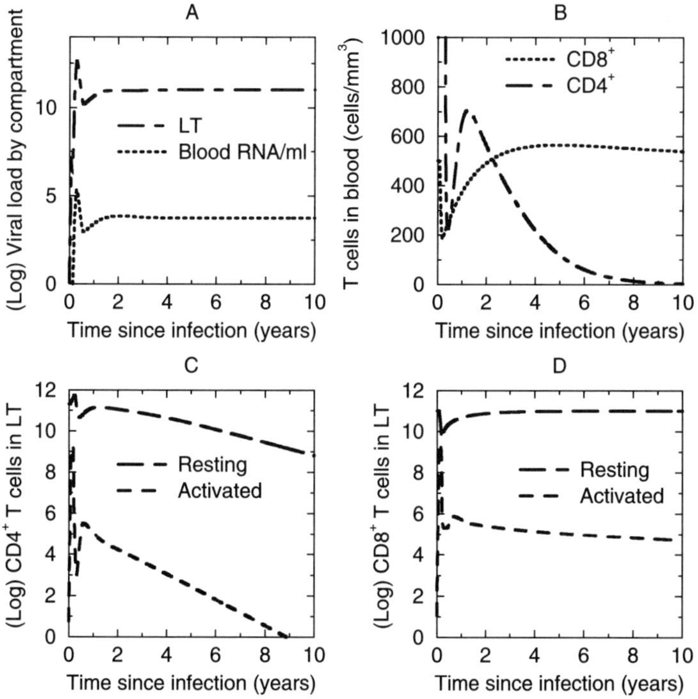

Fig. 3. **Three stages of typical HIV-1 infection.** Shown are simulations of early, chronic, and AIDS stages over a 10-year course of infection. Panel A: Virus levels in blood (dotted line) and LT (dot-dashed). Panel B: Total $CD4^+$ (dot-dashed) and $CD8^+$ (dotted) T cells in blood. Panel C: Resting (long dashed) and activated (short dashed) $CD4^+$ T cells in LT. Panel D: Resting (long dashed) and activated (short dashed) $CD8^+$ T cells in LT.

Table 4. Comparison of simulation at year 3 post-infection with clinical data.

Population	Model value	Clinical value	Reference
Resting $CD4^+$ T cells in LT	7×10^{10}	10^{10} cells	1
Resting $CD8^+$ T cells in LT	8×10^{10}	10^{10} cells	1
Infected cells in LT	4×10^{9}	10^{10} cells	1
Virions in LT	9×10^{10}	5×10^{10} virions	1
Activated $CD4^+$ T cells in blood	0.0000007%	0.12% of $CD4^+$ T cells	2
Activated $CD8^+$ T cells in blood	8.8%	0.1%–10% of $CD8^+$ T cells	3
Median DC in blood	10 cells/mL	7 cells/mL	4

[1]Haase, 1999.
[2]Pitcher *et al.*, 1999.
[3]Ogg *et al.*, 1998; Scott–Algara *et al.*, 2001.
[4]Barron *et al.*, 2003.

For comparison, we present clinical data for cell populations and virus in the chronic phase of infection with results from our model simulations (Table 4).

Progression to AIDS occurs for a number of reasons, and one key factor is likely the ultimate failure of the CTL response in controlling viral growth and spread. This may occur due to any of the reasons outlined in Section 1.4. With our model, we can observe decreased CTL levels with ongoing progression of infection, alterations in cell distributions, and a high number of activated cells with limited function.

We observe $CD4^+$ T-cell decline to AIDS-defining levels (<200 cells/mm^3 of blood) in the latter stage of infection. However, we do not observe an exponential rise in viral load in the blood in the end stage of infection. This could be due to several factors, among them the lack of target cells for infection once resting and activated $CD4^+$ T cells have declined to extremely low levels (Phillips, 1996). Homeostatic mechanisms may bring more uninfected target $CD4^+$ and $CD8^+$ T cells from the thymus during infection; however, we have not included this effect as we are solely examining CTL dynamics in the two compartments of blood and lymph tissues.

3.3. *HIV-1-specific CTL levels*

Measurement of HIV-1-specific $CD8^+$ T-cell responses varies between studies. Specificity for a single peptide is not indicative of the breadth of the CTL response. One study that measures specificity by flow cytometric detection of intracellular IFN-γ, found 1.4–22% of circulating $CD8^+$ T cells to be HIV-1-specific (Migueles and Connors, 2001). Gag, an HIV-1

Table 5. HIV-1-specific T cells in three stages in our model simulations.

Cell type	Acute (50 days)	Chronic (3 years)	AIDS (10 years)
$CD4^+$ helper T cells in blood	0.009%	0.0000007%	0.000000003%
$CD8^+$ cytotoxic T cells in blood	35%	8.8%	1.1%

protein, is most predominantly recognized by CTLs during asymptomatic HIV-1 infection (Johnson *et al.*, 1991; van Baalen *et al.*, 1993). A study that examined the frequency of CTL precursors in response to Gag found them to be in the range of 0.005–0.3% (Klein *et al.*, 1995). In another study, the gag-specific $CD4^+$ T-cell proportion ranged from 0.2–2% of total $CD4^+$ T cells with a mean of 1%. HIV-1-specific $CD8^+$ T cells ranged from 0.1–5% with a mean of 3% (Sester *et al.*, 2000). In that study, it was found that the frequency of HIV-1-specific $CD8^+$ T cells was typically higher than that of HIV-1-specific $CD4^+$ T cells. In general, as much as 10% of the $CD8^+$ T-cell population could be activated against HIV-1 antigens (Pantaleo *et al.*, 1997a). In Table 5, we present variations in specificity of $CD4^+$ and $CD8^+$ T cells concomitant with progression of infection based on our model simulations.

We see that there are extremely low percentages of activated $CD4^+$ T cells in blood. We need to examine LT dynamics (Fig. 3, Panel C) to observe the existence of $CD4^+$ T-helper cell activity, though it is undetectable in the blood compartment. CTLs however are greater in number than $CD4^+$ T helper cells, but also diminish throughout progression of infection (Fig. 3, Panel D).

3.4. *HIV-1-specific CTL action*

It is not known whether CTL killing or inhibition is more influential during infection, or whether or not a significant reduction in one or more of these mechanisms would have an impact on cell or viral dynamics. We vary those parameters that impact CTL action on viral infection and production (half-saturation constants on k_1 through k_4 and N). From these results (Table 6), we can see that infection via cell-cell interactions mediated by DCs (c_2 and c_4) has a larger impact than infection by free virus (c_1 and c_3) for both resting and activated $CD4^+$ T-cell infection. Our results show that a 100-fold change in inhibition of cell-mediated infection yields significant changes in disease progression, as compared to a 10,000-fold change

Table 6. Effects of variations in CTL activity on disease dynamics.

Parameter	Fold change	Results
Inhibition		
Viral infection of resting CD4+ T cells (c_1)	↓ 10,000	delayed acute stage
DC infection of resting CD4+ T cells (c_2)	↓ 100	↑ resting CD4+ T cells
Viral infection of activated CD4+ T cells (c_3)	↓ 10,000	no significant change
DC infection of activated CD4+ T cells (c_4)	↓ 100	delayed acute stage
Virus production by infected cells (c_5)	↓ 10	delayed acute stage
Killing		
Infected cells by CTLs (κ)	↑ 100	lower viral setpoint

on inhibition of virus infection necessary to significantly affect the results. Interestingly, changing the infection rate of activated CD4+ T cells by virus is not affected significantly by CTLs. Activated cells may not contribute significantly to the infected cell class due to their extremely short half-life. Additionally, the interaction between T helper cells and DCs could be far more influential in the infection process than the effect of free virus to T helper cells.

It is important to note that CTLs do not prevent infection, but only alter disease dynamics, i.e. the establishment of the acute stage of infection or the viral setpoint. Likely other cell types (such as those involved in innate immunity) or mutation in co-receptors (such as the CCR5Δ32 deletion) may play a role in the possible elimination of or resistance to initial virus introduced into the host.

3.5. *CD4+ T helper cells and dendritic cells work together to yield a successful CTL response*

One key hypothesis for the failing CTL response is the declining role of CD4+ T-cell help to CD8+ T-cell action. CD4+ T cells both activate and give instructional signals to CD8+ T cells during priming of CTLs. Low levels of help early in infection are attributed to a highly non-specific CD8+ T-cell response, which is ineffective for long-term immune control (McMichael and Rowland–Jones, 2001; Brodie *et al.*, 1999; Kalams and Walker, 1998). We explore the effects of both an increase or decrease in activation of CD4+ T cells into CD4+ T helper cells (parameter r in Eq. 1). Our simulations show that higher CD4+ T-cell numbers correlate with higher CTL levels (Fig. 4). Additionally, increased activation, although decreasing numbers of resting CD4+ and CD8+ T cells, results in a higher total

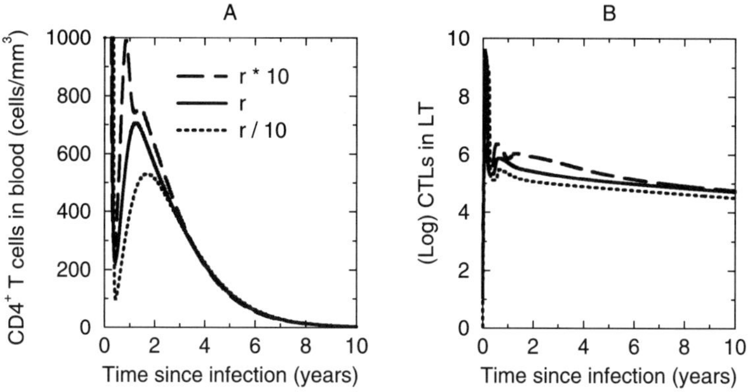

Fig. 4. **$CD4^+$ and $CD8^+$ T cell levels vary with activation rate, r.** Shown is a simulation of $CD4^+$ T cells in the blood (Panel A) and CTLs in the LT (Panel B) with respect to the rate of activation of resting $CD4^+$ T cells by MDCs in the LT (parameter r in Eq. (1)).

number of $CD4^+$ T cells in the blood (Fig. 4, Panel A) and a higher level of CTLs in the LT (Fig. 4, Panel B). This follows from our assumption that increased $CD4^+$ T helper cells activate DCs and $CD8^+$ T cells early enough to prevent further $CD4^+$ T-cell loss, as is observed during acute infection.

The interaction between resting $CD4^+$ T cells and MDCs in the LT can result in the development of $CD4^+$ T helper cells and dendritic cells with $CD8^+$ T-cell priming capability. Thus, the differentiation of $CD8^+$ T cells to HIV-1-specific CTLs can occur either through $CD8^+$ T cells coming into contact directly with $CD4^+$ T helper cells or through contact with dendritic cells. However, to distinguish between contact with any MDCs and those that have received activation signals from $CD4^+$ T cells, the equation for differentiation via dendritic cells includes augmentation by $CD4^+$ T-cell help (which saturates at f_3). To examine whether the dendritic cell-differentiation pathway or the direct $CD4^+$-$CD8^+$ T cell-activation pathway is more significant, we vary the differentiation parameters χ_D and χ_T, respectively, and examine effects to the total production of CTLs in the LT (Fig. 5). We find that differentiation of $CD8^+$ T cells by meeting $CD4^+$ T cells directly is more significant than by meeting MDCs (augmented by help), and speculate that dendritic cells may be more crucial in the activation of HIV-1-specific $CD4^+$ T helper cells than the activation of HIV-1-specific CTLs.

There is conflicting information as to numbers or proportions of blood dendritic cells in HIV-1-infected individuals. There is evidence that DCs

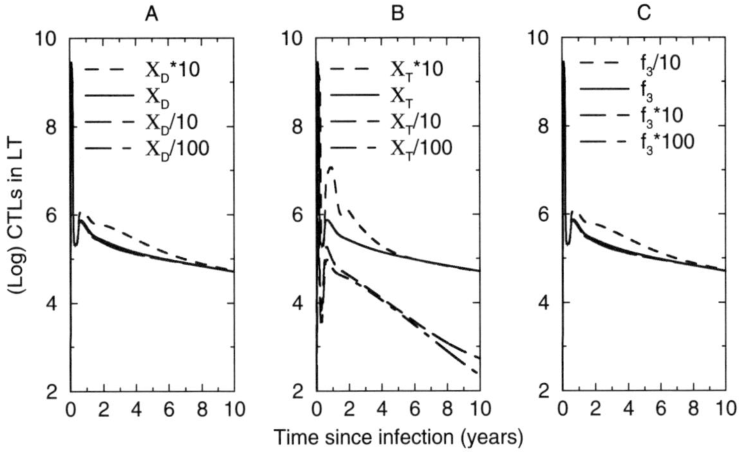

Fig. 5. **Success of CTL priming by MDCs and T helper cells.** Shown is a simulation of CTL levels in the LT based on changes in the differentiation rate of resting CD8$^+$ T cells by dendritic cells, χ_D, (Panel A), the differentiation rate of resting CD8$^+$ T cells by CD4$^+$ T helper cells, χ_T, (Panel B), and half-saturation for CD4$^+$ T-cell help to CD8$^+$ T-cell differentiation, f_3 (Panel C) from Eqs. 8 and 10. We vary these parameters over an order of magnitude range encompassing baseline values and observe effects to the time progression of CTLs.

are depleted in the blood early in HIV-1 infection, corresponding with an increase in blood viral levels (Donaghy *et al.*, 2001; Feldman *et al.*, 2001; Pacanowski *et al.*, 2001; Grassi *et al.*, 1999). This loss could be attributed to a down-regulation of DC markers in the blood, a decrease in new cells input from the bone marrow (Pacanowski *et al.*, 2001), enhanced DC deletion or death, or a rapid export of DCs into the lymph tissues (Cyster, 1999; Lore *et al.*, 2002). Support for migration of mature DCs to T-cell rich areas in the LT comes from studies of selective recruitment of DCs by chemokines to sites of infection (Dieu *et al.*, 1998; Foti *et al.*, 1999).

Our model does not account for differences between the plasmacytoid DC class and the myeloid DCs. These subsets are distinguished based on their phenotype and function during HIV-1 infection and these will be explored in future work. However, both subsets are depleted in the blood during infection to approximately half of their original values in therapy-naive patients (Barron *et al.*, 2003) and our model reflects that deficiency in the blood compartment (see Fig. 6, Panel A). This decrease in blood DCs has also been shown to have a strong correlation with viral load (Donaghy *et al.*, 2001). A rapid influx of DCs into LT is observed in acute infection, as

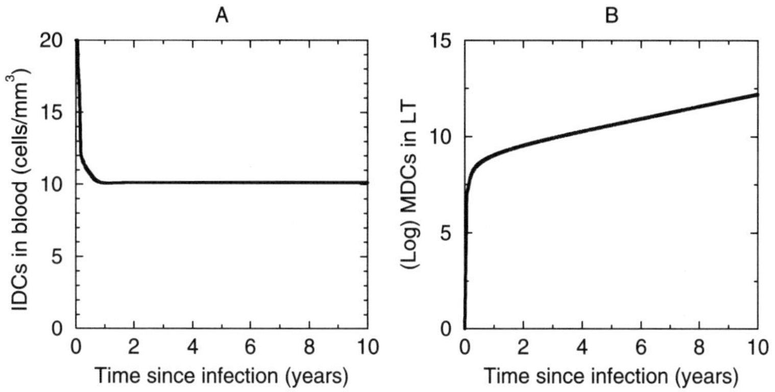

Fig. 6. **Dendritic cell dynamics.** Shown is a simulation of immature DCs (IDCs) in the blood (Panel A) and mature DCs (MDCs) in the LT (Panel B) over the course of a typical 10-year progression of infection.

shown in Fig. 6, Panel B and described in (Lore *et al.*, 2002). The specific mechanisms of DC trafficking into and out of the LT are considered in other work (in preparation).

In this work, we assume that the heightened state of immune activation during infection reported in several studies (Hazenberg *et al.*, 2003a,b; Leng *et al.*, 2001; Papagno *et al.*, 2004) can be captured by enhanced recruitment of DCs into the LT due to virus. Because dendritic cells are the primary activators of both resting $CD4^+$ and $CD8^+$ T cells in our model, this assumption mimics increased immune activation during the course of infection. To explore the role of priming, we study immune activation and activated T-cell levels in Fig. 7. We observe that for less activation (due to no recruitment of MDCs into LT with increasing viral levels), $CD4^+$ T cells in blood remain elevated, similar to that of a long-term non-progressor (Zhang *et al.*, 1997). With higher than baseline immune activation (excessive recruitment of MDCs into LT), there is a rapid decline in the $CD4^+$ T-cell count (Fig. 7, Panel B). Finally, even though activated T-cell levels fall greatly after the acute stage, our simulation shows their numbers remain steady in the LT when immune activation is low and fall more rapidly when immune activation is excessive (Fig. 7, Panel C). Some groups have suggested that HIV-specific CTLs might become exhausted through telomere shortening during constant divison resulting from chronic immune activation (Wodarz *et al.*, 1998; Effros *et al.*, 1996). Many studies have suggested that chronic immune activation may be a better predictor of disease progression than

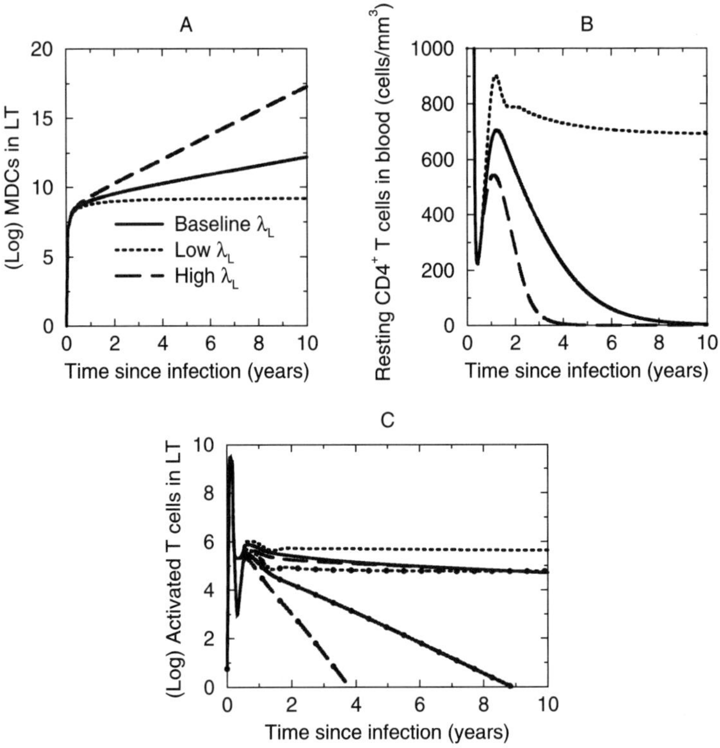

Fig. 7. **Changes in time to AIDS based on immune activation status.** Here, immune activation is captured by the rate of recruitment of DCs into the LT (parameter λ_L in Eq. (12)). We vary this parameter above (increased activation, $\lambda_L = 0.0105$) and below (decreased activation, $\lambda_L = 0.0035$) the baseline value ($\lambda_L = 0.007$) to observe effects. Panel A shows changes to levels of DCs in LT and Panel B shows changes to blood CD4$^+$ T cell counts for these three scenarios. In Panel C, we show levels of activated T cells in the LT (i.e. CD4$^+$ (filled circles) and CD8$^+$ T cells (lines only), respectively). We observe a more rapid decline in HIV-1-specific T helper cells and CTLs with increasing immune activation.

viral load (Simmonds *et al.*, 1991; Leng *et al.*, 2001; Roussanov *et al.*, 2000; Giorgi *et al.*, 1999) and our simulations concur (based on more rapid CD4$^+$ T-cell decline with increased immune activation).

4. Conclusions and Discussion

Several groups have considered the role of CTLs in HIV-1 infection using mathematical models (Wodarz *et al.*, 1998; Ribeiro *et al.*, 2002; Antia *et al.*, 2003; Wick and Self, 2002; Arnaout *et al.*, 2000; Fraser *et al.*, 2002), and others the role of $CD4^+$ T-cell help to CTL responses (Wodarz and Jansen, 2001; Wodarz, 2001). Our work expands on these ideas in many ways. First, we study lymphocyte circulation in health and infection with the use of a two-compartment model and are able to incorporate important dynamics of lymphocyte circulation as well as defects imposed on circulation by HIV-1 infection. Additionally, we include the specific roles of dendritic cells as antigen-presenting cells and their numbers and functional properties in antigen uptake as well as presentation to $CD4^+$ and $CD8^+$ T cells in blood and LT. We consider the total lymph system and use biologically relevant parameter estimations derived from literature, and our simulations compare with clinical and experimental data during various stages of disease progression.

Our simulations reflect the three-stage course of infection in a typical HIV-1 progressor with gradually declining $CD4^+$ T-cell levels throughout the chronic phase of infection (CASCADE Collaboration, 2003, and Fig. 3, Panel B). We use this positive control to test variations in $CD4^+$ T-cell help and DC activation to the successful differentiation of $CD8^+$ T cells to CTLs. In this context, we find that $CD4^+$ T-cell help is necessary for the priming of CTLs, and that interaction of $CD8^+$ T cells with both $CD4^+$ T cells and DCs is crucial for successful CTL development and actions during HIV-1 infection.

There is evidence that inadequately developed $CD8^+$ T cells lack the capacity to effectively control HIV-1 infection, especially as the infection progresses (Champagne *et al.*, 2001). It is thought that $CD4^+$ T-cell help may function in the memory $CD8^+$ T-cell response during recurring episodes of virus growth throughout the chronic phase of infection (Kaech and Ahmed, 2003; Janssen *et al.*, 2003; Sun and Bevan, 2003; Shedlock and Shen, 2003; Bourgeois *et al.*, 2002). However, in some experimental systems, $CD4^+$ T-cell help is also essential for the primary response (Wang and Livingstone, 2003). This model does not examine the development of a memory response. However, the development of activated CTLs does decline concommitant with the decline in $CD4^+$ T-cell help, and so we expect that there would also be a proportional decrease in memory $CD8^+$ T cells available for a recall response.

Due to high perforin expression early in infection that declines throughout progression of infection, it has been suggested that IL-2 activity is correlated with perforin expression (Zhang *et al.*, 2003). However, we tested the effect of reducing CTL killing of infected cells based on levels of $CD4^+$ T helper cells in the system (which are the main producers of IL-2) and found little to no variation in outcome (data not shown). This could be due to our assumption accounting only for HIV-1-specific CTLs, which have been shown to be generally deficient in perforin, and only a percentage successfully migrate to the LT for cytotoxic activities (Chen *et al.*, 2001). Therefore, as our model simulations show, it is important to consider the unique properties of HIV-1-specific CTLs so to be able to study their dysfunctions during progression of infection.

HIV-1 has evolved several properties to intervene in normal immune system functioning. For example, the HIV-1 protein Nef interferes with the antigen-presentation pathway by intersecting the CD40-CD40L pathway of dendritic cell activation (Pope, 2003; Andrieu *et al.*, 2001). This may result in impaired activation of dendritic cells that would render them inefficient at antigen presentation. However, dendritic cells and T cells are in close contact during infection, and the proximity of dendritic cell-associated virus to T cells will render T cells highly susceptible to cell-associated infection. Thus, activation levels of dendritic cells become important when determining under which circumstances dendritic cells result in either activation or in $CD4^+$ T-cell infection (Lutz and Schuler, 2002; McDonald *et al.*, 2003). Inadequate signaling may be deleterious to the system, and the role of dendritic cells in HIV-1 infection requires further study.

This work has examined the dynamics of CTLs in HIV-1 infection. Our simulations indicate that activation of $CD4^+$ T cells by DCs and activation of $CD8^+$ T cells by both $CD4^+$ T cells and DCs is crucial in infection. Thus, although CTLs cannot be the sole determinant of the outcome of infection, they are significant contributors to dynamics influencing progression of infection to AIDS.

Acknowledgments

This work was supported by National Institutes of Health Grant HL62119 and HL72682.

References

Albert M, Sauter B and Bhardwaj N (1998) Dendritic cells acquire antigen from apoptotic cells and induce class I-restricted CTLs, *Nature* **392**: 86–89.

Andersson J, Behbahani H and Lieberman J (1999) Perforin is not co-expressed with granzyme A within cytotoxic granules in CD8 T lymphocytes present in lymphoid tissue during chronic HIV infection, *AIDS* **13**: 1295–1303.

Andersson J, Kinloch S, Sonnerborg A, Nilsson J, Fehniger T, Spetz, A, Behbahani H, Goh L, McDade H, Gazzard B, Stellbrink H, Cooper D and Perrin L (2002) Low levels of perforin expression in CD8+ T lymphocyte granules in lymphoid tissue during acute immunodeficiency virus type 1 infection, *J Infect Dis* **185**: 1355–1358.

Andrieu M, Chassin D, Desoutter J, Bouchaert I, Baillet M, Hanau D, Guillet J and Hosmalin A (2001) Downregulation of major histocompatibility class I on human dendritic cells by HIV Nef impairs antigen presentation to HIV-specific CD8+ T lymphocytes, *AIDS Res Hum Retroviruses* **17**: 1365–1370.

Antia R, Bergstrom C, Pilyugin S, Kaech S and Ahmed R (2003) Models of CD8+ responses: 1. What is the antigen-independent proliferation program, *J Theor Biol* **221**: 585–598.

Appay V, Dunbar P, Callan M, Klenerman P, Gillespie G, Papagno L, Ogg G, King A, Lechner F, Spina C, Little S, Havlir D, Richman D, Gruener N, Pape G, Waters A, Easterbrook P, Salio M, Cerundolo V, McMichael A and Rowland–Jones S (2002) Memory CD8+ T cells vary in differentiation phenotype in different persistent virus infections, *Nat Med* **8**: 379–385.

Appay V, Nixon D, Donahoe S, Gillespie G, Dong T, King A, Ogg G, Spiegel H, Conlon C, Spina C, Havlir D, Richman D, Waters A, Easterbrook P, McMichael A and Rowland–Jones S (2000) HIV-specific $CD8^+$ T cells produce antiviral cytokines but are impaired in cytolytic function, *J Exp Med* **192**: 63–75.

Arnaout R, Nowak M and Wodarz D (2000) HIV-1 dynamics revisited: biphasic decay of cytotoxic T lymphocyte killing? *Proc R Soc Lond B* **267**: 1347–1354.

Auphan–Anezin N, Verdeil G and Schmitt–Verhulst AM (2003) Distinct thresholds for CD8 T cell activation lead to functional heterogeneity: CD8 T cell priming can occur independently of cell division, *J Immunol* **170**: 2442–2448.

Bajaria S, Webb G, Cloyd M and Kirschner D (2002) Dynamics of naive and memory $CD4^+$ T lymphocytes in HIV-1 disease progression, *J Acquir Immune Defic Syndr* **30**: 41–58.

Banchereau J and Steinman R (1998) Dendritic cells and the control of immunity, *Nature* **392**: 245–252.

Baribaud F, Pohlmann S and Doms R (2001) The role of DC-SIGN and DC-SIGNR in HIV and SIV attachment, infection, and transmission, *Virology* **286**: 1–6.

Barker E, Bossart K and Levy J (1998) Primary $CD8^+$ cells from HIV-infected individuals can suppress productive infection of macrophages independent of β-chemokines, *Proc Natl Acad Sci* **95**: 1725–1729.

Barron M, Blyveis N, Palmer B, MaWhinney S and Wilson C (2003) Influence of plasma viremia on defects in number and immunophenotype of blood dendritic cell subsets in human immunodeficiency virus 1-infected individuals, *J Infect Dis* **187**: 26–37.

Bennett S, Carbone F, Karamalis F, Flavell R, Miller J and Heath W (1998) Help for cytotoxic-T-cell responses is mediated by CD40 signalling, *Nature* **393**: 478.

Berke G (1995) The CTL's kiss of death, *Cell* **81**: 9–12.

Betts M, Ambrozak D, Douek D, Bonhoeffer S, Brenchley J, Casazza J, Koup R and Picker L (2001) Analysis of total human immunodeficiency virus (HIV)-specific CD4+ and CD8+ T-cell responses: relationship to viral load in untreated HIV infection, *J Virol* **75**: 11983–11991.

Blower S and Dowlatabadi H (1994) Sensitivity and uncertainty analysis of complex models of disease transmission: an HIV model, as an example, *Int Stat Rev* **62**: 229–243.

Borrow P, Lewicki H, Hahn BH, Shaw GM and Oldstone MB (1994) Virus-specific CD8+ cytotoxic T-lymphocyte activity associated with control of viremia in primary human immunodeficiency virus type 1 infection, *J Virol* **68**: 6103–6110.

Bottomly K (1999) T cells and dendritic cells get intimate, *Science* **283**: 1124–1125.

Bourgeois C, Rocha B and Tanchot C (2002) A role for CD40 expression on CD8+ T cells in the generation of CD8+ T cell memory, *Science* **297**: 2060–2063.

Bousso P and Robey E (2003) Dynamics of $CD8^+$ T cell priming by dendritic cells in intact lymph nodes, *Nat Immunol* **4**: 579–585.

Brodie S, Lewinsohn D, Patterson B, Jiyamapa D, Krieger J, Corey L, Greenberg P and Riddell S (1999) *In vivo* migration and function of transferred HIV-1-specific cytotoxic T cells, *Nat Med* **5**: 34–41.

Buseyne F, Le Gail S, Boccaccio C, Abastado JP, Lifson J, Arthur L, Riviere Y, Heard JM and Schwartz O (2001) MHC-1-restricted presentation of HIV-1 virion antigens without viral replication, *Nat Med* **7**: 344–349.

Caruso A, Licenziati S, Canaris A, Cantalamessa A, Fiorentini S, Ausenda S, Ricotta D, Dima F, Malacarne F, Balsari A and Turano A (1998) Contribution of CD4+, CD8+CD28+, and CD8+CD28-T cells to CD3+ lymphocyte homeostasis during the natural course of HIV-1 infection, *J Clin Invest* **101**: 137–144.

Caux C, Massacrier C, Vanbervliet B, Dubois B, Van Kooten C, Durand I and Bancherau J (1994) Activation of human dendritic cells through CD40 cross-linking, *J Exp Med* **180**: 1263.

Cavert W, Notermans D, Staskus K, Wietgrefe S, Zupancic M, Gebhard K, Henry K, Zhang Z, Mills E, McDade H, Schuwirth C, Goudsmit J, Danner S and Haase A (1997) Kinetics of response in lymphoid tissues to antiretroviral therapy of HIV-1 infection, *Science* **276**: 960–964.

Cella M, Engering A, Pinet V, Pieters J and Lanzavecchia A (1997) Inflammatory stimuli induce accumulation of MHC class II complexes on dendritic cells, *Nature* **388**: 782–787.

Cella M, Scheidegger D, Palmer–Lehmann K, Lane P, Lanzavecchia A and Alber G (1996) Ligation of CD40 on dendritic cells triggers production of high levels of interleukin-12 and enhances T cell stimulatory capacity: T-T help via APC activation, *J Exp Med* **184**: 747–752.

Champagne P, Ogg G, King A, Knabenhans C, Ellefsen K, Nobile M, Appay V, Paolo Rizzardi G, Fleury S, Lipp M, Forster R, Rowland–Jones S, Sekaly RP, McMichael A and Pantaleo G (2001) Skewed maturation of memory HIV-specific CD8 T lymphocytes, *Nature* **410**: 106–111.

Chang T, Francois F, Mosoian A and Klotman M (2003) CAF-mediated human immunodeficiency virus (HIV) type 1 transcriptional inhibition is distinct from alpha-defensin-1 HIV inhibition, *J Virol* **77**: 6777–6784.

Chen G, Shankar P, Lange C, Valdez H, Skolnik P, Wu L, Manjunath N and Lieberman J (2001) CD8 T cells specific for human immunodeficiency virus,

Epstein–Barr virus, and cytomegalovirus lack molecules for homing to lymphoid sites of infection, *Blood* **98**: 156–164.

Cocchi F, DeVico AL, Garzino–Deno A, Arya SK, Gallo RC and Lusso P (1995) Identification of RANTES, MIP-1α, and MIP-1β as the major HIV-suppressive factors produced by CD8$^+$ T cells, *Science* **270**: 1811–1815.

CASCADE Collaboration (2003) Differences in CD4 cell counts at seroconversion and decline among 5739 HIV-1-infected individuals with well-estimated dates of seroconversion, *J Acquir Immune Defic Syndr* **34**: 76–83.

Curtis B, Scharnowske S and Watson A (1992) Sequence and expression of a membrane-associated C-type lectin that exhibits CD4-independent binding of human immunodeficiency virus envelope glycoprotein gp120, *Proc Natl Acad Sci* **89**: 8356–8360.

Cyster J (1999) Chemokines and cell migration in secondary lymphoid organs, *Science* **286**: 2098–2102.

Dieu MC, Vanbervliet B, Vicari A, Bridon JM, Oldham E, Ait–Yahia S, Briere F, Zlotnik A, Lebecque S and Caux C (1998) Selective recruitment of immature and mature dendritic cells by distinct chemokines expressed in different anatomic sites, *J Exp Med* **188**: 373–386.

Donaghy H, Pozniak A, Gazzard B, Qazi N, Gilmour J, Gotch F and Patterson S (2001) Loss of blood CD11c$^+$ myeloid and CD11c$^-$ plasmacytoid dendritic cells in patients with HIV-1 infection correlates with HIV-1 RNA virus load, *Blood* **98**: 2574–2576.

Douek D (1998) Changes in thymic function with age and during the treatment of HIV infection, *Nature* **396**: 690–695.

Douek D, Brenchley J, Betts M, Ambrozak D, Hill B, Okamoto Y, Casazza J, Kuruppu J, Kunstman K, Wolinsky S, Grossmna Z, Dybul M, Oxenius A, Price D, Connors M and Koup R (2002) HIV preferentially infects HIV-specific CD4+ T cells, *Nature* **417**: 95–98.

Effros R, Allsopp R, Chiu C, Hausner M, Hirji K, Wang L, Harley C, Villeponteau B, West M and Giorgi J (1996) Shortened telomeres in the expanded CD28-CD8+ cell subset in HIV disease implicate replicative senescence in HIV pathogenesis, *AIDS* **10**: F17–22.

Ellefsen K, Harari A, Champagne P, Bart P, Sekaly R and Pantaleo G (2002) Distribution and functional analysis of memory antiviral CD8 T cell responses in HIV-1 and cytomegalovirus infections, *Eur J Immunol* **32**: 3756–3764.

Emilie D, Maillot M, Nicolas J, Fior R and Galanaud P (1992) Antagonistic effect of interferon-gamma on tat-induced transactivation of HIV long terminal repeat, *J Biol Chem* **267**: 20565–20570.

Engering A, van Vliet S, Geijtenbeek T and van Kooyk Y (2002) Subset of DC-SIGN$^+$ dendritic cells in human blood transmits HIV-1 to T lymphocytes, *Blood* **100**: 1780–1786.

Feldman S, Stein D, Amrute S, Denny T, Garcia Z, Kloser P, Sun Y, Megjugorac N and Fitzgerald–Bocarsly P (2001) Decreased interferon-alpha production in HIV-infected patients correlates with numerical and functional deficiencies in circulating type 2 dendritic cell precursors, *Clin Immunol* **101**: 201–210.

Ferbas J, Giorgi J and Amini S (2000) Antigen-specific production of RANTES, macrophage inflammatory protein (MIP)-1α, and (MIP)-1β *in vitro* is a correlate of reduced human immunodeficiency virus burden *in vivo*, *J Infect Dis* **182**: 1247–1250.

Folkvord J, Anderson D, Arya J, MaWhinney S and Connick E (2003) Microanatomic relationships between CD8$^+$ cells and HIV-1-producing cells in human lymphoid tissue *in vivo*, *J Acquir Immune Defic Syndr* **32**: 469–476.

Fong L, Mengozzi M, Abbey N, Herndier B and Engleman E (2002) Productive infection of plasmacytoid dendritic cells with human immunodeficiency virus type 1 is triggered by CD40 ligation, *J Virol* **76**: 11033–11041.

Forster R, Schubel A, Breitfeld D, Kremmer E, Renner–Muller I, Wolf E and Lipp M (1999) CCR7 coordinates the primary immune response by establishing functional microenvironments in secondary lymphoid organs, *Cell* **99**: 23–33.

Foti M, Granucci F, Aggujaro D, Liboi E, Luini W, Minardi S, Mantovani A, Sozzani S and Ricciardi–Castagnoli P (1999) Upon dendritic cell (DC) activation chemokines and chemokine receptor expression are rapidly regulated for recruitment and maintenence of DC at the inflammatory site, *Int Immunol* **11**: 979–986.

Fraser C, Ferguson N, de Wolf F, Ghani A, Garnett G and Anderson R (2002) Antigen-driven T-cell turnover, *J Theor Biol* **219**: 177–192.

Geijtenbeek T, Kwon D, Torensma R, van Vliet S, van Duijnhoven G, Middel J, Cornelissen I, Nottet H, KewalRamani V, Littman D, Figdor C and van Kooyk Y (2000) DC-SIGN, a dendritic cell specific HIV-1 binding protein that enhances trans-infection of T cells, *Cell* **100**: 587–597.

Gillespie G, Wils M, Appay V, O'Callaghan C, Murphy M, Smith N, Sissons P, Rowland–Jones S, Bell J and Moss P (2000) Functional heterogeneity and high frequencies of cytomegalovirus-specific CD8(+) T lymphocytes in healthy seropositive donors, *J Virol* **74**: 8140–8150.

Giorgi J, Hutlin L, McKeating J, Johnson T and Owens B (1999) Shorter survival in advanced human immunodeficiency virus type 1 infection is more closely associated with T lymphocyte activation than with plasma virus burden or virus chemokine coreceptor usage, *J Infect Dis* **179**: 859–870.

Gougeon M and Montagnier L (1993) Apoptosis in AIDS, *Science* **260**: 1269–1270.

Goulder P, Tang Y, Brander C, Betts M, Altfeld M, Annamalai K, Trocha A, He S, Rosenberg E, Ogg G, O'Callaghan C, Kalams S, McKinney R, Mayer K, Koup R, Pelton S, Burchett S, McIntosh K and Walker B (2000) Functionally inert HIV-specific cytotoxic T lymphocytes do not play a major role in chronically infected adults and children, *J Exp Med* **192**: 1819–1831.

Grabbe S, Kampgen E and Schuler G (2000) Dendritic cells: multi-lineal and multi-functional, *Immunol Today* **21**: 431–433.

Grassi F, Hosmalin A, McIlroy D, Calvez V, Debre P and Autran B (1999) Depletion in blood CD11c-positive dendritic cells from HIV-infected patients, *AIDS* **13**: 759–766.

Gummuluru S, Rogel M, Stamatatos L and Emerman M (2003) Binding of human immunodeficiency virus type 1 to immature dendritic cells can occur independently of DC-SIGN and mannose binding C-type lectin receptors via a cholesterol-dependent pathway, *J Virol* **77**: 12865–12874.

Haase A (1986) Pathogenesis of lentivirus infections, *Nature* **322**: 130–136.

Haase A (1999) Population biology of HIV-1 infection: Viral and $CD4^+$ T cell demographics and dynamics in lymphatic tissues, *Annu Rev Immunol* **17**: 625–656.

Hazenberg M, Otto S, Hamann D, Roos M, Schuitemaker H, de Boer R and Miedema F (2003a) Depletion of naive CD4 T cells by CXCR4-using HIV-1 variants occurs mainly through increased T-cell death and activation, *AIDS* **17**: 1419–1424.

Hazenberg M, Otto S, van Benthem B, Roos M, Coutinho R, Lange J, Hamann D, Prins M and Miedema F (2003b) Persistent immune activation in HIV-1 infection is associated with progression to AIDS, *AIDS* **17**: 1881–1888.

Hellerstein M, Hanley M, Cesar D, Siler S, Papagerorgopoulus C, Wieder E, Schmidt D, Hoh R, Neese R, Macallan S, Deeks S and McCune J (1999) Directly measured kinetics of circulating T lymphocytes in normal and HIV-1-infected humans, *Nat Med* **5**: 83.

Hislop A, Gudgeon N, Callan M, Fazou C, Hasegawa H, Salmon M and Rickinson A (2001) EBV-specific CD8+ T cell memory: relationships between epitope specificity, cell phenotype, and immediate effector function, *J Immunol* **167**: 2019–2029.

Ho D, Neumann A, Perelson A, Chen W, Leonard J and Markowitz M (1995) Rapid turnover of plasma virions and CD4 lymphocytes in HIV-1 infection, *Nature* **373**: 123–126.

Izmailova E, Bertley F, Huang Q, Makori N, Miller C, Young R and Aldovini A (2003) HIV-1 Tat reprograms immature dendritic cells to express chemoattractants for activated T cells and macrophages, *Nat Med* **9**: 191–197.

Janssen E, Lemmens E, Wolfe T, Christen U, von Herrath M and Schoenberger S (2003) CD4+ T cells are required for secondary expansion and memory in CD8+ T lymphocytes, *Nature* **421**: 852–856.

Jin X, Wills M, Sissons J and Carmichael A (1998) Progressive loss of IL-2-expandable HIV-1-specific cytotoxic T lymphocytes during asymptomatic HIV infection, *Eur J Immunol* **28**: 3564–3576.

Johnson R, Trocha A, Yang L, Mazzara G, Panicali D, Buchanan T and Walker B (1991) HIV-1 gag-specific cytotoxic T lymphocytes recognize multiple highly conserved epitopes. Fine specificity of the gag-specific response defined by using unstimulated peripheral blood mononuclear cell and cloned effector cells, *J Immunol* **147**: 1512–1521.

Kaech S and Ahmed R (2003) Immunology. CD8 T cells remember with a little help, *Science* **300**: 263–265.

Kalams S, Buchbinder S, Rosenberg E, Billingsley J, Colbert D, Jones N, Shea A, Trocha A and Walker B (1999) Association between virus-specific cytotoxic T-lymphocyte and helper responses in human immunodeficiency virus type 1 infection, *J Virol* **73**: 6715–6720.

Kalams S and Walker B (1998) The critical need for CD4 help in maintaining effective cytotoxic T lymphocyte responses, *J Exp Med* **188**: 2199–2204.

Katsikis P, Wunderlich E, Smith C and Herzenberg L (1995) Fas antigen stimulation induces marked apoptosis of T lymphocytes in human immunodeficiency virus-infected individuals, *J Exp Med* **181**: 2029–2036.

Klein M, van Baalen C, Holwerda A, Kerkhof Garde S, Bende R, Keet I, Eeftinck–Schattenkerk JKM, Osterhaus A, Schuitemaker H and Miedema F (1995) Kinetics of gag-specific cytotoxic T lymphocyte responses during the clinical course of HIV-1 infection: a longitudinal analysis of rapid progressors and long-term asymptomatics, *J Exp Med* **181**: 1365–1372.

Kostense S, Vandenberghe K, Joling J, Van Baarle D and Nanlohy N (2002) Persistent number of tetramer+ CD8(+) T cells, but loss of interferon-gamma+ HIV-specific T cells during progression to AIDS, *Blood* **99**: 2505–2511.

Koup R, Safrit J, Cao Y, Andrews C, McLeod G, Borkowsky W, Farthing C and Ho D (1994) Temporal association of cellular immune responses with the initial control of viremia in primary human immunodeficiency virus type 1 syndrome, *J Virol* **68**: 4650–4655.

Kovacs J, Lempicki R, Sidorov I, Adelsberger J and Herpin B (2001) Identification of dynamically distinct subpopulations of T lymphocytes that are differentially affected by HIV, *J Exp Med* **194**: 1731–1741.

Kwon D, Gregorio G, Bitton N, Hendrickson W and Littman D (2002) DC-SIGN-mediated internalization of HIV is required for trans-enhancement of T cell infection, *Immunity* **16**: 135–144.

Leng Q, Borkow G, Weisman Z, Stein M, Kalinkovich A and Bentwich Z (2001) Immune activation correlates better than HIV plasma viral load with CD4 T-cell decline during HIV infection, *J AIDS* **27**: 389–397.

Levy J, Mackiewicz C and Barker E (1996) Controlling HIV pathogenesis: the role of the noncytotoxic anti-HIV response of $CD8^+$ T cells, *Immunol Today* **17**: 217–224.

Lieberman J, Shankar P, Manjunath N and Andersson J (2001) Dressed to kill? A review of why antiviral CD8 T lymphocytes fail to prevent progressive immunodeficiency in HIV-1 infection, *Blood* **98**: 1667–1677.

Liu Y (2001) Dendritic cell subsets and lineages, and their functions in innate and adaptive immunity, *Cell* **106**: 259–262.

Lore K, Sonnerborg A, Brostrom C, Goh L, Perrin L, McDade H, Stellbrink H, Gazzard B, Weber R, Napolitano L, van Kooyk Y and Andersson J (2002) Accumulation of DC-SIGN+CD40+ dendritic cells with reduced CD80 and CD86 expression in lymphoid tissue during acute HIV-1 infection, *AIDS* **16**: 683–692.

Lutz M and Schuler G (2002) Immature, semi-mature and fully mature dendritic cells: which signals induce tolerance or immunity? *Trends Immunol* **23**: 445–449.

McCune J, Hanley M, Cesar D, Halvorsen R, Hoh R, Schmidt D, Wieder E, Deeks S, Siler S, Neese R and Hellerstein M (2000) Factors influencing T-cell turnover in HIV-1-seropositive patients, *J Clin Invest* **105**: R1.

McDonald D, Wu L, Bohks S, KewalRamani V, Unutmaz D and Hope T (2003) Recruitment of HIV and its receptors to dendritic cell-T cell junctions, *Science* **300**: 1295–1297.

McIlroy D, Autran B, Cheynier R, Wain–Hobson S, Clauvel J, Oksenhendler E, Debre P and Hosmalin A (1995) Infection frequency of dendritic cells and CD4+ T lymphocytes in spleens of human immunodeficiency virus-positive patients, *J Virol* **69**: 4737–4745.

McMichael A, Gotch F, Noble G and Beare P (1983) Cytotoxic T-cell immunity to influenza, *N Engl J Med* **309**: 13–17.

McMichael A and Rowland–Jones S (2001) Cellular immune responses to HIV, *Nature* **410**: 980–987.

Mellman I and Steinman R (2001) Dendritic cells: specialized and regulated antigen processing machines, *Cell* **106**: 255–258.

Meylan P, Guatelli J, Munis J, Richman D and Kornbluth R (1993) Mechanisms for the inhibition of HIV replication by interferons-α, -β, and -γ in primary human macrophages, *Virology* **193**: 138–148.

Migueles S and Connors M (2001) Frequency and function of HIV-specific CD8+ T cells, *Imm Lett* **79**: 141–150.

Moriuchi H, Moriuchi M, Combadiere C, Murphy P and Fauci A (1996) $CD8^+$ T-cell-derived soluble factor(s), but not β-chemokines RANTES, MIP-1α, and MIP-1β, suppress HIV-1 replication in monocyte/macrophages, *Proc Natl Acad Sci* **93**: 15341–15345.

Ogg G, Jin X, Bonhoeffer S, Dunbar P, Nowak M, Monard S, Segal J, Cao Y, Rowland–Jones S, Cerundolo V, Hurley A, Markowitz M, Ho D, Nixon D and McMichael A (1998) Quantitation of HIV-1-specific cytotoxic T lymphocytes and plasma load of viral RNA, *Science* **279**: 2103–2106.

Pacanowski J, Kahi S, Baillet M, Lebon P, Deveau C, Goujard C, Meyer L, Oksenhendler E, Sinet M and Hosmalin A (2001) Reduced blood $CD123^+$ (lymphoid) and $CD11c^+$ (myeloid) dendritic cell numbers in primary HIV-1 infection, *Blood* **98**: 3016–3021.

Pantaleo G, Demarest J, Schacker T, Vaccarezza M, Cohen O and Daucher M (1997a) The qualitative nature of the primary immune response to HIV infection is a prognosticator of disease progression independent of the initial level of plasma viremia, *Proc Natl Acad Sci USA* **94**: 254–258.

Pantaleo G, Graziosi C and Fauci A (1993) The immunopathogenesis of human immunodeficiency virus infection, *N Engl J Med* **328**: 327–335.

Pantaleo G, Soudeyns H, Demarest J, Vaccarezza M, Graziosi C, Paolucci S, Daucher M, Cohen O, Denis F, Biddison W, Sekaly R and Fauci A (1997b) Accumulation of human immunodeficiency virus-specific cytotoxic T lymphocytes away from the predominant site of virus replication during primary infection, *Eur J Immunol* **27**: 3166–3173.

Papagno L, Appay V, Sutton J, Rostron T, Gillespie G, Ogg G, King A, Makadzange A, Waters A, Balotta C, Vyakarnam A, Easterbrook P and Rowland–Jones S (2002) Comparison between HIV- and CMV-specific T cell responses in long-term HIV infected donors, *Clin Exp Immunol* **130**: 509–517.

Papagno L, Spina C, Marchant A, Salio M, Rufer N, Little S, Dong T, Chesney G, Waters A, Easterbrrok P, Dunbar P, Shepherd D, Cerundolo V, Emery V, Griffiths P, Conlon C, McMichael A, Richman D, Rowland–Jones S and

Appay V (2004) Immune activation and CD8+ T-cell differentiation towards senesence in HIV-1 infection, *PLoS Biol* **2**: Epub.

Patterson S, Rae A, Hockey N, Gilmour J and Gotch F (2001) Plasmacytoid dendritic cells are highly susceptible to human immunodeficiency virus type 1 infection and release infectious virus, *J Virol* **75**: 6710–6713.

Perelson A, Neumann A, Markowitz M, Leonard J and Ho D (1996) HIV-1 dynamics *in vivo*: clearance rate, infected cell life-span, and viral generation time, *Science* **271**: 1582–1586.

Phillips A (1996) Reduction of HIV concentration during acute infection: independence from a specific immune response, *Science* **271**: 497–499.

Phillips R, Rowland–Jones S, Nixon D, Gotch F, Edwards J, Ogunlesi A, Elvin J, Rothbard J, Bangham C and Rizza C (1991) Human immunodeficiency virus genetic variation that can escape cytotoxic T-cell recognition, *Nature* **354**: 453–459.

Pitcher C, Quittner C, Peterson D, Connors M, Koup R, Maino V and Picker L (1999) HIV-1-specific $CD4^{+}$ T cells are detectable in most individuals with active HIV-1 infection, but decline with prolonged viral suppression, *Nat Med* **5**: 518–525.

Pope M (2003) Nefarious abuse, *Nat Immunol* **4**: 729–730.

Quinn T (1997) Acute primary HIV infection, *JAMA* **278**: 58–62.

Ribeiro R, Mohri H, Ho D and Perelson A (2002) *In vivo* dynamics of T cell activation, proliferation, and death in HIV-1 infection: why are CD4+ but not CD8+ T cells depleted, *Proc Natl Acad Sci USA* **99**: 15572–15577.

Richman D (2000) Normal physiology and HIV pathophysiology of human T-cell dynamics, *J Clin Invest* **105**: 565–566.

Ridge J, Di Rosa F and Matzinger P (1998) A conditioned dendritic cell can be a temporal bridge between a $CD4^{+}$ T-helper and a T-killer cell, *Nature* **393**: 474–478.

Rosenberg Y, Anderson A and Pabst R (1998) HIV-induced decline in blood CD4/CD8 ratios: viral killing or altered lymphocyte trafficking? *Immunol Today* **19**: 10–16.

Roussanov B, Taylor J and Giorgi J (2000) Calculation and use of an HIV-1 disease progression score, *AIDS* **14**: 2715–2722.

Safrit J and Koup R (1995) Which immune responses control HIV replication? *Curr Opin Immunol* **7**: 456–461.

Sallusto F, Cella M, Danieli C and Lanzavecchia A (1995) Dendritic cells use macropinocytosis and the mannose receptor to concentrate macromolecules in the major histocompatibility complex class II compartment: downregulation by cytokines and bacterial products, *J Exp Med* **182**: 389–400.

Schacker T, Collier A, Hughes J, Shea T and Corey L (1996) Clinical and epidemiologic features of primary HIV infection, *Ann Intern Med* **125**: 257–264.

Schoenberger S, Toes R, van der Voort E, Offringa R and Melief C (1998) T-cell help for cytotoxic T lymphocytes is mediated by CD40-CD40L interactions, *Nature* **393**: 480–483.

Scott–Algara D, Buseyne F, Blanche S, Rouzioux C, Jouanne C, Romagne F and Riviere Y (2001) Frequency and phenotyping of human immunodeficiency virus

(HIV)-specific CD8+ T cells in HIV-infected children, using major histocompatibility complex class I peptide tetramers, *J Infect Dis* **183**: 1565–1573.
Sester M, Sester U, Kohler H, Schneider T, Deml L, Wagner R, Mueller–Lantzsch N, Pees H and Meyerhans A (2000) Rapid whole blood analysis of virus-specific CD4 and CD8 T cell responses in persistent HIV infection, *AIDS* **14**: 2653–2660.
Sharpe A and Freeman G (2002) The B7-CD28 superfamily, *Nat Rev Immunol* **2**: 116–126.
Shedlock D and Shen H (2003) Requirement for CD4 T cell help in generating functional CD8 T cell memory, *Science* **300**: 337–339.
Simmonds P, Beatson D, Cuthbert R, Watson H, Reynolds B, Peutherer JF, Parry JV, Ludlam LA and Steel CM (1991) Determinants of HIV disease progression: six-year longitudinal study in the Edinburgh haemophilia/HIV cohort, *Lancet* **338**: 1159–1163.
Sprent J (1973) Circulating T and B lymphocytes of the mouse, I. Migratory properties, *Cell Immunol* **7**: 40–59.
Sprent J and Basten A (1973) Circulating T and B lymphocytes of the mouse, II. Lifespan, *Cell Immunol* **7**: 10–39.
Stafford M, Corey L, Cao Y, Daar E, Ho D and Perelson A (2000) Modeling plasma virus concentration during primary HIV infection, *J Theor Biol* **203**: 285–301.
Sun J and Bevan M (2003) Defective CD8+ T cell memory following acute infection without CD4 T cell help, *Science* **300**: 339–342.
Teleshova N, Frank I and Pope M (2003) Immunodeficiency virus exploitation of dendritic cells in the early steps of infection, *J Leukoc Biol* **74**: 683–690.
Trimble L and Lieberman J (1998) Circulating CD8 T lymphocytes in human immunodeficiency virus-infected individuals have impaired function and downmodulate CD3 zeta, the signaling chain of the T-cell receptor complex, *Blood* **91**: 585–594.
van Baalen C, Klein M, Geretti A, Keet R, Miedema F, van Els C and Osterhaus A (1993) Selective *in vitro* expansion of HLA class I-restricted HIV-1 gag-specific CD8+ T cells: cytotoxic T-lymphocyte epitopes and precursor frequencies, *AIDS* **7**: 781–786.
van Baarle D, Kostense S, van Oers M, Hamann D and Miedema F (2002) Failing immune control as a result of impaired CD8+ T-cell maturation: CD27 might provide a clue, *Trends Immunol* **23**: 586–591.
van Essen D, Kikutani H and Gray D (1995) CD40 ligand-transduced costimulation of T cells in the development of helper function, *Nature* **378**: 620–623.
Wagner L, Yang O, Garcia–Zepeda E, Ge Y, Kalams S, Walker B, Pasternack M and Luster A (1998) β-chemokines are released from HIV-1-specific cytolytic T-cell granules complexed to proteoglycans, *Nature* **391**: 908–911.
Walker C, Moody D, Stites D and Levy J (1986) CD8+ lymphocytes can control HIV infection *in vitro* by suppressing virus replication, *Science* **234**: 1563–1566.
Wang J and Livingstone A (2003) Cutting edge: CD4+ T cell help can be essential for primary CD8+ T cell responses *in vivo*, *J Immunol* **171**: 6339–6343.

Webby R, Andreansky S, Stambas J, Rehg J, Webster R, Doherty P and Turner S (2003) Protection and compensation in the influenza virus-specific CD8+ T cell response, *Proc Natl Acad Sci USA* **100**: 7235–7240.

Wherry E, Blattman J, Murali–Krishna K, van der Most R and Ahmed R (2003) Viral persistence alters CD8 T-cell immunodominance and tissue distribution and results in distinct stages of functional impairment, *J Virol* **77**: 4911–4927.

Wick D and Self S (2002) What's the matter with HIV-directed killer T cells, *J Theor Biol* **219**: 19–31.

Wodarz D (2001) Helper-dependent vs. helper-independent CTL responses in HIV infection: implications for drug therapy and resistance, *J Theor Biol* **213**: 447–459.

Wodarz D and Jansen V (2001) The role of T cell help for anti-viral CTL responses, *J Theor Biol* **211**: 419–432.

Wodarz D, Klenerman P and Nowak M (1998) Dynamics of cytotoxic T-lymphocyte exhaustion, *Proc R Soc Lond B* **265**: 191–203.

Yang O, Kalams S, Rosenzweig M, Trocha A, Jones N, Koziel M, Walker B and Johnson R (1996) Efficient lysis of human immunodeficiency virus type-1-infected cells by cytotoxic T lymphocytes, *J Virol* **70**: 5799–5806.

Ye P, Kourtis A and Kirchner D (2002) The effects of different HIV type 1 strains of human thymic function, *AIDS Res Hum Retroviruses* **18**: 1239–1251.

Zajac A, Blattman J, Murali–Krishna K, Sourdive D, Suresh M, Altman J and Ahmed R (1998) Viral immune evasion due to persistence of activated T cells without effector function, *J Exp Med* **188**: 2205–2213.

Zhang D, Shankar P, Xu Z, Harnisch B, Chen G, Lange C, Lee S, Valdez H, Lederman M and Liberman J (2003) Most antiviral CD8 T cells during chronic viral infection do not express high levels of perforin and are not directly cytotoxic, *Blood* **101**: 226–235.

Zhang L, Huang Y, Yuan H, Chen B, Ip J and Ho D (1997) Genotypic and phenotypic characterization of long terminal repeat sequences from long-term survivors of human immunodeficiency virus type 1 infection, *J Virol* **71**: 5608–5613.

CHAPTER 11

IDENTIFIABILITY OF HIV/AIDS MODELS

Annah M. Jeffrey and Xiaohua Xia

There is a need to accurately estimate all viral, host cell and immune response specific parameters. Before such an extensive parameter estimation exercise can be carried out, an identifiability analysis needs to be done to investigate whether or not it is possible to determine all the parameters. If it is found to be possible, then the conditions or restrictions that apply need to be known beforehand. The issues to address are the variables to be measured, the minimal number of measurements for a complete determination of all parameters, the frequency and when, during the course of the viral infection, such measurements can be taken. It is important to know this in advance, especially where budgets are concerned. The measured variable combination and conditions that result in the least number of measurements or cost less could then be selected. All the foregoing will be investigated in this article using a well-established nonlinear system identifiability theory, where identifiability is a basic system property of whether all model parameters can be calculated from the measured system outputs and known inputs.

Keywords: HIV models, identifiability, minimal measurements, algebraic framework.

1. Introduction

1.1. *A control system perspective*

It is needless to say that the modelling approach has paved the way of theoretical research. In order to use the HIV/AIDS models as a tool for treatment decisions, it is necessary to determine all parameters of the models for individuals. Even though there are general observations that can be made from the model and its structure, it is only when the model is tailored to each patient's individual parameters that clear benefits in the treatment arise. After casting the HIV/AIDS models into a control system framework, many well-established control system techniques can be used to study various issues of HIV/AIDS modelling and control.

The available works on the engineering and related aspects of HIV/AIDS management generally involve sensor design (Grant and Xu, 2002), modelling the spread of infection in high risk populations, computer aided health care support systems (Beerenwinkel *et al.*, 2001), cellular level modelling of the interaction between the virus and the immune system (various), controllability and timing the initiation of therapy (Jeffrey, Xia and Craig, 2003a), feedback control of the viral load (Alvarez–Ramirez, Meraz and Velasco–Hermandez, 2000; Brandt and Chen, 2001), as well as optimal control and multi-drug therapy scheduling (Kirschner, Lenhart and Serbin, 1997; Wein, Zenios and Nowak, 1997; de Souza, Caetano and Yoneyama, 2000). Other model analysis works such as viral load time response under therapy and bifurcation are also available (Gumel, Twizell and Yu, 2000; Jeffrey and Xia, 2002). However, in total, the number of papers with a control system perspective is limited. There is scope and potential for more work to be done in order to further assist or help alleviate the burden imposed on health care givers by this wide spreading viral infection. To this end, this paper presents an identifiability analysis of some HIV/AIDS models. The intention is to provide more insight into the properties of these models, as well as help to formulate guidelines for and aid clinicians in measurement and test scheduling for the eventual estimation of the model parameters. The actual parameter estimation techniques or algorithms will however, not be dealt with here. For that, the reader is referred to Xia (2003), and Filter and Xia (2003).

1.2. *Use for parameter estimates*

If model parameter estimates can be obtained during the early stages of the HIV infection, they can be used to predict viral load set points, which are an important indicator of disease progression (Dewhurst, da Cruz and Whetter, 2000). In this case, if such estimates can be obtained, then it must be within a reasonably short period of time. From an HIV vaccination point of view, such estimates can be used to determine the vaccine efficacy. Estimates for model parameters are available in for example (Kirschner, Lenhart and Serbin, 1997; Nowak and May, 2000; Perelson *et al.*, 1996a; Ramratnam *et al.*, 1999). Not much effort however, has been put into *simultaneously* estimating all of these parameters. The available parameter estimates, especially for the compartmental models, are sparse and incomplete. There are indications though, that accuracy of estimates for some of the model parameters is increasing as more innovative approaches for their estimation are employed.

Another reason for the need for accurate parameter estimates is the fact that there are inter-individual variations in parameters (Muller, Maree and De Boer, 2001a; Little *et al.*, 1999). Furthermore, parameters are thought to vary from one stage to the next as the infection progresses (Kramer, 1999; Nowak and May, 2000). There is a possibility though, that what seems to be changes in model parameters with time could just be the effect of unmodelled dynamics or external disturbances. Obtaining an individuals parameters at different stages of the viral infection could therefore, settle this issue.

HIV drug pharmacokinetics, pharmacodynamics and adverse reactions are genetically predisposed (Pirmohamed and Back, 2001; Roden and George, 2002). Furthermore, the response to therapy, for example, the time to effectively suppress the viral load, are parameter dependent (Jeffrey, Xia and Craig, 2003b). Given this parameter dependence of the response to therapy, one can therefore consider exploiting inter-individual variations in parameters to individualize treatment and enhance the benefits of anti-retroviral therapy (Becker and Hoetelmans, 2002; Lindpainter, 2001). There will be a need then, for test measurements to be done over a short period of time.

1.3. *Layout of paper*

An identifiability analysis of some HIV/AIDS models will be presented in this paper. The intention is to provide more insight into the properties of these models, as well as help to formulate guidelines for and aid clinicians in measurement and test scheduling for the eventual estimation of the model parameters. The issues to address are the variables to be measured, the minimal number of measurements for a complete determination of all parameters, the frequency and when, during the course of the viral infection, such measurements can be taken.

The layout of the paper is as follows: Section 2 presents the models to be analyzed. Section 3 introduces the identifiability concept and the procedure to follow when applying this concept. Section 4 presents and discusses the analysis results of the HIV/AIDS models presented in Section 2, while Section 5 has the conclusions.

2. HIV/AIDS Models

Mathematical models that describe the host-pathogen interaction between the immune system and HIV should be able to explain the initial high rise in plasma viral load, its decline and settling to levels that are much lower than the peak viral load. The subsequent dramatic increase in the viral load during the later stage of the infection and timing of this increase should also be explained. Models have been developed and explained in, for example, Kirschner (1996), Perelson and Nelson (1999), Nowak and May (2000), Hraba, Dolezal and Celikovsky (1990), Hraba and Dolezal (1996), Tan and Wu (1998), Gumel, Shivakumar and Sahai (2001), de Souza (1999), Culshaw and Ruan (2000), Murray *et al.* (1998). Most of these models are deterministic and based on balancing population dynamics, while some are stochastic and take into account the random variations in HIV dynamics. Some of these models are single compartment and address the population dynamics of the virus and $CD4^+$ T cells in plasma, while others are multi compartment and take into consideration other cells in the body like macrophages that are as susceptible to the virus.

None of these models however, can completely exhibit all that is observed clinically and account for the full course of the infection. The main reason for the models' limitation is lack of a good understanding

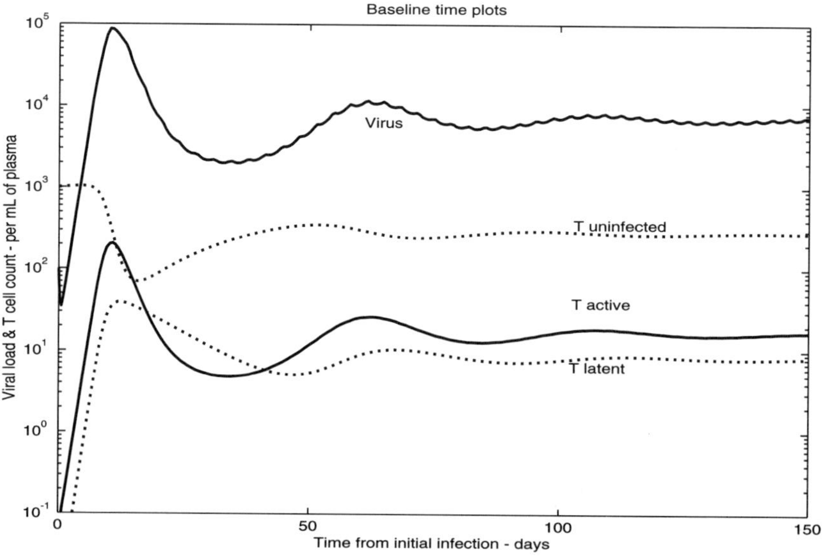

Fig. 1. Simulated viral load and $CD4^+$ T cell variation as the HIV infection progresses.

of the immunology of the human body against HIV. Biological systems tend to exhibit multi-compartmental interactions that are usually not well understood and as a result, cannot be accurately modelled mathematically. Another point to consider is that these models do not take into account other extenuating environmental, social and welfare factors that may affect the progression of the infection.

These models, however, do adequately explain the interaction of the virus and the immune system up to the clinical latency stage as illustrated in Fig. 1. In an attempt to account for the later or advanced stage of the infection, some model parameters are assumed to change as the infection progresses. These assumptions, though not clinically validated, do give a virus and CD4$^+$ cell profile that complies with clinical observations. Other suggestions are that the prolonged production and destruction of the CD4$^+$ lymphocytes ultimately results in an immune collapse. More recent studies have however, challenged these paradigms (Dewhurst, da Cruz and Whetter, 2000).

The following is a selected presentation of a class of HIV/AIDS models in Perelson and Nelson (1999).

2.1. *The basic 3-dimensional model*

The basic 3-dimensional (3D) model presented below in Eq. (1) is single compartment and shows the interaction between the virus and the CD4$^+$ T cells in blood.

$$\sum_{3D} = \begin{cases} \dot{T} &= s - dT - \beta TV, \\ \dot{T}^* &= \beta TV - \mu T^*, \\ \dot{V} &= pT^* - cV. \end{cases} \tag{1}$$

The state variables T, T^* and V are the plasma concentrations of the uninfected CD4$^+$ T cells, the actively infected CD4$^+$ T cells and the free virus particles, respectively. The first equation of (1) describes the population dynamics of the uninfected CD4$^+$ T cells. It shows that they are produced from a source at a rate s and die with a rate constant d.

Some authors assume that the source term s is constant. Other authors suggest that the source term depends on other variables of the system to best fit the clinical data, even though there is no direct medical evidence. Since HIV may be able to infect cells in the thymus and bone marrow and thus lead to a reduced production of new immunocompetent T cells, it is believed that the source term s is a decreasing function of the viral load. The following forms exist in literature: $s(v) = se^{(-\theta v)}$ in Perelson (1989);

$s(v) = \theta s/(\theta + v)$ in Perelson, Kirschner and De Boer (1993); Alvarez–Ramirez, Meraz and Velasco–Hermandez (2000); $s(v) = \theta_1 s + \theta_2 s/(B_s + v)$ in Kirschner and Webb (1996).

A proliferation rate term is sometimes added to the uninfected CD4$^+$ T cell dynamics. Since there are suggestions that the proliferation rate is density dependent with the rate of proliferation slowing as the T cell count gets high (Ho *et al.*, 1995), the most common form for proliferation is taken as the following logistic function (Perelson, Kirschner and De Boer, 1993):

$$\phi(T) = rT\left(1 - \frac{T}{T_{max}}\right), \tag{2}$$

with r as the proliferation rate constant and T_{max} is the T cell population density at which proliferation shuts off.

Uninfected CD4$^+$ T cells are infected by the virus at a rate that is proportional to the product of their abundance and the amount of free virus particles. The proportionality constant β is an indication of the effectiveness of the infection process and includes the rate at which virus particles find target cells, the rate of virus entry and probability of successful infection. The second equation of (1) describes the population dynamics of the actively infected CD4$^+$ T cells and shows that μ is their death rate constant. The third equation similarly describes the population dynamics of the free virus particles and it can be seen that an actively infected CD4$^+$ T cell produces infectious free virus particles with a rate constant p and c is the death rate constant for these virus particles.

2.2. *The latently infected 4-dimensional model*

There are additional reservoirs of virus. There is a pool of latently infected, resting CD4$^+$ T cells that is established early during primary HIV infection (Chun *et al.*, 1998). A quantitative image analysis technique was used to reveal viral burden in lymphoid tissue (Haase *et al.*, 1996), particularly on the surface of folicular dendritic cells (FDC). Perelson *et al.* observed that, after the rapid first phase of decay during the initial 1–2 weeks of antiretroviral therapy, plasma virus levels declined at a considerably slower rate (Perelson *et al.*, 1997). This second phase of viral decay was attributed to the turnover of a longer lived virus reservoir of infected cell population, which was determined to have a half-life of 1–4 weeks. This means that on average, it would take between $\frac{1}{2}$ and 3 years of perfectly effective antiretroviral therapy to eradicate the virus. This estimate generated a lot of enthusiasm and optimism in 1996 (Perelson *et al.*, 1996).

The so-called latently infected cell model presented in (3) takes note of the fact that not all CD4$^+$ T cells actively produce virus upon successful infection. This is reflected by dividing the infected cell pool into actively and latently infected cells.

$$\sum_{4D} = \begin{cases} \dot{T} &= s - dT - \beta TV, \\ \dot{T}_1 &= q_1\beta TV - kT_1 - \mu_1 T_1, \\ \dot{T}_2 &= q_2\beta TV + kT_1 - \mu_2 T_2, \\ \dot{V} &= pT_2 - cV. \end{cases} \tag{3}$$

In this 4-dimensional (4D) model, state variable T_2 represents the cells that actively produce virus, while the latently infected cells, denoted by T_1, harbor HIV proviral DNA and do not produce virus until they are activated. Parameter k is the activation rate constant. Parameters q_1 and q_2 are the probabilities that upon infection, a CD4$^+$ T cell will become either latent or actively produce virus.

2.3. *The extended 6-dimensional model*

While the latently infected cell model fits the patient data, it is not the only reasonable biological model. The source underlying the second phase kinetics might be the release of virus trapped in the lymphoid tissue. It could be linked to infected macrophage (Perelson *et al.*, 1997). Continued follow-up of persons who have remained on HAART for extended periods of time has provided strong evidence for the existence of a possible reservoir in long-lived CD4$^+$ memory T lymphocytes, a third phase of HIV decay observed during HAART. The kinetics of decay are extremely slow, and the half-life of the memory cell reservoir has been estimated at between 6 and 44 months. As a consequence, the predicted time for effective therapy required to fully eradicate the virus from the body ranges from 9 to 72 years. This suggested that a true virologic cure is unattainable using the conventional antiretroviral drugs. Table 1 is a summary of these important findings.

Table 1. Virus reservoir and life span.

Infected cell	Size	Half-life	Eradication
Active CD4	3×10^7	1 day	25 days
FDC	3×10^8–10^{11}	1–4 wks	0.5–2.8 yrs
Macrophage	?	1–4 wks	?
Memory CD4	10^5–10^6	6–44 months	9–72 yrs

Most of the figures can be found or deduced from Dewhurst, da Cruz and Whetter (2000).

Another totally different theory proposed by Grossman and colleagues says that the very slow rates of viral decay which occur following the initiation of HAART can be best explained if most of the virus is produced by cells infected after the commencement of treatment (Grossman *et al.*, 1999). These developments have brought to us challenges to face.

An upgrade of the 4D model is the extended 6-dimensional (6D) model given in (4), as it is apparent that there are other cells in the body besides CD4$^+$ T cells that are as susceptible to the virus. The release of the virus from these cells and other infected compartments has been shown to affect the virus kinetics in plasma (Muller, Maree and De Boer, 2001b). The particular cells of interest here are macrophages. These are large cells that live longer than the CD4$^+$ T cells and are chronic virus producers.

$$\sum_{6D} = \begin{cases} \dot{T} &= s_T - d_T T - \beta_T T V, \\ \dot{T}_1 &= q_1 \beta_T T V - k T_1 - \mu_1 T_1, \\ \dot{T}_2 &= q_2 \beta_T T V + k T_1 - \mu_2 T_2, \\ \dot{M} &= s_M - d_M M - \beta_M M V, \\ \dot{M}^* &= \beta_M M V - \delta M^*, \\ \dot{V} &= p_T T_2 + p_M M^* - c V. \end{cases} \tag{4}$$

M and M^* are the concentrations of the uninfected and infected macrophages respectively. The fourth equation of (4) shows that uninfected macrophages are produced from a source at a rate s_M, die with a rate constant d_M and are infected by the virus at a rate that is proportional to their abundance. Parameter β_M is an indication of the efficiency of the infection process. The fifth and sixth equations of (4) show that infected macrophages die with a rate constant δ and an infected macrophage cell produces virus with rate constant p_M. Table 2 summarizes the parameters that are in each model.

3. Identifiability Concepts

Observability (identifiability) is a basic system property of whether all state variables (all parameters) can be calculated from the measured output. The precise meaning of observability and identifiability is defined in (Conte, Moog and Perdon, 1999; Glad, 1997). Various works on nonlinear system identification can be found for example, in Tunali and Tarn (1987), Diop and Fliess (1991) and Ljung and Glad (1994).

Table 2. Parameters in each model.

	Parameter and description	3D	4D	6D
s, s_T	Source rate for uninfected CD4$^+$T cells	$\star$	$\star$	$\star$
d, d_T	Death rate for uninfected CD4$^+$ T cells	$\star$	$\star$	$\star$
β, β_T	Infection rate for CD4$^+$ T cells by virus	$\star$	$\star$	$\star$
μ, μ_2	Death rate for actively infected CD4$^+$ T cells	$\star$	$\star$	$\star$
μ_1	Death rate for latently infected CD4$^+$ T cells		$\star$	$\star$
k	Activation rate for latently infected CD4$^+$ T cells		$\star$	$\star$
q_1	Fraction of infected CD4$^+$ T cells that become latent		$\star$	$\star$
q_2	Fraction of infected CD4$^+$ T cells that become active		$\star$	$\star$
s_M	Source rate for uninfected macrophages			$\star$
d_M	Death rate for uninfected macrophages			$\star$
β_M	Infection rate for macrophages by virus			$\star$
δ	Death rate for infected macrophages			$\star$
p, p_T	Virus particle production rate per infected CD4$^+$ T cell	$\star$	$\star$	$\star$
p_M	Virus particle production rate per infected macrophage			$\star$
c	Death rate for virus particle	$\star$	$\star$	$\star$
Total number of parameters in model		6	10	15

*indicates the applicable parameter.
All parameters except s, s_T, and s_M are rate constants.

The identifiability concept applied here was presented and analyzed by Xia and Moog (2003). The concept is based on the practical requirement that if parameters can be expressed as functions of known quantities of the model, then it is possible to work in the algebraic framework. In essence, if as many such function expressions as there are unknown parameters can be generated, then it is possible to solve for all the unknown parameters. The identifiability characterizations presented here lend themselves to isolate the conditions under which the system parameters cannot be determined.

The following is the identifiability concept and the calculation procedure to follow when applying this concept to test a system's identifiability properties.

3.1. *Concepts*

Consider a nonlinear system,

$$\sum_{\theta} : \begin{cases} \dot{x} = f(x, \theta, u), & x(0, \theta) = x_0, \\ y = h(x, \theta, u), \end{cases} \tag{5}$$

where $x \in R^n$, $u \in R^m$ and $y \in R^p$ are the state, input and output variables of the system. Assume that

$$\operatorname{rank} \partial h(x, \theta, u)/\partial x = p. \tag{6}$$

θ is the parameter to be identified, and is assumed to belong to $\mathcal{P}$ which is an open subset of R^q. The functions $f(x, \theta, u)$ and $h(x, \theta, u)$ are meromorphic functions on a connected open subset $\mathcal{M}$ of R^n. Moreover, without loss of generality, x_0 is assumed to be independent of θ and not an equilibrium point of the system.

An input function $u(t)$: $[0, T] \to \mathcal{U}$, where $\mathcal{U}$ is an open subset of R^m, is called an admissible input (on $[0, T]$). For any initial condition x_0 and an admissible input $u(t)$ on $[0, T]$, there exists, on a possibly smaller time interval, $[0, \bar{T}]$, $\bar{T} \leq T$, a parameterized solution $x(t, \theta, x_0, u)$. The corresponding output is denoted by $y(t, \theta, x_0, u)$.

A classical definition of identifiability can be found in Tunali and Tarn (1987).

Definition 1: The system Σ_θ is said to be x_0-identifiable at θ through an admissible input u(on $[0, T]$) if there exists an open set $\mathcal{P}^0 \subset \mathcal{P}$ containing θ such that for any two θ_1, $\theta_2 \in \mathcal{P}^0$, $\theta_1 \neq \theta_2$, the solutions $x(t, \theta_1, x_0, u)$ and $x(t, \theta_2, x_0, u)$ exist on $[0, \epsilon]$, $0 < \epsilon \leq T$, and their corresponding outputs satisfy, on $t \in [0, \epsilon]$,

$$y(t, \theta_1, x_0, u) \neq y(t, \theta_2, x_0, u).$$

This property was termed in Tunali and Tarn (1987) as (instantaneously) locally strongly identifiability. We are more interested in a generic property of identifiability. This property was studied in Denis–Vidal, Joly–Blanchard and Noiret (2001) for polynomial systems. To introduce such a concept, we need a topology for the input function space. For any $T > 0$ and a positive integer N, the space $C^N[0, T]$ is the space of all functions on $[0, T]$ which have continuous differentiations up to the order N. A topology of the space $C^N[0, T]$ is the one associated with following well-defined norm: for $r(t) \in C^N[0, T]$,

$$\|r(t)\| = \sum_{i=0}^{N} \max_{t \in [0,T]} |r^{(i)}(t)|.$$

For any $T > 0$ and positive integer N, denote $C^N_{\mathcal{U}}[0, T]$ the set of all admissible inputs (on $[0, T]$) that have continuous derivatives up to the order N. The topology of $C^N_{\mathcal{U}}[0, T]$ is defined to be the m-fold product topology of $C^N[0,T]$. The topology of $C^N_{\mathcal{U}}[0,T] \times C^N_{\mathcal{U}}[0,T]$ is defined to be the product topology of $C^N_{\mathcal{U}}[0,T]$. The M-fold product of $C^N_{\mathcal{U}}[0,T]$ is denoted as $(C^N_{\mathcal{U}}[0,T])^M$.

Definition 2: The system Σ_θ is said to be structurally identifiable if there exist a $T > 0$, and a positive integer N, and open and dense subsets $\mathcal{M}^0 \subset \mathcal{M}, \mathcal{P}^0 \subset \mathcal{P}, \mathcal{U}^0 \subset C^N_{\mathcal{U}}[0,T]$ such that the system Σ_θ is x_0-identifiable at θ through u, for every $x_0 \in \mathcal{M}^0$, $\theta \in \mathcal{P}^0$ and $u \in \mathcal{U}^0$.

The structural identifiability is also interchangeably called geometrical identifiability in this paper, because Definition 2 is the generic version of the definition of Tunali and Tarn (1987). The structural identifiability is used to characterize the one-to-one property of the map from the parameters to the system output. The algebraic identifiability is about construction of parameters from algebraic equations of the system input and output. This concept was first employed in Glad (1997), and Ljung and Glad (1994) and later formally defined in Diop and Fliess (1991) in the differential algebraic framework.

Definition 3: The system Σ_θ is said to be algebraically identifiable if there exist a $T > 0$, positive integers N and k, and a meromorphic function Φ: $R^q \times R^{(k+1)m} \times R^{(k+1)p} \to R^q$ such that

$$\det \frac{\partial \Phi}{\partial \theta} \neq 0 \tag{7}$$

and

$$\Phi(\theta, u, \dot{u}, \ldots, u^{(k)}, y, \dot{y}, \ldots, y^{(k)}) = 0 \tag{8}$$

hold, on $[0,T]$, for all $(\theta, u, \dot{u}, \ldots, u^{(k)}, y, \dot{y}, \ldots, y^{(k)})$ where (θ, x_0, u) belong to an open and dense subset of $\mathcal{P} \times \mathcal{M} \times C^N_{\mathcal{U}}[0,T]$, and $\dot{u}, \ldots, u^{(k)}$ are the corresponding derivatives of u, and $y, \dot{y}, \ldots, y^{(k)}$ are the derivatives of the corresponding output $y(t, \theta, x_0, u)$.

Algebraic identifiability enables one to construct the parameters from solving algebraic equations depending only on the information of the input and output. As a matter of fact, under condition (7), one can solve locally an expression for θ from Eq. (8) by using the Implicit Function Theorem.

Sometimes an initial condition is known for a system. The information of the known initial state may provide additional help in determining the parameters. This phenomenon was recognized in Tunali and Tarn (1987), Glad (1997), and Ljung and Glad (1994).

Definition 4: The system Σ_θ is said to be identifiable with known initial conditions if there exist a positive integer k and a meromorphic function Φ: $R^q \times R^n \times R^{(k+1)m} \times R^{(k+1)p} \to R^q$ such that

$$\det \frac{\partial \Phi}{\partial \theta} \neq 0 \tag{9}$$

and

$$\Phi(\theta, x_0, u(0^+), \dot{u}(0^+), \ldots, u^{(k)}(0^+), y(0^+), \dot{y}(0^+), \ldots, y^{(k)}(0^+)) = 0 \tag{10}$$

hold for all $(\theta, x_0, u(0^+), \dot{u}(0^+), \ldots, u^{(k)}(0^+), y(0^+), \dot{y}(0^+), \ldots, y^{(k)}(0^+))$, where $(\theta, x_0,$ $u(0^+)$, $\dot{u}(0^+), \ldots, u^{(k)}(0^+)$ belong to an open and dense subset of $\mathcal{P} \times \mathcal{M} \times \mathcal{U}^{(k+1)m}$, and $y(0^+), \dot{y}(0^+), \ldots, y^{(k)}(0^+))$ are the derivatives of the corresponding output $y(t, \theta, x_0, u)$ evaluated at $t = 0^+$.

Again, under condition (9), one can solve locally an expression for θ from Eq. (10) by using the Implicit Function Theorem. The expression will depend on the known initial state, the input and the output.

3.2. *Characterizations*

We will give characterizations of the algebraic identifiability and structural identifiability in the linear algebraic framework of nonlinear systems (Conte, Moog and Perdon, 1999). The characterizations also lend themselves to isolate the initial conditions and inputs that are not persistently exciting, i.e. where the system parameters cannot be determined.

To recall the linear algebraic framework, let $\mathcal{K}$ be the field consisting of meromorphic functions of x, θ, u and finite derivatives of u, and define

$$E = \text{span}_{\mathcal{K}}\{\mathrm{d}\mathcal{K}\},$$

that is, a vector in E is a linear combination of a finite number of one-forms from $\mathrm{d}x, \mathrm{d}\theta, \mathrm{d}u, \mathrm{d}\dot{u}, \ldots, \mathrm{d}u^{(k)}, \ldots,$ with coefficients in $\mathcal{K}$. The vectors in E are called one-forms.

The differentiation of a function $\phi(x, \theta, u, \ldots, u^{(k)})$ along the dynamics of the system (5) is defined as

$$\dot{\phi} = \frac{\partial \phi}{\partial x} f(x, \theta, u) + \sum_{i=0}^{k} \frac{\partial \phi}{\partial u^{(i)}} u^{(i+1)},$$

and this operation can be extended to differential one-forms $\omega = \kappa_x \mathrm{d}x + \kappa_\theta \mathrm{d}\theta + \sum \eta_i \mathrm{d}u^{(i)} \in E$ as the following:

$$\dot{\omega} = \dot{\kappa}_x \mathrm{d}x + \dot{\kappa}_\theta \mathrm{d}\theta + \sum \dot{\eta}_i \mathrm{d}u^{(i)} + \kappa_x \mathrm{d}f(x, \theta, u) + \sum \eta_i \mathrm{d}u^{(i+1)}.$$

Table 3. Characterization of identifiability notions.

Algebraic identifiability	:	$\Theta \subset \mathcal{Y} + \mathcal{U}$
Identifiability with known initial conditions	:	$\Theta \subset \mathcal{X} + \mathcal{Y} + \mathcal{U}$
Geometric identifiability	:	$\Theta \subset \mathcal{X} + \mathcal{Y} + \mathcal{U}$

Note that $\dot{\omega} \in E$.

Denote

$$\begin{aligned}\mathcal{Y} &= \bigcup_{k=0}^{\infty} \operatorname{span}\{\mathrm{d}y, \mathrm{d}\dot{y}, \ldots, \mathrm{d}y^{(k)}\},\\ \mathcal{X} &= \operatorname{span}\{\mathrm{d}x\},\\ \mathcal{U} &= \bigcup_{k=0}^{\infty} \operatorname{span}\{\mathrm{d}u, \mathrm{d}\dot{u}, \ldots, \mathrm{d}u^{(k)}\},\\ \Theta &= \operatorname{span}\{\mathrm{d}\theta\},\end{aligned}$$

then the characterizations of the different notions of identifiability can be summarized in Table 3. For proof of these characterizations, the reader is referred to Xia and Moog (2003). It can be seen from the table that geometric identifiability is equivalent to identifiability with known initial conditions.

3.3. *Calculation*

Let us now investigate the computational issues for the determination of the parameters.

Note that $\mathcal{X} \cap (\mathcal{Y} + \mathcal{U})$ is the observation cospace of system (5) (Conte, Moog and Perdon, 1999), while $\mathcal{X} \cap (\mathcal{Y} + \Theta + \mathcal{U})$ can be interpreted as the observation cospace with parameter.

One checks the identifiability through calculating the input-output relations of the system. One way of doing this is first to eliminate x through observability properties of x.

Define, for system (5), the so-called observability indices. Let

$$\mathcal{F}_k := \mathcal{X} \cap \left(\operatorname{span}\{\mathrm{d}y, \mathrm{d}\dot{y}, \ldots, \mathrm{d}y^{(k-1)} + \mathcal{U} + \Theta\}\right)$$

for $k = 1, \ldots, n$. Consider the filtration

$$\mathcal{F}_1 \subset \mathcal{F}_2 \subset \cdots \subset \mathcal{F}_n.$$

Then, as done in Krener and Respondek (1985) for nonlinear systems without parameters and which are linearizable by output injections, define

$d_1 := \dim \mathcal{F}_1$ and $d_k := \dim \mathcal{F}_k - \dim \mathcal{F}_{k-1}$, for $k = 2, \ldots, n$. Let

$$k_i := \max\{k \mid d_k \geq i\}.$$

Then the list $\{k_1, k_2, \ldots, k_p\}$ is the list of observability indices and d_k represents the number of observability indices which are greater than or equal to k, for $k = 1, \ldots, n$. Reorder, if necessary, the output components such that

$$\begin{aligned} \operatorname{rank} \frac{\partial(y, \dot{y}, \ldots, y^{(n-1)}}{\partial x} &= \operatorname{rank} \partial(y_1, \dot{y}_1, \ldots, y_1^{(k_1-1)}, y_2, \ldots, y_2^{(k_2-1)}, \ldots, \\ &\quad y_p, \dot{y}_p, \ldots, y_p^{(k_p-1)})/\partial x \qquad (11) \\ &= k_1 + k_2 + \cdots + k_p. \qquad (12) \end{aligned}$$

Thanks to the assumption (6), the p observability indices are well defined. Compute

$$\begin{aligned} \mathrm{d}y_1 &= \xi_{11}\mathrm{d}x + \gamma_{11}\mathrm{d}\theta (\bmod \mathcal{U}), \\ &\vdots \\ \mathrm{d}y_p &= \xi_{p1}\mathrm{d}x + \gamma_{p1}\mathrm{d}\theta (\bmod \mathcal{U}). \end{aligned}$$

By assumption, $\operatorname{rank} \begin{bmatrix} \xi_{11} \\ \vdots \\ \xi_{p1} \end{bmatrix} = p$. And more generally, compute

$$\mathrm{d}y_i^{(j-1)} = \xi_{ij}\mathrm{d}x + \gamma_{ij}\mathrm{d}\theta (\bmod \mathcal{U}),$$

for $i = 1, \ldots, p$ and $j = 1, \ldots, k_i$. From (11), any ξ_{ij} can be written as a linear combination of $\{\xi_{11}, \ldots, \xi_{1k_1}, \ldots, \xi_{p1}, \ldots, \xi_{pk_p}\}$.

Higher order time derivatives $\mathrm{d}y_i^{(j)}$ can be computed and, from the implicit function theorem, $\mathrm{d}x$ can be substituted to obtain

$$\mathrm{d}y_i^{(j)} = \left(\sum_{r=1}^{p} \sum_{s=1}^{k-r} \eta_{rs} \mathrm{d}y_r^{(s-1)} \right) + \gamma_{i,j+1}\mathrm{d}\theta (\bmod \mathcal{U}).$$

Then, the system is geometrically identifiable if and only if there are integers k_i^*, for $i = 1, \ldots, p$, such that

$$\operatorname{rank} \begin{bmatrix} \gamma_{11} \\ \vdots \\ \gamma_{1k_1^*} \\ \gamma_{21} \\ \vdots \\ \gamma_{pk_p^*} \end{bmatrix} = q. \qquad (13)$$

The system is algebraically identifiable if and only if there exist p integers l_i^*, for $i = 1, \dots, p$, such that

$$\operatorname{rank} \begin{bmatrix} \gamma_{1(k_1+1)} \\ \vdots \\ \gamma_{1l_1^*} \\ \gamma_{2(k_2+1)} \\ \vdots \\ \gamma_{pl_p^*} \end{bmatrix} = q. \tag{14}$$

A trajectory such that (13) (or (14)) holds is called geometrically (or algebraically) persistently exciting.

4. Model Analysis

The success of any control system depends very much on the measurement subsystem. This is especially true in the modelling of HIV/AIDS dynamics. As a matter of fact, "the impetus for further modelling came with the development of rapid, sensitive and accurate methods of measuring the number of virus particles in the blood" (Perelson and Nelson, 1999). All variables in the models presented in this and Section 2 can essentially be measured, though some with less accuracy and high cost. However, variables that are routinely measured for deciding when to initiated therapy and for monitoring of patients on antiretroviral therapy are the viral load and the total CD4$^+$ T cell count T_{tot}, which is the sum of the uninfected and infected CD4$^+$ T cells. In settings where all variable measurements are obtainable, this will improve the identifiability properties of the system.

4.1. *Identifiability properties of the 3D model*

The identifiability properties of the basic 3D model (1) will be analyzed. Consider the viral load and uninfected CD4$^+$ T cell count as the measured outputs. Taking outputs as

$$y_1 = T, \quad y_2 = V; \quad y = (y_1, y_2),$$

compute

$$\dot{y}_1 = s - dy_1 - \beta y_1 y_2. \tag{15}$$

Thus, output y_1 has an observability index equal to 1 and three parameters that may be identified. Higher order derivatives yield

$$\ddot{y}_1 = -d\dot{y}_1 - \beta(y_1y_2)^{(1)}, \tag{16}$$

$$y_1^{(3)} = -d\ddot{y}_1 - \beta(y_1y_2)^{(2)}. \tag{17}$$

Now we have three equations (15), (16) and (17) with three unknown parameters. Parameters s, d and β can therefore, be computed from any persistently exciting trajectory $y(t)$ such that rank $\partial(\dot{y}_1, \ddot{y}_1, y_1^{(3)})/\partial(s, d, \beta) = 3$. That is, if

$$\text{rank}\begin{bmatrix} 1 & -y_1 & -y_1y_2 \\ 0 & -\dot{y}_1 & -(y_1y_2)^{(1)} \\ 0 & -\ddot{y}_1 & -(y_1y_2)^{(2)} \end{bmatrix} = 3.$$

The three equations (15), (16) and (17) can be solved to get a unique solution for s, d and β as functions of $y(t_o)$, $\dot{y}(t_o)$, $\ddot{y}(t_o)$ and $y_1^{(3)}(t_o)$ at some instant t_o. To cope with these order of derivatives, one can conclude that at least four measurements of the uninfected $CD4^+$ T cell count y_1 and at least three measurements of the viral load y_2, are needed for a complete first determination of these three parameters.

For the remaining 3 parameters, consider output y_2 and compute

$$\dot{y}_2 = pT^* - cy_2, \tag{18}$$

$$\ddot{y}_2 = \theta_1\dot{y}_2 + \theta_2 cy_2 + \theta_3 y_1y_2, \tag{19}$$

where

$$\theta_1 = -(\mu + c), \quad \theta_2 = -\mu c, \quad \theta_3 = p\beta.$$

Output y_2 has an observability index equal to 2 and higher order derivatives yield

$$y_2^{(3)} = \theta_1\ddot{y}_2 + \theta_2\dot{y}_2 + \theta_3(y_1y_2)^{(1)}, \tag{20}$$

$$y_2^{(4)} = \theta_1 y_2^{(3)} + \theta_2\ddot{y}_2 + \theta_3(y_1y_2)^{(2)}. \tag{21}$$

Again, we have three equations (19), (20) and (21) and three unknown parameters. "Parameters" θ_1, θ_2 and θ_3 can be similarly computed from any persistently exciting trajectory $y(t)$ such that rank $\partial(\ddot{y}_2, y_2^{(3)}, y_2^{(4)})/\partial(\theta_1, \theta_2, \theta_3) = 3$. That is, if

$$\text{rank}\begin{bmatrix} \dot{y}_2 & y_2 & y_1y_2 \\ \ddot{y}_2 & \dot{y}_2 & (y_1y_2)^{(1)} \\ y_2^{(3)} & \ddot{y}_2 & (y_1y_2)^{(2)} \end{bmatrix} = 3.$$

Table 4. Minimum measurements for 3D model.

Measured	Property	T_{tot}	T	T^*	V	Total
T, V	Algebraic	—	4	0	5	9
T_{tot}, V	Algebraic	4/5	—	—	5/4	9

— indicates that measurement is not applicable.

The map $\Theta = (\theta_1, \theta_2, \theta_3) = (-(\mu + c), -\mu c, p\beta)$ is one-to-one, hence the identifiability of $\theta_1, \theta_2, \theta_3$ is equivalent to that of μ, c, p. The three equations (19), (20) and (21) can be solved to get a unique solution for μ, c and p as functions of $y(t_o)$, $\dot{y}(t_o)$, $\ddot{y}(t_o), y_2^{(3)}(t_o)$ and $y_2^{(4)}(t_o)$ at some instant t_o, provided $\beta \neq 0$ and $c \neq \mu$. For the determination of these three parameters, at least three measurements of the uninfected CD4$^+$ T cell count y_1 and five measurements of the viral load y_2 are necessary. The system Σ_{3D} (1) is therefore algebraically identifiable and all six parameters can be computed from the measurements of the viral load and uninfected CD4$^+$ T cells. Overall, at least four measurements of the uninfected CD4$^+$ T cell count and five viral load measurements are required for a complete first estimate of all six parameters. Results are summarized in Table 4.

In current practice, discriminatory CD$_4^+$ T cell measurements are not readily attainable and the cell count that is actually measured is the total of the infected and uninfected CD4$^+$ T cells. Identifiability of this 3D model (1) has been analyzed with the viral load and the total CD4$^+$ T cell count as the measured outputs. Even though this results in the same total number of measurements, as indicated in Table 4, the eventual extraction of the parameters from the measured outputs will be more mathematically intensive.

It has been observed that for individuals in the asymptomatic stage of the infection and those on antiretroviral drugs, the infected CD4$^+$ T cell pool makes a very small percentage of the total CD4$^+$ T cell count (Embretson *et al.*, 1993; Janeway and Travers, 1997; Chun *et al.*, 1997). This means that in this case, the assumption that the uninfected CD4$^+$ T cell count is approximately equal to the total CD4$^+$ T cell count can be made. We have therefore opted to take the uninfected CD4$^+$ T cell count as the measured output instead of the total CD4$^+$ T cell count.

4.2. *When to take measurements*

If either one of the measured outputs y_1 or y_2 is constant, the higher order derivatives (16), (17), (20) and (21) in the previous section will be zero and parameter extraction from the measured outputs will not be possible. When one therefore, considers the variation of the model variables with time as depicted in Fig. 1, then one can see that the measured outputs are constant for the asymptomatic stage. It has been observed that for HIV-infected individuals, the viral load remains relatively constant during this long asymptomatic stage, while the $CD4^+$ T cell count slowly declines. This means that measurements should be taken during the acute infection stage and during the advanced stage of the HIV infection.

For individuals in the asymptomatic stage of the infection, one then needs to use antiretroviral drugs to perturb the quasi-steady state. When measurements are taken during the acute infection stage of the infection, then the assumption that the measured total $CD4^+$ T cell count is representative of the uninfected cell population does not hold. This will necessitate for discriminatory $CD4^+$ T cell measurements to be the standard practice, unless antiretroviral agents are again used, but in this case to reduce the proportion of infected cells in the total cell count.

4.3. *Identifiability with the use of antiretroviral agents*

The two classes of commonly used anti-retroviral agents are Reverse Transcriptase Inhibitors (RTI) and Protease Inhibitors (PI). Both agents work within the $CD4^+$ T cell because they do not prevent the virus from entering the cell. Model parameters that are affected by antiretroviral agents have been identified (McLeod and Hammer, 1992; Perelson and Nelson, 1999). RTIs reduce successful infection by reducing the value of β. Perfect inhibition, therefore, occurs when $\beta = 0$. In practice however, perfect inhibition is not attainable. PIs on the other hand, inhibit the protease enzyme so that the virus particles that are produced are mostly noninfectious. There are therefore two types of virus particles when protease inhibitors are used. The first type are the infectious virus particles that still continue to infect $CD4^+$ T cells and the other is the noninfectious type. Similarly, perfect inhibition occurs when all virus particles that are produced are non-infectious. Current therapies use a combination of Reverse Transcriptase and Protease Inhibitors and the combined therapy model can be presented as

$$\sum_{3Da} = \begin{cases} \dot{T} &= s - dT - u_{RT}\beta T V_I, \\ \dot{T}^* &= u_{RT}\beta T V_I - \mu T^*, \\ \dot{V}_I &= u_{PI} p T^* - c V_I, \\ \dot{V}_N &= (1 - u_{PI}) p T^* - c V_N. \end{cases} \tag{22}$$

$u_{RT} = 1 - \eta_{RT}$ and $u_{PI} = 1 - \eta_{PI}$ are the respective control efforts for the reverse transcriptase and protease inhibitors. η_{RT}, $0 \leq \eta_{RT} < 1$ is the combined effectiveness of all the reverse transcriptase inhibitors used and η_{PI}, $0 \leq \eta_{PI} < 1$ is the combined effectiveness of all the protease inhibitors used. State variables V_I and V_N are the plasma concentrations of the infectious and noninfectious virus particles respectively. It is assumed that both types of virus particles have the same death rate constant c.

For parameter identifiability under the use of antiretroviral agents, the effect of the drugs will affect the identification of the affected model parameters. In fact, for the model Σ_{3Da}, the identifiable parameters will be $u_{RT}\beta$ instead of β and $u_{PI}p$ instead of p. This means that the drug control effort or efficacy cannot be separated from the parameter it affects. Another point worth noting is that, current assays do not differentiate between infectious and noninfectious virus particles. That is, $V_{tot} = V_I + V_N$ is the measured viral load.

4.4. *Identifiability properties of the 4D model*

In this section, consider the latently infected cell model Σ_{4D} in Eq. (3). Taking the uninfected CD4$^+$ T cell count and the viral load as the measured outputs, then again,

$$y_1 = T, \quad y_2 = V.$$

Compute

$$\dot{y}_1 = s - dy_1 - \beta y_1 y_2. \tag{23}$$

The identifiability of parameters s, d and β has been presented in Section 4.1.

For the other parameters $(q_1, q_2, k, \mu_1, \mu_2, p, c)$, compute

$$\dot{y}_2 = pT_2 - cy_2, \tag{24}$$

$$\ddot{y}_2 = pq_2\beta y_1 y_2 + k_1 pT_1 - (\mu_2 + c)\dot{y}_2 - \mu_2 c y_2, \tag{25}$$

$$y_2^{(3)} = \theta_1 (y_1 y_2)^{(1)} + \theta_2 \ddot{y}_2 + \theta_3 \dot{y}_2 + \theta_4 y_1 y_2 + \theta_5 y_2, \tag{26}$$

where

$$\begin{aligned}
\theta_1 &= pq_2\beta, \\
\theta_2 &= -(\mu_1 + \mu_2 + c + k), \\
\theta_3 &= -(\mu_2 c + \mu_1\mu_2 + \mu_1 c + ck + k\mu_2), \\
\theta_4 &= kpq_1\beta + pq_2\beta(\mu_1 + k), \\
\theta_5 &= -c\mu_2(\mu_1 + k).
\end{aligned}$$

Thus output y_2 has an observability index equal to 3, and higher order derivatives of y_2 will just read as

$$y_2^{(i)} = \theta_1(y_1y_2)^{(i-2)} + \theta_2 y_2^{(i-1)} + \theta_3 y_2^{(i-2)} + \theta_4(y_1y_2)^{(i-3)} + \theta_5 y_2^{(i-3)}. \quad (27)$$

The five "parameters" $\theta_1, \ldots, \theta_5$ can be computed from any persistent exciting trajectory $y(t)$ such that rank $\partial(y_2^{(3)}, \ldots, y_2^{(7)})/\partial(\theta_1, \ldots, \theta_5) = 5$. That is, if

$$\text{rank}\begin{bmatrix} (y_1y_2)^{(1)} & \ddot{y}_2 & \dot{y}_2 & y_1y_2 & y_2 \\ (y_1y_2)^{(2)} & y_2^{(3)} & \ddot{y}_2 & (y_1y_2)^{(1)} & \dot{y}_2 \\ (y_1y_2)^{(3)} & y_2^{(4)} & y_2^{(3)} & (y_1y_2)^{(2)} & \ddot{y}_2 \\ (y_1y_2)^{(4)} & y_2^{(5)} & y_2^{(4)} & (y_1y_2)^{(3)} & y_2^{(3)} \\ (y_1y_2)^{(5)} & y_2^{(6)} & y_2^{(5)} & (y_1y_2)^{(4)} & y_2^{(4)} \end{bmatrix} = 5.$$

Thus, the system Σ_{4D} is not algebraically identifiable with the outputs taken as the uninfected $CD4^+$ T cell count and the viral load. Besides the identification of s, d and β, five of the seven parameters can be computed in terms of the measurements and two remaining parameters, if

$$q_1q_2\beta p(\mu_1 + k - \mu_2)(\mu_2 - c)(\mu_1 + k - c) \neq 0. \quad (28)$$

Some interesting conclusions can be drawn from (28):

(i) If $q_1 = 0$, then μ_1 and k cannot be separated from $\mu_1 + k$. Medically, this would mean that if the channel is cut off for infected cells to become latent, then one will not be able to tell from the viral load and the $CD4^+$ T cell count, how many latently infected cells die naturally and how many of them turn to actively infected cells.

(ii) If $p = 0$, then the actively infected $CD4^+$ T cells do not produce any virus. No information about these infected cells can therefore be extracted from measurements of the viral load.

(iii) If $\mu_1 + k = \mu_2$, then the latently infected cells and the actively infected cells disappear from their corresponding pools at the same rate. The dynamics of these two cell types will therefore be indistinguishable.

(iv) If $\mu_2 = c$, then the actively infected cells and the virus disappear at the same rate from their respective pools, hence it will not be possible to distinguish their dynamics.

(v) If $\mu_1 + k = c$, then similarly, the latently infected cells and the virus dynamics will be indistinguishable if they are cleared at the same rate from their respective pools.

Table 5. Identifiability of system Σ_{4D} with some known parameters.

Known parameters	Identifiability of remaining parameters
q_1, q_2	identifiable
q_1, μ_1	identifiable
q_1, k	identifiable
q_1, μ_2	not identifiable
q_1, p	identifiable
q_1, c	not identifiable
q_2, μ_1	identifiable
q_2, k	identifiable
q_2, μ_2	not identifiable
q_2, p	identifiable
q_2, c	not identifiable
μ_1, k	not identifiable
μ_1, μ_2	not identifiable
μ_1, p	identifiable
μ_1, c	not identifiable
k, μ_2	not identifiable
k, p	identifiable
k, c	not identifiable
μ_2, p	not identifiable
μ_2, c	not identifiable

It is noted that some of the parameters can be determined by other methods. For example, through the experiment of Chun *et al.* (1997), q_1 was found to be $q_1 = 0.01$. Janeway and Travers (1997) found that $q_2 = 0.02$. An exhaustive list of all cases where two of the model parameters are known is presented in Table 5.

The identifiability property of system Σ_{4D} (3) can be improved if the initial values of some other variables are known. In identifiability theory, this property is characterized by the so-called geometric identifiability, or equivalently, identifiability with known initial conditions.

To inspect the geometric identifiability of the seven remaining parameters $\{q_1, q_2, k, \mu_1, \mu_2, p, c\}$, introduce the notation $\Theta_1 = pq_2\beta$ and $\Theta_2 = kpq_1\beta$. Consider the new list of parameters $\{\Theta_1, \Theta_2, \mu_1, \mu_2, k, p, c\}$ and compute $\ddot{y}_2$, $y_2^{(3)}$ as

$$\ddot{y}_2 = \Theta_1 y_1 y_2 + kpT_1 - p\mu_2 T_2 - c\dot{y}_2, \tag{29}$$

$$\begin{aligned} y_2^{(3)} &= \Theta_1 (y_1 y_2)^{(1)} + \Theta_2 y_1 y_2 - \mu_2(\ddot{y}_2 + c\dot{y}_2) - c\ddot{y}_2 \\ &\quad - (\mu_1 + k)[\ddot{y}_2 - \Theta_1 y_1 y_2 + \mu_2(\dot{y}_2 + cy_2) - c\dot{y}_2]. \end{aligned} \tag{30}$$

Higher order derivatives are obtained by differentiating (30).

Introduce the notation

$$
\begin{aligned}
A &= y_1 y_2, \\
B &= (y_1 y_2)^{(1)} + (\mu_1 + k)(y_1 y_2), \\
C &= -[\ddot{y}_2 - \Theta_1 y_1 y_2 + \mu_2(\dot{y}_2 + c y_2) - c\dot{y}_2], \\
D &= -[(\mu_1 + k)(\dot{y}_2 + c y_2) + \ddot{y}_2 + c\dot{y}_2], \\
E &= -[y_2^{(2)} - \Theta_1 y_1 y_2 + \mu_2(\dot{y}_2 + c y_2) + c\dot{y}_2], \\
F &= -[(\mu_1 + k)(\mu_2 y_2 + \dot{y}_2) + \mu_2 \dot{y}_2 + \ddot{y}_2],
\end{aligned}
$$

and compute Γ_g as
$\partial(\dot{y}_2, \ddot{y}_2, y_2^{(3)}, y_2^{(4)}, y_2^{(5)}, y_2^{(6)}, y_2^{(7)})/\partial(\Theta_1, \Theta_2, \mu_1, \mu_2, k, p, c)$.
That is,

$$
\Gamma_g = \begin{bmatrix}
0 & 0 & 0 & 0 & 0 & T_2 & -y_2 \\
A & 0 & 0 & -pT_2 & pT_1 & [kT_1 - \mu_2 T_2] & -\dot{y}_2 \\
B & A & C & D & E & 0 & F \\
\dot{B} & \dot{A} & \dot{C} & \dot{D} & \dot{E} & 0 & \dot{F} \\
\ddot{B} & \ddot{A} & \ddot{C} & \ddot{D} & \ddot{E} & 0 & \ddot{F} \\
B^{(3)} & A^{(3)} & C^{(3)} & D^{(3)} & E^{(3)} & 0 & F^{(3)} \\
B^{(4)} & A^{(4)} & C^{(4)} & D^{(4)} & E^{(4)} & 0 & F^{(4)}
\end{bmatrix}.
$$

Then rank $\Gamma_g = 7$ if (28) holds. That is if,

$$q_1 q_2 \beta p(\mu_1 + k - \mu_2)(\mu_2 - c)(\mu_1 + k - c) \neq 0.$$

Thus, the system is geometrically identifiable and all the original parameters of Σ_{4D} are identifiable from the measurements of the viral load and uninfected $CD4^+$ T cell count if the initial values for both the actively and latently infected $CD4^+$ T cells for the individual are known. One therefore, needs to have a *comprehensive* test before measurements are taken. With the advancement of faster and cheaper measuring devices, it is envisaged that discriminatory $CD4^+$ T cell count measurements will soon be routine and available cost-effectively. The required minimum number of measurements for a complete first determination of all ten parameters are as summarized in Table 6.

Table 6. Minimum measurements for 4D model.

Measured	Property	T	T_1	T_2	V	Total
T, V	Geometric	6	1	1	8	16

4.5. *Identifiability properties of the 6D model*

An analysis of the 6D model Σ_{6D} in (4) will be presented for when different combinations of model variables are the measured outputs.

4.5.1. *Identifiability with viral load and uninfected CD4$^+$ T cell count measurements*

In the first instance, take outputs as the uninfected CD4$^+$ T cell count, T and the viral load V

$$y_1 = T, \quad y_2 = V.$$

Output y_1 has observability index equal to 1, while that of y_2 can be shown to be equal to 5. The resulting equations for outputs y_1 and y_2 are as given in (31) and (32), respectively.

$$\begin{aligned}
\dot{y}_1 &= s_T - d_T y_1 - \beta_T y_1 y_2, && (31)\\
y_2^{(5)} &= p_T q_2 \beta_T (y_1 y_2)^{(3)} + p_T \beta_T (\theta_2 + \delta q_2)(y_1 y_2)^{(2)} \\
&\quad + p_T \beta_T \delta \theta_2 (y_1 y_2)^{(1)} - (\theta_1 + c + \mu_2) y_2^{(4)} - (\theta_4 + (k + \mu_1)\theta_3) y_2^{(3)} \\
&\quad + (\theta_5 + s_M - (k + \mu_1)\theta_4)\ddot{y}_2 + (k + \mu_1)\theta_5 \dot{y}_2 \\
&\quad + s_M(\Sigma + (\psi_1 - \beta_M y_2)\ddot{y}_2) + \frac{\dot{\Sigma}}{\Sigma} - (d_M - \beta_M y_2)\Lambda, && (32)
\end{aligned}$$

where,

$$\begin{aligned}
\theta_1 &= k + \mu_1 + \delta, \\
\theta_2 &= k q_1 + (k + \mu_1) q_2, \\
\theta_3 &= \delta + c + \mu_2, \\
\theta_4 &= \delta c + \mu_2(\delta + c), \\
\theta_5 &= p_M \beta_M s_M - \delta c \mu_2, \\
\psi_1 &= \mu_2 - d_M - \beta_M y_2, \\
\Sigma &= \ddot{y}_2 + (\psi_1 - \beta_M y_2)\dot{y}_2, \\
\Lambda &= y_2^{(4)} - p_T q_2 \beta_T (y_1 y_2)^{(2)} - p_T \beta_T (\theta_2 + \delta q_2)(y_1 y_2)^{(1)} \\
&\quad - p_T \beta_T \delta \theta_2 y_1 y_2 + (k + \mu_1 + \theta_3) y_2^{(3)} + (\theta_4 + \theta_3(k + \mu_1))\ddot{y}_2 \\
&\quad - (\theta_5 + s_M - \theta_4(k + \mu_1))\dot{y}_2 - \theta_5(k + \mu_1) y_2 - \psi_1 s_M y_2.
\end{aligned}$$

However, the system is not algebraically identifiable. Besides the identification of s_T, d_T and β_T, only eight of the remaining twelve parameters can be computed in terms of the measured outputs and the other four parameters. The system can be shown to be geometrically identifiable, or identifiable with known initial conditions. One therefore, needs a comprehensive test to obtain initial measurements for both the actively and latently infected

CD4$^+$ T cells, as well as for the infected and uninfected macrophages. Even though this system is geometrically identifiable with measured outputs taken as the viral load and uninfected CD4$^+$ T cells, the required minimum number of measurements of these outputs is too high as outlined in Table 7. Attempts to obtain estimates of all 15 parameters of this model, with the viral load and uninfected CD4$^+$ T cell counts as the measured outputs will therefore, not be a practical approach when cost and patient discomfort are taken into consideration. One therefore needs to consider measuring something else or increasing the number of measured outputs.

4.5.2. *Identifiability with viral load and uninfected CD4$^+$ T cell count and macrophage measurements*

The identifiability property of system Σ_{6D} (4) can be improved by also measuring the uninfected macrophages. Then, taking the outputs as

$$y_1 = T, \quad y_2 = V, \quad y_3 = M, \quad \text{and} \quad \text{let } x_1 = y_1 y_2, \quad x_2 = y_2 y_3,$$

and compute

$$\dot{y}_1 = s_T - d_T y_1 - \beta_T x_1. \tag{33}$$

Then, s_T, d_T and β_T can therefore be computed from any persistently exciting trajectory $y(t)$ such that rank $\partial(\dot{y}_1, \ddot{y}_1, y_1^{(3)})/\partial(s_T, d_T, \beta_T) = 3$. That is, if

$$\text{rank} \begin{bmatrix} 1 & -y_1 & -x_1 \\ 0 & -\dot{y}_1 & -\dot{x}_1 \\ 0 & -\ddot{y}_1 & -\ddot{x}_1 \end{bmatrix} = 3.$$

For output y_3, compute

$$\dot{y}_3 = s_M - d_M y_3 - \beta_T x_2. \tag{34}$$

Similarly, s_M, d_M and β_M can be computed from any persistently exciting trajectory $y(t)$ such that rank $\partial(\dot{y}_3, \ddot{y}_3, y_3^{(3)})/\partial(s_M, d_M, \beta_M) = 3$. That is, if

$$\text{rank}\begin{bmatrix} 1 & -y_3 & -x_2 \\ 0 & -\dot{y}_3 & -\dot{x}_2 \\ 0 & -\ddot{y}_3 & -\ddot{x}_2 \end{bmatrix} = 3.$$

For the remaining nine parameters $(\mu_1, \mu_2, q_1, q_2, k, p_T, p_M, \delta, c)$, define

$$\begin{aligned}
\theta_1 &= k + \mu_1 + \delta, \\
\theta_2 &= kq_1 + (k + \mu_1)q_2, \\
\theta_3 &= \delta + c + \mu_2, \\
\theta_4 &= \delta c + \mu_2(\delta + c), \\
\theta_5 &= \delta c(\delta + c)(k + \mu_1), \\
\theta_6 &= c\mu_2(c + \mu_2)(k + \mu_1), \\
\theta_7 &= \mu_2\delta(\mu_2 + \delta)(k + \mu_1).
\end{aligned}$$

Then compute

$$\dot{y}_2 = p_T T_2 + p_M M^* - c y_2, \tag{35}$$

$$\begin{aligned}
\ddot{y}_2 &= q_2 p_T \beta_T x_1 + p_M \beta_M x_2 + p_T k T_1 \\
&\quad - (\delta + c)\dot{y}_2 - \delta c y_2 - p_T(\mu_2 - \delta)T_2,
\end{aligned} \tag{36}$$

$$\begin{aligned}
y_2^{(3)} &= q_2 p_T \beta_T \dot{x}_1 + p_M \beta_M \dot{x}_2 + p_T \beta_T (kq_1 + \delta q_2)x_1 \\
&\quad + p_M \beta_M \mu_2 x_2 - \theta_3 \ddot{y}_2 - \theta_4 \dot{y}_2 - \mu_2 \delta c y_2 \\
&\quad - p_T k(k + \mu_1 - \delta)T_1,
\end{aligned} \tag{37}$$

$$\begin{aligned}
y_2^{(4)} &= q_2 p_T \beta_T \ddot{x}_1 + p_M \beta_M \ddot{x}_2 + p_T \beta_T (\theta_2 + \delta q_2)\dot{x}_1 \\
&\quad + p_M \beta_M (k + \mu_1 + \mu_2)\dot{x}_2 + p_T \beta_T \delta \theta_2 x_1 \\
&\quad + p_M \beta_M \mu_2 (k + \mu_1) x_2 - (\theta_3 + k + \mu_1) y_2^{(3)} \\
&\quad - (\theta_4 + \theta_3(k + \mu_1))\ddot{y}_2 - (\mu_2 \delta c + \theta_4(k + \mu_1))\dot{y}_2 \\
&\quad - (k + \mu_1)\mu_2 \delta c y_2.
\end{aligned} \tag{38}$$

The remaining nine parameters can therefore be computed from any persistently exciting trajectory $y(t)$ such that rank $\partial(y_2^{(4)}, \ldots, y_2^{(12)})/\partial(\mu_1, \mu_2, q_1, q_2, k, p_T, p_M, \delta, c) = 9$. That is, if

$$\text{rank}\begin{bmatrix} \psi_1 & \psi_2 & \psi_3 & \cdots & \psi_8 & \psi_9 \\ \dot{\psi}_1 & \dot{\psi}_2 & \dot{\psi}_3 & \cdots & \dot{\psi}_8 & \dot{\psi}_9 \\ \vdots & \vdots & \vdots & \cdots & \vdots & \vdots \\ \psi_1^{(8)} & \psi_2^{(8)} & \psi_3^{(8)} & \cdots & \psi_8^{(8)} & \psi_9^{(8)} \end{bmatrix} = 9,$$

where,

$$\begin{aligned}
\psi_1 &= p_T\beta_T q_2(\dot{x}_1+\delta x_1)+p_M\beta_M(\dot{x}_2+\mu_2 x_2)\\
&\quad - y_2^{(3)} - \theta_3\ddot{y}_2 - \theta_4\dot{y}_2 - \mu_2\delta c y_2,\\
\psi_2 &= p_M\beta_M(\dot{x}_2+(k+\mu_1)x_2) - y_2^{(3)}\\
&\quad - (\theta_4+c)\ddot{y}_2 - \theta_5\dot{y}_2 - (k+\mu_1)\delta c y_2,\\
\psi_3 &= p_T\beta_T k(\dot{x}_1+\delta x_1),\\
\psi_4 &= p_T\beta_T(\ddot{x}_1+\theta_1\dot{x}_1+\delta(k+\mu_1)x_1),\\
\psi_5 &= p_T\beta_T(q_1+q_2)(\dot{x}_1+\delta x_1)+p_M\beta_M(\dot{x}_2+\mu_2 x_2)\\
&\quad - y_2^{(3)} - \theta_3\ddot{y}_2 - \theta_4\dot{y}_2 - \mu_2\delta c y_2,\\
\psi_6 &= \beta_T(q_2\ddot{x}_1+(\theta_3+\delta q_2)\dot{x}_1+\delta\theta_3 x_1),\\
\psi_7 &= \beta_M(\ddot{x}_2+\mu_2(k+\mu_1)x_2),\\
\psi_8 &= p_T\beta_T(q_2\dot{x}_1+\theta_3 x_1) - y_2^{(3)}\\
&\quad - (k+\mu_1+\delta+\mu_2)\ddot{y}_2 - \theta_6\dot{y}_2 - (k+\mu_1)\mu_2 c y_2,\\
\psi_9 &= -y_2^{(3)} - (\mu_2+\theta_1)\ddot{y}_2 - \theta_7\dot{y}_2 - (k+\mu_1)\mu_2\delta y_2.
\end{aligned}$$

System Σ_{6D} can be shown to be algebraically identifiable, and all fifteen parameters can be computed from measurements of the uninfected $CD4^+$ T cells, the viral load and uninfected macrophages, if

$$p_T\beta_T\beta_M k q_1 q_2(k-q_1)(k+\mu_1-\mu_2)\neq 0,$$

$$\mu_2\neq c \quad \text{and} \quad c\neq\delta.$$

Again, bearing in mind that the infected cells are a small portion of the total cell count, then in most instances where it is not possible to obtain discriminatory macrophage and $CD4^+$ T cell count measurements, one can take the total cell count as representative of the uninfected cells.

4.5.3. *Identifiability with viral load and discriminatory $CD4^+$ T cell count measurements*

Considering that macrophage measurements are currently more difficult to obtained compared to discriminatory $CD4^+$ T cell count measurements, one option would be to measure the actively infected $CD4^+$ T cells instead of the uninfected macrophages. Setting

$$y_1 = T,\quad y_2 = V,\quad y_4 = T_2,\quad \text{and}\quad x_1 = y_1 y_2,$$

then outputs y_1, y_2 and y_4 have observability indices $r_1 = 1$, $r_2 = 3$ and $r_4 = 2$, respectively. Identifiability of parameters s_T, d_T and β_T is as presented in Section 4.5.2. For identifiability of the remaining twelve parameters, define

$$\triangle_1 = \frac{\dot{y}_2}{y_2} - d_M - \beta_M y_2,$$
$$\triangle_2 = \ddot{y}_2 - p_T\dot{y}_4 + (\delta + c)\dot{y}_2 - p_T\delta y_4 + \delta c y_2,$$

and compute

$$\dot{y}_2 = p_T y_4 + p_M M^* - c y_2, \tag{39}$$
$$\ddot{y}_2 = p_T\dot{y}_4 + p_T\delta y_4 - (\delta + c)\dot{y}_2 - \delta c y_2 + p_M\beta_M y_2 M, \tag{40}$$
$$\begin{aligned} y_2^{(3)} &= p_T\ddot{y}_4 - (\delta + c + d_M)\ddot{y}_2 + p_T(\delta + d_M)\dot{y}_4 \\ &\quad - (\delta c + d_M(\delta + c))\dot{y}_2 + p_T d_M \delta y_4 \\ &\quad - (d_M\delta c - p_M\beta_M s_M)y_2 + \left(\frac{\dot{y}_2}{y_2} - \beta_M y_2\right)\triangle_2. \end{aligned} \tag{41}$$

Then the seven parameters $(p_T, p_M, \delta, c, s_M, d_M, \beta_M)$ in $y_2^{(3)}$ can be computed from any persistently exciting trajectory $y(t)$ such that rank $\partial(y_2^{(3)}, \ldots, y_2^{(9)})/\partial(p_T, p_M, \delta, c, s_M, d_M, \beta_M) = 7$. That is, if

$$\text{rank}\begin{bmatrix} v_1 & s_M v_6 & v_2 & v_3 & p_M v_6 & v_4 & v_5 \\ \dot{v}_1 & s_M\dot{v}_6 & \dot{v}_2 & \dot{v}_3 & p_M\dot{v}_6 & \dot{v}_4 & \dot{v}_5 \\ \vdots & \vdots & \vdots & \vdots & \vdots & \vdots & \vdots \\ v_1^{(6)} & s_M v_6^{(6)} & v_2^{(6)} & v_3^{(6)} & p_M v_6^{(6)} & v_4^{(6)} & v_5^{(6)} \end{bmatrix} = 7,$$

where,

$$\begin{aligned} v_1 &= \ddot{y}_4 + (\delta - \triangle_1)\dot{y}_4 - \delta\triangle_1 y_4, \\ v_2 &= -\ddot{y}_2 + p_T\dot{y}_4 - (c - \triangle_1)\dot{y}_2 - p_T\triangle_1 y_4 + \triangle_1 c y_2, \\ v_3 &= -\ddot{y}_2 - (\delta - \triangle_1)\dot{y}_2 + \triangle_1\delta y_2, \\ v_4 &= -\ddot{y}_2 + p_T\dot{y}_4 - (\delta + c)\dot{y}_2 + p_T\delta y_4 - \delta c y_2, \\ v_5 &= (p_M s_M + \triangle_2)y_2, \\ v_6 &= \beta_M y_2. \end{aligned}$$

However, the above matrix only has rank $= 6$, and therefore not all the seven parameters can be estimated. It can be shown that one needs prior knowledge of either s_M or p_M in order to determine the other five parameters.

For the still remaining parameters $(\mu_1, \mu_2, q_1, q_2, k)$, compute

$$\dot{y}_4 = q_2\beta_T x_1 + kT_1 - \mu_2 y_4, \tag{42}$$

$$\begin{aligned}\ddot{y}_4 &= q_2\beta_T \dot{x}_1 + \beta_T(kq_1 + (k+\mu_1)q_2)x_1 \\ &\quad - (k+\mu_1+\mu_2)\dot{y}_4 - \mu_2(k+\mu_1)y_4.\end{aligned} \tag{43}$$

The five parameters can be computed from any persistently exciting trajectory $y(t)$ such that rank $\partial(\ddot{y}_4, \ldots, y_4^{(6)})/\partial(\mu_1, \mu_2, k, q_1, q_2) = 5$. That is, if

$$\text{rank}\begin{bmatrix} \phi_1 & \phi_2 & \phi_1 + \beta_T q_1 x_1 & \beta_T k x_1 & \phi_3 \\ \dot{\phi}_1 & \dot{\phi}_2 & \dot{\phi}_1 + \beta_T q_1 \dot{x}_1 & \beta_T k \dot{x}_1 & \dot{\phi}_3 \\ \ddot{\phi}_1 & \ddot{\phi}_2 & \ddot{\phi}_1 + \beta_T q_1 \ddot{x}_1 & \beta_T k \ddot{x}_1 & \ddot{\phi}_3 \\ \phi_1^{(3)} & \phi_2^{(3)} & \phi_1^{(3)} + \beta_T q_1 x_1^{(3)} & \beta_T k x_1^{(3)} & \phi_3^{(3)} \\ \phi_1^{(4)} & \phi_2^{(4)} & \phi_1^{(4)} + \beta_T q_1 x_1^{(4)} & \beta_T k x_1^{(4)} & \phi_3^{(4)} \end{bmatrix} = 5,$$

where,

$$\begin{aligned}\phi_1 &= -\dot{y}_4 + \beta_T q_2 x_1 - \mu_2 y_4, \\ \phi_1 &= -\dot{y}_4 - (k+\mu_1)y_4, \\ \phi_3 &= \beta_T(\dot{x}_1 + (k+\mu_1)x_1.\end{aligned}$$

However, the above matrix has rank < 5, and therefore not all five remaining parameters can be estimated. One needs prior knowledge of either μ_1, q_1 or k in order to determine the other four parameters. The system is therefore not algebraically identifiable if the viral load, uninfected and actively infected $CD4^+$ T cell counts are the measured outputs.

To test for geometric identifiability of the remaining parameters, use (40) and (42) to generate a 7th equation for the y_2 dynamics and a 5th and 6th equation for the y_4 dynamics.

The system is geometrically identifiable if the viral load, the uninfected and actively infected $CD4^+$ T cells counts are the measured outputs. The initial measurements for the latently infected $CD4^+$ T cells, T_1 and uninfected macrophages, M will also be required.

For the output options considered in this section, measuring the actively infected $CD4^+$ T cells instead of the uninfected macrophage cells significantly reduces the number of required measurements, even though the system is no longer algebraically identifiable. This illustrates that improving the identifiability property of a system does not necessarily imply a reduction in the required number of measurements. More importantly, the point that is being illustrated here is that careful consideration of what needs to

Table 7. Minimum measurements for 6D model.

Measured	Property	T	T_1	T_2	M	M^*	V	Total
T, V	Geometric	15	1	1	1	1	17	36
T, M, V	Algebraic	11	0	0	11	0	13	35
T, T_2, V	Geometric	5	1	7	1	1	8	23

be measured is necessary. Table 7 summarizes the results for when various model variables are the measured outputs.

5. Conclusions

In this paper, a nonlinear system identifiability theory has been applied to analyze the identifiability properties of some HIV/AIDS models. The intention was to investigate the possibility of simultaneously estimating all the model parameters from measured system outputs, such as the viral load and $CD4^+$ T cell count. Other issues addressed include the minimum number of measurements for a complete first approximation of all parameters, the timing and conditions under which such measurements can be taken.

This information will be useful for the eventual parameter estimation, as well as for formulating guidelines for clinical practice.

References

Alvarez–Ramirez J, Meraz M and Velasco–Hermandez JX (2000) Feedback control of the chemotherapy of HIV, *Int J Bifurcat Chaos* **10**: 2207–2219.

Becker SL and Hoetelmans RM (2002) Exploiting pharmacokinetics to optimize antiretroviral therapy, http://www.medscape.com/viewprogram/703

Beerenwinkel N, Lengauer T, Selbig J, Schmidt B *et al.* (2001) Geno2pheno: interpreting genotypic HIV drug resistance tests, *IEEE Intell Syst Bio* Nov/Dec: 35–41.

Brandt ME and Chen G (2001) Feedback control of a biodynamical model of HIV-1, *IEEE Trans BME* **48**: 754–759.

Chun TW, Carruth L, Finzi D, Xhen X *et al.* (1997) Quantification of latent tissue reservoirs and total body viral load in HIV-1 infection, *Nature* **387**: 183–188.

Chun TW, Engel D, Berrey MM, Shea T, Corey L, Fauci AS (1998) Early establishment of a pool of latently infected, resting $CD4^+$ T cells during primary HIV-1 infection, *Proc Natl Acad Sci USA* **95**: 8869–8873.

Conte G, Moog CH and Perdon AM (1999) *Nonlinear Control Systems: An Algebraic Setting*, Springer Verlag, London.

Culshaw RV and Ruan S (2000) A delay-differential equation model of HIV infection of $CD4^+$ T-cells, *Math Biosci* **165**: 27–39.

de Souza FMC (1999) Modeling the dynamics of HIV-1 and CD4 and CD8 lymphocytes, *IEEE Eng Med Bio* Jan/Feb: 21–24.

de Souza JAMF, Caetano MAL and Yoneyama T (2000) Optimal control theory applied to the anti-viral treatment of AIDS, in *Proc IEEE Conference on Decision and Control*, pp. 4839–4844, Sydney, Australia.

Denis-Vidal L, Joly-Blanchard G and Noiret C (2001) Some effective approaches to check the identifiability of uncontrolled nonlinear systems, *Math Comput Simulat* **57**: 35–44.

Dewhurst S, da Cruz RLW and Whetter L (2000) Pathogenesis and treatment of HIV-1 infection: recent developments (Y2K update), *Front Biosci* **5**: 30–49.

Diop S and Fliess M (1991) Nonlinear observability, identifiability and persistent trajectories, *Proc 30th CDC*, Brighton, UK: 714–719.

Embretson J, Zupancic M, Beneke J, Till M, Wolinski S, Ribas JL *et al.* (1993) Analysis of HIV infected tissues by amplification and *in situ* hybridization reveals latent and permissive infections at single-cell resolution, *Proc Acad Sci* **90**: 357–361.

Filter R and Xia X (2003) A penalty function approach to HIV/AIDS model parameter estimation, 13th *IFAC Symp Syst Ident* 27–29 Aug. Rotterdam, Netherlands.

Glad ST (1997) Solvability of differential algebraic equations and inequalities: an algorithm, *Proc 4th ECC*, Bruxelles.

Grant SA and Xu J (2002) Investigation of an optical dual receptor method to detect HIV, *IEEE Sensors* **2**(5): 409–415.

Grossman Z, Polis M, Feinberg MB, Levi I, Jankelevich S, Yarchoan R *et al.* (1999) Ongoing dissemination during HAART, *Nat Med* **5**: 1099–1104.

Gumel AB, Shivakumar PN and Sahai BM (2001) A mathematical model for the dynamics of HIV-1 during the typical course of infection, *Nonlin Anal* **47**: 1773–1783.

Gumel AB, Twizell EH and Yu P (2000) Numerical and bifurcation analyses for a population model of HIV chemotherapy, *Math Comp Sim* **54**: 169–181.

Haase AT, Henry K, Zupancic M, Sedgewick G, Faust R *et al.* (1996) Quantitative image analysis of HIV-1 infection in lymphoid tissue, *Science* **274**: 985–989.

Ho DD, Newman AU, Perelson AS *et al.* (1995) Rapid turnover of plasma virions and CD4 lymphocytes in HIV-1 infection, *Nature* **273**: 123–126.

Hraba T, Dolezal J and Celikovsky S (1990) Model based analysis of $CD4^+$ lymphocyte dynamics in HIV infected individuals, *Immunbiol* **181**: 108–118.

Hraba T and Dolezal J (1996) A mathematical model and $CD4^+$ lymphocyte dynamics in HIV infection, *Emerg Infect Dis* **12**(4): 299–305.

Janeway CA and Travers P (1997) *Immunology: The Immune System in Health and Disease*, Garland, New York.

Jeffrey AM and Xia X (2002) Estimating the viral load response time after HIV chemotherapy, *IEEE Africon*, pp. 77–80, George, South Africa, 2–4 October.

Jeffrey AM, Xia X and Craig IK (2003a) When to initiate HIV therapy: a control theoretic approach, *IEEE Trans BME* **50**(11): 1213–1220.

Jeffrey AM, Xia X and Craig IK (2003b) On attaining maximal and durable suppresion of the viral load, *1st African Control Conference AFCON*, Cape Town, South Africa, 3–5 Dec.

Kirschner D (1996) Using mathematics to understand HIV immune dynamics, *Notices AMS* **43**(2): 191–202.

Kirschner D, Lenhart S and Serbin S (1997) Optimal control of the chemotherapy of HIV, *J Math Biol* **35**: 775–792.

Kirschner D and Webb GF (1996) A model for treatment strategy in the chemotherapy of AIDS, *Bull Math Biol* **58**: 367–390.

Kramer I (1999) Modeling the dynamical impact of HIV on the immune system: viral clearance, infection and AIDS, *Math Comp Mod* **29**: 95–112.

Krener AJ and Respondek W (1985) Nonlinear observers with linearizable error dynamics, *SIAM J Contr Optimiz* **23**: 197–216.

Lindpainter K (2001) Pharmacogenetics and the future of medical practice: conceptual considerations, *Pharmacogen J* **1**: 23–26.

Little SJ, McLean AR, Spina CA, Richman DD and Hvalir DV (1999) Viral dynamics of acute HIV-1 infection, *J Exp Med* **190**: 841–850.

Ljung L and Glad ST (1994) On global identifiability for arbitrary model parameterizations, *Automatica* **30**: 265–276.

McLeod GX and Hammer SM (1992) Zidovudine: 5 years later, *Ann Intern Med* **117**: 487–501.

Muller V, Maree AFM and De Boer RJ (2001a) Small variations in multiple parameters account for large variations in HIV-1 set-points: a novel modelling approach, *Proc R Soc Lond B Biol Sci* **268**: 235–242.

Muller V, Maree AFM and De Boer RJ (2001b) Release of virus from lymphoid tissue affects human immunodeficiency virus type 1 and hepatitis C virus kinetics in the blood, *J Virol* **75**: 2597–2603.

Murray JM, Kaufmann G, Kelleher AD and Cooper DA (1998) A model of primary HIV-1 infection, *Math Biosci* **154**: 57–85.

Nowak MA and May RM (2000) *Virus Dynamics: Mathematical Principles of Immunology and Virology*, Oxford University Press, New York.

Perelson AS (1989) Modelling the interaction of the immune system with HIV, in Castillo–Chavez C (ed.), *Mathematical and Statistical Approaches to AIDS Epidemiology*, Lecture Notes in Biomathematics, Vol. 83, pp. 350–370, Springer Verlag, New York.

Perelson AS, Kirschner D and De Boer R (1993) Dynamics of HIV infection of $CD4^+$ T cells, *Math Biosci* **114**: 81–125.

Perelson AS and Nelson PW (1999) Mathematical analysis of HIV-1 dynamics *in vivo*, *SIAM Rev* **41**: 3–44.

Perelson AS, Neumann AW, Markowitz M *et al.* (1996) HIV-1 dynamics *in vivo*: virion clearance rate, infected cell life-span and viral generation time, *Science* **271**: 1582–1586.

Perelson AS, Essunger P, Markowitz M and Ho DD (1996) How long should treatment be given if we had an antiretroviral regimen that completely blocked HIV replication? *XIth Intl Conf AIDS Abstracts*.

Perelson AS, Essunger P, Cao Y, Vesanen M, Hurley A, Saksela K, Markowitz M and Ho DD (1997) Decay characteristics of HIV-1 infected compartments during combination therapy, *Nature* **387**: 188–191.

Pirmohamed M and Back DJ (2001) The pharmacogenetics of HIV therapy, *Pharmacogen J Nature* **1**: 243–253.

Ramratnam B, Bonhoeffer S, Binley J, Hurley A *et al.* (1999) Rapid production and clearance of HIV-1 and hepatitis C virus assessed by large volume plasma apheresis, *Lancet* **354**: 1782–1785.

Roden DM and George AL Jr. (2002) The genetic basis of variability in drug responses, *Nat Rev Drug Disc* **1**: 37–44.

Tan W-Y and Wu H (1998) Stochastic modeling of the dynamics of $CD4^+$ T cell infection by HIV and some Monte Carlo studies, *Math Biosci* **147**: 173–205.

Tunali ET and Tarn TJ (1987) New results for identifiability of nonlinear systems, *IEEE Trans Auto Contr* **32**: 146–154.

Wein LM, Zenios SA and Nowak MA (1997) Dynamic multidrug therapies for HIV: a control theoretic approach, *J Theor Biol* **185**: 15–29.

Xia X (2003) Estimation of HIV/AIDS parameters, *Automatica* **39**: 1983–1988.

Xia X and Moog CH (2003) Identifiability of nonlinear systems with application to HIV/AIDS models, *IEEE Tran Auto Contr* **48**: 330–336.

CHAPTER 12

INFLUENCE OF DRUG PHARMACOKINETICS ON HIV PATHOGENESIS AND THERAPY

Narendra M. Dixit and Alan S. Perelson

Incorporation of descriptions of drug pharmacokinetics in models of viral dynamics is important in predicting the long-term outcome of therapy and in the analysis of short-term viral load decay from which crucial parameters that govern viral replication kinetics are determined. In this chapter, we discuss recent developments in this area, focusing on the role of drug pharmacokinetics in the suppression of viral load and the emergence of drug resistance in an effort to establish optimal treatment protocols for HIV infection.

Keywords: HIV infection, viral dynamics, antiretroviral therapy, drug pharmacokinetics, drug resistance.

1. Introduction

Current antiretroviral therapies rapidly reduce plasma viral loads in HIV infected individuals to below detection limits but fail to eradicate the virus despite prolonged treatment over several years. Often, drug resistant strains of HIV emerge rendering long-term therapy futile. Latently infected cells and follicular dendritic cells provide reservoirs for HIV from which it can resurface upon termination of therapy (Finzi and Siliciano, 1998; Hlavacek *et al.*, 1999). As a result, following the establishment of chronic infection eradication of HIV with current therapies has not been achieved. Yet, significant improvements in the conditions of HIV patients can be brought about by the optimization of current therapies.

Assuming long-term therapy is inevitable, what is then of interest is to minimize the amount of drug administered while maintaining viral loads below detection limits. Not only will this reduce treatment costs, making treatment more accessible in third world countries, but it will also curtail toxic side-effects commonly associated with antiretroviral therapy. In a recent study, a structured intermittent treatment strategy where patients were cyclically on therapy for 7 days and off therapy for 7 days successfully maintained viral loads below detection limits over the entire treatment period of ~1 year (Dybul *et al.*, 2001). Thus, the amount of drug administered was reduced to half of that in standard therapies, which in addition to commensurate benefits in treatment costs and toxic side effects allowed patients acceptable drug holidays. Whether the 7-day on-off structure is optimal, however, remains to be determined. A similar optimization problem is associated with the emergence of drug resistance, which is facilitated by low drug pressure. Drug-resistant strains often arise following noncompliance to therapy where missed doses result in low drug levels (Paterson *et al.*, 1999). Sustenance of drug susceptibility therefore calls for intense therapy, but is accompanied by enhanced side-effects. Dosage levels that ensure drug susceptibility while minimizing toxic side-effects remain to be established.

Standard models of viral dynamics cannot address problems of therapy optimization for they do not include descriptions of drug pharmacokinetics. In standard models (Perelson *et al.*, 1996; Stafford *et al.*, 2000; Nowak and May, 2000), drug efficacy is assumed to be constant, usually 100%, whereas drug concentrations wax and wane between doses. In particular, in intermittent treatment strategies or cases of noncompliance, drug concentrations and hence efficacies can drop significantly allowing the emergence of drug resistance and the resurgence of viral loads. Optimization of therapy therefore requires sophisticated models that explicitly incorporate variations in drug levels between doses and the consequent effects on viral dynamics and evolution.

Pharmacokinetic properties of antiretroviral drugs are well studied experimentally (for example, Buss and Cammack, 2001). Models have also been developed that successfully predict variations of drug concentrations in plasma following drug administration (Welling, 1986; Gabrielson and Weiner, 2000). *In vitro* studies provide links between drug concentration and efficacy (Buss and Cammack, 2001). At the same time, standard models of viral dynamics describe the evolution of viral load in infected individuals at fixed drug efficacies (Perelson *et al.*, 1996; Stafford *et al.*, 2000; Nowak and May, 2000; Perelson and Nelson, 2002). Combining these descriptions, attempts are now ongoing to develop models that capture viral load evolution as drug efficacies vary with time governed by the pharmacokinetic properties of the drugs employed and thereby predict outcomes of therapy (Austin *et al.*, 1998; Kepler and Perelson, 1998; Wahl and Nowak, 2000; Hurwitz and Schinazi, 2002; Huang *et al.*, 2003; Dixit and Perelson,

2004a). For instance, Austin *et al.* (1998) combine plasma pharmacokinetics of typical antiretroviral drugs with a basic model of viral dynamics to estimate the minimum dosage required to eradicate virus in the absence of drug resistant mutations. With simple models of drug resistance, Wahl and Nowak (2000) and Huang *et al.* (2003) have built on the Austin *et al.* (1998) description to determine the effects of adherence to prescribed dosing regimen on the evolution of viral load. These studies begin to provide valuable insights into the effects of drug pharmacokinetics on the within-host viral dynamics of HIV and establish guidelines for therapy optimization.

Models that combine drug pharmacokinetics and viral dynamics are also important in accurately analyzing short-term changes in viral load under therapy from which crucial parameters that govern the kinetics of viral replication *in vivo* are determined. For instance, viral loads remain at pre-treatment levels for 1–2 days following the onset of combination therapy before beginning to decline (Perelson *et al.*, 1996). This shoulder in viral load decay profiles arises as a combination of pharmacological and intracellular delays: An administered drug must be absorbed into circulation, transported to cells, and, in some cases, converted in cells into its therapeutically active form before it can influence viral replication. These processes contribute to the pharmacological delay in the shoulder. At the same time, alterations in viral replication due to drug action become evident as changes in plasma viral load only at the end of the ongoing viral replication cycle. For example, that drug action has aborted viral replication at the reverse transcription stage in an infected cell will be detected as a change in plasma viral load only after the time period required for the remaining steps in the replication cycle to occur, when contrary to expectation, no new virions will be released from the cell as a result of drug action. Thus the finite duration of the viral replication cycle introduces the second, intracellular contribution to the shoulder. Decoupling intracellular and pharmacological contributions to the shoulder, which is achieved by viral dynamics models that explicitly incorporate drug pharmacokinetics, can provide insights into the viral lifecycle (Dixit and Perelson, 2004a). Similarly, accurate estimates of the magnitude and the time dependence of the *in vivo* efficacy of drugs and the corresponding changes in viral load are best described by models that combine drug pharmacokinetics and viral dynamics.

As most antiretroviral drugs (with the exception of fusion inhibitors) act not in plasma but in cells, accurate estimation of drug efficacy requires knowledge of the intracellular pharmacokinetics of drugs. Dixit and

Perelson (2004a, b) recently developed a model to describe the intracellular pharmacokinetics of typical antiretroviral drugs and combined it with a model of viral dynamics. Analysis of the short-term evolution of viral load with their model yielded estimates of the *in vivo* efficacy of drugs and the duration of the viral replication cycle. Surprisingly, they found the *in vivo* efficacy of individual drugs to be significantly lower than estimates based on *in vitro* studies suggesting a reason for the failure of current therapies to fully block viral replication.

Despite limitations, these early attempts to integrate drug pharmacokinetics with viral dynamics have made significant advances in our understanding of HIV pathogenesis and our ability to predict outcomes of therapy. Clearly, much remains to be done, especially in terms of therapy optimization to minimize costs and side effects given the need for long-term treatment, which makes this an active area of ongoing research. In this chapter, we describe current modeling approaches that integrate drug pharmacokinetics and HIV dynamics, and discuss their implications on HIV pathogenesis and therapy.

2. Basic Pharmacokinetics Concepts — Plasma Pharmacokinetics

We begin with a simple description of the plasma pharmacokinetics of antiretroviral drugs. An antiretroviral drug is usually administered orally and is absorbed into systemic circulation through the gut. Following absorption, it is distributed to various parts of the body including cells, the locations of its therapeutic action. Absorbed drug is eliminated from circulation via the kidneys and the liver. Assuming first order processes for absorption and elimination, the time evolution of the drug concentration in plasma is determined by the following equations (Welling, 1986, Austin *et al.*, 1998, Dixit and Perelson, 2004a):

$$\frac{dM_g}{dt} = -k_a M_g \tag{1}$$

$$\frac{dM_b}{dt} = k_a M_g - k_e M_b. \tag{2}$$

Here, M_g is the mass of the drug in the gut and M_b that in blood. k_a and k_e are first order rate constants of absorption and elimination, respectively. Solving these equations with the initial conditions $M_g(0) = FD$ and $M_b(0) = 0$, where D is the mass of the drug administered in a dose, gives the time evolution of the concentration of the drug in blood, $C_b(t)$, following a single dose at $t = 0$:

$$C_b(t) = \frac{FD}{V_d}\frac{k_a}{k_e - k_a}\left[e^{-k_a t} - e^{-k_e t}\right], \tag{3}$$

where $V_d(= M_b/C_b)$ is the volume of distribution. F, known as the bioavailability, is the fraction of administered drug that is absorbed into circulation. Since orally administered drug passes through the liver before it reaches systemic circulation, a significant portion of the drug may be eliminated prior to entering the circulation. This loss arising from the so-called "first passage effect" is included in F.

According to Eq. (3), C_b is zero at the time of drug administration, rises to a maximum value C_{max} at $t = t_{max}$ and then decays to zero at long times $(t \to \infty)$. It can be shown that

$$t_{max} = \frac{\ln(k_a/k_e)}{k_a - k_e} \quad \text{and} \quad C_{max} = \frac{FD}{V_d}\left(\frac{k_e}{k_a}\right)^{\frac{k_e}{k_a - k_e}}. \tag{4}$$

The area under the concentration time curve,

$$AUC = \int_0^\infty C_b(t)\, dt = \frac{FD}{V_d k_e} \tag{5}$$

and the half-life of the drug, $t_{1/2} = \ln 2/k_e$. The overall clearance rate, $CL = k_e V_d$. These quantities, t_{max}, C_{max}, AUC, $t_{1/2}$, and CL are determined from experiments to characterize the pharmacokinetic properties of drugs in plasma.

To determine $C_b(t)$ when multiple doses are administered, Eq. (1) is generalized to include $j = 1, 2$, etc. additional doses given at a regular dosing interval, I_d, with the initial dose given at $t = 0$:

$$\frac{dM_g}{dt} = -k_a M_g + FD\sum_{j=1}^{\infty} \delta_D[t - jI_d] \tag{6}$$

Here, $\delta_D(x)$ is the Dirac delta function, equal to 1 when $x = 0$ and 0 otherwise. Solving Eq. (6) using the method of integrating factors yields

$$M_g(t) = M_g(0)e^{-k_a t} + e^{-k_a t}\int_0^t e^{k_a s} FD \sum_{j=1}^{\infty} \delta_D[s - jI_d]\, ds.$$

Combining this with the initial condition $M_g(0) = FD$ and the definition of the delta function we get

$$M_g(t) = FDe^{-k_a t}\left[1 + \sum_{j=1}^{N} e^{k_a j I_d}\right] = FDe^{-k_a t}\sum_{j=0}^{N} e^{k_a j I_d}.$$

Recognizing the latter summation to be a geometric series, the above expression is simplified to

$$M_g(t) = FDe^{-k_a t}\frac{e^{N_d k_a I_d} - 1}{e^{k_a I_d} - 1}, \tag{7}$$

where N is the greatest integer in t/I_d and $N_d = N+1$ is the total number of doses administered including the first dose at $t = 0$. Substituting Eq. (7) in Eq. (2), solving for M_b, and dividing by V_d yields

$$\begin{aligned} C_b(t) = \frac{FD}{V_d}\frac{k_a}{k_e - k_a}\left(\frac{e^{-k_e t}}{e^{k_a I_d} - 1}\right)\Bigg[1 - \left\{e^{(k_e - k_a)t}\left(1 - e^{N_d k_a I_d}\right)\right\} \\ + \left(e^{k_e I_d} - e^{k_a I_d}\right)\left(\frac{e^{(N_d - 1)k_e I_d} - 1}{e^{k_e I_d} - 1}\right) - e^{((N_d - 1)k_e + k_a)I_d}\Bigg]. \end{aligned} \tag{8}$$

According to Eq. (8), $C_b(t)$ oscillates between doses with the average plasma concentration in a dosing interval eventually reaching steady state. $C_b(t)$ during this steady state is obtained by letting $t \gg 1/k_a$ and $1/k_e$ in Eq. (8), which yields

$$C_b(t) = \frac{FD}{V_d}\frac{k_a}{k_e - k_a}\left[\frac{e^{(N_d I_d - t)k_a}}{e^{k_a I_d} - 1} - \frac{e^{(N_d I_d - t)k_e}}{e^{k_e I_d} - 1}\right], \tag{9}$$

where we note that $N_d I_d - t$ oscillates between 0 and I_d. The minimum value of C_b obtained by substituting $N_d I_d - t = 0$ or I_d is

$$C_{min}(t) = \frac{FD}{V_d}\frac{k_a}{k_e - k_a}\left[\frac{1}{e^{k_a I_d} - 1} - \frac{1}{e^{k_e I_d} - 1}\right]. \tag{10}$$

Generally, the instantaneous efficacy of a drug, $\varepsilon(t)$, is assumed to be a function of its plasma concentration (Welling, 1986; Austin *et al.*, 1998; Gabrielson and Weiner, 2000):

$$\varepsilon(t) = \frac{\varepsilon_{max}[C_b(t)]^n}{[IC_{50}]^n + [C_b(t)]^n}, \tag{11}$$

where IC_{50} is that value of the concentration at which the drug exhibits 50% of its maximum efficacy, ε_{max}. According to Eq. (11), ε is zero when no drug is present and reaches $\varepsilon_{max} \leq 1$ for large drug concentrations. A plot of ε versus C_b is sigmoidal in shape and the exponent n determines the steepness of the curve. For simplicity or in the absence of other information, both ε_{max} and n are assumed to be equal to 1. IC_{50} is usually determined from *in vitro* studies and employed to estimate efficacies *in vivo*.

As an illustration, we present in Fig. 1 the time evolution of the plasma concentration and efficacy of ritonavir, a potent antiretroviral drug, using

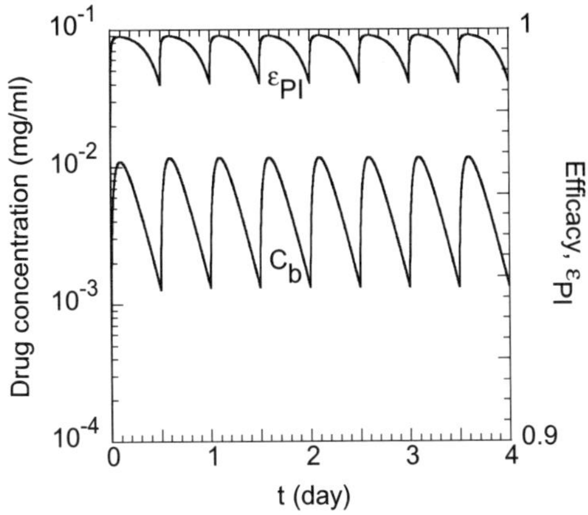

Fig. 1. Plasma concentrations (C_b) and efficacy (ε_{PI}) of ritonavir administered orally 600 mg bid, calculated using Eqs. (8) and (11), respectively, with the parameter values $D = 600$ mg, $I_d = 0.5$ day, $F = 1, V_d = 28{,}000\,\mathrm{l}, k_a = 14.64\ \mathrm{day}^{-1}, k_e = 6.86\ \mathrm{day}^{-1}$, and $IC_{50} = 1.7 \times 10^{-5}\,\mathrm{mg\,mL}^{-1}$ (geometric mean of reported values, Dixit and Perelson, 2004a).

Eqs. (8) and (11) for standard twice daily dosing. Note that C_b varies by nearly an order of magnitude between doses, yet stays sufficiently high compared to IC_{50} so that $\varepsilon \sim 1$ throughout (note the scale for ε). Given the time evolution of drug efficacy, of interest is to understand its influence on the evolution of plasma viral load in infected individuals. For this, we consider a standard model of HIV dynamics in the next section.

3. Viral Dynamics

A basic model of HIV dynamics considers the evolution of target cells, productively infected cells, and free virions according to the following equations (Perelson *et al.*, 1996; Stafford *et al.*, 2000; Nowak and May, 2000):

$$\frac{dT}{dt} = \lambda - dT - kTV \tag{12}$$

$$\frac{dT^*}{dt} = kTV - \delta T^* \tag{13}$$

$$\frac{dV}{dt} = N\delta T^* - cV. \tag{14}$$

Here, T is the population density of target CD4$^+$ T lymphocytes, which in the absence of infection are recruited into plasma at a rate λ and cleared with a first order rate constant d. Thus, a steady state population density of target cells, $T_{ss} = \lambda/d$, is achieved in the absence of infection. In the presence of virus, target cells are infected at a rate kV, where V is the viral load and k is a second order infection rate constant that accounts for the rate at which free virions encounter target cells and the fraction of these encounters that leads to successful infections. Productively infected cells, T^*, which are formed thus at the rate kVT, produce free virus at a rate $N\delta$, where N is the viral burst size and $1/\delta$ is the average lifetime of these cells. Free virus is cleared with a first order rate constant c.

Given the above model of HIV dynamics, whether a small invading population of free virions can establish infection within a host is determined by the basic reproductive ratio, R_0. R_0 is defined as the number of productively infected cells resulting from a single productively infected cell in a fully susceptible population of target cells (Nowak and May, 2000). If $R_0 > 1$, each productively infected cell gives rise to more than 1 productively infected cell in its lifetime and the infection spreads. If $R_0 < 1$, however, the number of productively infected cells gradually decreases and the infection dies out.

To derive an expression for R_0 based on the above model, we follow Ribeiro *et al.* (2004) and consider an invading viral population that gives rise to T_0^* productively infected cells at time $t = 0$ in a unit volume of a population of target cells at a number density, $T_{ss} = \lambda/d$, equal to that in the uninfected steady state. Productively infected cells die with a first order rate constant δ, so that at a later time t, $T^*(t) = T_0^* e^{-\delta t}$. The number density of free virions produced by these cells evolves according to Eq. (14). Substituting for $T^*(t)$ in Eq. (14) and integrating, we find $V(t) = N\delta T_0^*(e^{-\delta t} - e^{-ct})/(c - \delta)$, where we have employed the initial condition, $V(0) = 0$, as no new virions are present at $t = 0$. Since the rate at which virions infect target cells is kVT, the total number of target cells infected by these virions is $\int_0^\infty kV(t)T_{ss}dt = kNT_0^*\lambda/cd$, where the population of target cells is assumed to remain at its pretreatment steady-state. Thus, from one productively infected cell, $T_0^* = 1$, we obtain

$$R_0 = \frac{kN\lambda}{cd} \tag{15}$$

new productively infected cells.

That R_0 determines the establishment of infection is evident upon considering the steady-state solution of the above model obtained by equating the left hand sides of Eqs. (12)–(14) to zero:

$$T_{ss} = \frac{c}{kN}; \quad T_{ss}^* = \frac{\lambda}{\delta} - \frac{dc}{\delta kN}; \quad V_{ss} = \frac{\lambda N}{c} - \frac{d}{k} \tag{16}$$

Rewriting in terms of R_0 using Eq. (15) yields

$$T_{ss} = \frac{\lambda}{dR_0}; \quad T^*_{ss} = \frac{\lambda}{\delta}\left(1 - \frac{1}{R_0}\right); \quad V_{ss} = \frac{d}{k}(R_0 - 1). \tag{17}$$

It follows that infection is established when $R_0 > 1$ and cleared otherwise. How therapy perturbs the steady-state when $R_0 > 1$ and forces a decline in viral load is discussed next.

4. Impact of Therapy

Three classes of antiretroviral drugs are currently approved for use against HIV infection, *viz.*, reverse transcriptase inhibitors (RTIs), protease inhibitors (PIs), and fusion inhibitors. RTIs, which prevent successful infection of target cells, are of two kinds. Nucleoside RTIs act as chain terminators when incorporated into the DNA being produced in an infected cell, whereas non-nucleoside RTIs are small molecule inhibitors of the enzyme reverse transcriptase. PIs prevent the HIV protease enzyme from cleaving the polyprotein comprising reverse transcriptase, protease, integrase, and other matrix proteins, into functional units. The resulting virions are consequently rendered noninfectious. Fusion inhibitors block the entry of HIV into target cells by arresting the fusion of viral and cell membranes.

The basic model of viral dynamics described by Eqs. (12)–(14) is modified to account for drug action as follows. RTI action blocks infection and is therefore assumed to reduce the infection rate constant k by a factor $(1-\varepsilon_{RTI})$, where $0 < \varepsilon_{RTI} < 1$ is the efficacy of the RTI. PI action renders a fraction ε_{PI} of the virions produced noninfectious, where $0 < \varepsilon_{PI} < 1$ is the efficacy of the PI. Currently, no models explicitly account for fusion inhibitor action, but ignoring effects of intracellular delay (see below) fusion inhibitor action can be represented in a manner identical to that of an RTI since fusion inhibitors also block new infections. Under combined PI and RTI action, Eqs. (12)–(14) become (Perelson *et al.*, 1997; Perelson and Nelson, 2002):

$$\frac{dT}{dt} = \lambda - dT - (1-\varepsilon_{RTI})kTV_I \tag{18}$$

$$\frac{dT^*}{dt} = (1-\varepsilon_{RTI})kTV_I - \delta T^* \tag{19}$$

$$\frac{dV_I}{dt} = (1-\varepsilon_{PI})N\delta T^* - cV_I \tag{20}$$

$$\frac{dV_{NI}}{dt} = \varepsilon_{PI}N\delta T^* - cV_{NI}. \tag{21}$$

Here, V_I and V_{NI} are infectious and noninfectious virions, respectively. The measured viral load is $V = V_I + V_{NI}$. By calculating ε_{RTI} and ε_{PI} using Eq. (11), the above equations provide the necessary framework for combining drug pharmacokinetics and viral dynamics.

Several limitations exist in this description. First, the model does not include descriptions of intracellular delay, which affects drug action in subtle ways depending on the nature of the drug as discussed below. Ignoring intracellular delay, however, does not affect conclusions about the long-term impact of therapy as the delay is typically ~ 1 day whereas treatment proceeds over years. The second and more important drawback is that the model only includes a single population of target cells with an average lifetime of $1/\delta$ when infected and does not consider long-lived cells which are necessary to accurately describe HIV dynamics beyond 1–2 weeks (Perelson *et al.*, 1997). Also, the model ignores the emergence of drug resistance, which can be important in long-term therapy.

Notwithstanding, Austin *et al.* (1998) employ this minimal model to derive estimates of the dosage necessary to eradicate virus. For this purpose, they calculate R_0 in the presence of therapy, R_0^t, and argue that for therapy to be effective it must reduce R_0^t below unity. For monotherapy, assuming ε (which can be ε_{RTI} or ε_{PI}) does not vary with time, it can be shown following the above derivation that

$$R_0^t = (1 - \varepsilon)R_0, \tag{22}$$

where R_0 is the pretreatment value given by Eq. (15). To eradicate infection, $R_0^t < 1$, or $\varepsilon > \varepsilon_c$, where

$$\varepsilon_c = 1 - \frac{1}{R_0} \tag{23}$$

is called the critical efficacy. ε, however, is not constant and varies between doses according to its pharmacokinetic properties. Austin *et al.* (1998) then argue that therapy will be successful if the minimum value of ε is greater than ε_c. From the above estimates of C_{min}, the minimum drug concentration in plasma at steady state following multiple doses (Eq. (10)), we get for $\varepsilon_{min} > \varepsilon_c$ that the dose D should obey

$$D > \frac{R_0 V_d IC_{50}}{F} \frac{k_e - k_a}{k_a} \frac{1}{\left[\frac{1}{e^{k_a I_d} - 1} - \frac{1}{e^{k_e I_d} - 1}\right]}. \tag{24}$$

Equation (24) predicts the minimum dosage necessary for therapy to successfully clear infection.

With D that satisfies Eq. (24), solving Eqs. (18)–(21) with the pharmacokinetic parameters employed in Fig. 1 predicts that ritonavir monotherapy drives virus to extinction in ~2 months following the onset of therapy. Austin *et al.* (1998) present similar calculations under monotherapy with AZT and ddI and predict eradication of virus in ~1 year from the onset of monotherapy with either drug. Clearly, these predictions have not been borne out, as monotherapy is found to be unsuccessful in clearing infection.

Austin *et al.* (1998) estimate R_0 in the absence of therapy to be ~3 so that $\varepsilon_c \sim 2/3$. A recent survey finds R_0 to vary between ~1.2 and 30 across patients (Ribeiro *et al.*, 2004). Recognizing that $\varepsilon_c = 1 - T_{ss,inf}/T_{ss,uninf}$, where the latter ratio is the fraction of the target cell number density at steady state in an uninfected individual that survives in the chronically infected steady state, ε_c is estimated to be ~0.6 (Dixit and Perelson, 2004a). Also, the pharmacokinetic parameters employed by Austin *et al.* (1998) are the best estimates available ignoring inter-patient variation. The failure of monotherapy may therefore be attributed to the emergence of drug resistance, the effects of long-lived cells and reservoirs, and the possible low *in vivo* efficacy of drugs compared to *in vitro* estimates especially considering intracellular pharmacokinetics. We consider the emergence of drug resistance in the next section.

5. Drug Resistance and the Effect of Adherence

Several models have been developed to study the effect of drug pharmacokinetics on the emergence of drug resistance in HIV infection (Kepler and Perelson, 1998; Wahl and Nowak, 2000; Huang *et al.*, 2003). Wahl and Nowak (2000) assume the existence of wild-type and resistant species of the virus, the former fitter in the absence of therapy whereas the latter fitter in the presence of therapy. The resulting model of viral dynamics is written as follows:

$$\frac{dT}{dt} = \lambda - dT - (1-\varepsilon_1)k_1TT_1^* - (1-\varepsilon_2)k_2TT_2^* \tag{25}$$

$$\frac{dT_1^*}{dt} = (1-\varepsilon_1)k_1TT_1^* - \delta T_1^* \tag{26}$$

$$\frac{dT_2^*}{dt} = (1-\varepsilon_2)k_2TT_2^* - \delta T_2^* \tag{27}$$

where k_1 and k_2 are the infectivities of the wild type and resistant species respectively, with $k_1 > k_2$ so that the wild type is more infectious in the absence of therapy. (Note that here the infection rate is written as kTT^*

instead of kTV as a simplification following the assumption of rapid attainment of steady state between viral production and clearance (Eq. (14)); the rate constant k includes all the relevant parameters.) Cells T_1^* and T_2^* infected with wild-type and resistant virus, respectively, die with the same first order rate constant δ. The drug is assumed to be more effective against the wild type so that $\varepsilon_1 > \varepsilon_2$. The different drug susceptibilities are attributed to different values of IC_{50} and are determined using Eq. (11) where a simple approximation for the drug concentration is employed:

$$C_b(t) = \begin{cases} C_{min} + \dfrac{t}{t_{max}}(C_{max} - C_{min}), & 0 \le t \le t_{max} \\ C_{max}e^{-k_e(t-t_{max})}, & t_{max} \le t \le I_d. \end{cases} \tag{28}$$

According to this description, applicable for large t (steady-state), C_b rises linearly from C_{min} to C_{max} in time interval t_{max} following each dose, administered at regular intervals of time, I_d, and then decreases exponentially with the rate constant k_e.

Given this model of viral dynamics and drug pharmacokinetics, Wahl and Nowak (2000) analyze the emergence of the drug resistant strain of the virus as a function of the degree of adherence to therapy, p, defined as the fraction of prescribed doses taken. Since therapy proceeds over years and drug concentrations vary over a few hours to days, Wahl and Nowak (2000) discuss viral dynamics in terms of average susceptibilities, $\bar{s}_1(p)$ and $\bar{s}_2(p)$, of the two species, where $s_i = 1 - \varepsilon_i$ $(i = 1, 2)$. The averages depend not only on p but also on the pattern of adherence, for instance whether the average number of consecutive doses missed is large or small for a given p. Prior to therapy, the basic reproductive ratios of the two species, R_1 and R_2, obey $R_1 > R_2$ as the wild type is more virulent. Following the onset of therapy, both R_1 and R_2 decrease with p but R_1 decreases faster as the wild-type is more susceptible to the drug. Treatment is successful if it drives both R_1 and R_2 below 1. This is expected to happen at high degrees of adherence or high values of p.

For intermediate values of p several intriguing regimes arise in the Nowak and Wahl theory, which we show schematically in Fig. 2. For very small values of p, $R_1 > R_2 > 1$. In this case, treatment fails regardless of the emergence of drug resistance; the wild-type dominates the population. For higher values of p, since R_1 decreases faster than R_2 it may happen that $R_2 > R_1 > 1$. Treatment continues to fail but in this case, the resistant virus dominates the population whereas the wild-type is suppressed. However, since $R_1 > 1$ treatment would fail regardless of the emergence of drug

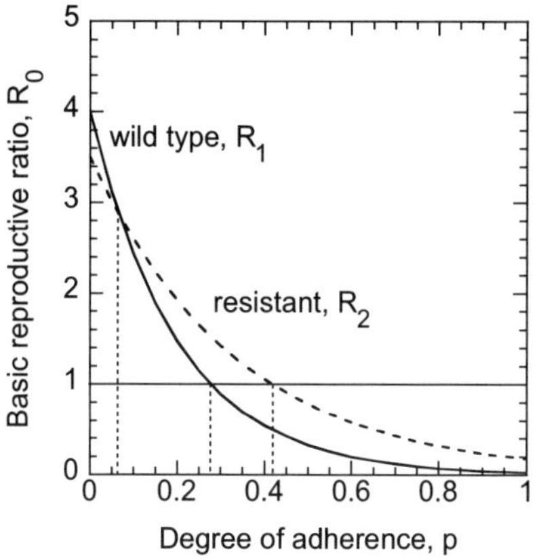

Fig. 2. A schematic representation of the possible dependence of the basic reproductive ratio (R_0) of the wild-type, R_1, (solid line) and drug-resistant virus, R_2, (dashed line) on the degree of adherence to drug therapy, p. The vertical dotted lines delineate various regimes discussed in the text.

resistance. For still higher values of p, R_1 may decrease further such that $R_2 > 1 > R_1$. For such values of p, therapy fails because of drug resistance. Therapy is predicted to eradicate the wild-type but not the resistant virus, which eventually dominates the population. Here, depending on the magnitude of R_2 and the initial abundance of the resistant virus, a long time may elapse before the resistant strain becomes detectable. During this period, the wild-type population decreases rapidly and virus may become undetectable for an extended period conveying the impression of successful therapy. Finally, for very large values of p, Nowak and Wahl predict that both R_1 and R_2 fall below 1, so that therapy should eradicate both wild type and resistant virus.

The precise values of p that determine these regimes depend on the relative susceptibilities of the two virus species to therapy, their basic reproductive ratios, and the patterns of adherence. Wahl and Nowak (2000) find that greater adherence is necessary for successful therapy as R_2 increases relative to R_1 and that combination therapy (with a 3-drug cocktail) is successful for lower degrees of adherence than monotherapy as simultaneous resistance to multiple drugs arises with a lower probability.

Wahl and Nowak (2000) also investigate the replicative advantage of the resistant virus as a function of the adherence pattern. For a given value of p, the adherence pattern is defined by the block size, which is the number of consecutive doses taken or missed each time a dose is taken or missed, respectively. For instance, if the block size is 5, then each time a dose is taken, 4 doses following it are also compulsorily taken, and if a dose is missed the following 4 doses are also missed.

The ratio $\bar{s}_2/\bar{s}_1$ is defined as the resistance advantage, as this ratio must be larger than the pretreatment ratio R_1/R_2 for the resistant virus to outcompete the wild-type. Interestingly, Wahl and Nowak (2000) find that the resistance advantage monotonically decreases with block size for any given value of p. To understand this, they argue that the resistant virus thrives under intermediate drug concentrations. At low drug concentrations, the wild-type continues to have an advantage, whereas at high drug concentrations, replication of both wild-type and resistant viruses is suppressed. When the block size is large, drug concentrations are high when doses are taken and low when missed. When the block size is small, concentrations tend to remain at intermediate levels allowing the resistant virus to grow.

In an earlier investigation, Kepler and Perelson (1998) have also predicted that drug resistant strains thrive at intermediate concentrations. They estimate the range of concentrations that is conducive to the emergence of drug resistance and find that a single compartment with a fixed drug concentration provides a narrow window of opportunity for the virus to develop drug resistance. They conclude from the prevalent failure of monotherapy that multiple compartments with different drug permeabilities probably exist in the body, for which they estimate a broader window of opportunity for the evolution of drug-resistant strains.

The calculations of Wahl and Nowak (2000) and Kepler and Perelson (1998) provide valuable insights into the evolution of drug resistance and the impact of adherence to prescribed dosing regimen. Similar arguments might underlie the success of the 7-day on-off short cycle therapy (Dybul *et al.*, 2001), presenting a possible framework for the identification of improved intermittent treatment protocols. Model predictions, however, await experimental validation.

Huang *et al.* (2003) employ a different model of the impact of drug pharmacokinetics on the emergence of drug resistance. Instead of a wild-type and a resistant virus population, they consider a single population of virus that becomes decreasingly susceptible to therapy with time as a result of the emergence of drug-resistant strains. They write

$$IC_{50}(t) = \begin{cases} I_0 + \dfrac{I_r - I_0}{t_r} t & 0 < t < t_r \\ I_r & t \geq t_t. \end{cases} \tag{29}$$

IC_{50} thus increases linearly from I_0 to I_r in time t_r following the onset of therapy and stays constant thereafter. The models of drug pharmacokinetics and viral dynamics employed are essentially identical to those described in Sections II and III. It follows that if ε remains greater than ε_c (defined above) at all times, therapy will eradicate virus regardless of drug resistance. Also, an efficacy ε_m is defined such that $dV_I/dt = 0$ when $\varepsilon = \varepsilon_m$. At this efficacy, viral loads cease to decrease due to therapy. Thus, for $\varepsilon < \varepsilon_m$ viral loads resurge during therapy. Huang *et al.* (2003) present the difference between the time when viral loads start to rise and when they become detectable (increase to 50 copies/mL). This difference is found in some cases to be as large as 6 months. When $\varepsilon_m < \varepsilon < \varepsilon_c$, viral loads do not resurge but virus is not eradicated. A low steady state viral load is achieved, which may resurge upon termination of therapy. For typical parameter values, Huang *et al.* (2003) estimate $\varepsilon_c \sim 0.7$ and $\varepsilon_m \sim 0.5$. From estimates of I_0, I_r, and t_r from *in vitro* studies, Huang *et al.* (2003) suggest that the dosage required to maintain $\varepsilon > \varepsilon_c$ can be identified to ensure viral eradication despite drug resistance.

Huang *et al.* (2003) also study the effect of adherence on viral load evolution in the absence of drug resistance ($t_r = 0$) and find that the tendency of viral load to resurge increases with block size for a given degree of adherence, p. As block size increases, drug concentrations drop during drug holidays to increasingly lower levels so that $\varepsilon < \varepsilon_c$ allowing viral loads to resurge. Although Huang *et al.* (2003) do not present calculations with drug resistance, it follows that incorporating the above description of IC_{50} variation with time will cause ε to decrease further, maintaining the trend of increasing viral resurgence with block size while allowing resistant virus to dominate increasingly. This prediction is in contrast with that of Wahl and Nowak (2000), where resistance advantage diminishes with block size, and raises an important question about how to model the emergence of drug resistance — as two distinct populations of wild-type and resistant virus with different but time invariant fitness values, or as a single population with a time dependent fitness.

Kepler and Perelson (1998) attempt to describe mechanistically the early evolution of a drug-resistant strain under therapy with constant drug efficacy. Given a mutation rate and the number of independent mutations necessary for drug resistance, they find that the waiting time for the successful evolution of drug-resistant virus, which in turn depends on the probability of the generation of desired mutations for resistance and the subsequent propagation of resistant virus, varies nonmonotonically with drug efficacy, exhibiting a minimum at intermediate drug concentrations. Further, this waiting time is drastically reduced when multiple compartments with different drug permeabilities are introduced. Resistant virus is produced with high probability in a compartment where intermediate drug concentrations exist, from which it can migrate to compartments with high drug concentrations where it enjoys a replication advantage over the wild-type and begin to dominate the population.

Building on such a model, and considering the precise mutations necessary for resistance to various drugs, the rate at which drug-resistant strains emerge can be estimated providing the necessary insight into the role of adherence on viral load evolution. Of importance in such studies is the accurate determination of drug efficacy and the corresponding selective advantage conferred on the resistant virus. Estimates of efficacy employed in extant models are inaccurate as they are based on plasma rather than intracellular pharmacokinetics of drugs. Dixit and Perelson (2004a) have recently developed a model of the intracellular pharmacokinetics of typical antiretroviral drugs, which provides more realistic estimates of drug efficacy. The model, which we discuss next, also includes rigorous descriptions of intracellular delay and therefore allows accurate analysis of short-term viral load evolution following therapy.

6. Intracellular Pharmacokinetics

Based on the plasma pharmacokinetics described in Section II, the intracellular concentrations of drugs can be determined by considering the transport of drugs across the cell membrane. At equilibrium, the intracellular concentration of the administered form of a drug, C_c, is related to the plasma concentration, C_b, through the partition coefficient, H, as $C_c^{eq} = (1 - f_b)HC_b^{eq}$, where f_b is the fraction of drug in plasma bound to proteins and hence which cannot transported into cells. Thus, depending on the magnitude of H and f_b, C_c^{eq} can be larger or smaller than C_b^{eq}. As C_b varies, the driving force for drug transport into cells varies in proportion to the difference $(1 - f_b)HC_b - C_c$. On the other hand, the driving force for transport from cells to plasma is proportional to C_c alone. The resulting time evolution of C_c is given by (Dixit and Perelson, 2004a):

$$\frac{dC_c}{dt} = k_{acell}C_x - k_{ecell}C_c \tag{30}$$

where

$$C_x = \begin{cases} (1-f_b)HC_b - C_c & \text{if } (1-f_b)HC_b - C_c > 0 \\ 0 & \text{otherwise} \end{cases} \tag{31}$$

with the initial condition $C_c(0) = 0$, where k_{acell} and k_{ecell} are the first order cellular drug absorption and elimination rate constants, respectively. Here, the mass of drug transported to cells is considered small so that C_b (given by Eq. (8)) is unaffected by variations in C_c.

For PIs, the administered form of the drug is therapeutically active in cells so that C_c determined by Eq. (30) provides an estimate of drug efficacy via Eq. (11) when C_b is replaced by C_c along with appropriate corrections for IC_{50}. RTIs, however, are phosphorylated in cells to their therapeutically active forms. Dixit and Perelson (2004a) model the phosphorylation of tenofovir DF, which starts out monophosphorylated and undergoes two additional phosphorylation steps in cells. The resulting reaction-transport scheme is written as:

$$\frac{dC_c}{dt} = k_{acell}C_x - k_{ecell}C_c - k_{1f}C_c + k_{1b}C_{cp} \tag{32}$$

$$\frac{dC_{cp}}{dt} = -k_{ecell}C_{cp} + k_{1f}C_c - k_{1b}C_{cp} - k_{2f}C_{cp} + k_{2b}C_{cpp} \tag{33}$$

$$\frac{dC_{cpp}}{dt} = -k_{ecell}C_{cpp} + k_{2f}C_{cp} - k_{2b}C_{cpp}, \tag{34}$$

where C_c, C_{cp}, and C_{cpp} are the intracellular concentrations of the native (monophosphorylated), diphosphorylated and triphosphorylated forms of the drug, and C_x is given by Eq. (31) above. The rate constants k_{1f}, k_{1b}, k_{2f}, and k_{2b} determine the rates of the phosphorylation reactions under the scheme:

$$C \underset{k_{1b}}{\overset{k_{1f}}{\rightleftharpoons}} CP \underset{k_{2b}}{\overset{k_{2f}}{\rightleftharpoons}} CPP, \tag{35}$$

where the native form of the drug, C, is not dephosphorylated. Solving the equations with the initial conditions $C_c(0) = C_{cp}(0) = C_{cpp}(0) = 0$ yields the time evolution of the intracellular concentration of the active form of the drug, C_{cpp}, which gives the time dependent efficacy when substituted for C_b in Eq. (11) with appropriate corrections for IC_{50}.

By obtaining estimates of the various rate constants from independent experiments, Dixit and Perelson (2004a) determine the efficacy of ritonavir and tenofovir DF for standard dosing regimens. The resulting drug concentrations and efficacies are shown in Figs. 3 and 4. Also shown is an

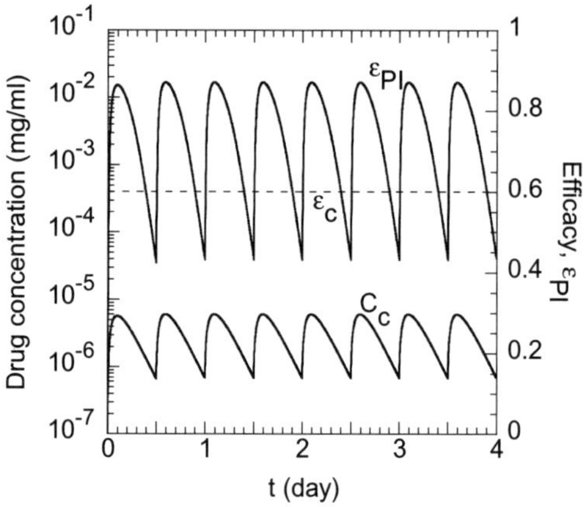

Fig. 3. Intracellular concentrations (C_c) and efficacy (ε_{PI}) of ritonavir administered orally 600 mg bid, calculated using Eqs. (30) and (11), respectively, with the parameter values $D = 600\,\text{mg}$, $I_d = 0.5$ day, $F = 1, V_d = 28{,}000\,\text{l}, k_a = 14.64\ \text{day}^{-1}$, $k_e = 6.86\ \text{day}^{-1}$, $f_b = 0.99, H = 0.052$, and $IC_{50} = 9 \times 10^{-7}\,\text{mg}\,\text{mL}^{-1}$. Rapid equilibrium across the cell membrane is assumed. The critical efficacy, $\varepsilon_c = 0.6$ (see text), is shown as a reference horizontal line.

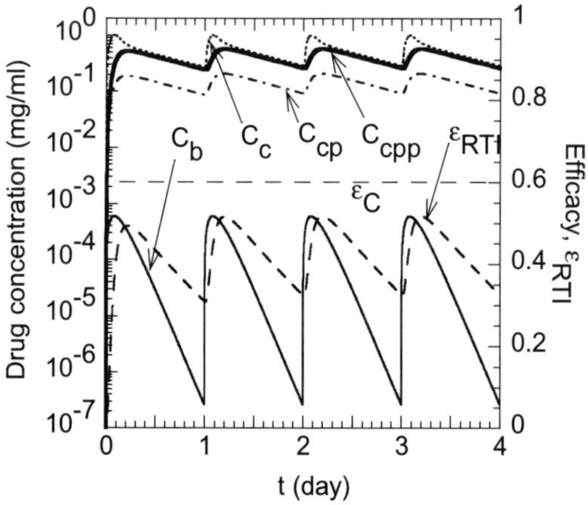

Fig. 4. Plasma (C_b) and intracellular concentrations of the native monophosphorylated (C_c), diphosphorylated (C_{cp}), and triphosphorylated (C_{cpp}) forms, and efficacy (ε_{RTI}) of tenofovir DF administered orally 300 mg daily, calculated using Eqs. (8), (32), (33), (34), and (11), respectively. Parameter values employed are $D = 300\,\text{mg}$, $I_d = 1$ day, $F = 0.39, V_d = 87,500\,\text{mL}, k_a = 14.64\ \text{day}^{-1}$, $k_e = 9.60\ \text{day}^{-1}$, $H = 1800$, $f_b = 0.07, k_{1f} = 9.6\ \text{day}^{-1}$, $k_{1b} = 30.3\ \text{day}^{-1}$, $k_{2f} = 270.7\ \text{day}^{-1}$, $k_{2b} = 95.5\ \text{day}^{-1}$, $k_{acell} = 24{,}000\ \text{day}^{-1}$, $k_{ecell} = 1.1\ \text{day}^{-1}$, and $IC_{50} = 0.54\,\text{mg}\,\text{mL}^{-1}$ (Dixit and Perelson, 2004a) The critical efficacy, $\varepsilon_c = 0.6$ (see text), is shown as a reference horizontal line.

estimate of the critical efficacy, $\varepsilon_c \sim 0.6$, determined from measurements of the steady state target cell density in chronically infected patients and typical estimates for uninfected individuals as described above (and in Dixit and Perelson, 2004a).

Note that due to the different partition coefficients (H) for the two cases, C_c is smaller than C_b for ritonavir, whereas C_c is higher than C_b for tenofovir DF. However, by converting reported *in vitro* plasma IC_{50} values to their corresponding intracellular values (Dixit and Perelson, 2004a), we find remarkably that $\varepsilon < \varepsilon_c$ at all times for tenofovir DF and for some duration between doses for ritonavir. Clearly, the efficacy based on intracellular concentrations of drugs is lower than that from plasma concentrations; the latter is >0.9 for ritonavir and much higher than ε_c throughout (Fig. 1). Monotherapy with either drug is therefore expected to fail in the long-term. Nevertheless, short-term viral load suppression is expected under monotherapy, which we describe next.

7. Viral Dynamics with Intracellular Pharmacokinetics — Short-term Therapy

Following the onset of therapy, viral loads remain at pretreatment levels for ~1 day and then decay nearly exponentially for ~1 week (Perelson *et al.*, 1996). To accurately analyze this early viral load evolution, models of viral dynamics must include descriptions of intracellular and pharmacological delays. Explicit incorporation of drug pharmacokinetics accounts for the latter part. We recognize that the intracellular contribution to the shoulder is given by the duration of the replication cycle from the stage affected by drug action to the end of the replication cycle and is therefore dependent on the nature of the drug (Dixit and Perelson, 2004a).

To account for this drug dependence, we define τ_1 and τ_2 to be the durations of the phases before and after the stage in the replication cycle affected by drug action, where $\tau = \tau_1 + \tau_2$ is the viral replication time. Then, modifying the description of Herz *et al.* (1996), we write (Dixit and Perelson, 2004a)

$$\frac{dT(t)}{dt} = \lambda - dT(t) - (1 - \varepsilon_{RTI}(t))kT(t-\tau_1)V(t-\tau_1)e^{-m\tau_1} \tag{36}$$

$$\frac{dT^*(t)}{dt} = (1 - \varepsilon_{RTI}(t-\tau_2))kT(t-\tau)V(t-\tau)e^{-m\tau} - \delta T^*(t) \tag{37}$$

$$\frac{dV(t)}{dt} = N\delta T^*(t) - cV(t) \tag{38}$$

for RTI monotherapy. The difference in this description from the basic model of viral dynamics arises in the infection terms (last term in Eq. (36) and first term in Eq. (37)). In the present description, cells infected by virus at time $t-\tau_1$ reach the reverse transcription stage at time t provided they survive the interval τ_1. The latter survival probability is $\exp(-m\tau_1)$, where m is the death rate of infected but not yet productive cells. In the surviving cells, drug action aborts reverse transcription with an efficacy $\varepsilon_{RTI}(t)$. Thus only a fraction $(1-\varepsilon_{RTI}(t))$ of the surviving cells are successfully infected, as quantified by the last term in Eq. 36. (Note that in the absence of drug, the delay τ_1 in the loss of infected cells vanishes as every infected cell is destined to produce virus. In this case, with $\tau_1 = 0$ and $\varepsilon_{RTI} = 0$, Eq. (36) reduces to the pretreatment situation of Eq. (12).) To become productively infected, however, the latter cells must survive the remaining interval, τ_2, of the replication cycle, which happens with the probability $\exp(-m\tau_2)$. Multiplying the last term in Eq. (36) with this survival probability gives the cells that become productively infected at time $t+\tau_2$. Replacing $t+\tau_2$ by t in the resulting expression gives the number of productively infected cells formed at time t, which is the first term in Eq. (37). We simplify these equations by recognizing that $\tau_1 \sim 0$ and $\tau_2 \sim \tau$ for RTIs (Dixit and Perelson, 2004a).

For PIs, the corresponding equations are

$$\frac{dT(t)}{dt} = \lambda - dT(t) - kT(t)V_I(t) \tag{39}$$

$$\frac{dT^*(t)}{dt} = kT(t-\tau)V_I(t-\tau)e^{-m\tau} - \delta T^*(t) \tag{40}$$

$$\frac{dV_I(t)}{dt} = N\delta T^*(t)(1-\varepsilon_{PI}(t-\tau_2)) - cV_I(t) \tag{41}$$

$$\frac{dV_{NI}(t)}{dt} = N\delta T^*(t)\varepsilon_{PI}(t-\tau_2) - cV_{NI}(t) \tag{42}$$

for which the complementary simplification $\tau_1 \sim \tau$ and $\tau_2 \sim 0$ applies (Dixit and Perelson, 2004a).

Solving these equations with the initial conditions given by the pre-treatment steady-state

$$T_{ss} = \frac{ce^{m\tau}}{kN}; \quad T^*_{ss} = \frac{\lambda}{\delta e^{m\tau}} - \frac{dc}{\delta kN}; \quad V_{ss} = \frac{\lambda N}{ce^{m\tau}} - \frac{d}{k} \tag{43}$$

for RTIs and

$$T_{ss} = \frac{ce^{m\tau}}{kN}; \quad T^*_{ss} = \frac{\lambda}{\delta e^{m\tau}} - \frac{dc}{\delta kN}; \quad V_{I,ss} = \frac{\lambda N}{ce^{m\tau}} - \frac{d}{k}; \quad V_{NI,ss} = 0 \tag{44}$$

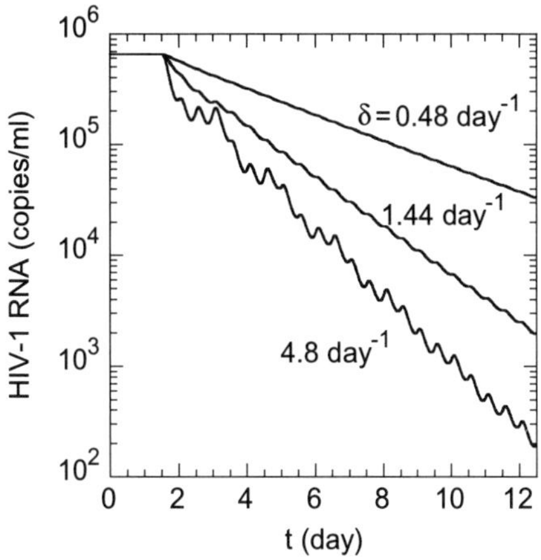

Fig. 5. Viral load decay under ritonavir monotherapy calculated for three values of the infected cell death rate, δ, by integrating Eqs. (18)–(21) with the initial conditions specified by Eq. (16). The values of the other parameters employed are $\lambda = 10^4\,\text{mL}^{-1}\,\text{day}^{-1}$, $d = m = 0.01\,\text{day}^{-1}$, $k = 2.4 \times 10^{-8}\,\text{mL day}^{-1}$, $c = 23\,\text{day}^{-1}$, $\tau = 1.5$ day, and $N = 2500$. The efficacy, ε_{PI}, is calculated as in Fig. 3.

for PIs, intriguing patterns of viral load decay result as a consequence of the subtle interplay between underlying viral dynamic and pharmacokinetic timescales. Typical profiles are shown in Fig. 5. For small values of the death rate of productively infected cells, δ, viral load decays nearly exponentially following a delay equal to the viral replication period, τ. For intermediate δ, mild oscillations are superimposed on the average exponential decay. The period of these oscillations equals the dosing interval, I_d, thus corresponding to oscillations in drug concentration and hence efficacy between doses. The overall decay rate increases monotonically with δ. Finally, for unrealistically large values of delta, superimposed on these oscillations are step decays at regular intervals of τ, the viral replication period.

Standard models of viral dynamics predict viral load to decay exponentially following the shoulder. The above calculations reveal that the exponential decay is only a special case; depending on the parameters, complex patterns of viral decay are possible. However, insufficient frequency of measurements coupled with large measurement errors precludes identification of these patterns from currently available viral load evolution data.

We applied the model to analyze data from 5 patients under ritonavir monotherapy and obtained estimates of the intracellular delay τ and the *in vivo* efficacy of ritonavir (Dixit and Perelson, 2004b). A typical fit is

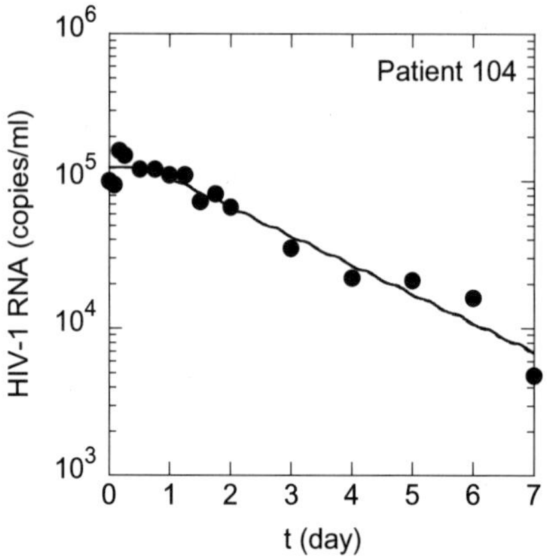

Fig. 6. Viral load decay in Patient 104 (Perelson *et al.*, 1996) (symbols) and best fit of the model (Eqs. (18)–(21)) to the data (line) (Dixit and Perelson, 2004a). The corresponding values of the adjustable parameters are $\delta = 0.74$ day^{-1}, $\tau = 0.66$ day, and $N = 1249$. The other parameter values are fixed to those in Fig. 6.

presented in Fig. 6. Accounting for known inter-patient variation in pharmacokinetic parameters and assuming the best estimate of $\delta = 1$ day^{-1}, we found $\tau \sim 1$ day and the average *in vivo* efficacy of ritonavir to be ~ 0.65. The estimate of $\tau \sim 1$ day is in agreement with previous estimates of the viral replication period and corresponds to a viral generation time of 2 days and ~180 replication cycles per year (Dixit and Perelson, 2004b). The average *in vivo* efficacy, however, is significantly lower than that expected from *in vitro* studies, the latter is ~0.9. This low *in vivo* efficacy together with the large replication rate of HIV facilitates the rapid emergence of drug resistance explaining the failure of monotherapy and perhaps combination therapy in combating HIV infection.

8. Conclusions

In this chapter, we have described models that combine pharmacokinetics of antiretroviral drugs and viral dynamics of HIV. Model predictions provide significant insights into HIV pathogenesis and the outcome of therapy establishing guidelines for identifying optimal treatment protocols. For instance, the models provide ways to estimates of the minimum dosage necessary to drive infection to extinction. Links between fitness of viral species and drug efficacy identify conditions under which drug-resistant strains of the virus emerge. Similarly, models that include intracellular pharmacokinetics of drugs and descriptions of intracellular delay provide estimates of viral replication times and the *in vivo* efficacy of drugs.

Extant models, however, are unable to predict the long-term outcome of therapy for several reasons. First, the descriptions of HIV dynamics employed here ignore long-lived cells and viral reservoirs, which are important in determining viral load evolution after 1–2 weeks of therapy. Models that include long-lived cells and viral reservoirs have been developed and applied to analyze viral load decay under combination therapy (Perelson *et al.*, 1997) but are yet to be combined with descriptions of drug pharmacokinetics.

Second, models that address drug resistance and the impact of adherence to therapy consider plasma pharmacokinetics of drugs and ignore intracellular delay. On the other hand, models that consider intracellular pharmacokinetics do not incorporate the emergence of drug resistance. Thus, combining the two approaches will provide a significantly improved description of viral dynamics under therapy. Third, models employ *ad hoc* descriptions of drug resistance either by assigning fitness values to competing species or by assuming variations in drug susceptibility based on *in vitro* studies and consequently make predictions difficult to reconcile.

Addressing these limitations will enhance the ability of extant models to accurately predict the long-term outcome of therapy and enable the identification of optimal treatment protocols that will minimize the prohibitive treatment costs and side-effects associated with current antiretroviral therapies.

Acknowledgments

This research is supported by the Department of Energy under contract W-7405-ENG-36 and by NIH grants RR06555 and AI28433. We thank Ruy M. Ribeiro for comments on the manuscript.

References

Austin DJ, White NJ and Anderson RM (1998) The dynamics of drug action on the within-host population growth of infectious agents: melding pharmacokinetics with pathogen population dynamics, *J Theor Biol* **194**: 313–339.

Buss N and Cammack N (2001) Measuring the effectiveness of antiretroviral agents, *Antiviral Therapy* **6**: 1–7.

Dixit NM and Perelson AS (2004a) Complex patterns of viral load decay under antiretroviral therapy: influence of pharmacokinetics and intracellular delay, *J Theor Biol* **226**: 95–109.

Dixit NM and Perelson AS (2004b) Estimates of intracellular delay and average drug efficacy from viral load data of HIV infected individuals under antiretroviral therapy, *Antiviral Therapy* in press.

Dybul M, Chun T-W, Yoder C, Hidalgo B, Belson M, Hertogs K, Larder B, Dewar RL, Fox CH, Hallahan CW, Justement JS, Migueles SA, Metcalf JA, Davey RT, Daucher M, Pandya P, Baseler M, Ward DJ and Fauci AS (2001) Short-cycle structured intermittent treatment of chronic HIV infection with highly active antiretroviral therapy: effects on virologic, immunologic, and toxicity parameters, *Proc Natl Acad Sci USA* **98**: 15161–15166.

Finzi D and Siliciano RF (1998) Viral dynamics in HIV-1 infection, *Cell* **93**: 665–671.

Gabrielson J and Weiner D (2000) Pharmacokinetic and pharmacodynamic data analysis: concepts and applications, *Swedish Pharmaceutical Press*, Stockholm.

Herz AVM, Bonhoeffer S, Anderson RM, May RM and Nowak MA (1996) Viral dynamics *in vivo*: limitations on estimates of intracellular delay and virus decay, *Proc Natl Acad Sci USA* **93**: 7247–7251.

Hlavacek WS, Wofsy C and Perelson AS (1999) Dissociation of HIV-1 from follicular dendritic cells during HAART: mathematical analysis, *Proc Natl Acad Sci USA* **96**: 14681–14686.

Huang Y, Rosenkranz S and Wu H (2003) Modeling HIV dynamics and antiviral responses with consideration of time-varying drug exposures, sensitivities and adherence, *Math Biosci* **184**: 165–186.

Hurwitz SJ and Schinazi RF (2002) Development of a pharmacodynamic model for HIV treatment with nucleoside reverse transcriptase and protease inhibitors, *Antiviral Res* **56**: 115–127.

Nowak MA and May RM (2000) Virus dynamics: mathematical principles of immunology and virology, Oxford University Press, Oxford.

Paterson D, Swindells S, Mohr J, Brester M, Vergis E, Squier C, Wagener M and Singh N (1999) How much adherence is enough? A prospective study of adherence to protease inhibitor therapy using MEMSCaps. In *Proceedings of the Sixth Conference on Retroviruses and Opportunistic Infections, Chicago, IL*, Abstract 092.

Perelson AS, Neumann AU, Markowitz M, Leonard JM and Ho DD (1996) HIV-1 dynamics *in vivo*: virion clearance rate, infected cell life-span, and viral generation time, *Science* **271**: 1582–1586.

Perelson AS, Essunger P, Cao Y, Vesanen M, Hurley A, Saksella K, Markowitz M and Ho DD (1997) Decay characteristics of HIV-1-infected compartments during combination therapy, *Nature* **387**: 188–191.

Perelson AS and Nelson PW (2002) Modeling viral infections, in Sneyd J (ed.) *An Introduction to Mathematical Modeling in Physiology, Cell Biology, and Immunology*, Amer. Math. Soc., Providence RI, pp 139–172.

Ribeiro RM, Dixit NM and Perelson AS, Modeling the *in vivo* growth rate of HIV: implications for vaccination, in Paton R and Leishman D (eds.) *Multi-disciplinary Approaches to Theory in Medicine*, Elsevier, in press.

Stafford MA, Corey L, Cao Y, Daar E, Ho DD and Perelson AS (2000) Modeling plasma virus concentration during primary HIV infection, *J Theor Biol* **203**: 285–301.

Wahl LM and Nowak MA (2000) Adherence and drug resistance: predictions for therapy outcome, *Proc R Soc Lond* B **267**: 835–843.

Welling PG (1986) *Pharmacokinetics: Processes and Mathematics*, American Chemical Society, Washington DC.

CHAPTER 13

A MODEL OF HIV-1 TREATMENT: THE LATENTLY INFECTED CD4$^+$ T CELLS BECOMES UNDETECTABLE

Karen O'Hara

The interactions between the HIV RNA and the CD4$^+$ T cells are well explained by a system of differential equations and predictions about growth of cell populations can be made. Reasonable parameter value estimation enabled meaningful analysis of system behavior, and Mathematica was used to simulate model dynamics. The model accounts for many observed behaviors, including the undetectability of the latent class of uninfected CD4$^+$ T cells. A better understanding of virus-immune dynamics are obtained, allowing for improved research on treatment.

1. Introduction

Ideas regarding the treatment of HIV-1 infections have changed greatly over the years. In the beginning, patients were treated with single antiretroviral drugs such as zidovudine (AZT), didanosine, zalcitabine, or stavudine. These drugs act against a retrovirus by preventing the reverse transcriptase from turning the viral RNA into DNA. HIV, however, is one of several viruses that replicate rapidly and carelessly. This means that they all have a high rate of mutation and these mutations cause different epitopes to be expressed in each viral generation. Antibody and cytotoxic T cell responses may be effective against one generation of virus, but less effective against subsequent generations (Bittner *et al.*, 1997; Nowak *et al.*, 1990). Because of this incomplete suppression of viral replication, or because of intolerance of infected individuals of the drugs, other treatments must be considered (Carr *et al.*, 1998).

Resistance to the antiretroviral drugs and the development of new protease inhibitors has led to the treatment regimen known as highly active antiretroviral therapy (HAART). A typical HAART regimen includes at least three antiretroviral drugs, one of which is usually a protease inhibitor

(Chun and Fauci, 1999). Studies have shown that HAART can significantly curtail viral replication — to less than 50 copies per mL of blood (Gulick *et al.*, 1997; Hammer *et al.*, 1997) and that this low level can be sustained over long periods of time (Gulick *et al.*, 1997; Hammer *et al.*, 1997). This has given rise to clinical investigations regarding the dynamics of virus replication, and early on, to the hope of eradication of the virus in some infected individuals (Nowak *et al.*, 1990; Perelson *et al.*, 1997; Wei *et al.*, 1995). Perelson *et al.* (1997), even went so far as to predict that a strong HAART regimen for 2–3 years could eradicate the infection. This model was based on the assumption that viral replication was completely suppressed, and that there were no other reservoirs of HIV anywhere in the body. Several studies have since shown, however, a pool of latently infected, resting $CD4^+$ T cells in the lymph system (Finzi *et al.*, 1997, 1999; Wong *et al.*, 1997). It has been shown that they can induce HIV replication either in the absence of HAART, or when HAART does not completely suppress viral replication (de Jong *et al.*, 1997).

Attempts to eradicate this latent pool, as well as the observation that the immune system does not fully rebound on its own, even when viral replication is suppressed below detection levels (Emery and Lane, 1997) have led to more recent advances in the treatment of HIV. Immune modulators are now being used in an attempt to increase the immune system's natural ability to ward off infection. The immune system regulates itself with cytokines, secreting more in response to foreign antigens, and less when there is no infection (Paul and Seder, 1994). Interleukin-2 (IL-2) was the first cytokine to be associated with the production of HIV (Kinter and Fauci, 1996). IL-2 regulates the proliferation and differentiation of several types of lymphocytes, including T cells, B cells, and natural killer (NK) cells (Arai *et al.*, 1990; Smith, 1988). Despite the fact that IL-2 is associated with toxicities such as flu-like fatigue, headache, pain and fever (Kovacs *et al.*, 1996), many studies have shown that the pool of $CD4^+$ T cells can be significantly increased with its use (Carr *et al.*, 1998; Chun *et al.*, 1999a; Jacobson *et al.*, 1996; de Jong *et al.*, 1997; Kovacs *et al.*, 1995, 1996). Most of the infected cells in the latent pool have what is called preintegration latency: they become infected, but only a partial degree of reverse transcription occurs and the viral DNA is unable to create new infections (Zack *et al.*, 1990). These cells can become activated, causing the reverse transcriptase to complete transcription, and the proviral DNA becomes integrated into the host cell DNA (Zack *et al.*, 1990). A very small percentage ($<0.1\%$) of the latent cells have what is called postintegration latency: the proviral DNA has already been integrated into the cell

DNA, but remains "transcriptionally silent" (Chun and Fauci, 1999; Chun *et al.*, 1997). In their 1998 study, Chun *et al.* (1998a) determined that this pool of latently infected, resting CD4$^+$ T cells becomes established so early in initial infection that even treatment with HAART as early as 10 days after primary infection did not prevent the pool from developing. The next step was to look at the ability of cytokines to stimulate, or activate, these latently infected, resting CD4$^+$ T cells. In the next study Chun and his group conducted (Chun *et al.*, 1998b), they found cytokines IL-2, IL-6 and TNF-α were able to activate these cells. Lastly, Chun *et al.* (1999a) studied individuals who had been taking HAART and whose plasma viral load was undetectable. When these patients were given IL-2, there was a decrease in the pool of latently infected, resting CD4$^+$ T cells. Using standard methods, they found replication competent virus in the pool of latent cells in all of the patients who were receiving HAART alone, but in 6 out of the 14 who were receiving HAART plus IL-2, they could not find any replication competent virus in the latent class. Additionally, when even more cells were examined, 3 of the 6 patients still had no replication competent virus in the latent class.

It is this last study that we will model in this paper. We will apply the treatment as discussed by Chun (1999a) to a model constructed from parts of models by Kirschner and Webb (1998, 2000).

2. Simulations

In order to simulate treatment, we used Mathematica. The equations used in the simulations are described in the Appendix. With IL-2 treatment, the patient comes in for regular treatments, typically approximately every 8 weeks. We used a 60-day cycle to simulate this. Initial values are set on the first day of treatment, and Mathematica provides an approximation for the solution over the next 60 days. The values of the five functions at the end of the first treatment are then used as the initial values for the second treatment. The number of treatment cycles is determined based on the data given by Chun *et al.* (1999a). For the patients not receiving IL-2, the simulation includes only one treatment cycle, for the length of the HAART treatment. Code for this is given in the Appendix.

We modeled four patients described in the Chun *et al.* (1999a) study. The patients were divided into two groups. Twelve patients received HAART alone, and 14 patients received intermittent IL-2 plus HAART. Of the 14 who received IL-2 plus HAART, 6 patients had no replication competent HIV in their resting CD4$^+$ T cells. All 12 receiving HAART

Table 1. Profiles of HIV-1 infected patients receiving IL-2 plus HAART or HAART alone (Chun *et al.*, 1999a). All measurement are taken per mm^3. Note that Patient 1 received 49 months of IL-2 and 20 months of HAART, Patient 10 received 30 months IL-2 and 21 HAART, Patient 20 received 16 months HAART, and Patient 22 received 21 months HAART.

Patient	CD4 count before IL-2	CD4 count before HAART	CD4 count at time of study	Plasma HIV before IL-2	Plasma HIV before HAART	Plasma HIV at time of study
1	675	1149	832	2.25	3.06	<0.05
10	602	581	647	2.417	3.927	<0.05
20	NA	446	732	NA	2.211	<0.05
22	NA	691	563	NA	14.400	<0.05

alone had detectable levels of replication competent HIV in their resting CD4$^+$ T cells. We modeled four of these patients, two from each group. We used the data given in Table 1 to define our patients. From personal correspondence with Dr. Chun, these data are the only available measurements. In order to model each patient, we needed to know when they began treatment. Since a typical patient will follow the disease progression in the basic model, we used this to initiate infection. Then, using the data, we chose the day for starting treatment based on the number of CD4$^+$ T cells the patient had before they began treatment and this becomes the initial value for the subsequent simulation. For example, Patient 1 had 675 CD4$^+$ T cells per mm^3. Following the treatment curve for the uninfected CD4$^+$ T cells from the basic model, we can see that a typical patient will have 675 CD4$^+$ T cells on approximately day 430. Thus we began treatment for Patient 1 on day 430, to simulate where he was in the normal disease progression when treatment was begun. As mentioned before, the patients in Dr. Chun's study were admitted to the hospital for a five-day intravenous administration of the IL-2. Since Dr. Chun did not have details available about individual treatment amounts, we had to determine the parameters used for each patient by reverse engineering in order for the graphs to match the data in Table 1.

We begin with Patient 1. This patient received 22 months of IL-2 and then 20 months of both IL-2 and HAART. Since the patient had 675 CD4$^+$ T cells per mm^3 before IL-2, we chose day 430 to begin IL-2 treatment. In the model, the patient had 674 CD4$^+$ T cells and 4.8 HIV RNA copies per mm^3. The patient received 11 IL-2 treatments, spaced 60 days apart.

Table 2. Parameters for Patient 1 IL-2 treatment.

Treatment cycle	1	2	3	4	5	6
a_1	0.08	0.08	0.07	0.07	0.06	0.05
a_2	0.4	0.4	0.4	0.4	0.5	0.5
b_1	1.0	1.0	1.0	1.0	1.0	1.0
b_2	0.1	0.1	0.1	0.1	0.1	0.1

Treatment cycle	7	8	9	10	11
a_1	0.05	0.04	0.03	0.03	0.02
a_2	0.5	0.5	0.5	0.5	0.5
b_1	1.0	1.0	1.0	1.0	1.0
b_2	0.1	0.1	0.1	0.1	0.1

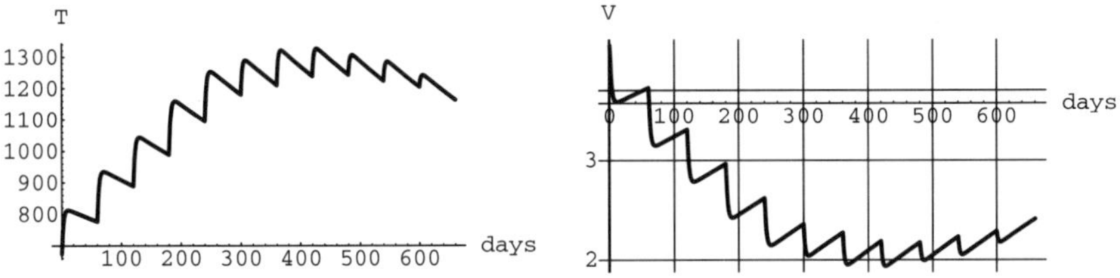

a. T cells during IL-2 treatment

b. HIV RNA during IL-2 treatment

Fig. 1. Patient 1 IL-2 treatment. Treatment parameters used are given in Table 2.

The treatment parameters are given in Table 2 and the graphs of $T(t)$ and $V(t)$ are shown in Fig. 1. After receiving 11 doses of IL-2, the model shows the patient with 1163 $CD4^+$ T cells and 2.4 HIV RNA copies per mm^3 which is consistent with the Chun data. We then added HAART to the regimen. We continued to space the IL-2 treatments 60 days apart, and include HAART. The patient continued this for 20 months. The treatment parameters are given in Table 3 and graphs for this simulation are given in Fig. 2. At the end of simulation, the model shows a $CD4^+$ T cell population of 1080.7 per mm^3, which is consistent with the data, a latently infected $CD4^+$ T cell population of 2.2×10^{-7} and an HIV RNA count of 1.2×10^{-23} which are both below detection level (10^{-4} cells per mm^3 for the latently infected $CD4^+$ T cells, and 0.05 copies per mm^3 for the virus). Thus for this patient, the IL-2 plus HAART treatment reduced both the virus and the latent class populations to below detectability.

Table 3. Parameters for Patient 1 HAART treatment.

Treatment cycle	12	13	14	15	16
a_1	0.02	0.02	0.02	0.02	0.02
a_2	0.5	0.5	0.5	0.5	0.5
b_1	1.0	1.0	1.0	1.0	1.0
b_2	0.1	0.1	0.1	0.1	0.1
c_1	0.29	0.28	0.28	0.28	0.27
c_2	0.29	0.28	0.28	0.28	0.27
c_3	0.29	0.28	0.28	0.28	0.27
Treatment cycle	17	18	19	20	21
a_1	0.01	0.01	0.01	0.01	0.01
a_2	0.5	0.5	0.5	0.5	0.5
b_1	1.0	1.0	1.0	1.0	1.0
b_2	0.1	0.1	0.1	0.1	0.1
c_1	0.27	0.26	0.26	0.25	0.25
c_2	0.27	0.26	0.26	0.25	0.25
c_3	0.27	0.26	0.26	0.25	0.25

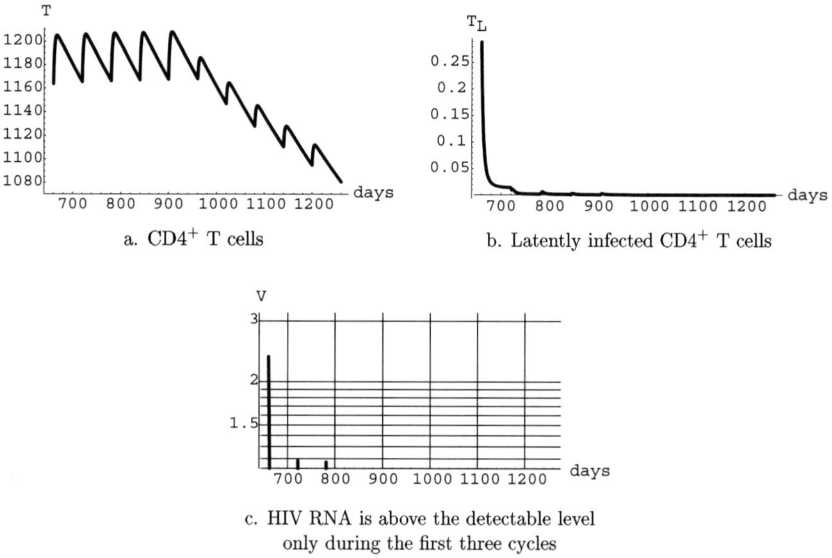

Fig. 2. Patient 1 HAART plus IL-2 treatment. The treatment parameters are given in Table 3.

Table 4. Parameters for Patient 10 IL-2 treatment.

Treatment cycle	1	2	3	4	5	6	7	8	9
a_1	0.05	0.05	0.05	0.04	0.04	0.03	0.03	0.03	0.03
a_2	0.4	0.4	0.4	0.4	0.5	0.5	0.5	0.5	0.5
b_1	1.0	1.0	1.0	1.0	1.0	1.0	1.0	1.0	1.0
b_2	0.1	0.1	0.1	0.1	0.1	0.1	0.1	0.1	0.1

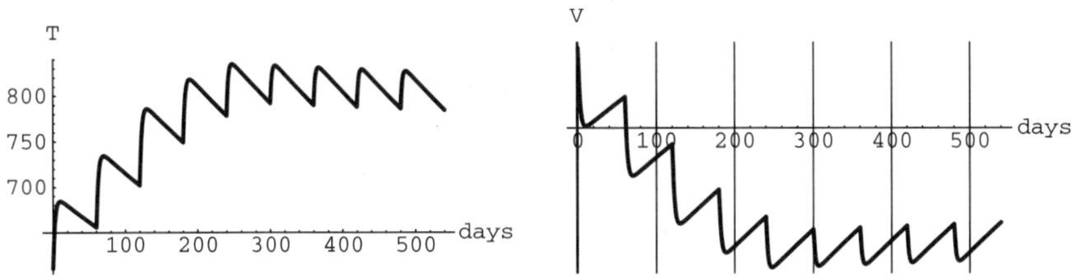

a. T cells during IL-2 treatment b. HIV RNA during IL-2 treatment

Fig. 3. Patient 10 IL-2 treatment. Treatment parameters used are given in Table 4.

Similarly, we modeled Patient 10. This patient had $CD4^+$ T cell count of 602 before treatment, so we began the first IL-2 day 550. The model shows a $CD4^+$ T cell population of 610 and an HIV RNA count of 5.4 per mm^3, which is consistent with the data. This patient received 18 months of IL-2. The treatment parameters are given in Table 4 and the graphs for the IL-2 treatment are given in Fig. 3. After 9 IL-2 treatments, we added HAART to Patient 10's regimen. At this point, the model shows a $CD4^+$ T cell population of 892 and an HIV RNA count of 3.988 per mm^3, which is consistent with the data. When HAART begins, we continue the cyclic IL-2, for 11 cycles. The graphs of this simulation are given in Fig. 4. At the end of treatment, the model shows a $CD4^+$ T cell population of 892 per mm^3, and a latently infected $CD4^+$ T cell population of 4.1×10^{-6} and an HIV RNA count of 1.2×10^{-14} per mm^3, both of which are well below the level of detectability. Again, IL-2 with HAART has reduced both the latently infected and the virus classes below detectable levels.

Next, we modeled Patient 20. This patient did not receive IL-2. At the beginning of HAART, Patient 20 had a $CD4^+$ T cell population of 446, so we began HAART at day 844. The graphs for this

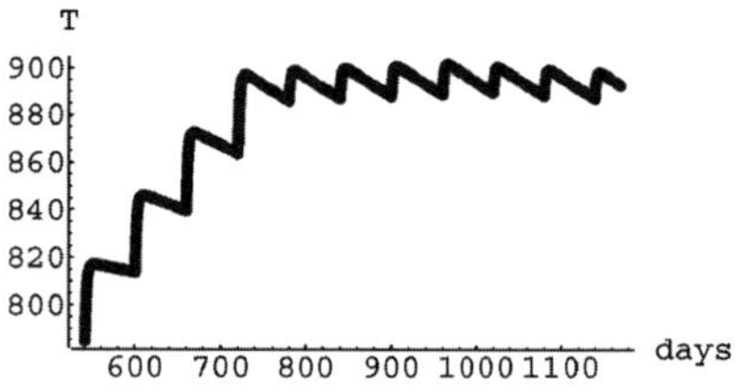

a. CD4$^+$ T cells

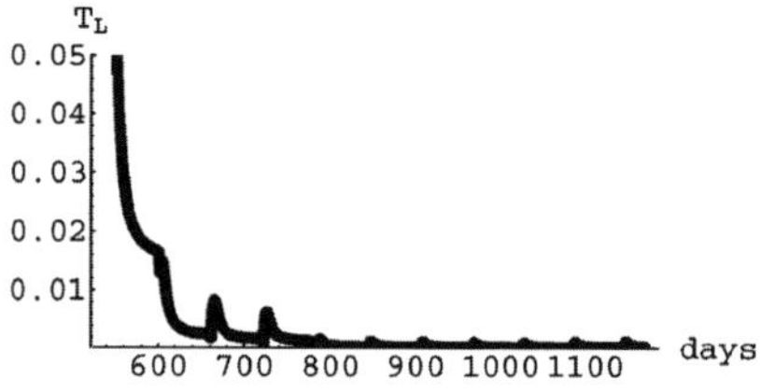

b. Latently infected CD4$^+$ T cells

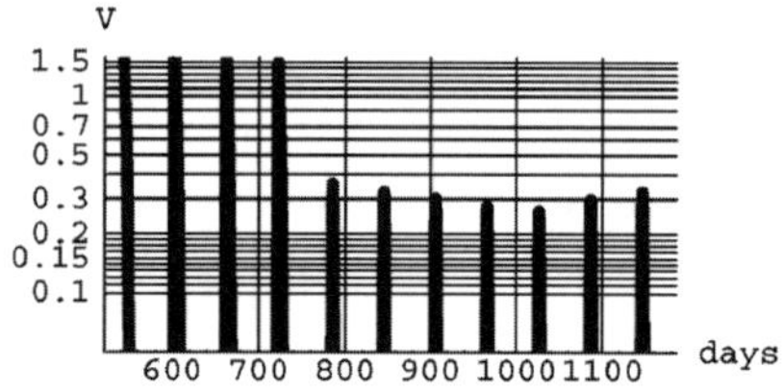

c. HIV RNA is above detectable level only at beginning of cycles

Fig. 4. Patient 10 HAART plus IL-2 treatment. The treatment parameters are given in Table 5.

Table 5. Parameters for Patient 10 HAART treatment.

Treatment cycle	10	11	12	13	14	15
a_1	0.02	0.02	0.02	0.02	0.01	0.01
a_2	0.5	0.5	0.5	0.5	0.6	0.6
b_1	1.0	1.0	1.0	1.0	1.0	1.0
b_2	0.1	0.1	0.1	0.1	0.1	0.1
c_1	2.0	2.0	2.0	2.0	2.0	2.0
c_2	0.2	0.2	0.2	0.2	0.2	0.2
c_3	0.2	0.2	0.2	0.19	0.19	0.19
Treatment cycle	16	17	18	19	20	
a_1	0.01	0.01	0.01	0.01	0.01	
a_2	0.6	0.6	0.7	0.7	0.7	
b_1	1.0	1.0	1.0	1.0	1.0	
b_2	0.1	0.1	0.1	0.1	0.1	
c_1	2.0	2.0	2.0	2.0	2.0	
c_2	0.2	0.2	0.2	0.2	0.2	
c_3	0.19	0.19	0.19	0.19	0.18	

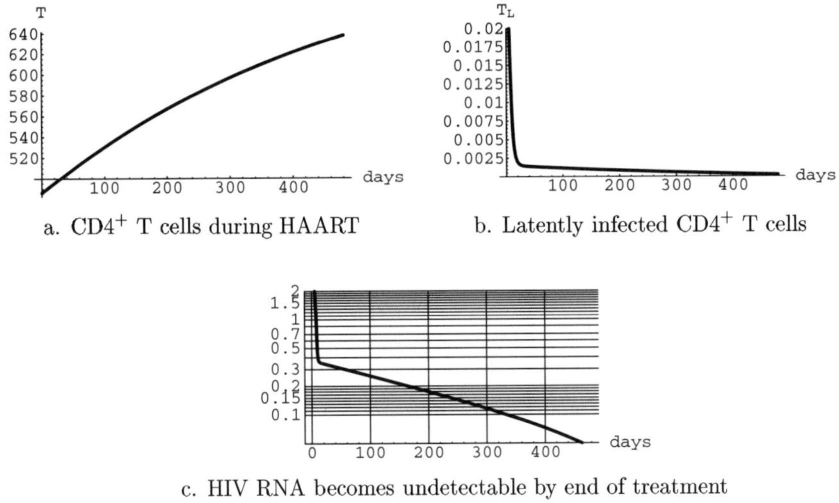

Fig. 5. Patient 20 HAART treatment. The treatment parameters are as follows: $c_1 = 2$, $c_2 = 0.2$, and $c_3 = 0.17$.

simulation are given in Fig. 5. The patient received HAART for 16 months. At the end of simulation, the model shows $T(t) \approx 637$ per mm^3, $T_P(t) \approx 2.5 \times 10^{-4}$ per mm^3, and $V(t) \approx 0.045$ per mm^3. Notice that for this patient, the latently infected CD4$^+$ T cell population is detectable, while the virus is not.

Last, we modeled Patient 22. This patient did not receive IL-2. At the beginning of HAART, Patient 22 had a CD4$^+$ T cell population of 691, so we began HAART at day 400. The graphs for this simulation are given in Fig. 6. The patient received HAART for 21 months. At the end of simulation, the model shows $T(t) \approx 769$ per mm^3, $T_P(t) \approx 1.2 \times 10^{-4}$ per mm^3, and $V(t) \approx 0.019$ per mm^3. Again, notice that for this patient, the latently infected CD4$^+$ T cell population is detectable, while the virus is not. Table 6 gives the data from the study and the corresponding model result for each patient. We see that the model, like all models, is imperfect. Since a model is not an exact solution, we must interpret it. For Patient 1, it is more accurate for the CD4$^+$ T cells, while for Patient 10 it is more accurate for the virus count. The patients on HAART alone have significant agreement with the CD4$^+$ T cells, while the virus counts are somewhat different between the model and the data. Since there is significant agreement, we can determine that it is a valid model.

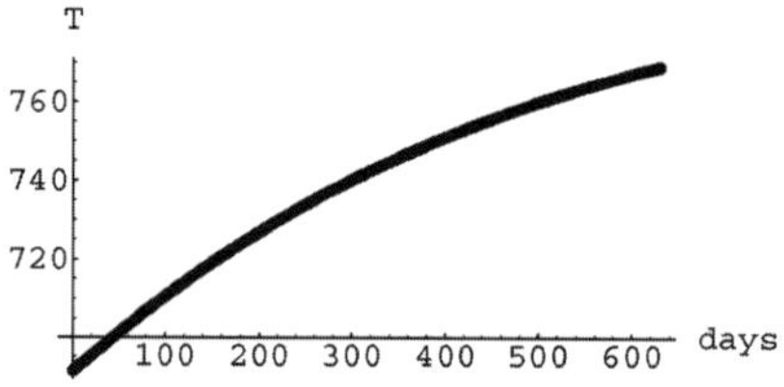

a. CD4$^+$ T cells

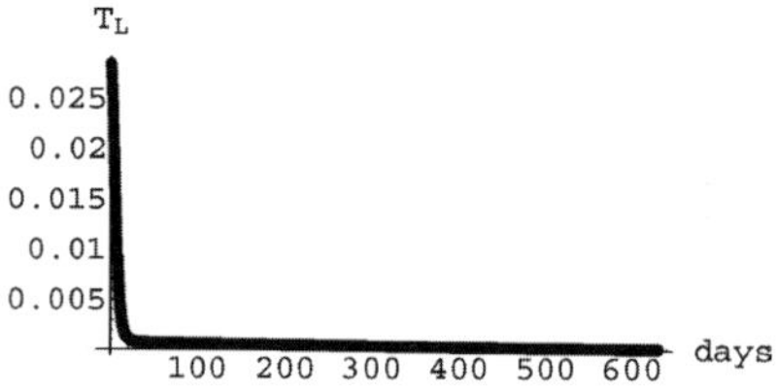

b. Latently infected CD4$^+$ T cells

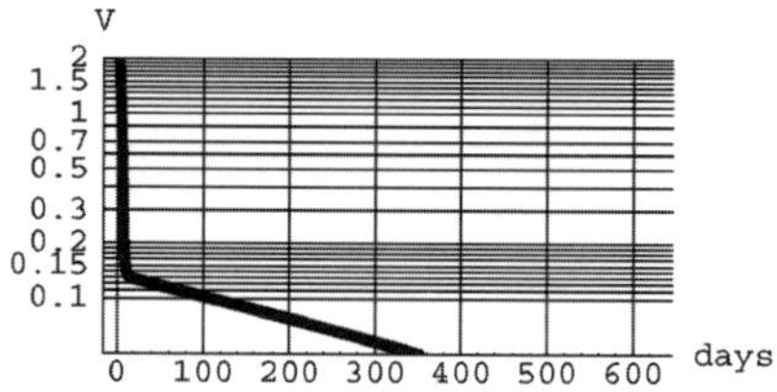

c. HIV RNA becomes undetectable by end of treatment

Fig. 6. Patient 22 HAART treatment. The treatment parameters are as follows: $c_1 = 2$, $c_2 = 0.23$, and $c_3 = 0.2$.

Table 6. Comparison of model results with study results.

Patient	CD4 count before HAART Study	CD4 count before HAART Model	CD4 count at time of study Study	CD4 count at time of study Model	Plasma HIV before HAART Study	Plasma HIV before HAART Model
1	1149	1163	832	1080	3.06	2.36
10	581	784	647	892	3.927	3.9
20	446	486	732	637	2.211	7.05
22	691	691	563	769	14.400	4.6

2.1. *Stopping treatment*

As patients do not always stay on the treatments indefinitely, we are interested in what happens to the populations when treatment is suspended, or stopped altogether. The data we have been modeling so far (Chun *et al.*, 1999a) did not stop treatment, but others have (Chun *et al.*, 2000; Davey *et al.*, 1999). It has generally been the case that no matter how low the levels of virus had become, or how long the virus had been undetectable, the virus almost immediately increases steeply, returning to levels approximately where it was when treatment began. We found this to be the case

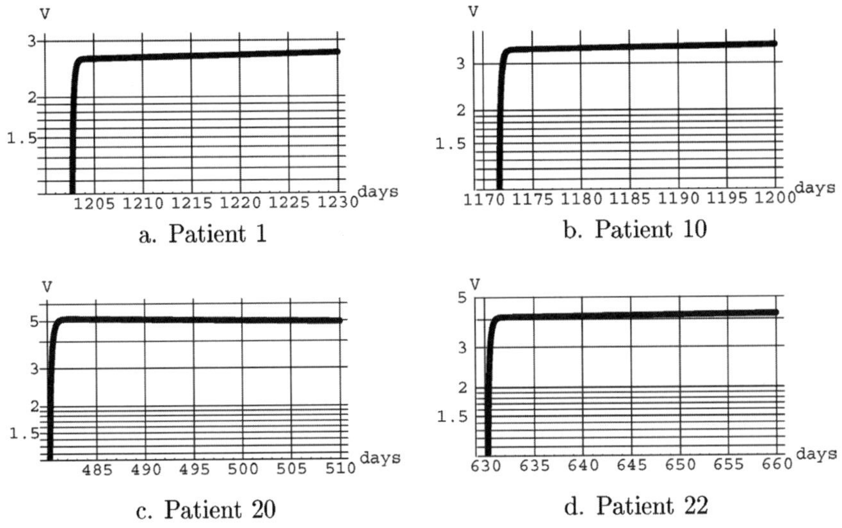

Fig. 7. Virus counts when all treatment was stopped.

with our simulations as well. Figure 7 has the graphs of all four patients when treatment was stopped.

3. Conclusions

We developed a model for the progression of HIV infection based on combining previously published models. When possible, well known parameters, such as the half-life of $CD4^+$ T cells (infected and uninfected), the normal T cell levels in an uninfected individual, the percentage of infected cells that fall into each class, and viral replication rates and lifespan have been used. Next, we determined reasonable functions to simulate treatment. These functions were based, in part, on previously published papers (Kirschner *et al.*, 2000; Kirschner and Webb, 1997, 1998). We ran simulations to model the treatment of patients in the study by Chun *et al.* (1999a). Using our mathematical model, we were able to reduce the latently infected class to below detectability. Unfortunately, when treatment is stopped, this class increases again to detectable levels.

3.1. *The next step*

There are still many questions about the treatment of HIV. Would a lower dose of IL-2 be as advantageous in rebuilding the $CD4^+$ T cell population with lower toxicity (Jacobson *et al.*, 1996)? Is it possible to eradicate the virus, or should we just aim for control (Ho, 1998)? There are groups working on vaccines. One such vaccine would be given to an HIV-positive person, then treatment would be halted. The hope is that the immune system would have been stimulated by the vaccine, and would be able to control the infection by itself (Smaglik, 2000). Clinical trials conducted on any of the above topics could be modeled mathematically, and determinations made about the feasibility of the treatments.

One area of particular interest is structured treatment interruptions (STI). Here, a patient would take medication, typically until undetectable levels of HIV have been maintained for a specified time period, the patient would have a break from the medications, for a specified time period. Studies have shown that the virus rebounds almost immediately, but only goes back to the pretreatment level, and stays steady. After a short break, the patient resumes treatment, and typically, the virus reverts to its undetectable level within 7–10 days. The hope in this type of treatment is that the patient can gain control over the infection, without the high levels of toxicity involved with long durations of treatment (Ortiz *et al.*, 1999; Lisziewicz *et al.*, 1999). In an attempt to see if our "patients" would also have undetectable virus, we took the simulations in Fig. 7 and restarted treatment (HAART only) at day 30. The results can be seen in Fig. 8. Notice that every single patient returned to their previously undetectable levels.

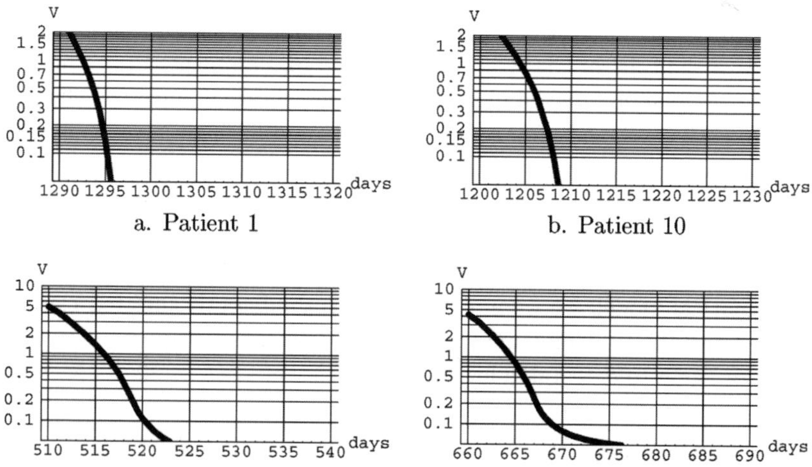

a. Patient 1

b. Patient 10

c. Patient 20

d. Patient 22

Fig. 8. Virus counts when HAART was resumed.

Again, with more clinical trials involving STI, data would be available that would enable us to model these treatments.

4. Appendix

Kirschner and Webb's model on immunotherapy (Kirschner and Webb, 1998) is based on two equations relating the uninfected CD4$^+$ T cells and the HIV virus. Their paper with Cloyd (Kirschner *et al.*, 2000) includes three types of infected CD4$^+$ T cells (productively, latently, and abortively infected), as well as the interaction between the lymph system and the plasma. It has been shown that the dynamics in the plasma closely mirror those in the lymph (Saag *et al.*, 1996), and since in practice, it is easier to take measurements of CD4$^+$ T cells and virus particles from the blood than from the lymph system, this model represents viral populations in the plasma. The model is a five compartment model designed to simulate the disease progression in these five classes with treatment using both HAART and IL-2. The variables used in this model are given in Table 7.

Some of the dynamic information in an HIV infected immune system is known, and this is incorporated in the model. Normal CD4$^+$ T cell counts are 800–1200 cells per mm^3 and their lifespan is approximately 200 days (Tough and Sprent, 1995). There is an initial period of acute viremia, after which the CD4$^+$ T cells decline from about 600–800 cells per mm^3 to 0 cells per mm^3 over a period of 7–12 years (Pennisi and Cohen, 1996). The CD4$^+$ T cells decline most rapidly early in infection (Phillips *et al.*, 1995), and the infected cells make up only about 4% or less of the total CD4$^+$ T cell population (Embretson *et al.*, 1993). The half life of a productively infected CD4$^+$ T cell has been found to be about 2 days (Wei *et al.*, 1995; Perelson *et al.*, 1996). The productively infected cells make up about 0.02% of the total CD4$^+$ T cell population, the latently infected cells about 0.01%, and the abortively infected cells about 0.5% (Chun *et al.*, 1997). The abortively infected CD4$^+$ T cells will convert to uninfected CD4$^+$ T cells

Table 7. Variables for the model. (Note: All measurements are per mm^3.)

T	=	the number of uninfected CD4$^+$ T cells
T_P	=	the number of productively infected CD4$^+$ T cells
T_L	=	the number of latently infected CD4$^+$ T cells
T_A	=	the number of abortively infected CD4$^+$ T cells
V	=	the number of HIV RNA copies

at a rate of about 0.2 per day (Zack *et al.*, 1990). During the asymptomatic phase, the viral replication rates remain relatively stable with a lifespan of about 7.2 hours outside a T cell (Wei *et al.*, 1995; Perelson *et al.*, 1996) and there is a dramatic increase in replication rates at the end of the symptomatic phase (Pennisi and Cohen, 1996). This model takes the uninfected $CD4^+$ T cell equation and the virus equation from Kirschner and Webb's immunology paper (Kirschner and Webb, 1998), and includes equations for each of the classes of infected T cells, modified from their paper with Cloyd (Kirschner *et al.*, 2000). The five equations are then

$$\frac{dT}{dt} = s(t) - \mu_T T(t) - \eta_1(t)\kappa V(t)T(t) + \nu(t)T(t) \tag{1a}$$

$$\frac{dT_P}{dt} = \eta_1(t)\sigma_P\kappa V(t)T(t) - \mu_{T_P}T_P(t) + \tau(t)\tau_P T_L(t) - \tau_L T_P(t) \tag{1b}$$

$$\frac{dT_L}{dt} = \sigma_L\kappa V(t)T(t) - \mu_{T_L}T_L(t) + \tau_L T_P(t) - \tau(t)\tau_P T_L(t) \tag{1c}$$

$$\frac{dT_A}{dt} = (r - \sigma_P - \sigma_L)\kappa V(t)T(t) - \mu_{T_A}T_A(t) \tag{1d}$$

$$\frac{dV}{dt} = -\kappa_V V(t)T(t) + \eta_2(t)\frac{l_s}{b_V + V(t)}V(t) \tag{1e}$$

with initial conditions

$$\begin{gathered} T(0) = T(t_1), \quad T_P(0) = T_P(t_2), \quad T_L(0) = T_L(t_2), \\ T_A(0) = T_A(t_2), \quad V(0) = V(t_2) \end{gathered} \tag{1f}$$

where t_2 is the time at which HAART is started.

We define $s_0 \equiv s_1 - \frac{s_2}{b(t)+V(t_1)}V(t_1)$ where t_1 is the time treatment begins. This measures the rate of influx of healthy $CD4^+$ T cells at time of treatment. Then we define $s(t) \equiv \min\left(s_0, s_1 - \frac{s_2}{b(t)+V(t)}V(t)\right)$. The second functions here represents the diminished capacity of the immune system to produce new T cells. So this term in Eq. (1a), has the desired effect; as treatment progresses, $V(t)$ will decrease, leaving s_0 the smaller of the two, and when treatment is stopped, the deterioration will once again take over. The normal death rate of $CD4^+$ T cells is μ_T, and the functions $\eta_1(t)$ and $\nu(t)$ are the functions necessary for treatment, as described below. In the third term, we assume that the rate of infection is governed by the mass action term $\kappa V(t)T(t)$, where κ is the infection rate.

In Eq. (1b), σ_P is the percentage of infections that result in a productively infected $CD4^+$ T cell and μ_{T_P} is the death rate of productively infected cells independent of the virus. The last two terms represent the exchange of cells between the latently and productively infected classes, where τ_P is the rate at which latent cells "activate" and become productive, τ_L the rate at which productive cells "shut down" and become latent, and $\tau(t)$ the function that modifies this exchange during treatment.

The third equation, (1c), represents the latently infected CD4$^+$ T cells, where σ_L is the percentage of new infections that result in latently infected CD4$^+$ T cells, μ_{T_L} is the death rate of those cells, and τ_P, τ_L, and $\tau(t)$ as above.

The fourth equation, (1d), represents the abortively infected CD4$^+$ T cells. Upon infection, a number of these cells almost immediately undergo apoptosis (Wang *et al.*, 1999) so for the purpose of this model, just disappear. The parameter r represents the proportion of infected cells that remain, the term $(r - \sigma_P - \sigma_L)V(t)T(t)$ represents the proportion of new infections that result in abortively infected CD4$^+$ T cells. The rate of conversion of these cells back to normal, uninfected CD4$^+$ T cells is μ_{T_A}.

In Eq. (1e), we have κ_V which represents the rate at which the immune system clears the virus, and $\frac{l_s}{b_V+V(t)}V(t)$ which represents the source of virus from the lymph system. This term, in which b_V is simply a saturation constant, is responsible for the steep increase in the virus population at the end of progression (Kirschner and Webb, 1997). The function $\eta_2(t)$ is a treatment function representing the reduction of this source during treatment.

Table 8 explains all of the parameters and the values used in simulation.

The patients in Chun's study (Chun *et al.*, 1999a) were admitted to the hospital every eight weeks for a five-day intravenous dose of IL-2. In our model, this is represented with the exponential function, $\nu(t) = a_1 e^{-a_2 t}$ in the $T'(t)$ Eq. (1a). This is an exponential function because the effect of the IL-2 will diminish over the 7 weeks between treatments.

One of the benefits attributed to IL-2 is the "purging" of the latently infected CD4$^+$ T cell class. We show this with a function $\tau(t) = b_1 e^{-b_2 t}$ in both the $T_P'(t)$ and $T_L'(t)$ equations. Again, it is exponential to give the diminishing effect. This function will be multiplied by the $\tau_P T_L(t)$ term in both equations, and has the effect of increasing the rate at which the latent cells become productive.

HAART, as described in the introduction, is a combination of drugs including at least one protease inhibitor, and a couple of reverse transcriptase inhibitors. The reverse transcriptase inhibitor will reduce the number of new infections. Thus the function $\eta_1(t) \equiv e^{-c_1 t}$, which is multiplied times κ

Table 8. Parameters and their values.

Parameter	Description	Simulation value	Reference
s_1	Source of healthy T cells from lymph system in absence of infection	$2.0/\text{mm}^3/\text{day}$	(Kirschner and Webb, 1997)
s_2	Reduction of s_1 due to infection	$1.6/\text{mm}^3/\text{day}$	(Kirschner and Webb, 1997)
b_T	Half-saturation constant of T cell source	$14.0/\text{mm}^3$	(Kirschner and Webb, 1997)
μ_T	Mortality rate of uninfected T cells	0.002/day	(Kirschner and Webb, 1998)
κ	Rate at which T cells are infected by virus	$0.00025\,\text{mm}^3/\text{day}$	(Kirschner and Webb, 1998)
σ_P	Percent of new infections that result in productively infected T cell	0.01	(Kirschner *et al.*, 2000)
μ_{T_P}	Mortality rate of productively infected T cells	0.25/day	(Kirschner *et al.*, 2000)
τ_P	Rate at which latently infected cells become activated	0.3/day	
τ_L	Rate at which productively infected cells shut down	0.009/day	
σ_L	Percent of new infections resulting in latently infected cells	0.01	(Kirschner *et al.*, 2000)
μ_{T_L}	Mortality rate of latently infected cells	0.003/day	(Kirschner *et al.*, 2000)
r	Proportion of infected cells that do not immediately die	0.5	(Kirschner *et al.*, 2000)
μ_{T_A}	Conversion rate of abortively infected to uninfected T cells	0.5/day	(Kirschner *et al.*, 2000)
κ_V	Rate at which T cells remove virus	$0.007\ \text{mm}^3/\text{day}$	(Kirschner and Webb, 1998)
l_s	Virus source from lymph system	$27.4/\text{mm}^3/\text{day}$	(Kirschner and Webb, 1998)
b_V	Half-saturation constant for virus source from the lymph system	$1/\text{mm}^3$	(Kirschner and Webb, 1998)

in both the $T'(t)$ and the $T'_P(t)$ equations. Since that term is negative in the $T'(t)$ equation and positive in the $T'_P(t)$ equation, it will have the effect of reducing the number of uninfected $CD4^+$ T cells lost to infection as well as reducing the number of new infections that appear in the productively infected class. The protease inhibitors will cause infected cells to not be able to produce new virus. This is shown in the model with the function

$\eta_2(t) \equiv \max(e^{-c_2 t}, c_3)$. This term is multiplied times the source from the lymph system, and has the effect of reducing the number of viruses sent in from the lymph system.

References

1. Arai K-I *et al.* (1990) Cytokines: coordinators of immune and inflammatory responses, *Ann Rev Biochem* **59**: 783–786.
2. Bittner B, Bonhoeffer S and Nowak MA (1997) Virus load and antigenic diversity, *Bull Math Biol* **59**(5): 881–896.
3. Carr A *et al.* (1998) Outpatient continuous intravenous Interleukin-2 or Subcutaneous, Polyethylene Glycol-modified Interleukin-2 in Human Immunodeficiency Virus-infected patients: A randomized, controlled, multicenter study, *J Infect Dis* **178**(4): 992–999.
4. Chun TW *et al.* (1998a) Early establishment of a pool of latently infected, resting $CD4^+$ T cells during primary HIV-1 infection, *Proc Natl Acad Sci USA* **95**(15): 8869–8873.
5. Chun TW *et al.* (1999a) Effect of Interleukin-2 on the pool of latently infected, resting $CD4^+$ T cells in HIV-1 infected patients receiving highly active anti-retroviral therapy, *Nature Med* **5**(6): 651–655.
6. Chun TW *et al.* (1998b) Induction of HIV-1 replication in latently infected $CD4^+$ T cells using a combination of cytokines, *J Exp Med* **188**(1): 83–91.
7. Chun TW and Fauci AS (1999b) Latent reservoirs of HIV: Obstacles to the eradication of the virus, *Proc Natl Acad Sci USA* **96**(20): 10958–10961.
8. Chun TW *et al.* (1997) Quantification of latent tissue reservoirs and total body viral load in HIV-1 infection, *Nature* **387**(6629): 183–188.
9. Chun TW *et al.* (2000) Relationship between pre-existing viral reservoirs and the re-emergence of plasma viremia after discontinuation of highly active anti-retroviral therapy, *Nature Med* **6**(7): 757–761.
10. Davey RT Jr *et al.* (1999) HIV-1 and T cell dynamics after interruption of highly active antiretroviral therapy (HAART) in patients with a history of sustained viral suppression, *Proc Natl Acad Sci* **96**(26): 1509–15114.
11. Embretson J *et al.* (1993) Massive covert infection of helper T lymphocytes and macrophages by HIV during the incubation period of AIDS, *Nature* **362**: 359–362.
12. Emery S and Lane HC (1997) Immune reconstitution in HIV infection, *Curr Opin Immunol* **9**(4): 568–572.
13. Finzi D *et al.* (1997) Identification of a reservoir for HIV-1 in patients on highly active antiretroviral therapy, *Science* **278**(5341): 1295–1300.
14. Finzi D *et al.* (1999) Latent infection of $CD4^+$ T cells provides a mechanism for lifelong persistence of HIV-1, even in patients on effective combination therapy, *Nature Med* **5**(5): 512–517.
15. Gulick R *et al.* (1997) Treatment with indinavir, zidovudine, and lamivudine in adults with human immunodeficiency virus infection and prior antiretroviral therapy, *New Engl J Med* **337**(11): 734–739.

16. Hammer SM *et al.* (1997) A controlled trial of two nucleoside analogues plus indinavir in persons with human immunodeficiency virus infection and CD4 cell counts of 200 per cubic millimeter or less, *New Engl J Med* **337**(11): 725–733.
17. Ho DD (1998) Toward HIV eradication or remission: The tasks ahead, *Science* **280**(5371): 1866–1867.
18. Jacobson EL, Pilaro F and Smith KA (1996) Rational interleukin 2 therapy for HIV positive individuals: Daily low doses enhance immune function without toxicity, *Proc Natl Acad Sci USA* **93**(19): 10405–10410.
19. de Jong MD *et al.* (1997) Overshoot of HIV-1 viremia after early discontinuation of antiretroviral therapy, *J Acq Imm Def Synd* **11**(11): F79–F84.
20. Kinter A and Fauci AS (1996) Interleukin-2 and Human Immunodeficiency Virus infection: Pathogenic mechanisms and potential for immunologic enhancement, *Immunol Res* **15**(1): 1–15.
21. Kirschner DE, Webb GF and Cloyd M (2000) A model of HIV-1 disease progression based on virus-induced lymph node homing and homing-induced apoptosis of CD4(+) lymphocytes, *J Acq Imm Def Synd* **24**(4): 352–362.
22. Kirschner DE and Webb GF (1998) Immunotherapy of HIV-1 infection, *J Biol Sys* **6**(1): 71–83.
23. Kirschner DE and Webb GF (1997) Resistance, remission, and qualitative differences in HIV chemotherapy, *Emerg Infect Dis* **3**(3): 273–283.
24. Kovacs JA *et al.* (1995) Increases in CD4 T lymphocytes with intermittent courses of interleukin-2 in patients with human immunodeficiency virus infection, *New Engl J Med* **332**(9): 567–575.
25. Kovacs JA *et al.* (1996) Controlled trial of Interleukin-2 infusions in patients infected with the human immunodeficiency virus, *New Engl J Med* **335**(18): 1350–1356.
26. Lisziewicz *et al.* (1999) Control of HIV despite the discontinuation of antiretroviral therapy, *New Engl J Med* **340**: 1683–1684.
27. Nowak MA, May RM and Anderson RM (1990) The evolutionary dynamics of HIV-1 quasispecies and the development of immunodeficiency disease, *J Acq Imm Def Synd* **4**(11): 1095–1103.
28. Ortiz *et al.* (1999) HIV specific immune responses in subjects who temporarily contain virus replication after discontinuation of highly active antiretroviral therapy, *J Clin Invest* **104**: R13–18.
29. Pantaleo G, Graziosi C and Fauci AS (1993) The immunopathogenesis of human immunodeficiency virus infection, *New Engl J Med* **328**(5): 327–335.
30. Paul WE and Seder RA (1994) Lymphocyte responses and cytokines, *Cell* **76**: 241–251.
31. Pennisi E and Cohen J (1996) Eradicating HIV from a patient: Not just a dream? *Science* **272**(5270): 1884.
32. Perelson AS *et al.* (1997) Decay characteristics of HIV-1-infected compartments during combination therapy, *Nature* **387**(6629): 188–191.
33. Perelson AS *et al.* (1996) HIV-1 dynamics *in vivo*: virion clearance rate, infected cell life-span, and viral generation time, *Science* **271**: 1582–1586.
34. Phillips AN *et al.* (1995) Antiviral therapy, *Nature* **375**(): 195.

35. Saag MS *et al.* (1996) HIV viral load markers in clinical practice, *Nature Medicine* **2**(6): 625–629.
36. Smaglik P (2000) Reservoirs dog AIDS therapy, *Nature* **405**(6784): 270–272.
37. Smith KA (1988) Interleukin-2: Inception, impact, and implications, *Science* **240**(4856): 112–113.
38. Tough DF and Sprent J (1995) Lifespan of lymphocytes, *Immunol Res* **14**(1): 1–12.
39. Wang LQ *et al.* (1999) A novel mechanism of CD4 lymphocyte depletion involves effects of HIV on resting lymphocytes induction of lymph node homing and apoptosis upon secondary signaling through homing receptors, *J Immunol* **162**(1): 268–276.
40. Wei XP *et al.* (1995) Viral dynamics in human immunodeficiency virus type 1 infection, *Nature* **272**: 117–122.
41. Wong JK *et al.* (1997) Recovery of replication-competent HIV despite prolonged suppression of plasma viremia, *Science* **278**(5341): 1291–1295.
42. Zack JA *et al.* (1990) HIV-1 entry into quiescent primary lymphocytes: molecular analysis reveals a labile, latent viral structure, *Cell* 213–222.

CHAPTER 14

A STATE SPACE MODEL FOR HIV PATHOGENESIS UNDER ANTI-VIRAL DRUGS AND APPLICATIONS

Wai-Yuan Tan, Ping Zhang and Xiaoping Xiong

In this chapter we have developed an individual-based state space model for HIV pathogenesis in HIV-infected individuals treated with various drug regimens including HAART. For monitoring effects of anti-viral drugs to search for optimal treatment regimens, we have used this model to develop a general procedure via multi-level Gibbs sampling to estimate the unknown parameters and the state variables. The unknown parameters include the infection rates, turn-over rates of HIV and CD4 T cells, and rates measuring effects of various drugs. The state variables include the numbers of infectious and non-infectious free HIV in blood over time. As an illustration, we have applied the method of this paper to the data of a patient from the St. Jude Children Hospital treated with various types of anti-retroviral drugs including HAART. The results indicated that for this individual, monotherapies using (Zidovudine or Abacavir or Stavudine) and 2 drugs therapies using (Zidovudine-Didanosine, or Abacavir-Stavudine, or Lamivudine-Zidovudine) were not very effective in suppressing HIV replication. On the other hand, the HAART regimens involving 2 NRTI drugs and 1 PI drug (Lamivudine-Zidovudine Nelfinavir, or Lamivudine-Stavudine-Indinavir) were very effective in suppressing HIV replication. In a few days it can bring down the infectious virus load from $\geq$146,000/mL to $\leq$1000/mL and stays at this low level during the whole treatment period under HAART. Also, using the total virus loads may be misleading and erroneous. It appeared that in some time periods, the number of infectious HIV can be very low ($\leq$1000/mL) indicating that the drugs were in fact effective but the total number of HIV were very high reaching $\geq$45,000/mL.

Keywords: Infectious free HIV; latently HIV-infected $CD4^{(+)}$ T cells; multi-level Gibbs sampling method; non-infectious free HIV; observation model; productively HIV-infected $CD4^{+}$ T cells; stochastic system model.

1. Introduction

To combat HIV, medical researchers and pharmaceutical companies have developed anti-viral drugs to treat HIV-infected individuals. These drugs include nucleoside reverse transcriptase inhibitors (NRTIs) such as Zidovudine (AZT) and Stavudine (D4T), non-nucleoside reverse transcriptase inhibitors (NNRTIs) such as Dilavirdine (Rescriptor) and Nevirapine (Viramune) and protease inhibitors (PIs) such as Indinavir (Crixivan) and Nelvinavir (Viracept). Recently, important breakthrough has been acheived through the combination of 3 drugs involving 2 NRTI and 1 PI, referred to as Highly Active Anti-Retroviral Therapy (HAART) (Bartlett and Moore, 1998). For monitoring HIV progression in HIV-infected individuals treated with anti-viral drugs to seek optimal treatment regiment, it is critical to estimate the effects of NRTIs, NNRTIs and PIs as well as the number of infectious RNA HIV and non-infectious RNA HIV over time. To achieve these objectives, in this paper we will develop an individual-based state space model based on some longitudinal data from a HIV-infected individual treated with various types of antiviral drugs. By using this state space model, we will then develop procedures via multi-level Gibbs sampling method to estimate the unknown parameters and state variables.

In Section 2, we will describe how to derive stochastic models for HIV pathogenesis at the cellular level in HIV-infected individuals by taking into account basic mechanisms of the HIV pathogenesis. By using the stochastic system model derived in Section 2, in Section 3, we will derive some state space models for HIV pathogenesis at the cellular level in HIV-infected individuals using data from the observed RNA virus counts over time. In Section 4, we will develop a general Bayesian procedure to estimate the unknown parameters and the state variables. As an application of our models, in Section 5, we will apply the results of our model to the data of a patient from the St. Jude Children's Research Hospital treated by various anti-viral drugs including HAART. Finally in Section 6, we will draw some conclusions and discuss some relevant issues regarding state space models.

2. A Dynamic Stochastic Model of HIV Pathogenesis Under HAART

In HIV pathogenesis, an unique feature is that free HIV infect mainly the $CD4^{(+)}$ T cells (T_4 cells) resulting in the depletion of these cells over time. When a resting T_4 cell is infected by a free HIV, it becomes a latently infected CD4$^{(+)}$ T cell (T_1 cell), which may either revert back to an uninfected T_4 cell or be activated at some time to become a productively HIV-infected CD4$^{(+)}$ T cell (T_2 cell). T_1 cells will not release free HIV until being activated to become T_2 cells. On the other hand, when a dividing T cell is infected by a free HIV, it becomes a T_2 cell that will release free HIV when it is activated to divide; when this happens the T_2 cell normally dies. These newly generated HIV bud out into the blood stream or body fluid and are free to infect new CD4 T cells. Under treatment by anti-viral drugs, because of the inhibition of protease by the PIs and/or the inhibition of reverse transcriptase by the NRTIs and/or NNRTIs during the activation and division of T_2 cells, one anticipates two types of free HIV in the blood and body fluid: The infectious HIV (V_1) and the non-infectious HIV (V_0); one also anticipate three types of CD4 T cells in the blood and body fluid: the normal uninfected CD4$^+$ T cells (T cells), the latently infected CD4$^+$ T cells, and the productively infected CD4$^+$ T cells.

To monitor the progression of HIV under treatment with HAART, one would need to derive stochastic models involving these cells and HIV and their interactions. We seek stochastic models which are complex enough to capture the most important biological features of HIV pathogenesis; yet are simple enough to permit efficient estimation of the unknown parameters and the prediction of state variables. One such model is a complex stochastic extension of a dynamic deterministic model proposed by Perelson *et al.* (1996) for HIV pathogenesis in HIV-infected individuals treated with anti-viral drugs. Under this model, it is assumed that the number of uninfected T_4 cells is a constant over the time interval and that the parameters are time-independent (i.e. time homogeneous). For short time intervals, this is justified by the observations that before the start of the treatment, the HIV pathogenesis is at a steady state condition and that the uninfected T_4 cells have a relatively long life span. (The average life span of uninfected CD4 T cells is more than 50 days; see Cohen *et al.* 1998). Mitter *et al.* (1998) have shown that even for a much longer period after treatment, this assumption has little impact on the total number of free HIV, thus providing further justification for this assumption. Because the time-span under treatment is quite long for most of the individuals and because over different time periods, different drugs and drug combinations are usually used for HIV-infected individuals treated with anti-viral drugs, in order for the above assumption to prevail, we partition the time-span under treatment into several non-overlapping sub-intervals each with a short time period. Then we may follow Perelson *et al.* (1996)

to assume that in each sub-interval, the number of uninfected T_4 cells is a constant and the parameters are time independent. Under this assumption, the state variables are the numbers of infectious and non-infectious HIV as well as the numbers of latently infected and productively infected CD4$^{(+)}$ T cells.

Let $\{T_i(t), i = 1, 2\}$ denote the numbers of $\{T_i, i = 1, 2\}$ cells at time t respectively and $\{V_j(t), j = 0, 1\}$ the numbers of $\{V_j, j = 0, 1\}$ free HIV at time t respectively. Then we are entertaining a 4-dimensional stochastic process $\underset{\sim}{X}(t) = \{T_i(t), i = 1, 2, V_j(t), j = 0, 1\}$. This is basically a 4-dimensional Markov process with discrete state space and continuous time and with time-dependent parameters. For this process, the traditional approach is too complicated to be of much use; we thus consider an alternative equivalent approach through stochastic equations proposed by Tan and Wu (1998).

Before we proceed, we illustrate how to model effects of the drugs NRTIs, NNRTIs, and PIs. For this, notice that under treatment by NRTIs and/or NNRTIs, the virus RNA cannot convert to virus DNA to complete the infection process although the HIV can enter the T cells and hence are removed from the V_1 virus pool. These drugs can also filter into T_2 cells to inhibit the generated reverse transcriptase during the activation and division of the T_2 cells, thus resulting in some non-infectious HIV due to the damaged reverse transcriptase in newly generated HIV. On the other hand, under treatment by PIs, because of the inhibition of protease by the PIs, most of the newly generated HIV are non-infectious, due to the lack of the envelop proteins. To model this stochastically, let $\xi_R(t)$ be the proportion of virus RNA $\rightarrow$ virus DNA blocked by the NRTIs at time t and let $\xi_P(t)$ be the proportion of non-infectious HIV from the newly generated HIV at time t, due to the inhibition of protease or the damage of reverse transcriptase by NRTIs or NNRTIs.

2.1. *The stochastic differential equations for the state variables*

To model the stochastic process $\underset{\sim}{X}(t)$ defined above, consider the time interval $[t, t+\Delta t)$. Then, by the Markov assumption, $\underset{\sim}{X}(t+\Delta t)$ derives from $\underset{\sim}{X}(t)$ through stochastic dynamic interactions between the HIV and the CD4$^{(+)}$ T cells during the interval $[t, t + \Delta t)$. From the basic biological mechanism as described in Tan (2000, Chapter 8), these stochastic interactions and transitions are characterized by the following stochastic variables:

$F(t) =$ Number of V_1 lost through entering CD4 T cells during $[t, t+\triangle t)$;

$F_i(t) =$ Number of T_i cells resulted from the infection of T cells by free HIV during $[t, t+\triangle t), i = 1, 2$;

$A(t) =$ Number of T_1 cells activated to become T_2 cells during $[t, t+\triangle t)$;

$R_j(t) =$ Number of non-infectious free HIV (i.e. V_0) generated by the death of the j-th T_2 cell during $[t, t+\triangle t)$;

$D_i(t) =$ Number of deaths of T_i cells during $[t, t+\triangle t), i = 1, 2$;

$D_{Vi}(t) =$ Number of free V_i HIV which have lost infectivity, or die, or have been removed during $[t, t+\triangle t)$;

$N(t) =$ Average number of free HIV released by a T_2 cell when it dies at time t. Because $N(t)$ is usually very large, in what follows we will assume that $N(t)$ is a deterministic function of t, unless otherwise stated.

Then, by the conservation law we have:

$$T_1(t+\triangle t) = T_1(t) + F_1(t) - A(t) - D_1(t), \tag{1}$$

$$T_2(t+\triangle t) = T_2(t) + F_2(t) + A(t) - D_2(t), \tag{2}$$

$$\begin{aligned} V_0(t+\triangle t) &= V_0(t) + \sum_{j=1}^{D_2(t)} R_j(t) - D_{V0}(t) \\ &= V_0(t) + R(t) - D_{V0}(t) \end{aligned} \tag{3}$$

$$\begin{aligned} V_1(t+\triangle t) &= V_1(t) + \sum_{j=1}^{D_2(t)} [N(t) - R_j(t)] - F(t) - D_{V1}(t) \\ &= V_1(t) + [D_2(t)N(t) - R(t)] - F(t) - D_{V1}(t). \end{aligned} \tag{4}$$

In the above equations, the variables in the right side are random variables which specify the random transition during the time interval $[t, t+\triangle t)$ from elements of $\boldsymbol{X}(t)$ to elements of $\boldsymbol{X}(t+\triangle t)$. Let $k(t)$ be the HIV infection rate of T_4 cells with $k_T(t) = k(t)T(t)$, $\alpha(t)$ the activation rate of $T_1 \to T_2$, $\mu_i(t)(i = 1, 2)$ the death rate of T_i cells $(i = 1, 2)$ and $\mu_{Vi}(t)$ the rate by which free V_i HIV are being removed, die, or have lost infectivity. Let $c(t)$ be the proportion of dividing uninfected T_4 cells at time t. Then the conditional probability distributions of the above variables given $\boldsymbol{X}(t)$ can be specified as follows:

$$
\begin{aligned}
&[F(t), D_{V1}(t)] \mid V_1(t) \sim Multinomial[V_1(t); k_T(t)\Delta t, \mu_{V1}(t)\Delta t];\\
&[F_1(t), F_2(t)] \mid F(t) \sim Multinomial\{F(t); [1-c(t)][1-\xi_R(t)], c(t)\\
&\qquad\qquad \times [1-\xi_R(t)]\};\\
&[A(t), D_1(t)] \mid T_1(t) \sim Multinomial[T_1(t); \alpha(t)\Delta t, \mu_1(t)\Delta t];\\
&D_2(t) \mid T_2(t) \sim Binomial[T_2(t); \mu_2(t)\Delta t];\\
&R_j(t) \sim Binomial[N(t); \xi_P(t)] \text{ independently for } j = 1, \ldots, D_2(t) \text{ if}\\
&\qquad D_2(t) > 0,\ (R_j(t) = 0 \text{ if } D_2(t) = 0);\\
&D_{V0}(t) \mid V_0(t) \sim Binomial[V_0(t); \mu_{V0}(t)\Delta t].
\end{aligned}
$$

Given $\boldsymbol{X}(t)$, conditionally $[A(t), D_1(t)], D_2(t), [F(t), D_{V1}(t)], D_2(t)$, and $D_{V0}(t)$ are independently distributed of one another; given $F(t)$, conditionally $[F_i(t),\ i = 1, 2]$ are independently distributed of other variables. Let $\varepsilon_i(t), i = 1, 2, 3, 4$ be defined by:

$$
\begin{aligned}
e_1(t)\Delta t &= \{F_1(t) - k_T(t)[1-\xi_R(t)][1-c(t)]V_1(t)\Delta t\}\\
&\quad - [A(t) - \alpha(t)T_1(t)\Delta t] - [D_1(t) - \mu_1(t)T_1(t)\Delta t],\\
e_2(t)\Delta t &= [F_2(t) - k_T(t)(1-\xi_R(t))c(t)V_1(t)\Delta t]\\
&\quad + [A(t) - \alpha(t)T_1(t)\Delta t] - [D_2(t) - \mu_2(t)T_2(t)\Delta t],\\
e_3(t)\triangle t &= \{[R(t) - N(t)\xi_P(t)T_2(t)\mu_2(t)\Delta t] - [D_{V0}(t) - \mu_{V0}(t)V_0(t)\Delta t]\}\\
\varepsilon_4(t)\triangle t &= [N(t)D_2(t) - R(t) - N(t)(1-\xi_P(t))T_2(t)\mu_2(t)\Delta t]\\
&\quad - [F(t) - k_T(t)V_1(t)\Delta t] - [D_{V1}(t) - \mu_{V1}(t)V_1(t)\Delta t].
\end{aligned}
$$

Then, Eqs. (1)–(4) lead to the following stochastic differential equations for $\{T_i(t), i = 1, 2, V_j(t), j = 0, 1\}$:

$$
\begin{aligned}
dT_1(t) &= T_1(t+\Delta t) - T_1(t) = \{k_T(t)[1-\xi_R(t)][1-c(t)]V_1(t)\\
&\quad - [\alpha(t) + \mu_1(t)]T_1(t)\}\Delta t + \varepsilon_1(t)\Delta t, \qquad (5)\\
dT_2(t) &= T_2(t+\Delta t) - T_2(t) = \{k_T(t)c(t)[1-\xi_R(t)]V_1(t)\\
&\quad + \alpha(t)T_1(t) - \mu_2(t)T_2(t)\}\Delta t + \varepsilon_2(t)\Delta t, \qquad (6)\\
dV_0(t) &= V_0(t+\Delta t) - V_0(t) = \{N(t)\xi_P(t)\mu_2(t)T_2(t)\\
&\quad - \mu_{V0}(t)V_0(t)\}\Delta t + \varepsilon_3(t)\Delta t, \qquad (7)\\
dV_1(t) &= V_1(t+\Delta t) - V_1(t) = \{N(t)[1-\xi_P(t)]\mu_2(t)T_2(t)\\
&\quad - k_T(t)V_1(t) - \mu_{V1}(t)V_1(t)\}\Delta t + \varepsilon_4(t)\Delta t, \qquad (8)
\end{aligned}
$$

In Eqs. (5)–(8), the random noises $\varepsilon_j(t)\Delta t, j = 1,2,3,4$ have expectation zero. Assuming that $T(t)$ is a deterministic function of t, the variances and covariances of these random variables are easily obtained as $\text{Cov}\{\varepsilon_i(t)\Delta t, \varepsilon_j(t)\Delta t\} = Q_{ij}(t)\Delta t + o(\Delta t)$, where:

$$Q_{11}(t) = [\alpha(t) + \mu_1(t)]\text{E}T_1(t) + k_T(t)[1 - \xi_R(t)][1 - c(t)]\text{E}V_1(t)],$$

$$Q_{12}(t) = -\alpha(t)\text{E}T_1(t),$$

$$Q_{13}(t) = 0,$$

$$Q_{14}(t) = -(1 - \xi_R(t))(1 - c(t))k_T(t)EV_1(t),$$

$$Q_{22}(t) = \text{E}\{k_T(t)c(t)[1 - \xi_R(t)]V_1(t) + \alpha(t)T_1(t) + \mu_2(t)T_2(t)\},$$

$$Q_{23}(t) = -\{N(t)\xi_P(t)\text{E}[\mu_2(t)T_2(t)]\},$$

$$Q_{24}(t) = -N(t)[1 - \xi_P(t)]\text{E}[\mu_2(t)T_2(t)] - [1 - \xi_R(t)]c(t)k_T(t)EV_1(t),$$

$$Q_{33}(t) = \text{E}\{N(t)\xi_P(t)[1 - \xi_P(t) + N(t)\xi_P(t)]\mu_2(t)T_2(t) + \mu_{V0}(t)V_0(t)\},$$

$$Q_{34}(t) = \xi_P(t)[1 - \xi_P(t)]N(t)[N(t) - 1]\mu_2(t)ET_2(t),$$

$$Q_{44}(t) = \text{E}\{N(t)[1 - \xi_P(t)][N(t)(1 - \xi_P(t)) + \xi_P(t)]\mu_2(t)T_2(t) + [\mu_{V1}(t) + k_T(t)]V_1(t)\}.$$

By using the formula $\text{Cov}(X,Y) = E\{\text{Cov}[(X,Y)|Z]\} + \text{Cov}[E(X|Z), E(Y|Z)]$, it can be shown easily that the random noises $\varepsilon_j(t)$ are uncorrelated with the state variables $T_i(t), i = 1,2$ and $V_j(t), = 0,1$. Since the random noises $\varepsilon_j(t)$ are random variables associated with the random transitions during the interval $[t, t+\Delta t)$, one may also assume that the random noises $\varepsilon_j(t)$ are uncorrelated with the random noises $\varepsilon_l(\tau)$ for all j and l if $t \neq \tau$.

2.2. *The probability distribution of the state variables*

Letting $\Delta t \sim 1$ corresponding to a small interval such as 0.1 day, then the state variables are $\boldsymbol{X} = \{\underset{\sim}{X}(t), t = 0,1,\ldots,t_M\}$ where t_M is the termination time of the study. Let Θ denote the collection of all parameters. Using the probability distributions given above, one can readily derive the conditional density $P\{\boldsymbol{X}(t+1)|\boldsymbol{X}(t),\Theta\}$ and hence

$$P\{\boldsymbol{X}|\Theta\} = P\{\boldsymbol{X}(0)|\Theta\}\prod_{t=1}^{t_M} P\{\boldsymbol{X}(t)|\boldsymbol{X}(t-1),\Theta\}, \tag{9}$$

where $\boldsymbol{X} = \{\boldsymbol{X}(t), t = 0,1,\ldots,t_M\}$.

In (9), $P\{\boldsymbol{X}(t+1)|\boldsymbol{X}(t),\Theta\}$ is given by:

$$
\begin{aligned}
P\{\boldsymbol{X}(t+1)|\boldsymbol{X}(t),\Theta\}
&= \sum_{i_1=0}^{T_1(t)} \sum_{i_2=0}^{T_1(t)-i_1} \binom{T_1(t)}{i_1, i_2} [\alpha(t)]^{i_1} [\mu_1(t)]^{i_2} [1-\alpha(t)-\mu_1(t)]^{T_1(t)-i_1-i_2} \\
&\times \sum_{i=0}^{T_2(t)} \binom{T_2(t)}{i} [\mu_2(t)]^{i} [1-\mu_2(t)]^{T_2(t)-i} \\
&\times \sum_{j=0}^{V_1(t)} \binom{V_1(t)}{j} [k_T(t)]^{j} h(i, i_1, i_2, j; t) g(i, j : t), \qquad (10)
\end{aligned}
$$

$$
\begin{aligned}
g(i,j;t) = \sum_{k=0}^{N(t)i} \binom{N(t)i}{k} \Big\{ & \xi_P(t)^k [1-\xi_P(t)]^{N(t)i-k} \\
& \times \binom{V_0(t)}{\zeta_1(k;t)} [\mu_{V_0}(t)]^{\zeta_1(k;t)} [1-\mu_{V_0}(t)]^{\zeta_2(k;t)} \\
& \times \binom{V_1(t)-j}{\eta_1(i,j,k;t)} [\mu_{V_1}(t)]^{\eta_1(i,j,k;t)} [1-k_T(t)-\mu_{V_1}(t)]^{\eta_2(i,j,k;t)} \\
& \times \prod_{r=1}^{2} \{\delta[\zeta_r(k;t)]\delta[\eta_r(i,j,k;t)]\} \Big\}, \qquad (11)
\end{aligned}
$$

where

$$
\begin{aligned}
\delta(x) &= 1 \quad \text{if } x \geq 0 \\
&= 0 \quad \text{if } x \leq 0, \qquad (12)
\end{aligned}
$$

$$
\zeta_1(k;t) = V_0(t) - V_0(t+1) + k, \quad \zeta_2(k;t) = V_0(t+1) - k,
$$

and

$$
\begin{aligned}
\eta_1(i,j,k;t) &= V_1(t) - V_1(t+1) + N(t)i - k - j, \\
\eta_2(i,j,k;t) &= V_1(t+1) - N(t)i + k,
\end{aligned} \qquad (13)
$$

$$
\begin{aligned}
h(i, i_1, i_2, j; t) = & \binom{j}{\rho_1(i_1, i_2; t), \rho_2(i_1, i; t)} \\
& \times \{[1-c(t)][1-\xi_R(t)]\}^{\rho_1(i_1,i_2;t)} \{c(t)[1-\xi_R(t)]\}^{\rho_2(i,i_1;t)} \\
& \times [\xi_R(t)]^{j-\rho_1(i_1,i_2;t)-\rho_2(i,i_1;t)} \delta\{\rho_1(i_1,i_2;t)\}\delta\{\rho_2(i,i_1;t)\},
\end{aligned} \qquad (14)
$$

with

$$\rho_1(i_1, i_2; t) = T_1(t+1) - T_1(t) + i_1 + i_2$$
$$\rho_2(i, i_1; t) = T_2(t+1) - T_2(t) - i_1 + i.$$

The probability distribution given by Eqs. (9)–(10) is very complicated. Alternatively, by using the stochastic equations given by (1)–(4) and the probability distributions in Section 2, one may readily generate Monte Carlo samples to study the stochastic behavior of the process of HIV pathogenesis. Also, by using these stochastic equations, one may derive differential equations for the mean numbers, the variances and covariances between the state variables as well as other higher cumulants. In particular, the differential equations for the mean numbers $\{u_i(t) = E[T_i(t)], i = 1, 2, u_{Vj}(t) = E[V_j(t)], j = 0, 1\}$ are given by,

$$\frac{d}{dt}u_1(t) = k_T(t)[1 - c(t)][1 - \xi_R(t)]u_{V_1}(t) - u_1(t)[\alpha(t) + \mu_1(t)], \quad (15)$$

$$\frac{d}{dt}u_2(t) = k_T(t)c(t)[1 - \xi_R(t)]u_{V_1}(t) + \alpha(t)u_1(t) - \mu_2(t)u_2(t), \quad (16)$$

$$\frac{d}{dt}u_{V_0}(t) = N(t)\xi_P(t)\mu_2(t)u_2(t) - \mu_{V_0}(t)u_{V_0}(t), \quad (17)$$

$$\frac{d}{dt}u_{V_1}(t) = N(t)[1 - \xi_P(t)]\mu_2(t)u_2(t) - k_T(t)u_{V_1}(t). \quad (18)$$

The above equations are derived by assuming that $T(t)$ is a deterministic function of t. In this case, these equations are the same equations for defining the deterministic model in which the $\{T_i(t), i = 1, 2, V_j(t), j = 0, 1\}$ are assumed as deterministic functions of time, ignoring completely the randomness of these variables. One may use these results to derive least square estimates of the parameters. Notice that when the parameters are independent of time, the above equations can easily be solved to give explicit equations for the state variables; however, the solution is extremely difficult, if not impossible, if the parameters are time dependent.

3. A State Space Model for HIV Pathogenesis

State space models of stochastic systems consist of two sub-models: The stochastic system model which is the stochastic model of the system and the observation model which is a statistical model based on available observed data from the system. They are advantageous over both the stochastic model and the statistical model when used alone since it combines information and advantages from both of these models; see Tan (2002, Chapter 8) for more details.

In this section we will develop a state space model for HIV pathogenesis under treatment by various drugs based on the observed number of RNA virus counts over time. For this state space model, the state variables of the state space model are $\underset{\sim}{X}(t) = \{T_{j+1}(t), V_j(t), j = 0, 1\}$. Thus the stochastic system model is represented by the stochastic equations (1)–(6). The observation model is given by the statistical model based on the observed number of RNA virus copies per mL of blood over time.

3.1. *The stochastic system model and the probability distribution of state variables*

For implementing the multi-level Gibbs sampling procedure described in Shephard (1994) and Tan (2002, Chapter 3) to estimate the unknown parameters, we define the un-observed state variables $\underset{\sim}{U}(t) = \{A(t), D_1(t), D_2(t), R(t), F(t)\}$ and put $\boldsymbol{U} = \{\underset{\sim}{U}(t), t = 0, 1, \ldots, t_M - 1\}$. Then, by using the distribution results in Section 3.2, one may readily generate $\boldsymbol{U}$ given $\boldsymbol{X}$ and generate $\boldsymbol{X}$ given $\boldsymbol{U}$. The conditional density of $\underset{\sim}{U}(t)$ given $\underset{\sim}{X}(t)$ is

$$\begin{aligned} P\{\underset{\sim}{U}(t)|\underset{\sim}{X}(t)\} &= C_1(t)[\alpha(t)]^{A(t)}[\mu_1(t)]^{D_1(t)}[1-\alpha(t)-\mu_1(t)]^{T_1(t)-A(t)-D_1(t)} \\ &\quad \times [\mu_2(t)]^{D_2(t)}[1-\mu_2(t)]^{T_2(t)-D_2(t)}[k_T(t)]^{F(t)}[1-k_T(t)]^{V_1(t)-F(t)} \\ &\quad \times [\xi_P(t)]^{R(t)}[1-\xi_P(t)]^{D_2(t)N-R(t)}, \end{aligned} \tag{19}$$

where $C_1(t) = \binom{T_1(t)}{A(t), D_1(t)}\binom{T_2(t)}{D_2(t)}\binom{V_1(t)}{F(t)}\binom{N(t)D_2(t)}{R(t)}$ and $R(t) = 0$ if $D_2(t) = 0$.

The conditional density of $\underset{\sim}{X}(t+1)$ given $\{\underset{\sim}{U}(t), \underset{\sim}{X}(t)\}$ is

$$\begin{aligned} P\{\underset{\sim}{X}(t+1)|\underset{\sim}{U}(t), \underset{\sim}{X}(t)\} &= \binom{F(t)}{F_1(t), F_2(t)}[(1-\xi_R(t))(1-c(t))]^{F_1(t)}[(1-\xi_R(t))c(t)]^{F_2(t)} \\ &\quad \times [\xi_R(t)]^{F(t)-F_1(t)-F_2(t)}\binom{V_0(t)}{B_1(t)}[\mu_{V_0}(t)]^{B_1(t)}[1-\mu_{V_0}(t)]^{V_0(t)-B_1(t)} \\ &\quad \times \binom{V_1(t)-F(t)}{B_2(t)}\left[\frac{\mu_{V_1}(t)}{1-k_T(t)}\right]^{B_2(t)}\left[1-\frac{\mu_{V_1}(t)}{1-k_T(t)}\right]^{V_1(t)-F(t)-B_2(t)}, \end{aligned} \tag{20}$$

where $F_1(t) = T_1(t+1) - T_1(t) + A(t) + D_1(t)$, $F_2(t) = T_2(t+1) - T_2(t) - A(t) + D_2(t)$, $B_1(t) = V_0(t) - V_0(t+1) + R(t)$, and $B_2(t) = V_1(t) - V_1(t+1) + [ND_2(t) - R(t)] - F(t)$.

The joint density of $\{\boldsymbol{X}, \boldsymbol{U}\}$ given Θ is

$$P\{\boldsymbol{X}, \boldsymbol{U}|\Theta\} = P\{\underset{\sim}{X}(0)|\Theta\} \prod_{t=1}^{t_M} P\{\underset{\sim}{X}(t)|\underset{\sim}{U}(t-1), \underset{\sim}{X}(t-1)\} \times P\{\underset{\sim}{U}(t-1)|\underset{\sim}{X}(t-1)\}. \tag{21}$$

3.2. *The observation model*

Let y_j be the observed total number of HIV RNA virus load at time $t_j, j = 1, \ldots, n$. Then the conditional mean of y_j given $V(t_j) = V_0(t_j) + V_1(t_j)$ and the conditional variance of y_j given $V(t_j)$ are given by $V(t_j)$ and $C_0 V(t_j)$ respectively (Tan and Xiang 1999, Tan 2000), where $C_0 > 0$. Because both y_j and $V(t_j)$ are very large and because the conditional variance of y_j is a function of the conditional mean of y_j, we seek a transformation of y_j such that the transformed variable is approximately normally distributed and that the conditional variance of the transformed variable given $V(t_j)$ is a constant approximately. Further, the transformation has to preserve the results $E[y_j|V(t_j)] = V(t_j)$ and $Var[y_j|V(t_j)] = C_0 V(t_j)$. One such transformation is the log transformation; see the discussion in Section 6 for the example. Thus, putting $z_j = \sqrt{V(t_j)}(\log y_j - \log V(t_j))$, then we assume that z_j is normally distributed with mean $\nu_z = -\frac{1}{2\sqrt{V(t_j)}}\sigma^2$ ($\nu_z \cong 0$ for large $V(t_j)$) and variance σ^2, independently for $j = 1, \ldots, n$. That is:

$$z_j \sim N\left\{-\frac{1}{2\sqrt{V(t_j)}}\sigma^2, \sigma^2\right\}, \quad \text{independently for } j = 1, \ldots, n. \tag{22}$$

Notice that using the distribution in Eq. (21), $E[y_j|V(t_j)] = E\{e^{\log y_j}\} = E\exp\left\{\frac{z_j}{\sqrt{V(t_j)}} + \log V(t_j)\right\} = V(t_j)\exp\left\{\frac{1}{\sqrt{V(t_j)}}\nu_z + \frac{\sigma^2}{2V(t_j)}\right\} = V(t_j)$. Similarly, $E[y_j^2|V(t_j)] = [V(t_j)]^2 e^{\frac{\sigma^2}{V(t_j)}}$ so that the conditional variance of y_j given $V(t_j)$ is $\text{Var}\{y_j|V(t_j)\} = V(t_j)^2\left[\exp\left(\frac{\sigma^2}{V(t_j)}\right) - 1\right] = \sigma^2 V(t_j)\left\{1 + \sum_{i=1}^{\infty}\frac{1}{(i+1)!}\left(\frac{\sigma^2}{V(t_j)}\right)^i\right\} \cong \sigma^2 V(t_j)$ for large $V(t_j)$. Notice also that $\sigma^2 \leq C_0$ although the difference is very small.

From Eq. (21), it follows that the y_j are independently distributed of one another and that the density of y_j is

$$f_Y\{y_j|X(t_j)\} = \frac{\sqrt{V(t_j)}}{y_j\sqrt{2\pi}\sigma} \exp\left[-\frac{V(t_j)}{2\sigma^2}\left(\log y_j - \log V(t_j) + \frac{\sigma^2}{2\sqrt{V(t_j)}}\right)^2\right]. \tag{23}$$

Let $\boldsymbol{Y} = \{y_j, j = 1, \ldots, n\}$. Then the joint density of $\{\boldsymbol{X}, \boldsymbol{U}, \boldsymbol{Y}\}$ given Θ is:

$$\begin{aligned} P\{\boldsymbol{X}, \boldsymbol{U}, \boldsymbol{Y}|\Theta\} &= P\{\underset{\sim}{X}(0)|\Theta\} \prod_{j=1}^{n} f_Y\{Y(j)|V(t_j)\} \prod_{t=t_{j-1}+1}^{t_j} \\ &\quad \times P\{\underset{\sim}{U}(t-1)|\underset{\sim}{X}(t-1)\}P\{\underset{\sim}{X}(t)|\underset{\sim}{X}(t-1), \underset{\sim}{U}(t-1)\}. \end{aligned} \tag{24}$$

In Section 4, these distribution results will be used to estimate the unknown parameters via multi-level Gibbs sampling procedures.

4. Estimation of Unknown Parameters and State Variables

To use the model to assess effects of drugs and to validate the model, one would need to estimate the unknown parameters and the state variables. For the above model, the important state variables are the numbers of infectious HIV and non-infectious HIV. The unknown parameters are $\{c(t), N(t), \mu_i(t), i = 1, 2, \mu_{Vi}(t), i = 0, 1, k_T(t), \xi_R(t), \xi_P(t), \alpha(t), \sigma^2\}$. As in Tan (2000, Chapter 8; 2002, Chapter 9), one may assume $\{N(t) = N, \mu_i(t) = \mu_i, i = 1, 2, \mu_{V_i}(t) = \mu_V, i = 0, 1, \alpha(t) = \alpha\}$. However, because different drugs with different dose levels are used over different time periods and because $k_T(t)$ is a product of the infection rate and the number of un-infected CD4$^{(+)}$ T cells, one cannot assume the parameters $\{\xi_R(t), \xi_P(t), k_T(t), c(t)\}$ as time independent.

Because different drugs and drug combinations are used over different time intervals in most of the cases (see Table 2), to estimate the time dependent parameters we partition the time interval $[0, t_M)$ into k non-overlapping sub-intervals $\{L_j = [s_{j-1}, s_j), j = 1, \ldots, k\}$ with $(s_0 = 0, s_k = t_M)$ and assume the above parameters as constants in each sub-interval. Notice that in each sub-interval, the same drugs are used throughout the interval and the number of un-infected CD4$^{(+)}$ T cells are assumed as a constant. As in Tan and Wu (1998), in this paper we will estimate N by least square method; however, we will estimate the other parameters by using multi-level Gibbs sampling method.

The multi-level Gibbs sampling procedure is a Gibbs sampling method applied to multi-variate data as proposed by Shephard (1994). The details of this procedure as applied to problems of AIDS and cancers have been given in Tan (2000, Chapter 8; 2002, Chapter 9).

Let $P\{\Theta\}$ be the prior density of the parameters $\Theta = \{\mu_1, \mu_2, \mu_V, \alpha, \sigma^2, \xi_R(t), \xi_P(t), k_T(t), c(t), t = 1, \ldots, k\}$.

For $i = 1, \ldots, k$ and $r = 1, 2$, put

$$\begin{aligned}
\bar{R}(i) &= \sum_{j=s_{i-1}+1}^{s_i} R(j), \quad \bar{D}_r(i) = \sum_{j=s_{i-1}+1}^{s_i} D_r(j);\\
\bar{F}(i) &= \sum_{j=s_{i-1}+1}^{s_i} F(j), \quad \bar{F}_r(i) = \sum_{j=s_{i-1}+1}^{s_i} F_r(j);\\
\bar{V}_r(i) &= \sum_{j=s_{i-1}+1}^{s_i} V_r(j), \quad \bar{A}(i) = \sum_{j=s_{i-1}+1}^{s_i} A(j);\\
\bar{B}_r(i) &= \sum_{j=s_{i-1}+1}^{s_i} B_r(j), \quad \bar{T}_r(i) = \sum_{j=s_{i-1}+1}^{s_i} T_r(j).
\end{aligned}$$

By using the densities given above, the conditional density of Θ given $\{\boldsymbol{Y}, \boldsymbol{X}, \boldsymbol{U}, N\}$ is

$$\begin{aligned}
P\{\Theta|\boldsymbol{X}, \boldsymbol{U}, \boldsymbol{Y}, N\} &\propto P\{\Theta\}\mu_1^{\bar{D}_1}(1-\mu_1)^{\bar{T}_1-\bar{D}_1-\bar{A}}\alpha^{\bar{A}}\\
&\times \left(1-\frac{\alpha}{1-\mu_1}\right)^{\bar{T}_1-\bar{A}-\bar{D}_1} \mu_2^{\bar{D}_2}(1-\mu_2)^{\bar{T}_2-\bar{D}_2}\mu_V^{\bar{B}}(1-\mu_V)^{\bar{V}-\bar{B}}\\
&\times \prod_{j=1}^{n}[\xi_R(j)]^{\bar{F}1(j)+\bar{F}2(j)}[1-\xi_R(j)]^{\bar{F}(j)-\bar{F}1(j)-\bar{F}2(j)}\\
&\times [c(j)]^{\bar{F}_2(j)}[1-c(j)]^{\bar{F}_1(j)}[\xi_P(j)]^{\bar{R}(j)}[1-\xi_P(j)]^{N\bar{D}_2(j)-\bar{R}(j)}\\
&\times \left(\frac{k_T(j)}{1-\mu_V}\right)^{\bar{F}(j)}\left(1-\frac{k_T(j)}{1-\mu_V}\right)^{\bar{V}_1(j)-\bar{F}(j)-\bar{B}_2(j)}, \qquad (25)
\end{aligned}$$

where $\bar{D}_r = \sum_{j=1}^{k} \bar{D}_r(j)$, $\bar{T}_r = \sum_{j=1}^{k} \bar{T}_r(j)$ for $r = 1, 2$; $\bar{V} = \sum_{r=0}^{1}\sum_{j=1}^{k} \bar{V}_r(j)$, $\bar{B} = \sum_{r=1}^{2}\sum_{j=1}^{k} \bar{B}_r(j)$; and $\hat{\sigma}^2 = \frac{1}{k}\sum_{j=1}^{k}\frac{1}{V(t_j)}[y_j - V(t_j)]^2$. To apply Eq. (25), one needs to specify the prior $P\{\Theta\}$. Because we have no prior information on Θ so that our knowledge about Θ is vague and imprecise, we follow Box and Tiao (1973) to assume a non-informative prior $P\{\Theta\} \propto \sigma^2$ but with $\sigma^2 < 100$ so that all parameters are bounded.

From the distribution in (25), one can easily derive the probability distribution of each of the parameters Θ given $\{\boldsymbol{X}, \boldsymbol{U}, N, \boldsymbol{Y}\}$. This shows that all parameters are identifiable if Θ is estimable; see Remark 1.

Remark 1: In the state space model, $\boldsymbol{X}$ and $\boldsymbol{U}$ are the state variables which are time dependent. The classical state space modeling approach (i.e. the Kalman filter) is to estimate and predict the state variables assuming the parameters as known. This is achieved through the stochastic system model, which provide information about the system via the mechanism of the system and the observation model which is a statistical model based on data. The number of data sets is very limited, far less than the number of state variables to be estimated; but due to additional information from the system besides information from data, this is made possible. Notice that the observation model is basically a statistical model which is the model used by statisticians to estimate parameters. (The information provided by the observation model is very limited and is not adequate even to estimate all parameters, not to mention state variables.) Now given the values of state variables (i.e. $\boldsymbol{X}, \boldsymbol{U}$), one can readily estimate the unknown parameters by using the stochastic system model and the observation model. Thus, in state space models, one can readily bring these two aspects together by Gibbs sampling method to estimate simultaneously the state variables and the unknown parameters. This is the approach used by Tan and Ye (2000a, b) to estimate the state variables (i.e. the numbers of S people, HIV infected people and AIDS cases) as well as the unknown parameters (the HIV infection incidence, the HIV incubation incidence, the death rates and the immigration rates) in the Swiss homosexual and IV drug population and in the San Francisco homosexual population. A formal proof of the convergence of this approach is given in Appendix B.

Using the above results, one can readily estimate simultaneously the state variables and the unknown parameters Θ given $(\boldsymbol{Y}, N)$ by using the multi-level Gibbs sampling procedure; see Remark 1. This procedure for estimating Θ and the state variables given $(\boldsymbol{Y}, N)$ is described in Tan (2000, Chapter 6; 2002, Chapter 9) and is given by the following loops:

1) Combining a large sample from $P\{\boldsymbol{U}, \boldsymbol{X}|\Theta, N\}$ with $P\{\boldsymbol{Y}|\Theta, \boldsymbol{X}\}$ through the weighted bootstrap method due to Smith and Gelfand (1992), we generate $\{\boldsymbol{U}, \boldsymbol{X}\}$ (denote the generated sample $\{\boldsymbol{U}^{(*)}, \boldsymbol{X}^{(*)}\}$) from $P\{\boldsymbol{U}, \boldsymbol{X}|\Theta, \boldsymbol{Y}\}$ although the latter density is unknown. An illustration of how to use the weighted bootstrap method is given in Appendix A.

2) On substituting $\{\boldsymbol{U}^{(*)}, \boldsymbol{X}^{(*)}\}$ which are generated by the above step, we generate Θ from the conditional density $P\{\Theta|\boldsymbol{X}^{(*)}.\boldsymbol{U}^{(*)}, \boldsymbol{Y}, N\}$ given by Eq. (25) above.
3) With Θ being generated from Step 2 above, go back to Step 1 and repeat the above [1]–[2] loop until convergence.

At convergence, one then generates a random sample of $\boldsymbol{X}, \boldsymbol{U}$ from the conditional distribution $P\{\boldsymbol{X}, \boldsymbol{U}|\boldsymbol{Y}, N\}$ of $\{\boldsymbol{X}, \boldsymbol{U}\}$ given $\{\boldsymbol{Y}, N\}$, independent of Θ and a random sample of Θ from the distribution $P\{\Theta|\boldsymbol{Y}, N\}$ of Θ given $\{\boldsymbol{Y}, N\}$, independent of $\{\boldsymbol{X}, \boldsymbol{U}\}$. Repeat these procedures we then generate a random sample of size n of $(\boldsymbol{X}, \boldsymbol{U})$ given $\{\boldsymbol{Y}, N\}$ and a random sample of size m of Θ given $\{\boldsymbol{Y}, N\}$. One may then use the sample means to derive the estimates of $(\boldsymbol{X}, \boldsymbol{U})$ and Θ and use the sample variances as the variances of these estimates. By extending the method give in Chapter 3 of Tan (2002), a proof of convergence is given in Appendix B. For implementing the above procedure, we use the least square method to estimate N.

5. An Illustrative Example

As an application of the above results, we consider a patient from St. Jude Children's Hospital treated by various drugs over a period of 10 years. For this patient, the RNA virus copies per mL over time are given in Table 1 and the history of drug treatment is given in Table 2. These data are longitudinal data from a single individual. For this individual, the treatment period is very long covering a period of more than 10 years and different drugs have been used over different time periods. The drug therapy include mono-therapy (Abacavir or Stavudine), combinations of 2 NRTIs (Zidovudine-Didanosine or Abacavir-Stavudine) and HAART involving 2 NRTIs and 1 PI (Lamivudine-Zidovudine-Nelfinavir or Lamivudine-Stavudine-Indinavir); in some intervals, treatments have been stopped.

To extract information from such data, to date models and methods for its analysis are nonexistent. In this chapter, we will apply the model and methods given in Sections 3–4 to analyze this data set to answer some important clinical questions and to compare effects of mono-therapy, combination therapy of 2 NRTIs and HAART regimens involving 2 NRTI and 1 PI as given in Table 2. To estimate the time dependent parameters, because different treatment regimens have been applied over different time periods for this individual, we partition the time period into k sub-intervals L_i as given in Table 3 and assume $\{\xi_R(t) = \xi_R(i), \xi_P(t) = \xi_P(i), k_T(t) = k_T(i), c(t) = c(i)\}$ for $t \in L_i$, $i = 1, \ldots, k$.

Table 1. The observed numbers of HIV RNA virus load of an HIV infected patient.

Days since HIV infection	Observed RNA Copies/mL	Predicted RNA Copies/mL ± Std error	Predicted infectious RNA copies/mL ± Std error	Residual based on log transformation	Pearson residual
2780	54614	53955 ± 14396	34169 ± 9099	1.3266	1.3300
2857	120000	120156 ± 61087	105259 ± 53378	−0.2080	−0.2110
2860	135000	135530 ± 128011	121719 ± 114794	−0.6734	−0.6749
2900	85000	84780 ± 19905	76082 ± 17832	0.3574	0.3542
2943	60000	60019 ± 13492	53891 ± 12145	−0.0320	−0.0364
2971	95500	96381 ± 21857	86445 ± 19614	−1.3331	−1.3304
2998	165000	162843 ± 113857	146259 ± 102070	2.4921	2.5059
3054	46000	46049 ± 21039	2142 ± 1068	−0.1021	−0.1071
3082	19000	18956 ± 11148	538 ± 407	0.1574	0.1498
3110	13000	12933 ± 6462	235 ± 146	0.2849	0.2762
3173	130000	130392 ± 137538	115738 ± 114860	−0.5067	−0.5089
3199	51000	51164 ± 12665	45415 ± 11224	−0.3357	−0.3399
3223	35000	34924 ± 10377	1802 ± 565	0.1962	0.1906
3255	28000	28174 ± 8488	1424 ± 442	−0.4811	−0.4860
3357	93000	93521 ± 25495	10434 ± 24286	−0.7974	−0.7987
3391	43000	43073 ± 74540	2989 ± 5300	−0.1599	−1.1649
3419	9100	9046 ± 6547	229 ± 226	0.2766	0.2662
3504	25000	25327 ± 15423	999 ± 714	−0.9629	−0.9633
3521	15000	14593 ± 11204	319 ± 305	1.5667	1.5795
3586	24000	23889 ± 23769	1566 ± 1681	0.3428	0.3367
3677	7600	7342 ± 25553	167 ± 272	1.3998	1.4115

Table 2. The history of drug treatment.

Days partition	Drug treatment	Inhibitor
[0, 1232)	No drugs	No inhibitor
[1232, 2860)	Zidovudine, Didanosine	NRTIs inhibitor
[2860, 2943)	Abacavir	NRTI inhibitor
[2943, 2986)	Abacavir, Stavudine	NRTIs inhibitor
[2986, 2998)	Stavudine	NRTI inhibitor
[2998, 3140)	Lamivudine, Nelfinavir, Zidovudine	NRTIs + PI
[3140, 3143)	No drugs	No inhibitor
[3143, 3199)	Lamivudine, Zidovudine	NRTIs inhibitor
[3199, 3670)	Lamivudine, Stavudine, Indinavir	NRTIs + PI

Table 3. The estimate of parameters.

Days after infection	$\{\xi_P(t)\}$ ± Std error	$\{\xi_R(t)\}$ ± Std error	$\{k_T(t)\}$ ± Std error	$\{c(t)\}$ ± Std error
[0, 1232)	0.0 ± 0.0	0.0 ±0.1094	1.95E−04 ±3.2569E−05	0.15 ±6.1112E−02
[1232, 2848)	0.365 ±5.1671E−03	0.9325 ±3.7591E−02	1.351E−03 ±2.2350E−04	0.11 ±6.1112E−02
[2848, 2857)	0.122 ±1.4781E−03	0.945 ±6.2663E−03	3.589E−03 ±1.6123E−04	0.3 ±4.6786E−02
[2857, 2860)	0.1 ±1.4416E−03	0.923 ±1.0110E−02	2.4998E−03 ±1.2899E−04	0.075 ±3.7143E−02
[2860, 2930)	0.1 ±1.8243E−03	0.965 ±5.3575E−03	2.498E−03 ±1.3715E−04	0.265 ±1.0143E−02
[2930, 2943)	0.1 ±2.1572E−03	0.975 ±4.8983E−03	2.883E−03 ±1.8031E−04	0.35 ±1.0754E−02
[2943, 2986)	0.1 ±2.2708E−03	0.917 ±1.6827E−03	1.15E−03 ±1.1310E−04	0.181 ±1.6308E−02
[2986, 2998)	0.1 ±1.5594E−03	0.8923 ±9.6893E−03	1.6268E−03 ±8.9671E−05	0.2097 ±5.2050E−02
[2998, 3069)	0.9475 ±1.0894E−03	0.9475 ±8.7712E−03	2.84985E−02 ±1.2414E−03	0.06 ±1.4300E−02
[3069, 3094)	0.9535 ±1.6391E−03	0.985 ±9.3496E−03	9.95E−02 ±8.0864E−03	0.08 ±4.0889E−02
[3094, 3140)	0.9648 ±2.2692E−03	0.9885 ±1.5167E−02	1.45E−01 ±1.8536E−02	0.145 ±6.2960E−02
[3140, 3143)	0.0 ±0.0	0.0 ±3.5893E−03	9.85E−04 ±5.8514E−05	0.5 ±1.0000E−06
[3143, 3184)	0.11 ±1.0941E−03	0.935 ±4.2779E−03	1.9348E−03 ±7.8796E−05	0.1 ±2.5248E−02
[3184, 3199)	0.11 ±1.0941E−03	0.9852 ±4.2779E−03	4.545E−03 ±2.0756E−04	0.35 ±9.0254E−03
[3199, 3246)	0.945 ±1.6396E−03	0.935 ±2.5882E−02	1.15E−02 ±1.2430E−03	0.2262 ±5.3771E−02
[3246, 3317)	0.945 ±3.0442E−03	0.955 ±0.1139	1.54E−02 ±3.8191E−03	0.315 ±0.1313
[3317, 3385)	0.885 ±1.3511E−03	0.915 ±2.2229E−02	1.5055E−02 ±1.5020E−03	0.1485 ±5.7825E−02
[3385, 3411)	0.925 ±1.7284E−03	0.935 ±5.5581E−02	9.835E−03 ±9.8474E−04	0.35 ±2.5822E−02
[3411, 3500)	0.955 ±2.6123E−03	0.985 ±2.7231E−02	6.295E−02 ±1.2816E−02	0.25 ±0.1202
[3500, 3517)	0.945 ±2.3064E−03	0.975 ±1.8764E−02	7.385E−02 ±8.4680E−03	0.135 ±6.6896E−02
[3517, 3579)	0.965 ±2.8792E−03	0.9885 ±2.2698E−02	5.805E−02 ±1.3318E−02	0.27 ±0.2085
[3579, 3670)	0.9255 ±2.6055E−03	0.9625 ±3.4497E−02	3.835E−02 ±4.7153E−03	0.215 ±2.2933E−02
[3670, 3680]	0.965 ±1.8456E−03	0.985 ±3.6004E−02	8.91E−01 ±9.4270E−03	0.35 ±6.3001E−02

Based on the above conditions and the procedures given in Section 4, we have developed a Fortran program to compute the estimates of the unknown parameters and the state variables. We have applied this program to the data given in Table 1 to estimate the parameters $\{\mu_i, i = 1, 2, \mu_{Vj}, j = 0, 1, N, c(i), \xi_R(i), \xi_P(i), k_T(i), i = 1, \ldots, k\}$ and the numbers of non-infectious HIV and infectious HIV per mL of blood. The estimates of the parameters are $\mu_1 = 0.005 \pm 1.9000E-02, \mu_2 = 0.5 \pm 1.0846e-02, \mu_{V_0} = 0.29 \pm 1.5081e{-}03, \mu_{V_1} = 0.3 \pm 2.1132e{-}03, \alpha = 0.05 \pm 6.9256e{-}02, \sigma^2 = 4.55 \pm 0.6526, N = 2000 \pm 0.3987$. The estimates of the time dependent parameters $\{\xi_R(i), \xi_P(i), k_T(i), c(i), i = 1, \ldots, k\}$ are given in Table 3. The estimates of the numbers of infectious HIV and non-infectious HIV per mL of blood over time are plotted in Fig. 1 (a)–(b) respectively. Given in the third column in Table 1 are the predicted total numbers of HIV by using the model and the Jackknife and cross-validation procedure given in Efron (1982). Plotted in Fig. 2 are the estimates of the numbers

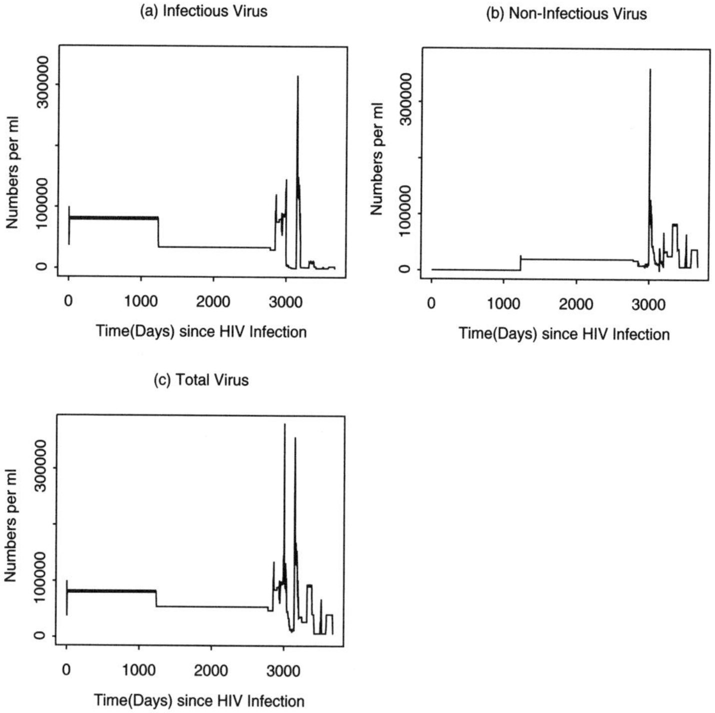

Fig. 1. Plots showing the numbers of infectious HIV, non-infectious HIV, and total number of HIV.

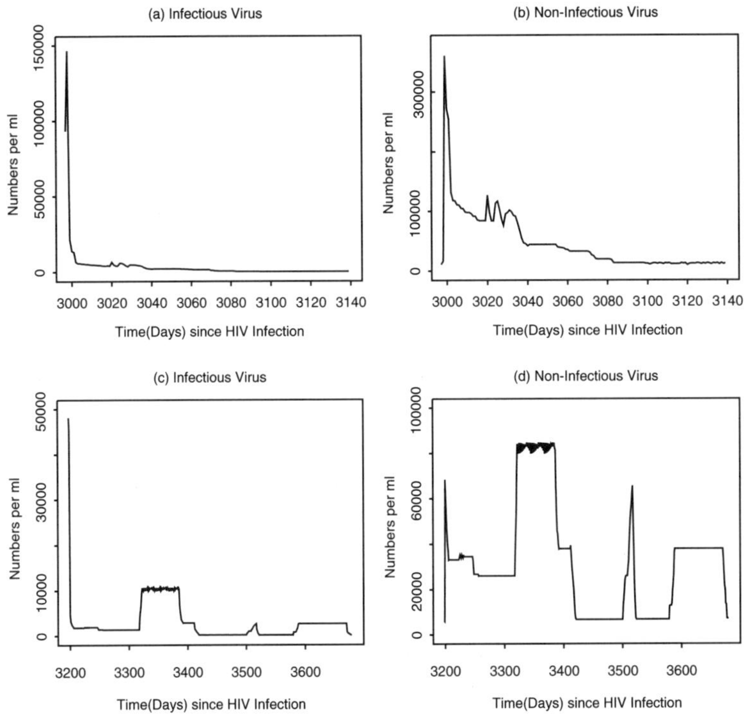

Fig. 2. Plots showing the numbers of infectious HIV, non-infectious HIV, under HAART during the time periods [2998, 3140) and [3199, 3680].

of infectious HIV and non-infectious HIV per mL of blood under HAART. From these estimates and the plots, we observed the following results:

$$\chi_1^2 = \sum_{j=1}^{25} \hat{\eta}_j^2 = 17.33, \quad \text{and} \quad \chi_2^2 = \sum_{j=1}^{25} \hat{\epsilon}_j^2 = 17.47.$$

1) The predicted total numbers of HIV viruses by using the model as given in Table 1 are very close to the observed numbers. This indicates that the fitting of the data by the model is extremely good; for more details, see the discussion in Section 6.
2) It appears that the estimates of $\{\mu_i, i = 1, 2, \mu_{Vj}, j = 0, 1\}$ are consistent with the estimates by Ho *et al.* (1995), indicating that both the productively HIV-infected CD4 T cells and the HIV are short lived. Similarly, the estimate of N is consistent with the estimate given by Tan and Wu (1998) and the estimate of $c(t)$ is close to the observation by Phillips (1996).

3) The estimates in Table 3 and the plots in Fig. 2 indicated that the HAART were quite effective in suppressing HIV replication. For example, during the period [2998, 3140) when the HAART consisting of Lamivudine and Zidovudine (NRTIs) and Nelfinavir (PI) were used, in about two weeks the number of infectious virus load were dropped from over 140,000/mL to below 2000/mL and stayed below 500/mL until the treatment were stopped at 3140 days since HIV infection. When the drugs were stopped at 3140 days since HIV infection, the infectious HIV virus load rebounded to over 115,000/mL in a few days.

4) The results in Table 3 and in Figure 2 indicated that in suppressing HIV duplication, there were no significant differences between the HAART consisting of (Lamivudine, Stavudine, Indinavir) and the HAART consisting of (Lamivuding, Zidovudine and Nelfinavir). Notice that during the period [3199, 3680) when the combination of (Lamivudine, Stavudine, Indinavir) were used, except during the period [3317, 3385) when the infectious virus loads fluctuated around 10,000/mL, in most time periods the infectious virus loads were below 2000/mL and in some time periods reaching below 300/mL.
5) From the estimates in Table 3 and in Fig. 1, we observed that the infectious HIV virus load under the regimen (Zidovudine, Didanosine) fluctuated between 50,000/mL and 60,000/mL whereas the infectious HIV virus load under the monotherapy regimen using abacavir or stavudine fluctuated between 70,000/mL and 90,000/mL. These results implied that the two drugs combinations using (Zidovudine, Didanosine) might be slightly more effective in suppressing HIV replication than the monotherapy using stavudine or abacavir. Comparing with the HAART, however, one might say that these regimens were not effective in suppressing HIV replication.
6) The estimates of the number of infectious HIV showed that under the combination regimen using (Zidovudine, Didanosine), the infectious HIV virus load fluctuated around 60,000/mL for about a year; then the infectious virus loads rebounded to the level before treatment, reaching more than 100,000/mL. This indicated that in about a year the HIV had developed drug resistance to both drugs.

6. Conclusions and Discussion

In this chapter, we have developed a state space model for the HIV pathogenesis under treatment by anti-viral drugs based on longitudinal data of RNA HIV virus copies over time. This is an individual-based model applicable to cases when the observed RNA HIV virus copies are available over time. Because different individuals have different genetic background, it is expected that different drugs and different treatment regimens are usually applied to different patients. This makes the individual-based model extremely useful and appropriate. Notice that by using mixed model to account for variations between individuals as in Wu *et al.* (1998), the model and methods in this chapter can easily be extended to cases where data on many patients are available.

To monitor the HIV progression in HIV-infected individuals treated by various drugs, it is important to estimate the number of infectious HIV and non-infectious HIV. In this chapter, by using the state space model and the multi-level Gibbs sampling method, we have developed procedures to estimate both the unknown parameters and the numbers of infectious HIV viruses and non-infectious HIV viruses per mL of blood. We have applied these procedures to the data of a patient treated with various antiviral drugs including HAART at St. Jude Children's Research Hospital in Memphis, TN. For this patient, the HAART regimens appeared to have suppressed HIV replication effectively although the numbers were over 10,000/mL at some time points. Besides these results, we also make the following observations:

1) In this chapter we have provided some examples indicating that in monitoring the disease status, using the total number of HIV to measure the success or failure of the drugs is very misleading and may be erroneous. For example, at 3054, 3082 and 3110 days since HIV infection, the total number of observed HIV copies per mL of blood are given by 46,000, 19,000, and 13,000 per mL of blood respectively. However, at these time points the estimated number of infectious HIV are 2142, 538, and 235 copies per mL of blood respectively, indicating the effectiveness of the treatment. These results suggest that for clinical applications and for providing guidelines for the medical doctors to follow, it is important to estimate both the numbers of infectious HIV and non-infectious HIV.
2) It has been reported in the literature that when the drugs are stopped, in 3–14 days the number of HIV rebounds, reaching the level before treatment; see Chun *et al.* (1999). In this chapter we have obtained similar results. For example, when the HAART regimen (Lamivudine, Zidovudine, Nelfinavir) were stopped at day 3140 since HIV infection, in a few days the infctious HIV virus load rebounded to more than 115,000/mL. When the drug combination regimen using (Lamivudine, Zidovudine) were used, this number were dropped to around 46,000/mL.

3) The estimates of infectious HIV virus-loads suggested that the HAART using (Lamivudine, Zidovudine, Nelfinavir) or using (Lamivudine, Stavudine, Indinavir) were effective in controlling HIV replication. Under HAART, the number of infectious HIV were dropped from 120,000/mL to below 1000/mL in about three weeks, reaching below 300/mL at some time points. On the other hand, mono-therapy using Abacavir or Stavudine and regimens using two NRTI drugs (Zidovudine-Didanosine, or abacavir-Stavudine or Lamivudine-Zidovudine) were not very effective.

To validate the model, we have used the Jackknife and cross-validation procedure as given in Efron (1982) to derive the predicted value of each observation. This procedure deletes one observation each time and use the remaining data (i.e. the data with the observation of question being deleted) and the model to derive the predicted value of the observed numbers of RNA HIV copies. The results are given in the third column of Table 1. The residuals based on the log transformation and the Pearson residuals $\left(\hat{\epsilon}_j = \frac{y_j - \hat{V}(t_j)}{\hat{\sigma}\sqrt{\hat{V}(t_j)}}\right)$ are given in column 4 in Table 1. The normal Q-Q plots of these residuals are given in Fig. 3. These results indicates that these residuals are approximately normally distributed with mean 0 and variance σ^2, thus justifying the normal assumption of the log transformation and the transformation $e_j = \{y_j - V(t_j)\}/\sqrt{V(t_j)}$.

While the models and the methods developed in this chapter are useful to estimate the unknown parameters and state variables and to monitor the HIV dynamics under HAART, some further researches are needed to address some important issues regarding the methods. Specifically, we need to do more research to answer the following questions:

1) To estimate the unknown parameters, we have proposed a method by combining the multi-level Gibbs sampling approach with the least square approach. The convergence of the multi-level Gibbs sampling procedure has been proved in Tan (2002, Chapter 3). In the application, the combination of the multi-level Gibbs sampling approach with the least square approach appears to converge. We conjecture that convergence will not be a problem here; but theoretical studies to prove the convergence of the procedure and to examine the efficiency of the method are definitely needed. This problem will be addressed in our future research.
2) In this chapter, we have partitioned the time period into non-overlapping sub-periods and assume that the parameters $\{c(t), \xi_R(t),\ \xi_P(t), k_T(t)\}$

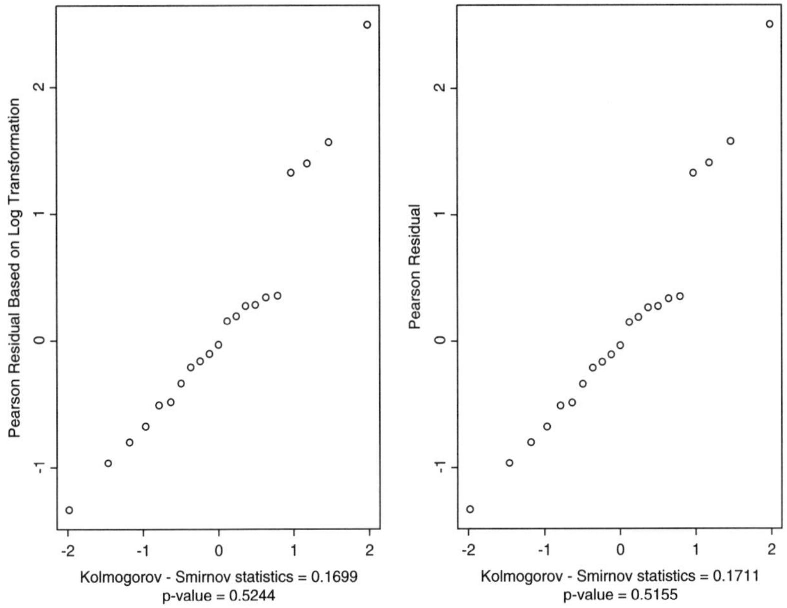

Fig. 3. Q-Q plots of the residuals and the p-values.

are constants in each sub-period. This partition is dictated by the treatment regimens and by the assumption that the number of non-infected $CD4^{(+)}$ T cells is a constant in each sub-period. In order for the later assumption to prevail, the sub-periods cannot be too long and more sub-periods are needed in intervals where the number of HIV changes wildly. In our future research, we will examine the impacts of this partitioning and look for optimal ways for selecting a partition.

In this chapter, we have proposed a state space model based on RNA HIV copies over time. In many occasions, data on $CD4^{(+)}$ T cell counts and $CD8^{(+)}$ T cell counts are also available over time. In these cases, it would be of considerable interest to use these data sets to derive state space models; this will be our further research topics. We will not go any further here.

References

1. Bartlett JG and Moore RD (1998) Improving HIV therapy. *Sci Amer* **279**: 84–93.
2. Besag J, Green P, Higdon D and Mengersen K (1995) Bayesian computation and stochastic systems (with discussion). *Statist Sci* **10**: 3–66.

3. Box GEP and Tiao GC (1973) *Bayesian Inference in Statistical Analysis.* Addison-Wesley, Reading MA.
4. Chun T-W, Davey RT, Engel D *et al.* (1999) Re-emergence of HIV after stopping therapy. *Nature* **401**: 874–875.
5. Cohen O, Weissman D and Fauci AS (1998) The immunopathogenesis of HIV infection. In: Paul WE (ed.) *Fundamental Immunology.* Chap-44, pp. 1511–1534 Fourth ed., Lippincott-Raven Publishers, Philadelphia.
6. Cowles MK and Carlin BP (1996) Markov chain Monte Carlo convergence diagnostics: a comparative review. *J Amer Statist Assoc* **91**: 883–904.
7. Efron B (1982) *The Jacknife, the Bootstrap and Other Resampling Plans.* SIAM, Philadelphia, PA.
8. Gelman A (1996) Inference and monitoring convergence. In: Gilks WR, Richardson S and Spiegelhalter DJ (eds.) *Markov Chain Monte Carlo in Practice, pp. 131–143,* Chapman and Hall, London.
9. Ho DD, Neumann AU, Perelson AS *et al.* (1995) Rapid turnover of plasma virus and CD4 lymphocytes in HIV-1 infection. *Nature* **373**: 123–126.
10. Kass RE, Carlin PR, Gelman A and Neal RM (1998) Markov chain Monte Carlo in practice: a roundatble discussion. *Amer Statist* **52**: 93–100.
11. Mitter JE, Sulzer B, Neumann AU and Perelson AS (1998) Influence of delayed viral production on viral dynamics in HIV-1 infected patients. *Math Biosci* **152**: 143–163.
12. Perelson AS, Neumann AU, Markowitz M *et al.* (1996) HIV-1 dynamics *in vivo*: Virion clearance rate, infected cell life-span, and viral generation time. *Science* **271**: 1582–1586.
13. Phillips AN (1996) Reduction of HIV concentration during acute infection: independence from a specific immune response. *Science* **271**: 497–499.
14. Roberts GO (1995) Convergence control methods for Markov chain Monte Carlo algorithms. *Statist Sci* **10**: 231–253.
15. Shephard N (1994) Partial non-Gaussian state space. *Biometrika* **81**: 115–131.
16. Smith AFM and Gelfand AE (1992) Bayesian statistics without tears: a sampling-resampling perspective. *Amer Statist* **46**: 84–88.
17. Tan WY (2000) *Stochastic Modeling of AIDS Epidemiology and HIV Pathogenesis.* World Scientific Publication Company, Singapore.
18. Tan WY (2002) *Stochastic Models with Applications to Genetics, Cancers, AIDS and other Biomedical Systems.* World Scientific Publication Company, Singapore.
19. Tan WY and Wu H (1998) Stochastic modeling of the dynamic of $CD4^{(+)}$ T cell infection by HIV and some Monte Carlo studies. *Math Biosci* **147**: 173–205.
20. Tan WY and Ye ZZ (2000) Estimation of HIV infection and HIV incubation via state space models. *Math Biosci* **167**: 31–50.
21. Tan WY and Ye ZZ (2000) Simultaneous estimation of HIV infection, HIV incubation, immigration rates and death rates as well as the numbers of susceptible people, infected people and AIDS cases. *Comm Statist (Theory & Methods)* **29**: 1059–1088.

22. Tan WY and Xiang Z (1999) The state space model of the HIV epidemic with variable infection. *J Statist Inf Plan* **78**: 71–87.
23. Wu H, Ding A and DeGruttola V (1998) Estimation of HIV dynamic parameters. *Statist Med* **17**: 2463–2485.

Appendix

Appendix A. Generating $(\boldsymbol{X}, \boldsymbol{U})$ from $P(\boldsymbol{X}, \boldsymbol{U}|\boldsymbol{Y}, \Theta)$ by the Weighted Bootstrap Method:
To illustrate the above, notice that

$$P(\boldsymbol{X}, \boldsymbol{U}|\boldsymbol{Y}, \Theta) \propto P(\boldsymbol{X}, \boldsymbol{U}|\Theta)P(Y|\boldsymbol{X}, \Theta)$$
$$\text{where } P(\boldsymbol{Y}|\boldsymbol{X}, \boldsymbol{U}, \Theta) = P(Y|\boldsymbol{X}, \Theta).$$

The weighted bootstrap method proposed by Smith and Gelfand (1992) to generate $(\boldsymbol{X}, \boldsymbol{U})$ from $P(\boldsymbol{X}, \boldsymbol{U}|\boldsymbol{Y}, \Theta)$ is then given by the following loops:

(a) Given Θ, generate a large sample $\{(\boldsymbol{X}_i, \boldsymbol{U}_i), i = 1, \ldots, N\}$ from $P(\boldsymbol{X}, \boldsymbol{U}|\Theta)$, where N is very large. (We take $N = 1\,000$.)

(b) Compute $(q_j, j = 1, \ldots, N\}$, where

$$q_j = P(\boldsymbol{Y}|\boldsymbol{X}_j, \Theta) \Big/ \left\{\sum_{i=1}^{N} P(\boldsymbol{X}_i, \boldsymbol{U}_i|\Theta)P(\boldsymbol{Y}|\boldsymbol{X}_i, \Theta)\right\}.$$

(c) Generate V from an uniform distribution $U(0,1)$. Let j^* be the integer satisfying $\sum_{j=1}^{j^*} q_j \le V < \sum_{j=1}^{j^*+1} q_j$. Then, when N is very large, $(\boldsymbol{X}_{j^*}, \boldsymbol{U}_{j^*})$ is a sample of size 1 from $P(\boldsymbol{X}, \boldsymbol{U}|\boldsymbol{Y}, \Theta)$ approximately.

Repeat the above loop n times, one may generate a sample of size n, $\{(\boldsymbol{X}_i, \boldsymbol{U}_i), i = 1, \ldots, n\}$ from $P(\boldsymbol{X}, \boldsymbol{U}|\boldsymbol{Y}, \Theta)$. The proof of this algorithm is given in Chapter 3 of Tan (2002). To implement the above procedure, one may use the distribution $P(\boldsymbol{X}, \boldsymbol{U}|\Theta)$ given in Eq. (21); or equivalently, one may use the stochastic Eqs. (1)–(4) to generate the sample $\{(\boldsymbol{X}_i, \boldsymbol{U}_i), i = 1, \ldots, N\}$.

Appendix B. Proof of the Multi-level Gibbs Sampling Procedures in Section 4:

To prove the convergence of the multi-level Gibbs sampling procedures in Section 4, put $\boldsymbol{Z} = (\boldsymbol{X}, \boldsymbol{U})$ and note that the joint density of $\{\boldsymbol{Z}, \Theta\}$ given

$\boldsymbol{Y}$ is:

$$P(\boldsymbol{Z}, \Theta|\boldsymbol{Y}) = P(\boldsymbol{Z}|\boldsymbol{Y})P(\Theta|\boldsymbol{Z}, \boldsymbol{Y}) = P(\Theta|\boldsymbol{Y})P(\boldsymbol{Z}|\Theta, \boldsymbol{Y}),$$

where

$$P(\boldsymbol{Z}|\boldsymbol{Y}) = \sum_{\Theta} P(\boldsymbol{Z}, \Theta|\boldsymbol{Y})$$
$$P(\Theta|\boldsymbol{Y}) = \sum_{Z} P(\boldsymbol{Z}, \Theta|\boldsymbol{Y}).$$

Denote the sample space of Θ and $\boldsymbol{Z}$ by $\Omega_{\Theta} = \{\Theta_i, i = 1, \ldots, M_1\}$ and $S_{\boldsymbol{Z}} = \{\boldsymbol{Z}_i, i = 1, \ldots, M_2\}$ respectively. Consider the Markov chain $\{\Theta(t), t = 1, \ldots,\}$ with state space Ω_{Θ} and with one-step transition probabilities given by

$$\begin{aligned} P(\Theta_i, \Theta_j; 1) &= P(\Theta(1) = \Theta_j|\Theta(0) = \Theta_i) \\ &= \sum_{Z} P(\Theta_j|\boldsymbol{Z}, \boldsymbol{Y})P(\boldsymbol{Z}|\Theta_i, \boldsymbol{Y}). \end{aligned} \tag{26}$$

Since $P(\Theta_i, \Theta_j; 1) > 0$ for all $\{\Theta_i \in \Omega_{\Theta}, \Theta_j \in \Omega_{\Theta}\}$ and since Ω_{Θ} is finite, the above Markov chain is aperiodic and irreducible; furthermore, the stationary distribution exists and is unique. We next show that the marginal distribution $P(\Theta|\boldsymbol{Y})$ is the stationary distribution of this chain. This follows readily from the result:

$$\begin{aligned} \sum_{\Theta_i} P(\Theta_i|\boldsymbol{Y})P(\Theta_i, \Theta_j; 1) &= \sum_{\Theta_i} P(\Theta_i|\boldsymbol{Y}) \sum_{Z} P(\Theta_j|\boldsymbol{Z}, \boldsymbol{Y})P(\boldsymbol{Z}|\Theta_i, \boldsymbol{Y}) \\ &= \sum_{Z} P(\Theta_j|\boldsymbol{Z}, \boldsymbol{Y}) \sum_{\Theta_i} P(\Theta_i|\boldsymbol{Y})P(\boldsymbol{Z}|\Theta_i, \boldsymbol{Y}) \\ &= \sum_{Z} P(\Theta_j|\boldsymbol{Z}, \boldsymbol{Y}) \sum_{\Theta_i} P(\Theta_i, \boldsymbol{Z}|\boldsymbol{Y}) \\ &= \sum_{Z} P(\Theta_j|\boldsymbol{Z}, \boldsymbol{Y})P(\boldsymbol{Z}|\boldsymbol{Y}) \\ &= \sum_{Z} P(\Theta_j, \boldsymbol{Z}|\boldsymbol{Y}) = P(\Theta_j|\boldsymbol{Y}). \end{aligned}$$

Similarly, we consider the Markov chain $\{\boldsymbol{Z}(t), t = 1, \ldots,\}$ with state space $S_{\boldsymbol{Z}}$ and with one-step transition probabilities given by

$$\begin{aligned} P(\boldsymbol{Z}_i, \boldsymbol{Z}_j \colon 1) &= P(\boldsymbol{Z}(1) = \boldsymbol{Z}_j|\boldsymbol{Z}(0) = \boldsymbol{Z}_i) \\ &= \sum_{\Theta} P(\boldsymbol{Z}_j|\Theta, \boldsymbol{Y})P(\Theta|\boldsymbol{Z}_i, \boldsymbol{Y}). \end{aligned} \tag{27}$$

Then, one can similarly show that the marginal distribution $P(\boldsymbol{Z}|\boldsymbol{Y})$ is the stationary distribution of this Markov chain. This proves the convergence of the multi-level Gibbs sampling procedures given in Section 4.

The above idea can readily be extended to bounded continuous random variables through embedded Markov chains; for details, see Chapter 4 in Tan (2002). Notice that for unbounded random variables, there is a possibility that the chain may not converge to a density function. This has been called the improper posterior in the literature. Thus, for unbounded random variables, it is important to monitor the convergence. Methods to monitor the convergence have been described in (Basag *et al.* 1995; Cowles and Carlin 1996; Gelman 1996; Kass *et al.* 1998; Roberts 1995).

CHAPTER 15

BAYESIAN ESTIMATION OF INDIVIDUAL PARAMETERS IN AN HIV DYNAMIC MODEL USING LONG-TERM VIRAL LOAD DATA

Yangxin Huang* and Hulin Wu*

There have been substantial interests in investigating HIV dynamics for understanding the pathogenesis of HIV-1 infection and antiviral treatment strategies. However, it is difficult to establish a relationship between pharmacokinetics (PK) and antiviral response due to too many confounding factors related to antiviral response during the treatment process. In this article, a mechanism-based dynamic model for HIV infection with intervention by antiretroviral therapies is proposed. In this model, we directly incorporate drug concentration, adherence and drug susceptibility into a function of treatment efficacy defined as an inhibition rate of virus replication. In order to focus our attention on estimating dynamic parameters for all subjects, we investigate a Bayesian approach under a framework of the hierarchical Bayesian (mixed-effects) model. The proposed methods and models not only can help to alleviate the difficulty in identifiability, but also can flexibly deal with sparse and unbalanced longitudinal data. The viral dynamic parameters estimated from the proposed method are, thus, more accurate since the variations in PK, adherence and drug resistance have been considered in the model.

Keywords: Hierarchical Bayesian models; HIV dynamics; MCMC; PK/PD modeling; parameter estimation; treatment efficacy.

1. Introduction

A virological marker, the number of human immunodeficiency virus type 1 (HIV-1) RNA copies in plasma (viral load), is currently used to evaluate anti-HIV therapies in AIDS clinical trials. Antiretroviral treatment of HIV-1-infected patients with highly active antiretroviral therapies (HAART),

*Department of Biostatistics and Computational Biology, School of Medicine and Dentistry, University of Rochester, 601 Elmwood Ave., Box 630, Rochester, New York 14642, USA. hwu@bst.rechester.edu

consisting of reverse transcriptase inhibitor (RTI) drugs and protease inhibitor (PI) drugs, results in several orders of magnitude reduction of viral load. The decay in viral load happens in a relatively short term, and it can be either sustained up to one year or may be followed by resurgence of virus within months. The resurgence of virus may be caused by drug resistance, noncompliance, pharmacokinetics problems and other factors during the treatment process.

Many HIV dynamic models have been proposed by AIDS researchers in the last decade to provide theoretical principles in guiding the development of treatment strategies for HIV-infected patients (Ding and Wu, 2000; Ho *et al.*, 1995; Huang *et al.*, 2003; Nowak *et al.*, 1995; Nowak *et al.*, 1997; Perelson *et al.*, 1996; Perelson *et al.*, 1997; Perelson and Nelson, 1999; Wu and Ding, 1999; Wei *et al.*, 1995). Most of the previous models are used to quantify short-term viral dynamics, but these models are rarely applied to model virological response for long-term treatment. In other words, these models were only used to fit the early segment of the viral load trajectory data and were limited to interpret long-term HIV dynamic data resulting from AIDS clinical trials since the viral load trajectory may change to different shapes in later stage due to drug resistance, non-compliance and other clinical factors. In this article, we put forward a set of relatively simple models to characterize long-term viral dynamics, which incorporate the factors associated with resurgence of viral load such as the intrinsic nonlinear HIV dynamics, drug exposure, compliance to treatment, and drug susceptibility. Thus, the models are flexible enough to quantify the long-term HIV dynamics.

Although a number of authors have investigated Bayesian methods to estimate population parameters in pharmacokinetic/pharmacodynamic (PK/PD) models (Gelfand *et al.*, 1990; Gelman *et al.*, 1996; Lunn *et al.*, 2002; Wakefield *et al.*, 1994; Wakefield 1996) or in viral dynamic models with fitting short-term viral load data (Han *et al.*, 2002; Putter *et al.*, 2002), there are relatively few papers to investigate the estimation of both population and individual dynamic parameters for long-term viral dynamics. In this article, a novel application of hierarchical Bayesian nonlinear (mixed-effects) models to long-term HIV dynamics described by a system of differential equations with time-varying coefficients will be presented. In this model, the imperfect treatment effect is taken into account by incorporating drug exposures and drug susceptibility, and complete long-term viral load data from a clinical study are used to fit our model. Thus, our estimates of viral dynamic parameters for individual subject are more accurate. We combine Bayesian approach and mixed-effects modeling approach to estimate dynamic parameters. Bayesian approach allows us to borrow appropriate prior information from the previous studies and the mixed-effects modeling method provides us a framework where both between-subject and within-subject variations are considered. The remaining sections are organized as follows. Section 2 introduces a simplified viral dynamic model with time-varying drug efficacy which incorporates the effect of pharmacokinetic (PK) variation, drug resistance and adherence. A hierarchical Bayesian modeling approach is proposed in estimating viral dynamic parameters in Section 3. Section 4 presents the results for individual dynamic parameter estimates by applying our method to an AIDS clinical trial study, and finally this article is concluded with some discussions in Section 5.

2. Mathematical Models for Long-Term HIV Dynamics

2.1. *Antiviral drug efficacy model*

Within the population of HIV virions in the human host, there is likely to be genetic diversity and corresponding diversity in susceptibility to the various antiretroviral (ARV) drugs. In recent treatment strategies, genotype or phenotype tests are conducted to determine the susceptibility of antiretroviral agents before a treatment regimen is selected. Here we use the phenotype marker IC_{50} (Molla *et al.*, 1996; Wainberg *et al.*, 1996), which represents the drug concentration necessary to inhibit viral replication by 50%, to quantify agent-specific drug susceptibility. Herein, we refer to this quantity as the median inhibitory concentration. To model the within-host changes over time in IC_{50} due to the emergence of new drug resistant mutations, we proposed the following function (Huang *et al.*, 2003):

$$IC_{50}(t) = \begin{cases} I_0 + \dfrac{I_r - I_0}{t_r}t & \text{for } 0 < t < t_r, \\ I_r & \text{for } t \geq t_r, \end{cases} \tag{1}$$

where I_0 and I_r are respective values of $IC_{50}(t)$ at baseline and time point t_r at which the resistant mutations dominate. Note that, here, we use a time-varying parameter $IC_{50}(t)$ to model the susceptibility (resistance) of a virus population with a mixture of quasispecies to the drugs in antiviral regimens. The make-up of these quasispecies may vary during treatment since drug resistant viral species may emerge if a sub-optimal dose of antiviral drugs is given. Thus, separate equations for quasispecies are not necessary under

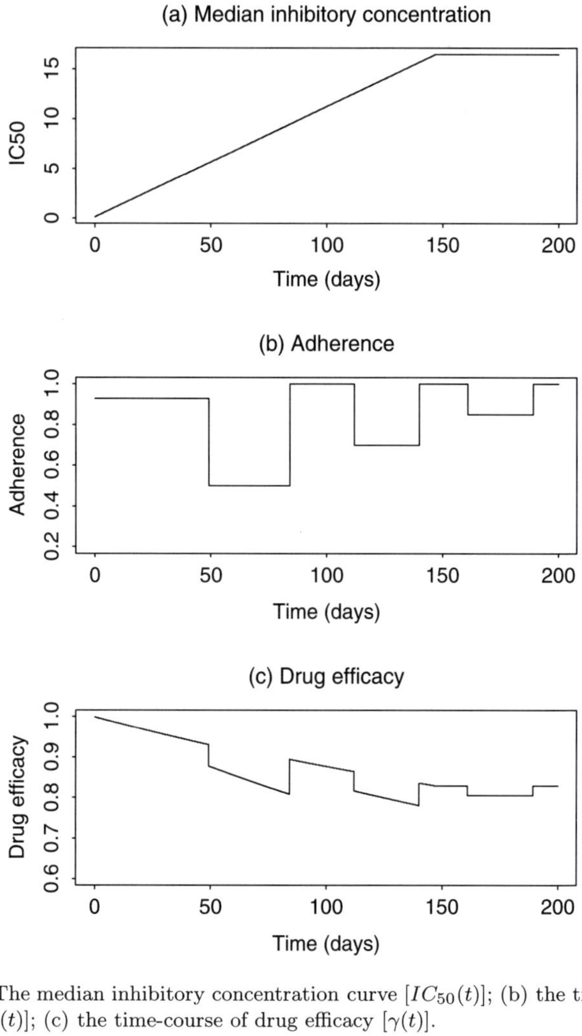

Fig. 1. (a) The median inhibitory concentration curve $[IC_{50}(t)]$; (b) the time-course of adherence $[A(t)]$; (c) the time-course of drug efficacy $[\gamma(t)]$.

our model setting. An example of such a function for a PI drug is plotted in Fig. 1(a). In clinical studies (e.g. A5055), it is common to measure IC_{50} values only at baseline and failure time (Molla *et al.*, 1996). Thus, this function may serve well as an approximation.

Poor adherence to a treatment regimen is one of the major causes for treatment failure (Besch, 1995; Ickovics and Meisler, 1997). Patients may occasionally miss doses, may misunderstand prescription instructions or may miss multiple consecutive doses for various reasons; these deviations from prescribed dosing affect drug exposure in predictable ways. We used the following model to represent adherence (Huang and Wu, 2004b),

$$A(t) = \begin{cases} 1 & \text{for } T_k < t \leq T_{k+1}, \quad \text{if all doses are taken in } [T_k, T_{k+1}], \\ R & \text{for } T_k < t \leq T_{k+1}, \quad \text{if 100 } R\% \text{ doses are taken in } [T_k, T_{k+1}], \end{cases} \tag{2}$$

where $0 \leq R < 1$ with R indicating adherence rate for drug (in our clinical study, we focus on the two PI drugs of the prescribed regimen: RTV and IDV). T_k denotes the adherence evaluation time at the kth clinical visit. As an example, Fig. 1(b) shows the effect of adherence over time for a PI drug.

In recent years, ARV drugs have been developed rapidly. Three types of ARV agents, nucleoside/non-nucleoside reverse transcriptase inhibitors (RTI) and protease inhibitors (PI) have been widely used in developed countries. The HAART has been proven to be extremely effective at reducing the amount of virus in the blood and tissues of infected patients. In viral dynamic studies by Ding and Wu (1999; 2000; 2001), Perelson and Nelson (1999), Wu and Ding (1999), and Wu *et al.* (1998), investigators assumed that the drug efficacy was constant over treatment time. However, drug efficacy may vary as the concentrations of ARV drugs and other factors (e.g. drug resistance) vary during treatment (Ding and Wu, 1999; Perelson and Nelson, 1999). Also, in practice, patient's viral load may rebound during treatment and the rebound may be associated with resistance to ARV therapy and other factors (Fitzgerald *et al.*, 2002). We incorporate drug resistance and drug exposures (pharmacokinetics and adherence) directly into the antiviral drug efficacy models based on principles of mass action, instead of modeling the diversity of virus mutations (Nowak and May, 2000) for which no clinical data are available. To model the relationship of drug exposure and resistance with antiviral drug efficacy, we used a modified $E_{\max}$ model (Sheiner, 1985) for the drug efficacy of single antiretroviral agent (Huang and Wu, 2004a),

$$\gamma(t) = \frac{C_{12h}A(t)}{\phi IC_{50}(t) + C_{12h}A(t)} = \frac{IQ(t)A(t)}{\phi + IQ(t)A(t)}, \tag{3}$$

where $IQ(t) = C_{12h}/IC_{50}(t)$ denotes the inhibitory quotient (IQ) (Hsu *et al.*, 2000; Kempf *et al.*, 2001); the IQ is the PK adjusted phenotypic susceptibility and has recently been shown to be related to antiviral responses (Hsu *et al.*, 2000; Kempf *et al.*, 2001). $\gamma(t)$, ranged from 0 to 1, indicates the antiviral drug efficacy (the inhibition rate of viral replication) in a viral dynamic (response) model (to be discussed below). C_{12h} is the drug concentration in plasma measured at 12 hours from doses taken. Note that C_{12h} could be replaced by other PK parameters such as area under the plasma concentration-time curve (AUC) or trough level of drug concentration ($C_{\min}$). Although $IC_{50}(t)$ can be measured by phenotype assays *in vitro*, it may not be equivalent to the $IC_{50}(t)$ *in vivo*. Parameter ϕ indicates a conversion factor between the two. This model is similar to the one used by Wahl and Nowak (2000). If $\gamma(t) = 1$, the drug is 100% effective, whereas if $\gamma(t) = 0$, the drug has no effect. Note that, if C_{12h}, $A(t)$ and $IC_{50}(t)$ can be observed from clinical study and ϕ can be estimated from clinical data, then time-varying drug efficacy $\gamma(t)$ can be estimated during the course of ARV treatment. An example of the time-course of the drug efficacy $\gamma(t)$ based on the model (3) with $\phi = 1$ and $C_{12h} = 80$ for a PI drug is shown in Fig. 1(c).

The model (3) is to quantify antiviral drug efficacy for a single drug from the RTI or PI class. However, HAART usually contains more than one RTI/PI drugs. We need to extend the model (3). For a regimen containing two agents within a class (for example, PI drugs), the combined drug efficacy of the two agents can be modeled as

$$\gamma(t) = \frac{IQ_1(t)A_1(t) + IQ_2(t)A_2(t)}{\phi + IQ_1(t)A_1(t) + IQ_2(t)A_2(t)}, \tag{4}$$

where $IQ_i(t) = C^i_{12h}/IC^i_{50}(t)$ $(i = 1, 2)$. Similarly, we can model the combined drug efficacy of an HAART regimen with both PI and RTI agents. Lack of adherence reduces the drug exposure, which can be quantified by Eq. (2). Thus it reduces drug efficacy based on the Eq. (3) or (4) and, in turn, can affect viral response.

2.2. *HIV dynamic model*

The aim of the basic dynamic model is to describe the population dynamics of HIV and its target cells in plasma. The mathematical details of this model have been presented elsewhere (Callaway and Perelson, 2002; Huang *et al.*, 2003; Nelson and Perelson, 2002; Nowak and May, 2000; Stafford *et al.*, 2000). In fact, we will need a trade-off between the model complexity and the parameter identifiability based on the measurements available from clinical trials. If a model has too many components, it may be difficult to analyze. Many of the variables in the model may not be measurable and parameters may not be identifiable. If a model is too simple, some important clinical factors (such as pharmacokinetics, adherence and drug resistance) and other biological mechanisms of HIV infection cannot be incorporated, although viral dynamic parameters can be identified and estimated. A good model should be simple enough to incorporate available clinical data in the analysis and to simulate actual trials, and complex enough to reflect the major biological mechanisms and components in HIV infection. In order to not only consider clinical factors and biological mechanisms of HIV infection, but also flexibly deal with data analysis for HIV dynamic models, an extended antiviral response model is proposed. We consider target uninfected cells (T), infected cells (T^*) that produce virus and free virus (V). This model differs from the previous models (Ding and Wu, 2000; Ho *et al.*, 1995; Nowak *et al.*, 1995; Nowak and May, 2000; Perelson *et al.*, 1996; Perelson *et al.*, 1997; Perelson and Nelson, 1999; Wu and Ding, 1999; Wei *et al.*, 1995) in that it includes a time-varying parameter $\gamma(t)$, which quantifies antiviral drug efficacy. The model is expressed as in terms of the following set of differential equations under the effect of ARV treatment,

$$\begin{aligned}\frac{d}{dt}T &= \lambda - \rho T - [1-\gamma(t)]kTV,\\ \frac{d}{dt}T^* &= [1-\gamma(t)]kTV - \delta T^*,\\ \frac{d}{dt}V &= N\delta T^* - cV,\end{aligned} \tag{5}$$

where λ represents the rate at which new T cells are created from sources within the body (such as the thymus), ρ is the death rate per T cell, k is the rate at which T cells become infected by virus, δ is the rate of death for infected cells, N is the number of new virions produced from each of infected cells during their life-time, and c is the clearance rate of free virions. The time-varying parameter $\gamma(t)$ is the time varying antiviral drug efficacy, as defined in the Eqs. (3) or (4).

If drug is not 100% effective (not perfect inhibition), the system of ordinary differential equations can not be solved analytically. The solutions to (5) then have to be evaluated numerically. Let $\beta = (\phi, c, \delta, \lambda, \rho, N, k)^T$ denote a set of parameters, and $\log_{10}(V_{ij}(\beta, t))$ denote the common logarithm of the numerical solution of $V(t)$ for the ith individual at time t_j, which is the quantity we measured in plasma and will be used for dynamic parameter estimation.

Similar to the analysis in Huang *et al.* (2003), it can be shown from (5) that if $\gamma(t) > e_c$ (e_c is a threshold of drug efficacy) for all t, where

$$e_c = 1 - \frac{c\rho}{kN\lambda}, \tag{6}$$

the virus will be eventually eradicated in theory. However, if $\gamma(t) < e_c$ (treatment is not potent enough) or if the potency falls below e_c before virus eradication (due to drug resistance, for example), viral load may rebound (see Huang *et al.*, 2003 for a detailed discussion). Thus, the efficacy threshold e_c may reflect the immune response of a patient for controlling virus replications by the patient's immune system. It therefore is important to estimate e_c for each patient based on clinical data. Since we consider the drug efficacy by incorporating pharmacokinetics, adherence and drug sensitivity data in our HIV dynamic model, viral dynamic parameters such as $c, \delta, \lambda, \rho, N, k$ as well as e_c can be better estimated from our model.

3. Statistical Inference Methods

In order to apply the proposed the long-term HIV dynamic model for modeling viral responses and estimating individual dynamic parameters using long-term viral load data from a clinical study, we need to resolve an important statistical problem, that is, how to conduct inference and handle the identifiability of model parameters. It is a great challenge to use the differential equation model (5) to fit clinical data and to predict antiviral response, because there are too many unknown parameters that need to be determined. We usually only have viral load data available from clinical trials ($CD4^+$ T cell count data are considered too noisy to be used for dynamic parameter estimation). In this case, we may not be able to identify all the unknown parameters in the model (5). To deal with the unidentifiability problem, mathematicians simply substitute some of the unknown parameters with their estimates from previous studies or literature (see Perelson *et al.*, 1997, for example). Here, we investigate a Bayesian approach to tackle this difficulty. A Bayesian analysis allows us to combine two forms of information, "prior knowledge" from the scientific literature and actual clinical data in the context of the antiviral response model. Neither source of information is complete. If prior knowledge were sufficient, then the clinical experiments would not have had to be done; but existing data alone are typically insufficient to obtain the reasonable estimation of parameters. We wish to fit the data using scientifically plausible parameter values, so that the analysis outputs distributions of parameter values that are consistent with both the data and the prior information.

Bayesian approaches offer many advantages (see Huang and Wu, 2004b, for the details). As indicated early, although a number of papers proposed the Bayesian approach to estimate population parameters in viral dynamic models using short-term viral load data, relatively, little work has been undertaken to investigate both population and individual dynamic parameter estimation for long-term viral dynamics. In this article, we offer a hierarchical Bayesian approach to estimate both population and individual parameters in viral dynamic model (5) using long-term viral load data for all subjects, but we mainly focus our attention on estimation of individual dynamic parameters. We use a hierarchical Bayesian (mixed-effects) model to incorporate the population prior for estimating posterior distribution for each individual using the Markov Chain Monte Carlo (MCMC) techniques. A description of the methodologies is given below.

3.1. *Hierarchical Bayesian modeling approach*

A hierarchical Bayesian (mixed-effects) model can be used by incorporating a prior at population level to estimate unknown dynamic parameters for individual subjects. We denote the number of subjects by n and the number of measurements on the ith subject by m_i. In the model (5), we are able to obtain measurements of viral load from AIDS clinical trials. For notational convenience, let $\boldsymbol{\mu} = (\log(\phi), \log(c), \log(\delta), \log(\lambda), \log(\rho), \log(N), \log(k))^T$, $\boldsymbol{\theta}_i = (\log(\phi_i), \log(c_i), \log(\delta_i), \log(\lambda_i), \log(\rho_i), \log(N_i), \log(k_i))^T$, $\boldsymbol{\Theta} = \{\boldsymbol{\theta}_i, i = 1, \ldots, n\}$, $\boldsymbol{\Theta}_{\{i\}} = \{\boldsymbol{\theta}_l, l \neq i\}$ and $\mathbf{Y} = \{y_{ij}, i = 1, \ldots, n;\ j = 1, \ldots, m_i\}$. Let $f_{ij}(\boldsymbol{\theta}_i, t_j)$ be the numerical solution of common logarithmic viral load $\log_{10}(V(t))$ to the differential equations (5) for the ith subject at time t_j. The repeated measurements of common logarithmic viral load for each subject, $y_{ij}(t)$, at treatment times t_j, $j = 1, 2, \ldots, m_i$, can be written as

$$y_{ij}(t_j) = f_{ij}(\boldsymbol{\theta}_i, t_j) + e_i(t_j), \quad i = 1, \ldots, n;\ j = 1, \ldots, m_i, \tag{7}$$

where $e_i(t)$ is a measurement error with mean zero. Then the hierarchical Bayesian nonlinear (mixed-effects) model can be written as following three stages (Davidian and Giltinan, 1995; Huang and Wu, 2004a).

Stage 1. Within-subject variation in common logarithmic viral load measurements:

$$\mathbf{y}_i = \mathbf{f}_i(\boldsymbol{\theta}_i) + \mathbf{e}_i, \quad [\mathbf{e}_i|\sigma^2, \boldsymbol{\theta}_i] \sim \mathcal{N}(\mathbf{0}, \sigma^2\mathbf{I}_{m_i}), \tag{8}$$

where $\mathbf{y}_i = (y_{i1}(t_1), \ldots, y_{im_i}(t_{m_i}))^T$, $\mathbf{e}_i = (e_i(t_1), \ldots, e_i(t_{m_i}))^T$, $\mathbf{f}_i(\boldsymbol{\theta}_i) = (f_{i1}(\boldsymbol{\theta}_i, t_1), \ldots, f_{im_i}(\boldsymbol{\theta}_i, t_{m_i}))^T$, and the bracket notation $[A|B]$ denotes the conditional distribution of A given B.

Stage 2. Between-subject variation:

$$\boldsymbol{\theta}_i = \boldsymbol{\mu} + \mathbf{b}_i, \quad [\mathbf{b}_i|\boldsymbol{\Sigma}] \sim \mathcal{N}(\mathbf{0}, \boldsymbol{\Sigma}). \tag{9}$$

Stage 3. Hyperprior distribution:

$$\sigma^{-2} \sim Ga(a, b), \quad \boldsymbol{\mu} \sim \mathcal{N}(\boldsymbol{\eta}, \boldsymbol{\Lambda}), \quad \boldsymbol{\Sigma}^{-1} \sim Wi(\boldsymbol{\Omega}, \nu), \tag{10}$$

where the mutually independent Gamma (Ga), Normal ($\mathcal{N}$) and Wishart (Wi) prior distributions are chosen to facilitate computations (Davidian and Giltinan, 1995; Gelfand *et al.*, 1990). Note that the parametrization of Gamma and Wishart distribution is such that $Ga(a, b)$ has mean ab and $Wi(\boldsymbol{\Omega}, \nu)$ has mean matrix $\nu\boldsymbol{\Omega}$. The parameters $a, b, \boldsymbol{\eta}, \boldsymbol{\Lambda}, \boldsymbol{\Omega}$ and ν that characterize the hyperprior distributions are known.

Following the studies by Davidian and Giltinan (1995), Gelfand *et al.* (1990) and Wakefield *et al.* (1994), we have shown (Huang and Wu, 2004a) from (8)–(10) that the full conditional distributions for the parameters σ^{-2}, $\boldsymbol{\mu}$ and $\boldsymbol{\Sigma}^{-1}$ can be obtained; the full conditional distribution of each $\boldsymbol{\theta}_i$, given the remaining parameters and the data, can not be calculated explicitly, but its density function is proportional to an exponential function (see Huang and Wu, 2004a).

The MCMC methods will be used to fit this three-stage hierarchical model (see Section 3.2 for details on the MCMC implementation). From this model-fitting, we expect to obtain the posterior distributions of unknown parameters ($\boldsymbol{\theta}_i, i = 1, \ldots, n$) for each individual. Under the normality assumption, it is enough to retain the posterior distributions for individuals, and the posterior mean is the individual parameter estimates which is our focus.

Note that the hierarchical model fitting may produce good estimates of population parameters, but it may not fit all individual data very well, in particular, when the between-subject variation is large (Carlin and Louis, 1996). This is because the objective of fitting the hierarchical model is to minimize overall risk of all subjects. Thus, some subjects whose data are away from the population mean may not get a very good fit. In addition, inaccurate measurement data may contribute to the poor fitting results for some of the subjects.

3.2. *MCMC implementation*

In the proposed Bayesian approach, we only need to specify the priors at population level. The population estimates of unknown parameters are easy to obtain from the previous studies or literature and usually the population estimates are accurate and reliable. The population prior information also helps to resolve the unidentifiability problem at individual levels.

It remains to specify the values of the hyper-parameters in the prior distributions (10). In principle, if we have reliable prior information for some of parameters, strong prior (smaller variance) may be used for these parameters. For the other parameters such as ϕ, we may not have enough prior information or we may intend to use the available clinical data to determine these parameters since they are critical to quantify between-subject variations in response; a non-informative prior (larger variance) may be given for these parameters. In particular, one usually chooses non-informative prior distributions for parameters of interest (Frost, 2001).

Based on the above discussion, the prior distribution for $\boldsymbol{\mu}$ was assumed to be $\mathcal{N}(\boldsymbol{\eta}, \boldsymbol{\Lambda})$ with $\boldsymbol{\Lambda}$ being diagonal matrix. Following the idea of Han *et al.* (2002), the prior construction for the parameters is detailed in Huang and Wu (2004a).

In order to implement our methodologies, we chose the values of the hyper-parameters (population priors) as follows:

$$
\begin{aligned}
a &= 4.5, \quad b = 9.0, \quad \nu = 8.0 \\
\boldsymbol{\eta} &= (1.5, 1.1, -1.0, 4.6, -2.5, 6.9, -11.0)^T \\
\boldsymbol{\Lambda} &= diag(1000.0, 0.0025, 0.0025, 0.0025, 0.0025, 0.0025, 0.001) \\
\boldsymbol{\Omega} &= diag(1.25, 2.5, 2.5, 2.0, 10.0, 2.0, 2.0)
\end{aligned}
$$

These values of the hyper-parameters were determined based on several studies (Han *et al.*, 2002; Ho *et al.*, 1995; Nowak and May, 2000; Perelson *et al.*, 1993, 1996, 1997, 1999; Stafford *et al.*, 2000; Wei *et al.*, 1995). Note that we selected a non-informative prior for $\log(\phi)$, while the strong informative priors are given for the other six parameters in order to adjust identifiability of parameters.

In the Bayesian approach, in addition to specifying the model for the observed data and prior distributions for the unknown parameters, inference concerning the unknown parameters is then based on their posterior distributions. In the above Bayesian modeling approaches, evaluation of the required posterior distributions in a closed-form is prohibitive. However, as indicated in Section 3.1, it is relatively straightforward to derive either the full conditional distributions for some parameters ($\sigma^{-2}, \boldsymbol{\mu}$ and $\boldsymbol{\Sigma}^{-1}$) or explicit expressions which are proportional to the full conditional distributions for other parameters ($\boldsymbol{\theta}_i,\ i = 1, \ldots, n$).

Bayesian techniques allow us to fit complex models using powerful MCMC techniques. MCMC methods are an established suite of methodologies that enable samples to be drawn from the target density of interest under the Bayesian modeling framework. The methods work by defining a Markov chain whose stationary density is equal to the target density. The chain is then simulated for a time deemed adequate for convergence to have occurred, and the samples are drawn from the simulated chain. These samples are samples from the target density of interest. To implement an MCMC algorithm, here Gibbs sampling step was used to update $\sigma^{-2}, \boldsymbol{\mu}$ and $\boldsymbol{\Sigma}^{-1}$, while we updated $\boldsymbol{\theta}_i$, $i = 1, \ldots, n$, using a Metropolis–Hastings (M-H) step with a reasonable proposal density (a normal density is used below).

For a more detailed discussion of the MCMC scheme and the convergence of MCMC methods, consisting of a series of Gibbs sampling and M-H algorithms, please refer to literature (Bennett *et al.*, 1996; Chib and Greenberg, 1995; Gamerman, 1997; Gelfand *et al.*, 1990; Gelman *et al.*, 1996; Lunn *et al.*, 2002; Smith and Roberts, 1993; Wakefield *et al.*, 1994, 1996). A description of the iterative MCMC algorithm and the choice of the proposal density are detailed in Huang and Wu (2004a). After we collect the final MCMC samples, we are interested in the posterior means or quantiles of functions of the parameters, such as credible intervals and posterior medians. Therefore, the sample generated during a run of the algorithm should adequately represent the posterior distribution of interest so that MCMC estimates of such quantities are approximately correct.

Following an informal check of convergence based on graphical techniques according to the suggestion of Gelfand and Smith (1990), for the three-stage hierarchical Bayesian modeling method, after an initial number of 30,000 burn-in iterations, every 5th MCMC sample was retained from the next 120,000 samples. Thus, we obtained 24,000 samples of targeted both population and individual posterior distributions of the unknown parameters.

4. Application to an AIDS Clinical Trial Study

4.1. *Clinical trials and data description*

We illustrate the proposed methodologies using an AIDS clinical study. This study was a Phase I/II, randomized, open-label, 24-week comparative study of the pharmacokinetic, tolerability and antiretroviral effects of two regimens of indinavir (IDV), ritonavir (RTV), plus two nucleoside analogue reverse transcriptase inhibitors (NRTIs) with HIV-1-infected subjects failing PI-containing antiretroviral therapies. The 44 subjects were randomly assigned to the two treatment arm A (IDV 800 mg q12h + RTV 200 mg q12h) and Arm B (IDV 400 mg q12h + RTV 400 mg q12h). Out of the 44 subjects (36 men and 8 women with median age 37 years old), the 42 subjects are included in the analysis. For the remaining two subjects, one was excluded from the analysis since the PK parameters can not be calculated and the other was excluded since PhenoSense HIV could not be completed on this subject due to an atypical genetic sequence that causes the viral genome to be cut by an enzyme used in the assay. Plasma HIV-1 RNA (viral load) measurements were taken at days 0, 7, 14, 28, 56, 84, 112, 140 and 168 of follow-up. The nucleic acid sequence-based amplification assay (NASBA) was used to measure plasma HIV-1 RNA, with a lower limit of quantification, 50 copies/mL. The HIV-1 RNA measures below this limit are not considered reliable, therefore we simply imputed such values as 25 copies/mL. Figure 2 shows the medians and 1st/3rd quantiles of $\log_{10}$(RNA) and the corresponding *P*-values from point-wise comparisons between Arm A and Arm B using the Wilcoxon Rank Sum test.

The data from the AIDS clinical study for pharmacokinetic parameters (C_{12h}), phenotype marker (baseline and failure $IC_{50}s$) and adherence were used in our modeling. Adherence data were determined from the percent-adherence of subjects by drug according to pill-count data.

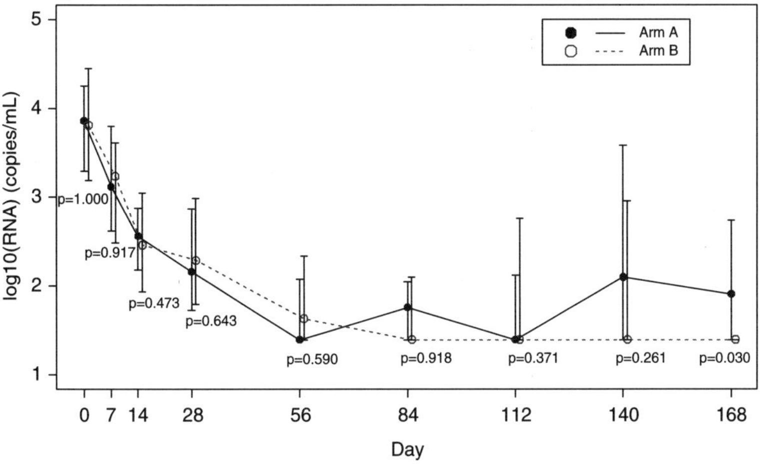

Fig. 2. Medians and 1st/3rd Quantiles of $\log_{10}$(RNA) and corresponding *P*-values from comparisons between the two Arms.

4.2. *Results of individual parameter estimates*

We fitted the model to the above data using the proposed Bayesian modeling approach. It was shown that the model provided a good fit to the observed data for most subjects. Figure 3 shows the individual fitted curve with observed viral load data in $\log_{10}$ scale, as well as the estimated drug efficacy ($\hat{\gamma}(t)$), observed $IC_{50}(t)$ and adherence for the three representative subjects. It is worth noting by comparing the plots of fitted curves and

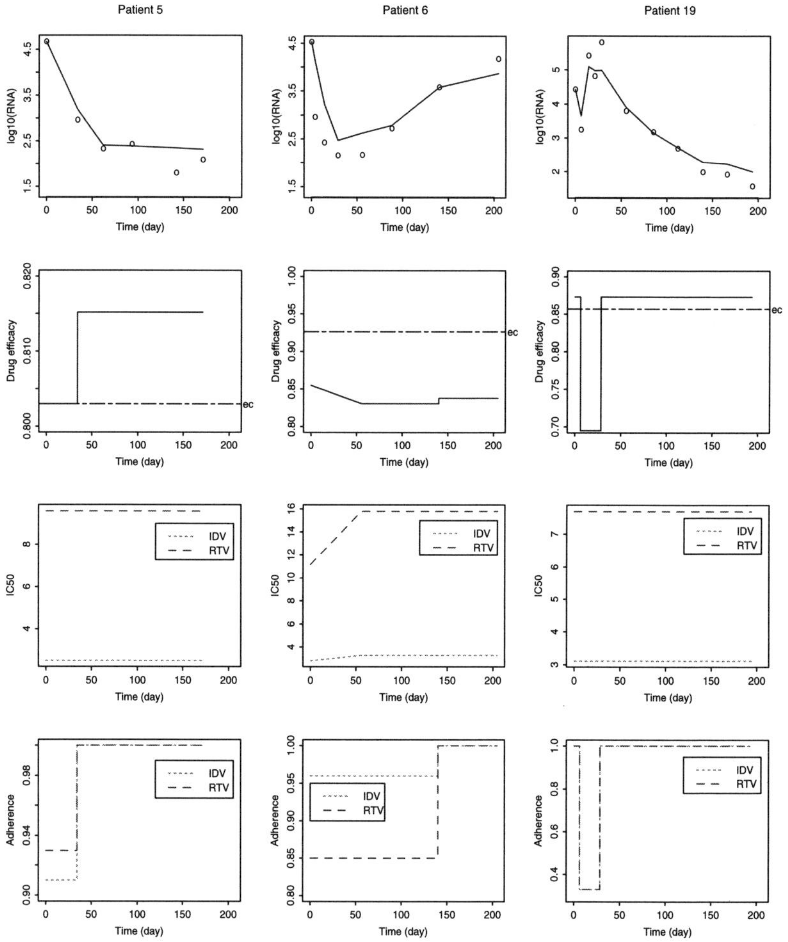

Fig. 3. Individual fitted curve with observed viral load measurements in $\log_{10}$ scale, drug efficacy with threshold (e_c), as well as $IC_{50}(t)$ and adherence of the two PI drugs for the three representative subjects.

estimated drug efficacy that, in general, if $\hat{\gamma}(t)$ falls below the threshold e_c, viral load rebounds, and in contrast, if $\hat{\gamma}(t)$ is above e_c, the corresponding viral load does not rebound, which is consistent with our theoretical analysis of the dynamic models (Huang and Wu, 2003). The results of estimated individual dynamic parameters are presented in the following.

Tables 1 and 2 present the results of estimated individual dynamic parameters and their summary statistics (minimum, median, 1st and

Table 1. The estimates of individual dynamic parameters and threshold of drug efficacy (e_c) in Arm A, where Min, 1st, Med, 3rd, Max, SD and CV = SD/Mean denote the minimum, first quantile, median, third quantile, maximum, standard deviation and coefficient of variation, respectively. Note that all entries of k_i (except for CV) are to be multiplied by 10^{-5}.

	ϕ_i	c_i	δ_i	λ_i	ρ_i	N_i	k_i	e_c
1	166.13	4.75	0.27	14.57	0.002	371.11	4.51	0.961
2	34.25	3.96	0.11	35.26	0.004	459.47	1.19	0.918
3	24.59	2.58	0.29	180.65	0.061	854.21	0.80	0.850
4	33.18	3.66	0.28	22.59	0.005	775.41	1.68	0.938
5	23.03	3.15	0.27	131.38	0.057	731.66	1.12	0.803
6	40.88	3.26	0.31	49.18	0.015	800.55	1.67	0.926
7	26.52	3.55	0.28	51.35	0.026	859.43	1.76	0.851
8	13.95	3.66	0.21	45.85	0.042	945.26	2.52	0.859
9	18.97	5.01	0.36	60.45	0.076	658.00	3.08	0.689
10	140.60	3.54	0.32	38.49	0.006	573.82	2.69	0.964
11	26.02	4.38	0.22	39.07	0.024	651.77	2.21	0.813
12	51.40	2.20	0.43	90.30	0.031	1174.79	1.37	0.953
13	21.87	4.07	0.14	35.42	0.009	597.47	1.68	0.897
14	19.96	7.91	0.09	8.16	0.003	437.62	5.09	0.759
15	25.39	6.16	0.09	16.38	0.002	422.53	2.20	0.919
16	27.03	4.50	0.16	25.85	0.009	574.37	2.36	0.884
17	53.18	3.84	0.07	18.23	0.001	318.56	1.17	0.743
18	23.05	5.80	0.21	30.42	0.018	510.93	2.99	0.776
19	44.18	2.48	0.98	328.17	0.331	1244.23	1.63	0.857
20	370.51	2.46	0.32	46.00	0.002	458.12	1.28	0.982
21	25.76	1.70	0.54	124.66	0.017	2231.72	0.33	0.969
22	23.20	5.66	0.12	17.79	0.007	466.35	4.01	0.821
Min	13.95	1.70	0.07	8.16	0.001	318.56	0.33	0.689
1st	23.09	3.18	0.14	23.41	0.004	461.19	1.30	0.815
Med	26.27	3.75	0.27	38.78	0.012	624.62	1.72	0.872
3rd	43.35	4.69	0.32	58.17	0.030	840.79	2.64	0.935
Max	370.51	7.91	0.98	328.17	0.331	2231.72	5.09	0.982
Mean	56.07	4.01	0.28	64.10	0.034	732.61	2.15	0.870
SD	79.85	1.47	0.20	73.23	0.070	415.36	1.20	0.082
CV(%)	142.40	36.52	71.17	114.24	204.980	56.70	55.78	9.422

Table 2. The estimates of individual dynamic parameters and threshold of drug efficacy (e_c) in Arm B, where Min, 1st, Med, 3rd, Max, SD and CV = SD/Mean denote the minimum, first quantile, median, third quantile, maximum, standard deviation and coefficient of variation, respectively. Note that all entries of k_i (except for CV) are to be multiplied by 10^{-5}.

	ϕ_i	c_i	δ_i	λ_i	ρ_i	N_i	k_i	e_c
23	44.72	1.56	0.27	120.99	0.009	1259.21	0.36	0.974
24	30.34	4.27	0.21	52.54	0.022	618.97	2.10	0.863
25	64.55	3.47	0.15	93.61	0.006	506.56	0.85	0.948
26	27.78	4.14	0.08	23.59	0.003	428.33	1.41	0.853
27	117.90	2.81	0.49	42.92	0.006	868.35	1.88	0.976
28	12.95	8.34	0.11	4.69	0.004	675.66	8.57	0.817
29	22.31	7.12	0.08	11.43	0.002	393.48	4.24	0.925
30	21.65	2.64	0.34	89.59	0.050	1178.90	1.37	0.909
31	12.21	4.09	0.23	44.87	0.066	910.45	2.41	0.726
32	26.17	2.75	0.25	152.18	0.052	621.82	0.90	0.832
33	38.30	4.43	0.12	32.70	0.003	411.15	1.38	0.928
34	22.40	4.25	0.32	31.17	0.044	841.16	4.31	0.835
35	12.33	4.46	0.25	65.36	0.088	755.11	2.24	0.645
36	34.33	2.60	0.15	38.12	0.003	870.19	0.63	0.963
37	11.14	6.09	0.15	9.43	0.012	834.44	4.68	0.801
38	36.92	5.64	0.04	8.60	0.001	317.38	1.43	0.757
39	43.14	5.50	0.13	29.43	0.004	370.89	2.02	0.900
40	13.81	5.96	0.22	25.59	0.032	685.57	4.17	0.739
41	25.29	5.24	0.39	25.52	0.070	938.71	5.18	0.704
42	21.07	3.91	0.24	88.89	0.085	741.77	2.49	0.798
Min	11.14	1.56	0.04	4.69	0.001	317.38	0.36	0.645
1st	19.26	3.31	0.13	25.04	0.004	487.01	1.38	0.788
Med	25.73	4.26	0.22	35.41	0.011	713.67	2.06	0.844
3rd	37.27	5.53	0.26	71.24	0.051	868.81	4.19	0.926
Max	117.90	8.34	0.49	152.18	0.088	1259.21	8.57	0.976
Mean	31.97	4.46	0.21	49.56	0.028	711.41	2.61	0.845
SD	24.29	1.66	0.11	40.26	0.030	261.34	2.01	0.097
CV(%)	76.00	37.26	54.03	81.22	108.077	36.74	76.32	11.438

3rd quartiles, mean, maximum, standard deviation (SD), and coefficient of variation (CV) of the estimated parameters) for Arm A and Arm B, respectively. We can see from the results in Tables 1 and 2 that the virological failure patient (viral load rebounds) corresponds to significant large efficacy parameter value (ϕ), which indicates a very low drug efficacy. Generally speaking, the virological success patients (plasma HIV-1 RNA level of less than 200 copies/mL) have significant higher death rate of free virions (c), but smaller efficacy parameter (ϕ), lower death rate of infected cells (δ), and lower drug efficacy threshold (e_c).

Table 3. A summary of the estimated dynamic parameters for 42 patients in the two arms, where Min, 1st, Med, 3rd, Max, SD and CV = SD/Mean denote the minimum, first quantile, median, third quantile, maximum, standard deviation and coefficient of variation, respectively. Note that all entries of k_i (except for CV) are to be multiplied by 10^{-5}.

	ϕ_i	c_i	δ_i	λ_i	ρ_i	N_i	k_i	e_c
Min	11.14	1.56	0.04	4.69	0.001	317.38	0.33	0.645
1st	21.98	3.18	0.13	24.08	0.004	461.19	1.37	0.802
Med	26.10	4.08	0.23	38.31	0.011	666.83	1.95	0.858
3rd	40.23	5.19	0.31	64.13	0.044	858.12	2.92	0.928
Max	370.51	8.34	0.98	328.17	0.331	2231.72	8.57	0.982
Mean	44.59	4.23	0.25	57.18	0.031	722.51	2.38	0.858
SD	60.73	1.56	0.16	59.60	0.054	346.60	1.63	0.089
CV(%)	136.17	36.90	66.92	104.23	173.435	47.97	68.59	10.378

Table 3 provides the summary statistics of the estimated dynamic parameters for the 42 subjects by combining the two Arms. It can be seen that the estimated parameter values of $\beta_i = (\phi_i, c_i, \delta_i, \lambda_i, \rho_i, N_i, k_i)^T$ are ranged with minimum $(11.14, 1.56, 0, 04, 4.69, 0.001, 317.38, 0.0000033)^T$ and maximum $(370.51, 8.34, 0.98, 328.17, 0.331, 2231.72, 0.0000857)^T$. We also can see from Table 3 for 42 patients that a relatively large between-subject variation in the seven viral dynamic parameters was observed (the CV ranges from 37% to 173%).

Boxplots and corresponding *P*-values from the Wilcoxon Rank Sum test for comparing the estimated dynamic parameters between Arm A and Arm B are displayed in Fig. 4. We can see that there was no significant difference in the estimated dynamic parameters between the two Arms.

5. Discussion

In this article, we established the relationship of virological response (viral load trajectory) with drug exposure (PK and adherence) and drug sensitivity (IC_{50}) via a viral dynamic model. We used PK, adherence and drug resistance data as well as long-term viral load data from an AIDS clinical study to fit our model. It was seen that the model fitting to the observed data for individual subjects was very well. The visual inspection suggests that the fitting for more than 90% of patients was reasonably good, and the number of patients with poor fitting was less than 10%. The poor fitting results for some of the patients may be caused by inaccurate measurements

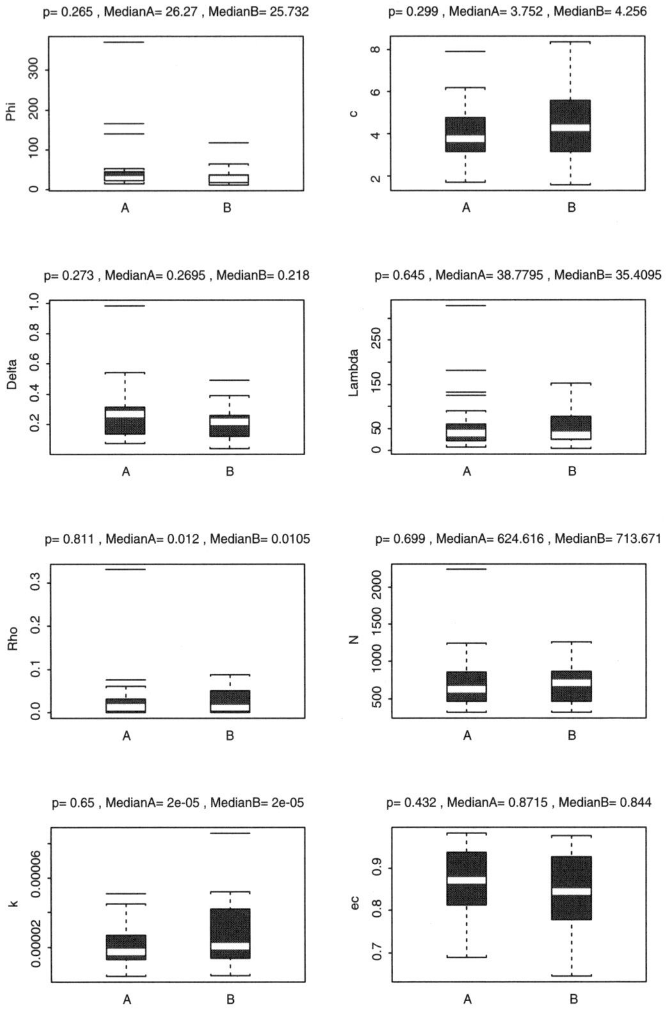

Fig. 4. Comparisons of estimated dynamic parameters between the two arms.

of drug exposure or adherence data. For example, the self-reported adherence data may not reflect the actual adherence of prescribed regimens for some patients.

Most of previous studies that assumed perfect drug effect (Ho *et al.*, 1995; Markowitz *et al.*, 2003; Nowak *et al.*, 1995; Perelson *et al.*, 1996; Perelson *et al.*, 1997; Perelson and Nelson, 1999; Wu *et al.*, 1998) or imperfect drug effect with constant (Ding and Wu, 2000; Perelson and Nelson, 1999;

Wu *et al.*, 1998) provided models to fit short-term viral load data for estimating dynamic parameters. These facts contributed to the limitations of those studies which might result in inaccurate results of dynamic parameter estimation. Compared with those studies, our model proposed in this article enjoys the following features: (i) time-varying drug efficacy during treatment; (ii) reasonable biological interpretation; (iii) incorporating drug concentration, adherence and resistance in the model; and (iv) good fit to the observed long-term viral load data or less frequent clinical data. Thus, based on this model, the results of estimated dynamic parameters should be more reliable and reasonable to interpret long-term HIV dynamics.

It was seen from Tables 1 and 2 that, although we used the same prior at a population level for all the patients, relatively large between-subject variations in parameters (ρ, λ) determined by the host and parameters (c, δ, N and k) decided by interaction between host and virus may be due to the highly diverse subject-to-subject viral concentration transients, drug exposure and resistance observed during the long-term treatment. This may indicate a large inhomogeneity in viral species and host characteristics, that may suggest the importance of individualized treatments. However, there was no significant difference in the dynamic parameter estimates between Arm A and Arm B, which is consistent with the primary analysis results showing no difference in responses and other aspects of the two regimens.

It is also interesting to note that large efficacy parameter values of ϕ, which indicate a very low drug efficacy, correspond to the virological failure patients (patients 1, 10, 20 and 27, for example).

From the results in Table 3, we can see that, over the 42 patients, the mean estimates are 4.23 and 0.25 for c and δ, respectively, which are the most important parameters in understanding viral dynamics. Compared with previous studies, our mean estimate of c (4.23) is greater than the previous mean estimates of c, 3.07 (Perelson *et al.*, 1996), 3.1 (Perelson and Nelson, 1999) and the population estimate of viral clearance rate, 2.81, with credible interval being (1.24, 6.49) obtained by Han *et al.* (2002). On the other hand, our mean estimate of δ (0.25) is less than the first-phase decay rate of 0.43 (Nowak, 1995), 0.49 (Perelson *et al.*, 1996) and 0.5 (Perelson and Nelson, 1999); our mean estimate of δ is also less than the mean value of δ, 0.37 (Klenerman *et al.*, 1996; Stafford *et al.*, 2000). In addition, in two separate studies by Markowitz *et al.* (2003) and Perelson *et al.* (1997), the mean values of 1.0 and 0.7 for δ were obtained by holding clearance rate c as constant of 23 and 3, respectively, and these two values are substantially greater than our mean estimate of 0.25 for δ. These differences may be due to the following reasons. The analysis of those studies assumed that viral replication was completely suppressed by the therapy or imperfect drug efficacy with the constant, and/or they used short-term viral load data to fit their models, while we used long-term viral load data to model fitting and the time-varying drug efficacy is incorporated in our model. Note that we are unable to validate our results for the other parameter estimates as no conclusive or comparable estimates have been published to date.

It is also important that we can estimate the treatment efficacy threshold (e_c). The efficacy threshold may represent how good the immune status of a patient is to control viral replication. If the efficacy threshold (e_c) for a patient is small, it may indicate that this patient's immune response to the virus is strong and a regimen with a mild potency can keep the virus on check for this patient. However, if the efficacy threshold (e_c) for a patient is high, it indicates that this patient needs a highly potent regimen to suppress the virus. Thus, the drug efficacy threshold (e_c) is important for individual patients.

In summary, the mechanism-based dynamic model is powerful and efficient to establish a relationship of antiviral response with drug exposure and drug susceptibility, although some biological assumptions have to be made. The dynamic parameters for individual subjects can be identified by borrowing information from prior population estimates using the developed Bayesian approach. The estimated dynamic parameters can help to understand the characteristics of the biological mechanisms of HIV dynamics, the pathogenesis of HIV-1 infection and the clinical roles in antiviral activities. In addition, the established models can be used to simulate antiviral responses of new antiviral agents and new treatment strategies.

Acknowledgments

This work was supported in part by National Institutes of Health (NIH) research grants RO1 AI052765 and RO1 AI055290.

References

1. Bennett JE, Racine–Poon A and Wakefield JC (1996) Markov chain Monte Carlo for nonlinear hierarchical models, in Gilks WR, Richardson S and Spiegelhalter DJ (eds), *Markov Chain Monte Carlo in Practice*, pp. 339–357, Chapman & Hall, London.
2. Besch CL (1995) Compliance in clinical trials. *AIDS* **9**: 1–10.

3. Bonhoeffer S, Lipsitch M and Levin BR (1997) Evaluating treatment protocols to prevent antibiotic resistance. *Proc Natl Acad Sci USA* **94**: 12106–12111.
4. Callaway DS and Perelson AS (2002) HIV-1 infection and low steady state viral loads. *Bull Math Biol* **64**: 29–64.
5. Carlin BP and Louis TA (1996) *Bayes and Empirical Bayes Methods for Data Analysis.* Chapman & Hall, London.
6. Chib S and Greenberg E (1995) Understanding the Metropolis–Hastings algorithm. *The American Statistician* **49**: 327–335.
7. Davidian M and Giltinan DM (1995) *Nonlinear Models for Repeated Measurement Data.* Chapman & Hall, London.
8. Ding AA and Wu H (1999) Relationships between antiviral treatment effects and biphasic viral decay rates in modeling HIV dynamics. *Math Biosci* **160**: 63–82.
9. Ding AA and Wu H (2000) A comparison study of models and fitting procedures for biphasic viral decay rates in viral dynamic models. *Biometrics* **56**: 16–23.
10. Ding AA and Wu H (2001) Assessing antiviral potency of anti-HIV therapies *in vivo* by comparing viral decay rates in viral dynamic models. *Biostatistics* **2**: 13–29.
11. Fitzgerald AP, DeGruttola VG and Vaida F (2002) Modelling HIV viral rebound using non-linear mixed effects models. *Stat Med* **21**: 2093–2108.
12. Frost SDW (2001) Bayesian modeling of viral dynamics and evolution. *AIDS Cyber J* **4**(2).
13. Gamerman D (1997) *Markov Chain Monte Carlo: Stochastic Simulation for Bayesian Inference.* Chapman & Hall, London.
14. Gelfand AE *et al.* (1990) Illustration of Bayesian inference in normal data models using Gibbs sampling. *J Amer Stat Assoc* **85**: 972–985.
15. Gelfand AE and Smith AFM (1990) Sampling-based approaches to calculating marginal densities. *J Amer Stat Assoc* **85**: 398–409.
16. Gelman A, Bois F and Jiang J (1996) Physiological pharmacokinetic analysis using population modeling and informative prior distributions. *J Amer Stat Assoc* **91**: 1400–1412.
17. Han C, Chaloner K and Perelson AS (2002) Bayesian analysis of a population HIV dynamic model. *Case Studies in Bayesian Statistics*, Vol. 6. Springer-Verlag, New York.
18. Ho DD, Neumann AU, Perelson AS *et al.* (1995) Rapid turnover of plasma virions and CD4 lymphocytes in HIV-1 infection. *Nature* **373**: 123–126.
19. Hsu A, Isaacson J *et al.* (2000) Trough concentrations-EC_{50} relationship as a predictor of viral response for ABT-378/ritonavir in treatment-experienced patients, in *40th Interscience Conference on Antimicrobial Agents and Chemotherapy.* San Francisco, CA, Poster session 171.
20. Huang Y, Rosenkranz SL and Wu H (2003) Modeling HIV dynamics and antiviral responses with consideration of time-varying drug exposures, sensitivities and adherence. *Math Biosci* **184**: 165–186.

21. Huang Y and Wu H (2004a) A Bayesian approach for estimating antiviral efficacy in HIV dynamic models. *J App Stat* (submitted).
22. Huang Y and Wu H (2004b) Mechanistic PK/PD modeling of antiretroviral therapies in AIDS clinical trials, in: *Biomedical Simulations Resource PK/PD*. Kluwer Academic Publishers, New York.
23. Ickovics JR and Meisler AW (1997) Adherence in AIDS clinical trial: a framework for clinical research and clinical care. *J Clin Epidemiol* **50**: 385–391.
24. Kempf DJ, Hsu A, Jiang P *et al.* (2001) Response to ritonavir intensification in indinavir recipients is highly correlated with virtual inhibitory quotient, in *8th Conference on Retroviruses and Opportunistic Infections*, Chicago, IL, Abstract 523.
25. Klenernam P, Phillips RE *et al.* (1996) Cytotoxic T lymphocytes and viral turnover in HIV type 1 infection. *Proc Natl Acad Sci USA* **93**: 15323–15328.
26. Lunn DJ, Best N *et al.* (2002) Bayesian analysis of population PK/PD models: general concepts and software. *J Pharmacokinet Pharmacodyn* **29**: 271–307.
27. Markowitz M, Louie M *et al.* (2003) A novel antiviral intervention results in more accurate assessment of human immunodeficiency virus type 1 replication dynamics and T-cell decay *in vivo*. *J Virol* **77**: 5037–5038.
28. Molla A *et al.* (1996) Ordered accumulation of mutations in HIV protease confers resistance to ritonavir. *Nature Med* **2**: 760–766.
29. Nelson PW and Perelson AS (2002) Mathematical analysis of delay differential equation models of HIV-1 infection. *Math Biosci* **179**: 73–94.
30. Nowak MA, Bonhoeffer S *et al.* (1995) HIV results in the frame. *Nature* **375**: 193.
31. Nowak MA, Bonhoeffer S, Shaw GM and May RM (1997) Anti-viral drug treatment: dynamics of resistance in free virus and infected cell populations. *J Theor Biol* **184**: 203–217.
32. Nowak MA and May RM (2000) *Virus Dynamics: Mathematical Principles of Immunology and Virology*. Oxford University Press, Oxford.
33. Perelson AS, Kirschener DE and Boer RD (1993) Dynamics of HIV infection of $CD4^+$ T cells. *Math Biosci* **114**: 81–125.
34. Perelson AS, Neumann AU, Markowitz M, Leonard JM and Ho DD (1996) HIV-1 dynamics *in vivo*: virion clearance rate, infected cell life-span, and viral generation time. *Science* **271**: 1582–1586.
35. Perelson AS, Essunger P *et al.* (1997) Decay characteristics of HIV-1-infected compartments during combination therapy. *Nature* **387**: 188–191.
36. Perelson AS and Nelson PW (1999) Mathematical analysis of HIV-1 dynamics *in vivo*. *SIAM Rev* **41**: 3–44.
37. Putter H, Heisterkamp SH, Lange JMA and De Wolf F (2002) A Bayesian approach to parameter estimation in HIV dynamical models. *Stat Med* **21**: 2199–2214.
38. Sheiner LB (1985) Modeling pharmacodynamics: parametric and nonparametric approaches, in Rowland M *et al.* (eds) *Variability in Drug Therapy: Description, Estimation, and Control*, pp. 139–152, Raven Press, New York.

39. Smith AFM and Roberts GO (1993) Bayesian computation via the Gibbs sampler and related Markov chain Monte Carlo methods. *J Roy Stat Soc B* **55**: 3–23.
40. Stafford MA *et al.* (2000) Modeling plasma virus concentration during primary HIV infection. *J Theor Biol* **203**: 285–301.
41. Wahl LM and Nowak MA (2000) Adherence and resistance: predictions for therapy outcome. *Proc Roy Soc Biol* **267**: 835–843.
42. Wainberg MA *et al.* (1996) Effectiveness of 3TC in HIV clinical trials may be due in part to the M184V substition in 3TC-resistant HIV-1 reverse transcriptase. *AIDS* **10**(suppl): S3–S10.
43. Wakefield JC, Smith AFM, Racine-Poon A and Gelfand AE (1994) Bayesian analysis of linear and non-linear population models using the Gibbs sampler. *App Stat* **43**: 201–221.
44. Wakefield JC (1996) The Bayesian approach to population Pharmacokinetic models. *J Amer Stat Assoc* **91**: 61–76.
45. Wei X, Ghosh SK *et al.* (1995) Viral dynamics in human immunodeficiency virus type 1 infection. *Nature* **373**: 117–122.
46. Wu H, Ding AA and de Gruttola V (1998) Estimation of HIV dynamic parameters. *Stat Med* **17**: 2463–2485.
47. Wu H and Ding AA (1999) Population HIV-1 dynamics *in vivo*: applicable models and inferential tools for virological data from AIDS clinical trials. *Biometrics* **55**: 410–418.

CHAPTER 16

WITHIN-HOST DYNAMICS AND TREATMENT OF HIV-1 INFECTION: UNANSWERED QUESTIONS AND CHALLENGES FOR COMPUTATIONAL BIOLOGISTS

John Mittler

Despite the tremendous progress that has been made in elucidating the molecular biology, virology, and immunology of HIV-1 infection, some of the most basic questions about how this virus causes disease remain unanswered. In this review, I outline seven problems relating to HIV-1 infection that might be addressed by computational biologists. Solutions to these problems would contribute significantly to our understanding of HIV-1 pathogenesis and improved treatments for HIV-1 infections.

Keywords: HIV-1; AIDS; mathematical modeling; pathogenesis; drug resistance; antiretroviral therapies.

1. Introduction

In the 20 years since the identification of HIV-1 as the agent responsible for AIDS, billions have been spent on basic research related to understanding the pathogenesis of this virus and on the development treatments and vaccines. These unprecedented expenditures have led to major advances in our understanding of the immunology and molecular biology of HIV-1 infection and to the development of highly suppressive combination drug therapies and candidate vaccines. In regions of the world where combination antiretroviral drug therapies are available, mortality rates of HIV-1 infected patients have dropped considerably (Pallela *et al.*, 1998). Extensive research in animal models and preliminary studies in human populations has renewed optimism that effective vaccines could be developed and introduced sometime within the next decade (Amara and Robinson, 2002; Bhardwaj and Walker, 2003).

Despite this progress, many of the most basic questions about the pathogenesis of this virus remained unanswered. In this review, I briefly outline seven questions concerning the pathogenesis and treatment of HIV-1 infection that may be of interest to quantitative biologists. These are questions that have confounded or confused HIV-1 researchers for which mathematical or computational approaches may be helpful. Some these questions are very hard to solve, while others are very basic ones that, one might think, would have been answered years ago. Others are questions for which many believe that answers already exist, and yet, can be shown to be quite complicated when approached using rigorous quantitative criteria.

1.1. *How much virus is in the body?*

At first glance this would appear to be a straightforward question in which the role of mathematics would be limited to adding up quantitative estimates for the amount of virus in different tissues. In fact, this has already been done. Using quantitative molecular techniques, virologists have been able to estimate the amount of HIV-1 RNA in plasma in patients in various stages of disease. Multiplying the amount of HIV-1 RNA in blood (typically between 1000 and 1,000,000 HIV-1 RNA copies per ml of plasma) by the volume of plasma and extracellular fluid has yielded estimates of 10^9 to 10^{10} virions in blood (Perelson *et al.*, 1996; Haase, 1999). This type of accounting can be extended to other tissues and organs, such as lymph nodes, spleen, and lung, by applying techniques such as *in situ* hybridization and immunohistochemical staining to biopsy and autopsy samples (Embertson *et al.*, 1993; Haase *et al.*, 1990; Pantaleo *et al.*, 1993). Current attempts to add up the amount of virus found in all infected tissues and organs, including follicular dendritic cells in the lymph nodes (cells that carry large amounts of virus on their surface), has led to estimates of 10^7 to 10^8 productively infected cells and 10^{10} to 10^{11} virions per body (Chun *et al.*, 1997a; Haase, 1999). Although there is still room for further refinements of these estimates (many of these virions are noninfectious), the overall message is clear: molecular-virological techniques for quantifying the amount of virus in tissues and organs suggest that there are tens of millions of HIV-infected cells and tens of billions of virions in a typical (untreated) HIV-1 infected patient.

This apparently straightforward answer gets shaken-up however, when viral populations are studied from a population genetics perspective. When quantifying populations, geneticists rely on a quantity known as the "effective population size" (designated as N_e) that indicates how the population compares to an idealized population (sometimes called a Wright–Fisher population) in which there is no selection, no migration, no subdivision, and no changes in the number of individuals over time (Hartl and Clark, 1989). A population that has an effective size of 10,000, for example, will have the same population genetic properties (e.g. genetic diversity) as an idealized population of size 10,000. Although there are some factors that can cause the effective size to be greater than the "census size" (raw count of individuals in the population), the majority of factors that distinguish real populations from idealized populations tend to lower effective population sizes. For example, the tendency of some individuals to produce significantly more offspring than others (a common occurrence in nature) will lower effective population size by reducing the amount of genetic diversity in the population relative to a population in which all individuals have an equal probability of producing offspring. Selection for a new trait within a population, likewise, will reduce the diversity of the population (relative to an idealized Wright–Fisher population) by removing individuals that do not carry that trait. Such departures can often reduce effective population sizes by a factor 2 or 3 relative to idealized populations.

When population geneticists have calculated the effective population size for HIV-1 within patients, they have obtained estimates that are many orders of magnitude lower than the census size — a result that has surprised and perplexed many in the HIV-1 community. This dilemma was first noted by Leigh Brown (1997), who used the theoretical relationship between diversity and population size $d = 2N_e\mu$, where d is the within host diversity of the virus and μ ($\sim 2.4 \times 10^{-5}$) is the viral mutation rate, to estimate the effective population size for HIV-1 within infected patients. Based on observed levels of diversity on the order of 5.6% or less, he obtained estimates for N_e ranging from 510 to 2100, roughly 100,000-fold lower than the estimated census sizes. Leigh Brown (1997) also applied Kuhner *et al.*'s (1999) maximum likelihood program "FLUCTUATE" to obtain an independent estimate of N_e. This method also gave estimates for the viral effective population size on the order of 1000 to 2000.

These population genetics estimates of population size are important because they can affect inferences that one might make about viral evolution. When effective population sizes are small, stochastic events become important (Crow and Kimura, 1970). According to population genetics theory, a mutation in a haploid population will put subject to neutral drift if the selection coefficient is less than $1/2N_e$. For $N_e \sim 1000$, this means that mutations that have small effects on fitness will be subject to genetic drift and will not rise in frequency in a predictable fashion. Small effective population sizes also influence the probability that any particular mutant will appear. Prior to Leigh Brown's estimate that $N_e = \sim 1000$, it had been suggested that due to the large census population size and high mutation rate for HIV-1 that variants with mutations at every single nucleotide in the genome will pre-exist in the body at any given time and that virtually no mutant could be considered to be completely neutral (Coffin, 1995). This idea was important because it suggested that the emergence of single-step drug resistance mutations would be virtually guaranteed in patients undergoing drug therapies in which a single mutation could render virus resistant to drug. However, if N_e is small (less than $1/\mu$) then there will be a large stochastic element to the emergence of escape mutants (Rouzine and Coffin, 1999). The wide variation in the rate of appearance of single-step mutants that confer resistance to the antiretroviral drug 3TC (Frost *et al.*, 2000) has lent support to Leigh Brown's assertion that stochastic processes play an important role in the emergence of drug resistance mutations.

Although Leigh Brown's estimate of $N_e \sim 1000$ has been confirmed by other researchers who have applied the similar methodologies to different data sets (Nijhuis *et al.*, 1998; Seo *et al.*, 2002; Shriner *et al.*, 2004), Rouzine and Coffin (1999) obtained a higher estimate for N_e using a different methodology. Their analyses focused on the frequency of the rarest of four possible haplotypes in patients in which viruses were polymorphic at two loci. Rouzine and Coffin first graphed the frequency of the rarest of four possible haplotypes in simulated populations with different values of $N\mu$. These simulations showed that the rarest haplotype will be seen very infrequently in populations with $N\mu < 10^3$; however, in their data sets they observed that all four haplotypes were present in most of the samples. By comparing the observed value for f_{rare} to the graphs of f_{rare} versus $N\mu$ from simulations, they estimated the effective population size to be $\sim 10^5$. Their estimate, therefore, is higher than the other estimates cited above, but still substantially lower than the census size.

These findings leave us with a technical question (why the methods of Leigh Brown (1997) and Rouzine and Coffin (1999) have yielded somewhat different estimates for N_e) and a more difficult conceptual question (why independent estimates for N_e differ are so much lower than the census size). A possible explanation to the technical question lies in the fact that nucleotide substitution rates are different in different regions of the genome. This variation in substitution rates has been described empirically using a gamma distribution with a shape parameter less than one; i.e. an L-shaped distribution in which most sites vary minimally and a few sites vary a lot. A recent re-analysis of the data used by Rouzine and Coffin has demonstrated that the sites that they studied in *env* and *pro* are among the most variable sites in these genes (Shriner *et al.*, 2004). To the extent that this variation represents variation in underlying neutral mutation rates, Shriner *et al.*'s analysis seems to resolve the discrepancy between Brown's estimate and Rouzine and Coffin's estimate: Shriner *et al.* estimated that the sites analyzed by Rouzine and Coffin had "mutation rates" 14-fold higher than the average site, a difference large enough to explain most of the discrepancy between the two estimates. Thus, the difference between these two estimates boils down to an issue of what is the correct mutation rate to plug into their respective formulas and graphs. Leigh Brown assumed that the overall diversity in Env reflected an average neutral mutation rate of 2.4×10^{-5} (Mansky and Temin, 1995; Mansky, 1996). The fact that a neutral mutation rate of 2.4×10^{-5} together with an average generation time of ~2 days (Perelson *et al.*, 1996) predicts an average rate of divergence in Env of ~1% per year (Shankarappa *et al.*, 1999) is consistent with Leigh Brown's neutral mutation rate; however, inferences about neutral mutation rates are tricky for genes, such as Env, that are thought to be under a combination of purifying selection and diversifying selection from HIV-1 specific antibodies — an assertion that is, itself, a matter of controversy (Leigh Brown, 1997; Richman *et al.*, 2003; Shriner *et al.*, 2004; Wei *et al.*, 2004).

While the analysis by Shriner *et al.* (2004) sheds light on the discrepancy between different estimates for N_e, it does not explain why N_e is so much lower than the census size. Hypotheses that have been proposed include the existence of genetic bottlenecks (Leigh Brown, 1997; Leigh Brown and Richman, 1997), population subdivision (Frost *et al.*, 2000), selection for new variants linked to sites under examination (Frost *et al.*, 2000; Rouzine *et al.*, 2001), variation in viral production rates between different infected cells (Brown, 1997; Brown and Richman, 1998), and the fact that most virions are noninfectious (Dimitrov *et al.*, 1993; Piatak *et al.*, 1993; Leigh Brown, 1997). However, no one to date has come up with a fully satisfying explanation for this large discrepancy. Future progress in this area may require the construction of more comprehensive stochastic models for the factors mentioned above and collection of detailed experimental data documenting the amount of selection, population subdivision, variation in viral production rates, and other factors that influence viral evolution *in vivo*.

1.2. *Can the progression to AIDS be modeled mathematically?*

Changes in viral loads and CD4 T-cell counts have distinctive dynamical features that present a challenge to mathematical biologists seeking to model HIV-1 infections. During a "typical" infection course in an untreated adult (Fig. 1), viral load rises rapidly during the first two to three weeks of "acute" infection, hitting a peak of $\sim 2 \times 10^6$ HIV-1 RNA copies per mL around day 21 post-infection before declining over a period of months to a patient-specific set-point of between 10^3 to 10^6 HIV-1 RNA copies per mL (Piatak *et al.*, 1993; Kaufmann *et al.*, 1998; Lindback *et al.*, 2000). During the acute phase of primary infection, patients may experience a transient flu-like illness characterized by swollen lymph nodes, fever, rash, and a transient decrease in peripheral blood CD4 "helper" T-cell counts from a typical pre-infection level of ~1000 cells per microliter of blood (Cooper *et al.*, 1988; Gaines *et al.*, 1988, 1990; Kinlock-de Loës *et al.*, 1993). As viral loads decline from the peak levels to their patient specific set-points, peripheral blood CD4 T-cell counts recover somewhat and the patient enters a period of clinical latency during which there are few clinical symptoms. During this period of symptom-free "chronic" infection, which lasts an average of ~10 years, viral loads gradually creep up and CD4 T-cell counts gradually decrease. However, these decreases in CD4+ T-cell counts do not become critical until they fall below ~200 cells per microliter, the threshold used by the CDC to define whether a patient has AIDS. As CD4 T-cell counts fall below 200 cells per microliter of blood, patients become increasingly vulnerable to a variety of opportunistic infections characteristic of an impaired

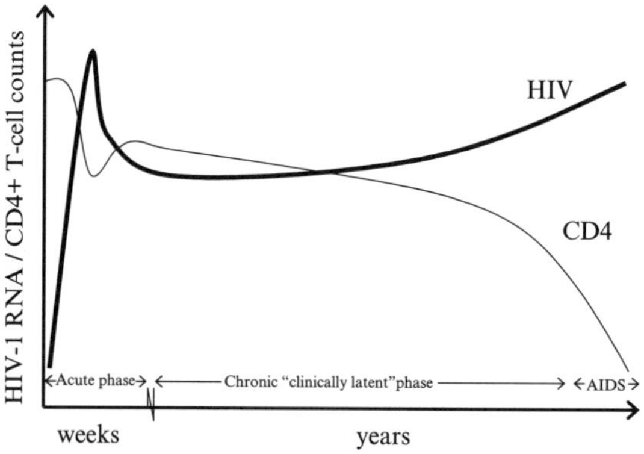

Fig. 1. Schematic illustrating "typical" HIV-1 infection dynamics *in vivo* in the absence of therapy.

CD4 T-helper cell response. Viral loads often rise to very high levels during this period, approaching or exceeding the very high levels seen during the acute phase of infection.

These changes in viral load and CD4 T-cell counts are interesting from a dynamical perspective because they resemble a classic predator-prey interaction in which viruses are the "predators" and the CD4 T-cells are the "prey." Unlike classic predator-prey interactions, however, this interaction has a twist: some of these CD4 target cells will assist other cells in removing of free virus and virally infected cells from the body. That is, HIV-specific CD4+ T-cells release cytokines and provide costimulatory signals via cellular receptors that stimulate HIV-1 specific B-cells to secrete antibodies that clear virus from the blood and HIV-1 specific CD8+ T-cells (CD8+ CTL) to release granzymes and perforins that kill HIV-1-infected cells in a contact dependent-fashion. Stimulated CD8+ T-cells also release cytokines that inhibit viral replication and chemokines that block HIV-1 from infecting new target cells by binding to the CCR5 co-receptor that this virus uses to enter the host cell. Some CD4+ T-cells can also act directly as cytotoxic lymphocytes (CD4+ CTL) that can kill HIV-infected cells directly. In other words, HIV-CD4 T-cell dynamics have the characteristic of a predator-prey system in which some of the prey prey upon the predators (or more precisely one in which some of the prey assist other cells in eliminating predators from the body).

Of course, as an interaction that involves cells within the body, as opposed to reproducing organisms, the dynamics depart from that of classic predator-prey models in several ways. To start with, CD4+ and CD8+ T-cells do not reproduce in the same way as free living organisms. These cells originate as CD4-CD8-precursors that migrate from the bone marrow to the thymus where they mature into functional lymphocytes that carry T-cell receptors that recognize foreign peptides presented on the surface of antigen presenting cells (cells bearing the MHC-II and MHC-I molecules, respectively). Although there are homeostatic signaling mechanisms in the body that tend to maintain total T-cell counts at a roughly constant level (Margolick *et al.*, 1995), the input of new CD4+ and CD8+ T-cells is usually modeled as occurring independently of T-cell densities. CD4+ and CD8+ T-cells can also expand in response to antigen: T-cells that recognize foreign antigens on the surface of antigen presenting cells populations will, in the presence of appropriate cell signaling events, undergo "clonal" expansions that greatly increase the number of T-cells that can respond to these foreign antigens. This is somewhat like a predator numerical response in ecological models for predator-prey interactions, but for a crucial difference: T-cell expansion *in vivo* is not limited by the availability of "food" for the cells. Given the appropriate stimulatory signals, T-cells will undergo several rounds of replication even in the absence of any further antigenic stimulation (Kaech *et al.*, 2001; von Stipdonk *et al.*, 2001).

With these considerations in mind, there have been a number of attempts to construct conceptual models for viral-T-cell interactions that mimic the dynamics summarized above (Cooper, 1986; Perelson *et al.*, 1989, 1993; Nowak *et al.*, 1991, 1992, 1995; McLean and Nowak, 1992; Essunger and Perelson, 1994; Bremermann, 1995; Mittler *et al.*, 1996; Stilianakis *et al.*, 1997; Kirschner *et al.*, 2000; Murray *et al.*, 1998; Wodarz and Nowak, 1999, 2000; Stafford *et al.*, 2000; Fraser *et al.*, 2001; Korthals Altes *et al.*, 2003, to list a few). These attempts incorporate, to varying degrees, mechanistically plausible terms for the complexities of the HIV-T-cell interactions, and are interesting because they provide conceptual frameworks by which HIV-1 dynamics may be understood. Most of these models are relatively simple, relying on a small number of differential equations (typically between 3 and 6).

1.2.1. *Potential for the construction of more realistic models (and problems in doing so)*

Researchers who are knowledgeable about the immunology and virology of HIV-1 infection could easily come up criticisms for most of the models presented above. By attempting to summarize HIV-1 dynamics with a small number of equations, much of the biology gets left behind. What is interesting and somewhat maddening, however, is that approximate, if not exact, mathematical solutions exist for many of simplifications made in the models above. The fact that most of these models do not distinguish between different classes of T-cells, for example, could be fixed by adding separate equations for naïve, memory, and effector T-cells (Essunger and Perelson, 1994). The problem is that a model that addressed all of the details that biologists are aware of would become large and cumbersome and would probably not help very much in understanding the fundamental forces that lead to the dynamics illustrated in Fig. 1.

The challenge, therefore, is to construct models that include mechanistically plausible terms for HIV-T-cell interactions that explain the general features of HIV-1 dynamics without becoming so unwieldy as to defy easily explanation. The ideal model would contain just enough equations to accurately mimic the overall dynamic while ignoring factors that have only minor effects on viral loads and CD4+ T-cell counts. It would also account for as much of the biology of infection as possible. A complete model would account for the following observations:

1) The existence of a primary infection viremia in which viral load doubles every day until around day 20 after which viral load decreases at a rate of ~0.4 per day for a couple of weeks before entering a slower rate of decline until it arrives at a patient-specific set point roughly 60–90 days after infection (Lindback *et al.*, 2000).
2) The transient dip in CD4+ T-cell counts that occurs during primary infection (Gaines *et al.*, 1988).
3) The existence of a wide range of relatively stable viral setpoints in HIV-1 infected patients (Piatak *et al.*, 1993).
4) The gradual increase in plasma HIV-1 RNA and gradual decline in CD4+ T-cell counts over a period of years.
5) Differences in the rate of decline of CD4+ cells in lymph nodes and in blood (Rosenberg *et al.*, 1993; Rosok *et al.*, 1996).
6) The eventual loss of T-cell homeostasis and decline of other populations of cells in the blood that precedes progression to AIDS (Margolick *et al.*, 1995).
7) The biphasic decline in HIV-1 plasma RNA following initiation of combination antiretroviral drugs: specifically, the observation that viral load remains roughly constant during the first day of treatment before declining at a rate of ~0.5 per day for ~10 days before declining at a rate of ~0.04 per day until viral load becomes undetectable (Perelson *et al.*, 1996, 1997; Notermanns *et al.*, 1998).
8) The increase in peripheral blood CD4+ T-cell counts that occurs following treatment (Ho *et al.*, 1995; Wei *et al.*, 1995).
9) The correlation between viral load and viral replication rate *in vitro* (Quinones–Mateu *et al.*, 2000).
10) The log-log relationships that have been observed between plasma viral load and the number of HIV-infected T-cells in lymph nodes (Hockett *et al.*, 1999).
11) The rapid increase in plasma viral SIV RNA in macaques treated with anti-CD8 antibodies (assuming that the same would be observed in HIV-1 infected humans) (Schmitz *et al.*, 1999; Jin *et al.*, 1999).
12) The ability of very early treatments (i.e., ones initiated during primary infection) to lower viral setpoints (Watson *et al.*, 1997; Rosenberg *et al.*, 2000).

13) Discordant responses wherein patients whose antiretroviral therapies fail to suppress viral replication due to the emergence of drug resistance viruses nevertheless experience immunological benefits (as quantified by peripheral blood CD4+ T-cell counts) (Kaufmann *et al.*, 1998; Perrin and Telenti, 1998; Piketty *et al.*, 1998).
14) Increases in viral diversity and divergence over time (Shankarappa *et al.*, 1999).
15) The ability of HIV-1 to mutate into forms, such as the so-called X4 variants, that can infect new types of cells (Tersmette *et al.*, 1989; Schuitemaker *et al.*, 1992; Koot *et al.*, 1993; Richman and Bozzette, 1994).
16) The ability of HIV-1 to mutate into forms that can evade HIV-specific immune responses (Borrow *et al.*, 1997; Evans *et al.*, 1999).
17) The correlation between steady-state viral loads and rate of progression to AIDS (Mellors *et al.*, 1996).
18) Existence of steady-state viral loads during reverse transcriptase monotherapies that are 10–100 fold lower than pre-treatment viral loads (Eron *et al.*, 1995). (An observation discussed in detail in Section 1.3 below.)

In addition to meeting these criteria (i.e. mimicking viral infection dynamics) and other factors not listed above, models need to make testable predictions so that they can be disproved in prospective experiments. What is particularly valued is the ability to make testable predictions that shed light on the fundamental forces that cause viral load to vary from person to person (Muller *et al.*, 2001; Bonhoeffer *et al.*, 2003). An understanding of these factors could help researchers develop new therapies, vaccines, and drug treatment regimens that specifically target those factors that cause viral loads to vary.

Having laid this out as a significant problem that remains unsolved, it is important to point out the potential dangers in pursuing this as a research problem. I have already touched on the multiple parameter problem. Modelers seeking to make models that match all of the requirements above face the danger that they will end up with a model with a huge number of parameters that could cause them to lose sight of the forest through the trees. It also makes it harder to create a model that is falsifiable. Modelers seeking to take up this challenge would be advised to make models that rely on small number of key parameters and that make testable predictions.

A second, more fundamental, danger lies in the assumption that HIV-1 dynamics can, in fact, be distilled down to a few key equations; i.e. that there are a small number of factors (e.g. variation in viral infectivity coupled with variation in the strength of HIV-specific CTL responses) that account of the majority of the 1000-fold variation in viral load observed in patients. As long as the majority of variation in viral load can be ascribed to a small number of factors, it becomes possible to create models that have a manageable number of equations and parameters. However, it is possible that there is no single factor that accounts for the majority of variation in viral load (Mueller *et al.*, 2001). It is conceivable, for example, that there are 18 factors that individually account for no more than a 1.5-fold variation in viral load, an amount that would be enough to account for 1000-fold variation among individuals (noting that $1.5^{18} = {\sim}1500$). Amongst the many biological factors that might affect viral load *in vivo* are (i) variation in thymic output levels (i.e. source term for target cells); (ii) variation in the inherent infectability of T-cells; (iii) variation in CTL densities; (iv) variation in T-helper cell responses; (v) variation in antibody responses; (vi) variation in the sensitivity of viruses to antibodies; (vii) variation in viral replication capacity; (viii) variation in viral tropism; (ix) variation in the concentration of β-chemokines (the natural ligands for the CCR5 co-receptor that HIV-1 relies upon to get into cells); (x) variation in CAF concentrations (a suppressive factor released by CD8+ T-cells); (xi) variation in interferon concentrations; (xii) variation in α-defensins concentrations (another type of suppressive factor); (xiii) variation in Th1-Th2 cytokine ratios; (xiv) variation in the ability of antigen presenting cells to process and present antigens; (xv) genetic variation in the CCR5 allele; (xvi) appearance of CTL and/or antibody escape mutants; (xvii) variation in the concentration of IL-16 (the natural ligand for CD4); (xviii) variation in inflammatory cytokine concentrations; and (xix) variation in cellular activation rates.

Despite this long list of potential factors, my bet is that one or two of them will ultimately be shown to have a dominant effect in the majority of patients; for example, variation in CTL responses and viral replication capacity might ultimately be shown to contribute to a 100-fold variation in viral setpoints, with all other factors combined contributing less than 10-fold to the viral setpoint in the average patient. Of course, this is just a bet (sorry, no monetary stakes taken), and it is entirely possible that HIV-1 dynamics cannot be modeled using a model that includes a small number of key equations. Having said this, it is worth pointing out that the question of whether there are a few dominant factors or many small factors that contribute to viral load variation is, itself, a question that mathematical and computational biologists could help to illuminate.

1.3. *The "knife-edge" problem*

Implicit in the discussion in Section 1.2 above is a certain tension between models that suppose that target cell limitations are the primary dynamical factor that regulates viral load *in vivo* and models that suppose that immunes responses, especially CD8+ T-cell responses, are the major dynamical factor that regulates viral load (de Boer and Perelson, 1998; Wick *et al.*, 2002). Although much of the experimental evidence that has been gathered over the years seems to point to CD8+ T-cell immune responses as the primary limiting factor (Borrow *et al.*, 1994; Schmitz *et al.*, 1999; Jin *et al.*, 1999), the most commonly invoked mathematical models for viral dynamics still invoke target-cell limitations as the primary limiting factor (Ding and Wu, 1999; Perelson and Nelson, 1999; Stafford *et al.*, 2000; Muller and Bonhoeffer, 2003). I review here a problem (first pointed out by Bonhoeffer *et al.*, 1997) that appears to falsify commonly used target-cell models for HIV-1 dynamics.

To illustrate the problem, it may be helpful to present a typical set of equations that posit that target-cell limitations as the primary limiting factor for HIV-1 infection. A commonly invoked target-cell limited model supposes that free virus, V, target cells, T, and infected cells, I, are related by the following equations:

$$dT/dt = s - mT - kVT \tag{1}$$

$$dI/dt = kVT - \delta I \tag{2}$$

$$dV/dt = pI - cV, \tag{3}$$

where s is the source of target cells, m is the death rate of target cells, k is the rate at which virus infects target cells, δ is the death rate of virally infected cells, p is the rate at which infected cells produce free virions, and c is the rate at which free virions are cleared from the body.

Under these assumptions the steady-state viral load is $V^* = ps/c\delta - m/k$. In an insightful analysis, Bonhoeffer *et al.* (1997) showed that this equilibrium has a peculiar sensitivity to changes in the viral infectivity parameter, k. For cases in which the product $k_{\text{crit}} = cm\delta/ps$ is much less than k, the steady-state viral load is relatively independent of k. As k

approaches k_{crit}, however, the steady state viral load rapidly drops to zero. For example, for $ps/c\delta = 10^4$ and $m = 0.01$ ($k_{\text{crit}} = 10^{-6}$), the steady state viral loads for $k = 10^{-3}$, 10^{-4}, 10^{-5}, 2.5×10^{-6}, and 10^{-6}, respectively, are 9990, 9900, 9000, 6000, and 0. In other words there is a wide range of parameter space over which k has very little effect on viral load, but once it approaches the threshold $k_{\text{crit}} = cm\delta/ps$, the resulting steady state becomes very sensitive to small changes in k.

This is a problem because viral loads in patients who have been treated with reverse transcriptase inhibitors have been observed to drop to new steady state levels that are anywhere from 10 to 100-fold lower the pretreatment steady state (Eron *et al.*, 1995). Both the effects of reverse transcriptase inhibitors and the accompanying genetic changes that confer resistance to these drugs are generally modeled by assuming a change in the infection rate constant k. In order for k to change in such a way that the steady-state viral load drops 10 to 100-fold, it would have to drop to a level that is just slightly above k_{crit}. Since there is no biological reason to expect that k has a lower bound that just happens to be slightly above k_{crit}, it becomes hard to explain the substantial reductions in viral load that have been observed in patients who have gone on RT inhibitor monotherapies using target-cell limited models along the lines of Eqs. (1)–(3). In other words, there is a very thin region of parameter space (a "knife-edge") under which target-cell limited models can explain these reductions.

In their exposition of this problem, Bonhoeffer *et al.* (1997) point out that this problem does not exist for commonly invoked immune control models. The steady-state viral load for the simple immune-control model, $dV/dt = pI - cV$, $dI/dt = kVT - \delta I - \delta_X IX$, $dX/dt = rI - mX$, (where X is the CTL response and target cell density T is assumed to be constant), for example, is $V^* = pm(kTp/c - \delta)/\delta_X rc$, a term that is linearly dependent on k and which can give a range of V^* values for different values of k. Although it too predicts a k_{crit} below which virus cannot persist, this model does not require that k fall within a very narrow range of parameter space in order for V^* to drop 100-fold.

Analyses by Bonhoeffer *et al.* (1997) and Callaway and Perelson (2002) demonstrate that this knife-edge problem exists for a number of straightforward variants of standard target cell limited models. Having said this, it should be pointed that this problem is partly a mathematical artifact arising from the equations that have been used to model target cell replenishment (as opposed to a biological problem). With a little bit of imagination, it is possible to concoct target cell-limited equations that do not suffer from this knife-edge problem. If one replaces Eq. (1) in Eqs. (1)–(3) with

$$\frac{dT}{dt} = \frac{s}{a+T} - m - kVT, \tag{4}$$

for example, the steady-state viral load becomes

$$V^* = \frac{p}{\delta c}\left\lfloor \frac{psk}{apk+\delta c} - m \right\rfloor, \tag{5}$$

an equation for which V^* is no longer inversely related to k, eliminating the knife-edge problem. To defend this equation for dT/dt, however, one would have the following assumptions: (i) that a set number of target cells are removed each day, perhaps due to some kind of filtering mechanism that works independently of T-cell densities; and (ii) that there is a somewhat unusual nonlinear homeostatic response that pushes the body to reach an equilibrium T-cell density defined by $T^* = s/m - a$ in the absence of virus. While it is conceivable that these conjectures are correct, it is not the first equation that most mathematical biologists would think of when it comes to modeling T-cell dynamics.

Another possible explanation comes from Callaway and Perelson (2002) who have pointed out that a model in which the death rate of infected cells, δ, is a function of the density of infected cells [i.e., $\delta(I) = \delta' I^{\omega}$] does not suffer this shortcoming. Although this hypothesis is broadly consistent with analyses indicating that viral load decay following treatment in HIV-1 infected children is density dependent (Holte *et al.*, 2001), the value for ω that gives the most reasonable behavior in Callaway and Perelson's model ($\omega = 0.01$) is substantially lower than that estimated by Holte *et al.* ($0.4 < \omega < 0.47$). Furthermore, in attempting to verify this explanation one needs to be careful to distinguish between density dependent death that is due to immune responses and density dependent death that is due to other causes. Density dependent death that is due to immune responses may take one back to immune control models in which the mechanism of control is enhanced killing of HIV-1 infected cells by HIV-1-specific CTL.

Callaway and Perelson (2002) also show that the knife-edge problem can be alleviated by postulating heterogeneity in the susceptibility of target cell populations to antiretroviral drugs. The basic idea of their model is that viral load may be maintained at a lower level due to continued production of viruses in compartments in which drugs fail to complete block viral replication. An interesting prediction of this model is that virological blips (i.e. transient viremias seen in combination therapy patients whose plasma viral loads have fallen below the level of detection) will be bigger in patients whose viral loads have dropped the most rapidly during the first few weeks of therapy. These analyses point to the importance of virological compartments and drug sanctuary sites as potentially important factors that allow viruses to persist in HIV-1 infected patients.

The potential explanations above suggest a need for a better understanding of factors that replenish target cell populations, factors that alter the rate of the decay of infected cells during therapy, and of the roles of virological compartments and drug sanctuary sites within the body (Section 1.7). In order to test these models, detailed collaborations between modelers and experimentalists will be required. For mathematically inclined readers who do not like the explanations above, but who still want to use target-cell limited models, the challenge is to come up with alternative explanations. Given the numerous complexities in the homeostatic response for T-cells and, in particular, the rate at which the primary target cells for HIV, i.e. activated CD4+ T-cells, are created and destroyed, there is still room for creative models that can solve the knife-edge problem using mechanistically plausible interaction terms.

1.4. *What role does viral diversity play in affecting the rate of progression to AIDS in untreated patients?*

The extraordinarily high mutation rate of HIV-1 is thought to play a major contributing role to the pathogenesis of HIV-1 infection. The combination of a large population size (at least as measured by the number of virally infected cells, see Section 1.1 above) combined with a short generation time (Perelson *et al.*, 1996) and a high mutation rate (Mansky and Temin, 1995; Mansky, 1996) allows this virus to mutate into forms that can escape from HIV-specific CTL and antibody responses (Borrow *et al.*, 1997; Evans *et al.*, 1999; Richman *et al.*, 2003; Wei *et al.*, 2003). What is not clear, however, is if the high mutation rate for HIV-1 *in vivo* is the primary factor that allows causes HIV-1 to be harmful to the host (as opposed to playing an ancillary role). In other words, is viral diversity, *per se*, a cause of AIDS?

The idea that viral diversity is a cause of AIDS has gotten a lot of attention due to some very interesting and clever analyses conducted by Nowak and colleagues in the 1990's that posited that there is "diversity threshold" above which the number of viral immune escape mutants exceeds the ability of the immune system to respond to all of the HIV-1 variants in the body (Nowak *et al.*, 1991, 1992, 1995; Nowak and May, 2000). In their model, there is an HIV-1 specific T-cell clone that can control each viral variant.

In the absence of mutation, this HIV-1 specific immune response is capable of controlling viral infection. In the presence of mutation, however, new variants that can escape these HIV-1 specific immune responses continually arise. As each new HIV-1 escape mutant arises, a corresponding T-cell clone that is capable of recognizing and eliminating the new variant arises. Thus, in their model, there is a race between the ability of existing T-cell clones to suppress viral replication and the ability of the virus to mutate to new forms that can escape existing HIV-specific immune responses. Their model makes the prediction that once the diversity of the viral population exceeds a critical diversity threshold that the virus will overwhelm the ability of new T-cell clones to control viral replication. This threshold exists because each viral variant is capable of infecting and killing any given T-cell clone, but T-cell clones can only eliminate viruses that carry CTL epitopes that are recognized by those T-cell clones. This asymmetry gives a rapidly mutating virus, such as HIV, the potential to overwhelm the immune response.

This model is interesting in that it is capable of mimicking the primary viremia, the shift to steady-state viral load, and then, in cases where viral diversity exceeds the diversity threshold, increasing viral loads and progression to AIDS. The model has the appeal that the basic forces outlined in this model have all been demonstrated to occur *in vivo*. Recent studies of HIV-1 infections in humans and simian immunodeficiency virus (SIV) infections in macaques have clearly shown that viruses can evolve to escape antibody responses (Richman *et al.*, 2003; Wei *et al.*, 2003). Studies of viral load change during primary infection, moreover, have shown that rebound in viral loads after the initial decline from the peak viremia is associated with the appearance of specific CTL escape mutants (Borrow *et al.*, 1997; Evans *et al.*, 1999). What has not been confirmed is the central claim that it is the ability of the virus to exceed a critical diversity threshold that causes patients to progress to AIDS.

The models of Nowak and colleagues have been said to have been undercut by a study by Wolinsky *et al.* (1996) that showed, contrary to the most straight-forward interpretation of Nowak's model, that viral diversity is, if anything, somewhat higher in slow progressors than it is in fast progressors. In an interesting correspondence in Science, Nowak *et al.* argued that this could be because some of these fast progressors had a very low diversity threshold. Their model does, in fact, predict that after HIV-1 has overwhelmed the immune response that diversity will decline as fast-growing variants take over the population in the absence of an effective immune response. Nowak and Bangham (1996), furthermore, have shown that if

CTLs from different patients differ from each other in their responsiveness to HIV-1 epitopes, one would expect a negative correlation between viral load and viral diversity when looking at a cross-sectional sample of patients.

The idea that viral diversity may play a role in HIV-1 pathogenesis has been further stimulated in recent years by a highly detailed study by Shankarappa *et al.* (1999) showing that diversity in *env* increases at a rate of around 1% per year in untreated intermediate progressors before hitting a peak a few years into infection. Paralleling the initial increase in diversity, Shankarappa observed an increased "divergence" (genetic distance of viruses in blood or PBMC from viruses that were obtained from the patient at the first time point). Unlike diversity, the divergence did not clearly hit a peak, though the rate of increase in divergence appeared to slow down 1–2 years after diversity peaked. Shankarappa *et al.* also observed that X4 mutants (cytopathic viruses that can infect naïve CD4+ T-cells) appeared around the time that diversity peaked. Interestingly, the appearance of X4 mutants in these patients was transitory: they went from undetectable to high frequency ~2 years after the diversity peak before falling back again. Shankarappa also noted a marked decline in CD4+ T-cell counts after the appearance of these X4 mutants creating an apparent link between increases in viral diversity and progression to AIDS.

While the results of Wolinsky *et al.* (1996) and Shankarappa *et al.* (1999) are intriguing, it hard to say for sure whether these studies refute or support the diversity threshold hypothesis as modeled by Nowak *et al.* because the viral diversity measure that really counts for these models is the diversity in CTL and antibody epitopes. Furthermore, the relationship between diversity and steady-state viral load under dynamical models is complex: as noted above, under some model conditions one will obtain a negative relationship between viral diversity and steady-state viral load (Bangham and Nowak, 1996). The increase and subsequent plateau in diversity observed in Shankarappa, meanwhile, may be simply be the product of neutral mutation in the context of a low effective population size. Under the neutral theory of evolution, the rate of increase of diversity after primary infection, which is usually characterized by low viral diversity, will be 2μ, while the equilibrium diversity in a haploid population will be $2N_e\mu$. Given estimates of $\mu = {\sim}3 \times 10^{-5}$, $N_e = {\sim}1000$, and a generation time ~2 days (Perelson *et al.*, 1996), one would expect diversity to increase at rate of ~1% per year before stabilizing at an equilibrium level of ~5.5% (which is about what was observed). However, since estimates of N_e are greatly influenced by viral diversity, which could be the product of non-neutral forces, one cannot take this agreement as proof of neutral evolution in HIV-1 *env*.

Thus, there is a great deal of work that still needs to be done to fully address the question of how important viral diversity is as a factor that allows HIV-1 to persist in the body and ultimately cause AIDS in the absence of treatment. Much of this work will depend on the availability of detailed immunological data on changes in the frequencies of HIV-specific CTL and virological data on the changes in CTL and antibody epitopes in the virus. The latter is an area where recent advances in sequencing create the potential for greatly improved models for HIV-T-cell interactions. Until recently, attempts to identify CTL escape mutants have been confined to individual genes. Over the last couple of years, however, researchers have increasingly been creating "whole genome" sequences of HIV-1 from individual patients. There is no reason (other than cost and the need to treat patients as CD4 counts decline) that studies along the lines of Shankarappa in which 20 or so viruses were sequenced at regular intervals could not be repeated using the entire HIV-1 genome. When combined with improvements in antibody and CTL detection, this should provide data needed to construct and test more sophisticated models for host-pathogen interactions, including variants of Nowak *et al.*'s antigenic diversity threshold model. While this is not something that will come easily or cheaply, this is an area where mathematical biologists, working in conjunction with virologists, immunologists, and population geneticists have the potential to make major breakthroughs over the next several years.

1.5. *Can drug-therapy failure be predicted in advance?*

Although combination antiretroviral therapies have greatly reduced the mortality rate due to HIV-1 infections in the USA and other developed countries, they are not effective in all patients. In most patients, viral load rapidly falls below the limits of detection (generally around 50 copies of HIV-1 RNA per mL of blood plasma). In many patients, however, viral load will either fail to decrease to undetectable levels or transitorily decrease to undetectable levels before rebounding. These indications of "virological failure" can be due to several causes, including poor adsorption, toxicities, failure of the patient to take drugs as prescribed (which may be related to toxicities), and the evolution of drug resistant variants in the viral population. Whatever the cause, continued viral replication in the presence of drug greatly increases the probability that viruses will become resistant to drug.

Evolution of resistance in the face of continued drug therapy is particularly pernicious in that this virus can evolve a series of compensatory mutations that reduce some of the fitness costs (i.e. reductions in replication capacity in the absence of drug) associated with drug resistance (Nijhuis *et al.*, 1999). For most of the currently available antiretroviral drugs, resistance is a sequential process: an initial mutation confers partial resistance to drug, while subsequent mutations increase the level of drug resistance or reduce the fitness costs associated with resistance to the drug (Molla *et al.*, 1996). The fact that drug resistance is a sequential process makes it imperative to determine as early as possible when antiretroviral therapies are failing.

In attempting to determine if an antiretroviral therapy is likely to fail there are two general things that one can look for: factors (such as baseline CD4 T-cell count) that can be quantified prior to the administration of drug, and factors (such as the rate of decline of HIV-1 RNA following therapy) that can only be quantified once therapy has been initiated. As one might expect, the presence of drug resistance mutants in the patient prior to therapy is a strong predictor of whether antiretroviral therapy will be effective (see Section 1.6 below). However, more subtle correlations have also been noted. Mueller *et al.* (1998), for example, fit the model $V(t) = V_0[ae^{-k_a t} + be^{-k_b t}]$ to viral decay data. They reported that high baseline viral loads, low baseline CD4+ T-cell counts, low drug concentrations, and a low k_b value correlated with high viral 12 weeks into therapy. While the importance of the k_b value was somewhat unexpected, their other findings are consistent with other studies of treatment efficacy (Mocroft *et al.*, 1998; Deeks *et al.*, 1999; Lucas *et al.*, 1999; and Powderly *et al.*, 1999).

In a retrospective study of a clinical trial in which patients were put on to nelfinavir monotherapy (Markowitz *et al.*, 1998), we were interested in determining to what extent the rate of decline of HIV-1 RNA during the first few weeks of therapy would correlate with longer-term reductions in viral load (Mittler *et al.*, 2001). This was a productive set of patients to examine because there was wide range of responses, with some patients managing to suppress viral load and others having strong virological rebounds. Using a quantity that we referred to as the "relative efficacy" of therapy [defined as $log(V(x)/V(0))$ for study patients divided by the equivalent ratio in combination therapy patients who had excellent responses to therapy], we looked at how early during the drug treatment could one detect signs that viral loads were likely to rebound. Comparing the reduction in viral load at day 56 to measures of relative efficacy based on differing values for x, we found no correlation between the rate of viral load decline during the first seven days of therapy and viral load reductions at day 56. By day 14, however, we could detect a statistically significant correlation between early treatment response and later viral load reductions. Even stronger correlations were observed between relative efficacy at day 21 and viral load reductions at day 56. This suggests that for nelfinavir monotherapy that early symptoms of virological failure could be detected as early as 14 days after the initiation of drug, though stronger correlations were observed after 21 days of therapy.

A similar study by Polis *et al.* (2001) on patients who treated with indinavir- and ritonavir-containing regimens, by contrast, found that viral declines over the first six days of treatment were predictive of viral load concentrations at weeks 4, 8, and 12 post-therapy. The earlier predictive value of viral load reduction in their study may have been due to better responses to therapy in their study. They found, in agreement with Mittler *et al.* (2001) that in patients who did not respond as well to therapy that there was no correlation between viral load decline during the first week of therapy and longer term virological responses. Together these studies suggests that viral loads be carefully monitored over the first few weeks of therapy to verify that viral loads are dropping at rates comparable to that which has been observed in "optimal therapy" patients whose viral loads decline rapidly.

While the above-cited studies give some guidance as to how to monitor treatment regimens, there is still room for improved models. In contrast to some of the other problems discussed in this review in which fundamental issues remained unresolved, this is an area where conscientious and detailed application of existing viral dynamics models (Fraser *et al.*, 2001; Wu *et al.*, 1999, 2003, 2004; Ding and Wu, 1999, 2001) combined with rigorous statistical analyses of drug concentration levels and records of adverse clinical events could lead to improved indicators of impending treatment failure. With the existence of numerous phase three studies of new antiretroviral regimens, there is plenty of data available for analysis for those with the inclination and statistical tools to integrate large amounts of real world clinical data.

1.6. *How can we improve genotypic algorithms for predicting drug resistance?*

Recent clinical trials in which patients who had experienced virological failures were randomly assigned to either standard of care or standard of cared supplemented with genotypic test results (GART study) have shown that provision of genotypic information greatly improves clinical outcomes (Baxter *et al.*, 2000; Durant *et al.*, 1999). This has helped to expand the use of commercial and noncommercial genotypic and phenotypic testing technologies. Phenotypic assays, such as Virologic's PhenoSense® assay and Virco's Antivirogram® assay, give a direct estimate of the IC_{50} value (concentration of drug at which viral replication is inhibited by 50%) but are expensive and time consuming. Genotypic assays are faster and less expensive but can be hard to interpret due to the large number of mutations that lead to full resistance to antiretroviral drugs (Schafer, 2000). Interpretation of genotypic results is also complicated because different mutations have different effects: primary mutations allow virus to grow in the presence of drugs, whereas secondary mutations either further increase the degree of resistance or function in an epistatic fashion to reduce the fitness costs associated with drug resistance mutations, or both.

These difficulties have led to the creation of a variety of genotypic interpretation algorithms to help researchers predict which drugs are likely to suppress viral replication *in vivo*. The most widely used algorithms are "knowledge-based" methods that attempt to predict IC_{50} values from sequence-IC_{50} information obtained from databases, such as the Stanford HIV Drug Resistance Database (http://hivdb.stanford.edu). Current knowledge-based prediction tools have drawn upon a number of different pattern matching algorithms, including various "rules-based" methods, support-vector-machines, decision trees, and neural networks [reviewed in Schmidt *et al.* (2002) and Sturmer *et al.* (2003)]. Current algorithms, however, have not done as well as could be expected on new datasets (Braun *et al.*, 2002; Zolopa *et al.*, 2002; De Luca *et al.*, 2003). In a recent test of methods, we have found that a new method that employed stepwise linear regression performed well against other algorithms in predicting IC_{50} values from genotypic information in both hold-one-out jackknifing experiments and in tests on an independent data set (Wang *et al.*, 2004).

While these algorithms can, with varying degrees of success, predict IC_{50} in *in vitro* tests, this is not the same as having high accuracy in predicting which mutations will actually appear *in vivo*. Although predictions based on genotypic data have been shown to reduce the rate of virological rebound in patients infected with drug resistant viruses, they do a poor job at predicting exactly which mutations will actually appear in patients (Quigg *et al.*, 2002). Although imperfections in predicting IC_{50} values from genotypic data may account for some of these poor predictions, it cannot be the entire explanation because raw IC_{50} values themselves often fail to predict which mutations will occur *in vivo* (Ives *et al.*, 1997; Rubsamen–Waigmann *et al.*, 1999). Thus, despite the existence of some fairly sophisticated "knowledge-based" algorithms, there is still a big gap between predictions from knowledge-based models and what actually happens *in vivo*.

To bridge this gap, new types of models that integrate new and existing data will have to be developed. These new models will need to take several different factors into account. First, models will need to take into account differences in mutation rates. Given estimates of a low effective population size for HIV-1 *in vivo*, it has been predicted that stochastic processes will have a big impact on which mutations will occur *in vivo*. Evidence that stochastic mutational forces are important comes from the study cited in Section 1.1 by Frost *et al.* (2000) that demonstrated wide variation in the rate of appearance of M184V mutations in patients on 3TC monotherapy. Frost *et al.* found that M184V mutants have higher fitness in the presence of drug than M184I mutants, and yet M184I mutants generally appear before M184V mutants. This appears to be because the particular nucleotide mutation required to create a M184I mutation occurs at about 4 times the rate of the nucleotide mutation needed to create a M184V mutation.

Second, models will need to account for reductions in intrinsic replication capacity that accompany changes in drug resistance (Goudsmit *et al.*, 1996; Croteau *et al.*, 1997; Zennou *et al.*, 1998; Erickson *et al.*, 1999; Barbour *et al.*, 2002). Intrinsic replication capacity can affect both the number of mutants that are present in the body when therapy is initiated and the fate of mutants during therapy. As mentioned above, many of the mutations that are associated with drug resistance are compensatory mutations that reduce the fitness costs associated with primary drug resistance mutations. Current databases only list changes in the IC_{50} values relative to a wild-type reference strain. In addition to numerous individual studies by academic researchers [e.g., Martinez-Picado *et al.* (1999) and Mammano *et al.* (2000)], there are now commercially available methods for estimating viral replication capacity (http://www.virologic.com). Addition of this type of data to publicly available databases could allow researchers to create new types of models that take viral replication capacity into account.

Third, there is a need for models that integrate pharmacokinetic data (Drusano *et al.*, 1993) with genotypic models for IC_{50} prediction [reviewed in Schmidt *et al.* (2002) and Sturmer *et al.* (2003)]. The peaks and troughs in drug concentrations in plasma over the course of a day have been quantified in great detail for all of the currently available antiretroviral drugs as part of the FDA mandated drug approval process. This information can be important to drug resistance prediction because it is during periods when drug concentration is low that drug resistant mutants are most likely to increase in frequency. Although there have been some nice papers showing how viral dynamic models and pharmacokinetic models can, in principle, be combined (Austin *et al.*, 1998; Wahl and Nowak, 2000; Huang *et al.*, 2003), there is still a great need for models that combine these features with detailed information on viral genotypes to create quantitative models for resistance to specific antiretroviral drugs.

Fourth, there is a need for models that can better account for epistatic interactions between different drug resistance mutations (Hoffman *et al.*, 2003; Ohtaka *et al.*, 2003). Given the large number of mutations that occur against both protease and RT inhibitors, it becomes difficult to estimate interaction terms using knowledge-based algorithms. A recent structural modeling paper by Shenderovich *et al.* (2003) has demonstrated good correlations between *in vitro* IC_{50} values and theoretical predictions of protease-inhibitor binding energies. Although the accuracies of such structural models are not as high as those from knowledge-based methods, these structural approaches have the ability, in theory, to predict interaction terms. As computers get faster and as molecular dynamics researchers get better at modeling protein-ligand interactions, it may become possible to integrate predictions from molecular models with knowledge-based approaches to create highly accurate algorithms for predicting drug resistance.

1.7. *How can we identify important viral reservoirs, sanctuary sites, and compartments within the body?*

1.7.1. *Identification and quantification of viral reservoirs*

Although combination antiretroviral drug therapies rapidly push plasma HIV-1 RNA concentrations below limits of detection (Gulick *et al.*, 1997; Hammer *et al.*, 1997; Perelson *et al.*, 1997), latently infected resting memory CD4+ T-cells can persist for years despite fully suppressive therapy (Chun *et al.*, 1997b, 1998; Finzi *et al.*, 1997; Wong *et al.*, 1997). Estimates for the half-life of these cells range from 6 months (Zhang *et al.*, 1999; Ramratnam *et al.*, 2000) to 44 months or longer (Finzi *et al.*, 1999; Blankson *et al.*, 2002). The latter estimate suggests that it would take 60 years or more for HIV-1 to be cleared from the body under current combination therapy regimens. Virus may also persist on follicular dendritic cells (Hlavacek *et al.*, 1999, 2000) and in monocytes/macrophage (Igarashi, 2001; Zhu *et al.*, 2002). These infected cells may represent an additional impediment to curing patients of their HIV-1 infections.

Understanding the persistence of latently infected cells following potent combination therapy is complicated by the possibility that these populations may be replenished due to transient or low-level viral replication during combination therapy. Gunthard *et al.* (1999), and Zhang *et al.* (1999) observed continued evolution of sequences in patients on therapy. Furthermore, Ramratnam *et al.* (2000) observed a negative correlation between the decay rate of latently infected cells and transient viremias ("blips" in viral load) in patients who had gone on potent combination therapy. The findings of Ramratnam *et al.* (2000) differed from those of Finzi *et al.* (1999), who did not observe slower declines in a subset of patients with virological blips. In Ramratnam *et al.*'s (2000) study, the average half-life of latently infected cells in patients with no blips was ~6 months, while little to no decay was observed in patients with two or more blips per year. This suggests that viral replication during transient viremias can reseed the latently infected cell reservoir, potentially undoing years of progress in reducing the density of latently infected cells. Indeed, a follow-up study by Ramratnam *et al.* (2003) has shown that intensifying therapy (adding additional drugs to the combination therapy regimen) can accelerate the rate of decline of latently infected cells in patients in whom latently infected cells were declining slowly.

The replenishment controversy is important because it holds out the possibility that latently infected cell populations could be driven to extinction following the introduction of better combination therapies. Even if latently infected cells cannot be eradicated, the ability to drive these cells to extremely low levels may push the density of spontaneously appearing drug resistant variants to such low levels that clinicians could move patients to maintenance regimens that include fewer antiretroviral drugs (Wein *et al.*, 1998). The calculus underlying these arguments is, of course, complex and one would need to be especially cautious in patients who have failed to completely suppress viral replication during previous attempts to administer antiretroviral therapies. Despite these concerns, the potential to push infected cell populations to extremely low levels brings hope that induction-maintenance regimens could be used to reduce long-term toxicities associated with current combination therapy regimens. Again, this is an area (closely related to the material in Section 1.6) in which quantitative biologists may be able to make contributions.

The current debate about the longevity of latently infected cells is based on data on the rate of decline in the *density* of latently infected cells. Recent studies suggest that the longevity of viral reservoirs can also be inferred from HIV-1 sequence data. Hermankova *et al.* (2001), Imamichi *et al.* (2001), Ruff *et al.* (2002), and Frenkel *et al.* (2003) have all observed reductions in the genetic "divergence" [genetic distance from the most recent common ancestor (MRCA) for viruses from that patient] of viruses obtained from patients following potent combination therapies. This contrasts with the more or less linear increase in genetic divergence that occurs in patients who do not go on effective combination therapies (Shankarappa *et al.*, 1999). A natural interpretation of these post-therapy divergence reductions is that viruses enter into latent reservoirs early during infection and increase in frequency (though not necessarily in absolute density) after antiretroviral therapy has reduced the density of actively replicating viral populations (Mullins, 2001; Ruff *et al.*, 2002; Nickle *et al.*, 2003).

Of particular interest is the possibility that these divergence reductions could be used to make quantitative deductions concerning the longevity of latently infected reservoirs. Ruff *et al.* (2002), for example, have calculated that if the latent reservoir had a half-life of six months, that "$> 98\%$ of the original latent reservoir sequences would be replaced in 3 years." The general thrust of this argument is correct: if latently infected cells turn over rapidly, then "older" sequences would eventually be replaced by "newer" sequences. However, their "$> 98\%$" figure should be interpreted cautiously. Their "$> 98\%$" figure makes the simplifying assumption that therapy is 100% effective and that the rate of input of viruses into latently infected reservoirs is constant over time prior to HAART. It is well known that viral loads climb to high levels during the acute phase of infection before decreasing to patient-specific viral setpoints which gradually increase as patients progress to AIDS. Antiviral therapies, furthermore, are not always 100% suppressive.

To do these calculations properly, models that account for changes in viral load and other factors that effect the creation and destruction of latently infected cells will need to constructed. In particular, it should be possible to use differential equation models that relate $V_m(t)$, the concentration of viruses with m mutations at time t in the actively replicating compartment, to $L_m(t)$, the number of viruses in latently infected cells with m mutations at time t. By combining pre-treatment viral loads, viral divergence data along the lines of that obtained in Shankarappa *et al.* (1999), and data on the divergence of viruses in latently infected cells, it should be possible to use models of this sort to estimate the death rate of latently infected cells in different compartments. Nickle *et al.* (2003) have used such an approach to show that pleural cavity in the lung may be latent reservoir for HIV-1. Models such as the one in Nickle *et al.* (2003) also have a theoretical appeal in that they combine models for viral dynamics with phylogenetic models for viral diversification and divergence.

1.7.2. *Distinguishing between latent reservoirs, compartments and drug sanctuary sites*

If we are to devise strategies for purging patients of HIV-1 infected cells following antiretroviral therapies, it will be essential to identify not only where virus comes from, but also why it persists in different tissues and organs. In addition to the possibility that the virus persists within a compartment because it contains long-lived infected cells, the virus may persist because antiretroviral drugs fail to penetrate that compartment (Kepler and Perelson, 1998) or because viruses continually enter that compartment from other organs or tissues in which viral replication is not suppressed.

Distinguishing between these latter hypotheses can be addressed through the rigorous application of phylogenetic models. If the virus is continually entering a particular compartment from other compartments, one should see intermingling of sequences from these sequences on a phylogenetic tree. If a compartment is isolated from other compartments, by contrast, the sequences should cluster in separate clades on a phylogenetic tree (Nickle *et al.*, 2003). If the virus persists in a compartment because it is a drug sanctuary site, one should see continued evidence of viral evolution over time (as opposed to the reduced divergences seen in latent reservoirs). Of course, drug sanctuary sites can also be identified by estimating drug concentrations and obtaining sequential estimates of the number of infected cells over time.

While the general strategy is easy to describe, the testing of these hypothesis requires a considerable amount of work. To begin with, it will require that experimentalists obtain multiple sequences from different tissues at different points in time. Even with these sequences, however, the phylogenetic and evolutionary interpretations are not easy. In addition to constructing correct models of evolution and constructing accurate phylogenetic trees, one must account for the effects that recombination and selection (both purifying and diversifying) have on these immigration and growth parameters. Thus, there are plenty of problems for phylogenetically inclined computational biologists to solve in attempting to determine various causes for the continued persistence of virus in patients on combination antiretroviral therapies.

Acknowledgments

The author acknowledges grant support from the National Institutes of Health (R01-HL72631, R03-AI55394, and R21-AI52063).

References

1. Amara RR, Robinson HL (2002) A new generation of HIV vaccines. *Trends Mol Med* **8**: 489–495.
2. Austin DJ, White NJ, Anderson RM (1998) The dynamics of drug action on the within-host population growth of infectious agents: melding pharmacokinetics with pathogen population dynamics. *J Theor Biol* **194**: 313–339.
3. Barbour JD, Wrin T, Grant RM, Martin JN, Segal MR, Petropoulos CJ, Deeks SG (2002) Evolution of phenotypic drug susceptibility and viral replication capacity during long-term virologic failure of protease inhibitor therapy in human immunodeficiency virus-infected adults. *J Virol* **76**: 11104–11112.
4. Baxter JD, Mayers DL, Wentworth DN, Neaton JD, Hoover ML, Winters MA, Mannheimer SB, Thompson MA, Abrams DI, Brizz BJ, Ioannidis JP, Merigan TC (2000) A randomized study of antiretroviral management based on plasma genotypic antiretroviral resistance testing in patients failing therapy. CPCRA 046 Study Team for the Terry Beirn Community Programs for Clinical Research on AIDS. *AIDS* **14**: F83–93.
5. Bhardwaj N, Walker BD (2003) Immunotherapy for AIDS virus infections: cautious optimism for cell-based vaccine. *Nat Med* **9**: 13–14.
6. Blankson JN, Persaud D, Siliciano RF (2002) The challenge of viral reservoirs in HIV-1 infection. *Annu Rev Med* **53**: 557–593.
7. Bonhoeffer S, Coffin JM, Nowak MA (1997) Human immunodeficiency virus drug therapy and virus load. *J Virol* **71**: 3275–3278.
8. Bonhoeffer S, Funk GA, Gunthard HF, Fischer M, Muller V (2003) Glancing behind virus load variation in HIV-1 infection. *Trends Microbiol* **11**: 499–504.
9. Borrow P, Lewicki H, Hahn BH, Shaw GM, Oldstone MB (1994) Virus-specific CD8+ cytotoxic T-lymphocyte activity associated with control of viremia in primary human immunodeficiency virus type 1 infection. *J Virol* **68**: 6103–6110.
10. Borrow P, Lewicki H, Wei X, Horwitz MS, Peffer N, Meyers H, Nelson JA, Gairin JE, Hahn BH, Oldstone MB, Shaw GM (1997) Antiviral pressure

exerted by HIV-1-specific cytotoxic T lymphocytes (CTLs) during primary infection demonstrated by rapid selection of CTL escape virus. *Nat Med* **3**: 205–211.
11. Braun P, Helm M, Ehret R, Schmidt B, Stürner K, Walter H, Hoehn C, Korn K, Knechten H (2002) Predictive value of different drug resistance interpretation systems in therapy management of HIV infected patients in daily routine. *Antivir Ther* **7**: S77.
12. Bucy RP, Hockett RD, Derdeyn CA, Saag MS, Squires K, Sillers M, Mitsuyasu RT, Kilby JM (1999) Initial increase in blood CD4(+) lymphocytes after HIV antiretroviral therapy reflects redistribution from lymphoid tissues. *J Clin Invest* **103**: 1391–1398.
13. Callaway DS, Perelson AS (2002) HIV-1 infection and low steady state viral loads. *Bull Math Biol* **64**: 29–64.
14. Carpenter CC, Cooper DA, Fischl MA, Gatell JM, Gazzard BG, Hammer SM, Hirsch MS, Jacobsen DM, Katzenstein DA, Montaner JS, Richman DD, Saag MS, Schechter M, Schooley RT, Thompson MA, Vella S, Yeni PG, Volberding PA (2000) Antiretroviral therapy in adults: updated recommendations of the International AIDS Society-USA Panel. *J Amer Med Assoc* **283**: 381–390.
15. Cooper DA, Gold J, Maclean P, Donovan B, Finlayson R, Barnes TG, Michelmore HM, Brooke P, Penny R (1985) Acute AIDS retrovirus infection. Definition of a clinical illness associated with seroconversion. *Lancet* **1**: 537–540.
16. Cooper LN (1986) Theory of an immune system retrovirus. *Proc Natl Acad Sci USA* **83**: 9159–9163.
17. Coffin JM (1995) HIV population dynamics *in vivo*: implications for genetic variation, pathogenesis, therapy. *Science* **267**: 483–489.
18. Croteau G, Doyon L, Thibeault D, McKercher G, Pilote L, Lamarre D (1997) Impaired fitness of human immunodeficiency virus type 1 variants with high-level resistance to protease inhibitors. *J Virol* **71**: 1089–1096.
19. Crow JF, Kimura M (1970) *An Introduction to Population Genetics Theory.* Harper and Row, New York, N.Y.
20. Chun TW, Carruth L, Finzi D, Shen X, Digiuseppe JA, Taylor H, Hermankova M, Chadwick K, Margolick J, Quinn TC, Kuo Y-H, Brookmeyer R, Zeiger MA, Barditch-Crovo P, Siliciano RF (1997a) Quantitation of latent tissue reservoirs and total body load in HIV-1 infection. *Nature* **387**: 183–188.
21. Chun TW, Engel D, Berrey MM, Shea T, Corey L, Fauci AS (1998) Early establishment of a pool of latently infected, resting CD4+ T cells during primary HIV-1 infection. *Proc Natl Acad Sci USA* **95**: 8869–8873.
22. Chun TW, Stuyver L, Mizell SB, Ehler LA, Mican JA, Baseler M, Lloyd AL, Nowak MA, Fauci AS (1997b) Presence of an inducible HIV-1 latent reservoir during highly active antiretroviral therapy. *Proc Natl Acad Sci USA* **94**: 13193–13197.
23. De Boer RJ, Perelson AS (1998) Target cell limited and immune control models of HIV infection: a comparison. *J Theor Biol* **190**: 201–214.

24. De Luca A, Cingolani A, Giambenedetto SD, Trotta MP, Baldini F, Rizzo MG, Bertoli A, Liuzzi G, Narciso P, Murri R, Ammassari A, Perno CF, Antinori A (2003) Variable prediction of antiretroviral treatment outcome by different systems for interpreting genotypic human immunodeficiency virus type 1 drug resistance. *J Infect Dis* **187**: 1934–1943.
25. Deeks SG, Hecht FM, Swanson M, Elbeik T, Loftus R, Cohen PT, Grant RM (1999) HIV RNA and CD4 cell count response to protease inhibitor therapy in an urban AIDS clinic: Response to both initial and salvage therapy. *AIDS* **13**: F35–43.
26. Dimitrov DS, Willey RL, Sato H, Chang LJ, Blumenthal R, Martin MA (1993) Quantitation of human immunodeficiency virus type 1 infection kinetics. *J Virol* **67**: 2182–2190.
27. Ding AA, Wu H (1999) Relationships between antiviral treatment effects and biphasic viral decay rates in modeling HIV dynamics. *Math Biosci* **160**: 63–82.
28. Ding AA, Wu H (2001) Assessing antiviral potency of anti-HIV therapies *in vivo* by comparing viral decay rates in viral dynamic models. *Biostatistics* **2**: 13–29.
29. Drusano GL (1993) Pharmacodynamics of antiretroviral chemotherapy. *Infect Control Hosp Epidemiol* **14**: 530–536.
30. Durant J, Clevenbergh P, Halfon P, Delgiudice P, Porsin S, Simonet P, Montagne N, Boucher CA, Schapiro JM, Dellamonica P (1999) Drug-resistance genotyping in HIV-1 therapy: the VIRADAPT randomised controlled trial. *Lancet* **353**: 2195–2199.
31. Embretson J, Zupancic M, Ribas JL, Burke A, Racz P, Tenner-Racz K, Haase AT (1993) Massive covert infection of helper T lymphocytes and macrophages by HIV during the incubation period of AIDS. *Nature* **362**: 359–362.
32. Erickson JW, Gulnik SV, Markowitz M (1999) Protease inhibitors: resistance, cross-resistance, fitness and the choice of initial and salvage therapies. *AIDS* **13** Suppl A: S189–204.
33. Eron JJ, Benoit SL, Jemsek J, MacArthur RD, Santana J, Quinn JB, Kuritzkes DR, Fallon MA, Rubin M (1995) Treatment with lamivudine, zidovudine, or both in HIV-positive patients with 200 to 500 CD4+ cells per cubic millimeter. North American HIV Working Party. *New Engl J Med* **333**: 1662–1669.
34. Essunger P, Perelson AS (1994) Modeling HIV infection of CD4+ T-cell subpopulations. *J Theor Biol* **170**: 367–391.
35. Evans DT, O'Connor DH, Jing P, Dzuris JL, Sidney J, da Silva J, Allen TM, Horton H, Venham JE, Rudersdorf RA, Vogel T, Pauza CD, Bontrop RE, DeMars R, Sette A, Hughes AL, Watkins DI (1999) Virus-specific cytotoxic T-lymphocyte responses select for amino-acid variation in simian immunodeficiency virus Env and Nef. *Nat Med* **5**: 1270–1276.
36. Finzi D, Blankson J, Siliciano JD, Margolick JB, Chadwick K, Pierson T, Smith K, Lisziewicz J, Lori F, Flexner C, Quinn TC, Chaisson RE, Rosenberg E, Walker B, Gange S, Gallant J, Siliciano RF (1999) Latent

infection of CD4+ T cells provides a mechanism for lifelong persistence of HIV-1, even in patients on effective combination therapy. *Nat Med* **5**: 512–517.
37. Finzi D, Hermankova M, Pierson T, Carruth LM, Buck C, Chaisson RE, Quinn TC, Chadwick K, Margolick J, Brookmeyer R, Gallant J, Markowitz M, Ho DD, Richman D, Siliciano RF (1997) Identification of a reservoir for HIV-1 in patients on highly active antiretroviral therapy. *Science* **278**: 1295–1300.
38. Fraser C, Ferguson NM, Anderson RM (2001) Quantification of intrinsic residual viral replication in treated HIV-infected patients. *Proc Natl Acad Sci USA* **98**: 15167–15172.
39. Frenkel LM, Wang Y, Learn GH, McKernan JL, Ellis GM, Mohan KM, Holte SE, De Vange SM, Pawluk DM, Melvin AJ, Lewis PF, Heath LM, Beck IA, Mahalanabis M, Naugler WE, Tobin NH, Mullins JI (2003) Multiple viral genetic analyses detect low-level human immunodeficiency virus type 1 replication during effective highly active antiretroviral therapy. *J Virol* **77**: 5721–5730.
40. Frost SD, Dumaurier MJ, Wain–Hobson S, Brown AJ (2001) Genetic drift and within-host metapopulation dynamics of HIV-1 infection. *Proc Natl Acad Sci USA* **98**: 6975–6980.
41. Frost SD, Nijhuis M, Schuurman R, Boucher CAB, Leigh Brown AJ (2000) Evolution of lamivudine resistance in human immunodeficiency virus type 1-infected individuals: the relative roles of drift and selection. *J Virol* **74**: 6262–6268.
42. Gaines H, von Sydow M, Pehrson PO, Lundbegh P (1988) Clinical picture of primary HIV infection presenting as a glandular-fever-like illness. *BMJ* **297**: 1363–1368.
43. Gaines H, von Sydow MA, von Stedingk LV, Biberfeld G, Bottiger B, Hansson LO, Lundbergh P, Sonnerborg AB, Wasserman J, Strannegaard OO (1990) Immunological changes in primary HIV-1 infection. *AIDS* **4**: 995–999.
44. Goudsmit J, de Ronde A, de Rooij E, de Boer R (1997) Broad spectrum of *in vivo* fitness of human immunodeficiency virus type 1 subpopulations differing at reverse transcriptase codons 41 and 215. *J Virol* **71**: 4479–4484.
45. Gulick RM, Mellors JW, Havlir D, Eron JJ, Gonzalez C, McMahon D, Richman DD, Valentine FT, Jonas L, Meibohm A, Emini EA, Chodakewitz JA (1997) Treatment with indinavir, zidovudine, and lamivudine in adults with human immunodeficiency virus infection and prior antiretroviral therapy. *New Engl J Med* **337**: 734–739.
46. Gunthard HF, Frost SDW, Leigh Brown AJ, Ignacio CC, Kee K, Perelson AS, Spina CA, Havlir DV, Hezareh M, Looney DJ, Richman DD, Wong JK (1999) Evolution of envelope sequences of human immunodeficiency virus type 1 in cellular reservoirs in the setting of potent antiviral therapy. *J Virol* **73**: 9404–9412.
47. Haase AT (1999) Population biology of HIV-1 infection: viral and CD4+ T cell demographics and dynamics in lymphatic tissues. *Annu Rev Immunol* **17**: 625–656.

48. Haase AT, Retzel EF, Staskus KA (1990) Amplification and detection of lentiviral DNA inside cells. *Proc Natl Acad Sci USA* **87**: 4971–7495.
49. Hammer SM, Squires KE, Hughes MD, Grimes JM, Demeter LM, Currier JS, Eron JJ, Feinberg JE, Balfour HH, Dayton LR, Chodakewitz JA, Fischl MA (1997) A controlled trial of two nucleoside analogues plus indinavir in persons with human immunodeficiency virus infection and CD4 cell counts of 200 per cubic millimeter or less. *New Engl J Med* **337**: 725–733.
50. Hartl DL, Clark AG (1989) *Principles of Population Genetics.* Sinauer Associates, Sunderland, Massachusetts.
51. Hermankova M, Ray SC, Ruff C, Powell-Davis M, Ingersoll R, D'Aquila RT, Quinn TC, Siliciano JD, Siliciano RF, Persaud D (2001) HIV-1 drug resistance profiles in children and adults with viral load of <50 copies/ml receiving combination therapy. *JAMA* **286**: 196–207.
52. Hlavacek WS, Stilianakis NI, Notermans DW, Danner SA, Perelson AS (2000) Influence of follicular dendritic cells on decay of HIV during antiretroviral therapy. *Proc Natl Acad Sci USA* **97**: 10966–10971.
53. Hlavacek WS, Wofsy C, Perelson AS (1999) Dissociation of HIV-1 from follicular dendritic cells during HAART: mathematical analysis. *Proc Natl Acad Sci USA* **96**: 14681–14686.
54. Hockett RD, Kilby JM, Derdeyn CA, Saag MS, Sillers M, Squires K, Chiz S, Nowak MA, Shaw GM, Bucy RP (1999) Constant mean viral copy number per infected cell in tissues regardless of high, low, or undetectable plasma HIV RNA. *J Exp Med* **189**: 1545–1554.
55. Hoffman NG, Schiffer CA, Swanstrom R (2003) Covariation of amino acid positions in HIV-1 protease. *Virology* **314**: 536–548.
56. Huang Y, Rosenkranz SL, Wu H (2003) Modeling HIV dynamics and antiviral response with consideration of time-varying drug exposures, adherence and phenotypic sensitivity. *Math Biosci* **184**: 165–186.
57. Igarashi T, Brown CR, Endo Y, Buckler–White A, Plishka R, Bischofberger N, Hirsch V, Martin MA (2001) Macrophage are the principal reservoir and sustain high virus loads in rhesus macaques after the depletion of CD4+ T cells by a highly pathogenic simian immunodeficiency virus/HIV type 1 chimera (SHIV): Implications for HIV-1 infections of humans. *Proc Natl Acad Sci USA* **98**: 658–863.
58. Imamichi H, Crandall KA, Natarajan V, Jiang MK, Dewar RL, Berg S, Gaddam A, Bosche M, Metcalf JA, Davey RT Jr, Lane HC (2001) Human immunodeficiency virus type 1 quasi species that rebound after discontinuation of highly active antiretroviral therapy are similar to the viral quasi species present before initiation of therapy. *J Infect Dis* **183**: 36–50.
59. Ives KJ, Jacobsen H, Galpin SA, Garaev MM, Dorrell L, Mous J, Bragman K, Weber JN (1997) Emergence of resistant variants of HIV *in vivo* during monotherapy with the proteinase inhibitor saquinavir. *J Antimicrob Chemother* **39**: 771–779.
60. Jin X, Bauer DE, Tuttleton SE, Lewin S, Gettie A, Blanchard J, Irwin CE, Safrit JT, Mittler J, Weinberger L, Kostrikis LG, Zhang L, Perelson AS,

Ho DD (1999) Dramatic rise in plasma viremia after CD8(+) T cell depletion in simian immunodeficiency virus-infected macaques. *J Exp Med* **189**: 991–998.
61. Kaufmann GR, Cunningham P, Kelleher AD, Zaunders J, Carr A, Vizzard J, Law M, Cooper DA (1998) Patterns of viral dynamics during primary human immunodeficiency virus type 1 infection. The Sydney Primary HIV Infection Study Group. *J Infect Dis* **178**: 1812–1815.
62. Kaufmann D, Pantaleo G, Sudre P, Telenti A (1998) CD4-cell count in HIV-1-infected individuals remaining viraemic with highly active antiretroviral therapy (HAART). Swiss cohort study. *Lancet* **351:** 723–724.
63. Kaech SM, Ahmed R (2001) Memory CD8+ T cell differentiation: initial antigen encounter triggers a developmental program in naive cells. *Nature Immunol* **2**: 415–422.
64. Kepler TB, Perelson AS (1998) Drug concentration heterogeneity facilitates the evolution of drug resistance. *Proc Natl Acad Sci USA* **95**: 11514–11519.
65. Kinloch-de Loes S, de Saussure P, Saurat JH, Stalder H, Hirschel B, Perrin LH (1993) Symptomatic primary infection due to human immunodeficiency virus type 1: review of 31 cases. *Clin Infect Dis* **17**: 59–65.
66. Kirschner D, Webb GF, Cloyd M (2000) Model of HIV-1 disease progression based on virus-induced lymph node homing and homing-induced apoptosis of CD4+ lymphocytes. *J Acquir Immune Defic Syndr* **24**: 352–362.
67. Korthals Altes H, Ribeiro RM, de Boer RJ (2003) The race between initial T-helper expansion and virus growth upon HIV infection influences polyclonality of the response and viral set-point. *Proc R Soc Lond B Biol Sci* **270**: 1349–1358.
68. Kuhner MK, Yamato J, Felsenstein J (1995) Estimating effective population size and mutation rate from sequence data using Metropolis-Hastings sampling. *Genetics* **140**: 1421–1430.
69. Leigh Brown AJ (1997) Analysis of HIV-1 env gene sequences reveals evidence for a low effective number in the viral population. *Proc Natl Acad Sci USA* **94**: 1862–1865.
70. Leigh Brown AJ, Richman DD (1997) HIV-1: gambling on the evolution of drug resistance? *Nat Med* **3**: 268–271.
71. Lindback S, Karlsson AC, Mittler J, Blaxhult A, Carlsson M, Briheim G, Sonnerborg A, Gaines H (2000) Viral dynamics in primary HIV-1 infection. Karolinska Institute Primary HIV Infection Study Group. *AIDS* **14**: 2283–2291.
72. Lucas GM, Chaisson RE, Moore RD (1999) Highly active antiretroviral therapy in a large urban clinic: Risk factors for virologic failure and adverse drug reactions. *Annals Intern Med* **131**: 81–87.
73. Mammano F, Trouplin V, Zennou V, Clavel F (2000) Retracing the evolutionary pathways of human immunodeficiency virus type 1 resistance to protease inhibitors: virus fitness in the absence and in the presence of drug. *J Virol* **74**: 8524–8531.

74. Mansky LM (1996) Forward mutation rate of human immunodeficiency virus type 1 in a T lymphoid cell line. *AIDS Res Hum Retroviruses* **12**: 307–314.
75. Mansky LM, Temin HM (1995) Lower *in vivo* mutation rate of human immunodeficiency virus type 1 than that predicted from the fidelity of purified reverse transcriptase. *J Virol* **69**: 5087–5094.
76. Margolick JB, Munoz A, Donnenberg AD, Park LP, Galai N, Giorgi JV, O'Gorman MR, Ferbas J (1995) Failure of T-cell homeostasis preceding AIDS in HIV-1 infection. The Multicenter AIDS Cohort Study. *Nat Med* **1**: 674–680.
77. Markowitz M, Conant M, Hurley A, Schluger R, Duran M, Peterkin J, Chapman S, Patick A, Hendricks A, Yuen GJ, Hoskins W, Clendeninn N, Ho DD (1998) A preliminary evaluation of nelfinavir mesylate, an inhibitor of human immunodeficiency virus (HIV)-1 protease, to treat HIV infection. *J Infect Dis* **177**: 1533–1540.
78. Martinez-Picado J, Savara AV, Sutton L, D'Aquila RT (1999) Replicative fitness of protease inhibitor-resistant mutants of human immunodeficiency virus type 1. *J Virol* **73**: 3744–3752.
79. McLean AR, Nowak MA (1992) Models of interactions between HIV and other pathogens. *J Theor Biol* **155**: 69–86.
80. Mellors JW, Rinaldo CR Jr, Gupta P, White RM, Todd JA, Kingsley LA (1996) Prognosis in HIV-1 infection predicted by the quantity of virus in plasma. *Science* **272**: 1167–1170.
81. Mittler JE, Essunger P, Markowitz M, Yuen GJ, Clendeninn N, Ho DD, Perelson AS (2001) Short-term measures of relative efficacy predict longer-term reductions in HIV-1 RNA following antiretroviral treatment. *Antimicrob Agents Chemo* **45**: 1438–1443.
82. Mittler JE, Levin BR, Antia R (1996) T-cell homeostasis, competition and drift: AIDS as HIV-Accelerated Senescence of the Immune Repertoire. *J Acquir Immune Defic Syndr Hum Retrovirol* **12**: 233–248.
83. Mocroft A, Gill MJ, Davidson W, Phillips AN (1998) Predictors of a viral response and subsequent virological treatment failure in patients with HIV starting a protease inhibitor. *AIDS* **12**: 2161–2167.
84. Molla A, Korneyeva M, Gao Q, Vasavanonda S, Schipper PJ, Mo HM, Markowitz M, Chernyavskiy T, Niu P, Lyons N, Hsu A, Granneman GR, Ho DD, Boucher CA, Leonard JM, Norbeck DW, Kempf DJ (1996) Ordered accumulation of mutations in HIV protease confers resistance to ritonavir. *Nat Med* **2**: 760–766.
85. Mueller BU, Zeichner SL, Kuznetsov VA, Heath–Chiozzi M, Pizzo PA, Dimitrov DS (1998) Individual prognoses of long-term responses to antiretroviral treatment based on virological, immunological and pharmacological parameters measured during the first week under therapy. *AIDS* **12**: F191–196.

86. Muller V, Bonhoeffer S (2003) Mathematical approaches in the study of viral kinetics and drug resistance in HIV-1 infection. *Curr Drug Targets Infect Dis* **3**: 329–344.
87. Muller V, Maree AF, De Boer RJ (2001) Small variations in multiple parameters account for wide variations in HIV-1 set-points: a novel modelling approach. *Proc R Soc Lond B Biol Sci* **268**: 235–242.
88. Mullins JI (2001) Compartments and reservoirs of HIV infection *in vivo*. Keystone Symposium on AIDS Pathogenesis. Keystone Symposia, Silverthorne, Colorado.
89. Murray JM, Kaufmann G, Kelleher AD, Cooper DA (1998) A model of primary HIV-1 infection. *Math Biosci* **154**: 57–85.
90. Nickle DC, Jensen MA, Shriner D, Brodie SJ, Frenkel LM, Mittler JE, Mullins JI (2003) Evolutionary indicators of human immunodeficiency virus type 1 reservoirs and compartments. *J Virol* **77**: 5540–5546.
91. Nijhuis M, Boucher CA, Schipper P, Leitner T, Schuurman R, Albert J (1998) Stochastic processes strongly influence HIV-1 evolution during suboptimal protease-inhibitor therapy. *Proc Natl Acad Sci USA* **95**: 14441–14446.
92. Nijhuis M, Schuurman R, de Jong D, Erickson J, Gustchina E, Albert J, Schipper P, Gulnik S, Boucher CA (1999) Increased fitness of drug resistant HIV-1 protease as a result of acquisition of compensatory mutations during suboptimal therapy. *AIDS* **13**: 2349–2359.
93. Notermans DW, Goudsmit J, Danner SA, de Wolf F, Perelson AS, Mittler J (1998) Rate of HIV-1 decline following antiretroviral therapy is related to viral load at baseline and drug regimen. *AIDS* **12**: 1483–1490.
94. Nowak MA (1992) Variability of HIV infections. *J Theor Biol* **7**: 1–20.
95. Nowak MA, Anderson RM, McLean AR, Wolfs TF, Goudsmit J, May RM (1991) Antigenic diversity thresholds and the development of AIDS. *Science* **254**: 963–969.
96. Nowak MA, Bangham CR (1996) Population dynamics of immune responses to persistent viruses. *Science* **272**: 74–79.
97. Nowak MA, May RM (2000). *Virus Dynamics. Mathematical Principles of Immunology and Virology.* Oxford University Press, Oxford.
98. Nowak MA, May RM, Phillips RE, Rowland-Jones S, Lalloo DG, McAdam S, Klenerman P, Koppe B, Sigmund K, Bangham CR, *et al.* (1995) Antigenic oscillations and shifting immunodominance in HIV-1 infections. *Nature* **375**: 606–611.
99. Ohtaka H, Schon A, Freire E (2003) Multidrug resistance to HIV-1 protease inhibition requires cooperative coupling between distal mutations. *Biochemistry* **42**: 13659–13666.
100. Palella FJ, Delaney KM, Moorman AC (1998) Declining morbidity and mortality among patients with advanced human immunodeficiency virus infection. *New Engl J Med* **338**: 853–860.
101. Pantaleo G, Graziosi C, Demarest JF, Butini L, Montroni M, Fox CH, Orenstein JM, Kotler DP, Fauci AS (1993) HIV infection is active and progressive in lymphoid tissue during the clinically latent stage of disease. *Nature* **362**: 355–358.

102. Perelson AS (1989) Modeling the interaction of the immune system with HIV, in Castillo–Chavez C (ed.) *Mathematical and Statistical Approaches to AIDS Epidemiology*, pp. 350–370, Springer–Verlag, New York.
103. Perelson AS, Essunger P, Cao Y, Vesanen M, Hurley A, Saksela K, Markowitz M, Ho DD (1997) Decay characteristics of HIV-1-infected compartments during combination therapy. *Nature* **387**: 188–191.
104. Perelson AS, Kirschner DE, de Boer R (1993) Dynamics of HIV infection of CD4+ T cells. *Math Biosci* **114**: 81–125.
105. Perelson AS, Nelson PW (1999) Mathematical analysis of HIV-1 dynamics *in vivo*. *SIAM Rev* **41**: 3–44.
106. Perelson AS, Neuman AU, Markowitz M, Leonard JM, Ho DD (1996) HIV-1 dynamics *in vivo*: virion clearance rate, infected cell life-span, and viral generation time. *Science* **271**: 1582–1586.
107. Perrin L, Telenti A (1998) HIV treatment failure: testing for HIV resistance in clinical practice. *Science* **280**: 1871–1873.
108. Piatak M, Saag MS, Yang LC, Clark SJ, Kappes JC, Luk KC, Hahn BH, Shaw GM, Lifson JD (1993) High levels of HIV-1 in plasma during all stages of infection determined by competitive PCR. *Science* **259**: 1749–1754.
109. Piketty C, Castiel P, Belec L, Batisse D, Si Mohamed A, Gilquin J, Gonzalez–Canali G, Jayle D, Karmochkine M, Weiss L, Aboulker JP, Kazatchkine MD (1998) Discrepant responses to triple combination antiretroviral therapy in advanced HIV disease. *AIDS* **12**: 745–750.
110. Polis MA, Sidorov IA, Yoder C, Jankelevich S, Metcalf J, Mueller BU, Dimitrov MA, Pizzo P, Yarchoan R, Dimitrov DS (2001) Correlation between reduction in plasma HIV-1 RNA concentration 1 week after start of antiretroviral treatment and longer-term efficacy. *Lancet* **358**: 1760–1765.
111. Powderly WG, Saag MS, Chapman S, Yu G, Quart B, Clendeninn NJ (1999) Predictors of optimal virological response to potent antiretroviral therapy. *AIDS* **13**: 1873–1880.
112. Quigg M, Frost SD, McDonagh S, Burns SM, Clutterbuck D, McMillan A, Leen CS, Brown AJ (2002) Association of antiretroviral resistance genotypes with response to therapy-comparison of three models. *Antivir Ther* **7**: 151–157.
113. Quinones-Mateu ME, Ball SC, Marozsan AJ, Torre VS, Albright JL, Vanham G, van Der Groen G, Colebunders RL, Arts EJ (2000) A dual infection/competition assay shows a correlation between *ex vivo* human-immunodeficiency virus type 1 fitness and disease progression. *J Virol* **74**: 9222–9233.
114. Ramratnam B, Mittler JE, Zhang L, Boden D, Hurley A, Fang F, Macken CA, Perelson AS, Markowitz M, Ho DD (2000) The decay of the latent reservoir of replication-competent HIV-1 is inversely correlated with the extent of residual viral replication during prolonged anti-retroviral therapy. *Nat Med* **6**: 82–85.
115. Ramratnam B, Ribeiro R, He T, Chung C, Simon V, Vanderhoeven J, Hurley A, Zhang L, Perelson AS, Ho DD, Markowitz M (2004) Intensification of antiretroviral therapy accelerates the decay of the HIV-1

latent reservoir and decreases, but does not eliminate, ongoing virus replication. *J Acquir Immune Defic Syndr* **35**: 33–37.
116. Richman DD, Wrin T, Little SJ, Petropoulos CJ (2003) Rapid evolution of the neutralizing antibody response to HIV type 1 infection. *Proc Natl Acad Sci USA* **100**: 4144–4149.
117. Rosenberg ES, Altfeld M, Poon SH, Phillips MN, Wilkes BM, Eldridge RL, Robbins GK, D'Aquila RT, Goulder PJ, Walker BD (2000) Immune control of HIV-1 after early treatment of acute infection. *Nature* **407**: 523–526.
118. Rosenberg YJ, Zack PM, White BD, Papermaster SF, Elkins WR, Eddy GA, Lewis MG (1993) Decline in the CD4+ lymphocyte population in the blood of SIV-infected macaques is not reflected in lymph nodes. *AIDS Res Hum Retroviruses* **9**: 639–646.
119. Rosok BI, Bostad L, Voltersvik P, Bjerknes R, Olofsson J, Asjo B, Brinchmann JE (1996) Reduced CD4 cell counts in blood do not reflect CD4 cell depletion in tonsillar tissue in asymptomatic HIV-1 infection. *AIDS* **10**: F35–38.
120. Rouzine IM, Coffin JM (1999) Linkage disequilibrium test implies a large effective population number for HIV *in vivo*. *Proc Natl Acad Sci USA* **96**: 10758–10763.
121. Rouzine IM, Rodrigo A, Coffin JM (2001) Transition between stochastic evolution and deterministic evolution in the presence of selection: general theory and application to virology. *Microbiol Mol Biol Rev* **65**: 151–185.
122. Rubsamen–Waigmann H, Huguenel E, Shah A, Paessens A, Ruoff HJ, von Briesen H, Immelmann A, Dietrich U, Wainberg MA (1999) Resistance mutations selected *in vivo* under therapy with anti-HIV drug HBY 097 differ from resistance pattern selected *in vitro*. *Antiviral Res* **42**: 15–24.
123. Ruff CT, Ray SC, Kwon P, Zinn R, Pendleton A, Hutton N, Ashworth R, Gange S, Quinn TC, Siliciano RF, Persaud D (2002) Persistence of wild-type virus and lack temporal structure in the latent reservoir for human immunodeficiency virus type 1 in pediatric patients with extensive antiretroviral exposure. *J Virol* **76**: 9481–9492.
124. Schmidt B, Walter H, Zeitler N, Korn K (2002) Genotypic drug resistance interpretation systems — the cutting edge of antiretroviral therapy. *AIDS Rev* **4**: 148–156.
125. Schmitz JE, Kuroda MJ, Santra S, Sasseville VG, Simon MA, Lifton MA, Racz P, Tenner–Racz K, Dalesandro M, Scallon BJ, Ghrayeb J, Forman MA, Montefiori DC, Rieber EP, Letvin NL, Reimann KA (1999) Control of viremia in simian immunodeficiency virus infection by CD8+ lymphocytes. *Science* **283**: 857–860.
126. Seo TK, Thorne JL, Hasegawa M, Kishino H (2002) Estimation of effective population size of HIV-1 within a host: a pseudomaximum-likelihood approach. *Genetics* **160**: 1283–1293.
127. Shafer RW, Kantor R, Gonzales MJ (2000) The genetic basis of HIV-1 resistance to reverse transcriptase and protease inhibitors. *AIDS Reviews* **2**: 211–228.

128. Shankarappa R, Margolick JB, Gange SJ, Rodrigo AG, Upchurch D, Farzadegan H, Gupta P, Rinaldo CR, Learn GH, He X, Huang XL, Mullins JI (1999) Consistent viral evolutionary changes associated with the progression of human immunodeficiency virus type 1 infection. *J Virol* **73**: 10489–10502.
129. Shenderovich MD, Kagan RM, Heseltine PN, Ramnarayan K (2003) Structure-based phenotyping predicts HIV-1 protease inhibitor resistance. *Protein Sci* **12**: 1706–1718.
130. Shriner D, Shankarappa R, Jensen MA, Nickle DC, Mittler JE, Margolick JB, Mullins JI (2004) Influence of random genetic drift on HIV-1 env evolution during chronic infection. *Genetics* **166**: 1155–1164.
131. Stafford MA, Corey L, Cao Y, Daar ES, Ho DD, Perelson AS (2000) Modeling plasma virus concentration during primary HIV infection. *J Theor Biol* **203**: 285–301.
132. Stilianakis NI, Dietz K, Schenzle D (1997) Analysis of a model for the pathogenesis of AIDS. *Math Biosci* **145**: 27–46.
133. Sturmer M, Doerr HW, Preiser W (2003) Variety of interpretation systems for human immunodeficiency virus type 1 genotyping: confirmatory information or additional confusion? *Curr Drug Targets Infect Disord* **3**: 373–382.
134. van Stipdonk MJB, Lemmens ED, Schoenberger SP (2001) Naive CTLs require a brief period of antigenic stimulation for clonal expansion and differentiation. *Nature Immunol* **2**: 423–429.
135. Wahl LM, Nowak MA (2000) Adherence and drug resistance: predictions for therapy outcome. *Proc Royal Soc Lond B Biol Sci* **267**: 835–843.
136. Wang K, Jenwitheesuk E, Samudrala R, Mittler J (2004) Simple linear model provides highly accurate genotypic predictions of HIV-1 drug resistance. *Antivir Ther* (in press).
137. Wei X, Decker JM, Wang S, Hui H, Kappes JC, Wu X, Salazar–Gonzalez JF, Salazar MG, Kilby JM, Saag MS, Komarova NL, Nowak MA, Hahn BH, Kwong PD, Shaw GM (2003) Antibody neutralization and escape by HIV-1. *Nature* **422**: 307–312.
138. Wei XP, Ghosh SK, Taylor ME, Johnson VA, Emini EA, Deutsch P, Lifson JD, Bonhoeffer S, Nowak MA, Hahn BH, Saag MS, Shaw GM (1995) Viral dynamics in human immunodeficiency virus type 1 infection. *Nature* **373**: 117–122.
139. Wein LM, Damato RM, Perelson AS (1998) Mathematical analysis of antiretroviral therapy aimed at HIV-1 eradication or maintenance of low viral loads. *J Theor Biol* **192**: 81–98.
140. Wick D, Self SG, Corey L (2002) Do scarce targets or T killers control primary HIV infection? *J Theor Biol* **214**: 209–214.
141. Wodarz D, Nowak MA (1999) Specific therapy regimes could lead to long-term immunological control of HIV. *Proc Natl Acad Sci USA* **96**: 14464–14469.
142. Wodarz D, Nowak MA (2000) Correlates of cytotoxic T-lymphocyte-mediated virus control: implications for immunosuppressive infections and their treatment. *Philos Trans R Soc Lond B Biol Sci* **355**: 1059–1070.

143. Wolinsky SM, Korber BT, Neumann AU, Daniels M, Kunstman KJ, Whetsell AJ, Furtado MR, Cao Y, Ho DD, Safrit JT (1996) Adaptive evolution of human immunodeficiency virus-type 1 during the natural course of infection. *Science* **272**: 537–542.
144. Wong JK, Hezareh M, Gunthard HF, Havlir DV, Ignacio CC, Spina CA, Richman DD (1997) Recovery of replication-competent HIV despite prolonged suppression of plasma viremia. *Science* **278**: 1291–1295.
145. Wu H, Kuritzkes DR, McClernon DR, Kessler H, Connick E, Landay A, Spear G, Heath-Chiozzi M, Rousseau F, Fox L, Spritzler J, Leonard JM, Lederman MM (1999) Characterization of viral dynamics in human immunodeficiency virus type 1-infected patients treated with combination antiretroviral therapy: relationships to host factors, cellular restoration, and virologic end points. *J Infect Dis* **179**: 799–807.
146. Wu H, Lathey J, Ruan P, Douglas SD, Spector SA, Lindsey J, Hughes MD, Rudy BJ, Flynn PM; PACTG 381 Team (2004) Relationship of Plasma HIV-1 RNA Dynamics to Baseline Factors and Virological Responses to Highly Active Antiretroviral Therapy in Adolescents (Aged 12–22 Years) Infected through High-Risk Behavior. *J Infect Dis* **189**: 593–601.
147. Wu H, Mellors J, Ruan P, McMahon D, Kelleher D, Lederman MM (2003) CNAA2004 Study Investigators. Viral dynamics and their relations to baseline factors and longer term virologic responses in treatment-naive HIV-1-infected patients receiving abacavir in combination with HIV-1 protease inhibitors. *J Acquir Immune Defic Syndr* **33**: 557–563.
148. Zennou V, Mammano F, Paulous S, Mathez D, Clavel F (1998) Loss of viral fitness associated with multiple Gag and Gag-Pol processing defects in human immunodeficiency virus type 1 variants selected for resistance to protease inhibitors *in vivo*. *J Virol* **72**: 3300–3306.
149. Zhang L, Ramratnam B, Tenner–Racz K, He Y, Vesanen M, Lewin S, Talal A, Racz P, Perelson AS, Korber BT, Markowitz M, Ho DD (1999) Quantifying residual HIV-1 replication in patients receiving combination antiretroviral therapy. *New Engl. J Med* **340**: 1605–1613.
150. Zhu T, Muthui D, Holte S, Nickle D, Feng F, Brodie S, Hwangbo Y, Mullins JI, Corey L (2002) Evidence for human immunodeficiency virus type 1 replication *in vivo* in CD14(+) monocytes and its potential role as a source of virus in patients on highly active antiretroviral therapy. *J Virol* **76**: 707–716.
151. Zolopa A, Lazzeroni L, Rinehart A, Kuritzkes D (2002) Accuracy, precision and consistency of expert HIV-1 genotype interpretation: an international comparison (The GUESS study). *Antivir Ther* **7**: S97.

CHAPTER 17

TREATMENT INTERRUPTIONS AND RESISTANCE: A REVIEW

Jane M Heffernan and Lindi M Wahl

In recent years, mathematical models have been used to examine the effects of adherence and/or structured treatment interruptions in HIV therapy and the resulting changes in population dynamics of HIV-1, the immune system and the emergence of drug resistance. These models have contributed to a better understanding of the effects of missed doses on the evolution of drug-resistant HIV strains. This chapter is a review of these studies; the methods and most important results are presented. We outline the significance of these results and discuss some of the drawbacks of the various modeling approaches.

Keywords: Adherence; structured treatment interruptions (STI); drug resistance.

1. Introduction

Poor adherence to antiretroviral treatment regimens has serious consequences for HIV infected patients, including failure to prevent viral replication and an increased risk of developing drug-resistant viral strains. The emergence of drug resistance is a major problem in the treatment of HIV infection. A large number of mathematical models have been developed to describe the population dynamics of HIV-1 including drug treatment and drug-resistant mutants (Nowak *et al.*, 1991; McLean *et al.*, 1991; McLean and Nowak, 1992; Frost and McLean, 1994; McLean and Frost, 1995; Loveday *et al.*, 1995; Wei *et al.*, 1995; Ho *et al.*, 1995; Perelson *et al.*, 1996; de Boer and Boucher, 1996; de Jong *et al.*, 1996; Kirschner and Webb, 1996; Stilianakis *et al.*, 1997; Nowak *et al.*, 1997; Bonhoeffer and Nowak, 1997; Perelson *et al.*, 1997; de Boer and Perelson, 1998; Wein *et al.*, 1998; Wodarz and Nowak, 1999; Nowak and May, 2000; Ribeiro and Bonhoeffer, 2000; Wodarz *et al.*, 2000; Wodarz, 2001; Perelson, 2002). Many address the epidemiology and emergence of drug resistance (Nowak *et al.*, 1991;

McLean and Nowak, 1992; Frost and McLean, 1994; Coffin, 1995; Austin *et al.*, 1997; Bonhoeffer *et al.*, 1997a; Davies, 1997; Bonhoeffer and Nowak, 1997; Levin *et al.*, 1997; Levy, 1997; Nowak *et al.*, 1997; Stilianakis *et al.*, 1997; Austin *et al.*, 1999; Austin and Anderson, 1999; Levin and Andreason, 1999; Lipsitch *et al.*, 2000). A handful of paper have focused on the interaction of changing drug concentrations with the population dynamics of a pathogen (Kepler and Perelson, 1998; Lipsitch and Levin, 1998; Austin *et al.*, 1998).

One drawback of previous studies in HIV drug resistance is that it has typically been assumed that therapy is taken continuously and that the infected individual strictly adheres to the drug regimen. In contrast, recent clinical work has shown that average adherence is only 70% (Bangsberg *et al.*, 2000; Paterson *et al.*, 2000; Arnsten *et al.*, 2001; McNabb *et al.*, 2001; Liu *et al.*, 2001; Gross *et al.*, 2001; Walsh *et al.*, 2002; Bangsberg *et al.*, 2003). Lack of adherence can drastically change the outcome of therapy, since missed doses allow the drug-resistant strain(s) in the system to become prevalent. This is possibly because new uninfected target cells have arisen during the period of therapy and have escaped infection due to the presence of antiretroviral drugs (McLean *et al.*, 1991), providing a pool of susceptible cells for the drug-resistant strain to infect. Considerations such as these have motivated both clinical and theoretical research into the effects of adherence on the emergence of drug resistance.

Meanwhile, attempts to reduce toxicity or boost the immune system have led to the idea of intermittent periods of therapy, usually referred to as "structured treatment interruptions" (STI). STI is a treatment strategy in which a patient on drug therapy discontinues all antiretroviral agents for defined periods of time while under careful medical observation. After each "non-treatment" period, antiretroviral therapy with the same or a new drug combination is started. Some clinical studies on the effects of STI have discovered that intermittent therapy and "drug holidays" can suppress the viral load for small periods of time. Other studies, however, found no prolonged suppression of the viral load (see Walmsley and Loutfy, 2002 for details). It appears that the STI approach to HIV therapy for a few carefully selected patients may allow for periods of time with control of HIV replication in the absence of drug therapy. This result has not, however, been consistently reproduced, leading to questions as to how early patients should begin treatment in order to gain benefit from STI, and how STI effects the emergence of drug-resistant viral strains in the system.

We will review some theoretical models of the effects of adherence (Section 1) and STI (Section 2) on the emergence of drug resistance in HIV. Each section will begin with a brief overview of recent clinical findings.

2. Adherence

It is widely acknowledged that lack of adherence to HIV regimens facilitates the emergence of drug resistance. The rapid replication and mutation rate of HIV (Coffin, 1992; Ho *et al.*, 1995) requires very high levels of adherence in order to achieve prolonged suppression of viral load, but maintaining these high levels of adherence may be difficult since drug regimens for HIV/AIDS are complex and the side effects are severe. It has been observed by the scientific and medical communities that, due to the severe side effects and the reduced quality of life, HIV/AIDS patients exhibit low adherence rates. Even when patients fully understand the ramifications of nonadherence to medications, adherence rates are not optimal. This may allow the emergence of a drug-resistant strain of the virus, which will complicate further therapy and also has the potential to be transmitted to others.

The development of more effective antiretroviral treatments has placed a new emphasis on studying the effects of adherence to treatment. With less effective regimens, viral replication was only partially suppressed, and resistance developed irrespective of adherence. In contrast, highly active antiretroviral therapy (HAART) regimens, which usually include two nucleoside reverse transcriptase inhibitors, non-nucleoside reverse transcriptase inhibitors and protease inhibitors, are capable of strongly suppressing HIV replication and thus suppressing the emergence of drug-resistant strains. Missed doses of medication, however, reduce the efficacy of viral suppression, allowing new viral replication and the development of drug-resistant mutations. In a study by Bangsberg *et al.* (2003) it was found that 23% of all drug resistance occurs in patients whose adherence rates are between 92 and 100%. Similarly, 50% of all drug resistance occurs in patients that have adherence rates of 79 to 100%. Thus, missing an occasional dose, even when overall adherence is high, can be significantly detrimental. Because of the critical importance of HIV antiretroviral adherence, HIV provides a compelling case for studying adherence and drug resistance.

When HAART was introduced, many hoped that it would completely eradicate the virus (Ho *et al.*, 1995; Perelson *et al.*, 1996), but it was soon observed that this was not the case since low levels of viral replication still existed in small reservoirs, maintaining the viral population even when viral loads are undetectable (WHO, 2003). Consequently, adherence to HAART must be almost perfect in order to achieve viral suppression over the long term. In an early study, Montaner *et al.* (1998) found that missing even a single dose of the drug regimen predicted treatment failure. Bangsberg *et al.* (2000) found that patients with adherence rates greater than 90% did not progress to AIDS, but 38% of those with adherence rates less than 50% progressed to AIDS as did 8% of the patients with adherence rates between 51 and 89%. In another study, Paterson *et al.* (2000) found that adherence rates less than 95% were associated with increased levels of viral resistance and opportunistic infections. In both of the latter studies patients who reported the same adherence rates might have had different patterns of adherence, thus changing the outcome of drug therapy.

Not only does nonadherence to HAART affect the individual taking the drugs, but it can also have important public health implications. It has been found in studies by Hecht *et al.* (1998), Boden *et al.* (1999), Little *et al.* (1999), Moeller *et al.* (2000) and Girardi (2003) that drug resistance can be transmitted to others during high risk activity. This, in turn, will effect drug therapy considerations for the newly infected individuals (WHO, 2003). In fact, Moeller *et al.* (2000) observed that up to 80% of isolates from newly infected people are resistant to at least one class of currently approved antiretroviral medications and 26% of these isolates are resistant to more than one class (WHO, 2003). Studies conducted by Blower and colleagues (2000, 2001, 2003) have determined that the current prevalence of resistance in HIV patients is high and that this will continue to rise even though the transmission of resistance is assumed to be low.

Although a number of studies have addressed the emergence of the drug-resistant strains under various treatment regimens, not much is known about how different rates and patterns of adherence affect drug resistance. A handful of theoretical studies have addressed this issue; in the sections to follow, we examine several of these studies in detail and address some ideas for future work.

2.1. *Wahl and Nowak, 2000*

Using an established model of viral replication, Wahl and Nowak (2000) addressed the effect of adherence on the evolution of resistance in HIV analytically, providing quantitative estimates of intermediate "danger zones" for particular drug regimens. They determined the conditions under which resistance dominates as a result of imperfect adherence for both single and triple drug therapy, and estimated the level of adherence needed to prevent the emergence or growth of drug-resistant strains.

Wahl and Nowak did this by simply looking at the average basic reproductive ratios of the drug-resistant mutant and drug-sensitive wild-type populations. In order to calculate the average basic reproductive ratios, a model that described the growth of the drug-sensitive and drug-resistant viral populations, the time-course of the drugs in the system and the effect of the drugs on both strains of the virus, subject to different adherence patterns was developed.

The equation governing drug concentration in plasma incorporated standard pharmacokinetics, assuming a linear rise in drug concentration followed by an exponential decay:

$$C(t) = \begin{cases} C(T) + \dfrac{t}{t_p}(C_{\max} - C(T)) & \text{for } 0 \le t \le t_p \\ C_{\max} e^{-\omega(t-t_p)} & \text{for } t_p \le t \le T \end{cases}, \tag{1}$$

where $C(t)$ is the plasma drug concentration at time t, T is the recommended dosing interval, $C_{\max}$ is the maximum drug concentration in plasma, t_p is the time to peak and $\omega = log(2)/T_{1/2}$ where $T_{1/2}$ is the serum half life. The authors noted that $C_{\max}$ must be the maximum concentration achieved in plasma after a number of successive doses such that the patient is in a "steady state" with respect to any gradual accumulation of the drug.

To model imperfect adherence, Wahl and Nowak first considered only the fraction of the prescribed doses of the drug which are taken, that is, the long term average of the number of doses. Equation (1) was used to describe the time course of the drug concentration if a drug was taken during a given interval. If the drug was not taken, it was assumed that the dose would decay from its current concentration with half life $T_{1/2}$. The pattern in which the doses were taken is considered in later sections of the paper.

An equation for the inhibition of viral replication at time t, $s(t)$, is then derived as a function of the drug concentration, $C(t)$, and the concentration of the drug which inhibits viral replication by 50%, IC_{50}:

$$s(t) = 1 - \frac{C(t)}{C(t) + \mathrm{IC}_{50}}. \tag{2}$$

Thus, s can take values between zero and one; when s is approximately equal to zero the drug completely inhibits the viral replication, whereas the drug has no effect when s is approximately one. Wahl and Nowak used an *in vitro* value of IC_{50} to estimate the *in vivo* value of s. As noted by the authors, this is a drawback of the model since the relationship between *in vitro* susceptibility of HIV to antiviral drugs and the *in vivo* inhibition of viral replication in humans has not been accurately established.

The susceptible cell and viral population dynamics were modeled using a well-known system developed by McLean and Nowak (1992) and Bonhoeffer and Nowak (1997). The model is as follows:

$$\begin{aligned}\dot{x} &= \lambda - dx - (s_1\beta_1 y_1 + s_2\beta_2 y_2)x\\ \dot{y}_1 &= (s_1\beta_1 x - a)y_1\\ \dot{y}_2 &= (s_2\beta_2 x - a)y_2,\end{aligned}$$

where x, y_1 and y_2 denote the populations of susceptible cells, infected cells with the drug sensitive virus and infected cells with the drug resistance strain of the virus respectively. The susceptible cells are produced at a constant rate λ from a pool of precursor cells, die at a rate d and become infected at rates $s_1\beta_1 y_1$ and $s_2\beta_2 y_2$ by sensitive and resistant virus respectively. Here β_1 and β_2 represent the infectivity of sensitive and resistant virus and $s_1(t)$ and $s_2(t)$ denote the inhibition of viral infectivity for the sensitive and resistant virus where the degree of inhibition is given by Eq. (2). Infected cells in both strains die at rate a.

Using the average values of s for the drug-sensitive strain and the drug-resistant strain, Wahl and Nowak derived an equation for the mean value of the basic reproductive ratio for each strain as a function of p, the long-term average of the fraction of doses taken:

$$\bar{R}_i(p) = \frac{\bar{s}_i(p)\beta_i\lambda}{ad},$$

where $i = 1, 2$ represents the drug-sensitive and drug-resistant strain respectively and $\bar{s}_i$ represents the mean value of s_i when adherence is p.

Wahl and Nowak used Eq. (1) for a given degree of adherence p to compute s_1 and s_2 using Eq. (2). When several drugs were taken simultaneously, the concentration time-course was computed for each drug and then the value of s was multiplied for each drug to give the total inhibition at each time point.

This model was used to simulate different adherence patterns to a given drug regimen, using single- or multi-drug therapy. Note that in this model, it was assumed that doses of drugs can only be taken at times corresponding to the dosing intervals. This was modeled using the Poisson distribution where for each new dosing interval the drug was taken with a probability equal to the long-term average adherence, p.

2.1.1. *Single-drug therapy*

When only a single drug is used in therapy, three separate regions of treatment outcome were identified [see Fig. (1)]:

1) The drug-sensitive virus dominates and treatment fails (diagonal hatching). Note that therapy fails in this region even in the absence of a drug-resistant strain.
2) Treatment fails because of the emergence of drug resistance (shaded region).
3) Treatment succeeds in eliminating both strains of the virus (solid white).

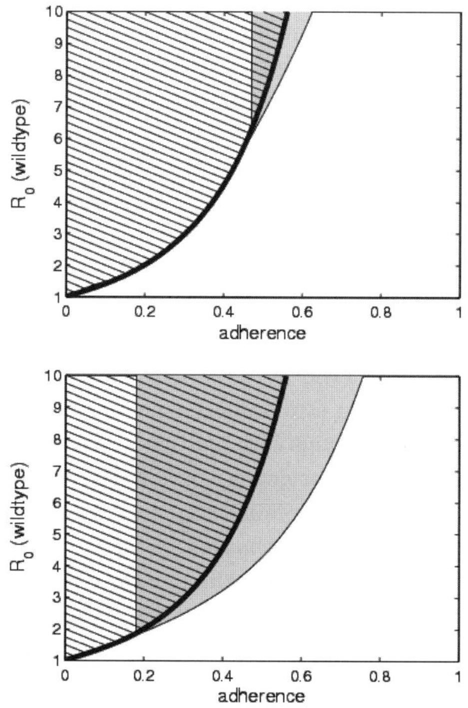

Fig. 1. Treatment outcome is shown for different values of drug-sensitive R_0, and for different levels of adherence, p. The upper and lower panels illustrate scenarios when the drug-resistant R_0 is 20% or 50% of the wild-type R_0 in the absence of therapy, respectively. See text for details.

Different dose patterns for the same value of p were also investigated. It was found that, in order for the resistant mutant to outcompete the wild-type, the "resistance advantage" (defined as the ratio of the mean value of the basic reproductive ratio for the resistant strain to the wild-type strain) must be greater than the value of R_1/R_2 in the pretreatment stage. They also found that doses which occur in blocks for the same value of p can reduce the advantage of resistance. This outcome has important implications for STIs.

Another important observation was made concerning the ratio of R_2 to R_1. Wahl and Nowak stated that if R_1 is higher than 2 or 3 and this ratio is higher than about 0.7, resistance is likely to dominate the system after treatment.

2.1.2. *Combination therapy*

The previous analysis was repeated for triple-drug therapy, which includes two nucleoside reverse transcriptase inhibitors and a single protease inhibitor. In this analysis the dosing interval and time-course of each drug were defined separately. The fraction of doses taken was also defined independently for each drug, and the effects of the drugs were combined to find the values of R_1 and R_i at each time for the seven possible drug-resistant strains — three are resistant to one of the three drugs, three are resistant to two drugs and one strain is resistant to all three of the drugs in the cocktail. Wahl and Nowak assumed the all of the drug-resistant drains had the same infectivity, β_2.

Once again using the ratio of R_2 to R_1, Wahl and Nowak identified regions where each resistant strain affects the outcome of drug treatment. It was observed that viral strains with multiple resistance have the ability to dominate the system for higher degrees of adherence, but these strains are less likely to exist. Overall, they found that treatment is most likely to fail by the emergence of a viral strain that is resistant to one of the three drugs since this is a precursor to obtaining the mutant strain that is resistant to all three drugs. It was also noted, not surprisingly, that treatment efficacy increased substantially when triple-drug therapy was used, in comparison to single-drug therapy since resistance was less likely to emerge and treatment was successful over a wider range of adherence levels.

An intriguing conclusion of this model is that Wahl and Nowak found that long-term adherence is the relevant predictor of therapy outcome i.e. missing a dose has a detrimental effect on the long-term antiviral effect, since R becomes briefly greater than one, but this will not determine the outcome of therapy in the long run. This means that the long-term average R_0, which is proportional to the long-term average of the antiviral effect s, can be used to predict the outcome of drug therapy. This has important implications for HIV research. First, the values of R_0 and s can be easily approximated using clinical data. Second, this goes against the intuitive notion that qualitative changes should occur if a dose is missed. The authors note, however, that the long term antiviral effect may differ for different dose taking patterns, even though the total adherence is the same.

This model also has some serious drawbacks. First, the model neglects cross-resistance between mutations. In clinical studies, cross-resistance has repeatedly been observed for drugs that are used together in drug cocktails, meaning that single mutations in the viral genome may confer a degree of resistance to both drugs. This is a clear topic for future work since studies of adherence and cross resistance will aid clinicians in designing treatment strategies.

Secondly, it was assumed that all the drug-resistant strains have the same infectivity, β_2. This is not necessarily true and future models should be modified accordingly.

Lastly, Wahl and Nowak used the *in vitro* IC_{50} value for each drug to provide an estimate of the *in vivo* value of s, as discussed previously.

Nonetheless, this paper offered a first step toward theoretical studies of the effects of adherence on the emergence of drug resistance in HIV.

2.2. *Phillips* et al., *2001*

Phillips *et al.* (2001) also studied the effects of drug adherence on the development of drug resistance. This study gives some insight into different patterns of adherence and which patterns are optimal to suppress the development of drug resistance.

Phillips *et al.* also modeled triple-drug therapy, defining the same eight populations as Wahl and Nowak. They also used the basic reproductive number of each strain to predict the outcome of the model. Phillips *et al.* developed a stochastic model, which is loosely based on ideas from previous work of McLean, Bonhoeffer, Kepler and colleagues (McLean *et al.*, 1991; McLean and Nowak, 1992; Bonhoeffer *et al.*, 1997; Kepler and Perelson, 1998). They used this model to illustrate how certain patterns of treatment lead to resistance development. Their model includes drugs A, B and C, a short-lived susceptible T cell population, eight viral subpopulations, and short-lived and latently infected cells infected by each of the viral subpopulations. They denote the viral subpopulations as V_0, V_a, V_b, V_c, V_{ab}, V_{ac}, V_{bc}, and V_{abc}, each of which has their own basic reproductive ratio, R_i, where the subscript denotes the drug(s) to which the virus subpopulation is immune. Note that this model assumes that the drug concentration is switched from being high to zero instantly from one day to the next. Please see Fig. (2) for a partial description of this model.

In order to determine the number of mutations incurred in one round of replication, Phillips *et al.* used a Binomial distribution when the population was small. They used a Poisson distribution to determine the number of cells created in each population in the next generation, using the value of the basic reproductive ratio for the population at that time. This model also includes the latently infected cell population. These cells are created with a small probability per new cell infection and are activated with a small probability per day.

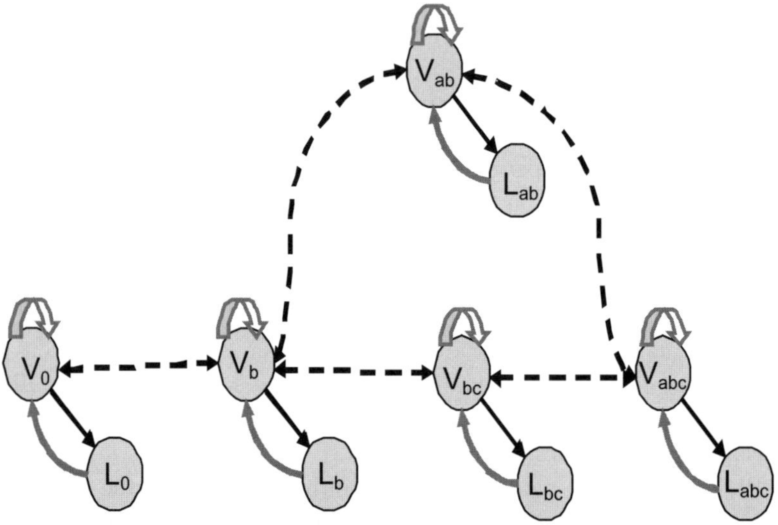

Fig. 2. The model studied by Phillips *et al.* incorporates cells infected or latently infected by viral subpopulations which are immune to no drugs (V_0 and L_0, respectively), or to some subset of drugs A, B and C. In total, eight viral subpopulations are possible; we illustrate only five for clarity, showing possible pathways that include resistance to drug B.

Phillips *et al.* assigned a value of the reproductive ratio to each viral strain, depending on drug exposure, the fitness cost of resistance mutations corresponding to drugs not in the regimen and availability of target cells. First, they assumed that there is a small fitness cost for each extra mutation which doesn't correspond to one of the drugs in use and reduced R further if one, two or three drugs for which the corresponding mutation was not present.

It was assumed that the system was in equilibrium before therapy.

Four different patterns of treatment adherence were studied:

1) Take drugs ABC for 2 days in every 3 days.
2) Take drugs ABC for 1 day in every 2 days.
3) Take drugs ABC for 50 days and then 50 days off.
4) Take drugs ABC for 200 days and then 200 days off.

For the first pattern, Phillips *et al.* found that completely different results can be obtained with different runs of the same model. This is due to the stochastic nature of the model. Some simulations resulted in the out-growth of subpopulation V_{abc} and some didn't. They found that the final outcome relied on the growth of the viral subpopulation V_{ab}, V_{bc} or V_{ac}. They note that in the simulations runs where V_{ab}, V_{bc} or V_{ac} existed at the start of therapy these populations may not survive to produce new virions since the value of R is only slightly greater than one. However, once the population of V_{ab}, V_{bc} or V_{ac} is substantial it will almost certainly continue to grow since random chance is unlikely to eliminate all infected cells. This leads to an eventual outgrowth of subpopulation V_{abc} since the mutation of V_{ab}, V_{bc} or V_{ac} to V_{abc} is certain and R_{abc} is substantially greater than one.

For other simulation runs subpopulations V_{ab}, V_{bc} and V_{ac} were empty at the time of start of therapy and no virions of these strains were created during therapy. Therefore, the viral subpopulation V_{abc} would never exist and will not overtake the system. Phillips *et al.* noted that due to the decline in V_a, V_b and V_c the probability of a viral strain with two mutations becomes very low after the first 50 days or so of therapy. They found that this outcome was actually more likely to occur than the case discussed previously. After 100 simulation runs, 15% of the simulations using the "2 days in every 3 days" adherence pattern had a subpopulation of V_{abc} present 1500 days after the start of therapy.

When instead the therapy is taken one day in every two days (pattern 2), it was found that the values of R_a, R_b and R_c are greater than one as well as the values of R_{ab}, R_{bc}, R_{ac} and R_{abc}. Thus, eventual outgrowth of V_{abc} is much more likely than for pattern 1. In fact, the V_{abc} population was present after 1500 days of therapy in all 100 simulations.

The next pattern of 50 days on drug ABC and 50 days off tested an intermittent therapy pattern. Using this pattern there was only a small chance of outgrowth of the V_{abc} population since the period off therapy allowed V_0 population to regrow to the size of the initial population. During this period the values of R for all strains with mutations were reduced to values less than one. During this period the values of R for all strains with mutations were reduced to values less than one. This halted any growth of V_{ab}, V_{bc}, V_{ac} and V_{abc}. Phillips *et al.* found that the results were similar for the adherence pattern of 200 days on and off therapy.

Phillips *et al.* found that their results were very sensitive to the parameter values. When they reduced the fitness cost of unnecessary mutations per mutation, there was a greater tendency for the viral population of ABC to grow. When the drug efficacy was increased, the triple-resistance viral population was less likely to outgrow. It was also observed that resistance was more likely to develop when the number of infected cells at steady state was increased.

A key contribution of this work is the observation of an increased risk of resistant infections during intervals when treatment has been recently interrupted. This is a result of the fact that during therapy there is an increased number of susceptible CD4 T-cells. Thus, just after therapy is interrupted, the replication rates of all the viral strains in the population are higher than they would be if the drugs were never taken. When this pool of susceptible cells decreases (i.e. they have been infected or have died), the basic reproductive ratios decrease to be less than one. Phillips *et al.* note that this presumably occurs when the viral load returns to pre-therapy levels. Thus, continually being on and off drugs for short periods can increase the risk of incurring mutations with drug resistance since the overall replicative capacity for strains with resistant mutations is greater than one.

Another key contribution of this paper is the finding that intermittent adherence can lead to resistance even though this model did not assume that there are periods of suboptimal drug concentration after interrupting therapy. This model assumes that drug concentration switches from being high to zero instantly from one day to the next.

In summary, this model suggested that different patterns of adherence or structured treatment interruptions affect the emergence of resistance in different ways. It was found that if therapy was given in blocks, the probability that the resistant population emerges is smaller than when the adherence pattern was stochastic. But, this seems to occur only when the viral load has returned to pre-therapy baseline levels before therapy is re-started. If this condition is not met, the viral subpopulation may grow. An extension to this study would be to explicitly consider cases where the viral load is not allowed to return to pre-therapy levels. This would help to determine the effect of viral load when therapy is re-started, an important clinical question.

One drawback of this model is that the effects of the latently infected cells were not quantified. It would be interesting to see which, if any, of the conclusions of the model would change if the effects of latently infected cell were neglected. Gaining some insight into the overall effect of the latently infected cells on the emergence of resistance in the system would help clinicians and researchers in developing long-term drug regimens. If the effect of the latently infected cells is negligible, this problem could be set aside until researchers have better quantified the emergence of resistance due only to infection of the active CD4 T-cells. If the effect is large, future studies of adherence and resistance should include the latently infected population.

Another drawback of the model is that it is assumed that only one mutation can occur at a time. Although, biologically, the probability of two mutations occurring at the same time is very small and for three mutations this probability is even smaller, these still occur at some frequency, particularly in large viral populations. This would affect the results of this model in that the time that it takes to incur these mutations would decrease and thus we might expect an earlier increase in the double- and triple-drug-resistant mutant subpopulations.

2.3. *Huang* et al., *2003*

Huang *et al.* (2003) compared the impact of perfect and imperfect adherence on the overall effect of a drug regimen. This model allowed for the derivation of threshold values of the overall treatment efficacy that determined the outcome of therapy. In this study, the relationship between the actual and detectable rebound failure times was also obtained and was found to be approximately linear. The actual rebound time is defined as time at which the change from positive to negative growth of the virus occurs and the detectable rebound time refers to the time at which the viral load rebounds to a detectable size. The model used was an established model of HIV dynamics including pharmacokinetic effects similar to those incorporated by Wahl and Nowak (2000).

Huang *et al.* defined the overall treatment effect of a regimen with both reverse transcriptase inhibitors and protease inhibitors as:

$$e(t) = 1 - [1 - \gamma(t)][1 - \eta(t)],$$

where $\gamma(t)$ and $\eta(t)$ are time varying parameters that represent the drug efficacies of protease inhibitors and reverse transcriptase inhibitors respectively. The parameters $\gamma(t)$ and $\eta(t)$ take the form:

$$\eta(t) = \frac{C(t)}{\phi \mathrm{IC}_{50}(t) + C(t)},$$

if the regimen contains a single drug from the reverse transcriptase inhibitor or protease inhibitor class. $C(t)$ represents the concentration of the drug in plasma [see Eqs. 1, 2 and 3 in Huang *et al.*], and IC_{50} is the concentration of the drug which inhibits viral replication by 50%.

Note that Huang *et al.* model drug resistance by allowing the value of IC_{50} to vary with time, in contrast with the model given by Wahl and Nowak [see Eq. (2)]. By allowing IC_{50} to vary with time, Huang *et al.* capture the changing sensitivity of a virus population with a mixture of quasispecies to the drugs in antiviral regimens. To model the within-host changes over time in IC_{50} (due to the emergence of new drug-resistant mutations), Huang *et al.* proposed the following:

$$\mathrm{IC}_{50}(t) = \begin{cases} I_0 + \dfrac{I_r - I_0}{t_r} & \text{for } 0 < t < t_r \\ I_r & \text{for } t \geq t_r \end{cases},$$

where I_0 and I_r are respective values of $\mathrm{IC}_{50}(t)$ at baseline and time point t_r at which the resistant mutations dominate. The drug efficacy for a drug regimen that consists of two agents within the same class is also derived (see Huang *et al.*, 2003 for details).

A system of differential equations then models the dynamics of uninfected cells, infected cells, infectious virions and non-infectious virions [see Eq. (7) in Huang *et al.*]. This system has two steady states: an uninfected steady-state and an endemically infected steady-state. Analysis of these steady-states gave rise to two threshold levels of efficacy, e_c and e_m, which delineate three possible outcomes for treatment:

1) If the treatment efficacy $e(t) > e_c$ for all t, the system converges to a stable uninfected steady-state and the virus is eventually eradicated.
2) If $e_m < e(t) < e_c$, the viral load will not rebound, but the virus will not be eradicated.
3) If $e(t)$ falls below e_m, the viral load will decline at the start of treatment but will eventually rebound at time $e(t) = e_m$.

Assuming that the system is in steady-state before treatment begins, the thresholds e_c and e_m are given by:

$$e_c = 1 - \frac{T_0}{T_1}$$
$$e_m = 1 - \frac{T_0}{T_m},$$

where T_0 represents the baseline uninfected T cell count before the initiation of the treatment, and T_1 denotes the recovered steady-state (uninfected) T cell count after the virus is eradicated (case 1 above). T_m gives the value of the uninfected cell population at the actual rebound time, i.e. when the change in the infectious virion and infected cell populations is zero. From these equations it is clear that if the baseline T cell count, T_0, is larger, the required treatment efficacy to reach a particular T cell count is lower. Thus, treatment should be initiated when patients have higher $CD4^+$ T cell counts as this will reduce the required efficacy of the treatment. Huang *et al.* note that this supports previous findings: that initiation of antiretroviral therapy should be based on the $CD4^+$ T cell counts; and that the $CD4^+$ T cell count, rather than the viral load, is predictive of treatment outcome.

By simulating perfect adherence to a triple drug regimen consisting of one protease inhibitor and two reverse transcriptase inhibitors, Huang *et al.* found that the overall treatment effect of the three-drug regimen falls below e_m because of the emergence of resistant mutations in the system. The drop in treatment efficacy is attributed to an increase in IC_{50} reflecting the emergence of drug resistance. The system then converges to the endemically infected steady-state. Eventually, the viral dynamic system converges to the endemically infected steady-state.

Huang *et al.* also simulated the effects of imperfect adherence on the dynamical system, but in these cases they did not consider the effects of drug resistance. Four patterns of drug adherence were considered; in each case the overall level of adherence was 50%:

1) Doses are missed randomly.
2) Every other dose is missed.
3) Three doses taken followed by three doses missed.
4) Five doses taken followed by five doses missed.

Huang *et al.* found that patterns 1 and 2 were the most effective. This is in contrast with the results found by Wahl and Nowak and by Phillips *et al.*, where "clumped" doses were found to be favorable over random patterns of adherence. The discrepancy may be due to the fact that the effect of missed doses was not included in the calculation of IC_{50} and thus the effect of the emergence of resistance was neglected.

Huang *et al.* used numerical techniques to study the relationship between the actual and detectable rebound times. Their results suggest that the actual rebound time may actually occur much earlier than the time at which the viral load exceeds the lower limit of detection. They found that the relationship between the actual and detectable rebound levels was approximately linear. Thus, the detectable rebound time could be used to approximate the actual rebound time in the drug regimen.

Although the effects of resistance were not considered in this model, the examination of threshold levels for treatment efficacy is an important contribution. A topic for future research would be to include the effects of the emergence of resistance in the IC_{50} derivation used with imperfect adherence. This will give valuable insight into variations in threshold levels with respect to changes in adherence rates and patterns. An understanding of these interdependencies could aid in treatment planning and in "recovery" after poor adherence, i.e. if a patient misses a dose, what is the change in these threshold levels and when should the next dose be taken in order to move these two threshold levels away from each other.

2.4. *Need for future studies in adherence and drug resistance*

The simulations and analytical studies reviewed here have allowed researchers to gain some insight into the dynamics of HIV and the evolution of resistance when drug regimens are not perfectly followed. These studies have offered some important conclusions for clinicians, and realistic values of the parameters were incorporated in order to facilitate clinical applicability. One conclusion of particular importance is that the relevant predictor to the outcome of therapy is the effective basic reproductive ratio of the virus in this system, and that this value can be easily calculated using experimental data. Another important conclusion for clinicians is that if therapy is given in blocks, the probability that the resistant population emerges is smaller than when the adherence pattern is stochastic. This can have important implications in clinical practice since, as these studies are refined, clinicians may be able to determine the optimal time for their patients to take the next dose if one or more doses are missed.

Although these models offer considerable insights into the effects of adherence on the emergence of resistance, they have a number of drawbacks as well. For example, the effects of cross resistance are not included in these papers although it is widely acknowledged that this is an important factor in deciding which drugs will be included in a drug regimen.

There is also a need to refine the models to include other phenomena which have been observed clinically. For example, recent experimental studies have seen signs of synergistic effects in HIV treatment (for example Buckheit *et al.*, 1999; Nitanda *et al.*, 2001). Synergism implies that two or more drugs have a stronger effect when combined than the sum of their individual effects.

Finally, perhaps the key factor to be determined in order to minimize the risks associated with missed doses is the optimal time and size of the next dose to be taken. A great deal more analytical work is needed to answer this key question, and to characterize the overall importance of adherence in HIV therapy.

3. Structured Treatment Interruption

Structured treatment interruptions (STI) or structured therapy interruptions are well-defined periods of time during which all antiretroviral therapy is discontinued. The idea behind STI is to use the patient's own viral load to boost HIV-1 specific immune responses. This process is similar to vaccination. If this occurs, the boosted immune response may help increase the rate of clearance of the virus during periods on and/or off drug treatment. It is thought that STI can also be used to reduce drug toxicities, allow reversion of drug-resistant mutations in highly drug-experienced patients, and help to maintain the viral load at a low level once the patient stops drug therapy altogether.

STI has been studied in many clinical cases. The most famous case of the benefits of STI is that of "The Berlin Patient" (Lisziewicz, 1999). After experiencing only two short treatment interruptions this patient's immune system was able to control the infection, and viral load has remained below detectable levels for more than four years after therapy was discontinued. Several other studies have reported viral suppression after intermittent compliance and subsequent drug therapy (Ortiz *et al.*, 1999; Lori *et al.*, 2000). It has also been observed that in some patients with chronic HIV infection, the interruption of antiretroviral drug therapy was able to boost cellular immune responses (Ortiz *et al.*, 1999; Papasavvas *et al.*, 2000). In a study by Dybul *et al.* (2001) patients followed a STI regimen of seven days on and seven days off drugs for 44 weeks. This study found that the $CD4^+$ T cells and the immune response were unaffected by the short breaks in treatment even though some patients exhibited several temporary increases in viral load. But, as mentioned in the introduction to this chapter, some studies of STI have found that patients did not achieve prolonged suppression of the viral load and some have found that it has not aided their patients at all (see Walmsley, 2002 for details). In fact, along with the potential benefits of STI, there can also be serious drawbacks, such as an increased risk of drug resistance.

There have been a number of mathematical investigations of STI, providing crucial insights relevant to STI strategies (see Frost, 2002 for review). Unfortunately, only a handful of these studies have explored mathematical models which include antiretroviral resistance. In this section we will review models that have contributed to the study of STI and the emergence of drug resistance.

3.1. *Kirschner and Webb, 1996*

Kirschner and Webb, in their paper "*A model for treatment strategy in the chemotherapy of AIDS*" (1996), suggested that the possibility of using multi-drug type treatment to combat resistance may be better facilitated by an intermittent treatment strategy i.e. periods on and off drugs. Kirschner and Webb are the first researchers to carry out a theoretical study of this topic.

Using a standard system of differential equations for HIV dynamics, Kirschner and Webb studied three different strategies of treatment using STIs:

1) 50 days on treatment followed by 50 days off treatment.
2) 25 days on treatment followed by 75 days off treatment.
3) 10 days on treatment followed by 90 days off treatment.

They found that the first pattern was the most effective in maintaining the T cell count, followed by the second pattern. The third strategy was the least effective of the three treatments. Although this model does not incorporate the effects of drug resistance it was concluded that the balanced intermittent treatment strategy (pattern 1) would merit further study.

3.2. *Dorman* et al., *2000*

Dorman *et al.* (2000) studied the effects of STI on the emergence of drug resistance. Their model predicts that partially drug-resistant mutant populations will rebound during treatment interruption and that this rebound may not be detected clinically since these resistant mutants have very small populations that are not significant relative to the wild-type. Dorman *et al.* also found that partially resistant mutants that exist in the viral population can decrease the efficacy and durability of drug therapy. These mutants can easily accumulate mutations to acquire full resistance, thus a substantial increase in these populations will affect the outcomes of drug therapy. The compelling conclusion of this study is that compared with continuous treatment, STIs can greatly increase the chance of mutation to full resistance.

The model derived by Dorman *et al.* includes compartments for the concentrations of uninfected cells, actively infected cells, latently infected cells, and infectious and noninfectious virus where the virus can be early state drug-resistant or drug-sensitive wild-type. This model also includes the inhibitory effects of reverse transcriptase inhibitors and protease inhibitors. In this model it is assumed that early-stage mutants are less fit than the wild-type viruses, not only when drugs are not in use, but also when drugs are introduced into the system.

Starting at pretreatment steady-state levels and varying both the amount and fitness of the early-state resistant virus, Dorman *et al.* studied the impact of STIs on small populations of partially resistant virus. It was found that the total viral load growth dynamics during the STI were the same regardless of the amount of early-stage virus and that the total viral load rebound was detectable in the early stages of the STI. In addition, the latently infected cell pool did not depend on the amount of early-stage virus present before the STI. In contrast with these results, the behavior of the mutant viral load was dependent on the initial size of this population. Although mutant viral load rebounded on the same time scale as the total viral load, the magnitude of the rebound depended on the size of the original mutant population which may not be at detectable levels. Using a small population of only 0.033 copies/mL of early-stage mutant virus at the start of STI Dorman *et al.* found that the body was exposed to 19 million infectious, early-stage resistant viruses over the course of a 100-day STI and the total exposure increased to 1.9 billion for an initial viral load of 33 copies/mL. Comparing these results to the effects of ongoing treatment they found that total exposure was over 1000 times greater.

Studying shorter STIs, Dorman *et al.*, found that they have similar impacts. The same initial value of 33 copies/mL, however, generates over 600 times fewer early-stage mutant viruses for a 30-day STI.

The effects of varying model parameters, and of allowing the inhibition of the drug to decay exponentially, were also studied. It was found that the qualitative behaviour was similar to that described above with few exceptions (see Dorman *et al.*, 2000).

There are a number of important conclusions suggested by this study. First, Dorman *et al.* found that there is an increased risk of the emergence of resistance during STI. Part of this increased risk is dependent on the latently infected cell pool. They found that the latent cell pool, which is partially made up of cells infected with resistant virus, is quickly replenished by the initial rise in viral load to levels greater than that before therapy was initiated. Thus, there is a threat that the partially resistant virus will decrease the efficacy of future treatment. Dorman *et al.* also concluded that because of the high mutation rate of HIV, it is probable that the early-stage resistant viruses that are able to replicate during STI will acquire further resistance. This effect is augmented by the fact that early-stage resistance viruses that are activated from the latent cell pool can also acquire further resistance.

Secondly, the observation that the rebound of the mutant viral load can be below detectable levels led to the question of how many viral clones would need to be sequenced in order for the population to be reliably detected. Dorman *et al.* found that in order to detect the population with probability α, the necessary number of clones is given by $\ln(\alpha)/\ln(1-b)$, where b is the fraction of mutants in the viral load. Conversely, if many samples are taken of a patient's viral load and some of them contain the early-stage resistance virus, the fraction of the viral load that is made up of the early-stage resistant virus can be estimated.

Thirdly, Dorman *et al.* observed that the initial dynamics of STI are dominated by parallel growth of the viral strains. The viral load and the early-stage resistant viral load rebound on the same time scale.

A drawback of this model is that it was assumed that the early-stage mutants were less fit than the wild-type virus, not only when drugs are not in use, but also when drugs are introduced into the system. This is not always the case since some early-stage mutants may have significant resistance under drug therapy, particularly when cross-resistance is taken into account. Future studies might include small numbers of early-stage mutants that have small advantages over the wild-type, in order to determine their effects on the evolution of resistance.

In both this study and in that of Kirschner and Webb (1996), the main disadvantage is that the immune response is not modeled. Since one motivation for STI is to potentially boost the immune response, models that do not include the immune response should not be used to dismiss the idea of STI. Instead, we need to consider models that include the immune response to explore how treatment interruptions affect the evolution of the precursor and effector cells.

3.3. *Bonhoeffer* et al., *2000*

Bonhoeffer *et al.* (2000), like Dorman *et al.* (2000), studied the effects of STI on the emergence of resistance, the replenishing of the latent cell pool and also compared STI to continuous treatment. Bonhoeffer *et al.*, however, used a model that includes the effector cells in the immune system. An effector cell in this model is a cytotoxic T cell, i.e., the component of the immune response which kills HIV-1 infected cells. This model suggests that STI can lead to transient or sustained control of the virus only if the increase in the effector cell population can offset the increase in susceptible target cells induced by therapy. Using this model it was found that the risks of inducing drug resistance during therapy were small as long as the virus population was held at levels considerably less than baseline before therapy, but that this risk increases considerably if the virus load is allowed to increase during drug-free periods to levels similar to or higher than this value. This was also observed for the replenishing of the latent cell pool during drug-free periods. The STI in this study was started at day 500 with a drug phase of 20 days on and 20 days off therapy.

The model proposed by Bonhoeffer *et al.* is as follows:

$$
\begin{aligned}
\frac{dT}{dt} &= s - d_T T - bTI \\
\frac{dI}{dt} &= bTI - d_I I - pIE \\
\frac{dE}{dt} &= cI - d_E E,
\end{aligned}
$$

where T, I and E denote the population densities of susceptible CD4 cells, active virus-producing CD4 cells and CTL effector cells respectively. The parameters are: the rate of immigration of susceptible cells, s; the infection rate, b; the killing rate, p; the stimulation of the effector cell proliferation, c; and d_T, d_I and d_E for the death rates of the corresponding cell populations. Using an extension of this model Bonhoeffer *et al.* demonstrated that STI may lead to an increase in CTL which may be able to control the virus even when therapy is halted. However, they also found that the viral load will increase and the CTL will decrease if the period in therapy is too long.

The number of newly infected cells during treatment compared to the number of infected cells at time zero, I_0, is given by:

$$J = \frac{1}{I_0} \int_0^{t_f} bTI\, dt,$$

where t_f is the time at the endpoint of the period of observation and the integral of bTI gives the number of newly infected cells during treatment. The value of J determines the probability of the emergence of the resistant mutant during therapy compared to the likelihood that it was present before the onset of therapy. If J is less than one, a resistant mutant is more likely to be present at the start of therapy than first produced during therapy. In this case, the origin of resistance is most likely due to the selective outgrowth of a resistant mutant that is exists when therapy is initiated. In contrast, if during STI the virus load grows to levels similar to baseline before therapy, J increases to a value greater than one. Bonhoeffer *et al.* note that, in this case, therapy interruption will dramatically increase the risk of treatment failure due to resistance.

Bonhoeffer *et al.* also studied the effects of STI on the latent cell pool. Assuming that the latently infected cell pool is in equilibrium before therapy is initiated, they found that the latently infected cells that are produced during therapy are very small in number compared to the number of latent cells that are cleared during therapy. This is because the population of infected cells decreases during therapy. Thus, Bonhoeffer *et al.* concluded that the latently infected cell population is likely to be refilled only if the load of latent cells has decreased by a larger factor than the free virus. Hence, STI may not lead to a repopulation of the latently infected cell pool as previously found by Dorman *et al.* Bonhoeffer *et al.* stated that this is of particular consideration in patients who have only been on continuous combination therapy for a short time.

The most important contribution of this paper is the inclusion of the immune response in the model, via effector cells. These cells have been ignored in many other models and as this study has shown, they have significant impact on the predictions of the analytical work. By including the effector cells in the model, Bonhoeffer *et al.* were able to determine events in the system that may increase the risk of acquiring mutations that confer drug resistance. These include situations in which the latently infected cell pool is likely to be refilled, and situations in which the viral load reaches a threshold level which significantly increases the risk of treatment failure due to resistance.

This study also provides some important results for clinicians by clarifying the major issues to be considered in determining when STI should be administered. Treatment success is heavily dependent on the pretreatment viral load, the probability of the emergence of the resistant mutant during therapy and the likelihood that it is present before therapy.

3.4. *Wodarz, 2001*

In another study that includes the immune response, Wodarz (2001) found that STI is less likely to induce resistance when better immunological control is established. Wodarz studied the CTL response in HIV infection using a mathematical model that included the dynamics of helper-dependent and helper-independent CTL in the system. He found that STI is beneficial in developing better CTL kinetics both on and off drug therapy, and also considered cases when drug resistance could evolve.

Wodarz developed a model that includes the dynamics of uninfected T-cells, infected T-cells, the helper-independent response (CTL) and the two types of helper-dependent responses (CTLp and CTLe). The CTLp is defined as the CD8 cells that have been activated in response to antigen and proliferate in the system at a significant rate. The model assumes that the CTLe do not have the capacity to proliferate at a significant rate but instead differentiate from CTLp. The model assumes that the CTLe do not have the capacity to proliferate at a significant rate but instead differentiate from CTLp. It is also assumed that the helper-dependent response is long lived in the absence of antigen and the helper-independent response is relatively short lived.

In this study, and in a previous study by Wodarz *et al.* (2000), it was found that interruptions in therapy can aid the CTL memory (CTLe) and therefore increase the probability that the immune system can control the infection of HIV even when therapy is stopped as long as the time on therapy is short. But, in the previous study, the development of resistant mutations in the system was not explicitly considered. In this study Wodarz investigates the conditions that are required in the CTL response for the drug-resistant mutant to grow and comments on how STI may affect this.

Assuming that the drug-resistant mutant exists before therapy is initiated, Wodarz found that there exists a threshold that determines whether the drug-resistant strain emerges. He observed that this threshold is dependent on the responsiveness of the CTL and the replication rate of the wild-type in the absence of the drug. If the helper-independent CTL response is small, the emergence of the drug-resistant virus is mainly determined by the basic reproductive ratio of the wild-type virus. If the responsiveness of the helper-independent CTL is stronger, then the drug-resistant strain can only persist if basic reproductive ratio of the resistant strain in the presence of the drug is larger than that of the wild-type strain in the absence of the drug. Wodarz notes that this is case since the helper-independent CTL counters the emergence of drug-resistant strains and the replication rate of the wild-type virus determines the level of the CTL at the start of treatment. Wodarz also notes that this condition is unlikely to occur *in vivo*, since resistance is probably costly to the virus in the presence of the CTL response. Thus, resistant strains present at any time before or during therapy are unlikely to grow as long as the CTL response is at approximately the equilibrium level.

Therefore, according to this study, it is unlikely that the resistant mutants will grow in the presence of efficient memory CTL. Unfortunately, as noted by the author, this conclusion has limited practical implications, since it is not realistic to assume that an efficient CTL memory response can control the infection. Finally, Wodarz suggests that STI is less likely to induce resistance the better the immunological control that is established.

Wodarz reaches the same conclusion as Bonhoeffer *et al.* (2000): that the emergence of the drug-resistant strains depends on the degree of the immune response. An important point made by Wodarz is that the immune response will become stronger if clearly defined treatment interruptions are used. That is, initiation and cessation of therapy must occur at certain levels of the viral load in order for the CTL memory to evolve and, thus, fight the virus in the absence of therapy altogether.

Wodarz also comments on the viral load dynamics during drug therapy that have been observed in many patients: during drug therapy the viral load will initially decrease to low levels, but this is followed by an increase to a new equilibrium. He stated that this could be attributed to a decline of the CTL response during therapy since growth of drug-resistant strains can only occur if the CTL has fallen below the threshold discussed in this study. This has important implications in the study of STI since it has been found that STI may induce the development of CTL responses and build up immunity to the virus. Therefore, STI could possibly decrease the chances of developing resistance since resistant populations are unlikely to grow in the presence of the CTL response.

3.5. *Walensky* et al., *2002*

Walensky *et al.* (2002) approached the study of STI and drug resistance in a different manner. Instead of solving a system of equations, Walensky *et al.* designed a computer simulation of primary HIV infection. They simulated a randomized clinical trial of patients diagnosed with primary HIV infection and used this to evaluate the clinical outcomes and life expectancy projections for three primary HIV infection treatment strategies:

1) Continuous antiretroviral therapy initiated at CD4 count < 350 cell/mm^3.
2) Continuous antiretroviral therapy initiated immediately on diagnosis of primary HIV infection.
3) Antiretroviral therapy initiated on diagnosis followed by STI.

Sensitivity analyses were conducted to examine the impact of potential variations in the short and long term consequences of immediate therapy and STIs, as well as reasonable variations in other model parameters such as the efficacy of the treatment, immunologic or CD4 benefit, the effect on later resistance to antiretroviral therapy and the possibility of more effective antiretroviral regimens in the future.

In this stochastic model, patients are considered one at a time and progress through the following health states: primary HIV infection, HIV infection, acute clinical event and death. Each patient's course is followed individually until death and then a new patient enters. When a patient enters the model, a random assignment of either immediate antiretroviral therapy, or antiretroviral therapy when the CD4 count is less than 350 cells/mm^3, is made. If the patient is assigned to immediate antiretroviral therapy, either continuous antiretroviral therapy or an antiretroviral induction period of one year is chosen randomly. If the antiretroviral induction period of one year is chosen, treatment is interrupted after that year if the HIV viral load is suppressed. The patient then continues an STI drug regimen for a year. In the STI phase the patient receives three cycles of the same drug combination with three months on therapy followed by one month off. If the patient's viral load was not suppressed after one year of antiretroviral therapy, therapy is discontinued until the CD4 count is less than 350 cells/mm^3 after which therapy is reinitiated.

The model tracked individual patient statistics regarding time spent in each health state, time on therapy, and length and quality of life using five assumptions, referred to as the base case (see Walensky *et al.*). At the end of the simulation summary statistics were calculated on a cohort that consisted of 1 million patients.

Walensky *et al.* found that STI was the most effective treatment strategy using the base case. They found that STI increases life expectancy to approximately 26 years after primary infection, as opposed to approximately 24 years for either of the other two strategies. When the effectiveness of STI was varied, it was found that STI was only needed to be effective in 20% of patients before it became the optimal strategy.

However, when the effects of resistance were included in the study, the life expectancy of the STI treated patient was reduced to approximately 24 years. This was done by assuming that the patient populations on STI have a bimodal distribution of life expectancy, where some fraction of patients develops resistance. In those who developed resistance it was assumed that the patients would experience a decrease in first-line therapy such that only 45% (as opposed to 70%) of patients achieved virologic suppression.

Through the examination of different parameter variations Walensky *et al.* found that, in the majority of cases, STI was still the optimal strategy. STI outcomes were not very sensitive to variations in the efficacy or duration of the treatment strategy. In contrast, STI was quite sensitive to variations in CD4 benefit during induction, and to the efficacy of subsequent antiretroviral therapy. It was also found that when more effective antiretroviral regimens were used (regimens which may be possible in future) STI remained optimal unless it resulted in a minimal decrease in antiretroviral efficacy. The authors note that currently, the outcome of drug therapy is largely dependent on patient adherence and that this will have to be overcome in order to reach antiviral efficacies as high as 90% that were studied in this paper.

An important conclusion of this work is that although the potential for antiretroviral resistance resulting from intermittent therapy during STI is uncertain, it is of critical concern. The immunologic benefit associated with immediate therapy and the potential for antiretroviral drug resistance due to STI have the most important impacts on the optimal strategy. In general, Walensky *et al.* found that if antiretroviral resistance occurs in response to STI and results in decreased antiretroviral efficacy, then STI is less favorable.

An important contribution of this model to studies on STI and the emergence of drug resistance is that although it is solely simulation-based, and therefore resting on entirely different assumptions than that of the analytical work in the field, it leads to the same conclusion as the previous studies: STI is not a beneficial strategy if resistant mutations exist in the system.

This model is beneficial for studies in clinical practices since individual patients, rather than population averages, were compared, yielding a range of possible outcomes for real patients undergoing therapy. This model also focused on the aspects of treatment that are important to patients undergoing therapy such as the clinical outcomes of the therapy and life expectancy projections for the STI regimen.

3.6. *Need for future studies in STI and drug resistance*

The studies described above have offered ground-breaking insights into STI and the emergence of drug resistance. It is clear that STI can increase the probability of the emergence of drug resistance and that STI is not the optimal strategy to be used if there is a chance that drug-resistant mutants exist in the system. The magnitude of the effect of STI on the emergence of the drug-resistant mutants, however, has not been clearly quantified. Some studies conclude that the effect is minimal while others predict a substantial effect. Clearly, more work is needed in this critical area.

These studies have also provided some useful results for clinicians, illustrating repeatedly that the outcome of STI strongly depends on pretreatment viral load, the probability of the emergence of resistant mutations during therapy and the likelihood that such a mutant is present before therapy. Using $\ln(\alpha)/\ln(1-b)$ given by Dorman *et al.*, clinicians can estimate the fraction of the virus population that is composed of the resistant virus simply by taking samples of the patient's viral load. Another significant implication is that early-stage resistant mutants that exist in the system and have no advantage over the wild-type during drug therapy can still lead to the eventual outgrowth of the resistant virus. It is important to note here that each of these studies has used realistic parameters whenever possible, in order to provide both qualitative and quantitative predictions.

Future studies should include models that compare the effects of STI on single drug and multi-drug treatment regimens. The studies presented here do not consider the different effects of these regimens. Although the efficacy of treatment is often varied, thus allowing the models to capture the increased overall effectiveness of a multi-drug regimen, the variable probabilities of emergent resistance under single- or multi-drug regimens are not considered.

Future studies in STI could also include investigation of relatively short STIs, such as those experienced by the Berlin patient. It would be interesting to see, for example, which parameter values would predict an outcome similar to this famous case. This would be of particular importance to the clinical understanding of the relationships between viral load, the length of the period on and off the drugs and the effective boost of the immune system. Similarly, the effects of STI during drug regimens that incorporate cross-resistance and/or synergism would be of interest.

A further point of interest is that although the latently infected cell population is included in several of these studies, the effects of this population on the emergence of drug-resistant virus in the models have not been isolated.

Finally, as noted by Phillips *et al.* (2001), few of these papers are easily understandable to clinicians. Although arguably very few clinicians in the field have the opportunity to keep abreast of even the primary clinical literature, it is still an important goal that future theoretical studies be made accessible to as wide a clinical audience as possible.

Acknowledgments

This work was supported by the Natural Sciences and Engineering Research Council of Canada

References

1. Arnsten JH, Demas PA, Farzadegan H, Grant RW, Gourevitch MN, Chang CJ, Buono D, Eckholdt H, Howard AA, Schoenbaum EE (2001) Antiretroviral therapy adherence and viral suppression in HIV-infected drug users: comparison of self-report and electronic monitoring. *Clin Infect Dis* **33**: 1417–1423.

2. Austin DJ, Kakehashi M, Anderson RM (1997) The transmission dynamics of antibiotic-resistant bacteria: the relationship between resistance in commensal organisms and antibiotic consumption. *Proc Roy Soc Lond B* **264**: 1629–1638.
3. Austin DJ, White NJ, Anderson RM (1998) The dynamics of drug action on the within-host population growth of infectious agents: melding pharmacokinetics with pathogen population dynamics. *J Theor Biol* **194**: 313–339.
4. Austin DJ, Anderson RM (1999) Studies of antibiotic resistance within the patient, hospitals and the community using simple mathematical models. *Phil Trans R Soc Lond B* **354**: 721–738.
5. Austin DJ, Kristinsson KG, Anderson RM (1999) The relationship between the volume of antimicrobial consumption in human communities and the frequency of resistance. *Proc Natl Acad Sci USA* **96**: 1152–1156.
6. Bangsberg DR, Hecht FM, Charlebois ED, Zolopa AR, Holodniy M, Sheiner L, Bamberger JD, Chesney MA, Moss M (2000) Adherence to protease inhibitors, HIV-1 viral load, and development of drug resistance in an indigent population. *AIDS* **14**: 357–366.
7. Bangsberg DR, Charlesbois ED, Grant RM, Holidniy M, Deeks SG, Perry S, Conroy KN, Clark R, Guzman D, Zolopa A, Moss A (2003) High levels of adherence do not prevent accumulation of HIV drug resistance mutations. *AIDS* **17**: 1925–1932.
8. Boden D, Hurley A, Zhang L, Cao Y, Guo Y, Jones E, Tsay J, Ip J, Farthing C, Limoli K, Parkin N, Markowitz M (1999) HIV-1 drug resistance in newly infected individuals. **282**: 1135–1141.
9. Bonhoeffer S, Nowak MA (1997) Pre-existence and emergence of drug resistance in HIV-1 infection. *Proc Roy Soc Lond B* **264**: 631–637.
10. Bonhoeffer S, Lipsitch M, Levin BR (1997) Evaluating treatment protocols to prevent antibiotic resistance. *Proc Natl Acad Sci USA* **94**: 12106–12111.
11. Bonhoeffer S, May RM, Shaw GM, Nowak MA (1997) Virus dynamics and drug therapy. *Proc Natl Acad Sci USA* **94**: 6971–6976.
12. Bonhoeffer S, Rembiszewski M, Ortiz GM, Nixon DF (2000) Risks and benefits of structured antiretroviral drug therapy interruptions in HIV-1 infection. *AIDS* **14**: 2313–2322.
13. Blower SM, Gershengorn HB, Grant RM (2000) A tale of two futures: HIV and antiretroviral therapy in San Francisco. *Science* **287**: 650–654.
14. Blower SM, Aschenback AN, Gershengorn HB, Kahn JO (2001) Predicting the unpredictable: Transmission of drug-resistant HIV. *Nature Med* **7**: 1016–1020.
15. Blower S, Schwartz EJ, Mills J (2003) Forecasting the future of HIV epidemics: the impact of antiretroviral therapies and imperfect vaccines. *AIDS Rev* **5**: 113–125.
16. Buckheit RW Jr, Russel JD, Pallansch LA, Driscoll JS. Anti-human immunodeficiency virus type 1 (HIV-1) activity of 2′-fluoro-2′,3′-dideoxyarabinosyladenine (F-ddA) used in combination with other mechanistically diverse inhibitors of HIV-1 replication.

17. Coffin JM (1992) Genetic diversity and evolution of retroviruses. *Curr Top Microbial Immunol* **176**: 143–164.
18. Coffin JM (1995) HIV population dynamics *in vivo*: implications for genetic variation, pathogenesis and therapy. *Science* **267**: 483–489.
19. Davies JE (1997) Origins, acquisition and dissemination of antibiotic resistance determinants. *Ciba Found Symp* **207**: 15–27.
20. de Boer RJ, Boucher CA (1996) Anti-CD4 therapy for AIDS suggested by mathematical models. *Proc R Soc Lond B* **263**: 899–905.
21. de Boer RJ, Perelson AS (1998) Target cell limited and immune control models of HIV infections: A comparison. *J Theor Biol* **190**: 201–214.
22. de Jong MD, Veenstra J, Stilianakis NI, Schuurman R, Lange JM, de Boer RJ, Boucher CA (1996) Host-parasite dynamics and outgrowth of virus containing a single K70R amino acid change in reverse transcriptase are responsible for the loss of HIV-1 RNA load suppression by zidovudine. *Proc Nat Acad Sci.*
23. Dorman KS, Kaplan AH, Kange K, Sinsheimer JS (2000) Mutation takes no vacation: can structured treatment interruptions increase the risk of drug-resistant HIV-1? *JAIDS* **25**: 398–402.
24. Dybul M, Chun T, Yoder C, Hidalgo B, Belson M, Hertogs K, Larder B, Dewar RL, Fox CH, Hallahan CW, Justement JS, Migueles SA, Metcalf JA, Davey RT, Daucher M, Pandya P, Baseler M, Ward DJ, Fauci AS (2001) Short-cycle structured intermittent treatment of chronic HIV infection with highly active antiretroviral therapy: Effects on virologic, immunologic, and toxicity parameters. *Proc Natl Acad Sci USA* **98**: 15161–15166.
25. Frost SDW (2002) Dynamics and evolution of HIV-1 during structured treatment interruptions. *AIDS Reviews* **4**: 119–127.
26. Frost SDW, McLean AR (1994) Germinal centre destruction as a major pathway of HIV pathogenesis. *JAIDS* **7**: 236–244.
27. Girardi E (2003) Epidemiological aspects of transmitted HIV drug resistance. *Scand J Infect Dis Suppl* **35**(106): 17–20.
28. Gross R, Bilker WB, Friedman HM, Strom BL (2001) Effect of adherence to newly initiated antiretroviral therapy on plasma viral load. *AIDS* **15**: 2109–2117.
29. Hecht FM, Grant RM, Petropoulos CJ, Dillon B, Chesney MA, Tian H, Hellmann NS, Bandrapalli NI, Digilio L, Branson B, Kahn JO (1998) Sexual transmission of an HIV-1 variant resistant to multiple reverse-transcriptase and protease inhibitors. *New Eng J Med* **339**: 307–311.
30. Ho DD, Neumann AU, Perelson AS, Chen W, Leonard JM, Markowitz M (1995) Rapid turnover of plasma virions and CD4 lymphocytes in HIV-1 infection. *Nature* **373**: 123–126.
31. Huang Y, Rosenkranz SL, Wu H (2003) Modeling HIV dynamics and antiviral response with consideration of time-varying drug exposures, adherence and phenotypic sensitivity. *Math Biosci* **184**: 165–186.
32. Kepler TB, Perelson AS (1998) Drug concentration heterogeneity facilitates the evolution of drug resistance. *Proc Natl Acad Sci USA* **95**: 11514–11419.

33. Kirschner DE, Webb GF (1996) A model for the treatment strategy in the chemotherapy of AIDS. *Bull Math Biol* **58**: 367–391.
34. Levin BR, Lipsitch M, Perrot V, Schrag S, Anita R, Simonsen L, Walker NM, Stewart FM (1997) The population genetics of antibiotic resistance. *Clin Infect Dis* **24** (Suppl. 1): S9–S16.
35. Levin SA, Andreason V (1999) Disease transmission dynamics and the evolution of antibiotic resistance in hospitals and communal settings. *Proc Natl Acad Sci USA* **96**: 800–801.
36. Levy SB (1997) Antibiotic resistance: an ecological imbalance. *Ciba Found Symp* **207**: 1–9.
37. Lipsitch M, Levin BR (1998) The population dynamics of tuberculosis chemotherapy: mathematical models of the roles of noncompliance and bacterial heterogeneity in the evolution of drug resistance. *Int J Tuberculosis Lung Dis* **2**: 187–199.
38. Lipsitch M, Bergstrom CT, Levin BR (2000) The epidemiology of antibiotic resistance in hostpitals: Paradoxes and prescriptions. *Proc Natl Acad Sci USA* **97**(4): 1938–1943.
39. Lisziewicz J, Rosenberg E, Lieberman J, Jessen H, Lopalco L, Siliciano R, Walker B, Lori F (1999) Control of HIV despite the discontinuation of antiretroviral therapy. *N Engl J Med* **340**: 1683–1684.
40. Little SI, Daar ES, D'Aquila RT, Keiser PH, Connick E, Whitcomb M, Hellmann NS, Petropoulos CJ, Sutton L, Pitt JA, Rosenberg ES, Koup RA, Walker B, Lori F (1999) Reduced antiretroviral drug susceptibility among patients with primary HIV infection. *J Amer Med Assoc* **282**: 1142–1149.
41. Liu H, Golin CE, Miller LG, Hays RD, Beck CK, Sanandaji S, Christian J, Maldonado T, Duran D, Kaplan AH, Wenger NS (2001) A comparison study of multiple measures of adherence to HIV protease inhibitors. *Ann Intern Med* **134**: 968–977.
42. Lori F, Maserati R, Foli A, Seminari E, Timpone J, Lisziewicz J (2000) Structured treatment interruptions to control HIV-1 infection. *Lancet* **355**: 287–288.
43. Loveday C, Kaye S, Tenant-Flowers M, Semple M, Ayliffe U, Weller IV, Tedder RS (1995) HIV-1 RNA serum-load and resistant viral genotypes during early zidovudine therapy. *Lancet* **345**: 820–824.
44. McLean AR, Emery VC, Webster A, Griffiths PD (1991) Population dynamics HIV within an individual after treatment with zidovudine. *AIDS* **5**: 485–489.
45. McLean AR, Frost SDW (1995) Ziduvidine and HIV: Mathematical models of within-host population dynamics. *Rev Med Virol* **5**: 141–147.
46. McLean AR, Nowak MA (1992) Competition between zidovudine-sensitive and zidovudine-resistant strains of HIV. *AIDS* **6**: 71–79.
47. McNabb J, Ross JW, Abriola K, Turley C, Nightingale CH, Nicolau DP (2001) Adherence to highly active antiretroviral therapy predicts virologic outcome at an inner-city human immunodeficiency virus clinic. *Clin Infect Dis* **33**: 700–705.

48. Moeller R (2000) HIV drug resistance. *J Amer Med Assoc* **284**: 169.
49. Montaner JS, Reiss P, Cooper D, Vella S, Harris M, Conway V, Wainberg MA, Smith D, Robinson P, Hall D, Myers M, Lange JM (1998) A randomized double-blind trial comparing combinations of nevirapine, didanosine, and zidovudine for HIV-infected patients: the INCAS trial. Italy, The Netherlands, Canada and Australia Study. *JAMA* **279**: 930–937.
50. Nitanda T, Wang X, Somekawa K, Yuasa S, Baba M. Three-drug combinations of emivirine and nucleoside reverse transcriptase inhibitors in vitro: long-term culture of HIV-1 infected cells and breakthrough viruses.
51. Nowak MA, Anderson RM, McLean AR, Wolfs TF, Goudsmit J, May RM (1991) Antigenic diversity thresholds and the development of AIDS. *Science* **254**: 963–969.
52. Nowak MA, Bonhoeffer S, Shaw GM, May RM (1997) Anti-viral drug treatment: Dynamics of resistance in free virus and infected cell populations. *J Theo Biol* **184**: 205–219.
53. Nowak MA, May RM (2000) *Virus Dynamics.* Oxford University Press.
54. Ortiz GM, Nixon DF, Trkola A, Binley J, Jin X, Bonhoeffer S, Kuebler PJ, Donahoe SM, Demoitie MA, Kakimoto WM, Ketas T, Clas B, Heymann JJ, Zhang L, Cao Y, Hurley A, Moore JP, Ho DD, Markowitz M (1999) HIV-1 specific immune responses in subjects who temporarily contain virus replication after discontinuation of highly active antiretroviral therapy. *J Clin Invest* **104**: R13–R18.
55. Papasavvas E, Ortiz GM, Gross R, Sun J, Moore EC, Heymann JJ, Moonis M, Sandberg JK, Drohan LA, Gallagher B, Shull J, Nixon DF, Kostman JR, Montaner LJ (2000) Enhancement of human immunodeficiency virus type 1-specific CD4 and CD8 T cell responses in chronically infected persons after temporary treatment interruption. *J Infect Dis* **182**: 766–755.
56. Paterson DL, Swindells S, Mohr J, Brester M, Vergis EN, Squier C, Wagener MM, Singh N (2000) Adherence to protease inhibitor therapy and outcomes in patients with HIV infection. *Ann Intern Med* **133**: 21–30.
57. Perelson AS, Neumann AU, Markowitz M, Leonard JM, Ho DD (1996) HIV-1 dynamics *in vivo*: virion clearance rate, infected cell lifespan, and viral generation time. *Science* **271**: 1582–1585.
58. Perelson AS, Essunger P, Cao Y, Vesanen M, Hurley A, Saksela K, Markowitz M, Ho DD (1997) Decay characteristics of HIV-1-infected compartments during combination therapy. *Nature* **387**: 188–191.
59. Perelson AS (2002) Modelling viral and immune system dynamics. *Nature* **2**: 28–36.
60. Phillips AN, Youle M, Johnson M, Loveday C (2001) Use of a stochastic model to develop understanding of the impact of different patterns of antiretroviral drug use on resistance development. *AIDS* **15**: 2211–2220.
61. Ribeiro RM, Bonhoeffer S (2000) Production of resistant HIV mutants during antiretroviral therapy. *Proc Natl Acad Sci USA* **97**: 7681–7686.
62. Stilianakis NI, Boucher CA, de Jong MD, Van Leeuwen R, Schuurman R, de Boer RJ (1997) Clinical data sets on human immunodeficiency virus type 1 reverse transcriptase resistant mutants explained by a mathematical model. *J Virol* **71**: 161–168.

63. Wahl LM, Nowak MA (2000) Adherence and drug resistance: predictions for therapy outcome. *Proc Roy Soc Lond B* **267**: 835–843.
64. Walensky RP, Goldie SJ, Sax PE, Weinstein MC, Paltiel AD, Kimmel AD, Seage GR 3rd, Losina E, Zhang H, Islam R, Freedberg KA (2002) Treatment for Primary HIV infection: Projecting outcomes of immediate, interrupted, or delayed therapy. *J Acq Immun Syn* **31**: 27–37.
65. Walmsley S, Loutfy M (2002) Can structured treatment interruptions (STIs) be used as a strategy to decrease total drug requirements and toxicity in HIV infection? *J Int Assoc Phy AIDS Care* **3**: 95–103.
66. Walsh JC, Mandalla S, Gazzard BG (2002) Responses to a 1 month self-report on adherence to antiretroviral therapy are consistent with electronic data and virological treatment outcome. *AIDS* **16**: 269–277.
67. Wei LM, Ghosh SK, Taylor ME, Johnson VA, Emini EA, Deutsch P, Lifson JD, Bonhoeffer S, Nowak MA, Hahn BH, Saag MS, Shaw GM (1995) Viral dynamics in HIV-1 infection. *Nature* **373**: 117–122.
68. Wein LM, D'Amato RM, Perelson AS (1998) Mathematical considerations of antiretroviral therapy aimed at HIV-1 eradication or maintenance of low viral loads. *J Theor Biol* **192**: 81–98.
69. Wodarz D, Nowak MA (1999) Specific therapy regimes could lead to long-term immunological control of HIV. *PNAS* **96**: 14464–14469.
70. Wodarz D, Page K, Arnaout R, Thomsen AR, Lifson JD, Nowak MA (2000) A new theory of cytotoxic T-lymphocyte memory: implications for HIV treatment. *Phil Trans R Soc Lond B* **355**: 329–343.
71. Wodarz D (2001) Helper-dependent vs. helper-independent CTL responses in HIV infection: Implications for drug therapy and resistance. *J Theor Biol* **213**: 447–459.
72. World Health Organization. Adherence to Long-Term Therapies: Evidence for Action. WHO Global Report. Geneva: WHO 2003.

CHAPTER 18

A BRANCHING PROCESS MODEL OF DRUG RESISTANT HIV

H. Zhou and K. S. Dorman

Drug resistance is one of the major barriers to the successful control of HIV infections. Despite numerous effective medications and increasingly tolerable regimens, resistance is a persistent and pervasive problem. Knowing when resistant mutants are present and where the mutants that cause treatment failure originate are important steps to understanding and ultimately defeating treatment failure. In this chapter, we explore a branching process model for the evolution of resistant mutants before and after treatment initiation. We find that simple resistant mutants with 1 or 2 mutations appear very early during the acute stage of infection, leaving only a small window of opportunity to treat before any resistant mutations are present. While mutants with single point mutations are much more likely to appear before treatment than from residual replication during treatment, double, triple and more complex mutants are often produced, on average, in greater numbers on treatment than before treatment. However, the amount of replication on treatment in this simple model is insufficient to explain most treatment failure, suggesting that most mutants, other than single mutants and some double mutants, neither pre-exist nor evolve in well-suppressed systems. It is much more likely that lack of replication control, due to faulty adherence or sanctuary replication, is the cause of complex resistance patterns. Additional research is needed to understand the evolution of resistance under such partially controlled replication.

Keywords: Branching process; stochastic model; drug resistance.

1. Introduction

Management of HIV infection has greatly improved since the introduction of antiretroviral (ARV) medications. Especially when used in combination, ARVs have transformed AIDS from a highly fatal disease into a chronic and somewhat manageable disease (Pomerantz *et al.*, 2003). Unfortunately, administration of only one drug, or monotherapy, may be the only feasible option in the developing world, and even with multidrug strategies, treatment may fail because of serious side effects, poor adherence to complex regimens, or emergence of resistant virus (Cohen, 2002).

When considering treatment options for patients, it is important to know whether drug resistant mutations exist and to what degree they dominate the infecting viral population. Under some circumstances, prescription of a potent monotherapy may be sufficient. Both phenotypic and genotypic assays have been proposed to test for the existence of HIV drug resistant strains. Unfortunately, current methods may not be sensitive enough to detect the minor variants such as early drug resistant mutants that can exist hidden in the viral quasispecies (Shafer, 2002). Mathematical models offer an opportunity to circumvent such practical limitations by providing a view of viral dynamics below the limits of detection. For example, Ribeiro *et al.* (Ribeiro *et al.*, 1998) used a quasispecies equation of viral dynamics to calculate the expected frequency of drug-resistant viruses in long term untreated patients. They calculated the equilibrium frequencies of resistant mutants with one, two, three or more point mutations given varying mutation and selection parameter values.

The resistant viruses that rebound during treatment failure may originate from one of two possible sources (Ribeiro and Bonhoeffer, 2000). The *pre-existence hypothesis* posits that treatment fails when pre-existing resistant strains are selected by the drug, while the *emergence hypothesis* holds that treatment failure is caused by resistant mutants that emerge *de novo* through residual virus replication during treatment. It is important to distinguish these two causes, because they could warrant different counteractive strategies. A combination of drugs defeated only with complex patterns of resistance mutations is needed if the treatment failure is caused by pre-existing mutants. On the other hand, increasing drug dosage and not regimen complexity may be sufficient if treatment is caused by resistant virus generated *de novo* during therapy. Ribeiro and Bonhoeffer (2000) compared the relative probabilities of these two hypotheses by calculating the number of infected cells present at the start of therapy and the number of cells newly infected during therapy. Both a deterministic model and a stochastic simulation suggested that treatment failure is most likely caused by the pre-existence of resistant mutants (Ribeiro and Bonhoeffer, 2000).

We formulate a stochastic branching process model to investigate the emergence of drug-resistant mutants prior to and after treatment. Two forms of the model are used for pre- and post-treatment environments. Before treatment, all mutant forms of the virus are modeled as a subcritical multitype branching process fed by continual immigration of mutation from the growing wild-type. After treatment, the wild-type and all mutant forms of the virus are modeled as a supercritical multitype branching process, with one mutant representing the viable, fully resistant virus capable of causing viral outgrowth and treatment failure. The abundance of mutants at any time prior to treatment is characterized by means, variances, or numerically calculated marginal distributions. The probability of no mutants present during the initial bursting stage of the acute infection is also characterized. To determine the two possible causes for treatment failure, we calculate the mean and variance of the number of resistant mutants that are generated *de novo* during drug therapy and compare it to the abundance of resistant mutants at equilibrium prior to treatment.

2. From the Biology to the Model

Point mutations at one or two sites may be enough to confer resistance to a single drug (de Jong *et al.*, 1996). Resistance to combination therapy, however, requires several point mutations. Suppose mutations at n sites in the HIV genome are associated with resistance to a particular drug therapy. The fully resistant mutant is achieved by accumulation of mutations at all n sites. Mutations at different sites may occur at different rates. For example, an amino acid change achieved by a single transition may occur more rapidly than one requiring a transversion or multiple DNA mutations. In addition, reverse transcriptase error rates may vary by genome position (Klarmann *et al.*, 1993; Ricchetti and Buc, 1990).

We define 2^n subpopulations of productively infected CD4 T cells (PITs) according to the number and locations of the point mutations in the genome of the inhabitant HIV virus. There are $\binom{n}{k}$ subpopulations of PITs infected by k-mutants, i.e. mutants having k mutations, with $0 \leq k \leq n$. Different subpopulations may be connected by mutational events. Mutation at one of the n sites within the genome of wild-type virus produces single mutants. Further mutation at one of the remaining sites produces double mutants, and so on. The k-mutants can also be converted into $(k-1)$ mutants by reversions to wild-type. For simplicity, we exclude the presumably rare possibility that two or more resistance-associated mutations occur simultaneously during one replication cycle. Since productively infected T cells contribute most of the plasma virus particles found in patients (Bonhoeffer *et al.*, 1997; Klenerman *et al.*, 1996) and free virus dynamics are faster compared to infected cell turnover (Perelson *et al.*, 1996), we do not consider a separate compartment of free virions, but simply assume that virion abundance is proportional to infected cell abundance.

To understand the stochastic dynamics of the viral mutants during untreated infection initiated with a drug-sensitive wild-type virus, we model the growth of PITs infected by wild-type virus deterministically and form a stochastic branching process model for the $2^n - 1$ subpopulations of PITs infected by HIV mutants. Mutants are assumed to have impaired fitness compared to wild-type virus in the absence of drugs (Coffin, 1995). In particular, we assume that, in the absence of drug and in the presence of competing wild-type, mutants cannot sustain growth and persist only because of renewal by mutation from wild-type. If some mutants are still able to establish infection in the absence of drugs and despite competition from wild-type, we can model these mutants deterministically according to the method used in Ribeiro *et al.* (1998). In branching process parlance, the $2^n - 1$ subpopulations of mutant-infected PITs constitute a subcritical multitype branching process, and this process is constantly refreshed by the immigration of mutants from the pool of wild-type virus. Since we exclude the possibility that mutations at two sites occur simultaneously, only single mutants immigrate. Back-mutation from single mutants to wild-type is ignored as negligible, allowing us to view immigration as independent of the branching process. Although we can safely assume that the stream of new immigrants is an independent Poisson process, the intensity of immigration depends directly on the population size of the surrounding PITs infected by wild-type virus, which is subject to growth and decline.

To model treated infection, we model all 2^n PIT populations, including wild-type PITs, as a supercritical multitype branching process. Now, in contrast to the pre-treatment scenario, wild-type-infected PITs and all k-mutant PITs with $k < n$ are alone incapable of establishing infection, while the n-mutant is the first mutant virus sufficiently resistant and fit to grow and expand its populations in the presence of drug. The process initiates with a number of each type of particle and has some probability of ending with the outgrowth of the n-mutant, also known as treatment failure.

Figure 1 shows an illustrative example system with point mutations at three sites, a, b and c. We will frequently refer to this example while developing and exploring the model. We define $2^3 = 8$ subpopulations of PITs. PIT_0 is the subpopulation of the PITs infected by the wild-type virus; PIT_a, PIT_b, and PIT_c by single mutants with one mutation at site a, b or c. The double mutants PIT_{ab}, PIT_{bc} and PIT_{ac} and the triple mutant PIT_{abc}

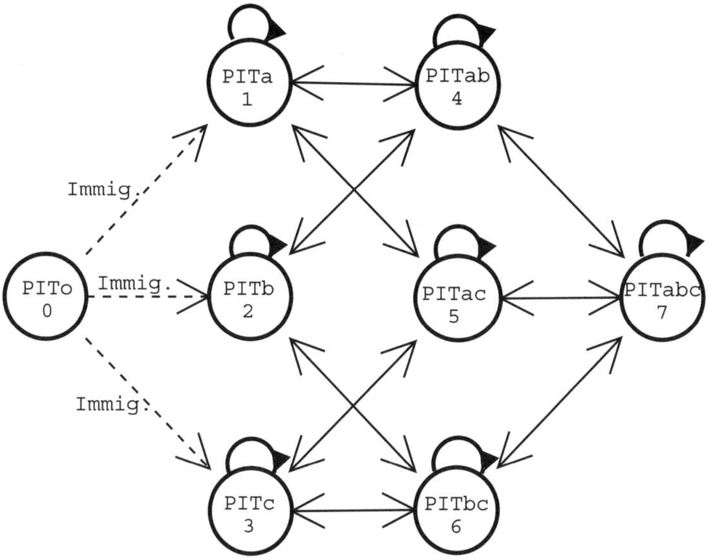

Fig. 1. A schematic illustration of the system with three resistance mutations resulting in eight populations of productively infected cells (PITs), starting with wild-type on the left and ending with the fully resistant triple mutant on the right.

are similarly identified. For later notational ease, these subpopulations are also subscripted 0 to 7 as in Fig. 1. In the absence of treatment, growth of PIT_0 is deterministic, while the counts of PITs infected by mutants, $\{\mathbf{Z}_t = (Z_{t1}, \ldots, Z_{t7}), t \geq 0\}$, constitute a subcritical branching process with seven types of particles, where Z_{ti} is the count of type i particle at time t. In the presence of treatment, $\{\mathbf{Z}_t = (Z_{t0}, \ldots, Z_{t7}), t \geq 0\}$ includes all 8 particle types and constitutes a supercritical branching process. Time 0 is the start of infection in the first case or the start of treatment in the second case, unless otherwise specified. The mutation events connecting these subpopulations are illustrated by arrows in the figure. Mutation rates at sites a, b and c are m_a, m_b and m_c per base per replication cycle.

The growth of the PIT subpopulations is determined by the effective reproductive numbers (Anderson and May, 1991), R_i in the absence of treatment and R'_i in the presence of treatment for type i particles, and the PIT death rate C. Here, the reproductive number is defined as the average number of secondarily infected cells produced by a type i infected cell under the given conditions. R_i and R'_i reflect the replicative capacity (fitness) of the inhabitant virus within the type i cells, but is also a function of the environment, for example the abundance of susceptible cells and host immunity. By the subcriticality assumption, mutant viruses are expected to have an effective reproductive number less than 1 in the presence of wild-type virus and no treatment. Additionally, R_i' are assumed less than 1 for all $0 \leq i < 2^n$, i.e. all but the n-mutant. The clearance rate of PIT cells, C, can be estimated from the half-life of productively infected cells upon treatment with potent ARVs (Coffin, 1995; Perelson *et al.*, 1996; Ho *et al.*, 1995; Wei *et al.*, 1995). We suppose that the population of mutant-infected PITs $\sum_i Z_{ti}$ remains small relative to the whole PIT plus susceptible cell population and host immunity is approximately constant during at least the initial rise of resistant mutants, so the replicative capacities R_i, R_i' and the clearance rate C do not change with time. In other words, we assume the process is time homogeneous.

3. From the Model to the Mathematics

With the model framework in hand, we now delve into the mathematical details. In constructing a branching process model, there are two primary tasks: first, to define the rate at which particles die and reproduce; and second, to define the probability model of reproduction, i.e. how many and what type of descendents each particle will produce.

In the standard branching process formulation, reproduction occurs coincident with the death of the parent particle. Because the host cell normally does not die with the release of a each progeny virus, our branching process is constructed by viewing the mother cell as dying at each reproduction event and being replaced by an identical substitute. Reproduction continues until one of the substitutes dies an ordinary death. Thus, type i particles have an exponentially distributed lifetime with rate $\omega_i = C(1 + R_i)$, where R_i may be decorated as R_i' if treatment is in place. The reproduction probability model is conveniently summarized in terms of a progeny generating function. The progeny generating function for type 1 particles is

$$P_1(\mathbf{s}) = \frac{C}{C(1+R_1)} + \frac{m_b C R_1}{C(1+R_1)} s_1 s_4 + \frac{m_c C R_1}{C(1+R_1)} s_1 s_5 + \frac{(1-m_b-m_c) C R_1}{C(1+R_1)} s_1^2,$$

since death with no reproduction occurs with probability $\frac{C}{C(1+R_1)}$, while death/rebirth with production of a type 4 particle occurs with probability $\frac{m_b C R_1}{C(1+R_1)}$, and so forth. With simplification, the complete set of progeny generating functions for the 3-mutant system in the absence of treatment is

$$
\begin{aligned}
P_1(\mathbf{s}) &= \frac{1}{1+R_1} + \frac{R_1 s_1}{1+R_1}[m_b s_4 + m_c s_5 + (1-m_b-m_c)s_1] \\
P_2(\mathbf{s}) &= \frac{1}{1+R_2} + \frac{R_2 s_2}{1+R_2}[m_a s_4 + m_c s_6 + (1-m_a-m_c)s_2], \\
P_3(\mathbf{s}) &= \frac{1}{1+R_3} + \frac{R_3 s_3}{1+R_3}[m_a s_5 + m_b s_6 + (1-m_a-m_b)s_3], \\
P_4(\mathbf{s}) &= \frac{1}{1+R_4} + \frac{R_4 s_4}{1+R_4}[m_b s_1 + m_a s_2 + m_c s_7 \\
&\quad + (1-m_a-m_b-m_c)s_4], \qquad (1) \\
P_5(\mathbf{s}) &= \frac{1}{1+R_5} + \frac{R_5 s_5}{1+R_5}[m_c s_1 + m_a s_3 + m_b s_7 + (1-m_a-m_b-m_c)s_5], \\
P_6(\mathbf{s}) &= \frac{1}{1+R_6} + \frac{R_6 s_6}{1+R_6}[m_c s_2 + m_b s_3 + m_a s_7 + (1-m_a-m_b-m_c)s_6], \\
P_7(\mathbf{s}) &= \frac{1}{1+R_7} + \frac{R_7 s_7}{1+R_7}[m_c s_4 + m_b s_5 + m_a s_6 + (1-m_a-m_b-m_c)s_7].
\end{aligned}
$$

Study of the process $\{\mathbf{Z}_t, t \geq 0\}$ is through a set of probability generating functions

$$Q_i(t,\mathbf{s}) = E(\mathbf{s}^{\mathbf{Z}_t} | \mathbf{Z}_0 = \mathbf{u}_i) = \sum_{\mathbf{z}_t \geq \mathbf{0}} p_{i\mathbf{z}_t} \mathbf{s}^{\mathbf{z}_t} = \sum_{\mathbf{z}_t \geq \mathbf{0}} p_{i\mathbf{z}_t} s_1^{z_{t1}} \cdots s_r^{z_{tr}},$$

for $i = 1, \ldots, r$, where the number of particle types $r = 2^n - 1$ off treatment or $r = 2^n$ on treatment. These functions $Q_i(t,\mathbf{s})$ satisfy the system of backward differential equations

$$\frac{\partial}{\partial t} Q_i(t,\mathbf{s}) = -\omega_i Q_i(t,\mathbf{s}) + \omega_i P_i\left[Q_1(t,\mathbf{s}), \ldots, Q_r(t,\mathbf{s})\right]. \qquad (2)$$

Calculating $f_{ij} = \frac{\partial}{\partial s_j} P_i(\mathbf{1})$ and $\Omega_{ij} = \omega_i(f_{ij} - \delta_{ij})$, where δ_{ij} is 1 when $i = j$ and 0 otherwise, yields the branching process matrix

$$\Omega = C \times \begin{bmatrix} - & 0 & 0 & m_b R_1 & m_c R_1 & 0 & 0 \\ 0 & - & 0 & m_a R_2 & 0 & m_c R_2 & 0 \\ 0 & 0 & - & 0 & m_a R_3 & m_b R_3 & 0 \\ m_b R_4 & m_a R_4 & 0 & - & 0 & 0 & m_c R_4 \\ m_c R_5 & 0 & m_a R_5 & 0 & - & 0 & m_b R_5 \\ 0 & m_c R_6 & m_b R_6 & 0 & 0 & - & m_a R_6 \\ 0 & 0 & 0 & m_c R_7 & m_b R_7 & m_a R_7 & - \end{bmatrix}, \qquad (3)$$

where the diagonal entry at row i is set such that row i of Ω sums to $C(R_i - 1)$.

In the absence of treatment, PIT_0 is modeled deterministically according to the observed clinical course of untreated HIV infection (Clark, 1991; Daar *et al.*, 1991; Henrard *et al.*, 1995; Schacker *et al.*, 1998). The acute phase of HIV infection lasts only a few weeks, with blood viral load and infected CD4 T cell count increasing dramatically and then declining by two or more orders of magnitude to a quasi-stable equilibrium. The quasi-equilibrium slowly decays during approximately 10 years of clinical latency in an untreated patient, and when the CD4 T cells of the body are sufficiently depleted (< 200 cells per μl), AIDS develops. The pattern during acute infection has been modeled using a basic population model for HIV primary infection (Bonhoeffer *et al.*, 1997; Nowak *et al.*, 1996). The solid line in Fig. 2 is the numerical solution of these equations for the number

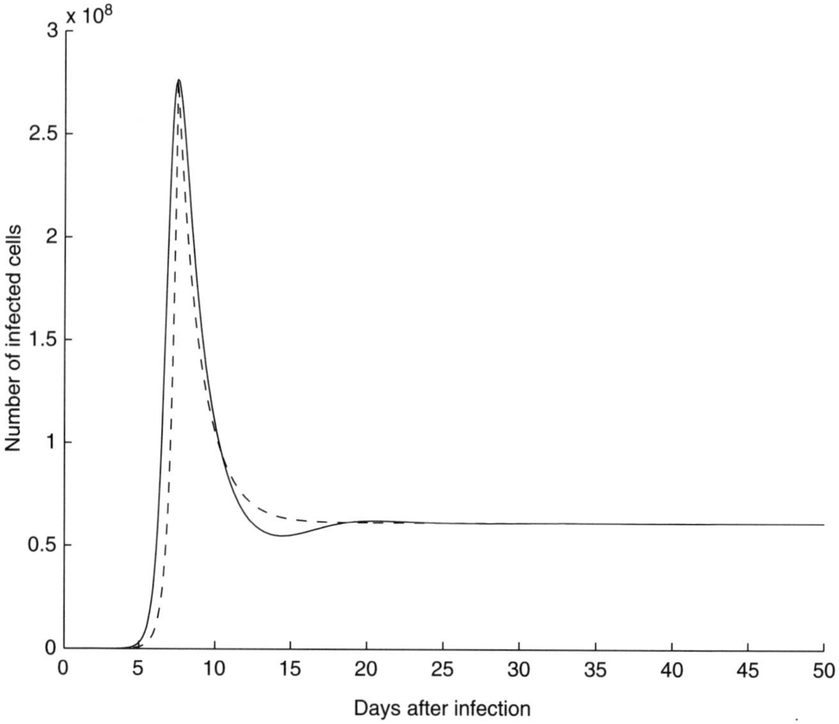

Fig. 2. The dynamics of PIT_0 during primary infection calculated by the basic mathematical model (solid line) and approximated by exponential functions (dashed line). Parameter settings are $\lambda = 5 \times 10^7, d = 0.1, a = 0.7, \beta = 5 \times 10^{-10}, k = 100, b = 5$, using the notation of Bonhoeffer *et al.* (1997).

of infected CD4 cells during primary infection, using typical parameter settings.

When immigration rates are exponential or linear combinations of exponential functions, analytic solutions for many quantities of interest are available (Dorman *et al.*, 2004; Zhou, 2003). So to capture these PIT_0 dynamics for use in the branching process model, we divide the growth of PIT_0 into two stages and approximate the numerical solution using exponential functions. Let n_0 be the initial inoculum of PIT_0, n_{peak} the peak viral load, n_{eq} the equilibrium viral load, and T_{peak} the time PIT_0 reaches its peak. In the *early acute stage*, the population of PIT_0 explodes exponentially from n_0 to n_{peak} with rate $\beta_I = (\log n_{\text{peak}} - \log n_0)/T_{\text{peak}}$. In the *late acute and chronic stages*, the population declines from n_{peak} to n_{eq} exponentially according to $(n_{\text{peak}} - n_{\text{eq}})e^{\beta_{II}t} + n_{\text{eq}}$, where β_{II} is negative and can be set by fitting the curve to clinical data or the numerical solution of the basic mathematical model (Bonhoeffer *et al.*, 1997). As comparison, the fitted approximation is shown as a dashed line in Fig. 2. In this description, we have ignored the gradual decline in PIT_0 during the chronic, or latent, stage, assuming instead that an equilibrium is achieved.

Generally, suppose immigration rates of type i particles are of the form $\eta_i(t) = \alpha_{i1}e^{\beta_{i1}t} + \cdots + \alpha_{iD}e^{\beta_{iD}t} = \sum_{d=1}^{D} \alpha_{id}e^{\beta_{id}t}$, then the mean $\mathbf{m}(t)$ of $\mathbf{Z}_t$ at time t can be calculated according to

$$\mathbf{m}(t) = \mathbf{m}(0)e^{t\Omega} + \mathbf{1}\sum_{i=1}^{r} \mathbf{u}_i^* \mathbf{u}_i \sum_{d=1}^{D} \alpha_{id}(\Omega - \beta_{id}\mathbf{I})^{-1}(e^{t\Omega} - e^{t\beta_{id}\mathbf{I}}), \tag{4}$$

where $\mathbf{I}$ is the $r \times r$ identity matrix, $\mathbf{1}$ is a vector of 1's, and u_i is the elementary vector of 0's everywhere but in the ith position where there is a 1. The * notation represents the transpose operator. The variance of $\mathbf{Z}_t$ is given by

$$\text{Var}(\mathbf{Z}_t) = \sum_{i=1}^{r} n_i \mathbf{V}_i(t) + \sum_{i=1}^{r} \mathbf{W}_i(t), \tag{5}$$

where $n_i = Z_{0i}$ is the number of type i particles present at time $t = 0$, $\mathbf{V}_i(t)$ is the variance of the immigration-free process initiated with one particle of type i, i.e. $\mathbf{Z}_0 = \mathbf{u}_i$, and $\mathbf{W}_i(t)$ is the variance of the the process with immigration of type i particles only and initiated with no particles, i.e. $\mathbf{Z}_0 = \mathbf{0}$. Formulae for both $\mathbf{V}_i(t)$ and $\mathbf{W}_i(t)$ are derived in Dorman *et al.* (2004) and are repeated in the appendix for completeness.

In addition, if we add another probability generating function for the immigration-only process, call it $R(t, \mathbf{s})$, then the probability generating function of the full branching process with immigration is given by the overall probability generating function

$$Q_{bp}(t,\mathbf{s}) = R(t,\mathbf{s})Q(t,\mathbf{s})^{\mathbf{n}} = \left(e^{-\sum_{i=1}^{r} H_i(t,\mathbf{s})}\right)\prod_{i=1}^{r} Q_i(t,\mathbf{s})^{n_i}, \tag{6}$$

where

$$H_i(t,\mathbf{s}) = \int_0^t \eta_i(\tau)[1 - Q_i(t-\tau,\mathbf{s})]\,d\tau.$$

The above integral equation for $H_i(t,\mathbf{s})$ is not particularly useful, but can be differentiated to produce

$$\frac{\partial}{\partial t}H_i(t,\mathbf{s}) = \beta_i H_i(t,\mathbf{s}) + \alpha_i[1 - Q_i(t,\mathbf{s})], \tag{7}$$

when $\eta_i(t) = \alpha_i e^{\beta_i t}$. Similar equations are possible when the immigration rate is a sum of exponentials. These differential equations supplement the backward system of equations (2) when immigration is present.

4. Modeling the Early Acute Stage

We take typical parameter values, listed in Table 1, and investigate the dynamics that generate viral mutants during the early acute stage up to the initial peak (Fig. 2). The means, variances, and marginal distributions of different mutants produced during this stage are calculated. For the details of calculation, we refer the reader to Dorman *et al.* (2004). We also examine the probability $e_i(t)$ that no type i mutants are present, and the probability $e(t)$ that no mutants of any type are present in the body at time t after first infection. Finally, the sensitivity of the results to uncertainty in the parameters settings is examined.

Table 1. Typical parameter settings.

Parameters	Notation	Value	Unit	Reference
Cell clearance	C	0.7	/day	(Markowitz *et al.*, 2003)
Mutation	m_a, m_b, m_c	0.00003	/base/cycle	(Mansky and Temin, 1995)
Reproductive	R_1, R_2, R_3	0.95		
Numbers	R_4, R_5, R_6	0.90		
	R_7	0.85		
PIT_0 inoculum	n_0	10		
PIT_0 peak	n_{peak}	5×10^8		(Haase, 1999)
PIT_0 equilibrium	n_{eq}	4×10^7		(Haase, 1999)
Time to peak	T_{peak}	10	days	(Reimann *et al.*, 1994)

4.1. *An example with typical parameter settings*

In this example, we assume that the mutation rates at all three sites are the same, i.e., $m_a = m_b = m_c = 3 \times 10^{-5}$ per base per replication cycle, and the fitness of mutants decreases with the number of point mutations, e.g., $R_1 = R_2 = R_3 = 0.95$, $R_4 = R_5 = R_6 = 0.90$ and $R_7 = 0.85$. These settings are by no means a practical restriction and should be adapted to fit the drug and mutant combination of interest.

First we determine the immigration rates of single mutants during the early acute stage. Suppose the infection starts with an inoculum of $n_0 = 10$ PIT_0s and reaches the peak load, $n_{\text{peak}} = 5 \times 10^8$, at $T_{\text{peak}} = 10$ days after infection. Then $\beta_I = [\log(5 \times 10^8) - \log 10]/10 \approx 1.77$. At time t, there are $n_0 e^{\beta_I t}$ PIT_0s, so the rate of immigration for particles of type 1 (PIT_a) is $\eta_1(t) = m_a R_0 C n_0 e^{\beta_I t}$, where R_0 is the expected number of secondary PITs produced from a single PIT_0 particle. Note the constants β_I and R_0 are related by the identity $\beta_I = (R_0 - 1)C$ since reproduction must be adjusted for death to compute the net rate of exponential growth. Thus, $\eta_1(t) = m_a(\beta_I + C)n_0 e^{\beta_I t}$. Calculation of the immigration intensity for particles 2 (PIT_b) and 3 (PIT_c) is similar.

Formulae (4) and (5) offer analytical means $\mathrm{E}(\mathbf{Z}_t)$ and variances $\mathrm{Var}(\mathbf{Z}_t)$ of the counts of particles 1–7 given immigration of single mutants from the pool PIT_0. Figure 3 plots the mean numbers of single mutants (particles 1, 2, 3), double mutants (particles 4, 5, 6) and the triple mutant (particle 7) against time as solid lines. The symmetry of the parameter settings means all mutants with the same number of mutations behave identically. The mean plus/minus the standard deviation (SD) are plotted as dotted lines. The deterministic growth of PIT_0 is plotted as comparison in the upper, left quadrant. Numerical values at days 0, 2, 4, 6, 8, and the peak day 10 are listed in Table 2. During this early acute stage, single mutants rise to the order of 10^4 in concert with the exponential growth of PIT_0, while the presence of double and triple mutants remains uncertain.

Inference based on the mean and variance may be biased if the distribution is skewed, as is clearly the case for double and triple mutants. To get a clearer picture, we also obtain the marginal distributions of the number of mutants, using numerical methods described in Dorman *et al.* (2004). Distributions of single mutants at days 6, 8 and of double and triple mutants at days 6, 8, and 10 are shown in Fig. 4. To verify the accuracy of the numerical method, the means and standard deviations calculated from the marginal distributions are displayed in Fig. 4 and compare favorably with

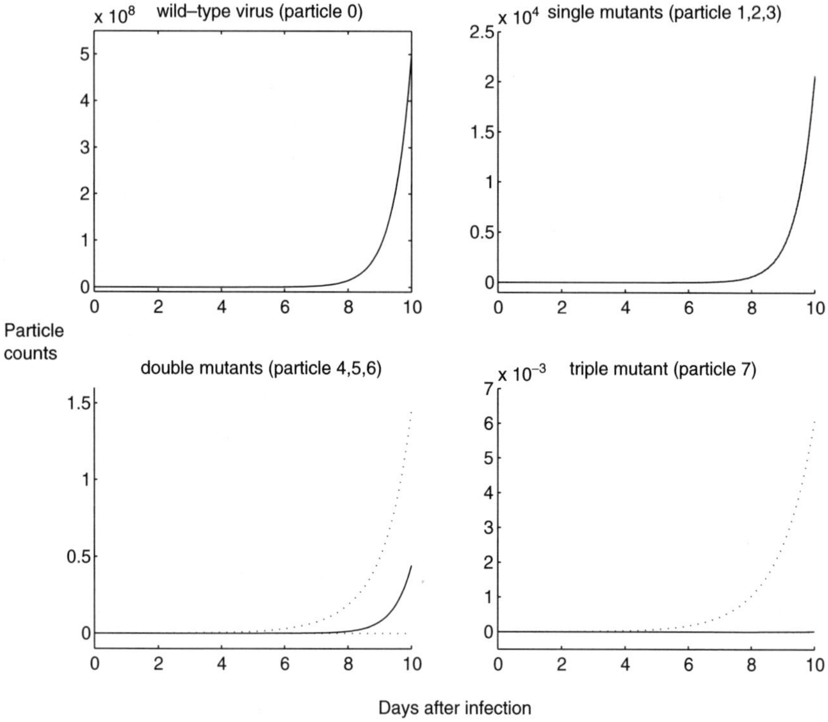

Fig. 3. Mean ± SD of single mutants, double mutants and triple mutants in the early acute stage using parameter values in Table 1.

Table 2. Mean ± SD of the number of mutants after early acute stage.

Day	Single mutants	Double mutants	Triple mutant
0	0.00 ± 0.00	0.00 ± 0.00	0.00 ± 0.00
2	0.01 ± 0.15	0.00 ± 0.00	0.00 ± 0.00
4	0.49 ± 0.92	0.00 ± 0.00	0.00 ± 0.00
6	17 ± 5.4	0.00 ± 0.03	0.00 ± 0.00
8	592 ± 32	0.01 ± 0.17	0.00 ± 0.00
10	20517 ± 188	0.44 ± 1.01	0.00 ± 0.01

the analytical values in Table 2. The distribution of single mutants at viral load peak on day 10 is not computed because numerical accuracy decreases when the mean is large. By day 6, a few to a few dozen single mutants are already present in $> 99\%$ of the patients. The presence of double mutants

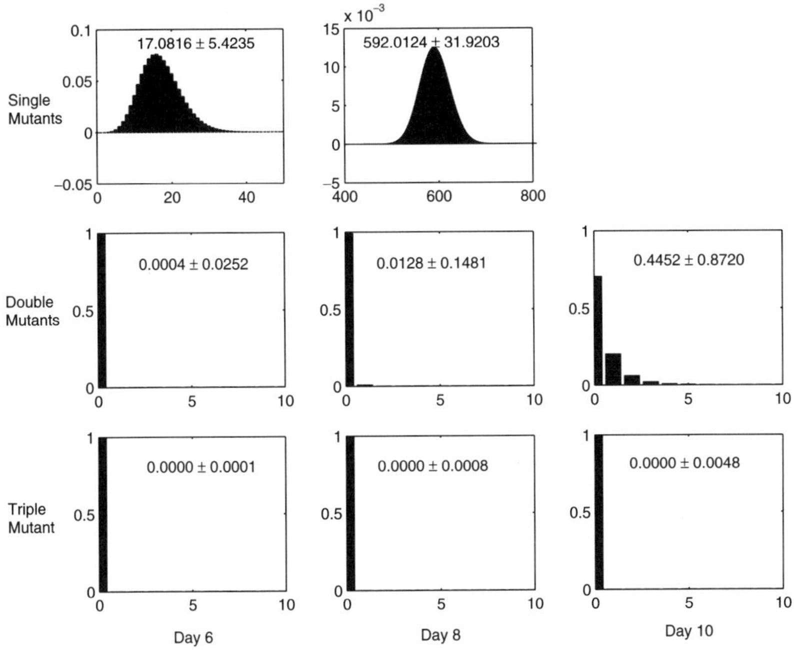

Fig. 4. Marginal distributions of single mutants at days 6, 8 and marginal distributions of double and triple mutants at days 6, 8 and 10 using parameter values in Table 1. Means and standard deviations calculated from the distributions are also displayed.

remains a matter of chance during this stage, however, by the peak time T_{peak}, double mutants are expected in a substantial fraction of patients. Throughout the early acute stage, the distribution of the triple mutant is concentrated on 0.

The probability of the presence or absence of mutant particles is probably the most important prediction of the model. This information is included in the marginal distributions, which are obtained through repeated integrations of the backward differential equations (Dorman *et al.*, 2004). The same information can be obtained more rapidly and with less error using a single integration. Note that the probability of no type i particles at time t $e_i(t)$ is $\Pr(Z_{ti} = 0) = Q_{bp}(t, \mathbf{1}_i)$, where $\mathbf{1}_i$ is the vector with all entries 1 except the ith entry where there is a 0. $Q_{bp}(t, \mathbf{s})$ is defined in Eq. (6). The quantity $e_i(t)$ can be obtained via numerical integration of Eqs. (2) and (7), evaluated at $\mathbf{s} = \mathbf{1}_i$ (Zhou, 2003). The $e_i(t)$ for $i = 1, 4, 7$ are plotted in Fig. 5 and numerical values for $i = 1, 4, 7$ and $t = 0, 2, 4, 6, 8, 10$ are recorded in Table 3. It is clear from Fig. 5 that single

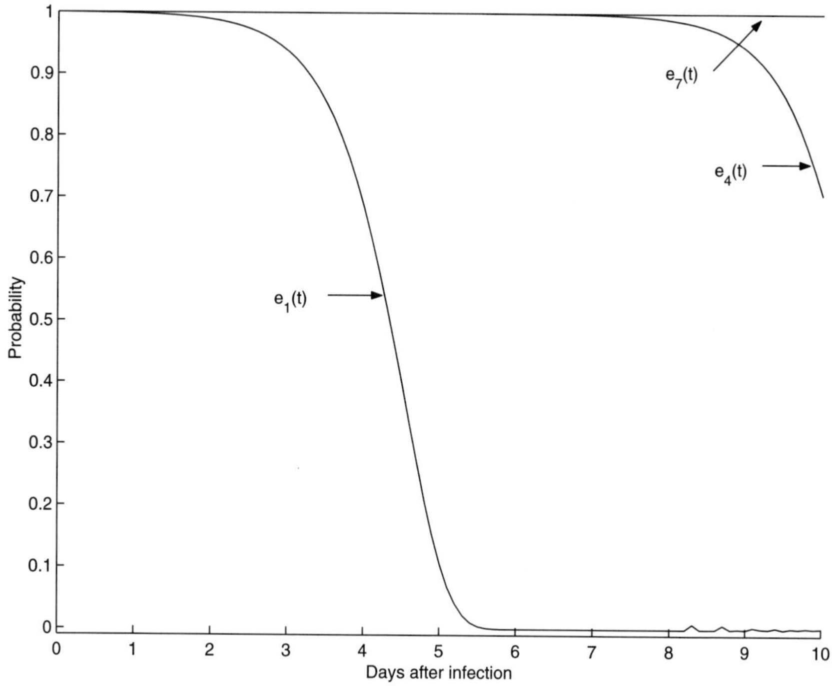

Fig. 5. The probability of no type i particle, $e_i(t)$, for single ($i = 1$), double ($i = 4$) and triple ($i = 7$) mutants during the early acute stage using parameter values in Table 1.

Table 3. The probability of no mutants during the early acute stage.

Day	Single mutants	Double mutants	Triple mutant
0	1.00	1.00	1.00
2	0.99	1.00	1.00
4	0.68	1.00	1.00
6	0.00	1.00	1.00
8	0.00	0.99	1.00
10	0.00	0.71	1.00

The probability of no type i mutants $e_i(t)$, where $i = 1$, 2, or 3 for single mutants, $i = 4$, 5, or 6 for double mutants and $i = 7$ for the triple mutant on days 0, 2, 4, 6, 8, 10 in the early acute stage using parameter values in Table 1.

mutants appear rapidly. They will be present in more than 90% of patients just 5 days after infection. The appearance of more complex mutants is delayed, but double mutants nevertheless appear a mere 6 days after the single mutants.

The time period in early infection during which it is *completely* safe to administer any effective monotherapy rather than combination therapy is the time during which no single mutants appear. Let the probability that there are *no* mutants of any type at time t be $e(t)$. Intuitively, $e(t) = Q_{bp}(t, \mathbf{0})$ will decrease monotonically. By setting $\mathbf{s} = \mathbf{0}$, numerically integrating the backward system of ordinary differential equations for the process with immigration (Eq. 2 supplemented by Eq. 7), and using the product formula Eq. (6), we calculate $e(t)$ during the early acute stage (data not shown). The probability of no mutant particles declines rapidly from 0.95 to 0.05 during days 2.3–4.5. Only treatments initiated in the first 2.3 days after infection will catch more than 95% of patients with no resistance mutants. In contrast, after 4.5 days, mutants are present in over 95% of patients and caution must be taken when instituting treatment, especially monotherapy.

4.2. *Sensitivity to the parameter settings*

So far, we have explored the dynamics for generation of mutants in the early acute stage using the typical parameter settings in Table 1, including rather arbitrary effective reproductive numbers. It is important to evaluate the sensitivity of the results to different parameter settings. The growth of wild-type PIT_0 and mutant fitness levels are expected to vary because of inter-patient variability, while mutation rates are expected to vary for different sites and viruses. We vary the values of different parameters and report the mean $\pm$ SD of the number of mutants present at end of the early acute stage and the period of time (window of closing opportunity) during which the probability of no mutants $e(t)$ declines from 0.95 to 0.05. An "upper bound" of the number of mutants expected during the early acute stage is also obtained by using extreme, mutant-favoring parameters.

Different patients may start with different inoculum sizes n_0 of PIT_0 at the start of infection and reach different peak loads n_{peak} of PIT_0 at different times T_{peak}. With other parameters in Table 1 fixed, we vary the starting inoculum n_0 of PIT_0 and record the impact in Table 4. Similarly, fixing n_0 and T_{peak}, we record the outcome for varying n_{peak} in Table 5. According to the simulation of stochastic models (Kamina *et al.*, 2001;

Table 4. The effect of varying the initial inoculum size n_0.

n_0	Single	Double	Triple	Window (days)
1	19894 ± 181	0.38 ± 0.91	0.00 ± 0.01	3.1–5.1
10	20517 ± 188	0.44 ± 1.01	0.00 ± 0.01	2.3–4.5
100	21323 ± 197	0.53 ± 1.14	0.00 ± 0.01	1.2–3.7
500	22041 ± 206	0.61 ± 1.26	0.00 ± 0.01	0.5–2.9
1000	22403 ± 210	0.65 ± 1.32	0.00 ± 0.01	0.3–2.6

The mean $\pm$ SD of the number of different mutant types is reported along with the window during which the probability of no mutants $e(t)$ declines from 0.05 to 0.95. Here, the inoculum size n_0 is varied while all other parameters are set as in Table 1.

Table 5. The effect of varying the peak size n_{peak}.

n_{peak} ($\times 10^8$)	Single	Double	Triple	Window
0.1	441 ± 29	0.01 ± 0.18	0.00 ± 0.00	2.9–5.8
0.5	2132 ± 62	0.05 ± 0.36	0.00 ± 0.00	2.6–5.2
1	4211 ± 87	0.10 ± 0.49	0.00 ± 0.00	2.5–4.9
5	20517 ± 188	0.44 ± 1.01	0.00 ± 0.00	2.3–4.5
10	40627 ± 262	0.85 ± 1.38	0.00 ± 0.00	2.2–4.3

Table 6. The effect of varying the peak time T_{peak}.

T_{peak} (day)	Single	Double	Triple	Window
5	17786 ± 156	0.20 ± 0.57	0.00 ± 0.00	1.2–2.2
10	20517 ± 188	0.44 ± 1.01	0.00 ± 0.01	2.3–4.5
15	23197 ± 219	0.74 ± 1.46	0.00 ± 0.01	3.4–6.7
20	25825 ± 248	1.08 ± 1.92	0.00 ± 0.02	4.5–9.0
25	28404 ± 277	1.45 ± 2.38	0.00 ± 0.02	5.5–11.2
30	30935 ± 305	1.87 ± 2.85	0.00 ± 0.03	6.6–13.3

Tuckwell and Le Corfec, 1998), the time T_{peak} to reach peak load, ranges from 19 days to 27 days. Observation of experimental SIV infection shows T_{peak} to be 7 to 21 days (Reimann *et al.*, 1994; Sodora *et al.*, 1998). Fixing n_0 and n_{peak}, the results for varying T_{peak} are given in Table 6.

While the starting inoculum has relatively small influence on the mean counts of mutants, the mutants appear much more rapidly as the starting

Table 7. The effect of varying the mutation rate.

m	Single	Double	Triple	Window
0.00003	20517 ± 188	0.44 ± 1.01	0.00 ± 0.01	2.3–4.5
0.00004	27356 ± 217	0.79 ± 1.35	0.00 ± 0.01	2.1–4.3

The mutation rates $m = m_a = m_b = m_c$ are set to 3×10^{-5} or 4×10^{-5} while all other parameters are set as in Table 1.

inoculum increases (Table 4). On the contrary, the peak load n_{peak} of PIT_0 has greater influence on the count of mutants than on the timing of their appearance (Table 5). The larger the peak load, the more mutants are generated during the early acute stage. The same n_0 and n_{peak} but different T_{peak} should stretch the process in time, as evidenced in Table 6. With larger T_{peak}, more mutants accumulate at end of the early acute stage, but the window of closing opportunity starts later and lasts longer (Table 6).

Large mutation rates will accelerate the generation of mutants. With the parameters other than mutation rates set as in Table 1, the mean $\pm$ SD of mutants at day $T_{\text{peak}} = 10$ and extinction windows with $m_a = m_b = m_c = 3 \times 10^{-5}$ (Mansky and Temin, 1995) and $m_a = m_b = m_c = 4 \times 10^{-5}$ (Manksy, 1996) are recorded in Table 7. With the larger mutation rates, the mean number of single mutants at the end of the early acute stage increases by thousands and the window of closing opportunity shifts forward in time by hours. In addition, different mutation rates at different sites will favor some mutants more than others. When $m_a = 4 \times 10^{-5}$ and $m_b = m_c = 3 \times 10^{-5}$, a higher mutation rate at site a promotes the single mutant PIT_a and as a consequence, the double mutants PIT_{ab}, PIT_{ac}. Expected levels are 27357 ± 217 for the single mutant PIT_a vs. 20517 ± 188 for the single mutants PIT_b and PIT_c and 0.59 ± 1.17 for the double mutants PIT_{ab} and PIT_{ac} vs. 0.44 ± 1.01 for the double mutant PIT_{bc}.

The fitness profile of mutants is an important determinant of the process. In the former examples, we assume a simple inverse relationship between the number of point mutations and mutant fitness. However, the fitness profile of mutants may be quite variable (Mammano *et al.*, 2000). In Table 8, we record mean $\pm$ SD of different mutants at day $T_{\text{peak}} = 10$ for three fitness profiles. The first row is the previous case where mutants have sequentially decreasing fitness. The second row emulates the scenario when all intermediate mutants have high fitness. On day 10, there is only a minor increase in the expected number of single and double mutants. We also compute the marginal distributions of double and triple mutants at

Table 8. The effect of variable fitness profiles.

R_s	R_d	Single	Double	Triple	Window
0.95	0.90	20517 ± 188	0.44 ± 1.01	0.00 ± 0.01	2.3–4.5
0.99	0.99	20840 ± 192	0.49 ± 1.11	0.00 ± 0.01	2.3–4.5
0.50	0.50	17473 ± 150	0.17 ± 0.51	0.00 ± 0.00	2.3–4.5

Here $R_1 = R_2 = R_3 = R_s$ and $R_4 = R_5 = R_6 = R_d$ and $R_7 = 0.85$. Other parameters are set as in Table 1.

day 10 (data not shown). There is a 31% chance for each type of double mutant (PIT_{ab}, PIT_{ac} or PIT_{bc}) to be present at day 10, slightly different from the original case (29%). The triple mutant PIT_{abc} is still unlikely to appear. For the last case, all intermediate mutants have low fitness and thus pose a barrier to generation of triple mutants. Now there is only a 14% chance of having one or more of each double mutant present by day 10. The three profiles in Table 8 have little effect on the time evolution of the process as witnessed by the unchanging window of closing opportunity.

So far, we have explored the sensitivity of results to a specific parameter by varying its value while other parameters remain fixed. To get an "upper bound" on how many mutants could be generated during the early acute stage, we consider an extreme situation favorable for generation of mutants. The average number of each single mutant that immigrates during the early acute stage is computed by integrating its immigration rate

$$\int_0^{T_{\text{peak}}} mn_0(\beta_I + C)e^{\beta_I t}\, dt$$
$$= m\left(1 + \frac{CT_{\text{peak}}}{\log n_{\text{peak}} - \log n_0}\right)(n_{\text{peak}} - n_0). \tag{8}$$

To maximize the number of immigrants while retaining biologically reasonable values for the parameters, we set $n_0 = 1000$, $n_{\text{peak}} = 10 \times 10^8$, $T_{\text{peak}} = 30, m_a = m_b = m_c = 4 \times 10^5, R_i = 0.99$, for $1 \leq i \leq 7$. Mean $\pm$ SDs of the counts of mutants at day $T_{\text{peak}} = 30$ are 99280 ± 624 for single mutants, 12 ± 9 for double mutants, and 0.00 ± 0.14 for the triple mutant. As shown in Fig. 6, the probability of no mutants $e(t)$ declines from 0.95 to 0.05 during days 0.4–6.4. Marginal distributions of double and triple mutants at day 30 are also obtained (Fig. 7). Under these settings, there is a nearly 100% probability of 1 to 40 double mutants at the end of the early acute stage, but there is still only a slightly larger than 0.1% chance of a triple mutant.

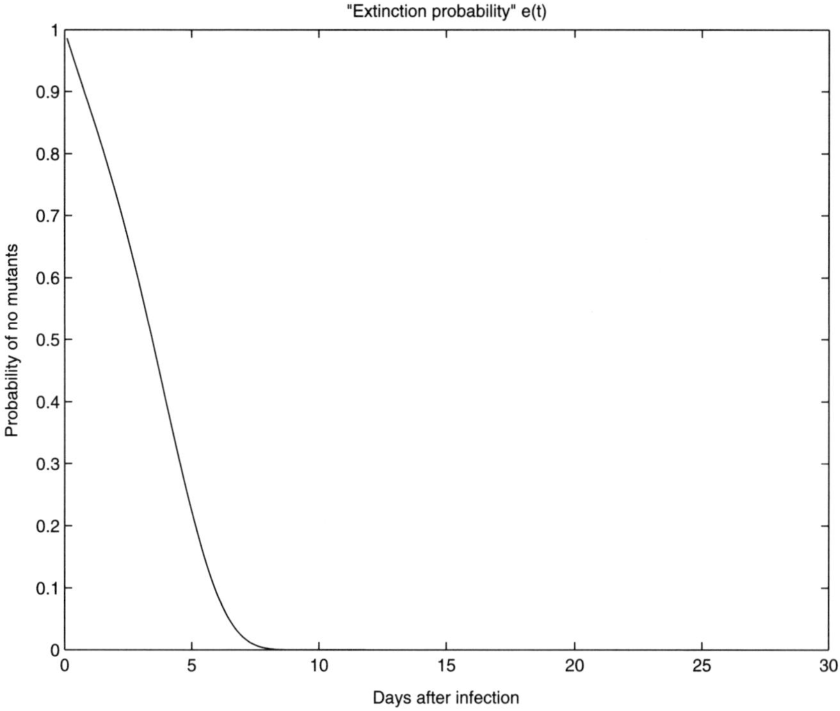

Fig. 6. Time evolution of the probability of no mutants $e(t)$ for $n_0 = 1000$, $n_{\text{peak}} = 10 \times 10^8, T_{\text{peak}} = 30, m_a = m_b = m_c = 4 \times 10^5$, and $R_i = 0.99, 1 \leq i \leq 7$. Other parameters are as in Table 1.

5. Modeling the Late Acute and Chronic Stages

Following the dramatic increase in PIT_0 during the early acute stage comes a equally dramatic decrease during the late acute stage. This decrease is caused by a decline in the number of available susceptible CD4 T cells and/or induction of host immunity (Phillips, 1996). The PIT_0 levels then approaches a new quasi-stable equilibrium that will persist more or less constantly for several years. Most drug therapy is initiated well after the early acute stage. Knowledge of the levels of mutants in the chronic stage is essential in guidance of treatment.

5.1. *An example with typical parameter settings*

Out of necessity, we have assumed that during the late acute and chronic stage the number of PIT_0, say $n(t)$, declines from the peak load n_{peak} to

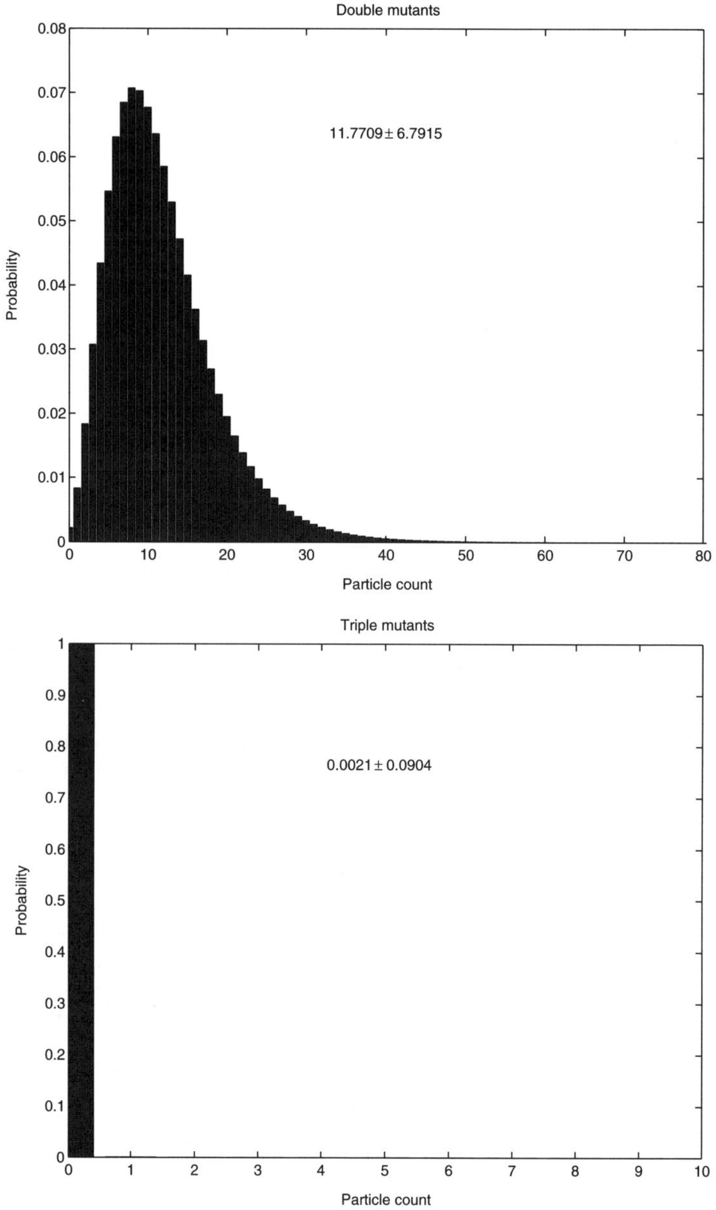

Fig. 7. Marginal distributions of double mutants (upper graph) and triple mutant (lower graph) at day 30 for mutant-favoring settings $n_0 = 1000$, $n_{\text{peak}} = 10 \times 10^8, T_{\text{peak}} = 30$, $m_a = m_b = m_c = 4 \times 10^5$, and $R_i = 0.99, 1 \leq i \leq 7$. Other parameters are as in Table 1.

equilibrium load n_{eq} according to $n(t) = (n_{\text{peak}} - n_{\text{eq}})e^{\beta_{II}t} + n_{\text{eq}}$, where $\beta_{II} < 0$ (see Fig. 2). Since we will focus on the equilibrium achieved during the chronic stage, the effect of this approximation is minimal. We set $t = 0$ at T_{peak} and $t > 0$ is the time since T_{peak}. In this context, the counts of mutants $\{\mathbf{Z}_t = [Z_{t1}, \ldots, Z_{t7}], t \geq 0\}$ still constitute a subcritical multitype branching process, with the single mutants immigrating from the PIT_0 pool with rates $m(\beta_{II}+C)(n_{\text{peak}}-n_{\text{eq}})e^{\beta_{II}t}+mn_{\text{eq}}C$. Here, m is equal to m_a, m_b or m_c for single mutants PIT_a, PIT_b or PIT_c. Note the immigration rates are linear combinations of exponential functions and converge to constants $m_a n_{\text{eq}} C$, $m_b n_{\text{eq}} C$ and $m_c n_{\text{eq}} C$ as $t \to \infty$.

For this subcritical process with immigration rates converging to constants, a stochastic equilibrium will be achieved between immigration (mutation) and extinction (selection). Let $\eta(t)$ be the vector of immigration rates at time t. As indicated above, $\eta(t)$ converges to the equilibrium immigration rates

$$\eta = [m_a n_{\text{eq}} C, m_b n_{\text{eq}} C, m_c n_{\text{eq}} C, 0, 0, 0, 0]$$

as $t \to \infty$. Let

$$\mathbf{m}(t) = [m_1(t), \ldots, m_r(t)]$$

be the vector of expected particle counts at time t. The forward equations for means can be expressed in vector-matrix form as

$$\frac{d}{dt}\mathbf{m}(t) = \mathbf{m}(t)\Omega + \eta(t).$$

Setting this equation to 0 and solving, we obtain the mean at equilibrium as

$$\mathbf{m} = [m_1, \ldots, m_7] = -\eta\Omega^{-1}.$$

The exact solution of this linear system is complicated to write down, but if we neglect back-mutations, we obtain an elegant approximation

$$m_1 = \frac{m_a n_{\text{eq}}}{1 - (1 - m_b - m_c)R_1} \approx \frac{m_a n_{\text{eq}}}{1 - R_1},$$

$$m_2 = \frac{m_b n_{\text{eq}}}{1 - (1 - m_a - m_c)R_2} \approx \frac{m_b n_{\text{eq}}}{1 - R_2},$$

$$m_3 = \frac{m_c n_{\text{eq}}}{1 - (1 - m_b - m_c)R_3} \approx \frac{m_c n_{\text{eq}}}{1 - R_3},$$

for the single mutants,

$$m_4 = \frac{m_a m_b n_{\text{eq}}}{1-(1-m_a-m_b-m_c)R_4}\left(\frac{R_1}{1-R_1}+\frac{R_2}{1-R_2}\right)$$
$$\approx \frac{m_a m_b n_{\text{eq}}}{1-R_4}\left(\frac{R_1}{1-R_1}+\frac{R_2}{1-R_2}\right),$$
$$m_5 = \frac{m_a m_c n_{\text{eq}}}{1-(1-m_a-m_b-m_c)R_5}\left(\frac{R_1}{1-R_1}+\frac{R_3}{1-R_3}\right)$$
$$\approx \frac{m_a m_c n_{\text{eq}}}{1-R_5}\left(\frac{R_1}{1-R_1}+\frac{R_3}{1-R_3}\right),$$
$$m_6 = \frac{m_b m_c n_{\text{eq}}}{1-(1-m_a-m_b-m_c)R_6}\left(\frac{R_2}{1-R_3}+\frac{R_3}{1-R_3}\right)$$
$$\approx \frac{m_b m_c n_{\text{eq}}}{1-R_6}\left(\frac{R_2}{1-R_3}+\frac{R_3}{1-R_3}\right),$$

for the double mutants, and

$$m_7 = \frac{m_4 m_c R_4 + m_5 m_b R_5 + m_6 m_a R_6}{1-(1-m_a-m_b-m_c)R_7}$$
$$\approx \frac{m_a m_b m_c n_{\text{eq}}}{1-R_7}\left[\frac{R_4}{1-R_4}\left(\frac{R_1}{1-R_1}+\frac{R_2}{1-R_2}\right)\right.$$
$$\left.+\frac{R_5}{1-R_5}\left(\frac{R_1}{1-R_1}+\frac{R_3}{1-R_3}\right)+\frac{R_6}{1-R_6}\left(\frac{R_2}{1-R_2}+\frac{R_3}{1-R_3}\right)\right],$$

for the triple mutant. The additional approximations made in these equations are valid when mutation rates are much less than 1. This approximate solution has a clear interpretation. The average number of single mutants at equilibrium is proportional to the mutation rate at the associated site and the equilibrium levels of wild-type infected cells n_{eq}, and is inversely related to 1 minus its effective reproductive number, which can be considered a measure of the selective force. The frequency of double and triple mutants is proportional to the mutation rates at the associated sites and related to the effective reproductive numbers of the mutant and its predecessor mutants. Formulae for variance/covariance of these counts at equilibrium can also be obtained through limits (Dorman *et al.*, 2004; Zhou, 2003) but are too complicated to include here.

We are interested in the expected frequency of the triple mutant relative to the wild-type virus at equilibrium, i.e. m_7/n_{eq}, a quantity that

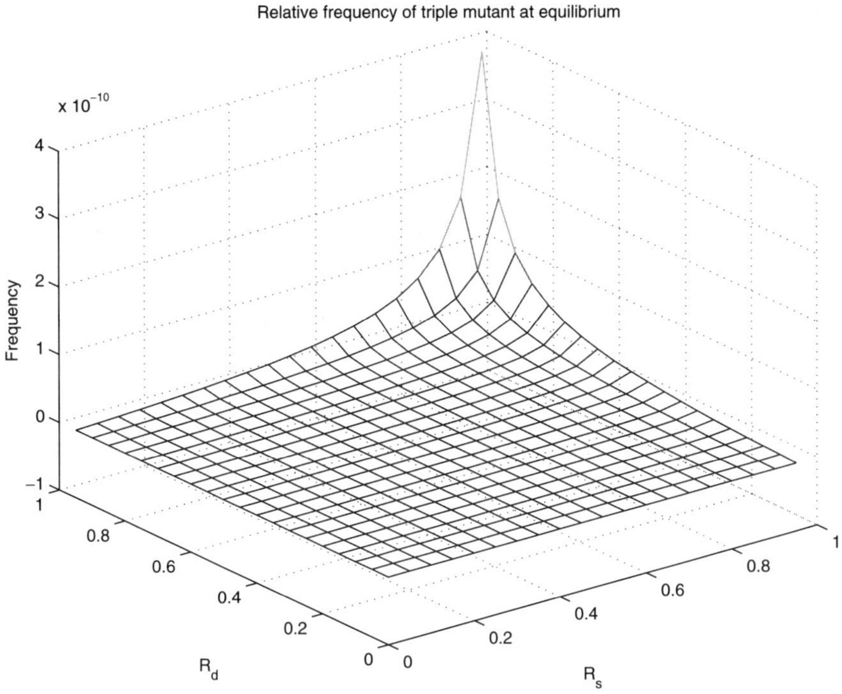

Fig. 8. Relative frequency of triple mutant at equilibrium and its dependence on the effective reproductive numbers of single mutants and double mutants. Lower graph is the contour plot of the relative frequency. The contour elevation levels are $1 \times 10^{-14}, 10^{-13}, 10^{-12}, 10^{-11}, 10^{-10}$ and 10^{-9} from lower left to upper right.

determines whether resistance testing can detect the mutant. We calculate the relative frequency under different combinations of effective reproductive numbers of single mutants, $R_s = R_1 = R_2 = R_3$, and double mutants, $R_d = R_4 = R_5 = R_6$ (Fig. 8). Here we assume $m_a = m_b = m_c = 3 \times 10^{-5}$ and $R_7 = 0.85$. Variability in these two values will only shift the surface upward or downward. Not surprisingly, low values of both R_s and R_d block the generation of the triple mutant and the relative frequency is near 0; only high values of both R_s and R_d allow relatively high frequencies of triple mutant. When $R_s = R_d = R_7 \approx 1$, the relative frequency reaches 0.037, but even this is not easily detectable.

Calculation of the mean and variance of the equilibrium counts is readily expanded to mutants that differ from wild-type virus by arbitrary n point mutations. Elegant, approximate formulae are obtained if we neglect back-mutation, assume all mutants have the same effective reproductive number

$R_i = R, 1 \leq i \leq n$, and all sites have the same mutation rate $m \ll 1$. Then, the mean is

$$\frac{n! n_{\text{eq}}}{R} \left(\frac{mR}{1-R} \right)^n \tag{9}$$

and the variance is

$$\frac{n! n_{\text{eq}}}{R(1-R)} \left(\frac{mR}{1-R} \right)^n . \tag{10}$$

Note, the means, variances, and distributions of particles at equilibrium are the result of interaction between mutation (immigration) and selection (fitness) and unrelated to the starting conditions.

As a numerical example, we set the parameter values as in Table 1. The mean $\pm$ SDs of mutants at equilibrium are 23973 ± 692 for single mutants, 14 ± 15 for double mutants, and 0.01 ± 0.29 for the triple mutant. The rise in the equilibrium counts from those shown in Table 2 at the end of the early acute phase indicates that the number of mutants continues to rise despite

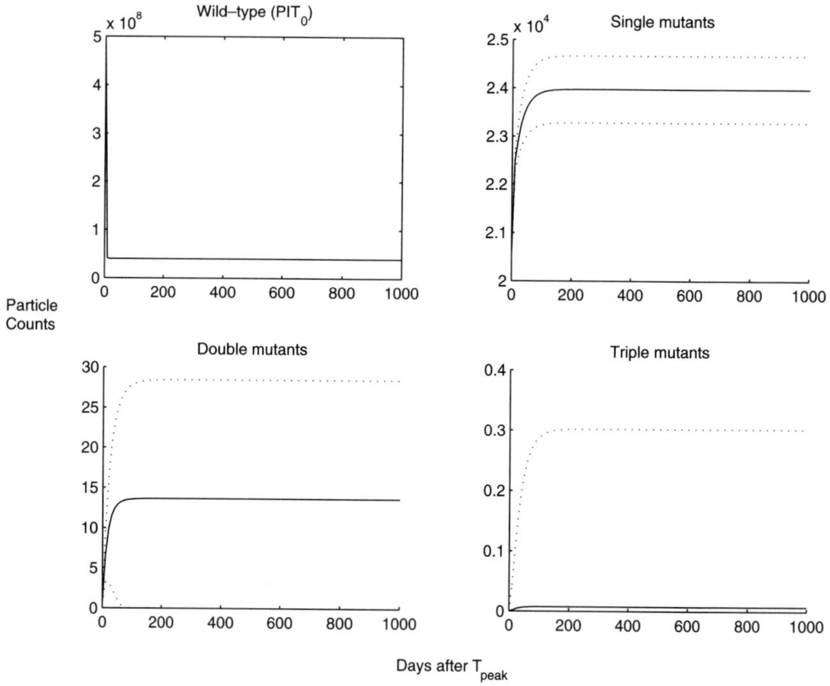

Fig. 9. Mean $\pm$ SD of mutants during the late acute and chronic stage. Parameters are set as in Table 1.

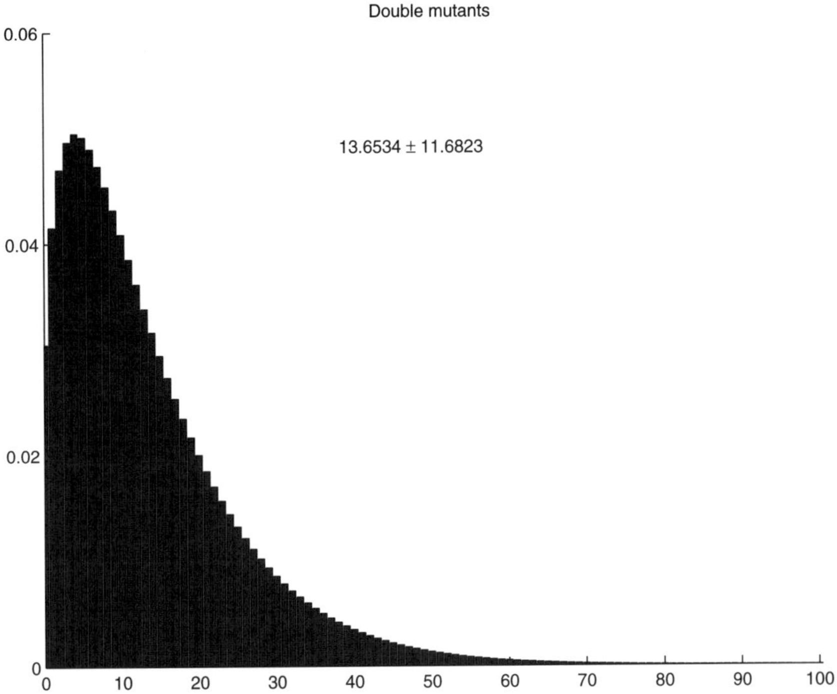

Fig. 10. Marginal distribution of the number of double mutants 1000 days after T_{peak}. Parameters are set as in Table 1.

the decline in viral and PIT load during the late acute phase. Taking the means at day 10 from the early acute stage (Table 2) as starting counts of mutants, the time evolution of mean ± SDs of mutants in the late acute and chronic stages are plotted in Fig. 9. The deterministic growth of PIT_0 is also plotted as comparison. We stress that the calculation is conditional on the specific starting conditions in Table 2. Variability in particle counts at the start of the late acute and chronic stage is overlooked for mathematical convenience. As a result, initial variability is underestimated, but since the process converges to an equilibrium independent of starting conditions, the approximation becomes more reasonable as time progresses.

The distributions of double and triple mutants 1000 days after T_{peak}, which can be assumed to be near equilibrium, are shown in Fig. 10. We observe that the triple mutant is still unlikely (∼0.0025 probability) to exist in a patient prior to the treatment. The double mutants are present in about 97% of patients with counts ranging from 1–100.

5.2. *Sensitivity to the parameter settings*

As comparison, we repeat the analysis with parameter values favoring mutants. In particular, parameters are set as before except that the mutation rates $m_a = m_b = m_c = 4 \times 10^{-5}$ and reproductive numbers $R_i = 0.99$, $1 \leq i \leq 7$, are increased. Mean ± SDs at equilibrium under these setting are $158,753 \pm 3977$ for single mutants, 1243 ± 430 for double mutants, and 15 ± 50 for the triple mutant. The high fitness of mutants not only increases the equilibrium counts of mutants by one to two orders of magnitude, but also affects the timing of mutant appearance. The process takes much longer to reach the higher equilibrium, achieving it only after approximately 2000 days (data not shown). Figure 11 shows the marginal distribution of the number of triple mutants 2000 days after T_{peak}. Approximately 50% of patients are expected to have one or more triple mutants present.

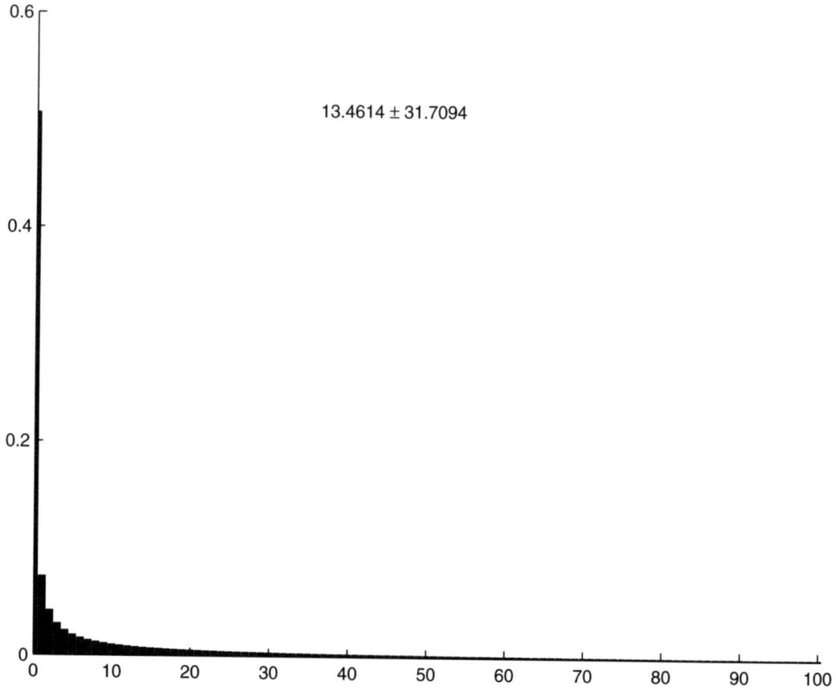

Fig. 11. Effect of high mutation rates and high fitness of mutants on the equilibrium numbers of triple mutant. $m_a = m_b = m_c = 4 \times 10^{-5}$ and $R_i = 0.99$, $1 \leq i \leq 7$. Other parameters are set as in Table 1. Calculations made at 2000 days after initiation of the late acute stage.

6. Modeling Resistance Evolution during Therapy

Treatment failure may be caused either by the existence of resistant mutants in the patient's virus population before the start of therapy or by resistant mutants newly produced by residual replication during therapy. In this section, we compare the likelihood of these two causes in the branching process framework. We make direct comparison between the abundance of resistant mutants prior to treatment and the total number of resistant mutants newly produced during treatment.

While previously we have used the effective reproductive numbers, R_i, for particles in the absence of drugs, we now turn to the effective reproductive numbers, R'_i, for particles in the presence of drug treatment. Again, we stress that the effective reproductive numbers are a combined property of host, virus and drug. In particular, the host environment is quite different pre- and post-treatment. Susceptible cell counts rise substantially on treatment, leading to an apparent enhancement of fitness for all mutants capable of any, even feable, replication in the presence of drugs. In reality, the mutants are expected to be intrinsically less fit both on and off treatment, i.e., they produce, on average, fewer offspring per infection than a typical wild-type virus operating in a comparable environment absent the drug. Because the values of effective reproductive numbers, R'_i, depend on the susceptible cell counts, they undoubtedly change with time. However, by necessity, it is assumed that on the time-scale we are considering, the abundance of susceptible cells and host immunity are constant, so effective reproductive numbers on this time-scale do not change. That is, the process is time-homogeneous. The implications of this time-homogeneity assumption are discussed in the conclusions.

6.1. *Single mutant model*

We begin with a simple model that contains only wild-type virus and one resistant mutant with a single point mutation at site a (Fig. 12(a)). The time evolution of the counts of wild-type virus and mutant during drug therapy constitutes a two-type branching process. The mutation rate is m_a.

Suppose the system achieves equilibrium prior to the drug therapy. Then there are n_{eq} PIT_0 in the body. PIT_a is in balance between mutation and selection. As a trivial case of Eq. (9), the mean number of PIT_a at equilibrium is given by $m_a n_{\text{eq}}/(1 - R_1)$. The variance is given by $m_a n_{\text{eq}}/(1 - R_1)^2$ (Zhou, 2003).

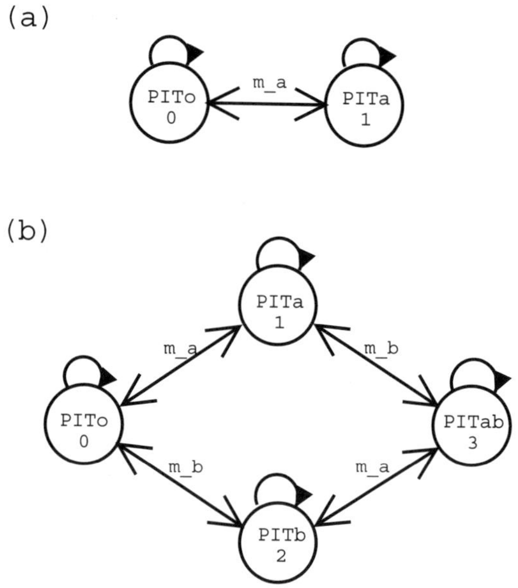

Fig. 12. A schematic illustration of the system with (a) one and (b) two point mutations between sensitive wild-type virus and resistant mutant virus.

To compute the number of mutant PIT_a produced *de novo* during treatment, we construct a branching process and assume there are no PIT_a present at $t = 0$, the start of treatment. Once drug therapy is started, the effective reproductive number of wild-type sensitive virus R_0' is driven below one. If the therapy is so successful that $R_0' = 0$, then we are in the happy state that no new cells will be infected and hence no resistant virus can emerge *de novo*. When $0 < R_0' < 1$, the population of PIT_0 will be driven to extinction but some residual replication will continue. The *de novo* production of resistant mutants during therapy is possible so long as there is residual replication.

It is trivial to set up the progeny generating functions, $P_0(\mathbf{s})$ and $P_1(\mathbf{s})$, of the two particles as in Eq. (1). Taking partial derivatives of progeny generating functions, we get the matrix of progeny means

$$\mathbf{F} = \begin{bmatrix} \frac{(2-m_a)R_0'}{1+R_0'} & \frac{m_a R_0'}{1+R_0'} \\ 0 & \frac{2R_1'}{1+R_1'} \end{bmatrix},$$

where the entry $f_{ij} = \frac{\partial}{\partial s_j} P_i(\mathbf{1}), 0 \leq i, j \leq 1$, is the mean number of daughter particles of type j produced by a particle of type i. Using Eq. (3), the

Ω matrix of the branching process is

$$\mathbf{\Omega} = C \times \begin{bmatrix} (1-m_a)R'_0 - 1 & m_a R'_0 \\ 0 & R'_1 - 1 \end{bmatrix}.$$

Note the process is super-critical since $R'_1 > 1$. In order to account for the total number of PIT_a generated *de novo* from wild-type sensitive virus PIT_0, we set $R'_1 = 0$, forcing the process to be subcritical. Following the derivations in (Dorman *et al.*, 2004) and under the reasonable assumption that $m_a \ll 1$, the matrix of average total descendents is

$$\mathbf{A} = (\mathbf{I} - \mathbf{F})^{-1} = \begin{bmatrix} \frac{1+R'_0}{1-(1-m_a)R'_0} & \frac{m_a R'_0}{[1-(1-m_a)R'_0](1-R'_1)(1+R'_1)} \\ 0 & \frac{1+R'_1}{1-R'_1} \end{bmatrix}$$
$$\approx \begin{bmatrix} \frac{1+R'_0}{1-R'_0} & \frac{m_a R'_0}{1-R'_0} \\ 0 & 1 \end{bmatrix},$$

where $A_{ij}, 0 \leq i,j \leq 1$, has the interpretation that when the subcritical branching process starts with a single particle of type i, an average A_{ij} particles of type j will be produced. Further work produces the variance/covariance matrices of these counts (Dorman *et al.*, 2004).

$$\mathbf{B}_0 \approx \begin{bmatrix} \frac{4R'_0(1+R'_0)}{(1-R'_0)^3} & \frac{m_a R'_0(1+R'_0)^2}{(1-R'_0)^3} \\ \frac{m_a R'_0(1+R'_0)^2}{(1-R'_0)^3} & \frac{m_a R'_0}{1-R'_0} \end{bmatrix}$$

for the descendents of type 0 particles and $\mathbf{B}_1 = \mathbf{0}$ for descendents of type 1 particles.

Suppose at the start of treatment, there are n_{eq} wild-type PIT_0 but no PIT_a. Then upon initiation of treatment, around $n_{\text{eq}} m_a R'_0/(1-R'_0) \pm \sqrt{n_{\text{eq}} m_a R'_0/(1-R'_0)}$ mutant PIT_a will be generated *de novo* from PIT_0 during the decline of population PIT_0. As comparison, $n_{\text{eq}} m_a/(1-R_1) \pm \sqrt{n_{\text{eq}} m_a}/(1-R_1)$ PIT_a are expected at the start of treatment from Eqs. (9) and (10). For example, if $m_a = 3 \times 10^{-5}$ (Mansky and Temin, 1995) and $n_{\text{eq}} = 4 \times 10^7$ (Haase, 1999), the abundance of mutant prior to treatment and the number of mutants generated *de novo* with different values of R_1 and R'_0 are listed in Table 9.

We are interested in the ratio, Θ, of the number of PIT_a produced *de novo* from wild-type PIT_0 during therapy to the number of the PIT_a

Table 9. The number of mutants present at or produced after treatment initiation.

R_1	0	0.2	0.4	0.6	0.8	1
Before	1200 ± 35	1500 ± 43	2000 ± 58	3000 ± 87	6000 ± 173	∞
R'_0	0	0.2	0.4	0.6	0.8	1
During	0 ± 0	300 ± 17	800 ± 28	1800 ± 42	4800 ± 69	∞

present at the start of therapy. The mean and variance of Θ can be approximated by the delta method (Casella and Berger, 2001).

$$\mathrm{E}(\Theta) \approx \frac{R'_0(1-R_1)}{(1-R'_0)},$$

$$\mathrm{Var}(\Theta) \approx \frac{R'_0(1-R_1)^2}{m_a n_{\mathrm{eq}}(1-R'_0)^2}.$$

$\mathrm{E}(\Theta)$, plotted in Fig. 13, depends only on the fitness of mutant before treatment (R_1) and the fitness of wild-type mutant in presence of drugs (R'_0). When the wild-type virus is highly sensitive to the drug therapy (small R'_0) and the mutant has high fitness in the absence of drug (large R_1), the ratio Θ is much smaller than 1, implying that treatment failure is much more likely to be caused by pre-existing resistance than the resistance produced *de novo* during therapy.

6.1.1. *n-mutant example*

Now we expand the model to consider resistant mutants that differ from wild-type virus by two point mutations (Fig. 12(b)). Let m_a and m_b be the mutation rates at two resistance-associated sites a and b. Let R'_i, $0 \le i \le 3$, be the effective reproductive number for type i particles during drug therapy. Still ignoring back-mutation, the progeny generating functions are otherwise set as in Eq. (1). We set $R'_3 = 0$ in order to compute the total number of particle 3 (PIT_{ab}) generated *de novo*. The mean progeny matrix is

$$\mathbf{F} = \begin{bmatrix} \frac{R'_0(2-m_a-m_b)}{1+R'_0} & \frac{m_a R'_0}{1+R'_0} & \frac{m_b R'_0}{1+R'_0} & 0 \\ 0 & \frac{R'_1(2-m_b)}{1+R'_1} & 0 & \frac{m_b R'_1}{1+R'_1} \\ 0 & 0 & \frac{R'_2(2-m_a)}{1+R'_2} & \frac{m_a R'_2}{1+R'_2} \\ 0 & 0 & 0 & 0 \end{bmatrix}$$

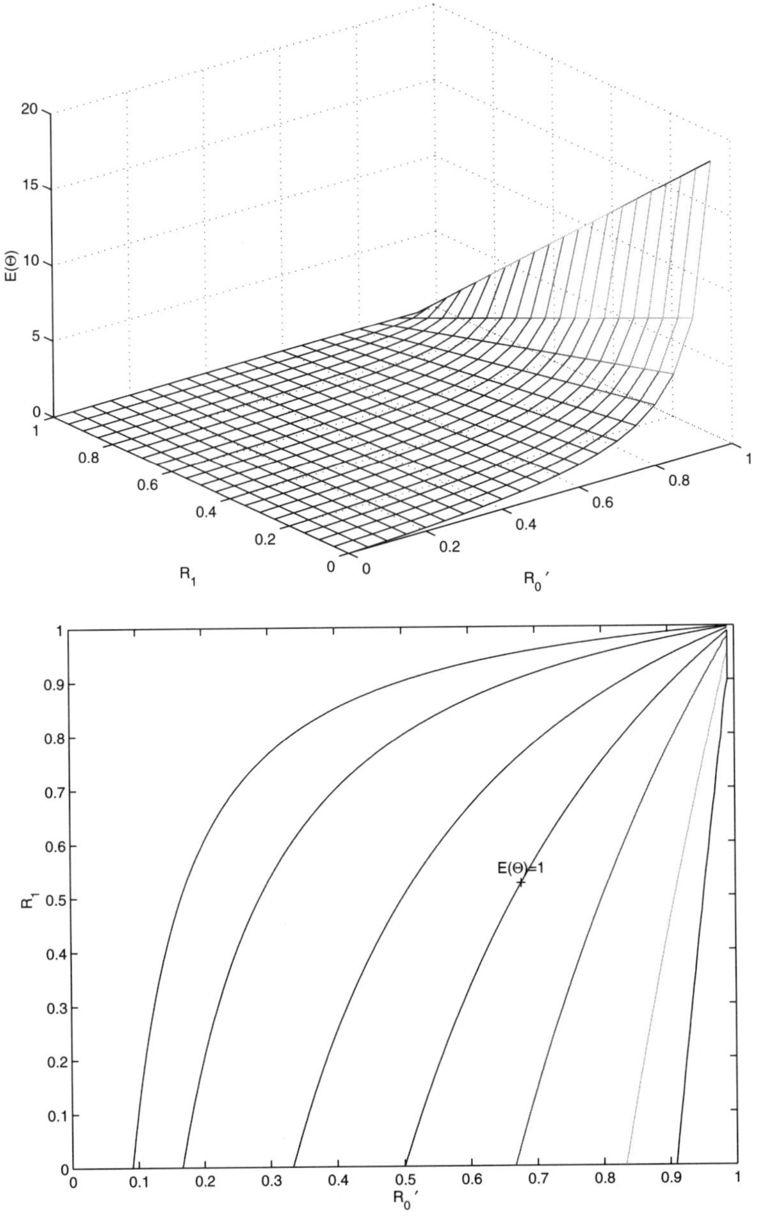

Fig. 13. The upper figure shows the dependence of the average ratio E(Θ) on the effective reproductive number R_1 of PIT_a before treatment and the effective reproductive number R'_0 of wild-type PIT_0 during treatment. The lower figure is the contour plot of E(Θ). The contour plot elevation levels are $0.1, 0.2, 0.5, 1, 2, 5$, and 10 from left to right.

through which we obtain the matrix for average total descendents

$$\mathbf{A} = (\mathbf{I} - \mathbf{F})^{-1}$$
$$\approx \begin{bmatrix} \frac{1+R_0'}{1-R_0'} & \frac{m_a R_0'(1+R_1')}{(1-R_0')(1-R_1')} & \frac{m_b R_0'(1+R_2')}{(1-R_0')(1-R_2')} & \frac{m_a m_b R_0'}{1-R_0'}\left(\frac{R_1'}{1-R_1'} + \frac{R_2'}{1-R_2'}\right) \\ 0 & \frac{1+R_1'}{1-R_1'} & 0 & \frac{m_b R_1'}{1-R_1'} \\ 0 & 0 & \frac{1+R_2'}{1-R_2'} & \frac{m_a R_2'}{1-R_2'} \\ 0 & 0 & 0 & 1 \end{bmatrix}.$$

The variance of these counts are complicated to write, but again if we assume that the mutation rate for each site is the same and much smaller than unity, $m_a = m_b = m \ll 1$, and that all intermediate mutants have the same effective reproductive number as wild-type virus during treatment, $R_0' = R_1' = R_2' = R'$, then the variance of the number of resistant mutants PIT_{ab} descendent to a single ancestor PIT_0, PIT_a or PIT_b is $B_0(3,3) = 2m^2R'^2/(1-R')^2$ and $B_1(3,3) = B_2(3,3) = mR'/(1-R')$ respectively.

Now we consider how many PIT_{ab} are present at the start of treatment. Let $\eta = [m_aCn_{\text{eq}}, m_bCn_{\text{eq}}, 0]$ be the vector of immigration rates at equilibrium. Then the mean counts of PIT_a, PIT_b and PIT_{ab} at equilibrium are

$$-\eta \times \Omega^{-1} \approx \left[\frac{m_a n_{\text{eq}}}{1-R_1}, \frac{m_b n_{\text{eq}}}{1-R_2}, \frac{m_a m_b n_{\text{eq}}}{1-R_3}\left(\frac{R_1}{1-R_1} + \frac{R_2}{1-R_2}\right)\right]$$

with variance/covariance matrix

$$\begin{bmatrix} \frac{mn_{\text{eq}}}{(1-R)^2} & 0 & \frac{m^2 n_{\text{eq}} R}{2(1-R)^3} \\ 0 & \frac{mn_{\text{eq}}}{(1-R)^2} & \frac{m^2 n_{\text{eq}} R}{2(1-R)^3} \\ \frac{m^2 n_{\text{eq}} R}{2(1-R)^3} & \frac{m^2 n_{\text{eq}} R}{2(1-R)^3} & \frac{2m^2 n_{\text{eq}} R}{(1-R)^3} \end{bmatrix}$$

if $m_a = m_b = m \ll 1$ and $R_1 = R_2 = R_3 = R$.

The model can be expanded to consider resistant mutants that differ from wild-type virus by n point mutations. For simplicity, we assume that the mutation rates for each site are the same, m, and much less than unity, all mutants have the same effective reproductive number, R, prior to treatment and all intermediate mutants have the same effective reproductive number, R', as wild-type virus during treatment. Then means and variances of the number of different mutants existing prior to treatment and the means and variances of the number of *de novo* produced resistant n-mutants descendent to a pre-existent wild-type virus or a pre-existent intermediate k-mutant can be calculated. For example,

$$\frac{k!n_{\mathrm{eq}}}{R}\left(\frac{mR}{1-R}\right)^k \pm \sqrt{\frac{k!n_{\mathrm{eq}}}{R(1-R)}\left(\frac{mR}{1-R}\right)^k}$$

k-mutant exists at the start of treatment and each of the pre-existing k-mutant will finally produce about

$$(n-k)!\left(\frac{mR'}{1-R'}\right)^{n-k} \pm \sqrt{(n-k)!\left(\frac{mR'}{1-R'}\right)^{n-k}}$$

resistant n-mutants during treatment. Note for each class of k-point mutants, there are $\binom{n}{k}$ strains. Thus, by the delta method, the mean of the total number of resistant n-mutants produced *de novo* during treatment from pre-existing wild-type and intermediate mutants is approximately

$$n_{\mathrm{eq}}n!\left(\frac{mR'}{1-R'}\right)^n + \frac{n!n_{\mathrm{eq}}m^n}{R}\sum_{k=1}^{n-1}\left(\frac{R}{1-R}\right)^k\left(\frac{R'}{1-R'}\right)^{n-k}. \tag{11}$$

Dividing Eq. (11) by the average number of resistant mutants present prior to treatment, $(n!n_{\mathrm{eq}}/R)[mR/(1-R)]^n$, we get an estimate (by the delta method) of the mean of the ratio of resistant mutants produced during therapy to their abundance at the start of therapy

$$\mathrm{E}(\Theta_n) \approx \begin{cases} \frac{1-R}{R-R'}\left[R' - R\left(\frac{R'(1-R)}{(1-R')R}\right)^n\right] & \text{if } R \neq R' \\ n-1+R & \text{if } R = R' \end{cases}$$

$\mathrm{E}(\Theta_n)$ for $n = 1, 2, 5, 10$ is plotted in Fig. 14. $\mathrm{E}(\Theta_n)$ depends only on the fitness of mutants prior to treatment, R, the fitness of the sensitive wild-type and intermediate mutant virus during treatment, R', and n. It is independent of the mutation rate m and the clearance rate of productively infected cells C. When $R' > R$, $\mathrm{E}(\Theta_n)$ increases geometrically with n; when $R' < R$, $\mathrm{E}(\Theta_n)$ decreases geometrically with n. From the graph, for $n \geq 2$, resistant mutants are more likely to be generated (for the first time) *de novo* during therapy than to pre-exist when $R' > R$ and even for a small region where $R' \leq R$. The region of (R, R') space over which $\mathrm{E}(\Theta_n) > 1$ increases as the number of required resistance mutations n grows.

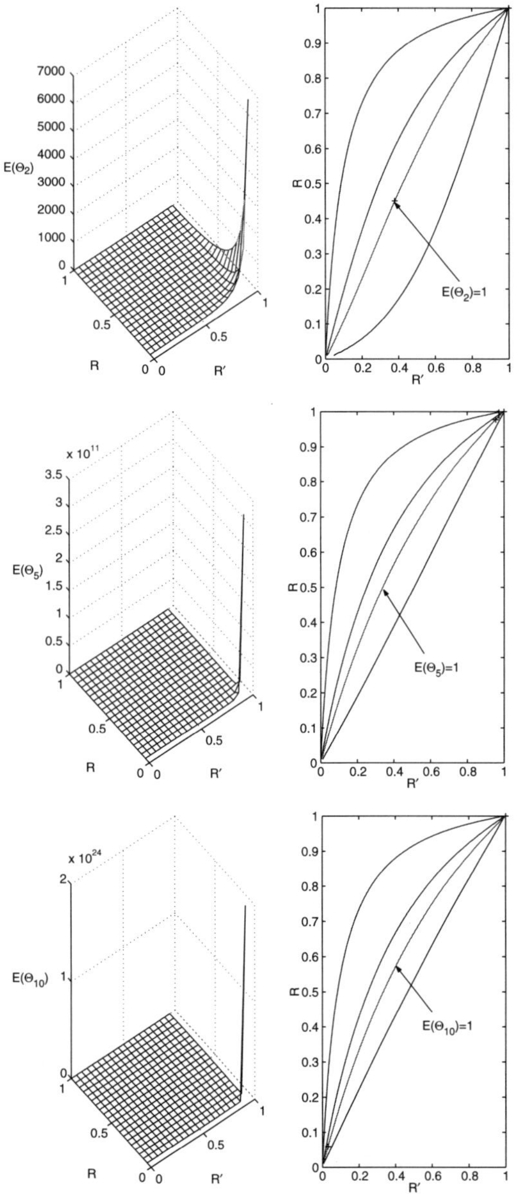

Fig. 14. Left column shows dependence of the average ratios $\mathrm{E}(\Theta_n)$ on the effective reproductive number R of all mutants before treatment and the effective reproductive number R' of wild-type and intermediate mutant virus during treatment for $n = 1, 5, 10$. Right column shows contour plots of $\mathrm{E}(\Theta_n)$ for $n = 1, 5, 10$. The contour plot elevation levels are 0.1, 0.5, 1, and 5 from left to right.

7. Conclusions

The success of drug therapy depends to a large extent on whether or not resistant virus is present in patients prior to treatment. Using a multitype continuous-time branching process model for the mutant populations, we track the timing and variability in the appearance of mutants from the start of acute infection through the equilibrium state during disease latency, and after initiation of treatment.

During the early acute stage, or the first few weeks after the start of infection, the number of single mutants increases exponentially as the surrounding PIT_0 population grows. At the end of the early acute stage, single mutants are well represented on the order 10^2–10^4. The timing and abundance of single mutants depends greatly on the dynamics of PIT_0. Double mutants may or may not be present. The chance that any double mutant is present at the end of the acute stage is approximately 0.29 (Fig. 4) in the typical numeric example of Table 1 and just shy of 1 in a setting extremely favorable for the generation of mutants. The triple and more complex mutants have little chance to occur during this stage as shown in a variety of conditions. The probability, $e(t)$, that no mutants are present drops rapidly in a matter of days. The timing of this period is determined by various factors, with for example, a larger inoculum of infecting virus indicative of rapid resistance. In all cases, the probability of no mutants will certainly be less than 0.05 before the end of the acute stage.

In the late acute and chronic stages, the amount of wild-type PIT_0 and mutant viruses converges to an approximate equilibrium. The equilibrium frequency of mutants depends on the mutation rates at the mutated sites and the fitness of all intermediate mutants. For the numeric example in Table 1, we expect to find $23,973 \pm 693$ single mutants, 14 ± 15 double mutants and ~ 0 triple mutant at equilibrium. The chance that no double mutant is present at equilibrium is small, approximately 0.03. The chance that any triple mutant is present at equilibrium is consistently near 0. The counts of mutants at equilibrium increase with higher mutation rates and mutant fitness. In the numeric example in Fig. 11, triple mutants are expected in about 50% of patients at equilibrium. However the rate of convergence to equilibrium is slowed with higher mutant fitness, and even in the most mutant-favorable conditions at final equilibrium, the relative frequency of the triple mutant remains low and difficult to detect (Fig. 8).

The model provides some practical lessons for clinical treatment. Not surprisingly, the pre-existence of resistant mutants depends on the number of point mutations required to confer full resistance. For monotherapy, to which one or two point mutations can confer resistance, the early high abundance of single mutants could cause rapid selection of resistant virus and lead to therapy failure. Only monotherapies starting in the first few days or hours can be absolutely assured of success by the total lack of resistant mutations. Resistance to two or three drugs may require several point mutations. If resistance requires three or more point mutations, resistant mutants have virtually zero probability of appearing prior to treatment. Under these conditions, increasing the efficacy of drug (for example by increasing the dosage) is an effective way to control the virus.

Comparing the number of resistant mutants generated *de novo* during treatment and the pre-existence of resistant mutants, it is possible to identify the likely source of resistant viruses that cause treatment failure. Again the answer depends on the number of point mutations required for resistance. For resistance conferred by single mutants, treatment failures are more likely to be caused by pre-existing resistant mutants when the wild-type post-treatment fitness $R_0' < 0.5$, while for resistance requiring more than two point mutations, mutants are increasingly likely to evolve in the presence of medication. Our model, however, predicts very low treatment failure rates when resistance requires 3 or more mutations. Such mutants do not exist prior to treatment and rarely appear after treatment. It is likely, therefore, that treatment failure in patients has another source, such as poor adherence or residual replication in sanctuaries. More work is needed to understand their impact on resistance development. In this light, it is interesting to note that very effective treatment is also likely to produce conditions most favorable for mutants by raising the level of susceptible cells and, hence, the mutant reproductive numbers on treatment R_i'. While this effect is no risk when residual replication is completely blocked, it could be a major factor when there is hidden replication, for example in a drug sanctuary.

For ease of calculation, we have assumed that the effective reproductive numbers of wild-type virus and mutants are constant during treatment, i.e., the process is time-homogeneous, an assumption that may not hold in reality. For instance, with the decline of wild-type virus upon treatment, the susceptible cell count may increase substantially and improve the fitness (effective reproductive number) of all viruses in the system. If we assume that the starting conditions are applicable throughout treatment, we obtain a best case for the patients. Vice versa, if we use the on-treatment equilibrium conditions, we obtain the worst case scenario for patients. Thus, with more precise data on fitness during treatment, we can make predictions that are bounded from above and below.

We have not included latently infected cells in our model. These cells do not produce virus, but are notoriously tenacious. When a patient goes on effective treatment, the virus enters a race to survive. The wild-type and incompletely resistant mutants are destined to extinction; there are a limited number of replications cycles left to generate a sufficiently fit resistant virus that can persist in the face of treatment. Latent cells prolong the survival of the virus by delaying the choice between viral eradiation or viral outgrowth (Saksena and Potter, 2003).

One of the difficulties in applying the model is the need for relative fitness measures in the presence of drug. While fitness measurements in the absence of drug are readily available, very few studies have measured fitness quantitatively in the presence of drug (Mammano *et al.*, 2000). Measurements are usually taken relative to a wild-type comparison virus, but wild-type is often incapable of measurable growth in the presence of medication. These models require several critical quantities to make relevant predictions. First, in the absence of a drug, a measure of the absolute fitness of wild-type virus and the relative fitness of all mutants are required. In the presence of a drug, a measure of the absolute fitness of a replication-competent virus and the fitness of all mutants relative to the reference mutant virus are needed. Unfortunately, closer examination makes even these data insufficient. Mammano *et al.* (2000) show that the relative fitness of mutants changes dramatically with varying drug concentrations. The complete picture of resistance evolution is very complex indeed.

Appendix

The variance of the branching process with immigration given in Eq. (5) is a sum involving $\mathbf{V}_i(t)$, the variance of the process initiated with one particle of type i and no immigration and $\mathbf{W}_i(t)$, the variance of the process initiated with no particles but fed by immigration of type i particles. Here, we provide formulae for these components of the total variance.

These equations are best summarized by relying on the Vec and Kronecker $\otimes$ operators. The Vec operator stacks the successive columns of a matrix to form a column vector. If $\mathbf{A} = (a_{ij})$ is an $k \times l$ matrix and $\mathbf{B} = (b_{ij})$ is an $m \times n$ matrix, then the Kronecker product $\mathbf{A} \otimes B$ is the $km \times ln$ block matrix

$$\mathbf{A} \otimes \mathbf{B} = \begin{pmatrix} a_{11}\mathbf{B} & \cdots & a_{1l}\mathbf{B} \\ \vdots & \ddots & \vdots \\ a_{k1}\mathbf{B} & \cdots & a_{kl}\mathbf{B} \end{pmatrix}.$$

So, if $\text{Vec}[\mathbf{V}(t)]$ is obtained by stacking vectors $\text{Vec}[V_i(t)]$ and $\text{Vec}\,(C)$ is obtained by stacking vectors $\text{Vec}(C_i)$, then the variance of the process without immigration is given by

$$\text{Vec}[\mathbf{V}(t)] = \int_0^t e^{\tau\Omega} \otimes e^{(t-\tau)\Omega^*} \otimes e^{(t-\tau)\Omega^*}\, d\tau \text{Vec}(\mathbf{C}).$$

Matrix $\mathbf{C}_i$ is defined as

$$\mathbf{C}_i = \frac{d}{dt}\mathbf{V}_i(0) = \omega_i(\mathbf{G}_i + \mathbf{u}_i^*\mathbf{u}_i + \mathbf{f}_i^*\mathbf{f}_i - \mathbf{u}_i^*\mathbf{f}_i - \mathbf{f}_i^*\mathbf{u}_i).$$

where $\mathbf{G}_i$ is the covariance matrix of the progeny distribution

$$\mathbf{G}_i = \mathbf{H}_i + \text{diag}(\mathbf{f}_i) - \mathbf{f}_i^*\mathbf{f}_i. \tag{12}$$

obtained from the progeny means

$$f_{ij} = \frac{\partial}{\partial s_j} P_i(\mathbf{1})$$

and the progeny Hessian matrix

$$(\mathbf{H}_i)_{jk} = \frac{\partial^2 P_i(\mathbf{1})}{\partial s_j \partial s_k},$$

for $1 \leq i, j, k \leq r$.

Finally, the variance of the process fed only by immigration of type i particles is given by

$$\text{Vec}[\mathbf{W}_i(t)] = \int_0^t \eta_i(\tau)\{\text{Vec}[\mathbf{V}_i(t-\tau)] + e^{(t-\tau)\Omega^*} \otimes e^{(t-\tau)\Omega^*}\text{Vec}(\mathbf{u}_i^*\mathbf{u}_i)\}\, d\tau.$$

The integrals in the above formulae are easily computed when immigration is exponential or sums of exponential functions.

References

Anderson RM and May RM (1991) *Infectious Diseases in Humans: Dynamics and Control.* Clarendon, Oxford.

Bonhoeffer S, May RM, Shaw GM and Nowak MA (1997) Models of virus dynamics and drug therapy. *Proc Natl Acad Sci USA* **94**: 6971–6976.

Casella G and Berger RL (2001) *Statistical Inference*, pp. 244, Belmont, California.

Clark SJ, Saag MS, Decker WD, Campbell–Hill S, Roberson JL, Veldkamp PJ, Kappes JC, Hahn BH and Shaw GM (1991) High titers of cytopathic virus in plasma of patients with symptomatic primary HIV-1 infection. *New Eng J Med* **324**: 954–960.

Coffin JM (1995) HIV population dynamics *in vivo*: implications for genetic variation, pathogenesis, and therapy. *Science* **267**: 482–489.

Cohen J (2002) Confronting the limits of success. *Science* **296**: 2320–2324.

Daar ES, Moudgil T, Meyer RD and Ho DD (1991) Transient high levels of viremia in patients with primary human immunodeficiency virus type 1 infection. *New Eng J Med* **324**: 961–964.

Dorman KS, Sinsheimer JS and Lange K (2004) In the garden of branching processes. *SIAM Rev*, in press.

Finzi D and Blankson J and Siliciano JD and Margolick JB, Chadwick K, Pierson T, Smith K, Lisziewicz J, Lori F, Flexner C, Quinn TC, Chaisson RE, Rosenberg E, Walker B, Gange S, Gallant J and Siliciano RF (1999) Latent infection of CD4+ T cells provides a mechanism for lifelong persistence of HIV-1, even in patients on effective combination therapy. *Nat Med* **5**: 512–517.

Ho DD, Neumann AU, Perelson AS, Chen W, Leonard JM and Markowitz M (1995) Rapid turnover of plasma virions and CD4 lymphocytes in HIV-1 infection. *Nature* **373**: 123–126.

de Jong MD, Veenstra J, Stilianakis NI, Schuurman R, Lange JM, de Boer RJ and Boucher CA (1996) Host-parasite dynamics and outgrowth of virus containing a single K70R amino acid change in reverse transcriptase are responsible for the loss of human immunodeficiency virus type 1 RNA load suppression by zidovudine. *Proc Natl Acad Sci USA* **93**: 5501–5506.

Haase AT (1999) Population biology of HIV-1 infection: viral and CD4+ T cell demographics and dynamics in lymphatic tissues. *Ann Rev Immunol* **17**: 625–656.

Henrard DR, Daar E, Farzadegan H, Clark SJ, Phillips J, Shaw GM and Busch MP (1995) Virologic and immunologic characterization of symptomatic and asymptomatic primary HIV-1 infection. *J Acquir Immune Defic Syndr Hum Retrovirol* **9**: 305–310.

Kamina A, Makuch RW and Zhao H (2001) A stochastic modeling of early HIV-1 population dynamics. *Math Biosci* **170**: 187–198.

Klarmann GJ, Schauber CA and Preston BD (1993) Template-directed pausing of DNA synthesis by HIV-1 reverse transcriptase during polymerization of HIV-1 sequences *in vitro*. *J Biol Chem* **268**: 9793–9802.

Klenerman P, Phillips RE, Rinaldo CR, Wahl LM, Ogg G, May RM, McMichael AJ and Nowak MA (1996) Cytotoxic T lymphocytes and viral turnover in HIV type 1 infection. *Proc Natl Acad Sci USA* **93**: 15323–15328.

Mammano F, Trouplin V, Zennou V and Clavel F (2000) Retracing the evolutionary pathways of human immunodeficiency virus type 1 resistance to protease inhibitors: virus fitness in the absence and in the presence of drugs. *J Virol* **74**: 8524–8531.

Mansky LM and Temin HM (1995) Lower *in vivo* mutation rate of human immunodeficiency virus type 1 than that predicted from the fidelity of purified reverse transcriptase. *J Virol* **29**: 5087–5094.

Mansky LM (1996) Forward mutation rate of human immunodeficiency virus type 1 in a T lymphoid cell line. *AIDS Res Hum Retrovir* **12**: 307–314.

Markowitz M, Louie M, Hurley A, Sun E, Mascio MD, Perelson AS and Ho DD (2003) A novel antiviral intervention results in more accurate assessment of human immunodeficiency virus type 1 replication dynamics and T-cell decay *in vivo*. *J Virol* **77**: 5037–5038.

Nowak MA, Anderson RM, Boerlijst MC, Bonhoeffer S, May RM and McMichael AJ (1996) HIV evolution and disease progression. *Science* **274**: 1008–1011.

Perelson AS, Neumann AU, Markowitz M, Leonard JM and Ho DD (1996). HIV-1 dynamics *in vivo*: virion clearance rate, infected cell life-span, and viral generation time. *Science* **271**: 1582–1586.

Phillips AN (1996) Reduction of HIV concentration during acute infection: independence from a specific immune response. *Science* **271**: 497–499.

Pomerantz RJ and Horn DL (2003) Twenty years of therapy for HIV-1 infection. *Nat Med* **9**: 867–873.

Reimann KA, Tenner-Racz K, Racz P, Montefiori DC, Yasutomi Y, Lin W, Ransil BJ and Letvin NL (1994) Immunopathogenic events in acute infection of rhesus monkeys with simian immunodeficiency virus of macaques. *J Virol* **68**: 2362–2370.

Ribeiro RM, Bonhoeffer S and Nowak M (1998) The frequency of resistant mutant virus before antiviral therapy. *AIDS* **12**: 461–465.

Ribeiro RM and Bonhoeffer S (2000) Production of resistant HIV mutants during antiretroviral therapy. *Proc Natl Acad Sci USA* **97**: 7681–7685.

Ricchetti M and Buc H (1990) Reverse transcriptases and genomic variability: the accuracy of DNA replication is enzyme specific and sequence dependent. *EMBO J* **9**: 1583–1593.

Saksena NK and Potter SJ (2003) Reservoirs of HIV-1 *in vivo*: implications for antiretroviral therapy. *AIDS Rev* **5**: 3–18.

Schacker TW, Hughes JP, Shea T, Coombs RW and Corey L (1998) Biological and virologic characteristics of primary HIV infection. *Ann Intern Med* **128**: 613–620.

Shafer RW (2002) Genotypic testing for human immunodeficiency virus type 1 drug resistance. *Clin Microbiol Rev* **15**: 247–177.

Sodora DL, Lee F, Dailey PJ and Marx PA (1998) A genetic and viral load analysis of the simian immunodeficiency virus during the acute phase in macaques inoculated by the vaginal route. *AIDS Res Hum Retrovir* **14**: 171–181.

Tsai CC, Follis KE, Grant RF, Nolte RE, Wu H and Benveniste RE (1993) Infectivity and pathogenesis of titered dosages of simian immunodeficiency virus experimentally inoculated into longtailed macaques (*Macaca fascicularis*). *Lab Anim Sci* **43**: 411–416.

Tuckwell HC and Le Corfec E (1998) A stochastic model for early HIV-1 population dynamics. *J Theor Biol* **195**: 451–463.

Wei X, Ghosh SK, Taylor ME, Johnson VA, Emini EA, Deutsch P, Lifson JD, Bonhoeffer S, Nowak MA, Hahn BH, Saag MS and Shaw GM (1995) Viral dynamics in human immunodeficiency virus type 1 infection. *Nature* **373**: 117–122.

Zhou H (2003). Branching process models for HIV-1 drug resistant mutants, M. Sci. Thesis, Interdepartmental Program in Bioinformatics and Computational Biology, Iowa State Univ., 106p.

CHAPTER 19

A BAYESIAN APPROACH FOR ASSESSING DRUG RESISTANCE IN HIV INFECTION USING VIRAL LOAD

Hua Liang, Waiyuan Tan and Xiaoping Xiong

In this chapter, we have assessed the time to development of drug resistance in HIV-infected individuals treated with antiviral drugs by using longitudinal viral load HIV-1 counts. Through log transformed data of HIV virus counts over time, we have assumed a linear changing-point model and developed procedures to estimate the unknown parameters by using the Bayesian approach. We have applied the method and procedure to the data generated by the ACTG 315 involving treatment by the drug combination (3TC, AZT and Ritonavir). Our analysis showed that the mean time to the first changing point (i.e. the time the macrophage and long-lived cells began to release HIV particles) was around 15 days whereas the time to development of drug resistance by HIV was around 75 days. The Bayesian HPD intervals for these changing points are given by $(8.7, 21.3)$ and $(42, 108)$ respectively. This analysis indicated that if we use the combination of three drugs involving 2 NRTI inhibitors (3TC and AZT) and 1 PI inhibitor (Ritonavir) to treat HIV-infected individuals, in about two and half months it would be beneficial to change drugs to avoid the problem of drug resistance.

Keywords: AIDS clinical trial; Gibbs sampler; HIV dynamics; longitudinal data; random change points; nonlinear mixed-effects models.

1. Introduction

It has been well documented that plasma HIV RNA level (viral load) is a very effective predictor for clinical outcomes in HIV-infected individuals (Saag *et al.*, 1996; Mellors *et al.*, 1995, 1996), and has thus become a primary surrogate marker in most AIDS clinical trials. To assess effects of drugs and to monitor the progression of the disease in HIV-infected individuals treated with anti-retroviral drugs, it appears necessary to study the HIV dynamics during treatment through HIV viral loads and how the dynamics being related to the changing numbers of HIV virus load over time.

To illustrate the basic HIV dynamics in HIV-infected individuals, consider an HIV-infected individual treated with highly active antiretroviral therapy (HAART) involving 2 nucleoside reverse transcriptase inhibitors (NRTI) such as 3TC and AZT and 1 protease inhibitor (PI) such as ritonavir. Then, it has been shown that within the first week, the total number of HIV counts per mL of blood can be approximated by an exponential function. It follows that in the first week, the log of the HIV virus load decreases linearly and sharply with large negative slope (Ho *et al.*, 1995; Wei *et al.*, 1995; see Tan, 2000, Chapter 8, for other publications). After the first week, however, the picture is very different and show multi-phases of decline and increase. In log scale, within the first 2–3 months after the first week, the picture is again linear but the curve is very flat due presumably to the release of HIV by macrophage and other long-lived cells from the lymph nodes (Perslosn *et al.*, 1997; Tan, 2000, Chapter 8); at some time between the 3rd month and the 8th month for some individuals, the virus load increases sharply reaching the level before the treatment, due presumably to the development of drug resistance of HIV to the drugs. By using a comprehensive stochastic model of HIV pathogenesis involving flow of HIV from the lymph nodes and development of drug resistance, Tan (2000, Chapter 8) has demonstrated through computer simulation how this changes and outcomes are predicted by the model. To illustrate these points, we give in Fig. 1 a scatter plot of viral load (in $\log_{10}$ scale) at different treatment times from an AIDS clinical study conducted by the AIDS Clinical Trials Group (ACTG 315). In this study, 48 HIV-1 infected patients were treated with potent antiviral therapy consisting of ritonavir, 3TC and AZT (see Lederman *et al.*, 1998 and Wu *et al.*, 1999 for more details on this study). Viral load was monitored simultaneously at treatment days 0, 2, 7, 10, 14, 28, 56, 84, 168, and 336.

From Fig. 1, we observe that the viral load of most patients decline rapidly within the first two weeks with large negative slope; after two weeks, however, the trend of decline decreases and becomes flat. At about 10 weeks, the viral loads rebound mostly upward to the end of treatment. From this, it is logical to assume that the effect of treatment can be divided into three stages: rapid decline, slow decline, and rebound. We are interested in the rate of declines/rebound and the time of changing points. From the biological perspective, one may link the time of first changing point as the time when the macrophage or other long-lived HIV-infected cells from

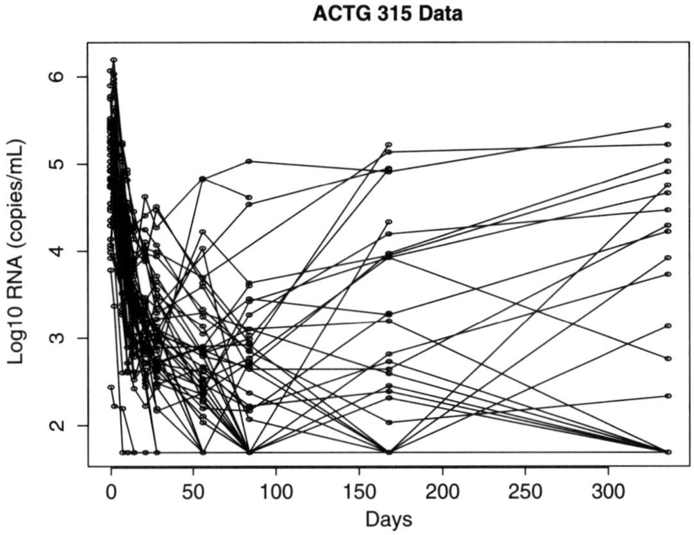

Fig. 1. Viral load data for ACTG 315 study.

lymphoid tissues release free HIV to the blood; one may link the second time of changing point as the time when HIV have developed resistance to the drugs. These time points have important clinical implications because this is the time to change drugs to avoid the problem of drug resistance.

For better understanding of the HIV pathogenesis and for better treatment management and care of AIDS patients, it is important to identify when the patients' viral load or CD4 cell counts decline, and when to change declining trend, and rebound. Lange *et al.* (1992) considered a fully Bayesian analysis of CD4 cell counts to assess HIV infection. Kiuchi *et al.* (1995) used a similar approach to examine change points in the series of T4 counts prior AIDS. Putter *et al.* (2002) and Han *et al.* (2002) have considered a population HIV dynamic model using the Bayesian approach. But none of these studies have ever attempted to estimate the time to drug resistance. By using data from the clinical trial ACTG 315, in this paper we will proceed to develop a Bayesian procedure to estimate the time of these changing points.

In this chapter, we propose a three-segment model with random change-point to describe viral load trajectory data. We are concerned with the decline/rebound rates and the location of changing points. This chapter is organized as follows. In Section 2, we propose our model structure, prior distributions and conditional posterior distributions. In Section 3, we describe the multi-level Gibbs sampling procedures to estimate the unknown parameters and the times of changing points. In Section 4, we present the analysis and results for the ACTG 315 data by using the models and methods given in previous sections. Finally, in Section 5, we present some conclusions and some discussions.

2. The Model

When monitoring the disease progression in HIV-infected individuals, the data are usually the number of RNA virus copies over different time points. As illustrated in Section 1, in log scale these data sets are best described by a piecewise linear model with at most two unknown changing points. The first changing point is the time at which some HIV are released by macrophage or other long-lived cells from the lymphoid tissues (Perelson *et al.*, 1997). The second changing point is the time at which the HIV in the HIV-infected individual develops resistance to the drugs. For those individuals in which the HIV have not yet developed resistance to the drugs, the log of the RNA virus copies per mL then showed a two-segment linear curve with one changing point; the first linear segment is a sharp, linear declining curve with large negative slope showing a rapid decline of the HIV numbers due to the inhibition effect of the drugs whereas the second linear segment is a less rapid linearly declining-curve with small negative slope due to additional new release of HIV of macrophage from lymphoid tissues (Perelson *et al.*, 1997). When the HIV in the HIV-infected individual has developed drug resistance, then one would expect a 3-piece linear model with 2 changing points; the third linear segment is a linear ascending curve with positive slope and the second changing point is the time when the HIV develop resistance to the drugs. A typical picture describing this situation is shown in Fig. 2.

To describe the model, suppose that there are n HIV-infected individuals treated by the same drug combination. For the ith individual, let t_{Li} and t_{Ri} denote the time of the first and second changing points respectively. Assuming a mixed model to account for the variation between individuals, then $\{t_{Li}, i = 1, \ldots, n\}$ is a random sample from an unknown density with mean t_L and $\{t_{Ri}, i = 1, \ldots, n\}$ is a random sample from an unknown density with mean t_R. It is generally believed that t_L occurs at about two weeks (Perelson *et al.*, 1997) and t_R occurs some time between months 3 and 11. (Tan, 2000, Chapter 8). For the ith individual, the 3 linear segments

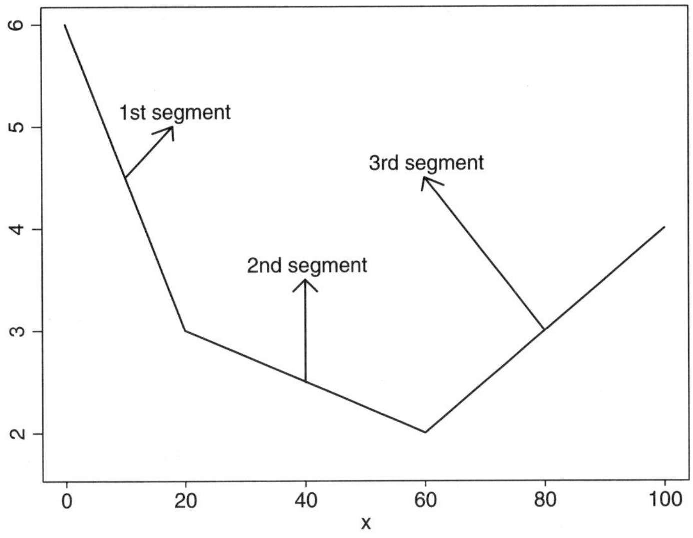

Fig. 2. Illustrative plot of three segments used to explore viral load trajectory.

are separated by t_{Li} and t_{Ri}. For this individual, the first line segment, $\eta_{0i} + \eta_{1i}t$, shows a rapid decline in $\log(RNA)$ after treatment. The second line segment, $\eta_{2i} + \eta_{3i}t$, shows a less rapid decline after some unknown time point t_{Li} due to the release of HIV from macrophage or other long-lived cells from the lymphoid tissues. The third segment, $\eta_{4i} + \eta_{5i}t$, shows an increase of the $\log(RNA)$ after some unknown time point t_{Ri}, because of the drug resistance. Notice that the first two segments agree at t_{Li}, while the last two segments agree at t_{Ri}. Denote by $\Delta_{1i} = \eta_{3i} - \eta_{1i}$ and $\Delta_{2i} = \eta_{5i} - \eta_{3i}$. Then, by combining these three line segments, we can express the piecewise linear model by the following equation:

$$y_{ij} = \eta_{0i} + \eta_{1i}t_{ij} + \Delta_{1i}(t_{ij} - t_{Li})_+ + \Delta_{2i}(t_{ij} - t_{Ri})_+ + \varepsilon_{ij}, \tag{1}$$

where $s_+ = \max(s, 0)$ and ε_{ij} is the random measurement error for measuring y_{ij}.

It is assumed that the ε_{ij}'s are independently distributed normal random variables with means 0 and variance σ_i^2 independently of $\{\eta_{ri}, r = 0, 1, \Delta_{ui}, u = 1, 2, t_{Li}, t_{Ri}\}$ for all $i = 1, \ldots, n$. That is, $\varepsilon_{ij} \sim N(0, \sigma_i^2)$, independently distributed of the random variables η_{0i}, η_{1i}, Δ_{1i}, Δ_{2i}, t_{Li}, and t_{Ri} for all $i = 1, \ldots, n$.

In the above model, because of variation between individuals, the $\{\eta_{ji}, j = 0, 1\}$ and the $\{\Delta_{ri}, r = 1, 2\}$ are random variables. As in Wu *et al.*

(1999), we use a mixed-effects model to account for variation between individuals. Also, following Lange *et al.* (1992) and Kiuchi *et al.* (1995), we will model these variables by using first-order linear equations:

$$\begin{aligned}\eta_{0i} &= \alpha_0 + \beta_{0i}\\ \eta_{1i} &= \alpha_1 + \beta_{1i}\\ \Delta_{1i} &= \alpha_2 + \beta_{2i}\\ \Delta_{2i} &= \alpha_3 + \beta_{3i}\end{aligned}$$

In the above equations, the α's denote population effects, and the β's individual effects to account for variations between individuals. Put $\boldsymbol{\beta}_i = (\beta_{0i}, \beta_{1i}, \beta_{2i}, \beta_{3i})^{\mathrm{T}}$. As in Lange *et al.* (1992) and Kiuchi *et al.* (1995), we assume that the $\boldsymbol{\beta}_i$'s are independently and identically distributed as a Gaussian vector with means 0 and unknown covariance matrix $\mathbf{V}$, i.e., $\boldsymbol{\beta}_i \sim N(0, \mathbf{V})$, independently.

2.1. *The likelihood function*

In the above model, the $\{\mathbf{t}_{LR,i} = (t_{Li}, t_{Ri})^{\mathrm{T}}, \boldsymbol{\beta}_i\}$ are random variables. The unknown parameters are $\{\boldsymbol{\alpha} = (\alpha_0, \alpha_1, \alpha_2, \alpha_3)^{\mathrm{T}}, \mathbf{V}, \sigma_i^2\}$. Let $\boldsymbol{\beta}$ denote the collection of all $\boldsymbol{\beta}_i$, $\boldsymbol{\sigma}^2$ the collection of all σ_i^2, and $\mathbf{t}_{LR}$ the collection of all $\mathbf{t}_{LR,i} = (t_{Li}, t_{Ri})^{\mathrm{T}}$. Let $N(y|\mu, \sigma^2)$ denote the density of a normal distribution with mean μ and variance σ^2 and put $\mathbf{y} = (y_{ij}, i = 1, \ldots, n, j = 1, \ldots, n_i)^{\mathrm{T}}$. Then the conditional likelihood given $(\boldsymbol{\alpha}, \boldsymbol{\beta}, \mathbf{t}_{LR})$ is

$$\mathcal{L}(\mathbf{y}|\boldsymbol{\alpha}, \boldsymbol{\beta}, \mathbf{t}_{LR}) = \prod_{i=1}^{n}\prod_{j=1}^{m_i} N\left(y_{ij}|\mu_{ij}, \sigma_i^2\right),$$

where μ_{ij} is the right-hand side of (1) apart from ε_{ij}.

Let $p(\mathbf{t}_{LR,i})$ denote the probability density function of $\mathbf{t}_{LR,i}$. Then the joint density of $\{\mathbf{y}, \boldsymbol{\beta}, \mathbf{t}_{LR}\}$ given the parameters $\{\boldsymbol{\alpha}, \mathbf{V}, \boldsymbol{\sigma}^2\}$ is:

$$\mathcal{P}\left(\mathbf{y}, \boldsymbol{\beta}, \mathbf{t}_{LR}|\boldsymbol{\alpha}, \mathbf{V}, \boldsymbol{\sigma}^2\right) = \prod_{i=1}^{n}\prod_{j=1}^{m_i} N\left(y_{ij}|\mu_{ij}, \sigma_i^2\right)\prod_{i=1}^{n}\{N(\boldsymbol{\beta}_i|0, \mathbf{V})p(\mathbf{t}_{LR,i})\}.$$

In the above distribution, $p(\mathbf{t}_{LR,i})$ is a density of discrete random variables. To specify this density, we assume that the t_{Li}'s are independently distributed of the t_{Ri}'s. Because we have very little information about t_{Li} except that the expected value of t_{Li} probably occurs at about two weeks (Perelson *et al.*, 1997), we assume that the t_{Li} is a discrete uniform random variable on the set $\{8, 9, \ldots, 18\}$; similarly, because we have no information about t_{Ri} except that the plots of the data seemed to suggest that its expected value probably occurs between 60 days and 90 days, we assume that t_{Ri} is a discrete uniform random variable on the set $\{60, 65, \ldots, 90\}$.

2.2. *The prior distribution*

For the prior distribution of the parameters $\{\boldsymbol{\alpha}, \boldsymbol{\sigma}^2, \mathbf{V}\}$, we assume *a priori* that these parameters are independently distributed of one another. As in Lange *et al.* (1992), we assume that $\boldsymbol{\alpha}$ is a normal vector with prior mean μ_α and prior covariance matrix D_α. For specifying these hyperparameters, we will adopt an empirical Bayesian approach by using some data to estimate its values; for details, see Section 4. For the prior distribution of σ_i^2's, it is assumed that the σ_i^2's are independently and identically distributed as an inverted gamma distribution $IG(\lambda_1, \lambda_2)$ with $\lambda_1 = 10$ and $\lambda_2 = 0.06$. For the prior distribution of $\mathbf{V}$, we assume that the $\mathbf{V}$ is distributed as an inverted Wishart random matrix. Denote by $S \sim W(\Sigma, f)$ that the symmetric random matrix S is distributed as a Wishart distribution with matrix Σ and degrees of freedom f. Then, $\mathbf{V}^{-1} \sim W((\rho\boldsymbol{\Gamma})^{-1}, \rho)$. In $\boldsymbol{\Gamma}$, the left upper 2×2 submatrix is initiated by the covariance matrix of random term from modeling (9); the right-bottom 2×2 submatrix is an identity matrix and the other elements are zeros. We also take $\rho = 2$.

Let $N(\boldsymbol{\alpha}|\mu_\alpha, D_\alpha)$ denote the density of $\boldsymbol{\alpha}$, $IG(z|f_1, f_2)$ the density of an inverted Gamma distribution with parameters (f_1, f_2) and $W(\mathbf{V^{-1}}|(\rho\boldsymbol{\Gamma})^{-\mathbf{1}}, \rho)$ the density of $\mathbf{V^{-1}}$. Using the above prior distributions, the joint density of $\{\mathbf{y}, \boldsymbol{\alpha}, \boldsymbol{\beta}, \mathbf{t}_{LR}, \boldsymbol{\sigma}^2, \mathbf{V}\}$ is:

$$\begin{aligned}\mathcal{P}(\mathbf{y}, \boldsymbol{\alpha}, \boldsymbol{\beta}, \mathbf{t}_{LR}, \boldsymbol{\sigma}^2, \mathbf{V}) &= \prod_{i=1}^{n}\prod_{j=1}^{m_i} N\big(y_{ij}|\mu_{ij}, \sigma_i^2\big) N(\boldsymbol{\alpha}|\mu_\alpha, \mathbf{D}_\alpha)\\ &\quad \times \prod_{i=1}^{n} N(\boldsymbol{\beta}_i|0, \mathbf{V}) \prod_{i=1}^{n} IG\big(\sigma_i^2|\lambda_1, \lambda_2\big)\\ &\quad \times \prod_{i=1}^{n} p(\mathbf{t}_{LR,i}) \mathbf{W}(\mathbf{V}^{-1}|(\rho\boldsymbol{\Gamma})^{-1}, \rho). \end{aligned} \tag{2}$$

2.3. *The posterior distributions*

From the joint density given in Eq. (2), one may readily derive the conditional posterior distributions of the parameters and the conditional distributions of $\{\boldsymbol{\beta}_i, \mathbf{t}_{LR,i}\}$. For this purpose, we introduce the following notations: $X_{ij} = \{1, t_{ij}, (t_{ij} - t_{Li})_+, (t_{ij} - t_{Ri})_+\}$, $\mathbf{X}_i = (X_{i1}^{\mathrm{T}}, \dots, X_{im_i}^{\mathrm{T}})^{\mathrm{T}}$, $\boldsymbol{\varepsilon}_i = (\varepsilon_{i1}, \dots, \varepsilon_{im_i})^{\mathrm{T}}$, $\mathbf{y}_i = (y_{i1}, \dots, y_{im_i})^{\mathrm{T}}$. Then (1) can be described as

$$y_{ij} = X_{ij}(\boldsymbol{\alpha} + \boldsymbol{\beta}_i) + \varepsilon_{ij} \quad \text{or} \quad \mathbf{y}_i = \mathbf{X}_i\boldsymbol{\alpha} + \mathbf{X}_i\boldsymbol{\beta}_i + \boldsymbol{\varepsilon}_i$$

Then, with $\boldsymbol{\varepsilon}_i = \mathbf{y}_i - (\mathbf{X}_i\boldsymbol{\alpha} + \mathbf{X}_i\boldsymbol{\beta}_i)$,

$$\mathcal{L}(\mathbf{y}|\boldsymbol{\alpha}, \boldsymbol{\beta}, \mathbf{t}_{LR}, \boldsymbol{\sigma}^2) \propto \exp\left\{-\sum_{i=1}^{n} \frac{\boldsymbol{\varepsilon}_i{}^{\mathrm{T}}\boldsymbol{\varepsilon}_i}{2\sigma_i^2}\right\}.$$

One can readily derive the posterior distributions of the parameters by direct calculation. For implementing the Gibbs sampling method to estimate the unknown parameters and $\{\boldsymbol{\beta}_i, \mathbf{t}_{LR,i}\}$, we summarize these distributions as in the following:

(a) Denote $\mathbf{r}_{\alpha,i} = \mathbf{y}_i - \mathbf{X}_i\boldsymbol{\beta}_i$. Then the conditional posterior distribution of $\boldsymbol{\alpha}$ given $\{\mathbf{y}, \mathbf{t}_{LR}, \boldsymbol{\beta}, \mathbf{V}, \boldsymbol{\sigma}^2\}$ is a Gaussian random vector with means $\hat{\boldsymbol{\alpha}}$ and covariance matrix $\hat{\Sigma}_\alpha$, where,

$$\hat{\boldsymbol{\alpha}} = \left(\sum_{i=1}^{n} \frac{\mathbf{X}_i^{\mathrm{T}}\mathbf{X}_i}{\sigma_i^2} + \mathbf{D}_\alpha^{-1}\right)^{-1} \left(\sum_{i=1}^{n} \frac{\mathbf{X}_i^{\mathrm{T}}\mathbf{r}_{\alpha,i}}{\sigma_i^2} + \mathbf{D}_\alpha^{-1}\mu_\alpha\right),$$

and

$$\hat{\Sigma}_\alpha = \left(\sum_{i=1}^{n} \frac{\mathbf{X}_i^{\mathrm{T}}\mathbf{X}_i}{\sigma_i^2} + \mathbf{D}_\alpha^{-1}\right)^{-1}.$$

That is,

$$\boldsymbol{\alpha}|\{\mathbf{y}, \mathbf{t}_{LR}, \boldsymbol{\beta}, \mathbf{V}, \boldsymbol{\sigma}^2\} \sim N(\hat{\boldsymbol{\alpha}}, \hat{\Sigma}_\alpha). \tag{3}$$

(b) Denote $\mathbf{r}_{\beta,i} = \mathbf{y}_i - \mathbf{X}_i\boldsymbol{\alpha}$. Then,

$$\boldsymbol{\beta}_i|\{\mathbf{y}, \mathbf{t}_{LR}, \mathbf{V}, \boldsymbol{\alpha}, \boldsymbol{\sigma}^2\} \sim N(\hat{\boldsymbol{\beta}}_i, \hat{\Sigma}_\beta), \tag{4}$$

where

$$\hat{\boldsymbol{\beta}}_i = \left(\frac{\mathbf{X}_i^{\mathrm{T}}\mathbf{X}_i}{\sigma_i^2} + \mathbf{V}^{-1}\right)^{-1} \left(\frac{\mathbf{X}_i^{\mathrm{T}}\mathbf{r}_{\beta,i}}{\sigma_i^2} + \mathbf{V}^{-1}\mu_\alpha\right),$$

and

$$\hat{\Sigma}_\beta = \left(\frac{\mathbf{X}_i^{\mathrm{T}}\mathbf{X}_i}{\sigma_i^2} + \mathbf{V}^{-1}\right)^{-1}.$$

(c) Denote $\mathbf{r}_{\sigma^2,i} = \mathbf{y}_i - \mathbf{X}_i\boldsymbol{\alpha} - \mathbf{X}_i\boldsymbol{\beta}_i$. Then

$$\sigma_i^2|\{\mathbf{y}_i, \boldsymbol{\alpha}, \boldsymbol{\beta}_i, \mathbf{V}, \mathbf{t}_{LR}\} \sim IG\left\{\lambda_1 + \frac{m_i}{2}, \left(\frac{1}{\lambda_2} + \frac{1}{2}\mathbf{r}_{\sigma^2,i}^{\mathrm{T}}\mathbf{r}_{\sigma^2,i}\right)^{-1}\right\}. \tag{5}$$

(d) Denote $\mathbf{a}_i = \boldsymbol{\alpha} + \boldsymbol{\beta}_i - \mu_\alpha$. Then,

$$\mathbf{V}^{-1}|\{\boldsymbol{\alpha}, \boldsymbol{\beta}_i, \boldsymbol{\sigma}, \mathbf{t}_L, \mathbf{t}_R\} \sim W\left\{\left(\sum_{i=1}^{n} \mathbf{a}_i\mathbf{a}_i{}^{\mathrm{T}} + \rho\boldsymbol{\Gamma}\right)^{-1}, n+\rho\right\}. \quad (6)$$

(e) Denote the exponent of exp in the expression $\prod_{j=1}^{m_i} N(y_{ij}|\mu_{ij}, \sigma_i^2)$ from the expression of $\mathcal{L}(\mathbf{y}|\boldsymbol{\alpha}, \boldsymbol{\beta}, \mathbf{t}_{LR})$ as $\mathcal{G}$. Then the conditional distribution of $\mathbf{t}_{Li}$ given $\{\mathbf{y}, \mathbf{t}_R, \boldsymbol{\alpha}, \boldsymbol{\beta}_i, \boldsymbol{\sigma}^2, \mathbf{V}\}$ is

$$\mathbf{t}_{Li}|\{\mathbf{y}, \mathbf{t}_{Ri}, \boldsymbol{\alpha}, \boldsymbol{\beta}_i, \boldsymbol{\sigma}^2, \mathbf{V}\} \sim \frac{\exp\{-\mathcal{G}(\mathbf{t}_{Li})\}p(\mathbf{t}_{Li})}{\sum_{\mathbf{t}_{Lk}} \exp\{-\mathcal{G}(\mathbf{t}_{Lk})\}p(\mathbf{t}_{Lk})}. \quad (7)$$

(f) In the exactly same way as (e), we obtain the conditional distribution of $\mathbf{t}_{Ri}$ given $\{\mathbf{y}, \mathbf{t}_L, \boldsymbol{\alpha}, \boldsymbol{\beta}_i, \boldsymbol{\sigma}^2, \mathbf{V}\}$ as:

$$\mathbf{t}_{Ri}|\{\mathbf{y}, \mathbf{t}_{Li}, \boldsymbol{\alpha}, \boldsymbol{\beta}_i, \boldsymbol{\sigma}^2, \mathbf{V}\} \sim \frac{\exp\{-\mathcal{G}(\mathbf{t}_{Li})\}p(\mathbf{t}_{Ri})}{\sum_{\mathbf{t}_{Lk}} \exp\{-\mathcal{G}(\mathbf{t}_{Rk})\}p(\mathbf{t}_{Rk})}. \quad (8)$$

3. The Gibbs Sampling Procedure

In the Bayesian approach, one derives the Bayesian estimates via the posterior means and derives the HPD intervals of the parameters from the marginal posterior distributions. With the conditional posterior distributions given by Eqs. (3)–(8), one may generate random samples from the marginal distributions via the multi-level Gibbs sampling method. General theories of these procedures are given by Sheppard (1994), Liu and Chen (1998) and Kitagawa (1998). The detailed procedures and its proofs have been given in Chapter 3 of Tan (2002). This is an iterative algorithm by sampling from the conditional distributions alternatively and sequentially with updated parameter values. At convergence, this then gives samples from the marginal distributions respectively. For the model in this chapter, each cycle in this algorithm loops through the following steps:

Step 1. Starting with initial values $\boldsymbol{\Theta}_0^\alpha = (\sigma_0^{-2}, \boldsymbol{\beta}_0, \mathbf{V}_0, \mathbf{t}_{L0}, \mathbf{t}_{R0})^{\mathrm{T}}$, one draws a sample from the density $f_\alpha(\boldsymbol{\alpha}|\mathbf{y}, \boldsymbol{\Theta}_0^\alpha)$ given by (3). Denote the sample value of $\boldsymbol{\alpha}$ by $\boldsymbol{\alpha}_1$.

Step 2. With $\boldsymbol{\Theta}_0^\beta = (\sigma_0^{-2}, \boldsymbol{\alpha}_1, \mathbf{V}_0, \mathbf{t}_{L0}, \mathbf{t}_{R0})^{\mathrm{T}}$, one draws a sample from the density $f_\beta(\boldsymbol{\beta}_i|\mathbf{y}, \boldsymbol{\Theta}_0^\beta)$ given by (4). Denote the sample value of $\boldsymbol{\beta}_i$ by $\boldsymbol{\beta}_{i1}, i = 1, \ldots, n$ and put $\boldsymbol{\beta}_1 = \{\boldsymbol{\beta}_i, i = 1, \ldots, n\}^{\mathrm{T}}$.

Step 3. With $\boldsymbol{\Theta}_0^{\sigma^2} = (\boldsymbol{\alpha}_1, \boldsymbol{\beta}_1, \mathbf{V}_0, \mathbf{t}_{L0}, \mathbf{t}_{R0})^{\mathrm{T}}$, one draws a sample of σ_i^2 from the density $f_\sigma(\sigma_i^2|\mathbf{y}, \boldsymbol{\Theta}_0^{\sigma^2})$ given by (5). Denote the sample value of σ_i^2 by $\sigma_{i1}, i = 1, \ldots, n$ and put $\boldsymbol{\sigma}_1 = \{\sigma_{i1}, i = 1, \ldots, n\}^{\mathrm{T}}$.

Step 4. With $\mathbf{\Theta}_0^{(V)} = (\boldsymbol{\sigma}_1, \boldsymbol{\alpha}_1, \boldsymbol{\beta}_1, \mathbf{t}_{L0}, \mathbf{t}_{R0})^{\mathrm{T}}$, one draws a sample of $\mathbf{V}$ from the density $f_V(\mathbf{V}^{-1}|\mathbf{y}, \mathbf{\Theta}_0^{(V)})$ given by (6). Denote the sample value of $\mathbf{V}$ by $\mathbf{V}_1$.

Step 5. With $\mathbf{\Theta}_0^{(L)} = (\boldsymbol{\sigma}_1, \boldsymbol{\alpha}_1, \boldsymbol{\beta}_1, \mathbf{t}_{R0})^{\mathrm{T}}$, one draws a sample of $\mathbf{t}_{Li}$ from the density $f_L(\mathbf{t}_{Li}|\mathbf{y}, \mathbf{\Theta}_0^{(L)})$ given by (7). Denote the sample value of $\mathbf{t}_{Li}$ by $\mathbf{t}_{Li,1}$ and put $\mathbf{t}_{L1} = \{\mathbf{t}_{Li,1}, i = 1, \ldots, n\}^{\mathrm{T}}$.

Step 6. With $\mathbf{\Theta}_0^{(R)} = (\boldsymbol{\sigma}_1, \boldsymbol{\alpha}_1, \boldsymbol{\beta}_1, \mathbf{t}_{L1})^{\mathrm{T}}$, one draws a sample of $\mathbf{t}_{Ri}$ from the density $f_R(\mathbf{t}_{Ri}|\mathbf{y}, \mathbf{\Theta}_0^{(R)})$ given by (8). Denote the sample value of $\mathbf{t}_{Ri}$ by $\mathbf{t}_{Ri,1}$ and put $\mathbf{t}_{R1} = \{\mathbf{t}_{Ri,1}, i = 1, \ldots, n\}^{\mathrm{T}}$.

Step 7. With $\mathbf{\Theta}_0^{\alpha}$ in Step 1 replaced by $\mathbf{\Theta}_0^{\alpha} = (\sigma_1^{-2}, \boldsymbol{\beta}_1, \mathbf{V}_1, \mathbf{t}_{L1}, \mathbf{t}_{R1})^{\mathrm{T}}$, repeat the above cycle until convergence is reached.

As shown in Chapter 3 of Tan (2002), at convergence one generates a sample of size one from the marginal distribution of each parameter and of $\{\mathbf{t}_L, \mathbf{t}_R\}$ respectively. By repeating the above procedures, one generates a random sample of size m from the marginal posterior distribution for each parameter and for $\{\mathbf{t}_L, \mathbf{t}_R\}$ respectively. Then one may compute the sample means and the sample variances and covariances for these parameters. These estimates are optimal in the sense that under squared loss function, the posterior mean values minimize the Bayesian risk function.

4. Analysis of the ACTG 315 Data

Using data from the ACTG 315 clinical trial on HAART, in this section we will derive a full Bayesian analysis to estimate the times to changing points and to drug resistance. In this analysis, we will adopt an empirical Bayesian approach to estimate the prior means and prior variances of the parameters $\boldsymbol{\alpha}$. Because the first change point generally occurs about two weeks, we use the first two weeks data to conduct linear mixed effects modeling via

$$y_{ij} = \eta_{0i} + \eta_{1i} t_{ij} + \varepsilon_{ij}. \tag{9}$$

The estimated values of α_0 and α_1 given here are taken as the means of our prior distributions of α_0 and α_1. The prior distribution of α_2 and α_3 are assumed as both of $N(\bullet|0,1)$. The covariance matrix of α_0 and α_1 obtained from Eq. (9) is taken as the left-upper submatrix of $\mathbf{D}$. The random estimated values are taken as the prior distributions of β_{0i} and β_{1i}.

With the posterior distributions given in Section 2 and with the Gibbs sampling procedure in Section 3, one can readily develop a full Bayesian

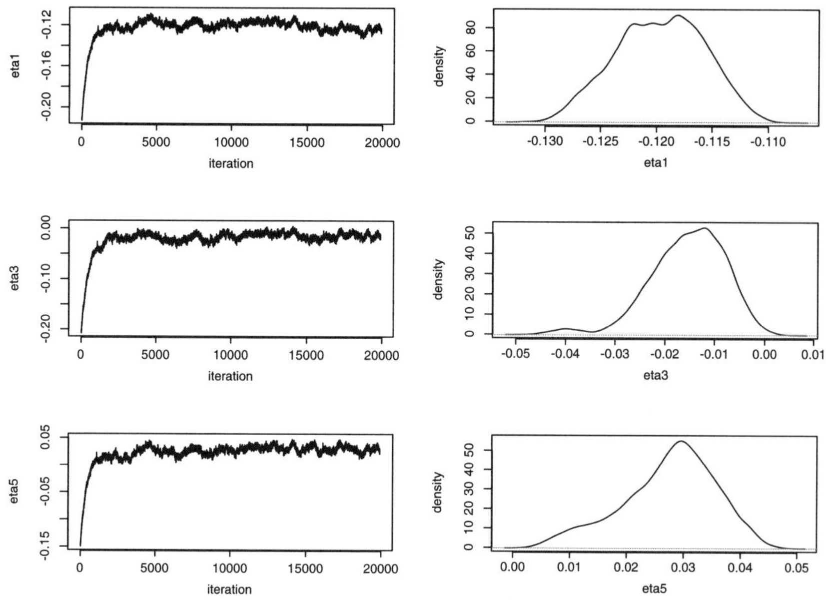

Fig. 3. Diagnostic plots. Left panel: the number of MCMC iterations and posterior means. Right panel: the densities of the posterior means.

analysis to estimate the relevant parameters and the times to changing points. To implement the MCMC sampling scheme we follow the method proposed by Raftery and Lewis (1992). The procedure works as follows: After an initial number of 1000 burn-in iterations, every 10th MCMC sample is retained from the next 200,000 samples. Thus, we obtain 20,000 samples of targeted posterior distributions of the unknown parameters. The stability of the posterior means is checked informally by examining the graphics of the runs. Figure 3 shows the number of MCMC iterations and convergence diagnostics.

Figure 4 shows us a population trend of the viral load trajectory with $\eta_1 = -0.12(0.005)$, $\eta_3 = -0.011(0.006)$, and $\eta_5 = 0.003(0.003)$, $t_L = 15$ and $t_R = 75$. Denote the time period from entry time to 15 days as the first stage and the time period from 15 days to 75 days as the second stage. Figure 4 then implies that the decline rate at the beginning stage is about 10 times of the decline rate of the second stage. After 75 days, the viral load trajectory rebounds slightly.

To characterize the patterns of viral dynamics, we select 4 subjects and examine the trajectories of their viral load. Given in Table 1 are the

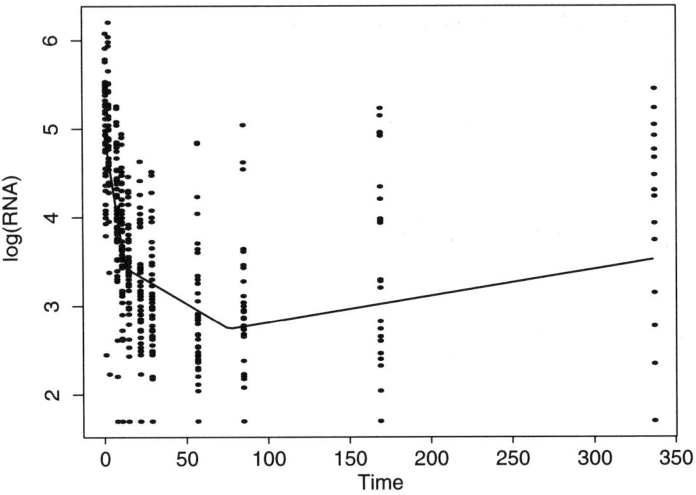

Fig. 4. The estimated population mean curve obtained by using ACTG 315 data, and Bayesian approach. The observed values are indicated by plus.

Table 1. Estimates of the parameters of 4 subjects.

ID	η_0	η_1	η_3	η_5	t_L	t_R
8	4.947	−0.12	−0.003	−0.005	14	75
10	5.38	−0.112	−0.017	0.009	15	76
30	4.741	−0.118	−0.019	0	15	76
48	4.838	−0.144	0.032	0.002	14	76

estimates of the parameters for these 4 subjects. Given in Fig. 5 are the plots of fitted values for these 4 subjects based on the three-segment model and the Bayesian approach. These results indicated that the individual estimates of η_1, t_L, and t_R are quite similar to the corresponding ones of the population estimates respectively. However, the individual estimates of η_3 and η_5 varied widely so that the individual estimates may totally different from the ones of the population. Figure 5 showed that the viral load trajectory of patient 10 is comparable to that of the population but with a stronger rebound in the later stage; but the patterns of viral load trajectory in subjects 8 and 20 are very different from that of the population in that in the second segment, their viral loads continued declining (patient 8) or flat (patient 20) instead of rebounded like the population pattern, suggesting that drug resistance in these patients had not yet been developed.

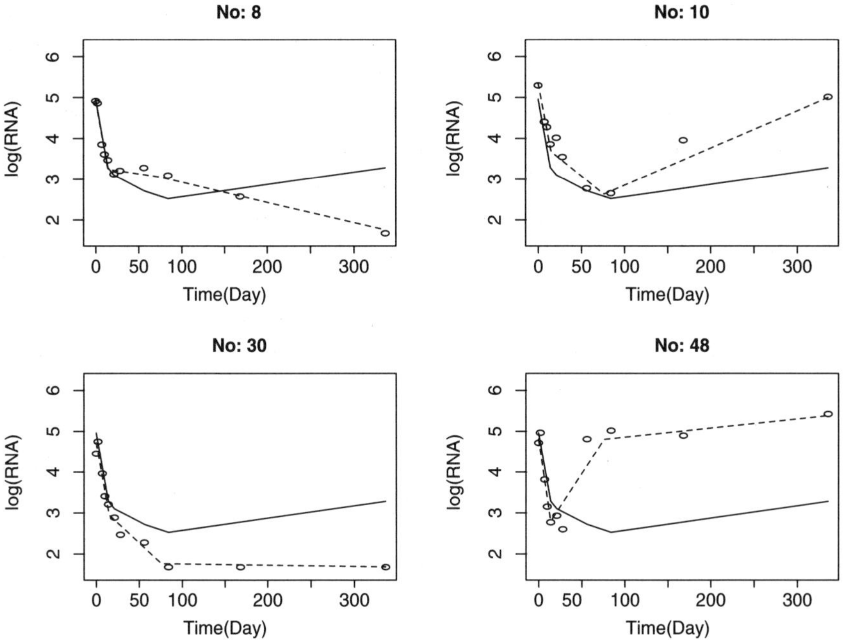

Fig. 5. Profiles of four arbitrarily selected patients for ACTG 315 data. The dotted and solid lines are estimated individual and population curves. The observed values are indicated by the circle signs.

Figure 5 also showed large difference between individual viral load trajectory in patient 48 and that from the population; for this patient, the viral loads rapidly declined in the first stage, then rapidly rebounded in the second segment and kept flatly rebound in the third segment to the end of the treatment. These results indicated that for the clinicians to manage treatment and care of AIDS patients, it is necessary to study the viral load trajectory of each individual. Because the HIV pathogenesis is a stochastic dynamic process, this calls for individual-based models to assess effects of treatment and drug resistance.

5. Conclusion and Discussion

To treat HIV-infected individuals with anti-retroviral drugs, a major obstacle is the development of drug resistance. In order to avoid the problem of drug resistance, it is therefore of considerable interest to estimate the time at which some resistance to the drugs have developed by the HIV. To answer this question, in this paper we have developed a full Bayesian approach by assuming a three-segments linear model with two unknown changing points for the log of RNA viral load. In this model, the first changing point is the time when the macrophage or other long-lived cells release HIV to the blood from lymph nodes (Perelson *et al.*, 1997); the second changing point is the time when the HIV develop resistance to the drugs (Tan, 2000, Chapter 8). This statistical model is motivated and hence dictated by the dynamic models of HIV pathogenesis under treatment and by the observed and simulated data on HIV pathogenesis (Perelson *et al.*, 1997; Tan, 2000, Chapter 8). The Bayesian approach is useful because it can incorporate prior information from empirical study or other related or previous studies. This is important because current clinical data are not enough to identify all viral dynamic parameters.

In this chapter, we have applied the above Bayesian procedure to analyze the ACTG 315 data. This data treated patients with HAART involving 3TC and AZT (NRTI) and ritonavir (PI). For each patient, the data gave RNA viral load per mL of blood at different time points after treatment. Using this data set and the log transformation, we have estimated the expected time to the first changing point as 15 days after treatment and the time to development of drug resistance as 75 days after treatment. The Bayesian HPD intervals for these changing points are $(8.7, 21.3)$ and $(42, 108)$ respectively. For the 3-line segments, the estimate of the slope are given by -0.12, -0.011 and 0.003 respectively. The variances of these estimates are given by 0.005, 0.006 and 0.003 respectively. From this estimates, it is clear that most patients have very much the same slope for the first segment; but the opposite is true for the other 2 segments. Observing the plots of log of RNA load, one may explain this by noting that (i) for some patients, the second segment is very short or hardly noticeable, while for other patients, the second segment is long; and (ii) while drug resistance have developed in some patients (about 21 patients), in other patients drug resistance had not been developed or hardly noticeable. Thus, for the second and third segments, there are considerable variation between individuals.

While we have estimated the time to drug resistance by using a Bayesian approach, in treating HIV-infected individuals with anti-retroviral drugs, many important problems can not be answered by this or any other statistical models. In particular, to monitor the disease progression, to assess efficacy of the drugs and to search for optimal treatment protocols, it is important to estimate effects of different drugs and how these effects are affected by many risk factors such as CD4 T cell counts and CD8 T cell counts; also it is important to estimate the numbers of infectious as well as non-infectious virus loads. Because the drug efficacy depends on pharmacokinetics and the bio-availability of the drugs, it is also of considerable interest to study how the pharmacokinetics affects effects of treatment and drug resistance. Because the HIV pathogenesis is a stochastic dynamic process, this calls for stochastic dynamic models and state space models. Using the ACTG 315 data, we are currently developing stochastic dynamic model and state space models to answer these questions. We will not go any further here.

Acknowledgments

Liang and Xiong's research was partially supported by the American Lebanese Syrian Associated Charities (ALSAC). Liang's research was also partially supported by a grant from the National Institute of Allergy and Infectious Diseases (R01 AI062247-01).

References

1. Han C, Chaloner K and Perelson AS (2002) Bayesian analysis of a population HIV dynamic model, in *Case Studies in Bayesian Statistics.* Vol. 6. Springer-Verlag, New York.
2. Ho DD, Neumann AU, Perelson AS, Chen W, Leonard JM and Makowitz M (1995) Rapid turnover of plasma virus and CD4 lymphocytes in HIV-1 infection, *Nature* **373**: 123–126.
3. Kitagawa G (1998) A self organizing state space model, *J Am Stat Assoc* **93**: 1203–1215.
4. Kiuchi AS, Hartigan JA, Holford TR, Rubinstein P and Stevens CE (1995) Change points in the series of T4 counts prior to AIDS, *Biometrics* **51**: 236–248.
5. Lange N, Carlin BP and Gelfand AE (1992) Hierarchical Bayes models for the progression of HIV infection using longitudinal CD4 T-cell numbers (with discussion), *J Am Stat Assoc* **87**: 615–632.
6. Lederman MM, Connick E, Landay A, Kuritzkes DR, Spritzler J, St Clair M, Kotzin BL, Fox L, Chiozzi MH, Leonard JM, Rousseau F, Wade M, D'Arc Roe J, Martinez A and Kessier H (1998) Immunological responses associated with 12 weeks of combination antiretroviral therapy consisting of Zidovudine, Lamivudine, and Ritonavir: results of AIDS clinical trial group protocol 315, *J Infect Diseases* **178**:70–79.
7. Liu JS and Chen R (1998) Sequential Monte Carlo method for dynamic systems, *J Am Stat Assoc* **93**: 1032–1044.
8. Mellors JW, Kingsley LA, Rinaldo CR *et al.* (1995) Quantitation of HIV-1 RNA in plasma predicts outcome after seroconversion, *Ann Inter Med* **122**: 573–579.

9. Mellors JW, Rinaldo CR, Gupta P *et al.* (1996) Prognosis in HIV-1 infection predicted by the quantity of virus in plasma, *Science* **272**: 1167–1170.
10. Perelson AS, Essunger O, Cao YZ, Vesanen M, Hurley A, Saksela K, Markowitz M and Ho DD (1997) Decay characteristics of HIV infected compartments during combination therapy, *Nature* **387**: 188–191.
11. Putter H, Heisterkamp SH, Lange JMA and De Wolf F (2002) A Bayesian approach to parameter estimation in HIV dynamical models, *Stat Med* **21**: 2199–2214.
12. Raftery AE and Lewis S (1992) How many iterations in the Gibbs sample? in Bernardo J, Berger J, Dawid A and Smith A (eds) *Bayesian Statistics 4*, pp. 763–773. Oxford University Press, Oxford.
13. Saag MS, Holodniy M, Kuritzkes DR, O'Brien WA, Coombs R, Poscher ME, Jacobsen DM, Shaw GM, Richman DD and Volberding PA (1996) HIV viral load markers in clinical practice, *Nat Med* **2**: 625–629.
14. Shephard N (1994) Partial non-Gaussian state space model, *Biometrika* **81**: 115–131.
15. Tan WY (2000) *Stochastic Modeling of AIDS Epidemiology and HIV Pathogenesis.* World Scientific, New Jersey.
16. Tan WY (2002) *Stochastic Models with Applications to Genetics, Cancers, AIDS and other Biomedical Systems.* World Scientific, New Jersey.
17. Wei X, Ghosh SK, Taylor ME, Johnson VA, Emini EA, Deutsch P, Lifson JD, Bonhoeffer S, Nowak MA, Hahn BH, Saag MS and Shaw GM (1995) Viral dynamics in human immunodeficiency virus type 1 infection, *Nature* **373**: 117–122.
18. Wu H and Ding A (1999) Population HIV-1 dynamics *in vivo*: applicable models and inferential tools for virological data from AIDS clinical trials, *Biometrics* **55**: 410–418.
19. Wu H, Kuritzkes DR, McClernon DR *et al.* (1999) Characterization of viral dynamics in human immunodeficiency virus type 1-infected patients treated with combination antiretroviral therapy: relationships to host factors, cellular restoration and virological endpoints, *J Infect Diseases* **179**: 799–807.

CHAPTER 20

ESTIMATING HIV INCIDENCE FROM A CROSS-SECTIONAL SURVEY WITH THE LESS SENSITIVE ASSAY

Robert H. Byers, Jr., Dale J. Hu and Robert S. Janssen

The use of a less sensitive or detuned test for recent human immunodeficiency virus (HIV) infection requires some modification to the standard incidence estimator. We use a stochastic simulation to demonstrate the details of using a detuned test to estimate incidence. We calculate the probabilities of the outcomes and give the result of the simulation as an example. Two estimators are compared and recommendations are made for calculating confidence intervals. We also show that the window period should be chosen to be half the collection period in order to minimize bias in the incidence estimator.

Keywords: HIV; incidence; less sensitive assay; detuned assay.

1. Introduction

Estimation of incidence is an important component for surveillance and monitoring the global HIV pandemic in different populations around the world. In addition, once medication that prolonged the lives of infected persons became widely available, identification of persons with early HIV infection became increasingly important. Early identification allows us to focus prevention efforts and to improve opportunities for early HIV therapy and prevention of opportunistic infections. However, identification of recently infected persons and accurate estimation of incidence are extremely difficult and have principally relied on the prospective testing and longitudinal follow-up of cohorts of individuals at risk. Besides the obvious logistical challenges of such studies, estimates of incidence based on prospective cohorts are biased for a number of reasons (Brookmeyer *et al.*, 1987).

Brookmeyer and Quinn (1995) suggested that the time from infection to appearance of p24 antigen could be used to estimate incidence rates. This approach is based on early diagnostic tests to identify HIV p24 antigen before the appearance of detectable antibodies to HIV such that someone who has detectable p24 antigen but is still HIV seronegative is assumed to be recently infected. A disadvantage of this approach is that large samples or high incidence would be needed because the time before the appearance of p24 antigen is short.

There is a time after infection with human immunodeficiency virus (HIV) when antibodies to the virus are not detectable. The appearance of detectable antibodies is termed "seroconversion". Observing that recent HIV tests can detect antibody sooner than the earlier tests, Janssen *et al.* (1998) have suggested creating a test that is deliberately less sensitive to make the time of seroconversion later than on a standard test. We define the time from seroconversion on the sensitive test to seroconversion on the less sensitive test as the "window period". By "seroconversion on the less-sensitive test", we mean that the standardized optical density (SOD) measured by the test rises above a pre-specified cut-off. The window period varies from person to person, but its mean can be adjusted by varying the SOD cut-off. If a specimen from an HIV-infected person tests positive on a sensitive test and negative (below the SOD cut-off) on the less sensitive test, you could conclude that the person probably was infected within a known time window. This allows the estimation of incidence using a cross-sectional survey.

Reported HIV-1 incidence can vary greatly in different population groups. In North America, the incidence of HIV-1 during the 1990s has been extremely low (less than 0.01/100 person-years) among low risk groups such as blood donors (Janssen *et al.*, 1998; Glynn *et al.*, 1991). High incidence has been reported among men who have sex with men (MSM) and injection drug users (IDU). For example, the incidence of HIV-1 among MSM who participated in a hepatitis B vaccine trial in San Francisco was very low (0.3/100 person-years) in 1978 during the early part of the epidemic, but increased to almost 20/100 person-years by 1982 before declining to low levels by 1987 (Hessol *et al.*, 1989). Similarly, the incidence among IDU has decreased in the 1990s in large urban areas such as New York City to under 3/100 person-years (Des Jarlais *et al.*, 2000). In contrast, an explosive epidemic of HIV-1 was documented among IDU in Bangkok, Thailand during the late 1980s where incidence was estimated to be as high as 20 to almost 60 per 100 person-years (Weniger *et al.*, 1991).

To understand the issues involved in estimating incidence with a cross-sectional survey using the less sensitive test, we constructed two stochastic simulations of an epidemic. The first epidemic has a low incidence of 3.3%, and the second of a high incidence of 15.4%.

Section 2 describes the possible outcomes of a survey using the less-sensitive assay. In Section 3, we derive the probabilities of the various outcomes of the test. In Section 4, we motivate the use of a previously described estimator, and use the theory in Section 3 to deduce a new estimator for incidence. Section 5 describes two stochastic simulations of a survey which are used to compare the two estimators. The first simulation is for an epidemic with a low incidence of 3.3%, and the second has a high incidence of 15.4%. We present the results of the simulations and show that they agree with the theory. In Section 6, we discuss the variability of two estimates of incidence and recommend a preferred estimator and method for calculating confidence intervals.

2. Outcomes of a Survey with the Less Sensitive Assay

We imagine that N people are tested during a testing period that starts on a day designated as τ_0 and ends at day τ_1. The tests are uniformly distributed throughout the year. The tested people can be classified into mutually exclusive and exhaustive categories according to their seroconversion dates (s), their test dates (t), the lengths of their window periods (ω), and the beginning and ending dates of the test period (τ_0 and τ_1, respectively). Seroconversion dates can be classified into 3 categories: $s < \tau_0$ (prevalent), $\tau_0 \leq s \leq \tau_1$ (incident), and $s > \tau_1$ (negative). We assume that everyone with $s < t$ will test positive on the standard HIV test and we will call them "HIV+". It is possible for someone who seroconverts in the testing period or later to have a negative HIV test if $s > t$; we will call them "HIV−". (Since we cannot distinguish between those who seroconvert after testing and those who are never infected, we put them in the same category.) If $s + \omega \leq t$, the less-sensitive test will be positive and the person will be classified as "LS+". If $s + \omega > t$, the less sensitive test will be negative and the person will be classified as "LS−" and considered recently infected. The counts can be arrayed in the following table:

Table 1. Possible outcomes for tests.

	$s < \tau_0$	$\tau_0 \leq s \leq \tau_1$	$s > \tau_1$
$s < t$, $s + \omega > t$, HIV+, LS−	n_{11}	n_{12}	0
$s < t$, $s + \omega \leq t$, HIV+, LS+	n_{21}	n_{22}	0
$s > t$, HIV−	0	n_{32}	n_{33}

If we could observe the individual entries in this table, estimation of incidence would be easy. But after a cross-sectional survey, we will know the sums of the rows of this table. We will not know the individual entries nor the sums of the columns.

3. Probabilities of the Outcomes

We assume that seroconversion day s has a probability density $h(s)$ with cumulative distribution function (CDF) $H(s)$. The time from seroconversion to the day a less sensitive test would be positive has a density $g(s)$ with CDF $G(s)$. The distribution of a sum of random variables is a convolution, so the probabilities for the outcomes are:

$$P(n_{11}) = \frac{\int_{\tau_0}^{\tau_1} \int_0^{\tau_0} h(s)(1 - G(t-s))\, ds\, dt}{\tau_1 - \tau_0} \tag{1}$$

$$P(n_{12}) = \frac{\int_{\tau_0}^{\tau_1} \int_{\tau_0}^{t} h(s)\,(1 - G(t-s))\, ds\, dt}{\tau_1 - \tau_0} \tag{2}$$

$$P(n_{21}) = \frac{\int_{\tau_0}^{\tau_1} \int_0^{\tau_0} h(s)\, G(t-s)\, ds\, dt}{\tau_1 - \tau_0} \tag{3}$$

$$P(n_{22}) = \frac{\int_{\tau_0}^{\tau_1} \int_{\tau_0}^{t} h(s)\, G(t-s)\, ds\, dt}{\tau_1 - \tau_0} \tag{4}$$

$$P(n_{32}) = \frac{\int_{\tau_0}^{\tau_1} (s - \tau_0)\, h(s)\, ds}{\tau_1 - \tau_0} \tag{5}$$

$$P(n_{33}) = 1 - H(\tau_1) \tag{6}$$

4. Estimation of Incidence

An estimator for incidence from Table 1 is:

$$I = \frac{n_{12} + n_{22} + n_{32}}{n_{12} + n_{22} + n_{32} + n_{33}} \tag{7}$$

We could use this as an estimator of the population incidence if we knew the sums of the columns of Table 1. In reality, we only know the sums of the rows. Now the expected value of the first row is

$$\mathcal{E}(n_{11}+n_{12}) = N\frac{\int_{\tau_0}^{\tau_1}\int_0^t h(s)(1-G(t-s))\,ds\,dt}{\tau_1-\tau_0} \tag{8}$$

We assume that incidence is fairly constant over the testing interval so that we can apply the mean value theorem:

$$h(s) \approx \frac{H(\tau_1)-H(\tau_0)}{\tau_1-\tau_0}. \tag{9}$$

Now Eq. (8) becomes

$$\begin{aligned}\mathcal{E}(n_{11}+n_{12}) &\approx N\frac{H(\tau_1)-H(\tau_0)}{\tau_1-\tau_0}\frac{\int_{\tau_0}^{\tau_1}\int_0^t (1-G(t-s))\,ds\,dt}{\tau_1-\tau_0}\\ &= \frac{\bar{\omega}}{\tau_1-\tau_0}[H(\tau_1)-H(\tau_0)]N,\end{aligned} \tag{10}$$

where $\bar{\omega}$ is the mean window period.

The expected value of the numerator of incidence (7) is

$$\mathcal{E}(n_{12}+n_{22}+n_{32}) = [H(\tau_1)-H(\tau_0)]N \tag{11}$$

Substituting (11) into (7) and applying (10) to the numerator gives

$$\hat{I}_1 = \frac{(\frac{\tau_1-\tau_0}{\bar{\omega}})(n_{11}+n_{12})}{n_{12}+n_{22}+n_{32}+n_{33}}. \tag{12}$$

We are tempted to apply the same approximation to the denominator, but we do not know n_{33}. We do know $n_{32}+n_{33}$, so we need an adjustment for $n_{12}+n_{22}$. From (2) and (4), we have

$$\begin{aligned}\mathcal{E}(n_{12}+n_{22}) &= \frac{N}{\tau_1-\tau_0}\int_{\tau_0}^{\tau_1}\int_{\tau_0}^{t} h(s)\,ds\,dt\\ &= \frac{N}{\tau_1-\tau_0}\int_{\tau_0}^{\tau_1}[H(t)-H(\tau_0)]\,dt.\end{aligned} \tag{13}$$

Since $H(t)$ is nearly linear, and from (10)

$$\mathcal{E}(n_{12}+n_{22}) \approx N\frac{H(\tau_1)-H(\tau_0)}{2} \approx \mathcal{E}(n_{11}+n_{12})\frac{\tau_1-\tau_0}{2\bar{\omega}}. \tag{14}$$

This suggests an estimator of the form

$$\hat{I}_2 = \frac{(\frac{\tau_1 - \tau_0}{\bar{\omega}})(n_{11} + n_{12})}{(\frac{\tau_1 - \tau_0}{2\bar{\omega}})(n_{11} + n_{12}) + n_{32} + n_{33}}. \quad (15)$$

If $\bar{\omega}$ is chosen to be half the testing interval, the multiplier in the denominator is 1. This is the estimator used by Janssen *et al.* (1998):

$$\hat{I}_1 = \frac{(\frac{\tau_1 - \tau_0}{\bar{\omega}})(n_{11} + n_{12})}{n_{11} + n_{12} + n_{32} + n_{33}}. \quad (16)$$

The multiplier $(\frac{\tau_1 - \tau_0}{\bar{\omega}})$ can be thought of as an inflation factor that estimates the number of seroconverters from the number LS−.

5. Results of the Simulations

The theoretical probability for each cell in Table 1 can be calculated using the assumed probability densities. We used a probability distribution based on a stochastic simulation of the HIV epidemic given in Tan and Byers (1993). The probability of seroconversion is given by

$$P(s \leq S) = H(S) = \left(1 + e^{-\frac{(log(S) - \mu)}{\sigma}}\right)^{-\lambda}, \quad (17)$$

with $\mu = 8.567$, $\sigma = 0.094$, and $\lambda = 0.225$. Time is measured in days, with time zero arbitrarily set seventeen years before the testing interval.

We modeled distribution of the window period as a generalization of the log-logistic distribution (Pack *et al.*, 1990):

$$P(\omega \leq w) = G(w) = 1 - \left(1 + \tau e^{\frac{(log(w) - \nu)}{\rho}}\right)^{-\frac{1}{\tau}}, \quad (18)$$

where $\nu = 5.023$, $\rho = 0.225$, $\tau = 0.805$. These parameters are estimated using the data described in Parekh *et al.* (2001), and give a mean window period of 158 days. The dimensions of μ and ν are log(days). The corresponding probability density functions are $h(s)$ and $g(w)$, respectively.

The probability of a test date is given by:

$$P(t \leq T) = \frac{T - \tau_0}{\tau_1 - \tau_0} \quad (19)$$

and $\tau_0 = 1460$, $\tau_1 = 1825$.

The probabilities were calculated using numerical integration in *Mathematica®* (Wolfram, 1991). Here are the results:

Table 2. Probabilities for outcomes.

	$s < \tau_0$	$\tau_0 \leq s \leq \tau_1$	$s > \tau_1$	Total
$s < t$, LS−, $s + \omega > t$	0.003	0.010	0.0	0.013
$s < t$, LS+, $s + \omega \leq t$	0.044	0.005	0.0	0.049
$s > t$, HIV−	0.0	0.017	0.920	0.937
Total	0.047	0.033	0.920	1.000

Since we know the underlying mechanism of the simulated epidemic, we can calculate the true incidence for comparison with other estimators.

$$I = \frac{H(\tau_1) - H(\tau_0)}{1 - H(\tau_0)} = 0.0346 \tag{20}$$

Using the theoretical values from Table 2, the two estimators give:

$$\hat{I}_1 = \frac{0.0132241}{0.0132241 + 0.9377087} \times \frac{365}{158} = 0.0322,$$

and

$$\hat{I}_2 = \frac{0.0132241 \times \frac{365}{158}}{\frac{365}{2 \times 158} \times 0.0132241 + 0.9377087} = 0.0321.$$

We repeated the simulation 1000 times with the following results. The cells contain the mean counts for 2500 tested each time. These agree very well with the theoretical percentages (Table 2).

Table 3. Mean counts from 1000 repetitions.

	$s < \tau_0$	$\tau_0 \leq s \leq \tau_1$	$s > \tau_1$
$s + \omega > t$, LS−	7.084	25.876	0.0
$s + \omega \leq t$, LS+	109.053	12.969	0.0
$s > t$, HIV−	0.0	43.821	2301.197

Table 4. Percents from mean counts.

	$s < \tau_0$	$\tau_0 \leq s \leq \tau_1$	$s > \tau_1$
$s + \omega > t$, LS−	0.00283	0.01035	0.0
$s + \omega \leq t$, LS+	0.04362	0.00519	0.0
$s > t$, HIV−	0.0	0.01752	0.920498

The mean incidence for the 1000 simulations is 0.0346 with a standard deviation of 0.00370. For $\hat{I}_1$ the mean and standard deviation are 0.0318 and 0.00549; for $\hat{I}_2$ they are 0.03219 and 0.00552. Figure 1 shows a nonparametric density estimate of the probability densities of the two statistics.

The difference between these estimators is small when incidence is low. What about relatively high incidence? We simulated 1000 more epidemics, this time with a theoretical incidence of 15%. To do this we changed the parameters of the seroconversion distribution to $\mu = 8.57$, $\sigma = 0.094$, and $\lambda = 0.225$, and set $\tau_0 = 3465$ and $\tau_1 = 3830$. We started with 2500 people at risk and allowed all of them to eventually seroconvert. The expected values of the outcomes are in Table 5.

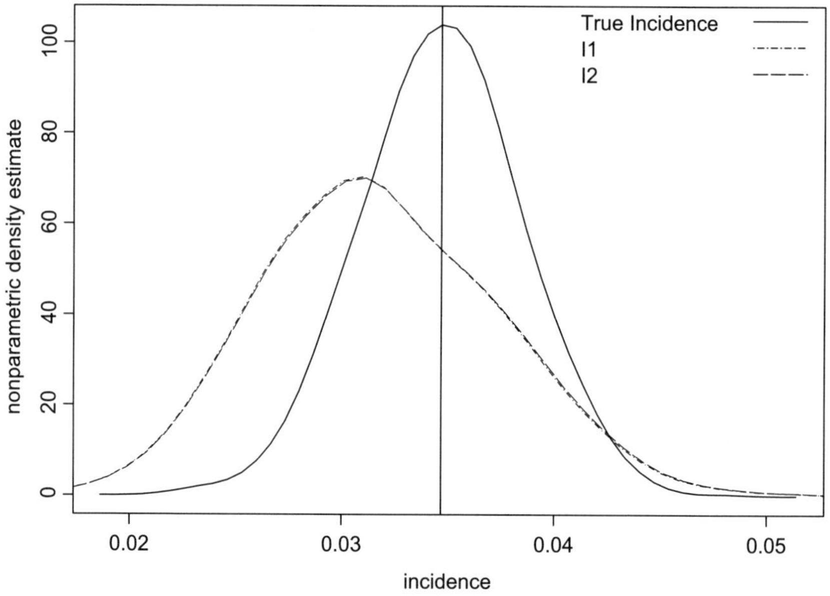

Fig. 1. Estimates when incidence is low (3.46%).

Table 5. Expected values for high incidence simulation.

	$s < \tau_0$	$\tau_0 \leq s \leq \tau_1$	$s > \tau_1$
$s + \omega > t$, LS−	24.263	78.524	0.0
$s + \omega \leq t$, LS+	894.893	40.698	0.0
$s > t$, HIV−	0.0	124.491	1337.13

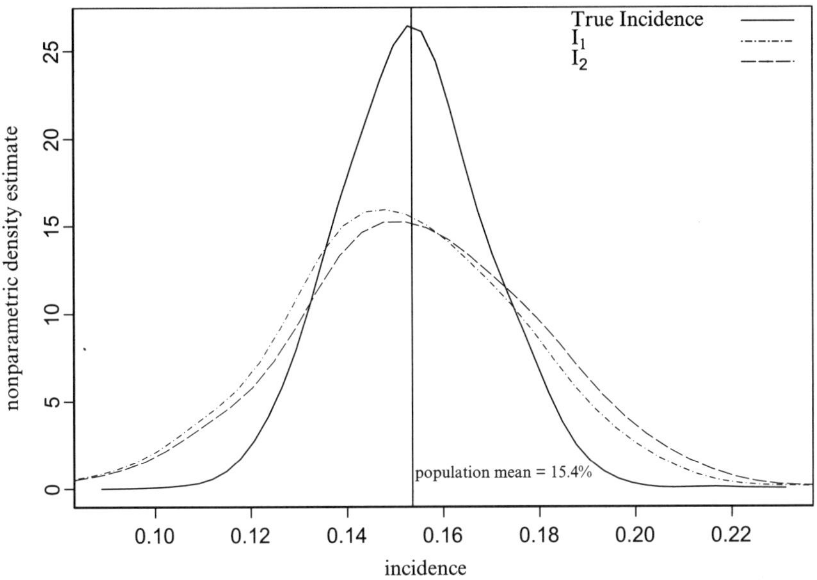

Fig. 2. Estimates when incidence is high (15.4%).

The means from the simulation are: true incidence $I = 0.1541$; $\hat{I}_1 = 0.1497$; $\hat{I}_2 = 0.1511$; with standard deviations of 0.00958, 0.01373, and 0.01401, respectively. The probability densities are plotted in Fig. 2.

6. Confidence Intervals

Satten suggested a method of calculating confidence intervals for $\hat{I}_1$ incidence based on a probability inequality due to Bonferroni. This takes into account the variability in the mean window period and the number of LS−tests. To calculate the Bonferroni limits, we use 97.5% confidence limits on the mean window period. These are $\omega_- = 152$, and $\omega_+ = 163$, respectively; these limits are from a bootstrap estimate using the Parekh data. We assume that the numerator ($n_{ls} = n_{11} + n_{12}$) has a Poisson distribution and calculate the 97.5% limits of that. Call these n_- and n_+. The approximate 95% confidence limits are then

$$\left(\frac{365\, n_-}{N\, \omega_+}, \frac{365\, n_+}{N\, \omega_-} \right), \tag{21}$$

where N is the number detuned plus the number negative.

Another way to calculate confidence intervals was motivated by observing in the simulations that the estimator $\hat{I}_2$ is symmetrically distributed with skewness nearly zero and kurtosis nearly 3. A normal approximation could be used if an estimate of the variance of $\hat{I}_1$ were available. Using the conditional variance formula and Taylor series approximations:

$$V(\hat{I}_1) \approx \frac{n_{ls}}{\bar{\omega}^2}\left(\frac{\tau_1 - \tau_0}{n_{11}+n_{12}+n_{32}+n_{33}}\right)^2 = \hat{I}_1^2/n_{ls}. \tag{22}$$

In the 15% simulation the observed variance of 1000 estimates was 0.00003012. Equation (22) gives 0.00003037. A confidence interval with approximately 95% coverage is:

$$\hat{I} \pm 1.96\,\hat{I}/\sqrt{n_{ls}}. \tag{23}$$

where 1.96 is the 97.5% quantile of the standard normal distribution. In a simulation of 1000 samples from a population with annual incidence of 3.1% we found the coverage to be 96.6%; the Bonferroni gave 98.8%.

7. Bias

The $\hat{I}_2$ estimator is slightly biased. The bias is given by

$$\begin{aligned}\hat{B} &= I - \hat{I}_2 \\ &= \frac{n_{12}+n_{22}+n_{32}}{n_{12}+n_{22}+n_{32}+n_{33}} - \frac{(\frac{\tau_1-\tau_0}{\bar{\omega}})(n_{11}+n_{12})}{(\frac{\tau_1-\tau_0}{2\bar{\omega}})(n_{11}+n_{12})+n_{32}+n_{33}}.\end{aligned} \tag{24}$$

Using the assumption that testing is done uniformly over the collection period and that seroconversion occurs uniformly over the collection period, we can substitute $n_{12}+n_{22}+n_{32}$ for $2(n_{11}+n_{22})$ in the numerator of $\hat{I}_2$. This is because about half the seroconverters will be tested after they seroconvert, i.e., $n_{32} \approx n_{12}+n_{22}$. This gives

$$\hat{B} = \frac{-I((n_{22}-n_{11})\Delta t + (\Delta t - 2\bar{\omega})(n_{32}+n_{33}))}{((n_{11}+n_{12})\Delta t + 2\bar{\omega}(n_{32}+n_{33}))}, \tag{25}$$

where $\Delta\tau = \tau_1 - \tau_0$.

From Eq. (14) we see that, if $2\bar{\omega} = \Delta\tau$, then $\mathcal{E}(n_{12}+n_{22}) \approx \mathcal{E}(n_{11}+n_{12})$, so n_{11} will on average be close to n_{22} and $\hat{B}$ will be zero. Based on this observation, we recommend that the optical density should be chosen to set the mean window period to one-half the collection interval. This should give a bias of zero. (This has the trivial benefit that the correction $(\frac{\tau_1-\tau_0}{2\bar{\omega}})$ in the denominator of $\hat{I}_2$ can be neglected.)

8. Discussion

The basic strategy for estimating incidence with a less sensitive assay is to apply an inflation factor derived from the probability that a seroconverter will test HIV positive and less sensitive negative. This factor is approximately the length of the test interval divided by the mean window period, assuming incidence is constant over the testing interval. We have considered two ways to do this. The first $(\hat{I}_1)$ is to calculate incidence using the number LS− and then multiply the incidence by the inflation factor. The second $(\hat{I}_2)$ is to adjust the number of LS− in the numerator and the denominator. We have shown that the correct factor in the denominator is half the adjustment for the numerator.

In order to minimize the bias in the estimator, we have shown that the optical density cut-off should be set so that the mean window period is half the collection interval. This makes the denominator adjustment equal to 1 and the two estimators are the same.

The concepts of sensitivity, specificity and positive predictive value are used to measure the accuracy of a biological test. We have been unable to develop definitions of these terms that are sensible in the context of the less-sensitive assay.

The standard definition of sensitivity is the proportion of tests that are positive among people who actually have a disease. In this case, we are trying to determine if someone is within $\bar{\omega}$ days of seroconversion, so we might try as a definition: "the proportion of people whose window period is less than average among those whose SOD is less than the cut-off: $P(t - s < \hat{\omega}|SOD < c)$". In this view, someone whose window period is less than average, i.e. $(SOD > c$ and $t - s < \hat{\omega})$, is considered an error. But window periods vary: 60% of the population will have windows less than the mean no matter what cut-off is chosen. It's perfectly natural that some proportion of those will have SOD greater than the cut-off when they are tested. This definition depends on the time of testing, which is independent of the accuracy of the test.

For purposes of estimating incidence, all that is needed is knowledge of the distribution of the window period, which is determined by the optical density cut-off. The concepts of sensitivity and specificity are unnecessary in context of testing with the less-sensitive assay.

References

1. Brookmeyer R and Gail MH (1987) Biases in prevalent cohorts. *Biometrics* **43**(4): 739–750.
2. Brookmeyer R and Quinn TC (1995) Estimation of current human immunodeficiency virus incidence rates from a cross-sectional survey using early diagnostic tests. *Amer J Epidemiol* **141**: 166–172.
3. Janssen RS, Satten GA, Stramer SL, Rawal BD *et al.* (1998) New testing strategy to detect early HIV-1 infection for use in incidence estimates and for clinical and prevention purposes. *J Amer Med Assoc* **280**(1): 42–48.
4. Glynn SA, Kleinman SH, Schrieber GB, Busch MP, Wright DJ, Smith JW *et al.* (2000) Trends in incidence and prevalence of major transfusion-transmissible viral infections in U.S. blood donors, 1991 to 1996. *J Amer Med Assoc* **284**: 229–235.
5. Hessol NA, Lifson AR, O'Malley PM, Doll LS, Jaffe HW and Rutherford GW (1989) Prevalence, incidence and progression of human immunodeficiency virus infection in homosexual and bisexual men in hepatitis B vaccine trials, 1978–1988. *Am J Epidemiol* **130**: 1167–1175.
6. Des Jarlais DC, Marmor M, Friedmann P, Titus S, Aviles E, Deren S *et al.* (2000) HIV incidence among injection drug users in New York City, 1992–1997: evidence for a declining epidemic. *Am J Public Health* **90**: 352–359.
7. Weniger BG, Limpakarnjanarat K, Ungchusak K, Thanprasertsuk S, Choopanya K, Vanichseni S *et al.* (1991) The epidemiology of HIV infection and AIDS in Thailand. *AIDS* **5**: S71–S85.
8. Tan W-Y and Byers RH Jr. (1993) A stochastic model of the HIV epidemic and HIV infection distribution in a homosexual population. *Math Biosci* **113**: 115–143.
9. Pack SE and Morgan BJT (1990) A mixture model for interval-censored time-to-response quantal assay data. *Biometrics* **46**(3): 749–757.
10. Parekh B, Kennedy MS, Dobbs T, Pau C-P, Byers RH, Green T *et al.* (2001) Quantitative detection of increasing HIV-1 antibodies following seroconversion: A simple assay for detecting recent HIV infection and estimating incidence. *AIDS Res Hum Retroviruses* **18**(4): 295–397.
11. Wolfram S (1991) *The Mathematica Book*, 4th ed. New York.

CHAPTER 21

DESIGN OF POPULATION STUDIES OF HIV DYNAMICS

Cong Han and Kathryn Chaloner

This chapter reviews design issues for studies of HIV dynamics that use a nonlinear mixed-effects model. Both the choice of sampling times and also the trade-off between the number of sampling times and the number of subjects are discussed. A method based on a first-order approximation, used for similar design issues in pharmacokinetics and pharmacodynamics, is reviewed and discussed. Limitations of this method are discussed and other methods are described, including methods based on the exact calculation of the Fisher information matrix and Bayesian methods, which, although computationally intensive, provide alternatives. A computationally intensive Bayesian method is recommended.

Keywords: Bayesian methods; Fisher information; Gaussian quadrature; Markov chain Monte Carlo; pseudo-mixed-effects; mixed-effects; random effects.

1. Introduction

HIV viral load refers to the concentration of circulating virion-associated HIV RNA in plasma. HIV/AIDS clinical trials that evaluate antiviral therapies often use changes in HIV viral load over time as one of the outcome variables to measure treatment effectiveness. Studies of HIV dynamics examine repeated measurements, over time, of HIV viral load, to understand the mechanism of HIV replication and antiviral therapies.

When a study is conducted in more than one subject, the subjects are usually heterogeneous. Such situations arise in population studies of virus dynamics (Wu, Ding, and de Gruttola, 1998). With such data, application of the nonlinear mixed-effects models is appropriate. Such models stipulate an underlying distribution of the parameters for each subject. Each subject's parameters, or random effects, are a realization from this distribution of parameters. Each observed response is a realization from the conditional distribution of the data, conditioning on the individual's parameter. Hence, in the nonlinear mixed-effects model approach, both the similarity and the difference across the subjects are modeled. Application of this approach to HIV dynamics can be found in Wu and Ding (1999). Han, Chaloner and Perelson (2002) and Putter *et al.* (2002) analyze population HIV dynamics data using a Bayesian nonlinear mixed-effects model.

An important design issue in studies of HIV dynamics is the choice of a set of time points at which blood samples should be drawn and viral load should be measured. Different choices of time points lead to different efficiency in the estimation of HIV dynamic parameters. Experimental design for nonlinear mixed-effects models, however, has been a challenging statistical problem.

In this paper, previous work in the design for nonlinear mixed-effects models will be reviewed, some new results will be given, and recent developments will be summarized. The materials will be presented as they relate to studies of HIV dynamics.

2. Design via Simulation

Wu and Ding (2002) adopt a simulation approach to the design issue in HIV dynamics. Their work is briefly reviewed in this section.

Under antiviral treatment, after an initial "shoulder" of about 2 days, the trajectory of viral load can be described approximately by a random-effects nonlinear regression model (Ding and Wu, 1999). That is, let t_c denote the time of the end of the shoulder, then for all $t \geq t_c$, the viral load $V(t)$ can be approximated by an bi-exponential model, $P_1 \exp(-d_1 t) + P_2 \exp(-d_2 t)$. Hence, a possible statistical model for viral load data, taking into account the fact that substantial individual differences in viral dynamics may exist, is as follows:

1. For individual i, the parameter θ_i follows a multivariate normal distribution; across individuals, the parameters are independent and identically distributed (*iid*).
2. Given the individual parameter $\theta_i = (P_{1i}\ P_{2i}\ d_{1i}\ d_{2i})^T$, and assuming that exactly the same sampling schedule is to be used for each subject, the logarithm of the jth measurement of the ith individual is
$$y_{ij} = \log[P_{1i} \exp(-d_{1i} t_j) + P_{2i} \exp(-d_{2i} t_j)] + \varepsilon_{ij}, \tag{1}$$
with $\varepsilon_{ij} \overset{iid}{\sim} N(0,\ \sigma^2)$.

The design problem is the choice and number of the t_j's and the number of subjects. Wu and Ding (2002) explore the design issue via simulation. In particular, they consider a set of four designs that are regarded as "practical", and simulate data for two different treatment groups according to the four different designs. The model under which the data are generated is slightly different from (1): it has an extra compartment to represent the shoulder:

$$y_{ij} = \log[P_{0i}\exp(-d_{0i}t_j) + P_{1i}\exp(-d_{1i}t_j) + P_{2i}\exp(-d_{2i}t_j)] + \varepsilon_{ij}, \quad (2)$$

with the obvious modification in the between-subject variation.

For each simulation run, Wu and Ding fit the bi-exponential model, model (1), instead of model (2), to the data, and predictions for the individual parameters and some functions of them are obtained. Non-parametric tests are performed on functions of individual parameters and then the empirical powers of the tests are used to evaluate the designs. Based on the parameter values they use, they recommend that first, weekly measurements be taken during the first month of treatment, second, the parameter d_1 and the viral load change in the first week be used as markers of drug potency, and third, their approach be taken when designs are investigated in other studies.

Wu and Ding (2002) argue that formal optimality criteria (such as D-optimality) are inappropriate since first, these (local) optimality criteria usually result in no more design points than the unknown parameters so that model validation is impossible, and second, they may lead to infeasible designs.

The following issues, however, should also be observed:

1. Bayesian optimal designs often result in more design points than unknowns (Chaloner and Larntz, 1989; Chaloner and Verdinelli, 1995).
2. Even if the optimal design is infeasible, it may still serve as a benchmark against which feasible designs can be evaluated.
3. The traditional design criteria such as D- or A-optimality can still be used when the set of designs is finite.

3. A First-order Approximation

Assume that there are n subjects from whom data are obtained, and m observations are obtained from each subject at time points t_j, $j = 1, \ldots, m$; then also assume that the random effects form a p-vector, that the measurement errors are independent and identically distributed from a univariate normal distribution, and that the random effects are independent and identically distributed from a multivariate normal distribution. The jth observation on the ith subject is then

$$y_{ij}|\theta_i, \sigma^2 \overset{ind}{\sim} N\left(r(\theta_i, t_j), \sigma^2\right), \quad \theta_i \overset{iid}{\sim} N_p(\mu, \Sigma). \tag{3}$$

It is assumed that Σ is positive definite. The parameters μ, vechΣ, and σ^2 are called population parameters, because they do not vary across individuals.

One simple example of a mixed-effects model is

$$\begin{aligned} y_{ij}|\theta_i, \sigma^2 &\overset{ind}{\sim} N(\exp(\theta_i t_j), \sigma^2), \\ \theta_i|\mu, \tau^2 &\overset{iid}{\sim} N(\mu, \tau^2). \end{aligned} \tag{4}$$

In this example, the population parameters are μ, τ^2, and σ^2.

A major difficulty with nonlinear mixed-effects models is the lack of analytic expression of the marginal likelihood function $L(\mu, \Sigma, \sigma^2; \mathbf{y})$; and hence a related problem for design is the lack of analytic expression of the expected Fisher information matrix. For optimal design problems, evaluation of the expected Fisher information matrix is usually essential, since the classical optimal design criteria are usually functions of the information matrix (see, for example, Silvey, 1980).

For likelihood-based estimation, the likelihood can be numerically computed by using adaptive Gaussian quadrature (Pinheiro and Bates, 1995, 2000; Wolfinger, 2000). To facilitate the computation of the information matrix and overcome this difficulty for design, Mentré, Mallet, and Baccar (1997) suggest the use of a first-order Taylor expansion of the mean function around μ. In particular,

$$\begin{aligned} y_{ij}|\theta_i &= r(\theta_i, t_j) + \varepsilon_{ij} \\ &\approx s(\mu, t_j) + (\theta_i - \mu)'\nabla r(\mu, t_j) + \varepsilon_{ij}, \quad \varepsilon_{ij} \overset{iid}{\sim} N(0, \sigma^2), \end{aligned} \tag{5}$$

where ∇ denotes the gradient with respect to μ.

Mentré, Mallet, and Baccar treat σ^2 as known; the approach is extended in Retout, Duffull, and Mentré (2001) to treat σ^2 as unknown, and in Retout and Mentré (2003a, 2003b) to allow for heteroscedastic ε_{ij}'s. See also Fedorov, Gagnon and Leonov (2002).

Model (5) is called a pseudo-mixed-effects model. This model has a likelihood identical to the first-order approximation to the likelihood of model (3) as implemented in SAS (Wolfinger, 2000). The use of the first-order approximation for parameter estimation was proposed by Sheiner and Beal (1980) and refined by Vonesh and Carter (1992). Note that the two models are identical if the mean function is linear in the random effects.

The use of this approximation, however, is only appropriate if the variances of the random effects are small, or the nonlinearity is mild (Jones and Wang, 1999; Jones *et al.*, 1999). The performance of the "optimal" designs identified using the approximation needs further investigation. In addition, with a fixed number of measurements per subject, the approximation-based estimator has been shown to be inconsistent for model (4) (Demidenko, 1997) and a slight variation of it (Ramos and Pantula, 1995).

The derivation of the Fisher information matrix for model (5) is outlined below.

Define a vector-valued function $\mathbf{f}: \mathbb{R}^p \to \mathbb{R}^m$ by $\mathbf{f}(\theta) = (r(\theta, t_1) \cdots r(\theta, t_m))^T$. For $j = 1, \ldots, p$, let $\mathbf{g}_j(\theta) = \partial \mathbf{f}(\theta)/\partial \theta_j$. Then, the Jacobian matrix of $\mathbf{f}$ is

$$\mathbf{Df}(\theta) = \begin{pmatrix} \nabla^T r(\theta, t_1) \\ \vdots \\ \nabla^T r(\theta, t_m) \end{pmatrix} = \Big(\mathbf{g}_1(\theta) \cdots \mathbf{g}_p(\theta) \Big).$$

So, the vector of the responses of the ith subject is

$$\begin{aligned} \mathbf{Y}_i &= \begin{pmatrix} r(\mu, t_1) \\ \vdots \\ r(\mu, t_m) \end{pmatrix} + \begin{pmatrix} \nabla^T r(\mu, t_1) \\ \vdots \\ \nabla^T r(\mu, t_m) \end{pmatrix} (\theta_i - \mu) + \begin{pmatrix} \varepsilon_{i1} \\ \vdots \\ \varepsilon_{im} \end{pmatrix} \\ &= \mathbf{f}(\mu) + \mathbf{Df}(\mu)(\theta_i - \mu) + \varepsilon_i, \quad \varepsilon_i \overset{iid}{\sim} N_m(0, \sigma^2 \mathbf{I}_m). \end{aligned}$$

Note that $\mathbf{Df}(\mu)(\theta_i - \mu) \overset{iid}{\sim} N_m(0, \mathbf{Df}(\mu)\Sigma\mathbf{D}^T\mathbf{f}(\mu))$; thus, define $\eta_i = \mathbf{Df}(\mu)(\theta_i - \mu) + \varepsilon_i$, then $\eta_i \overset{iid}{\sim} N_m(0, \mathbf{Df}(\mu)\Sigma\mathbf{D}^T\mathbf{f}(\mu) + \sigma^2\mathbf{I}_m)$, $i = 1, \ldots, n$. Let $\mathbf{V} = \mathbf{Df}(\mu)\Sigma\mathbf{D}^T\mathbf{f}(\mu) + \sigma^2\mathbf{I}_m$, the model now becomes

$$\mathbf{Y}_i = \mathbf{f}(\mu) + \eta_i, \quad \eta_i \overset{iid}{\sim} N_m(0, \mathbf{V}).$$

Assume that the elements of Σ do not depend on any other parameters and denote $\Omega = (\mu^T, \mathrm{vech}(\Sigma), \sigma^2)^T$. Then, the log-likelihood function based on subject i is

$$\begin{aligned} l(\Omega; \mathbf{Y}_i) = &-\frac{m}{2}\log(2\pi) - \frac{1}{2}\log\det\mathbf{V} \\ &-\frac{1}{2}\mathrm{tr}\left[\mathbf{V}^{-1}(\mathbf{Y}_i - \mathbf{f}(\mu))(\mathbf{Y}_i - \mathbf{f}(\mu))^T\right]. \end{aligned}$$

After much algebra, it can be shown that the Fisher information matrix M, with (j,k)th elements $m_{jk} = \mathrm{E}\left[-\partial^2 l(\Omega; \mathbf{Y}_i)/\partial\omega_j\partial\omega_k\right]$, can be expressed as:

$$M = (m_{jk}) = \left(\frac{\partial \mathbf{f}^T(\mu)}{\partial \omega_j} \mathbf{V}^{-1} \frac{\partial \mathbf{f}(\mu)}{\partial \omega_k} + \frac{1}{2} \mathrm{tr} \left(\mathbf{V}^{-1} \frac{\partial \mathbf{V}}{\partial \omega_j} \mathbf{V}^{-1} \frac{\partial \mathbf{V}}{\partial \omega_k} \right) \right). \tag{6}$$

Note that all the components in this expression depend on the design, that is, the time points at which the observations are taken; and hence, the determinant of M also depends on the design in general.

Note also that if the elements of θ_i are assumed independent in the population, that is, if $\Sigma = \mathrm{diag}(\tau_1^2 \cdots \tau_p^2)$, then $\Omega = (\mu^T, \tau_1^2, \ldots, \tau_p^2, \sigma^2)^T$, and the Fisher information matrix is of dimension $(2p+1) \times (2p+1)$.

Calculation of the partial derivatives of $\mathbf{f}$ and $\mathbf{V}$ is involved in determining the elements of M. The expressions for these partial derivatives are given below. Detailed derivation can be found in Han (2002, Secs. 4.7 and 4.8).

Let μ_k be the kth element of μ and let

$$\frac{\partial}{\partial \mu_k} \mathbf{Df}(\mu) = \left(\frac{\partial}{\partial \mu_k} \mathbf{g}_1(\mu) \cdots \frac{\partial}{\partial \mu_k} \mathbf{g}_p(\mu) \right) = \left(\mathbf{h}_1^{(k)}(\mu) \cdots \mathbf{h}_p^{(k)}(\mu) \right).$$

Then,

$$\frac{\partial}{\partial \omega_j} \mathbf{f}(\mu) = \begin{cases} \mathbf{g}_j(\mu), & \text{if } j \in \{1, \ldots, p\}, \\ 0, & \text{otherwise;} \end{cases}$$

and

$$\frac{\partial}{\partial \omega_k} \mathbf{V} = \begin{cases} \sum_{j=1}^p \Big[\tau_j^2 \Big(\mathbf{h}_j^{(k)}(\mu) \mathbf{g}_j^T(\mu) \\ \qquad + \mathbf{g}_j(\mu) \mathbf{h}_j^{(k)T}(\mu) \Big) \Big], & \text{if } k \in \{1, \ldots, p\}, \\ \mathbf{g}_{k-p}(\mu) \mathbf{g}_{k-p}^T(\mu), & \text{if } k \in \{p+1, \ldots, 2p\}, \\ \mathbf{I}_m, & \text{if } k = 2p+1. \end{cases}$$

Note that the results presented here are different from those given in Retout, Duffull, and Mentré (2001). The difference is caused by their assumption that $\mathbf{V}$ does not depend on μ. This assumption is usually invalid, unless the random effects are linear (see, for example, Mentré, Mallet, and Baccar, 1997). Han (2002) and Retout and Mentré (2003a, 2003b) give an expression for M without this independence assumption, which is identical to (6).

4. Design under the Approximation

In this section the design issues in HIV dynamics are investigated using the first-order approximation introduced in the previous section. With repeated measurements of HIV RNA level, measurements obtained from the same subject are not independent, and so, the information matrix cannot be expressed as a (weighted) sum of the information matrices at each time point. As a consequence, exact designs, rather than approximate designs, are found here. The following definitions are needed to facilitate the discussion.

Let Ξ_m be the set of designs that are supported at m (not necessarily distinct) points in a compact set denoted $\mathscr{T}$ and take 1 observation at each point, and let $M(\xi)$ be the Fisher information matrix induced by design $\xi \in \Xi_m$. Note that $M(\xi)$ can be computed using (6). The analogue of the D-optimality criterion (see, for example, Silvey, 1980) maybe of interest, which will be called ϕ_m^D-optimality, under which $\log\det M(\xi)$ is to be maximized over Ξ_m. Since it is sometimes desirable to compare designs with different numbers of measurements per individual, ϕ^D-optimality is introduced, which, for a model with k unknowns, is defined as $\log\det M(\xi) - k\log m$. This can be informally interpreted as an indication of "information per measurement".

Other population optimality criteria, for example, those analogous to D_A- and A-optimality (Silvey, 1980), can be defined similarly.

In what follows, a ϕ_m^D-optimal design will refer to the design that maximizes the ϕ_m^D-optimality for a pseudo-mixed-effects model. To better distinguish from the optimal designs for the individual models, the term "population ϕ_m^D-optimal design" will also be used.

Closed form results for designs have not been found and the examples are all numerical results.

4.1. *An analog of D-optimality, with examples*

In all the following examples, the number of subjects is denoted by n, and the total number of measurements is denoted by N. These examples are approximate models for HIV dynamics (see, Wu and Ding, 1999).

Example 1 Let $r(\theta_i, t_j) = \log[\exp(\theta_{i1}) + \exp(\theta_{i2} - \theta_{i3}t_j)]$, and assume measurement errors are independent and identically distributed $N(0, \sigma^2)$ while θ_i's are independent and identically distributed $N_3(\mu, \Sigma)$. Using the numerical results of Wu and Ding (1999), let $\mu = (5.530\ 11.023\ 0.216)^T$, $\Sigma = \text{diag}(1.100^2\ 1.284^2\ 0.035^2)^T$, and $\sigma^2 = 0.381^2$. Note that, if the observations are to be taken between 2 and 84 days, the D-optimal design for individual parameter estimation at μ puts 1/3 mass at each of 2, 20.39, and 84 days (see, Theorem 4 of Han and Chaloner, 2003).

To examine, informally, how well the pseudo-mixed-effects model approximates the mixed-effects model in this example, twenty sets of individual parameters (random effects) are drawn from their distribution. Figure 1 shows the expected response (that is, excluding measurement errors) for these twenty realizations under the two different models. The

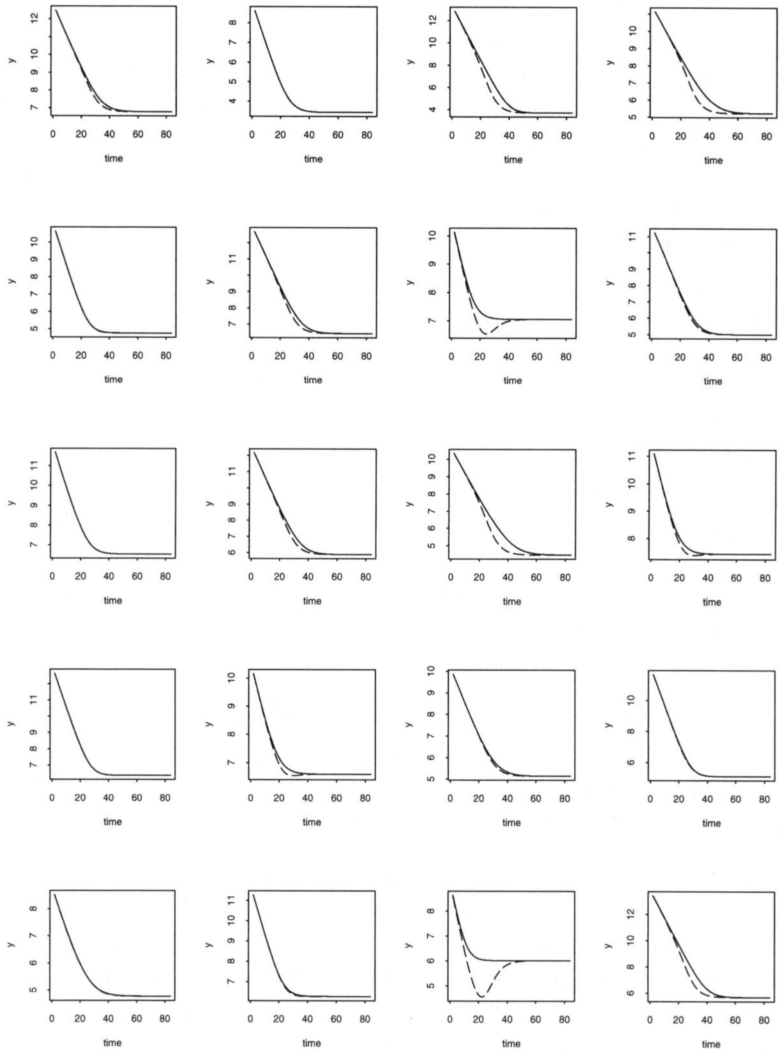

Fig. 1. Realizations of individual models in Example 1; the solid lines are for the mixed-effects models and the dashed lines for the pseudo-mixed-effect models.

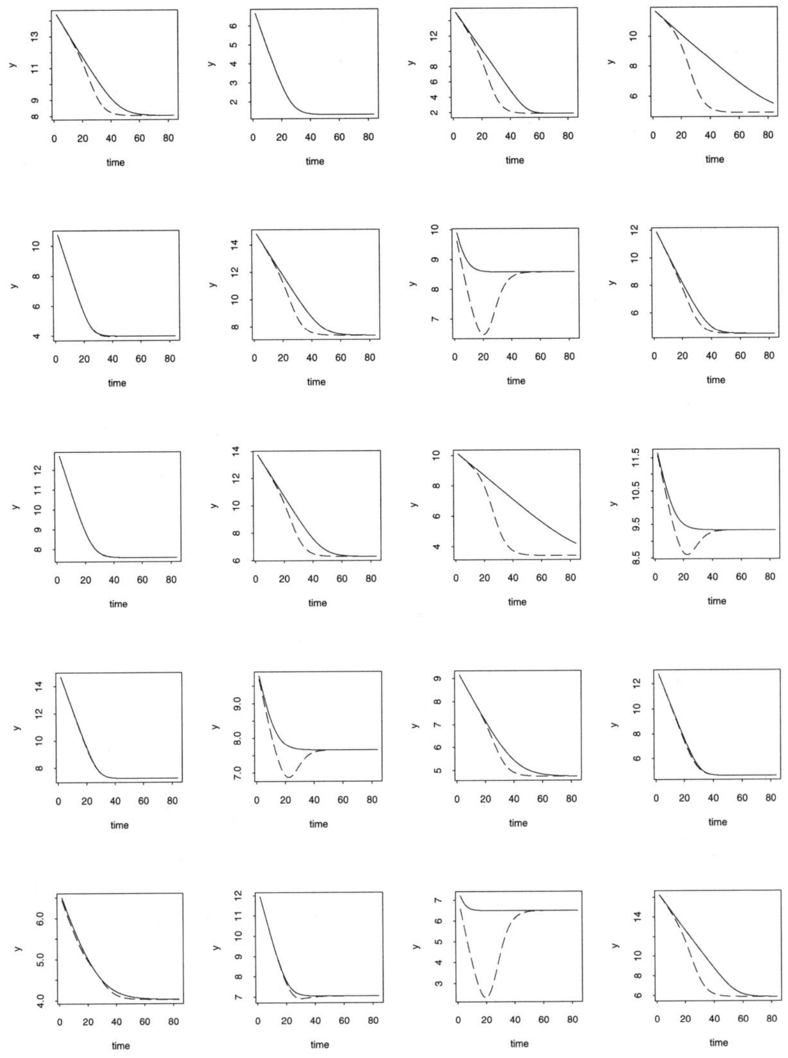

Fig. 2. Realizations of individual models with variances quadrupled; the solid lines are for the mixed-effects models and the dashed lines for the pseudo-mixed-effect models.

approximation is usually reasonable. If the variances of the random effects are quadrupled, however, the approximation is rather poor (Fig. 2).

Suppose again that the design space is $\mathscr{T} = [2, 84]$. The ϕ_3^D-optimal design is supported at 2, 22.17, and 29.40. Using the ϕ_3^D-criterion (logarithm of the determinant of the population Fisher information matrix), the criterion is 14.53 for the ϕ_3^D-optimal design and 13.93 for the three-point individual D-optimal design; thus, to achieve the same ϕ_3^D-criterion, the individual design needs a sample size that is $\exp[(14.53-13.93)/7] \approx 1.09$ times the sample size of the ϕ_3^D-design.

On the other hand, if the individual D-criterion is used, with the assumption that one (instead of one-third) measurement is taken at each time point, then the logarithm of the determinant of the information matrix of the individual design is -27.87 while that of the population ϕ_3^D-optimal design is -29.29; to achieve the same individual criterion, the population design needs a sample size that is about $\exp[(-27.87+29.29)/7] = 1.22$ times the sample size of the individual design.

The population ϕ_4^D-optimal design is supported at 2, 19.7, 26.3, and 84, with a ϕ_4^D-criterion value 18.06. Note that if the number of subjects using a population ϕ_3^D-optimal design is increased to 4/3 as large as in a population ϕ_4^D-optimal population design, its ϕ_4^D-criterion value would become $14.53 + 7\log(4/3) \approx 16.54$.

Alternatively, it can be observed that the ϕ^D-criterion value is $14.53 - 7\log 3 = 6.84$ for the ϕ_3^D-optimal design and $18.06 - 7\log 4 = 8.36$ for the ϕ_4^D-optimal design; hence taking 4 measurements in n patients is more ϕ^D-efficient than taking 3 measurements in 4n/3 patients. It can be calculated that to match the efficiency of the population ϕ_4^D-optimal design administered in n subjects, about $1.66 \times n$ subjects are needed in the population ϕ_3^D-optimal design; implementing such a design would take $1.66/(4/3) \approx 1.24$ times as many measurements as in the population ϕ_4^D-optimal design.

The population ϕ_5^D-, ϕ_6^D-, ϕ_7^D-, and ϕ_8^D-optimal designs can be found in a similar fashion. Table 1 summarizes these designs, and gives the numbers

Table 1. ϕ_m^D-optimal designs for Example 1.

m	Support points	Log-det	n	N	$\lceil n \rceil$	N^*
3	2 22.17 29.40	14.53	38.03	114.10	39	117
4	2 19.70 26.30 84	18.06	22.96	91.84	23	92
5	2 16.90 25.76 25.76 84	19.15	19.65	98.24	20	100
6	2 2 16.77 25.57 25.57 84	19.94	17.57	105.43	18	108
7	2 2 16.90 26.05 26.05 84 84	20.50	16.21	113.48	17	119
8	2 2 17.14 17.14 26.60 26.60 84 84	20.96	15.13	121.03	16	128

Table 2. ϕ_m^D-optimal designs for Example 2.

m	Support points	Log-det	n	N	$\lceil n \rceil$	N^*
4	2 12.00 48.60 84	31.66	23.32	93.27	24	96
5	2 7.35 12.57 24.00 84	35.31	15.54	77.68	16	80
6	2 7.37 12.69 24.39 24.39 84	36.24	14.02	84.11	15	90
7	2 7.34 12.77 12.77 24.12 24.12 84	37.01	12.86	90.04	13	91
8	2 7.35 12.77 12.77 24.61 24.61 84 84	37.42	12.30	98.38	13	104

of subjects, n, and the total number of measurements, $N = mn$, that are needed to achieve an information matrix whose determinant is exp(40). Also tabulated are $\lceil n \rceil$, the smallest integer no less than n, and $N^* = m\lceil n \rceil$. Among these designs, the one with 4 measurements per subject is the best one.

Example 2 As another example, let $r(\theta_i, t_j) = \log[\exp(\theta_{i1} - \theta_{i3}t_j) + \exp(\theta_{i2} - \theta_{i4}t_j)]$, and assume measurement errors are *iid* $N(0, \sigma^2)$ while θ_i's are *iid* $N_4(\mu, \Sigma)$. Using the numerical results of Wu and Ding (1999) again, let $\mu = (12.142\, 7.624\, 0.442\, 0.032)^T$, $\Sigma = \text{diag}(1.397^2\, 1.545^2\, 0.137^2\, 0.015^2)^T$, and $\sigma^2 = 0.267^2$. With the design space being $\mathscr{T} = [2, 84]$, the D-optimal design for individual parameter estimation at μ puts mass 1/4 at 2, 9.199, 20.88, and 84. With the same design space, the population ϕ_m^D designs are obtained and listed in Table 2, along with the numbers of subjects and measurements that are needed to achieve an information matrix whose determinant is exp(60). Among these designs, the one with 5 measurements per subject is the best one.

4.2. *Alternative criteria, with examples*

The derivation in the previous subsections assumes that the information matrix is $2p + 1$ by $2p + 1$, and the previous examples focus on the ϕ_m^D-optimality criterion. Alternative criteria may also be of interest. For example, if the variance of the measurement errors, σ^2, is assumed known, then the information matrix becomes $2p$ by $2p$; if in addition, the covariance matrix of the random effects is assumed known, then the information matrix becomes p by p.

Furthermore, when the information matrix is $2p+1$ by $2p+1$ or $2p$ by $2p$, interest may only be focused on estimating the mean vector of the random effects. In such cases, the proper optimality criterion is the $\phi_m^{D_s}$-optimality with $s = p$ (that is, the $\phi_m^{D_A}$-optimality with $A^T = (\mathbf{I}_s\ 0)$), defined as follows.

Formally, let the $2p+1$ by $2p+1$ information matrix M be partitioned into

$$M = \begin{pmatrix} M_{11} & M_{12} & M_{13} \\ M_{21} & M_{22} & M_{23} \\ M_{31} & M_{32} & M_{33} \end{pmatrix},$$

where M_{11}, M_{12}, and M_{22} are p by p matrices, M_{13} and M_{23} are p-vectors, M_{33} is a scalar, and $M_{ji} = M_{ij}^T$, $i, j = 1, 2, 3$. Then the aforementioned $2p$ by $2p$ and p by p information matrices are

$$\begin{pmatrix} M_{11} & M_{12} \\ M_{21} & M_{22} \end{pmatrix}$$

and M_{11}, respectively. Also, depending on whether or not σ^2 is known, the $\phi_m^{D_s}$- (or $\phi_m^{D_A}$-) optimality seeks to minimize

$$\log\det\left[(\mathbf{I}_p\ 0\ 0)\begin{pmatrix} M_{11} & M_{12} & M_{13} \\ M_{21} & M_{22} & M_{23} \\ M_{31} & M_{32} & M_{33} \end{pmatrix}^{-}\begin{pmatrix} \mathbf{I}_p \\ 0 \\ 0 \end{pmatrix}\right]$$

or

$$\log\det\left[(\mathbf{I}_p\ 0)\begin{pmatrix} M_{11} & M_{12} \\ M_{21} & M_{22} \end{pmatrix}^{-}\begin{pmatrix} \mathbf{I}_p \\ 0 \end{pmatrix}\right]$$

over Ξ_m, where the 0 matrices/vectors are such that the partitions are conformable and the matrix multiplications are defined.

Example 3 This example compares two criteria:

1. criterion a, maximizing $\log\det M_{11}$ for fixed m; this corresponds to the population D-optimality when Σ and σ^2 are known.
2. criterion b, maximizing

$$-\log\det\left[(\mathbf{I}_p\ 0\ 0)\begin{pmatrix} M_{11} & M_{12} & M_{13} \\ M_{21} & M_{22} & M_{23} \\ M_{31} & M_{32} & M_{33} \end{pmatrix}^{-}\begin{pmatrix} \mathbf{I}_p \\ 0 \\ 0 \end{pmatrix}\right].$$

First, suppose the model and the parameter values are the same as in Example 1. Under criterion a, the designs in Table 3 are obtained, where the number of subjects n is the one required to achieve an information matrix whose determinant is exp(15).

Table 3. Optimal designs under criterion a, using the model in Example 1.

m	Support points	Criterion	n	N	$\lceil n \rceil$	N^*
3	2 25.67 84	6.27	18.33	54.99	19	57
4	2 26.11 26.11 84	6.64	16.24	64.96	17	68
5	2 14.77 26.32 26.32 84	6.87	15.02	75.11	16	80
6	2 14.76 26.81 26.81 84 84	7.04	14.19	85.12	15	90
7	2 14.36 26.88 26.88 26.88 84 84	7.24	13.30	93.07	14	98
8	2 14.38 27.19 27.19 27.19 84 84 84	7.33	12.91	103.31	13	104

Table 4. Optimal designs under criterion b, using the model in Example 1.

m	Support points	Criterion	n	N	$\lceil n \rceil$	N^*
3	2 26.05 84	6.04	19.80	59.41	20	60
4	2 16.42 26.69 84	6.40	17.58	70.34	18	72
5	2 15.21 26.50 26.50 84	6.71	15.84	79.21	16	80
6	2 15.32 27.02 27.02 84 84	6.87	15.03	90.19	16	96
7	2 15.13 27.10 27.10 27.10 84 84	7.04	14.19	99.29	15	105
8	2 2 15.13 15.13 26.95 26.95 84 84	7.16	13.63	109.03	14	112

Table 4 summarizes the results under criterion b.

It appears that under these two different criteria, the optimal 4-, 7-, and 8-point designs are quite different. It is, therefore, interesting to examine the efficiency of the optimal m-point design under criterion a, when the criterion actually employed is b, relative to the optimal m-point optimal design under criterion b; and vice versa.

In particular, for any given m, let ψ_{aa} denote the value of criterion a of the optimal design under criterion a, ψ_{ab} the value of criterion b of the optimal design under criterion a, ψ_{ba} the value of criterion a of the optimal design under criterion b, and ψ_{bb} denote the value of criterion b of the optimal design under criterion b. Then, the efficiency of the optimal design under criterion a, when the criterion actually employed is b, relative to the optimal design under criterion b, is defined as $eff_{a|b} = \exp[(\psi_{ab} - \psi_{bb})/p]$. Similarly, $eff_{b|a}$ is defined as $\exp[(\psi_{ba} - \psi_{aa})/p]$.

Table 5. Relative efficiencies, using the model in Example 1.

m	ψ_{aa}	ψ_{ba}	ψ_{ab}	ψ_{bb}	$eff_{a\|b}$	$eff_{b\|a}$
3	6.2743	6.2708	6.0392	6.0427	0.9988	0.9989
4	6.6375	6.5307	6.3807	6.3991	0.9939	0.9650
5	6.8715	6.8706	6.7109	6.7121	0.9996	0.9997
6	7.0431	7.0417	6.8679	6.8697	0.9994	0.9995
7	7.2377	7.1710	7.0403	7.0435	0.9989	0.9780
8	7.3251	7.2744	7.1174	7.1635	0.9847	0.9832

Table 6. Optimal designs under criterion a, using the model in Example 2.

m	Support points	Criterion	n	N	$\lceil n \rceil$	N^*
4	2.39 12.41 25.13 84	12.45	23.00	92.00	23	92
5	2 6.95 12.55 25.08 84	12.94	20.40	101.97	21	105
6	2 7.08 12.61 12.61 24.65 84	13.22	18.99	113.95	19	114
7	2 7.14 12.88 12.88 25.18 25.18 84	13.43	18.06	126.39	19	133
8	2 7.08 11.82 13.45 13.45 25.18 25.18 84	13.58	17.39	139.13	18	144

Then, Table 5 can be formulated.

Now, suppose the model and the parameter values are the same as in Example 2. Under criterion a, Table 6 summarizes the optimal designs. The number of subjects, n, is the one required to achieve an information matrix whose determinant is exp(25).

Table 7 corresponds to criterion b, and Table 8 summarizes the efficiencies.

Note that under the "assumption" that $\mathbf{V}$ does not depend on μ and hence that the Fisher information matrix is block-diagonal (Retout, Duffull, and Mentré, 2001), criteria a and b would result in exactly the same optimal designs. Example 3 clearly shows that the optimal designs are different under different criteria. However, for the particular parameter values that are used in this example, optimal designs under one criterion also seem to be highly efficient under the other.

Table 7. Optimal designs under criterion b, using the model in Example 2.

m	Support points	Criterion	n	N	$\lceil n \rceil$	N^*
4	2 12.27 24.69 84	11.77	27.33	109.33	28	112
5	2 6.91 12.66 25.02 84	12.72	21.55	107.75	22	110
6	2 7.08 12.73 12.73 24.62 84	12.98	20.18	121.08	21	126
7	2 7.13 13.01 13.01 25.13 25.13 84	13.17	19.25	134.78	20	140
8	2 7.03 7.03 12.79 12.79 25.09 25.09 84	13.37	18.30	146.38	19	152

Table 8. Relative efficiencies, using the model in Example 2.

m	ψ_{aa}	ψ_{ba}	ψ_{ab}	ψ_{bb}	$eff_{a\|b}$	$eff_{b\|a}$
4	12.4582	12.4533	11.7397	11.7679	0.9930	0.9988
5	12.9389	12.9376	12.7174	12.7186	0.9997	0.9997
6	13.2240	13.2226	12.9797	12.9811	0.9996	0.9997
7	13.4260	13.4247	13.1676	13.1691	0.9996	0.9997
8	13.5761	13.5727	13.3052	13.3728	0.9832	0.9992

5. Effect of Approximation on Optimality Criteria

It has been shown that with fixed number of measurements per subject, the estimator based on the first-order approximation (5) is inconsistent for model (4) (Demidenko, 1997) and a slight variation of it (Ramos and Pantula, 1995). When an "optimal" design is found using this approximation, but the estimation is carried out using the true nonlinear mixed-effects model (3), the performance of this design needs investigation.

Note that when the approximation (5) is to be avoided, one can proceed as follows.

Let $f(\mathbf{y}_i)$ denote the marginal density of $\mathbf{y}_i = (y_{i1} \cdots y_{im})^T$, the vector of observations from ith subject. Let $\nabla f(\mathbf{y}_i)$ denote the gradient of $f(\mathbf{y}_i)$ with respect to the population parameters. The gradient of the log-likelihood function, that is, the score function, is then $\nabla \log f(\mathbf{y}_i) = \nabla f(\mathbf{y}_i)/f(\mathbf{y}_i)$. Using "$\square$" to denote the outer product, the information matrix is then

$$\sum_{i=1}^{n} \mathrm{Cov}_{\mathbf{y}_i} \frac{\nabla f(\mathbf{y}_i)}{f(\mathbf{y}_i)} = \sum_{i=1}^{n} \left\{ \mathrm{E}_{\mathbf{y}_i} \frac{\nabla f(\mathbf{y}_i) \square \nabla f(\mathbf{y}_i)}{f^2(\mathbf{y}_i)} - \left[\mathrm{E}_{\mathbf{y}_i} \frac{\nabla f(\mathbf{y}_i)}{f(\mathbf{y}_i)} \right] \square \left[\mathrm{E}_{\mathbf{y}_i} \frac{\nabla f(\mathbf{y}_i)}{f(\mathbf{y}_i)} \right] \right\}.$$

Under regularity conditions similar to Theorem 9 of Lehmann (1986, Section 2.7),

$$\mathrm{E}_{\mathbf{y}_i} \frac{\nabla f(\mathbf{y}_i)}{f(\mathbf{y}_i)} = \int \nabla f(\mathbf{y}_i)\, d\mathbf{y}_i = \nabla \int f(\mathbf{y}_i)\, d\mathbf{y}_i = 0,$$

and the above simplifies to

$$\sum_{i=1}^{n} \mathrm{E}_{\mathbf{y}_i} \frac{\nabla f(\mathbf{y}_i) \square \nabla f(\mathbf{y}_i)}{f^2(\mathbf{y}_i)}.$$

Let $p(\mathbf{y}_i|\theta_i, \sigma^2)$ denote the conditional density of $\mathbf{y}_i$, and $q(\theta_i|\mu, \Sigma)$, the density of θ_i. Then,

$$f(\mathbf{y}_i) = \int p(\mathbf{y}_i|\theta_i, \sigma^2) q(\theta_i|\mu, \Sigma)\, d\theta_i;$$

under the same regularity conditions,

$$\begin{aligned} \nabla f(\mathbf{y}_i) &= \nabla \int p(\mathbf{y}_i|\theta_i, \sigma^2) q(\theta_i|\mu, \Sigma)\, d\theta_i \\ &= \int \nabla \left[p(\mathbf{y}_i|\theta_i, \sigma^2) q(\theta_i|\mu, \Sigma) \right] d\theta_i. \end{aligned}$$

The nonlinear mixed-effects model (3) satisfies the regularity conditions (see, Chapter 4 of Han, 2002). Hence, for any $\mathbf{y}_i$, both $f(\mathbf{y}_i)$ and $\mathbf{g}(\mathbf{y}_i)$ can be evaluated using Gaussian quadrature, provided that the dimension of θ_i is not high. This is referred to as an "inner" integration since it is to be evaluated for a certain $\mathbf{y}_i$. The dimension of the inner integration is equal to the number of random effects (that is, p).

When the conditional distribution of data given the random effects is continuous, and when the number of design points is small, for example, less than or equal to 4, the expectation with respect to the marginal distribution of $\mathbf{y}_i$ can also be evaluated using Gaussian quadrature:

$$\mathrm{E}_{\mathbf{y}_i} \frac{\nabla f(\mathbf{y}_i) \square \nabla f(\mathbf{y}_i)}{f^2(\mathbf{y}_i)} = \int \frac{\nabla f(\mathbf{y}_i) \square \nabla f(\mathbf{y}_i)}{f(\mathbf{y}_i)}\, d\mathbf{y}_i. \tag{7}$$

This is an "outer" integration, and has dimension the same as the number of design points, that is, the number of observations per individual. The implementation details for evaluating (7) can be found in Han (2002, Section 4.3).

Han (2002, Section 4.3.3) compares the optimality criteria based on the true nonlinear mixed-effects model and the first-order approximation for the following model:

$$\begin{aligned} y_{ij}|\theta_i &\overset{ind}{\sim} N(e^{-\theta_i t_j}, \sigma^2), \\ \theta_i &\overset{iid}{\sim} N(\mu, \tau^2), \end{aligned} \tag{8}$$

where $\mu = 1.5$ and $\sigma^2 = 0.05$; two different values for τ^2 where considered: 0.03 and 0.09. The designs evaluated are two-point designs; each point belongs to the equally spaced partition of $[0.1, 2.5]$ with mesh 0.1.

The outcome of interest includes the D-optimality criterion, the logarithm of the determinant of the 3×3 Fisher information matrix, and a D_s-optimality criterion, the negative logarithm of the $(1,1)$ element of the inverse of the Fisher information matrix. Let M denote the Fisher information matrix, and m^{11}, the $(1,1)$ element of its inverse; then the two criteria are $\log\det M$ and $-\log m^{11}$, respectively. Also, assuming that σ^2 and τ^2 are both known, then the information would reduce to the $(1,1)$ element of the 3×3 information matrix for the model with σ^2 and τ^2 both unknown; so, the logarithm of the $(1,1)$ element of M, $\log m_{11}$, is also examined.

The information matrices and the optimal design criteria are computed based on both the first-order expansion of the mean function and the true nonlinear mixed-effects model. Figures 3 and 4 show the criteria based on the true mixed-effects model and their difference with the criteria based on the pseudo-mixed-effects model (the former minus the latter). In each plot, the higher surface is the criteria based on the true mixed-effects model, and the lower one represents the difference. While among the candidate designs it appears that the "optimal" designs coincide with or without the first-order approximation, it can be seen that the difference can be very substantial relative to the criteria based on the true mixed-effects model. This is especially true with the large τ^2, 0.09.

Merlé and Tod (2001) use nested loops of stochastic simulation to evaluate the true information of the nonlinear mixed-effects model and compared the optimality criteria of candidate designs with and without first-order approximation. Using a population pharmacokinetic model and a population pharmacodynamic model, and based on ϕ_m^D-optimality (called D-optimality in Merlé and Tod, 2001), their numerical examples suggest that with or without the first-order approximation, the optimal designs are the same; the values of the criterion function, however, vary between the true model and the approximate model. The authors caution that such discrepancies can lead to problems when computing design efficiency.

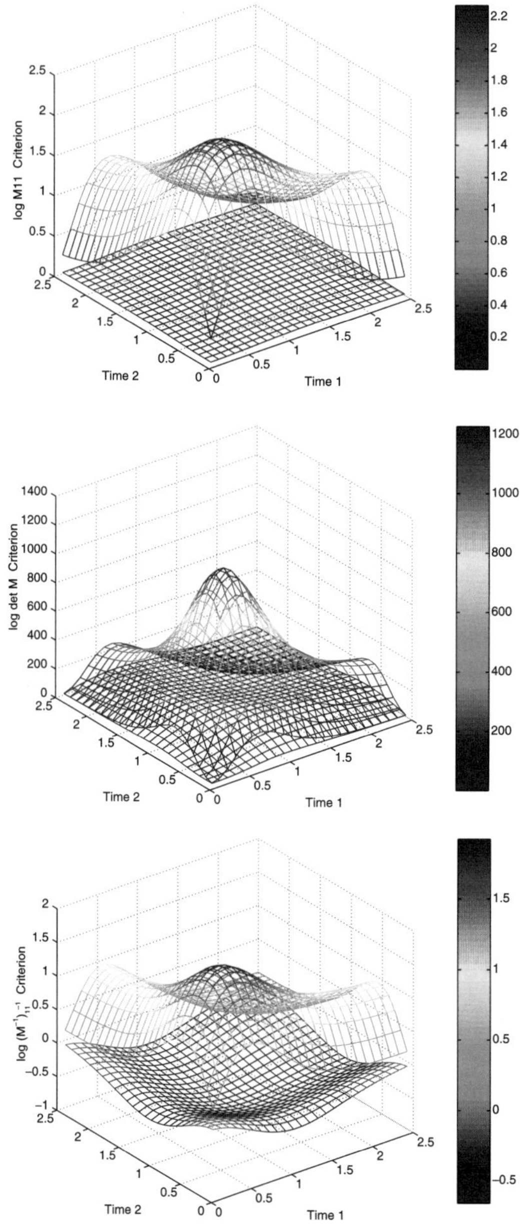

Fig. 3. Criteria based on the true mixed-effects model (the higher surfaces) and their difference (the lower surfaces) with the criteria based on the pseudo-mixed-effects model of 2-point designs for model (8). The criteria are $\log m_{11}$ (top), $\log \det M$ (middle), and $-\log m^{11}$ (bottom). $\mu = 1.5$, $\sigma^2 = 0.05$, $\tau^2 = 0.03$. Time 1 and Time 2 are the two design points.

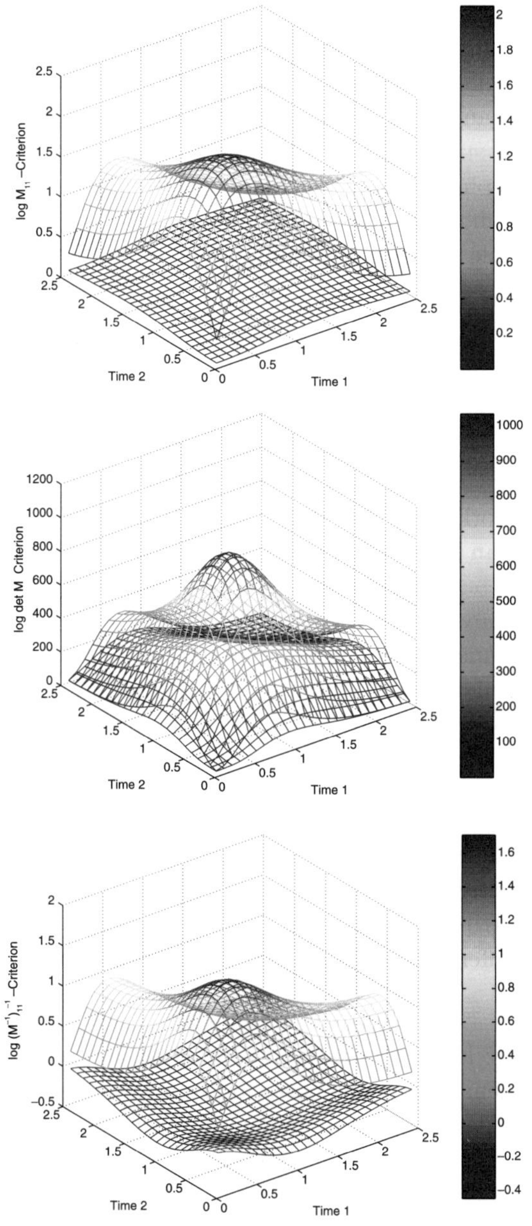

Fig. 4. Criteria based on the true mixed-effects model (the higher surfaces) and their difference (the lower surfaces) with the criteria based on the pseudo-mixed-effects model of 2-point designs for model (8). The criteria are $\log m_{11}$ (top), $\log\det M$ (middle), and $-\log m^{11}$ (bottom). $\mu = 1.5$, $\sigma^2 = 0.05$, $\tau^2 = 0.09$. Time 1 and Time 2 are the two design points.

Whether via nested stochastic simulation or via nested Gaussian quadrature, the direct evaluation of the Fisher information matrix for the nonlinear mixed-effects model is very computationally intensive. Both Merlé and Tod (2001) and Han (2002) provide examples where the number of observations per subject is low (≤ 4). In practice, virus dynamic experiments, as well as pharmacokinetic and pharmacodynamic studies, are likely to involve much more observations per subject. In such situations, it does not appear feasible to use either technique to evaluate the Fisher information matrix.

6. Bayesian Design

Designing an experiment can be formulated as a decision of choosing the values of the predictor(s) so as to maximize an expected utility or minimize a risk. Han and Chaloner (2004) study design of HIV experiments in a Bayesian framework. They adopt the assumption that the data will be analyzed in a Bayesian paradigm and at the design stage, a prior distribution is used which may (but does not necessarily) differ from the prior distribution used for data analysis (see, for example, Lindley and Singpurwalla, 1991; Etzioni and Kadane, 1993; and Tsai and Chaloner, 2001).

Han and Chaloner (2004) assume that in addition to the nonlinear mixed-effects model (3), the following prior distribution, π, is used in data analysis:

$$\sigma^{-2} \sim G(\alpha, \eta), \quad \mu \sim N(\eta, \Lambda), \quad \Sigma^{-1} \sim W(\Omega, \nu), \tag{9}$$

where σ^2, μ, and Σ^{-1} are mutually independent, α and η are positive, Λ and Ω are positive definite, and $\nu \geq p$. The parameterization of gamma and Wishart distributions follows Carlin and Louis (1996, Appendix A).

A possibly different prior distribution ω is used for design:

$$\sigma^{-2} \sim G(\gamma, \delta), \quad \mu \sim N(\zeta, \mathbf{K}), \quad \Sigma^{-1} \sim W(\Psi, \chi). \tag{10}$$

If a quadratic loss for parameter estimation is used, then solving the estimation problem leads to the posterior expectation as a Bayes estimator; the loss for the design problem is then a weighted sum of the posterior variances.

Han and Chaloner have shown that for (3) and (9), the posterior risk for estimation of μ always exists. When the same prior distribution is used for both design and inference, they have proven that the preposterior risk for design exists. If the prior distribution used in design is different from that used for inference, sufficient conditions have been established for existence of the preposterior risk for design. These conditions can be roughly described as ω being as informative as, or more informative than π.

Using this approach, an HIV dynamic model of Perelson *et al.* (1996), and the prior distribution constructed by Han, Chaloner and Perelson (2002), a set of eight candidate designs, including the original design used in Perelson *et al.*, are compared. Details can be found in Han and Chaloner (2004) and Han (2002, Chapter 6). It should be noted that this approach involves generating samples from the marginal distribution of the data under each design and implementing a Markov chain Monte Carlo algorithm for each generated sample of data. It is therefore also time consuming.

Bayesian design problems for population pharmacokinetic models have been studied by Stroud, Müller and Rosner (2001), who assume the same prior distribution for design and analysis, and use the posterior precision of the area under the time-concentration curve as the utility. While the method can be applied to nonlinear mixed-effects models in general, the existence of preposterior expected utility is not obvious under their utility function.

7. Summary

Design issues for population HIV dynamics, or more generally, for nonlinear mixed-effects models, represent a challenging problem. Wu and Ding (2002) use simulation to compare different designs. This can be time consuming, and may encounter convergence problems for some of the generated datasets. Evaluation of the Fisher information is relatively easy when a nonlinear mixed-effects model is linearized using the first-order approximation (5), but such evaluation may be of limited use since Merlé and Tod (2001) and Han (2002) show that the Fisher information for the approximate model (5) can be quite different from the Fisher information for the true model (3). Directly evaluating the Fisher information for the true model (3) is not only time consuming, but may also suffer from underflow/overflow in high dimemsions. Han and Chaloner (2004) provide an attractive alternative to the design problem in population HIV dynamics, although their approach, involving Markov chain Monte Carlo nested within Monte Carlo simulation, is time consuming as well.

Acknowledgements

This work was supported in part by a grant from the National Institutes of Health to the Great Lakes Regional Center for AIDS Research (1P30 CA79458), a Doctoral Dissertation Fellowship from the University of Minnesota Graduate School and a grant from the National Security Agency.

References

1. Carlin BP and Louis TA (1996) *Bayes and Empirical Bayes Methods for Data Analysis*, Chapman & Hall, London.
2. Chaloner K and Larntz K (1989) Optimal Bayesian design applied to logistic regression experiments. *J Stat Plan Infer* **21**: 191–208.
3. Chaloner K and Verdinelli I (1995) Bayesian experimental design: A review. *Stat Sci* **10**: 273–304.
4. Demidenko E (1997) Asymptotic properties of nonlinear mixed-effects models, in: Gregoire TG, Brillinger DR, Diggle PJ, Russek–Cohen E, Warren WG and Wolfinger RD (eds.) *Modelling Longitudinal and Spatially Correlated Data: Methods, Applications, and Future Directions*, pp. 49–62, Springer-Verlag, New York.
5. Ding AA and Wu H (1999) Relationships between antiviral treatment effects and biphasic viral decay rates in modeling HIV dynamics. *Math Biosci* **160**: 63–82.
6. Etzioni R and Kadane JB (1993) Optimal experimental design for another's analysis. *J Amer Stat Assoc* **88**: 1404–1411.
7. Fedorov VV, Gagnon RC and Leonov SL (2002) Design of experiments with unknown parameters in variance. *App Stoch Model Bus* **18**: 207–218.
8. Han C (2002) *Optimal Designs for Nonlinear Regression Models with Applications to HIV Dynamic Studies*, PhD Dissertation in Biostatistics, University of Minnesota, Minneapolis.
9. Han C and Chaloner K (2003) *D*- and *c*-optimal designs for exponential regression models used in viral dynamics and other applications. *J Stat Plan Infer* **115**: 585–601.
10. Han C and Chaloner K (2004) Bayesian experimental design for nonlinear mixed-effects models with application to HIV dynamics. *Biometrics* **60**: 25–33.
11. Han C, Chaloner K and Perelson AS (2002) Bayesian analysis of a population HIV dynamic model, in: Gatsonis C, Carriquiry A, Gelman A, Higdon D, Kass R, Pauler D and Verdinelli I (eds.) *Case Studies in Bayesian Statistics Volume VI*, pp. 223–237, Springer-Verlag, New York.
12. Jones B and Wang J (1999) Constructing optimal designs for fitting pharmacokinetic models. *Stat Comput* **9**: 209–218.
13. Jones B, Wang J, Jarvis P and Byrom W (1999) Design of cross-over trials for pharmacokinetic studies. *J Stat Plan Infer* **78**: 307–316.
14. Lehmann EL (1986) *Testing Statistical Hypotheses (2nd ed.)*, Wiley, New York.
15. Lindley DV and Singpurwalla ND (1991) On the evidence needed to reach agreed action between adversaries, with application to acceptance sampling. *J Amer Stat Assoc* **86**: 933–937.
16. Mentré F, Mallet A and Baccar D (1997) Optimal design in random-effects regression models. *Biometrika* **84**: 429–442.
17. Merlé Y and Tod M (2001) Impact of pharmacokinetic-pharmacodynamic model linearization on the accuracy of population information matrix and optimal design. *J Pharmacokinet Pharmacodynam* **28**: 363–288.

18. Perelson AS, Neumann AU, Markowitz M, Leonard JM and Ho DD (1996) HIV-1 dynamics *in vivo*: Virion clearance rate, infected cell life-span, and viral generation time. *Science* **271**: 1582–1586.
19. Pinheiro JC and Bates DM (1995) Approximations to the log-likelihood function in the nonlinear mixed-effects model. *J Comput Graph Stat* **4**: 12–35.
20. Pinheiro JC and Bates DM (2000) *Mixed-Effects Models in S and S-PLUS.* Springer-Verlag, New York.
21. Putter H, Heisterkamp SH, Lange JMA and de Wolf F (2002) A Bayesian approach to parameter estimation in HIV dynamical models. *Stat Med* **21**: 2199–2214.
22. Ramos RQ and Pantula SG (1995) Estimation of nonlinear random coefficient models. *Stat Prob Lett* **24**: 49–56.
23. Retout S, Duffull S and Mentré F (2001) Development and implementation of the population Fisher information matrix for the evaluation of population pharmacokinetic designs. *Comp Prog Methods Biomed* **65**: 141–151.
24. Retout S and Mentré F (2003a) Further development of the Fisher information matrix in nonlinear mixed effects models with evaluation in population pharmacokinetics. *J Biop Stat* **13**: 209–227.
25. Retout S and Mentré F (2003b) Optimization of individual and population designs using Splus. *J Pharmacokinet Pharmacodynam* **30**: 417–443.
26. Sheiner LB and Beal SL (1980) Evaluation of methods for estimating population pharmacokinetic parameters. I. Michaelis-Menten model: Routine clinical pharmacokinetic data. *J Pharmacokinet Biop* **8**: 553–571.
27. Silvey SD (1980) *Optimal Design: An Introduction to the Theory of Parameter Estimation*, Chapman and Hall, London.
28. Stroud JR, Müller P and Rosner GL (2001) Optimal sampling times in population pharmacokinetic studies. *App Stat* **50**: 345–359.
29. Tsai C-P and Chaloner K (2001) *Bayesian Experimental Design for the Analysis of another Bayesian (or of a Frequentist).* Technical Report, School of Statistics, University of Minnesota, Minneapolis.
30. Vonesh EF and Carter RL (1992) Mixed-effects nonlinear regression for unbalanced repeated measures. *Biometrics* **48**: 1–17.
31. Wolfinger RD (2000) *Fitting Nonlinear Mixed Models with the New NLMIXED Procedure.* http://www.sas.com/usergroups/sugi/sugi24/sipapers/p287-24.pdf.
32. Wu H and Ding AA (2002) Design of viral dynamic studies for efficiently assessing potency of anti-HIV therapies in AIDS clinical trials. *Biomet J* **44**: 175–196.
33. Wu H and Ding AA (1999) Population HIV-1 dynamics *in vivo*: Applicable models and inferential tools for virological data from AIDS clinical trials. *Biometrics* **55**: 410–418.
34. Wu H, Ding AA and de Gruttola V (1998) Estimation of HIV dynamic parameters. *Stat Med* **17**: 2463–2485.

CHAPTER 22

STATISTICAL ESTIMATION, INFERENCE AND HYPOTHESIS TESTING OF PARAMETERS IN ORDINARY DIFFERENTIAL EQUATIONS MODELS OF HIV DYNAMICS

Sarah Holte

This chapter provides a brief summary of the use of mathematical models and associated statistical analysis which have contributed to research on the pathogenesis of HIV infection and etiology. A review of some associated statistical methods is included with an example hypothesis testing using a mathematical model.

Keywords: Differential equations; hypothesis testing.

1. Introduction

The central theme of this article is to describe the integrated use of *mathematical models* in a *statistical* framework to address *clinical* questions about the etiology and pathogenesis of HIV infection. Most analysis of data from clinical trials and epidemiology studies uncover important empirical correlations without shedding light on the underlying biological mechanisms. *In vitro* experiments often identify these mechanisms, but do not always explain the bigger picture of why such a mechanism has evolved in the larger context of population biology. Mathematical models provide a powerful tool for the analysis of data from both clinical and experimental studies. They provide a means by which the mechanistic insights gained in the laboratory can be applied to clinical data from a population of infected individuals. However, without formal statistical methods that can be used to analyze data from clinical and experimental research, they remain academic exercises in the realm of theory.

Recent work by Costantino *et al.* (1997) provides an example of the type of interplay between mathematicians, statisticians, and experimental scientists that has resulted in the development of a validated model in mathematical biology. In that work, the researchers considered the population dynamics of flour beetles. The population dynamics were modeled; the model carefully analyzed for bifurcation profiles of parameters that could be perturbed in growth experiments; the experiments conducted, controlling the parameters as specified by the model; and statistical analysis performed to recover the model parameters from the data. Many of the model predictions turned out to be accurate, including the observation of various limit cycles and even chaotic dynamics. The success of this project is well known in the mathematical biology community, and relies on the successful interplay between researchers in mathematics, statistics, and biology.

Numerous groups have used mathematical models to explore and hypothesize possible mechanisms for the dynamics of HIV infection. However, the research with most impact is that which utilizes mathematical models in combination with formal statistical tools and clinical or experimental data. This article provides a summary of mathematical models which have been used to analyze clinical data. There are numerous theoretical models which offer significant insight into the possible mechanisms of HIV infection, but in this article we will primarily restrict our attention to those studies where mathematical models have been used for specific data analysis. We include a section with a brief summary of some of the relevant statistical methods, and conclude with an example which demonstrates the use of data to "reject the null hypothesis" that a linear system of differential equations accurately characterizes viral decay after treatment.

2. Examples of Mathematical Models in HIV Research

To date, some of the most successful uses of mathematical models in research on the pathogensis of HIV are described in Ho *et al.* (1995), Perelson *et al.* (1996), and Wei *et al.* (1995). These studies examined the kinetics of virus and $CD4^+$ T-cells populations using mathematical models with data from patients undergoing highly active antiretroviral therapy (HAART) and showed a very rapid and constant turnover of the viral and infected cell populations in all individuals studied. Whereas researchers had previously assumed that the stable viral and $CD4^+$ T-cell concentrations seen during the period of clinical latency of chronic HIV infection were due to the absence of any significant viral replication, these studies indicated that both the viral and infected cell populations were turning over rapidly and continuously. This information regarding HIV dynamics has enhanced basic understanding of HIV pathogenesis and has impacted clinical approaches to antiretroviral therapy. Further work by Perelson *et al.* (1997) describes a model with an additional compartment and uncovers a second population of "long-lived" infected cells contributing to the population of viral RNA. This model and associated analysis of data have been used to explain the biphasic decay profiles of viral RNA observed in most patients on treatment. These studies introduced a new era of collaboration between basic/clinical scientists and quantitative researchers in HIV science, and since these reports, many groups have used mathematical models to estimate decay rates and half-lives for infected cell populations for both children (Luzuriaga *et al.*, 1999; Melvin *et al.*, 1999) and adults (Neumann *et al.*, 1999; Notermans *et al.*, 1998; Wu and Kuritzkes *et al.*, 1999).

The linear differential equation models used in Ho *et al.* (1995), Perelson *et al.* (1996, 1997), and Wei *et al.* (1995) are approximations to more realistic nonlinear models for viral and infected cell decay, and as such are only applicable over short periods of time, probably on the order of days or weeks. While these linear models have been extremely useful in characterizing short-term dynamics of HIV infection after therapy, the use of these models to estimate time to eradication of virus from an individual is probably not valid. For example, the authors of Notermans *et al.* (1998) and Perelson *et al.* (1997) use a linear differential equations model and data collected on participants up to twelve weeks post-treatment to obtain estimates of the decay of the long-lived infected cell populations. Using these estimates and a linear differential equations model, the predicted time to viral eradication is two to three years. These predictions rely on the assumption that the linear approximation to the dynamics of the infected cell populations are appropriate over the time interval of two to three years. If the production or decay of infected cells varies during this time period, this assumption is invalid. In reality, it is now clear that infected cell populations are not eradicated during the time predicted (Finzi *et al.*, 1999; Ramratnam *et al.*, 2000; Zhang *et al.*, 1999) indicating that the linear differential equations models did not provide accurate long-term predictions.

Researchers have also used simple exponential decay models to describe the decay of the other infected cell populations. The authors of Finzi *et al.* (1999) and Zhang *et al.* (1999) assess the decay of latently infected cells using a linear decay model and make predictions based on this model about the time to eradication of the latently infected pool of infected cells. In Finzi *et al.* (1999), an estimate of 73 years required for eradication of the latently infected cell pool is obtained, and in Zhang *et al.* (1999), the eradication time is estimated as 7–10 years. Both studies based their estimates on data obtained within the first 3 years of treatment but rely on the assumption of constant decay (a linear differential equations model) through the entire course of time to eradication. Thus, the same concerns about the validity of the use of a linear model over long periods of time apply here. Differences between these two studies in the estimates of time to eradication of the latently infected cell compartment could be due to the different time periods over which the data were collected; and due to differences in decay of the latently infected cell compartment over these different time periods.

A few authors have raised the question of whether or not these linear mathematical models have adequately described the decay of compartments relevant to HIV infection dynamics. In Furtado *et al.* (1999), the authors measure decay of proviral DNA and viral mRNA. They found that proviral DNA "plateaus" after an initial decrease, and suggest that infected cell populations may not be decaying at a simple exponential rate. Furthermore, they suggest that there are not yet sufficient data to allow estimates of the duration of therapy required to eradicate infection. The authors of Bonhoeffer *et al.* (1997) offer arguments to show that more complex nonlinear models are needed to accurately describe long-term viral decay. In Luzuriaga *et al.* (1999), the authors note that the biphasic pattern which has been attributed to two populations of infected cells could be the result of exponential decay of a single population of infected cells with decreasing exponent over time. This phenomena is well known in population biology, and is often referred to as density-dependent decay (Hoppensteadt *et al.*, 1982). In Muller *et al.* (2002), the authors describe a model for decelerating decay of latently infected cells which postulates that residual viral production after many years of successful therapy is due to the decelerating decay of this compartment.

Many more complex models, including models with compartments for immune response, have been used to theorize about possible mechanisms for disease pathogenesis and viral decay, but they have not been rigorously analyzed or tested with clinical or experimental data. These models provide valuable insights into "what-if" scenarios, and offer alternative explanations for a variety of biological mechanisms in HIV infection and the associated immune response. However, in spite of this extensive literature, there remain several significant opportunities for mathematical modeling in combination with formal statistical analysis of data to contribute to ongoing research on HIV infection and associated immune response.

3. Background on Statistical Methods for Longitudinal HIV Dynamics Data

In order to successfully utilize mathematical models as a tool in the study of the pathogenesis of HIV infection and etiology, formal statistical tools are needed for analysis of data. For such an analysis, there are two models involved: the mathematical model, often in the form of systems of ordinary differential equations, which describes the mean value or expectation of the data over time; and the statistical model, which describes the variability and correlation in the residuals errors after fitting the expected mathematical model to the data. In this section, we will provide a summary of statistical models for HIV dynamics data and describe three of the most commonly used frequentist methods. There are Bayesian methods which have been quite successful in the analysis of dynamics data, but we will not discuss them further in this article. An excellent review of both frequentist and Bayesian methods for the analysis of dynamics data can be found in Wu *et al.* (2004).

The general form for mathematical models which describe dynamics in HIV research for a single individual or experimental unit is specified by the s-dimensional system of differential equations where the compartments represent various states of interest, such as viral RNA, infected cell populations, or various measures of immune response.

$$\frac{d\boldsymbol{X}}{dt} = f(\boldsymbol{X}, \boldsymbol{\beta}), \quad \boldsymbol{X}(0) = \boldsymbol{X}_0. \tag{1}$$

The solutions to this system of equations, $\boldsymbol{X}(t, \boldsymbol{\beta}, \boldsymbol{X}_0)$, considered as a function of the parameter $\boldsymbol{\beta}$ and initial conditions $\boldsymbol{X}_0$, determine expected trajectories that the model compartments could potentially traverse and serve as a vector-valued conditional (on the parameters $\boldsymbol{\beta}$) mean function for that data. To simplify notation, we include $\boldsymbol{X}_0$ in the vector of parameters $\boldsymbol{\beta}$. Data are collected at times $t_1, \ldots, t_n$ from a single individual or experimental unit. Assuming that data are collected on each state of the system at different points in time, we specify the statistical model for each observation time as follows: Let $\boldsymbol{Y}_j = (Y_{j1}, \ldots, Y_{js})^T$ be the $(s \times 1)$ vector of observations at time t_j on the s states of the system of ordinary differential equations (ODE) (1). The statistical model for the data $\boldsymbol{Y}_j$ can be expressed as:

$$\boldsymbol{Y}_j = \boldsymbol{X}(t_j, \boldsymbol{\beta}) + \boldsymbol{\epsilon}_j, \tag{2}$$

where $\boldsymbol{\epsilon}_j$ is a random variable representing deviations of the observed data from the dynamic model, $\boldsymbol{X}(t_j, \boldsymbol{\beta})$, due to measurement error, sampling variation, within-individual local physiological "fluctuations" or unmodeled dynamics not represented in (1).

Assuming that $E(\boldsymbol{\epsilon}_j|\boldsymbol{\beta}) = 0$ our statistical model (2) implies that the conditional mean at time t_j is $E(\boldsymbol{Y}_j|\boldsymbol{\beta}) = \boldsymbol{X}(t_j, \boldsymbol{\beta})$. To complete specification of the statistical model, we assume that observations from different compartments are independent conditional on the mean structure. In most cases, we assume that the variation in observations at a given state is constant over time and is denoted by $\text{var}(\epsilon_{jk}) = \sigma_{Y_k}$, $k = 1, \dots, s$. It is possible that the variances for different compartments in the model are different. Variance-stabilizing transforms, such as log transform, will often be used to achieve the constant variance assumption. Adjustment for heteroscedacity is also available in software packages such as the gnls package in Splus (Pinheiro *et al.*, 2000). Conditions (1) and (2) define the complete mathematical and statistical model for an individual patient or experiment with samples collected on s states over times $t_1, \dots, t_n$.

In all but the very simplest cases, the statistical models for HIV dynamics are nonlinear in the parameters of interest, $\boldsymbol{\beta}$, and thus require some type of nonlinear regression method for statistical analysis. Parameter estimation and inference for data from a single individual measured on a single compartment can generally be carried out in a relatively straightforward manner using nonlinear least-squares regression and profile likelihood methods (Venzon *et al.*, 1988). Estimation of $\boldsymbol{\beta}$ requires minimization of the sum of squared differences between the observed data and model mean. The form of the quantity to be minimized is:

$$min_{\boldsymbol{\beta}} \sum_{j}^{n} \left(X(t_j, \boldsymbol{\beta}) - Y_j\right)^2,$$

where $X(t_j, \boldsymbol{\beta})$ is either the analytical or numerical solution to (1). If the errors, $\boldsymbol{\epsilon}_j$ are normally distributed, this is also the maximum likelihood estimate for $\boldsymbol{\beta}$.

Similarly, when observations are available from more than one compartment of the system, multivariate techniques are available, and generally require some type of weighting of observations in different compartments by the variance of that compartment. In general, analytic solutions for the mathematical model for the mean structure are not available, and numerical solutions must be used in combination with nonlinear regression techniques. In comparison to linear regression analysis, this estimation problem is significantly more complex, and requires attention to issues such as convergence of the estimation algorithm and identifiability of the parameters of interest from the data at hand.

In most HIV dynamics studies, data are generally available from more than one individual. Although the functional and conditional (on the dynamics parameters) distributional form of the mathematical and statistical models for the data may remain the same for different subjects, i.e (1) and (2) apply to all individuals in an aggregate population, the times at which data are collected, the compartments which are observed, and the specific parameter values and initial conditions will likely vary from individual to individual.

With these considerations, (2) can be easily modified to describe data from a population of individuals. In this case, we explicitly model individual heterogeneity by viewing the parameter $\boldsymbol{\beta}$ as a random variable that varies across individuals, with mean, variance, and covariance to be estimated from aggregate population data. This variation can be described by either formulating regression-type models for specific components of $\boldsymbol{\beta}$ as a function of observed covariates (fixed effects), by formulating random effects models for specific components of $\boldsymbol{\beta}$ (random effects) or by a combination of both (mixed effects). To simplify notation, for the remainder of the article we will assume the (1) has only a single state variable. We can drop the index k indicating state, and will add an index i for each individual in the population of m individuals.

For random effects type models, we assume that the population model is described by

$$\boldsymbol{\beta}_i \sim F, \tag{3}$$

where F is a distribution in some admissible class. The mean of $\boldsymbol{\beta}_i$ under F, $E_F(\boldsymbol{\beta}_i) = \boldsymbol{\beta}$, represents the average or marginal parameter values across the population of interest. The covariance matrix, $\text{var}(\boldsymbol{\beta}_i) = \boldsymbol{\theta}$ describes heterogeneity in the dynamic parameters across the population of interest. We use $\boldsymbol{\theta}$ to denote the collection of parameters that define $\text{var}(\boldsymbol{\beta}_i)$. Together, the individual model described by (1) and (2) and the population model (3) describe a multivariate ODE-based nonlinear hierarchical random effects model for the observed data from m individual drawn from a population of interest. An introduction to models of this type can be found in Davidian *et al.* (1995).

One natural approach for analysis of repeated measures from an aggregate population of individuals is the "two-stage" approach. Many studies that aim to estimate model parameters in HIV infection dynamics have used this method for assessing the multivariate distribution of parameters across a population. With this method, data from each individual are analyzed separately and inference is conducted on the estimated parameters (Ho *et al.*, 1995; Luzuriaga *et al.*, 1999; Notermans *et al.*, 1998; Perelson *et al.*, 1997). Drawbacks to the two-stage method include the necessity of obtaining complete and comprehensive longitudinal series of measurements on all patients in a study.

Once a nonlinear hierarchy of the form (1), (2) and (3) is specified, in principle, the parameters of interest, $\boldsymbol{\beta}$ and $\boldsymbol{\Sigma}(\boldsymbol{\theta})$ may be estimated using standard inferential principles as an alternative to the two stage approach. Assuming that $\boldsymbol{\epsilon}_{ij}$ have a conditional probability density, e.g. normal or log normal, then $\boldsymbol{Y}_i = (\boldsymbol{Y}_{i1}^T, \ldots, \boldsymbol{Y}_{in_i}^T)$ have a conditional density $p(\boldsymbol{Y}_i|\boldsymbol{\beta})$ so that the marginal density of $\boldsymbol{Y}_i$ satisfies

$$p(\boldsymbol{Y}_i|\boldsymbol{\beta}, \boldsymbol{\Sigma}) = \int p(\boldsymbol{Y}_i|\boldsymbol{\beta}_i)\, dF(\boldsymbol{\beta}_i|\boldsymbol{\beta}, \boldsymbol{\Sigma}) \tag{4}$$

Using (4) and a specification for F, e.g. multivariate normal, inference may (in principle) be carried out by maximizing the likelihood for $\boldsymbol{\beta}$ and $\boldsymbol{\Sigma}$,

$$\prod_{i=1}^{m} p(\boldsymbol{Y}_i; \boldsymbol{\beta}, \Sigma)$$

based on the data vectors $\boldsymbol{Y}_i$.

However, in practice, even when an analytic solution for $\boldsymbol{X}(t, \boldsymbol{\beta})$ is available, challenges associated with maximization of the likelihood for $\boldsymbol{\beta}$ and $\boldsymbol{\Sigma}$ are significant, requiring the evaluation of the multi-dimensional integral in (4) for each of the m terms in the likelihood (where m is the number of individuals in the aggregate population). This integration is often intractable, even under parametric distributional assumptions for both $p(\boldsymbol{Y}_i|\boldsymbol{\beta})$ and F.

Other methods of estimation and inference in nonlinear random effects models include using closed-form approximations to the integrals in the marginal density (4). Software for implementing this approach is readily available and includes the FORTRAN program NONMEM (Beal *et al.*, 1998), the nlme() macro in Splus (Pinheiro *et al.*, 2000), and the SAS PROC NLMIXED based on the nlinmix SAS macro (Wolfinger *et al.*, 1995). Most of this software are set up so that an analytic form of the system of ODE's is required, something which will not be the case for many of the models we explore. The nlme() macro in Splus can be used with a numerical differential equation solver. A variety of HIV dynamics data have been successfully described using this approach (Ding *et al.*, 1999; Wu *et al.*, 1996, 1998, 1999, 1999).

In some cases, the primary goal of the analysis of longitudinal data which are modeled as (1), (2), and (3) is the estimation, inference, or hypothesis tests of the marginal parameters $\boldsymbol{\beta}$. In this case, we can ignore estimation of the variance/covariance for $\boldsymbol{\beta}_i$ and model the residual errors, ϵ_{ij}, to account for correlations in data from the same individual. It is generally assumed that the time points are sufficiently intermittent, and that the mean structure adequately describes temporal trends, so any additional correlation attributable to local fluctuations over time is negligible, however this component of correlation can be modeled with a time varying stochastic process (Diggle *et al.*, 2002). Available software for nonlinear marginal analysis include the gnls package in S-plus (Pinheiro *et al.*, 2000). These types of analyses are appropriate when differences in population level values are to be compared, for example if the goal of a study is to determine the effect of some type of treatment on viral or cell decay rates.

This section provides a brief overview of some of the statistical methods available for working with HIV dynamics data from an individual or population of individuals. The three frequentist methods described for dealing with aggregated data; the two-phase approach, marginal analysis using nonlinear regression techniques and nonlinear random effects methods, are all well developed in theory, but suffer various inadequacies. In addition, very little has been done in the area of study design for longitudinal studies where differential equations model parameters are the outcome of interest, although a few references are available (Marschner *et al.*, 1998; Wu *et al.*, 2000, 2004). Thus, there remains extensive opportunities for statisticians and numerical analysts to contribute to HIV dynamics research.

4. Hypothesis Testing for Viral Decay after Start of HAART

In linear systems of differential equations the decline (or growth) of a population is exponential, and depends on the size of the population in a linear way, i.e. the decay in the population is proportional (via a constant) to the size of the population. This model is referred to as the constant decay model. In a density dependent decay model, this proportionality is variable: the per capita decay rate depends on the size of the population, resulting in a model described by nonlinear differential equations which govern the population dynamics. Density dependant decay models are often used in population biology (Hoppensteadt *et al.*, 1982) as an alternative to long-term exponential growth or decay. In cell populations, density dependent homeostatic mechanisms are described for lymphocyte populations in mice (Tanchot *et al.*, 1995) and humans (Davey *et al.*, 1999; Haase *et al.*, 1999; Hockett *et al.*, 1999). In HIV research, time dependent decay of a single infected cell compartment was suggested as a possible alternative explanation for the biphasic decay pattern observed in HIV-1 decline by Arnaout *et al.* (2000) and Luzuriaga *et al.* (1999).

In this section, we provide an example of formal hypothesis testing of a model with the following null hypothesis:

H_0: *A linear differential equations model with two compartments of infected cells is sufficient to describe biphasic viral decay dynamics after treatment.*

and the alternative hypothesis:

H_a: *Density dependent decay more accurately describe the viral decay profile in the data analyzed.*

Our model and analysis can also be used to determine if in addition to density dependent decay, more than one population of infected cells is contributing to the overall population of viral RNA. The authors of Arnaout *et al.* (2000) and Luzuriaga *et al.* (1999) suggest the following null hypothesis: *Biphasic viral RNA decay data are the result of density dependent decay of a single population of infected cells, rather than the sum of virus from two populations of infected cells which both follow a simple exponential decay trajectory.* The alternative hypothesis is that *at least two populations of infected cells contribute to the overall viral RNA population, even in the presence of density dependent decay.*

The model used to test these hypothesis is very similar to the one proposed by Perelson *et al.* (1997), with the addition of a single parameter that will be used to test the null hypothesis of linear decay of two infected cell compartments. In fact, the model in the null hypotheses is equivalent to Perelson's model. In this model X represents a population of infected cells which produce large amounts of virus and die quickly, e.g. activated $CD4^+$ T-cells. The compartment Y represents a population of infected cells (possibly macrophages) which produce less virus but have a longer life expectancy, and V is the population of HIV viral RNA. The parameters δ, μ, and c represent the linear decay rates of the productively infected cells, longer-lived infected cells and viral RNA, respectively; p_x and p_y represent the contribution to the population of viral RNA by productively and longer-lived infected cells respectively. Finally, the parameter r measures the degree to which the data deviate from a biphasic linear decay model for viral decay after (fully efficacious) treatment.

$$\frac{dX}{dt} = -\delta X^r \tag{5}$$

$$\frac{dY}{dt} = -\mu Y^r \tag{6}$$

$$\frac{dV}{dt} = p_x X + p_y Y - cV \tag{7}$$

Note that the right-hand sides of equations (5) and (6) can be written as $(\delta X^{r-1})\, X$ and $(\mu Y^{r-1})\, Y$ respectively so that the per capita decay rate for the short-lived infected cells is δX^{r-1} and for long-lived infected cells is μY^{r-1}. If r is not equal to one, the per capita decay rate of these two populations depends on the density of the decaying population. If the parameter r is equal to one, this model reduces to the model presented in Perelson *et al.* (1999), which we refer to as the constant decay model for viral decay after initiation with HAART. Thus the primary hypotheses of interest is $r = 1$ (simple exponential decay of two infected cell compartments), vs. the alternative hypothesis $r > 1$ (density dependent decay). The other relevant hypothesis test is $p_x = 0$ or $p_y = 0$ (only one population of infected cells contributing to the total observed viral load) vs. $p_x > 0$ and $p_y > 0$ (at least two populations of infected cells contribute to the total observed viral load).

To perform our hypothesis test we used data collected on children after they start treatment. These data have been previously analyzed (Melvin *et al.*, 1999) for biphasic linear decay using Perelson's model. Details pertaining to data collection and methods can be found in Melvin *et al.* (1999). Using data from children was important for this analysis, since the time between treatment start until the patient achieves an undetectable viral load is much longer in children than in adults. The time during which viral load is detectable in adults who start treatment is short enough so that the constant decay model is likely to capture all the relevant information in the data, since over shorter periods of time, linear models are good approximations of nonlinear models. Table 1 provides baseline viral loads for the children included in our analysis.

Since not all parameters can be estimated from the data, we assume $c = 3$ (Ho *et al.*, 1995; Perelson *et al.*, 1996, 1997). To assess the robustness of this assumption, we fit all the models with $c = 5, 10, 20$ and 30 and found no qualitative differences in our results. To accurately estimate

Table 1. Baseline characteristics of participants.

	HIV-1 plasma per mL	HIV-1 plasma total body	Proviral DNA per mm^3	Proviral DNA total body
Child 1	$1.14 * 10^5$	$3.94 * 10^9$	521	$5.58 * 10^7$
Child 2	$3.28 * 10^6$	$5.54 * 10^{10}$	1930	$3.81 * 10^8$
Child 3	$9.96 * 10^4$	$2.65 * 10^9$	148	$1.21 * 10^7$
Child 4	$2.17 * 10^5$	$5.76 * 10^9$	529	$1.63 * 10^8$
Child 5	$1.97 * 10^6$	$1.92 * 10^{10}$	382	$3.73 * 10^7$
Child 6	$1.56 * 10^6$	$5.19 * 10^9$	236	$1.46 * 10^7$
Median	$8.89 * 10^5$	$5.48 * 10^9$	452	$4.66 * 10^7$

density dependent decay, we must work with total body values of plasma viremia at each time point, which we estimated using our measurements of viral RNA per mL of plasma adjusted using each individual's weight and age at the time the measurement was obtained (Barone *et al.*, 1996). In addition, we must have an estimate of the initial size of the total body populations of short-lived and long-lived infected cell populations, x_0 and y_0, which we estimated from proviral DNA measurements obtained at the time treatment was started. Our measurements of proviral DNA were not distinguished by cell type or replication competence, so we considered a number of possible distributions of proviral DNA between short-lived and long-lived virus producing cells. Based on data in Chun *et al.* (1997a, 1997b), Derdeyn *et al.* (1999), Haase *et al.* (1999), and Rosok *et al.* (1997), it seems likely that not more than 5% of the total proviral load in both peripheral blood and lymphoid tissue is involved in active viral replication. The remainder is most likely transcriptionally silent replication competent provirus or defective or incomplete provirus (Haase *et al.*, 1999). Previous estimates (Melvin *et al.*, 1999; Perelson *et al.*, 1997) suggest that approximately 95% of the viral RNA measured in plasma is produced by infected cells with a short life-expectancy with the remainder having a longer life-expectancy and possibly producing less infectious virus. We used this estimates to guide our choices of x_0 and y_0 based on observed proviral DNA at start of treatment.

In our primary analysis, we assume that 5% of the total measured provirus is replication competent, and that 80% of the provirus involved in replication composes the population of short-lived virus producing infected cells, with the remaining 20% in the population of long-lived virus producing infected cells. The estimates, confidence intervals, and figures we report are based on these assumptions. To assess the robustness of these assumptions, we fit the model allowing the percentage of provirus that are involved in viral production to vary from 3 to 10%, and the distribution of this provirus in the pool with shorter life expectancy to vary from 70 to 95%. We also varied the assumption of the percentage of total body populations of viral RNA and infected cells populations measured in blood and plasma, assuming that 1% of the peripheral measurements of viral RNA and proviral DNA represent the total body loads. We concluded that the qualitative results were robust to assumptions, based on this sensitivity analysis.

Our estimates of time to eradication of the short-lived infected and long-lived infected cell populations under the density dependent decay model were obtained by using equations (8) and (9) to determine the time required for the initial populations (estimated as the median of the initial total body short-lived and long-lived infected cell populations) to decay to one infected cell. For the constant decay model, we used first-order exponential trajectories for the infected cell populations to determine the time to decay to one infected cell.

We fit the model to all six children simultaneously and performed careful diagnostics to determine that within subject correlations did not affect our inference. To obtain the model solutions for plasma viremia, short-lived infected cells, and long-lived infected cells, we solved equations (5) and (6) under the assumption that $r \neq 1$ to obtain

$$X(t) = \left[\delta(r-1)t + x_0^{1-r}\right]^{\frac{1}{1-r}} \tag{8}$$

$$Y(t) = \left[\mu(r-1)t + y_0^{1-r}\right]^{\frac{1}{1-r}} \tag{9}$$

and substituted these solutions into equation (7) to obtain

$$\frac{dV}{dt} = p_x\left[\delta(r-1)t + x_0^{1-r}\right]^{\frac{1}{1-r}} + p_y\left[\mu(r-1)t + y_0^{1-r}\right]^{\frac{1}{1-r}} - cV$$

Using a numerical differential equation solver, we used the numerical solution for $V(t, r, \delta, \mu, p_x, p_y)$ and nonlinear least squares to obtain parameters estimates. Confidence intervals were calculated using profile likelihood. The estimates of r, p_x, and p_y from the density dependent and constant decay model and associated 95% confidence intervals are shown in Table 2. The confidence interval for the parameter r does not contain one, indicating that density-dependent decay plays a role in the decline of infected cell populations after initiation of treatment. In other words, the assumption of simple exponential decay for both short-live and long-lived infected cells is violated for the data we analyzed. The estimates of p_x and p_y, the viral

Table 2. Parameter estimates from constant and density dependent decay models with 95% confidence intervals.

	r	p_x	p_y
		Density dependent decay	
Estimate	1.435	431.5	22.7
95% CI	(1.396, 1.479)	(252.2, 902.7)	(10.0–49.5)
		Constant decay	
Estimate	1	253.1	60.0
95% CI	NA	(107.7, 543.0)	(27.5, 132.9)

production rates for short-lived and long-lived infected cells, are both significantly greater than zero, indicating that at least two populations of infected cells are contributing to the viral pool in the presence of density dependent decay. Finally, our estimates of both δ and μ (data not shown, see Holte *et al.* (2001, 2002)), the decay coefficients for the short-lived and long-lived infected cell populations, are both significantly different from zero, suggesting that there is an overall decline in both infected cell populations. Both the estimated density dependent and constant decay model trajectories along with the observed plasma viremia data are shown in Fig. 1(A).

The model based trajectories and predicted time to eradication for the productively infected and long-lived infected cell populations under the density dependent decay and constant decay models using parameter estimates obtained from the analysis of the aggregate data are shown in Figs. 1(B) and (C). These estimates indicate that the time to eradication for both cell populations is much longer under the density dependent decay model than under the constant decay model. The density dependent decay model predicts years to eradication of the short-lived infected cell population, rather than weeks or months as predicted by the constant decay model. Furthermore, the density dependent decay model predicts that long-lived infected cells may not be eradicated within an infected individuals lifetime in contrast to predictions of 3–5 years to eradication under the constant decay model.

The conclusion should not be made that the density dependant decay model is correct, or that the associated estimates of time to viral eradication are likely to be accurate. A more reasonable conclusion is that the constant decay model does not accurately represent long-term dynamics of viral decay after HAART, in other words, we have rejected the null hypothesis that the constant decay model is appropriate for the data at

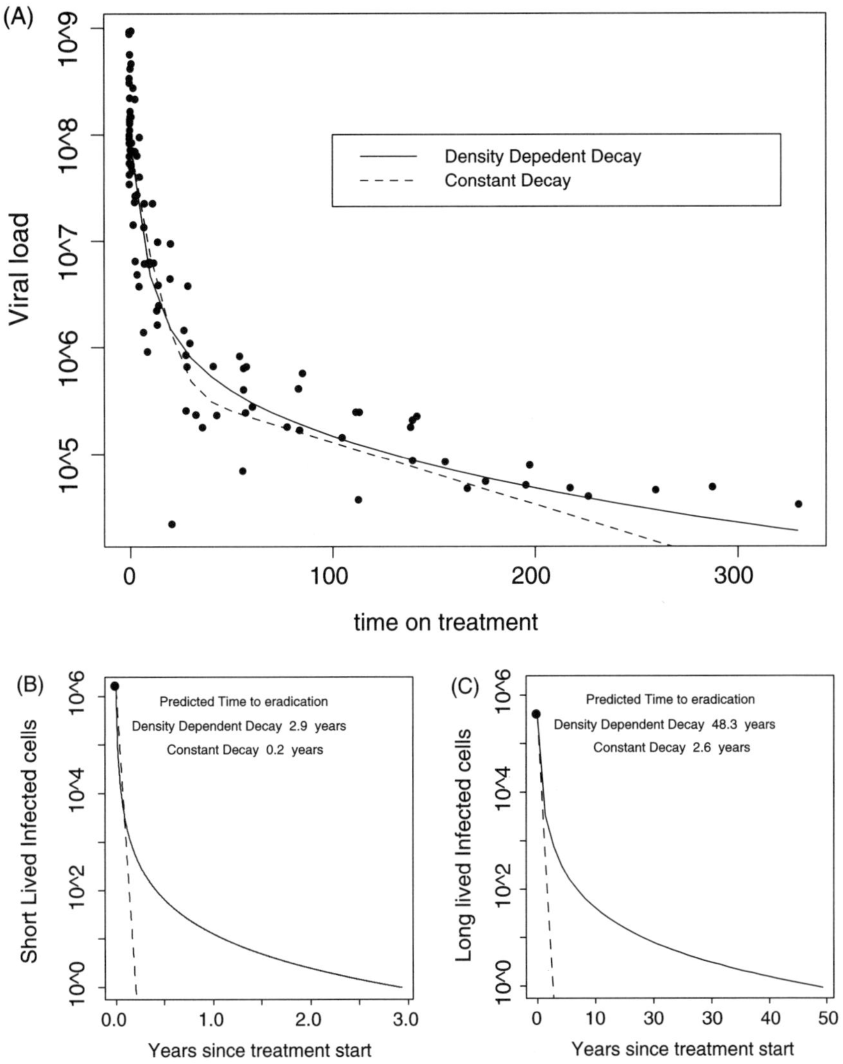

Fig. 1. Total body plasma HIV-1 RNA by months since start of treatment for children with fitted trajectories from the density dependent and constant decay models Panel (A). Theoretical trajectories of the total body short-lived infected cell population by months since start of treatment (Panel B) and trajectories of the total body long-lived infected cell population by years since start of treatment (Panel C). For all panels, solid lines indicate fitted model trajectories from the density dependent decay model and dashed lines indicate fitted model trajectories from the constant decay model.

hand. However, the density dependant decay model predictions are in closer agreement with recent data on persistence of viral replication in successfully treated individuals and can explain why viral rebound is often observed quite quickly when therapy is discontinued (Davey *et al.*, 1999) or why viral replication continues to be detected after year of successful therapy (Chun and Fauci *et al.*, 1999; Zhu *et al.*, 2002). The earliest estimates of decay rates of infected cell populations which produced the estimates of time to eradication of infected cells (Notermans *et al.*, 1998; Perelson *et al.*, 1997) are based on the assumption that the per capita rates of decay are constant over the entire time course required for eradication. Our analysis shows that per capita decay rates are not consistant and vary according to the size of the decaying population. These results imply that investigators should continue to measure infected cell decay rates over different periods of time, since it is likely that these decay rates are decelerating over time. In addition, any intervention that would prevent this deceleration of decay rates, such as a theraputic vaccine, which might increase the probability of total viral eradication.

5. Discussion

HIV is an immensely complicated disease, with many questions remaining regarding its pathogenic properties and etiology, even after 25 years of intense research. It is also a disease that has benefitted from the use of mathematical models; both applied models used with clinical and experimental data and theoretical models which suggest directions for future research. This chapter provides a summary of some of the mathematical modeling of HIV viral dynamics accomplished to date, with a description of statistical models and methods appropriate for analyzing data with these mathematical models. What has been missing from this overall successful marriage of quantitative and experimental/clinical science is formal model comparison. We propose that in order to make further significant contributions to HIV research, the models used should be treated as formal hypotheses to be tested. A basic model, pared down to the simplest terms, can be treated as a null hypothesis, with more elaborate extensions viewed as alternative hypothesis. Standard hypothesis testing in any standard statistical analysis offers a null hypothesis and an alternative hypothesis. The goal of statistical analysis is to determine whether the data are sufficient to reject the null hypothesis in favor of the alternative hypothesis. Failure to do so does not necessarily imply that the null hypothesis is correct, only that the data are not sufficient to reject it in favor of the alternative hypothesis. In the same way, "null models" can be rejected based on available data, and analysis of this type can serve as a first step in choosing a model most appropriate for the data under consideration. As the types of data available for analysis become more complex, so will the models which describe it, and consideration of appropriate models will be an important factor in successful modeling research in HIV.

References

Arnaout RA, Nowak MA and Wodarz D (2000) HIV-1 dynamics revisited: biphasic decay or cytotoxic T lymphocyte killing? *Proc Roy Soc Lond B* **267**: 1347–1354.

Barone MA (ed.) (1996) *The Harriet Lane Handbook. A Manual for Pediatric House Officers*, 14th ed., St. Louis, MO: Mosby-Year Book.

Beal SL and Sheiner LM (1998) *NONMEM User's Guides*, Version V, NONMEM Project Group, University of California, San Fransisco.

Bonhoeffer S, Coffin JM and Nowak MA (1997) Human immunodeficiency virus drug therapy and virus load, *J Virol* **71**: 3275–3278.

Bonhoeffer S, Rembiszewski M, Ortiz GM and Nixon DF (2000) Risks and benefits of structured antiretroviral drug therapy interruptions in HIV-1 infection, *AIDS* **14**: 2313–2322.

Chun T-W, Stuyver L, Mizell SB *et al.* (1997) Presence of an inducible HIV-1 latent reservoir during highly active antiretroviral therapy, *Proc Natl Acad Sci* **94**: 13193–13197.

Chun TW, Carruth L, Finzi D *et al.* (1997a) Quantification of latent tissue reservoirs and total body viral load in HIV-1 infection, *Nature* **387**: 183–188.

Chun T-W *et al.* (1997b) Presence of an inducible HIV-1 latent reservoir during highly active antiretroviral therapy, *Proc Natl Acad Sci* **94**: 13193–13197.

Chun T-W, Enge D, Berrey MM *et al.* (1998) Early establishment of a pool of latently infected, resting CD4+ T-cells during primary HIV-1 infection, *Proc Natl Acad Sci* **95**: 8869–8873.

Chun T-W, Davey RT, Engel D, Lane HC and Fauci AS (1999) Re-emergence of HIV after stopping therapy, *Nature* **401**: 874–875.

Chun T-W and Fauci AS (1999) Latent reservoirs of HIV: Obstactles to the eradication of virus, *Proc Natl Acad Sci* **96**: 10958–10961.

Costantino RF, Desharnais RA, Cushing JM and Dennis B (1997) Chaotic dynamics in an insect population, *Science* **275**: 389–391.

Dalod M *et al.* (1999) Broad, intense anti-human immunodeficiency virus (HIV) ex vivo CD8+ responses in HIV type 1-infected patients: Comparison with anti-epstein-barr virus responses and changes during antiretroviral therapy, *J Virol* **73**: 7108–7116.

Davey RT, Bhat N, Yoder C *et al.* (1999) HIV-1 and T-cell dynamics after interruption of highly active antiretroviral therapy (HAART) in patients with a history of sustained viral suppression, *Proc Natl Acad Sci* **96**: 15109–15114.

Davidian M and Giltinan DM (1995) *Nonlinear Models for Repeated Measurement Data*, Chapman and Hall/CRC, Boca Raton, FL.

Derdeyn CA, Kilby JM, Miralles GD *et al.* (1999) Evaluation of distinct blood lymphocyte populations in human immunodeficiency virus type-1-infected subjects in the absence or presence of effective therapy, *J Infect Dis* **180**: 1851–1862.

Diggle PJ, Heagerty P, Liang K-Y and Zeger SL (2002) *Analysis of Longitudinal Data*, Oxford University Press, Oxford.

Ding AA and Wu H (1999) Statistical tests in nonlinear mixed-effects models with application to potency comparison of anti-HIV therapies, in *ASA Proceedings of the Biopharmaceutical Section, American Statistical Association* , pp. 58–63 Alexandria, VA.

Finzi D, Blankson J *et al.* (1999) Latent infection of CD4+ T-cells provides a mechanism for lifeling persistence of HIV-1, even in patients on effective combination therapy, *Nat Med* **5**(5): 512–517.

Furtado MR, Callaway DS, Phair JP *et al.* (1999) Persistence of HIV-1 transcription in peripheral blood mononuclear cells in patients receiving potent antiretroviral therapy, *New Eng J Med* **340**(21): 1614–1622.

Gray CM *et al.* (1999) Frequency of class I HLA-restricted anti-HIV CD8+ T-cells in individual receiving highly active antiretroviral therapy (HAART), *J Immunol* **162**: 1780–1788.

Haase AT (1999) Population biology of HIV-1 infection: viral and CD4+ T-cell demographics and dynamics in Llymphatic tissues, *Ann Rev Immunol* **17**: 625–656.

Ho DD, Neumann AU, Perelson AS *et al.* (1995) Rapid turnover of plasma virions and CD4 lymphocytes in HIV-1 infection, *Nature* **373**: 123–126.

Hockett RD, Kibly JM, Derdeyn CA *et al.* (1999) Constant mean viral copy number per infected cell in tissues regardless of high, low, or undetectable plasma HIV RNA, *J Exp Med* **189**: 1545–1554.

Holte S, Melvin A, Mullins J and Frenkel L (2001) Density dependent decay in HIV-1 dynamics after HAART, in *Eighth Conference on Retroviruses and Opportunistic Infections*, Abstract Number 394.

Holte S, Melvin A, Mullins J and Frenkel L (2002) Density dependent decay in HIV-1 dynamics after HAART. Preprint.

Hoppensteadt FC (1982) *Mathematical Methods of Population Biology*, Cambridge University Press, New York.

Kalams SA *et al.* (1999) Levels of human immunodeficiency virus type 1-specific cytotoxic T-lymophocyte effector and memory responses decline after suppression of viremia with highly active antiretrovrial therapy, *J Virol* **73**: 6721–6728.

Klenerman P, Phillips R, Rinaldo CR *et al.* (1996) Cytotoxic T lymphocytes and viral turnover in HIV type 1 infection, *Proc Natl Acad Sci USA* **93**: 15323–15328.

Luzuriaga K, Wu H, McManus M *et al.* (1999) Dynamics of human immunodeficiency virus type 1 replication in vertically infected infants, *J Virol* **73**(1): 362–367.

Marschner I (1998) Design of HIV viral dynamics studies, *Stat Med* **17**: 2421–2434.

Melvin AJ, Rodrigo AG, Mohan KM *et al.* (1999) HIV-1 dynamics in children, *J AIDS* **20:** 468–473.

Muller V, Vigueras-Gomez JF and Bonhoeffer S (2002) Decelerating decay of latently infected cells during prolonged therapy for human immunodeficiency virus type 1 infection, *J Virol* **76**(17): 8963–8965.

Neumann AU, Tubiana R, Calvez V *et al.* (1999) HIV-1 rebond during interruption of highly active antiretroviral therapy has no deleterious effect on reinitiated treatment, *AIDS* **13**: 677–683.

Notermans DW, Goudsmit J, Danner SA *et al.* (1998) Rate of HIV-1 decline following antiretroviral therapy is related to viral load at baseline and drug regimen, *AIDS* **12**: 1483–1490.

Ogg GS *et al.* Decay kinetics of human immunodeficiency virus-specific effector cytotoxic T lymphocytes after combination antiretroviral therapy, *J Virol* **73**: 797–800.

Palumbo PE, Kwok S, Waters S *et al.* (1995) Viral measurement by polymerase chain reaction-based assays in human immunodeficiency virus-infected infants, *J Pediat* **126**: 592–595.

Perelson AS, Neumann AU, Markowitz M, Leonard JM and Ho DD (1996) HIV-1 dynamics *in vivo*: virion clearance rate, infected cell life-span, and viral generation time, *Science* **271**: 1582–1586.

Perelson AS, Essunger P *et al.* (1997) Decay characteristics of HIV-1 infected compartments during combination therapy, *Nature* **387**: 188–191.

Pinheiro JC and Bates DM (2000) *Mixed Effects Models in S and S-PLUS*. Springer-Verlag, New York.

Pitcher C, Quittner C, Peterson D *et al.* (1999) HIV-1-Specific CD4+ T-cells are detectable in most individuals with active HIV-1 infection, but decline with prolonged viral suppression, *Nat Med* **5**: 518–525.

Press WH, Teukolsky SA, Vetterline WT and Flannery BP (1986) *Numerical Recipes in Fortran*. Cambridge University Press, New York.

Ramratnam B, Mittler JE, Zhang L *et al.* (2000) The decay of the latent reservoir of replication-competent HIV-1 is inversely correlated with the extent of residual viral replication during prolonged anti-retroviral therapy, *Nat Med* **6**(1): 82–85.

Reibnegger RR *et al.* (1989) Stability analysis of simple models for immune cells interacting with normal pathogens and immune system retroviruses, *Proc Natl Acad Sci USA* **86**: 2026–2030.

Rosok B, Brinchmann JE, Voltersvik P *et al.* (1997) Correlates of latent and productive HIV type-1 infection in tonsillar CD4+ T cells, *Proc Natl Acad Sci USA* **94**: 9332–9336.

Speigel HM *et al.* (1999) Changes in frequency of HIV-1 specific cytotoxic T-cell precursors and circulating effectors after combination antiretroviral therapy in children, *J Infect Dis* **180**: 359–368.

Tanchot C and Rocha B (1995) The peripheral T cell repertoire: independent homeostatic regulation of virgin and activated CD8+ T cell pools, *Eur J Immunol* **25**: 2127–2136.

Tetali S, Abrams E, Bakshi S *et al.* (1996) Virus load as a marker of disease progression in HIV-infected children, *AIDS Res Hum Retrov* **12**: 669–675.

Venzon DJ and Moolgavkar SH (1988) A method for computing profile-likelihood-based confidence intervals, *App Stat* **37**: 87–94.

Wakefield J, Smith AFM, Racine-Poon A and Gelfand AE (1994) Bayesian analysis of linear and non-linear populations models by using the Gibbs sampler, *App Stat* **43**: 201–221.

Wakefield J (1996) The Bayesian analysis of population pharmacokentic models. *J Amer Stat Assoc* **91**: 62–75.

Wei X, Ghosh SK, Taylor ME *et al.* (1995) Viral dynamics of HIV-1 infection, *Nature* **373**: 117–122.

Wolfinger R (1995) *The NLINMIX macro.* SAS Institute, Cary, North Carolina.

Wu H and Tan WY (1996) Modeling and monitoring the progression of HIV infection using nonlinear Kalman filter, in *ASA Proceedings of the Biometrics Section*, American Statistical Association, pp. 372–377, Alexandria, VA.

Wu H, Ding A and DeGruttola V (1998) Estimation of HIV dynamic parameters, *Stat Med* **17**: 463–485.

Wu H, Zhang B and Spritzler J (1999) Modeling and analyzing cellular kinetics for HIV-infected patients treated with potent antiviral therapies in AIDS clinical trials, in *ASA Proceedings of the Biometrics Section*, American Statistical Association, pp. 123–128, Alexandria, VA.

Wu H and Ding A (1999) Population HIV-1 dynamics *in vivo*: Applicable models and inferential tools for virological data from AIDS clinical trials, *Biometrics* **55**: 410–418.

Wu H, Kuritzkes DR, McClernon DR *et al.* (1999) Characterization of viral dynamics in human immunodeficiency virus type 1-infected patients treated with combination antiretroviral therapy: relationships to host factors, cellular restoration and virlogic end points, *J Infect Dis* **179**: 799–807.

Wu H and Ding A (2000) Design of viral dynamic studies for efficiently assessing anti-HIV therapies in AIDS clinical trials, *Biometrical J* **2**: 175–196.

Wu H (2004) Statistical methods for HIV dynamics studies in AIDS clinical trial, *Stat Method Med Res*, in press.

Zhang L, Ramratnam B, Tenner-Racz K *et al.* (1999) Quantifying residual HIV-1 replication in patients receiving combination antiretroviral therapy, *New Eng J Med* **340**(21): 1605–1613.

Zhu T, Holte S, Chang Y *et al.* (2002) Evidence for human immunodefieciency virus type 1 replication *in vivo* in CD14+ monocytes and its potential role as a source of virus in patients on highly active antiretroviral therapy, *J Virol* **76**(2): 707–716.

CHAPTER 23

CONVERGENCE TO AN ENDEMIC STATIONARY DISTRIBUTION IN A CLASS OF STOCHASTIC MODELS OF HIV/AIDS IN HOMOSEXUAL POPULATIONS

Charles J. Mode

Department of Mathematics
Drexel University
Philadelphia, PA 19104

In recent years, the incidence of infection with HIV among young homosexual men has been on the rise in developed countries. It is, therefore, of interest to study stochastic models of HIV/AIDS in which HIV infections become endemic in a population of homosexuals. Within a stochastic paradigm, the idea of an endemic equilibrium corresponds to the mathematical concept of a stochastic process converging in distribution to a stationary distribution. In Mode and Sleeman (2002), it was conjectured that in a Monte Carlo simulation experiment convergence to a stationary distribution of a Markov chain was being observed. In this paper, a proof of this conjecture is constructed under a reasonable sufficient condition within an abstract framework that resembles a general subcritical branching process. The stationary distribution turns out to be a mixture of conditionally independent Poisson densities and the mixing measure is that underlying the population process. It is often difficult to decide in a Monte Carlo simulation experiment whether convergence to a stationary distribution is actually being observed, which sometime leads to controversies. Knowing that convergence to a stationary distribution does indeed occur, under a plausible sufficient in the class of models considered in the paper, will provide a firm mathematical basis for avoiding controversies. In the literature on stochastic models of HIV/AIDS, the class of models considered in this paper is often referred to as chain multinomial models.

Keywords: HIV/AIDS; homosexual population; convergence in distribution; probability generating functions; Markov chains; convergence of embedded deterministic model; chain multinomial models.

1. Introduction

In Mode and Sleeman (2002), it was conjectured that convergence to a stationary distribution was being observed in a Monte Carlo simulation experiment with a model designed to study an epidemic of HIV/AIDS in a population of homosexuals. At first glance it may seem to some that whether convergence occurred in the experiment would be obvious. If the type of convergence were point-wise to a set of points, as is the case in deterministic models, then it would have been possible to decide from the experiment whether convergence had indeed occurred. However, unlike point-wise convergence, when convergence is in the sense of distribution, it will usually not be clear as to whether convergence has occurred, because variability among the sample of realizations of the process will still be observed even in the limit as $t \to \infty$.

The particular focus of attention in this experiment was the study of the role that recurrent rare events may play in the transmission of a virus from a species that had evolved resistance to it to a susceptible species. It was also conjectured that a simple example would be useful in suggesting approaches to proving the conjecture. After a deeper examination of the problems involved in constructing a proof of the conjecture, however, it was found that it was actually easier to work with an abstract framework rather than a simpler model. Although abstract frameworks are actually implied by the models considered in Chapters 10, 11, and 12 of the book (Mode and Sleeman, 2000), they overlooked, because the thought processes involved in abstract mathematical thinking are quite different from those exercised in the book, where the construction of Monte Carlo simulation algorithms and writing software for their computer implementation was emphasized. Accordingly, the purpose of this paper is to present an abstract framework within which reasonable sufficient conditions may be found to prove that convergence to a stationary distribution may have actually been observed in the experiment.

Before proceeding to the presentation of the results, it is desirable that connections be made with other recently published papers on stochastic models of epidemics. One of these papers is that of Nasell (2002) in which attention was focused on approximations of quasi-stationary distributions and times to extinction in stochastic versions of SI, SIS, SIR and SIRS models. Another recent paper on a related subject was that of Ball and Neal (2002), where a general model for stochastic SIR epidemic with two levels of mixing was studied. Among many other things, these later authors used multitype branching process approximations to an epidemic model in derivations of threshold conditions. In both these papers, the basic class of stochastic process underlying their models was a Markov jump process in continuous time. In this paper, however, the abstract framework defined and analyzed in what follows belongs to a class of processes that are closely related to multitype branching processes with immigration. More specifically, they are related to sub-critical multitype general branching processes with immigration and state or density dependence. The book (Mode, 1970), as well as other literature on branching processes, may be consulted for further details on the non-density dependent, multitype Galton–Watson process with immigration. An interesting book by Tan (2000) also contains discussions of stochastic models of HIV/AIDS that are related to the class of model considered in this paper but differ significantly in details.

The structure of the paper is as follows. In Section 2, a discrete time Markov chain model for the life cycle of individuals with a finite state space, whose transition matrix is dependent on the state of the population at any time, is introduced, and in Section 3, it is shown that this model for individuals, along with a distribution for recruits entering the population at any time, determine a transition matrix for the population process, which is also a discrete time Markov chain with an infinite state space. Section 3 also concludes in a brief description of a probability space, which is needed to accommodate convergence in distribution and whose measure is determined by the Markov transition matrix of the population process. In Section 4, a formula for the probability generating function of the one-dimensional distributions of the population process at any time t is derived under the assumption that the number of recruits entering the population at any time follows a Poisson distribution. In Section 5, it is shown under a sufficient condition that the population process converges to a stationary distribution as $t \to \infty$, which is a mixture of conditionally independent Poisson densities. Moreover, the mixing measure is that for the probability space underlying the population process. Finally, in Section 6 it is shown that, under the sufficient condition used in proving the population process converges to a stationary distribution, the embedded deterministic model also converges to a limit as $t \to \infty$, as was observed for the parameters point consider in a computer experiment reported in Mode and Sleeman (2002) as well as in other experiments.

The results of the paper will apply to the models in Chapter 10 of Mode and Sleeman (2002) and it seems likely that modifications of the techniques used in this paper may also be applied to the models of Chapters 11 and 12 of this book, where epidemics of sexually transmitted diseases, with the formation and dissolution of partnerships in homosexual or heterosexual populations, are considered.

2. Individuals, Population and Population Process

In a model of the life cycle of individuals, let $\mathfrak{S}_1$ denote the set of $r_1 \geq 1$ absorbing states and let $\mathfrak{S}_2$ denote the set of $r_2 \geq 1$ transient states, which represent types of individuals in a population. By definition, the state space for the life cycle model for individuals is the set

$$\mathfrak{S} = \mathfrak{S}_1 \cup \mathfrak{S}_2, \tag{2.1}$$

containing $r = r_1 + r_2$ states. By way of a concrete example, in Chapter 10 of Mode and Sleeman (2000), in a model for an epidemic of HIV/AIDS in a population of homosexual males, the set of transient states was chosen at the set of ordered pairs

$$\mathfrak{S}_2 = ((j,k) \mid j = 1,2,\ldots,m_1; k = 0,1,2,\ldots,n_1), \tag{2.2}$$

where j denotes a behavioral class defined in terms of the number sexual contacts per unit time, $k = 0$ denotes a susceptible person and $k = 1,2,\ldots,n_1$ denotes a person in a stage of HIV disease. In one class of models considered in Chapter 10, there were $r_1 = 2$ absorbing states, denoting background death and deaths due to disease, four stages of disease expressed in terms of the concentration of $CD4^+$ cells per unit volume, and $m_1 = 3$ behavioral classes. Thus, in this model, because the state $k = 0$ was a transient state, the number of transient states was $r_2 = 3 \times 5 = 15$. In this paper, attention will be focused on the general classes of models considered in Chapter 10 of the book cited above (Mode and sleeman, 2000).

Let

$$\mathbb{N}_1^+ = (n \mid n = 0,1,2,\ldots) \tag{2.3}$$

denote the set of non-negative integers, and let $\mathbb{N}_{r_2}^+$ denote the Cartesian product of the set $\mathbb{N}_1^+$ with itself r_2 times. Every point $\mathbf{x} \in \mathbb{N}_{r_2}^+$ is a vector

$$\mathbf{x} = \left(x_1, x_2, \ldots, x_{r_2} \mid x_k \in \mathbb{N}_1^+ \text{ for } k = 1,2,\ldots,r_2\right). \tag{2.4}$$

The state of the population at any time is, by definition, a vector $\mathbf{x}$ of non-negative integers.

Just as in the branching process, the structure of the models flows from a model for the life cycle of individuals after they become sexually mature. By construction, individuals in a population evolve according to a discrete time parameter Markov chain, whose one-step transition matrix $\mathbf{\Pi}(\mathbf{x})$ depends on the state $\mathbf{x}$ of the population at any time. This matrix has the partitioned form

$$\mathbf{\Pi}(\mathbf{x}) = \begin{bmatrix} \mathbf{I}_{r_1} & \mathbf{O} \\ \mathbf{\Pi}_{21}(\mathbf{x}) & \mathbf{\Pi}_{22}(\mathbf{x}) \end{bmatrix}, \tag{2.5}$$

where $\mathbf{I}_{r_1}$ is an identity matrix of order $r_1, \mathbf{O}$ is a $r_1 \times r_2$ matrix of zeros, $\mathbf{\Pi}_{21}(\mathbf{x})$ is a $r_2 \times r_1$ matrix of conditional probabilities governing transitions from the set $\mathfrak{S}_2$ of transient states to the set $\mathfrak{S}_1$ of absorbing states and $\mathbf{\Pi}_{22}(\mathbf{x})$ is a $r_2 \times r_2$ matrix of conditional probabilities governing transitions among the states of the set $\mathfrak{S}_2$ of transient states. Elements of the latter two matrices will be denoted by $\pi_{ij}(\mathbf{x})$. For example,

$$\mathbf{\Pi}_{22}(\mathbf{x}) = (\pi_{ij}(\mathbf{x})). \tag{2.6}$$

A detailed account of the algorithms used to compute numerical values of the elements of the matrix $\mathbf{\Pi}(\mathbf{x})$ for every $\mathbf{x} \in \mathbb{N}_1^+$, using the classical theory of competing risks, may be found in the book cited previously (Mode and Sleeman, 2000). In particular, Chapter 10 of this book contains an account of these algorithms for the case of a one-sex model of an epidemic of HIV/AIDS in a population of homosexuals considered in this paper.

With regard to the evolution of the population, the matrix $\mathbf{\Pi}(\mathbf{x})$ for individuals, along with a distribution of recruits entering the population at any time, determines a Markov chain with state space $\mathbb{N}_{r_2}^+$ and a stationary matrix of transition probabilities. When computing Monte Carlo realizations of the population process, it is assumed that the discrete time set for the evolution of the process is the lattice

$$\mathbb{T}_h = (t \mid t = kh \text{ for } k = 0, 1, 2, \ldots), \tag{2.7}$$

where $h > 0$ is the length of the time unit used in an experiment. For the sake of brevity, however, any element of $\mathbb{T}_h$ will be denoted by t. Because the discrete time process may be viewed as an approximation to a process in continuous time, in the software implementing any model in the class under consideration, h is a parameter that may be varied so as to test empirically changes among Monte Carlo realizations of the process as h varies from small to larger values.

Given an initial state vector $X(0) = \mathbf{x}_0$ for the population at time $t = 0$, let the vector $\mathbf{X}(t)$ denote the state of the population at time $t = 1, 2, \ldots$, and let $\mathbf{X}_R(t)$, $t = 1, 2, \ldots$, represent a sequence of independent random variables with values in $\mathbb{N}_{r_2}^+$ with the common probability density function $(p.d.f.)$

$$g_R(\mathbf{x}) \tag{2.8}$$

for $\mathbf{x} \in \mathbb{N}^+_{r_2}$, governing the number of recruits that enter the population during per unit time. Furthermore, if at time t the population is in state $\mathbf{x}$, let $P(\mathbf{x}, \mathbf{y})$ denote the conditional probability that at time $t+1$ the population is in state $\mathbf{y}$. In symbols,

$$P(\mathbf{x}, \mathbf{y}) = P\left[\mathbf{X}(t+1) = \mathbf{y} \mid \mathbf{X}(t) = \mathbf{x}\right] \tag{2.9}$$

for all $t = 0, 1, 2, \ldots$ and vectors $\mathbf{x}$ and $\mathbf{y}$ in $\mathbb{N}^+_{r_2}$. In the next section, it will be shown that the matrix in (2.5) along with the density for recruits in (2.8) determine a transition matrix $\mathbf{P}$ for the population process with elements in (2.9).

3. Determination of Transition Matrix for Population Process

Probability generating functions $(p.g.f.'s)$ will play a basic role in the derivation of the results presented in this and subsequent sections. By way of illustration, let $\mathbf{z} = (z_1, z_2, \ldots, z_{r_2})$ be any vector of complex or real numbers such that $\mid z_k \mid \leq 1$ for all $k = 1, 2, \ldots, r_2$. In what follows, the z-variables in a $p.g.f.$ may carry one or more subscripts. Then, by definition, the $p.g.f.$ of the probability distribution for recruits is

$$G_R(\mathbf{z}) = \sum_{\mathbf{x}} g_R(\mathbf{x}) \prod_{k=1}^{r_2} z_k^{x_k}, \tag{3.1}$$

where the sum runs over all vectors $\mathbf{x} = (x_1, x_2, \ldots, x_{r_2}) \in \mathbb{N}^+_{r_2}$. Similarly, let

$$(\pi_{ij}(\mathbf{x}) \mid j \in \mathfrak{S}) \tag{3.2}$$

be the vector of conditional probabilities in the row of the matrix in (2.5) corresponding to transient state $i \in \mathfrak{S}_2$. Then, by definition, the $p.g.f.$ for this row of probabilities for state $\mathbf{x}$ of the population is

$$H_i(\mathbf{x};\, z_{i1}, z_{i2}, \ldots, z_{ir_2}) = \sum_j \pi_{ij}(\mathbf{x}) z_{ij}, \tag{3.3}$$

where the sum runs over all states $j \in \mathfrak{S}$.

The Monte Carlo algorithm for computing realizations of the process entails the computation of samples from a set of multinomial distributions corresponding to each transient state $i \in \mathfrak{S}_2$. For example, suppose that at time t the state vector for the population has the value

$$\mathbf{X}(t) = \mathbf{x} = (x_1, x_2, \ldots, x_{r_2}) \in \mathbb{N}^{+}_{r_2}. \tag{3.4}$$

By assumption, given the state $\mathbf{x}$ of the population at time t, during a time interval t to $t+1$, individuals in transient state $i \in \mathfrak{S}_2$ at time t undergo transitions independently of some state $j \in \mathfrak{S}$ by time $t+1$ according to the multinomial probabilities in (3.2). Let the random variable $Y_{ij}(t+1)$ denote the number of individuals among the x_i in transient state i at time t who undergo the transition $i \to j$ during the time interval t to $t+1$. Then, for all $i \in \mathfrak{S}_2$

$$\sum_j Y_{ij}(t+1) = x_i, \tag{3.5}$$

where the sum runs over all states $j \in \mathfrak{S}$. According to a well-known property of the multinomial distribution, the *p.g.f.* of the summands on the left in (3.5) is

$$(H_i(\mathbf{x};\, z_{i1}, z_{i2}, \ldots, z_{ir_2}))^{x_i} \tag{3.6}$$

for every transient state $i \in \mathfrak{S}_2$.

Given the state vector of $\mathbf{x}$ of the population at time t, let the random variable $W_j(t+1)$ denote the number of individuals who made a transition to transient state $j \in \mathfrak{S}_2$ by time $t+1$. Then,

$$W_j(t+1) = \sum_i Y_{ij}(t+1), \tag{3.7}$$

where the sum runs over all $i \in \mathfrak{S}_2$. Let the

$$\mathbf{W}(t+1) = (W_1(t+1), W_2(t+1), \ldots, W_{r_1}(t+1)) \tag{3.8}$$

denote a vector random variable whose elements are given in (3.7), and let the $r_2 \times 1$ vector $\mathbf{X}_R(t+1)$ denote the number of recruits of each type who enter the population during the time interval t to $t+1$. Then, given the state $\mathbf{X}(t) = \mathbf{x}$ of the population at time t, the recursive algorithm for computing Monte Carlo realizations of the population process takes the form

$$\mathbf{X}(t+1) = \mathbf{X}_R(t+1) + \mathbf{W}(t+1), \tag{3.9}$$

for $t = 0, 1, 2, \ldots$.

To begin a simulation experiment, the initial vector $\mathbf{X}(0) = \mathbf{x}_0$ is assigned arbitrarily. It will also be assumed that the collections of vector-valued random variables

$$\{\mathbf{X}_R(t) \mid t \in \mathbb{T}_h\} \quad \text{and} \quad \{\mathbf{X}(t) \mid t \in \mathbb{T}_h\} \tag{3.10}$$

are independent. This means that any finite subsets selected from either collection are independent in the probabilistic sense. When writing software to implement a model in the class under consider, these assumption of conditional independence are incorporated in the programs.

Let

$$G(\mathbf{x}, \mathbf{z}) = \sum_{\mathbf{y}} P(\mathbf{x}, \mathbf{y}) \prod_{k=1}^{r_2} z_k^{y_k} \tag{3.11}$$

denote the *p.g.f.* for the row $\mathbf{x}$ in the matrix $\mathbf{P}$ for the population process corresponding to state $\mathbf{x} \in \mathbb{N}_{r_2}^+$, where the sum runs over all $\mathbf{y} = (y_1, y_2, \ldots, y_{r_2}) \in \mathbb{N}_{r_2}^+$. To show that the *p.g.f's* in (3.1) and (3.3) determine the *p.g.f.* in (3.11), it suffices to consider the case of $r_1 = 1$ absorbing state and $r_2 = 2$ transient states in the life cycle model for individuals. Let 0 denote the absorbing state and 1, 2 the transient states. Then the *p.g.f.* in (3.3) takes the simple form

$$(H_i(\mathbf{x};\, z_{i0}, z_{i1}, z_{i2}))^{x_i} \tag{3.12}$$

for $i = 1, 2$.

By assumption, given $\mathbf{X}(t) = \mathbf{x} = (x_1, x_2)$ individuals evolve independently. Therefore, the joint *p.g.f.* for state transitions of all individuals is

$$\prod_{i=1}^{2} (H_i(\mathbf{x};\, z_{i0}, z_{i1}, z_{i2}))^{x_i}. \tag{3.13}$$

To obtain the marginal *p.g.f.* for those individuals in transient states 1 and 2, set $z_{i0} = 1$ for $i = 1, 2$. Then, to obtain the *p.g.f.* for the random vector sum in (3.8) and (3.9) set $z_{i1} = z_1$ and $z_{i2} = z_2$ for $i = 1, 2$. Then, under assumptions of conditional independence stated in this section, it follows that

$$G(\mathbf{x}, \mathbf{z}) = G_R(\mathbf{z}) \prod_{i=1}^{2} (H_i(\mathbf{x};\, 1, z_1, z_2))^{x_i}. \tag{3.14}$$

It is clear that the argument may easily be extended to the cases such that $r_1 > 1$ and $r_2 \geq 2$.

In principle, by expanding the *p.g.f.* on the right in (3.14) in an infinite series, the transition probability $P(\mathbf{x}, \mathbf{y})$ may be obtained by deriving the coefficient of

$$z_1^{y_1} z_2^{y_2} \tag{3.15}$$

in the series expansion for every $\mathbf{x} \in \mathbb{N}_{r2}^+$ and $\mathbf{y} = (y_1,y_2) \in \mathbb{N}_{r2}^+$. Needless to say, this exercise would not be particularly useful from either the symbolic or computational points of view, but, nevertheless, having a proof of the theoretical existence of the transition matrix for the population process is a necessary step towards understanding the results presented in the remainder of this paper.

As an aid to understanding the results in the next section, it will be helpful to define a probability space underlying the population process, which is Markov chain with stationary transition matrix

$$\mathbf{P} = (P(\mathbf{x}, \mathbf{y})). \tag{3.16}$$

Let $\mathbf{\Omega}$ be the set of all functions with domain $\mathbb{T}_h$ and range $\mathbb{N}_{r2}^+$. More precisely, each $\omega \in \mathbf{\Omega}$ is a sequence

$$\left(\omega = (\mathbf{x}_0, \mathbf{x}_1, \mathbf{x}_2, \ldots) \mid \mathbf{x}_t \in \mathbb{N}_{r2}^+ \text{ for all } t \in \mathbb{T}_h\right), \tag{3.17}$$

Let $\mathfrak{A}$ be a σ-algebra of subsets of $\mathbf{\Omega}$. Then, as is well known from the foundations of mathematical probability, given the transition matrix in (3.16), $\mathfrak{A}$ may be chosen in such a way that a probability measure $\mathbb{P}$ may be defined on it such that $0 \leq \mathbb{P}[A] \leq 1$ for all $A \in \mathfrak{A}$ and $\mathbb{P}[\mathbf{\Omega}] = 1$. Such books as Chung (1960), as well as more recent books, may be consulted for technical details on the construction of a probability space underlying a Markov Chain, which, by definition, is the triple $(\mathbf{\Omega}, \mathfrak{A}, \mathbb{P})$. Briefly, this construction involves a basic theorem due to Kolmogorov, which is widely used in books on stochastic processes. From now on, whenever the symbol $\mathbb{P}$ appears in a derivation, it should be understood that such phrases as convergence in probability or in distribution will be with respect to the probability space $(\mathbf{\Omega}, \mathfrak{A}, \mathbb{P})$.

It is helpful to point out that the assumption of conditional independence used in this a subsequent sections is one that is common to all chain multinomial models. Also implicit in the use of chain multinomial models is the assumption that the evolution of all individuals among transient states and into absorbing states are conditionally independent, given the state $\mathbf{x}$ of the population at any time. It is in this sense that the assumptions underlying the class of models considered in this paper are similar to those used in branching processes, where it is commonly assumed that individuals evolve independently according to some life cycle model.

4. One Dimensional Distributions of the Population Process

Throughout this section, a random sum of multinomial indicators will play a basic role in the derivation of the results. Let $p_i \geq 0,\ i = 1, 2, \ldots, n$, be probabilities of disjoint events such the $p_1 + p_2 + \cdots + p_n = 1$, and let the random vector $\xi = (\xi_1, \xi_2, \ldots, \xi_n)$ be such that if the ith event occurs, then $\xi_i = 1$ and $\xi_j = 0$ for $j \neq i$. It is in this sense that the value of random vector ξ indicates which among the n events has occurred. By way of an illustrative example, in terms of probabilities for the case $n = 3$,

$$P\left[\xi = (0, 0, 1)\right] = p_3. \tag{4.1}$$

A similar statement holds for the other possible values of ξ; namely $(1, 0, 0)$ and $(0, 1, 0)$. In the general case, the *p.g.f.* of the multinomial indicator ξ is, by definition,

$$E\left[\prod_{i=1}^{n} z_i^{\xi_i}\right] = p_1 z_1 + p_2 z_2 + \cdots + p_n z_n. \tag{4.2}$$

Let X denote a Poisson random variable with parameter $\lambda > 0$ that takes values in $\mathbb{N}_1^+$. Given that $X = x \in \mathbb{N}_1^+$, let $\xi_k,\ k = 1, 2, \ldots$ be a conditionally independent sequence of multinomial indicators. The random sum of multinomial indicators that will play a basic role in this section is a vector random variable of the form

$$\mathbf{Y} = (Y_1, Y_2, \ldots, Y_n) = \sum_{k=1}^{\mathbf{X}} \xi_k. \tag{4.3}$$

By the assumption of conditional independence, the conditional *p.g.f.* of $\mathbf{Y}$, given $X = x$, is

$$E\left[\prod_{j=1}^{n} z_j^{Y_j} | X = x\right] = \left(p_1 z_1 + p_2 z_2 + \cdots + p_n z_n\right)^x. \tag{4.4}$$

Therefore, the unconditional *p.g.f.* of the random vector $\mathbf{Y}$ is

$$\begin{aligned} E\left[\prod_{j=1}^{n} z_j^{Y_j}\right] &= E\left[E\left[\prod_{j=1}^{n} z_j^{Y_j} | X = x\right]\right] \\ &= \sum_{x=0}^{\infty} e^{-\lambda} \frac{\lambda^x \left(p_1 z_1 + p_2 z_2 + \cdots + p_n z_n\right)^x}{x!}. \end{aligned} \tag{4.5}$$

It can be shown that, because $p_1 + p_2 + \cdots + p_n = 1$, the infinite series in (4.5) has the closed form

$$E\left[\prod_{j=1}^{n} z_j^{Y_j}\right] = \exp\left[\sum_{j-1}^{n} \lambda p_j(z_j - 1)\right] = \prod_{j=1}^{n} \exp\left[\lambda p_j(z_j - 1)\right]. \qquad (4.6)$$

From this expression it can be seen that the elements of the random vector $\mathbf{Y} = (Y_1, Y_2, \ldots, Y_n)$ have independent Poisson distributions with parameters λp_j for $j = 1, 2, \ldots, n$.

Throughout chapters 10 of the book (Mode and Sleeman, 2000), a Poisson distribution with a *p.g.f.* of the form

$$G_R(\mathbf{z}) = \prod_{i=1}^{r_2} \exp\left[\lambda \varphi_i(z_i - 1)\right] \qquad (4.7)$$

for recruits was used, where φ_i is the probability a recruit is of type i. From now on, this distribution for recruits will be in force in this and subsequent sections.

In the previous section, the evolution of the population was viewed from cross-sectional perspective. More specifically, this cross-sectional perspective was used in the derivation of the recursive equation (3.9). A useful alternative way of viewing the evolution of the population is to consider a longitudinal perspective in which individuals are classified by the time they entered the population as recruits. To this end, let the random vector $\mathbf{X}(s,t)$ be the state vector at time t for those individuals who entered the population as recruits at time $s \leq t$. By definition, $\mathbf{X}(t,t) = \mathbf{X}_R(t)$, the vector of recruits entering the population at time $t > 0$. Then, for every $t > 0$, the population state vector may be represented in the form

$$\mathbf{X}(t) = \sum_{s=0}^{t} \mathbf{X}(t-s, t). \qquad (4.8)$$

Except for the cases, $\mathbf{X}(0,t)$ and $\mathbf{X}(t,t)$, the other vectors in the sum on the right are such that each element is a random sum of multinomial indicators of form (4.3), with probability vectors depending on the transition matrix for individuals in (2.5). For the vector $\mathbf{X}(0,t)$, each element is a fixed sum of the elements of the state vector $\mathbf{x}_0 = (x_{01}, x_{02}, \ldots, x_{0r_2})$ of the initial population.

As a first step in deriving a formula for the one dimensional distributions of the process, consider a particular realization of the population process up to time $t \geq 1$, which has the form

$$\omega(0,t) = (\mathbf{x}(0), \mathbf{x}(1), \mathbf{x}(2), \ldots, \mathbf{x}(t)), \tag{4.9}$$

where $\mathbf{x}(0) = \mathbf{x}_0$, the initial vector. Then, given this realization, consider the evolution of those recruits who enter the population at time $s < t$. The times at which these individuals may undergo transitions among the states of the state space for individuals $\mathfrak{S}$ are $k = s+1, \ldots, t$. A matrix product defined by

$$\mathbf{\Pi}_{22}^{(s,t)}(\omega(0,t)) = \prod_{k=s+1}^{t} \mathbf{\Pi}_{22}(\mathbf{x}_k), \tag{4.10}$$

where $\mathbf{\Pi}_{22}(\mathbf{x}_k)$ is the submatrix in (2.5), governing transitions among the set of transient states $\mathfrak{S}_2$ in the life cycle model for individuals when the state of the population is $\mathbf{x}_k$, will play a basic role in the derivation of a formula for the one dimensional distributions of the population process. Briefly as an aside, if the branching perspective were being emphasized, the matrix in (4.10) would be the first moment matrix in a sub-critical multitype density dependent general branching process that had evolved for $t - s$ generations. For every $\omega(0,t)$, let $\mathbf{\Pi}_{22}^{(0,0)}(\omega(0,t)) = \mathbf{I}_{r_2}$, a $r_2 \times r_2$ identity matrix, and define a matrix-valued random variable $\mathbf{M}(t;\omega(0,t))$ as the sum

$$\mathbf{M}(t;\omega(0,t)) = \sum_{s=0}^{t-1} \mathbf{\Pi}_{22}^{(t-s,t)}(\omega(0,\ t)) \tag{4.11}$$

for $t \geq 1$.

As a first step towards describing the elements of the vector $\mathbf{X}(s,t)$ in terms of random sums, as in (4.10) define the matrix product by

$$\mathbf{\Pi}^{(s,t)}(\omega(0,t)) = \prod_{k=s+1}^{t} \mathbf{\Pi}(\mathbf{x}_k), \tag{4.12}$$

where $\mathbf{\Pi}(\mathbf{x}_k)$ is the partitioned matrix for individuals in (1.4). Then, this matrix also has the partitioned form

$$\mathbf{\Pi}^{(s,t)}(\omega(0,t)) = \begin{bmatrix} \mathbf{I}_{r_r} & \mathbf{0} \\ \mathbf{\Pi}_{21}^{(s,t)}(\omega(0,t)) & \mathbf{\Pi}_{22}^{(s,t)}(\omega(0,t)) \end{bmatrix}. \tag{4.13}$$

The submatrix $\mathbf{\Pi}_{21}^{(s,t)}(\omega(0,t))$ is a complicated sum of products, but, fortunately, there is no need to exhibit its precise form here. Let

$$\left(\pi_{ij}^{(s,t)}(\omega(0,t)) \mid j \in \mathfrak{S}\right) \tag{4.14}$$

be a vector of multinomial probabilities corresponding to the transient state $i \in \mathfrak{S}_2$ in the matrix in (4.13).

By proceeding analogously to (3.5) in Section 3, let the random variable $Y_{ij}(s,t)$ be the number of individuals in state $j \in \mathfrak{S}$ at time t who entered the population as recruits of type $i \in \mathfrak{S}_2$ at time $s < t$. According to (4.7), the number of recruits of type i entering the population at time s follows a Poisson distribution with parameter $\lambda\varphi_i$. Hence, by (4.6), the joint *p.g.f.* of the random variables $Y_{ij}(s,t)$, $j \in \mathfrak{S}$, is

$$\exp\left[\sum_j \lambda\varphi_i\pi_{ij}^{(s,t)}(\omega(0,t))(z_j - 1)\right], \tag{4.15}$$

where the sum runs over all $j \in \mathfrak{S}$. Let the random variable $X_j(s,t)$ denote the jth component of the vector random variable $\mathbf{X}(s,t)$. Then,

$$X_j(s,t) = \sum_{i \in \mathfrak{S}_2} Y_{ij}(s,t) \tag{4.16}$$

for every $j \in \mathfrak{S}_2$. Given the representation in (4.16), the next step is to derive the *p.g.f.* of the vector random variable $\mathbf{X}(s,t)$.

By setting $z_j = 1$ for all $j \in \mathfrak{S}_1$, the set of absorbing states for the life cycle model for individuals, it follows from (4.15) that the marginal *p.g.f.* of the random variables $Y_{ij}(s,t)$ for $j \in \mathfrak{S}_2$ is

$$\exp\left[\sum_{j \in \mathfrak{S}_2} \lambda\varphi_i\pi_{ij}^{(s,t)}(\omega(0,t))(z_j - 1)\right]. \tag{4.17}$$

Under the assumption that, given a realization of the population process $\omega(0,t)$ up to time $t \geq 1$, the components of $\mathbf{X}(s,t)$ are conditionally independent, it follows that the conditional *p.g.f.* of the vector random variable $\mathbf{X}(s,t)$ is

$$\prod_{i \in \mathfrak{S}_2} \exp\left[\sum_{j \in \mathfrak{S}_2} \lambda\varphi_i\pi_{ij}^{(s,t)}(\omega)(z_j - 1)\right] = \exp\left[\sum_{i \in \mathfrak{S}_2}\sum_{j \in \mathfrak{S}_2} \lambda\varphi_i\pi_{ij}^{(s,t)}(\omega)(z_j - 1)\right]. \tag{4.18}$$

It will be helpful in what follows to represent the function on the right is a succinct vector-matrix form. Let μ_R denote a $1 \times r_2$ row vector with elements $\lambda\varphi_i$, $\mathbf{z}$ be a $r_2 \times 1$ column vector with elements z_i, and $\mathbf{1}$ be a $r_2 \times 1$ column vector with constant element 1. Then, the conditional *p.g.f.*

of the vector $\mathbf{X}(s,t)$, given $\omega(0t)$, has the succinct form

$$E\left[\prod_{j=1}^{r_2} z_i^{X_j(s,t)} | \omega(0,t)\right] = \exp\left[\mu_R \mathbf{\Pi}_{22}^{(s,t)}(\omega(0,t))(\mathbf{z}-\mathbf{1})\right]. \tag{4.19}$$

From the assumption that all individuals undergo state transitions in the model for individuals independently, given a realization $\omega(0,t)$, it follows that the vector summands in (4.8) are conditionally independent given $\omega(0,t)$. Therefore, by (4.19), the conditional *p.g.f.* of the random vectors in the sum in (4.8) up to $t-1$, given $\omega(0,t)$, is

$$\prod_{s=0}^{t-1} \exp\left[\mu_R \mathbf{\Pi}_{22}^{(t-s,t)}(\omega(0,t))(\mathbf{z}-\mathbf{1})\right] = \exp\left[\mu_R \mathbf{M}(t;\omega(0,t))(\mathbf{z}-\mathbf{1})\right], \tag{4.20}$$

where $\mathbf{M}(t;\omega(0,t))$ is the sum in (4.11). Furthermore, the conditional *p.g.f.* of the vector $\mathbf{X}(0,t)$, given $\omega(0,t)$, is

$$G^{(0,t)}(\mathbf{z} \mid \omega(0,t)) = \prod_{i\in\mathfrak{S}_2}\left(\sum_{j\in\mathfrak{S}_1} \pi_{ij}^{(0,t)}(\omega(0,t)) + \sum_{j\in\mathfrak{S}_2} \pi_{ij}^{(0,t)}(\omega(0,t)) z_j\right)^{x_{oi}}, \tag{4.21}$$

where z_j for $j \in \mathfrak{S}_2$ are the elements of the vector $\mathbf{z}$. Observe the principles used in the derivation of (4.21) are the same as those in the derivation of (3.14). Therefore, the conditional *p.g.f.* of the state vector $\mathbf{X}(t)$ in (4.8), given $\omega(0,t)$, is

$$\exp\left[\mu_R \mathbf{M}(t;\omega(0,t))(\mathbf{z}-\mathbf{1})\right] G^{(0,t)}(\mathbf{z} \mid \omega(0,t)). \tag{4.22}$$

Let $G(t,\mathbf{z})$ denote the unconditional *p.g.f.* of the random vector $\mathbf{X}(t)$. To derive a formula for this function, a formula for the probability of the realization $\omega(0,t)$ of the population process will be needed. According to the assumption that the population process is a Markov chain, it follows that the formula for this probability has the form

$$P\left[\omega(0,t)\right] = \prod_{k=1}^{t} P\left(\mathbf{x}(k-1), \mathbf{x}(k)\right), \tag{4.23}$$

where $P\left(\mathbf{x}(k-1, \mathbf{x}(k)\right)$ are elements of the transition matrix in (3.16) for every $k \geq 1$. Therefore, the unconditional *p.g.f.* $G(t,\mathbf{z})$ has the form

$$G(t,\mathbf{z}) = \sum_{\omega(0,t)} \exp\left[\mu_R \mathbf{M}(t;\omega(0,t))(\mathbf{z}-\mathbf{1})\right] G^{(0,t)}(\mathbf{z} \mid \omega(0,t)) P\left[\omega(0,t)\right], \tag{4.24}$$

where the sum run over all possible realizations of the population process up to time $t \geq 1$. Observe that this sum has infinitely many terms, because with the initial vector $\mathbf{x}(0)$ fixed the vector $\mathbf{x}(k)$ runs over all elements in the set $\mathbb{N}_{r_2}^{+}$ for every $k \geq 1$.

In terms of the abstract probability space $(\boldsymbol{\Omega}, \mathfrak{A}, \mathbb{P})$ introduced in Section 3, each realization $\omega(0,t)$ may be viewed as the basis for a cylinder set in the σ-algebra $\mathfrak{A}$. Consequently, the infinite sum in (4.24) may be represented as the abstract integral

$$G(t, \mathbf{z}) = \int_{\boldsymbol{\Omega}} \exp\left[\mu_R \mathbf{M}_{22}(t;\omega)(\mathbf{z} - \mathbf{1})\right] G^{(0,t)}(\mathbf{z} \mid \omega) \mathbb{P}(d\omega) \tag{4.25}$$

on this probability space for every $t \geq 1$. In principle, this formula determines the finite dimensional distributions of the population process for every $t \geq 1$. Observe that the exponential function in the integrand is actually the *p.g.f.* of conditionally independent Poisson random variable, and in the next section, it will be shown that the function $G^{(0,t)}(\mathbf{z} \mid \omega)$ converges in probability to 1 as $t \uparrow \infty$ under a plausible condition that will described in the next section. It should also be remarked that formula (4.25) is of interest in its own right, because it shows that for any $t \geq 1$ a sample of realizations of the population process has a unconditional distribution determined by the *p.g.f.* in (4.25). This observation will be of significance when interpreting and analyzing a Monte Carlo simulation sample of realizations of the population process.

5. Convergence to a Stationary Distribution

In order to simplify the notation in a proof that the population process does indeed converge to a stationary distribution under a plausible sufficient condition, it will be helpful to introduce the idea of a matrix norm. Let $\mathbf{A} = (a_{ij})$ denote a square or rectangular matrix of real or complex numbers. Then, the norm of this matrix is defined by

$$\|\mathbf{A}\| = \max_{\mathbf{i}} \sum_j |a_{ij}|. \tag{5.1}$$

If $\mathbf{a}$ is either a row or column vector with elements a_i, then the norm of $\mathbf{a}$ is defined by $\|\mathbf{a}\| = \mathbf{max_i}|a_i|$. Furthermore, if the elements of the matrix $\mathbf{A}$ are random variables, then, by definition, the expectation of the random matrix $\mathbf{A}$

$$E[\mathbf{A}] = (E[a_{ij}]). \tag{5.2}$$

An inequality that will be used in this section is that if $\mathbf{A}$ is a random matrix, then it can be shown that

$$\|E[\mathbf{A}]\| \leq E[\|\mathbf{A}\|]. \tag{5.3}$$

Another useful property of the matrix norm defined in (5.1) is that if $\mathbf{A}$ and $\mathbf{B}$ are matrices such that the product $\mathbf{AB}$ is well defined, then

$$\|\mathbf{AB}\| \leq \|\mathbf{A}\| \times \|\mathbf{B}\|. \tag{5.4}$$

A sufficient condition for the population process to converge to a stationary distribution may be succinctly stated in terms of this norm.

Condition A. Let $\mathbf{\Pi}_{22}(\mathbf{x})$ for the submatrix in (2.5), governing transitions of individuals among the transient states in the set $\mathfrak{S}_2$ when the population state vector is $\mathbf{x}$. In order to use this condition to also prove that the embedded deterministic model also converges point wise to a limit, it will be helpful to define the condition in terms of vectors of non-negative real numbers of dimension $r_2 \geq 1$. By definition, the set of non-negative real numbers is

$$\mathbb{R}_1^+ = (x \mid x \geq 0)\,. \tag{5.5}$$

Let $\mathbb{R}_{r_2}^+$ denote the set of all r_2 dimensional vectors of non-negative real numbers. Then suppose there is a number $\delta > 0$ such that

$$\sup_{\mathbf{x}\in\mathbb{R}_{r_2}^+} \|\mathbf{\Pi}_{22}(\mathbf{x})\| \leq \delta < 1. \tag{5.6}$$

Observe that because $\mathbb{N}_{r_2}^+ \subset \mathbb{R}_{r_2}^+$, it follows that

$$\sup_{\mathbf{x}\in\mathbb{N}_{r_2}^+} \|\mathbf{\Pi}_{22}(\mathbf{x})\| \leq \delta < 1. \tag{5.7}$$

To illustrate by an example that condition $\mathbf{A}$ is plausible for the class of models of sexually transmitted diseases considered in Chapter 10 of (Mode and Sleeman, 2000), it will suffice to consider an illustrative simple example in which there is one behavioral class, $m_1 = 1$, and one stage, $n_1 = 1$, of disease. In this case, the set of transient states for individuals would be $\mathfrak{S}_2 = \{0, 1\}$, where 0 and 1 denote, respectively, a susceptible and infected person. With regard to set of two absorbing states $\mathfrak{S}_1$, let d_1 denote a case in which an individual dies from a background risk that is common to all individuals in the population and let d_2 denote a case in which an infected individual dies from disease. For this model, the matrix of latent risks has the partitioned form

$$\mathbf{\Theta} = [\mathbf{\Theta}_{12}, \mathbf{\Theta}_{22}]\,, \tag{5.8}$$

where each of the submatrices is 2×2.

In particular for this model, the submatrix $\mathbf{\Theta}_{12}$, which governs rates of transition from transient to absorbing states, has the form

$$\mathbf{\Theta}_{12} = \begin{bmatrix} \mu_0 & 0 \\ \mu_0 & \mu_1 \end{bmatrix}, \tag{5.9}$$

where $\mu_0 > 0$ is a back ground risk of death and $\mu_1 > 0$ in an incremental risk of death when a person is infected. The submatrix $\mathbf{\Theta}_{22}$ has the form

$$\mathbf{\Theta}_{22} = \begin{bmatrix} 0 & \lambda_1 Q(\mathbf{x}) \\ 0 & 0 \end{bmatrix}, \tag{5.10}$$

where λ_1 is the expected number of sexual partners per unit time and $Q(\mathbf{x})$ is the probability per unit time that a susceptible person becomes infected by a infected partner in transient state 2. Chapter 10 of the book cited above (Mode and Sleeman, 2000) may be consulted for a detailed derivation of the function $Q(\mathbf{x})$, which satisfies in condition $0 \leq Q(\mathbf{x}) \leq \mathbf{1}$ for all $\mathbf{x} \in \mathbb{R}_2^+$. The second row of zeros in the matrix in (5.10) indicates that once a person is infected, he or she remains infected for the rest of his or her life. Hence, the risk of a transition from the infected state 1 to the susceptible state 0 is zero. According to the second row of the matrix in (5.9), however, there are positive risk that an infected person will die from either a background risk common to individuals in the population or a risk due to disease.

To illustrate the plausibility of condition **A** for this simple model as well as more complex models, let $\mathbf{\Theta}(\mathbf{x}) = (\theta_{ij}(\mathbf{x}))$ be a $r_2 \times r$ matrix of non-negative latent risks when the state vector of the population is $\mathbf{x}$. Then, the total risk function for transient state $i \in \mathfrak{S}_2$ is, by definition,

$$\theta_i(\mathbf{x}) = \sum_{j \in \mathfrak{S}} \theta_{ij}(\mathbf{x}). \tag{5.11}$$

As illustrated by many examples in Chapter 10 of Mode and Sleeman's book, using the classical theory of competing risks, it can be shown that elements $\pi_{ij}(\mathbf{x})$ in the $r_2 \times r$ partitioned transition matrix $\mathbf{\Pi}_2(\mathbf{x}) = (\mathbf{\Pi_{21}}(\mathbf{x}), \mathbf{\Pi_{22}}(\mathbf{x}))$ for individuals, when the state vector of the population is $\mathbf{x}$, have the following forms. If $i = j$, then

$$\pi_{ii}(\mathbf{x}) = \exp\left[-\theta_i(\mathbf{x})\right], \tag{5.12}$$

but if $i \neq j$, then

$$\pi_{ij}(\mathbf{x}) = \left(1 - \exp\left[-\theta_i(\mathbf{x})\right]\right) \frac{\theta_{ij}(\mathbf{x})}{\theta_i(\mathbf{x})} \tag{5.13}$$

for $\theta_i(\mathbf{x}) \neq 0$.

For the simple model under discussion, the total risk function for transient state 0 is

$$\theta_0(\mathbf{x}) = \mu_0 + \lambda Q(\mathbf{x}), \tag{5.14}$$

and that for the infected state 1 is

$$\theta_1(\mathbf{x}) = \mu_0 + \mu_1. \tag{5.15}$$

Therefore, by applying formulas (5.12) and (5.13), it follows that the 2×2 submatrix $\mathbf{\Pi_{21}}(\mathbf{x})$ has the form

$$\mathbf{\Pi_{21}}(\mathbf{x}) = \begin{bmatrix} (1 - \exp[-\theta_0(\mathbf{x})]) \dfrac{\mu_0}{\theta_0(\mathbf{x})} & 0 \\ (1 - \exp[-\theta_1(\mathbf{x})]) \dfrac{\mu_0}{\theta_1(\mathbf{x})} & (1 - \exp[-\theta_1(\mathbf{x})]) \dfrac{\mu_1}{\theta_1(\mathbf{x})} \end{bmatrix}, \tag{5.16}$$

and the 2×2 submatrix $\mathbf{\Pi_{22}}(\mathbf{x})$ has the form

$$\mathbf{\Pi_{22}}(\mathbf{x}) = \begin{bmatrix} \exp[-\theta_0(\mathbf{x})] & (1 - \exp[-\theta_0(\mathbf{x})]) \dfrac{\lambda Q(\mathbf{x})}{\theta_0(\mathbf{x})} \\ 0 & \exp[-\theta_1(\mathbf{x})] \end{bmatrix}. \tag{5.17}$$

For any state vector $\mathbf{x}$ of the population, the norm of this matrix is

$$\|\mathbf{\Pi_{22}}(\mathbf{x})\| = \max\left(\exp[-\theta_0(\mathbf{x})] + (1 - \exp[-\theta_0(\mathbf{x})]) \frac{\lambda Q(\mathbf{x})}{\theta_0(\mathbf{x})}, \exp[-\theta_1(\mathbf{x})]\right) \tag{5.18}$$

From (5.15), it can be seen that $\theta_1(\mathbf{x})$ is constant for all $\mathbf{x} \in \mathbb{R}_2^+$. Therefore, $\exp[-\theta_1(\mathbf{x})] < 1$ for all $\mathbf{x} \in \mathbb{R}_2^+$. Next consider the function

$$f(\mathbf{x}) = \exp[-\theta_0(\mathbf{x})] + (1 - \exp[-\theta_0(\mathbf{x})]) \frac{\lambda Q(\mathbf{x})}{\theta_0(\mathbf{x})}, \tag{5.19}$$

which is defined for all $\mathbf{x} \in \mathbb{R}_2^+$. This function must also have the property $f(\mathbf{x}) < 1$ for all $\mathbf{x} \in \mathbb{R}_2^+$. This follows from the condition that, by construction, each row in the transition matrix $\mathbf{\Pi}_2(\mathbf{x})$ sums to one for all $\mathbf{x} \in \mathbb{R}_2^+$. In particular, the sum of the elements in the row of this matrix corresponding to transient state 0 has the form

$$(1 - \exp[-\theta_0(\mathbf{x})]) \frac{\mu_0}{\theta_0(\mathbf{x})} + f(\mathbf{x}) = \mathbf{1} \tag{5.20}$$

for all $\mathbf{x} \in \mathbb{R}_2^+$. Therefore, if there is a $\mathbf{x} \in \mathbb{R}_2^+$ such that $f(\mathbf{x}) \geq 1$, condition would be violated. Thus, $f(\mathbf{x}) < 1$ for all $\mathbf{x} \in \mathbb{R}_2^+$.

It can be seen from the many examples in Chapter 10 of Mode and Sleeman's book that $Q(\mathbf{x})$ is a continuous function of $\mathbf{x} \in \mathbb{R}_2^+$. Hence, $f(\mathbf{x})$

is a continuous function of $\mathbf{x} \in \mathbb{R}_2^+$. Therefore, if a maximum of the function is attained for some $\mathbf{x}_{\max} \in \mathbb{R}_2^+$, then $f(\mathbf{x}_{\max}) = \kappa < 1$ so that the inequality in (5.6) of condition **A** would be satisfied, where $\delta = \max(\kappa, \exp[-\theta_1(\mathbf{x})])$. It should be pointed out that the argument just outlined is not a rigorous proof that condition **A** is satisfied for the illustrative model under consideration, but it seems plausible that this condition will be satisfied for not only the simple model just discussed but also the class of models described in Chapter 10 of the book cited above (Mode and Sleeman, 2000). In what follows, functions such as $\mathbf{M}(t;\omega(0,t))$ in (4.11) will be written in the simpler form $\mathbf{M}(t;\omega)$, because $\omega(0,t)$ may be viewed as a point $\omega \in \mathbf{\Omega}$ whose first $t+1$ elements are $\omega(0,t)$.

Lemma 5.1. If condition **A** is satisfied, then the random vector $\mathbf{X}(0,t)$ converges to $\mathbf{0}$, a vector of zeroes, in probability as $t \uparrow \infty$, and there is a random $r_2 \times r_2$ random matrix $\mathbf{\Xi}(\omega)$ such that the limit

$$\mathbf{\Xi}(\omega) = \lim_{t\uparrow\infty} \mathbf{M}(t;\omega) \tag{5.21}$$

holds with probability one, where the matrix $\mathbf{M}(t;\omega)$ is defined in (4.11).

Proof: To prove the first assertion, because all the elements of the random vector are ≥ 0, it follows that by the Markov inequality that for every $\epsilon > 0$

$$\mathbb{P}[X_j(0,t) > \epsilon] \leq \frac{E[X_j(0,t)]}{\epsilon}$$

for all $j \in \mathfrak{S}_2$. Therefore,

$$\begin{aligned}\max_{j\in\mathfrak{S}_2} \mathbb{P}[X_j(0,t) > \epsilon] &\leq \frac{1}{\epsilon} \max_{j\in\mathfrak{S}_2} E[X_j(0,t)] \\ &= \frac{1}{\epsilon}\|E[\mathbf{X}(0,t)]\| = \frac{1}{\epsilon}E\|\mathbf{x}_0\mathbf{\Pi}_{22}^{(0,t)}(\omega)\| \leq \frac{1}{\epsilon}\|\mathbf{x}_0\|\delta^t.\end{aligned}$$

Assertion one follows by letting $t \uparrow \infty$.

To prove the second assertion, observe from (4.11) that

$$\left\|\sum_{s=0}^{t-1} \mathbf{\Pi}_{22}^{(t-s,t)}(\omega)\right\| \leq \sum_{s=0}^{t-1} \|\mathbf{\Pi}_{22}^{(t-s,t)}(\omega)\| \leq \sum_{s=0}^{t-1} \delta^t \leq \frac{1}{1-\delta}. \tag{5.22}$$

for all $t \geq 1$. By a Weierstrass type convergence theorem, it follows that the limit

$$\mathbf{\Xi}(\omega) = (\gamma_{ij}(\omega)) = \lim_{t\uparrow\infty} \mathbf{M}(t;\omega) = \lim_{t\uparrow\infty} \sum_{s=0}^{t-1} \mathbf{\Pi}_{22}^{(t-s,t)}(\omega) \tag{5.23}$$

holds for all $\omega \in \mathbf{\Omega}$ and thus convergence is with probability $\mathbb{P}[\mathbf{\Omega}] = 1$.

By way of preparation for the main result of this section, let

$$\eta_j(\omega) = \sum_{i \in \mathfrak{S}_2} \lambda \varphi_i \gamma_{ij}(\omega) \tag{5.24}$$

for $j \in \mathfrak{S}_2$.

Theorem 5.1. If condition **A** is satisfied and $\|\mathbf{z}\| \leq \mathbf{1}$, then the sequence of random vectors

$$\{\mathbf{X}(t) \mid t = 1, 2, \ldots\}$$

converges in distribution for a random vector $\mathbf{X}_{STA}$ whose probability density function is the mixture

$$\pi_{STA}(\mathbf{y}) = \int_{\mathbf{\Omega}} \prod_{j \in \mathfrak{S}_2} \exp\left[-\eta_j(\omega)\right] \frac{(\eta_j(\omega))^{y_j}}{y_j!} \mathbb{P}(d\omega) \tag{5.25}$$

for every $\mathbf{y} = (y_1, y_2, \ldots, y_{r_2}) \in \mathbb{N}^+_{r_2}$.

Proof: Because the vector $\mathbf{X}(0,t) \to \mathbf{0}$ in probability as $t \uparrow \infty$, it suffices to derive the conditional *p.g.f.* of the sum of random vectors

$$\sum_{s=0}^{t-1} \mathbf{X}(t-s, t, \omega), \tag{5.26}$$

given $\omega \in \mathbf{\Omega}$, see (4.8) and (4.11), converges in distribution. According to (4.25), the *p.g.f.* of the random vector $\mathbf{X}(t)$ has the form

$$G(t, \mathbf{z}) = \int_{\mathbf{\Omega}} \exp\left[\mu_R \mathbf{M}_{22}(t; \omega)\left(\mathbf{z} - \mathbf{1}\right)\right] G^{(0,t)}(\mathbf{z} \mid \omega) \mathbb{P}(d\omega). \tag{5.27}$$

But by Lemma 5.1, the random vector $\mathbf{X}(0,t)$ converges in probability to a zero vector 0 as $t \uparrow \infty$. Therefore, the conditional *p.g.f.* of this random vector $G^{(0,t)}(\mathbf{z} \mid \omega)$ converges in probability to 1 as $t \uparrow \infty$ for all vectors $\mathbf{z}$ such that $\|\mathbf{z}\| \leq \mathbf{1}$. Moreover, if $\|\mathbf{z}\| \leq \mathbf{1}$ then

$$\|\exp\left[\mu_R \mathbf{M}_{22}(t; \omega)\left(\mathbf{z} - \mathbf{1}\right)\right] G^{(0,t)}(\mathbf{z} \mid \omega)\| \leq 1 \tag{5.28}$$

for all $\omega \in \mathbf{\Omega}$ and all $t \geq 1$. By letting $t \uparrow \infty$ and applying the dominated convergence theorem in (5.27), it follows from (5.28) that the unconditional *p.g.f.* of the random vector $\mathbf{X}(t)$ converges to the limit

$$\lim_{t \uparrow \infty} G(t, \mathbf{z}) = \int_{\mathbf{\Omega}} \exp\left[\mu_R \Xi(\omega)(\mathbf{z} - \mathbf{1})\right] \mathbb{P}(d\omega) \tag{5.29}$$

for every vector $\mathbf{z}$ such that $\|\mathbf{z}\| \leq \mathbf{1}$. But, the *p.g.f.* in (5.29) is that for the mixture of conditionally independent Poisson densities in (5.25), which completes the proof of the theorem.

From (5.29), it also follows that the expectation of the stationary vector $\mathbf{X}_{STA}$ has the form

$$E\left[\mathbf{X}_{STA}\right]=\int_{\mathbf{\Omega}}\mu_R\Xi(\omega)\mathbb{P}(d\omega). \tag{5.30}$$

It is also possible to set down a formula for the covariance matrix of the stationary vector $\mathbf{X}_{STA}$ but no further details will be pursued here.

In general, the abstract form the probability density function $\pi_{STA}(\mathbf{y})$ in (5.25) is not computationally tractable but both this expectation vector and covariance matrix may be estimated by Monte Carlo simulation methods. There is, however, a case in which the density in (5.25) has a simple form. For if the transition matrix $\mathbf{\Pi}_{22}(\mathbf{x})$ for individuals is constant and does not depend on the population state vector $\mathbf{x}$, then this density takes the simple form

$$\pi_{STA}(\mathbf{y})=\prod_{j\in\mathfrak{S}_2}\exp\left[-\eta_j\right]\frac{(\eta_j)^{y_j}}{y_j!} \tag{5.31}$$

for every $\mathbf{y}=(y_1,y_2,\ldots,y_{r_2})\in\mathbb{N}^+_{r_2}$, where

$$\eta=(\eta_1,\eta_2,\ldots,\eta_{r_2})=\mu_R\left(\mathbf{I}_{r_2}-\mathbf{\Pi}_{22}\right)^{-1} \tag{5.32}$$

When the matrix $\mathbf{\Pi}_{22}$ is not too large to be accommodated in a computer, the computation of the inverse matrix on the right will be a straight forward exercise at any point in the parameter space of a model.

If a reader is interesting in more general treatments of Markov chains that converge to a stationary distribution, the books by Chung (1960), and Karlin and Taylor (1981) may be consulted.

In any reasonable formulation of a stochastic model for a population of human beings, the model should satisfy the condition that every individual should die eventually with probability one. It is, therefore, of interest to note that condition **A** implies this condition. For any transient state $i\in\mathfrak{S}_2$, let

$$\left(\pi_{ij}^{(s,t)}(\omega)\mid j\in\mathfrak{S}\right)$$

be the ith row in the product matrix in (4.10) for the realization ω of the population process. Then, for every $s\geq 1$

$$\sum_{j\in\mathfrak{S}_1}\pi_{ij}^{(s,t)}(\omega)+\sum_{j\in\mathfrak{S}_2}\pi_{ij}^{(s,t)}(\omega)=1 \tag{5.33}$$

for all $i\in\mathfrak{S}_2$, $t\geq 1$ and $\omega\in\mathbf{\Omega}$. But, if condition **A** is satisfied, then

$$\lim_{t\uparrow\infty}\sum_{j\in\mathfrak{S}_2}\pi_{ij}^{(s,t)}(\omega)=0 \tag{5.34}$$

for all $s \geq 1$, $i \in \mathfrak{S}_2$ and $\omega \in \boldsymbol{\Omega}$. Therefore, it follows from (5.33) and (5.34) that

$$\lim_{t\uparrow\infty} \sum_{j\in\mathfrak{S}_1} \pi_{ij}^{(s,t)}(\omega) = 1 \tag{5.35}$$

for all $s \geq 1$, $i \in \mathfrak{S}_2$ and $\omega \in \boldsymbol{\Omega}$. Therefore, all individuals die with eventually with probability one.

6. Convergence of Embedded Deterministic Model to a Limit

In chapters 10, 11 and 12 of the book (Mode and Sleeman, 2000) and elsewhere, procedures for embedding deterministic models in stochastic population processes are described. This embedding procedure gives rise to a vector-valued non-linear difference equation of the form

$$\hat{\mathbf{X}}(t) = \mu_R + \hat{\mathbf{X}}(t-1)\boldsymbol{\Pi}_{22}(\hat{\mathbf{X}}(t-1)), \tag{6.1}$$

for $t \geq 1$, where $\hat{\mathbf{X}}(0) = \mathbf{x}_0$, the state vector for the initial population. The symbol $\hat{\mathbf{X}}(t)$ is used to indicate, in a certain sense explained in Mode and Sleeman (2000) and elsewhere, that $\hat{\mathbf{X}}(t)$ is an estimate of the random vector, $\mathbf{X}(t)$ for the state of the population at time t. For the computer experiment reported in Mode and Sleeman (2002), it was observed empirically that the limit

$$\lim_{t\uparrow\infty} \hat{\mathbf{X}}(t) = \hat{\mathbf{x}} \tag{6.2}$$

did exist at the set of parameters considered in the experiment. This type of convergence was also observed in other computer experiments not reported in the latter paper. In this section, it will be proven that convergence in (6.2) does indeed occur when condition **A** is satisfied. As in Section 5, let $\mathbb{R}_{r_2}^+$ be the r_2-fold Cartesian product of $\mathbb{R}_1^+$ with itself.

Theorem 6.1. If condition **A** is satisfied and the elements of matrix $\boldsymbol{\Pi}_{22}(\mathbf{x})$ are continuous functions of the vector $\mathbf{x} \in \mathbb{R}_{r_2}^+$, then the limit in (6.2) exists and satisfies the non-linear equation

$$\hat{\mathbf{x}} = \mu_R + \hat{\mathbf{x}}\boldsymbol{\Pi}_{22}(\hat{\mathbf{x}}). \tag{6.3}$$

Proof: Let

$$\hat{\omega} = (\mathbf{x}_0, \hat{\mathbf{x}}_1, \hat{\mathbf{x}}_2, \ldots)$$

be a sequence of points in $\mathbb{R}_{r_2}^{+}$ computed recursively from (6.1). Just as in (4.12), let

$$\mathbf{\Pi}_{22}^{(s,t)}(\hat{\omega}) \tag{6.4}$$

be a product of matrices evaluated at the points $\hat{\mathbf{x}}_k$ for $k = s+1, \ldots, t$. Then, by either iterating Equation (6.3) or by using the cohort representation of $\mathbf{X}(t)$ in (4.8), it can be seen that

$$\hat{\mathbf{X}}(t) = \mu_R \left(\sum_{s=0}^{t-1} \mathbf{\Pi}_{22}^{(t-s,t)}(\hat{\omega}) \right) + \mathbf{x}_0 \mathbf{\Pi}_{22}^{(0,t)}(\hat{\omega}). \tag{6.5}$$

If condition **A** is satisfied, then

$$\left\| \mathbf{x}_0 \mathbf{\Pi}_{22}^{(0,t)}(\hat{\omega}) \right\| \leq \|\mathbf{x}_0\| \delta^t.$$

Therefore, the last term on the right in (6.4) converges to $\mathbf{0}$, a vector of zeros, as $t \uparrow \infty$. Furthermore, proceeding as in (5.22), it can be seen that

$$\left\| \sum_{s=0}^{t-1} \mathbf{\Pi}_{22}^{(t-s,t)}(\hat{\omega}) \right\| \leq \sum_{s=0}^{t-1} \left\| \mathbf{\Pi}_{22}^{(t-s,t)}(\hat{\omega}) \right\| \leq \frac{1}{1-\delta} \tag{6.6}$$

for all $t \geq 1$. Therefore, the limit in (6.2) exists and

$$\lim_{t \uparrow \infty} \hat{\mathbf{X}}(t) = \hat{\mathbf{x}} = \mu_R \left(\lim_{t \uparrow \infty} \sum_{s=0}^{t-1} \mathbf{\Pi}_{22}^{(t-s,t)}(\hat{\omega}) \right). \tag{6.7}$$

That Equation (6.3) is satisfied follows by letting $t \uparrow \infty$ in (6.1) and using the continuity of the elements of the matrix $\mathbf{\Pi}_{22}(\mathbf{x})$. This completes the proof of the theorem.

Equation (6.3) may have more than one solution so that the question arises as to which solution is the limit in (6.7). An approach to finding answers to this question in Mode and Sleeman (2000) was to let $h \downarrow 0$, where h is the length of the time interval in $\mathbb{T}_h$, and derive a set of nonlinear differential equations from the difference equations in (6.1). Then, by deriving the Jacobian matrix for this system of differential equations and testing for stability at a disease free equilibrium some insights as to which solution of (6.3) is the limit in (6.7). If the Jacobian matrix was stable at this equilibrium, then it was empirically observed in numerous computer experiments that the disease free equilibrium was the limit point in (6.7). This limit point had the property that all coordinates corresponding to stages of disease were zero. On the other hand, if the Jacobian matrix evaluated at a disease free equilibrium was not stable, then the limit in (6.7) was a point with positive elements, indicating an infectious disease had been established in a population. In a Monte Carlo simulation experiment reported Mode and Sleeman (2002), the Jacobian matrix evaluated as a disease free equilibrium was not stable and it was observed that this endemic equilibrium point was between the 25th and 75th quantile trajectories as computed from a simulated random sample of realizations of the population process. Because these trajectories were relatively constant, it seems likely that convergence of the population process to the stationary distribution

$$\left(\pi_{STA}(\mathbf{y}) \mid \mathbf{y} \in \mathbb{N}_{r_r}^{+}\right) \tag{6.8}$$

was being observed and that the quantile trajectories were based approximately on a random sample from the distribution in (6.8).

References

Ball F and Neal P (2002) A general model for stochastic SIR epidemics with two levels of mixing, *Math Biosci* **180**: 73–102.

Chung KL (1960) *Markov Chains with Stationary Transition Probabilities*, Springer, Berlin, Gottingen, Heidelberg.

Karlin S and Taylor HM (1981) *A Second Course in Stochastic Processes*, Academic Press, Boston, New York, London, Tokyo.

Mode CJ and Sleeman CK (2002) An algorithmic synthesis of the deterministic and stochastic paradigms via computer intensive methods, *Math Biosci* **180**: 115–126.

Mode CJ and Sleeman CK (2000) *Stochastic Processes in Epidemiology, HIV/AIDS, Other Infectious Diseases and Computers*, World Scientific, Singapore.

Mode CJ (1971) *Multitype Branching Process — Theory and Applications*, American Elsevier, New York.

Nasell I (2002) Stochastic models of some epidemic infections, *Math Biosci* **179**: 1–19.

Tan WY (2000) *Stochastic Modeling of AIDS Epidemiology and HIV Pathogenesis*, World Scientific, Singapore.

INDEX